THE PROBABILITY LIFESAVER

A PRINCETON LIFESAVER STUDY GUIDE

The Calculus Lifesaver: All the Tools You Need to Excel at Calculus by Adrian Banner
The Real Analysis Lifesaver: All the Tools You Need to Understand Proofs by Raffi Grinberg
The Probability Lifesaver: All the Tools You Need to Understand Chance by Steven J. Miller

The PROBABILITY LIFESAVER

All the tools you need to understand chance

STEVEN J. MILLER

PRINCETON UNIVERSITY PRESS
Princeton and Oxford

MATLAB® is a registered trademark of The Math Works Inc. and is used with
permission. The MathWorks does not warrant the accuracy of the text in
this book. This book's use of a MATLAB® related products does not constitute
an endorsement or sponsorship by the MathWorks of a particular pedactogical
approach or a particular use of MATLAB software.

Library of Congress Cataloging-in-Publication Data

Names: Miller, Steven J., 1974–
Title: The probability lifesaver : all the tools you need to understand
chance / Steven J. Miller.
Description: Princeton : Princeton University Press, [2017] | Series: A
Princeton lifesaver study guide | Includes bibliographical references and index.
Identifiers: LCCN 2016040785| ISBN 9780691149547 (hardcover : alk. paper) |
ISBN 9780691149554 (pbk. : alk. paper)
Subjects: LCSH: Probabilities. | Chance. | Games of chance (Mathematics) |
Random variables.
Classification: LCC QA273 .M55185 2017 | DDC 519.2-dc23 LC record
available at https://lccn.loc.gov/2016040785

British Library Cataloging-in-Publication Data is available

This book has been composed in Times New Roman with Stencil and Avant Garde

Printed on acid-free paper. ∞

Typeset by Nova Techset Pvt Ltd, Bangalore, India
Printed in the United States of America

1 3 5 7 9 10 8 6 4 2

CONTENTS

NOTE TO READERS

Welcome to *The Probability Lifesaver*. My goal is to write a book introducing students to the material through lots of worked out examples and code, and to have lots of conversations about not just why equations and theorems are true, but why they have the form they do. In a sense, this is a sequel to Adrian Banner's successful *The Calculus Lifesaver*. In addition to many worked out problems, there are frequent explanations of proofs of theorems, with great emphasis placed on discussing why certain arguments are natural and why we should expect certain forms for the answers. Knowing why something is true, and how someone thought to prove it, makes it more likely for you to use it properly and discover new relations yourself. The book highlights at great lengths the methods and techniques behind proofs, as these will be useful for more than just a probability class. See, for example, the extensive entries in the index on proof techniques, or the discussion on Markov's inequality in §17.1. There are also frequent examples of computer code to investigate probabilities. This is the twenty-first century; if you cannot write simple code you are at a competitive disadvantage. Writing short programs helps us check our math in situations where we can get a closed form solution; more importantly, it allows us to estimate the answer in situations where the analysis is very involved and nice solutions may be hard to obtain (if possible at all!).

The book is designed to be used either as a supplement to any standard probability book, or as the primary textbook. The first part of the book, comprising six chapters, is an introduction to probability. The first chapter is meant to introduce many of the themes through fun problems; we'll encounter many of the key ideas of the subject which we'll see again and again. The next chapter then gives the basic probability laws, followed by a chapter with examples. This way students get to real problems in the subject quickly, and are not overloaded with the development of the theory. After this examples chapter we have another theoretical chapter, followed by two more examples loaded chapters (which of course do introduce some theory to tackle these problems).

The next part is the core of most courses, introducing random variables. It starts with a review of useful techniques, and then goes through the "standard" techniques to study them.

Specific, special distributions are the focus of Part III. There are many more distributions that can be added, but a line has to be drawn somewhere. There's a nice mix of continuous and discrete, and after reading these chapters you'll be ready to deal with whatever new distributions you meet.

The next part is on convergence theorems. As this is meant to supplement or serve as a first course, we don't get into as much detail as possible, but we do prove

Markov's inequality, Chebyshev's theorem, the Weak and Strong Laws of Large Numbers, Stirling's formula, and the Central Limit Theorem (CLT). The last is a particularly important topic. As such, we give a lot of detail here and in an appendix, as the needed techniques are of interest in their own right; for those interest in more see the online resources (which include an advanced chapter on complex analysis and the CLT).

The last part is a hodgepodge of material to give the reader and instructor some flexibility. We start with a chapter on hypothesis testing, as many classes are a combined probability and statistics course. We then do difference equations, continuing a theme from Chapter 1. I really like the Method of Least Squares. This is more statistics, but it's a nice application of linear algebra and multivariable calculus, and assuming independent Gaussian distribution of errors we get a chi-square distribution, which makes it a nice fit in a probability course. We touch upon some famous problems and give a quick guide to coding (there's a more extensive introduction to programming in the online supplemental notes). In the twenty-first century you absolutely *must* be able to do basic coding. First, it's a great way to check your answers and find missing factors. Second, if you can code you can get a feel for the answer, and that might help you in guessing the correct solution. Finally, though, often there *is no simple closed form solution*, and we have no choice but to resort to simulation to estimate the probability. This then connects nicely with the first part of this section, hypothesis testing: if we have a conjectured answer, do our simulations support it? Analyzing simulations and data are central in modern science, and I strongly urge you to continue with a statistics course (or, even better, courses!).

Finally, there are *very* extensive appendixes. This is deliberate. A lot of people struggle with probability because of issues with material and techniques from previous courses, especially in proving theorems. This is why the first appendix on proof techniques is so long and detailed. Next is a quick review of needed analysis results, followed by one on countable and uncountable sets; in mathematics the greatest difficulties are when we encounter infinities, and the purpose here is to give a quick introduction to some occurrences of the infinite in probability. We then end the appendices by briefly touching on how complex analysis arises in probability, in particular, in what is needed to make our proofs of the Central Limit Theorem rigorous. While this is an advanced appendix, it's well worth the time as mastering it will give you a great sense of what comes next in mathematics, as well as hopefully help you appreciate the beauty and complexity of the subject.

There is a lot of additional material I'd love to include, but the book is already quite long with all the details; fortunately they're freely available on the Web and I encourage you to consider them. Just go to

http://press.princeton.edu/titles/11041.html

for a wealth of resources, including all my previous courses (with videos of all lectures and additional comments from each day).

Returning to the supplemental material, the first is a set of practice calculus problems and solutions. Doing the problems is a great way of testing how well you know the material we'll need. There are also some advanced topics that are beyond many typical first courses, but are accessible and thus great supplements. Next is the Change of Variable formula. As many students forget almost all of their Multivariable Calculus, it's useful to have this material easily available online. Then comes the distribution of longest runs. I've always loved that topic, and it illustrates some powerful techniques. Next is the Median Theorem. Though the Central Limit Theorem deservedly sits at the pinnacle of a course, there are times its conditions are not met and thus the Median Theorem has an important role. Finally, there is the Central Limit Theorem itself. In a first course we can only prove it in special cases, which begs the question of what is needed for a full proof. Our purpose here is to introduce you to some complex analysis, a wonderful topic in its own right, and both get a sense of the proof and a motivation for continuing your mathematical journey forward.

Enjoy!

HOW TO USE THIS BOOK _____

This book was written to help you learn and explore probability. You can use this as a supplement to almost any first text on the subject, or as a stand-alone introduction to the material (instructors wishing to use this as a textbook can e-mail me at the addresses at the end of this section for a partial solution key to exercises and exams). As you'll see in your studies, probability is a vast subject with numerous applications, techniques, and methods. This is both exciting and intimidating. It's exciting as you'll see so many strange connections and seemingly difficult problems fall from learning the right way to look at things, but it's also intimidating as there is so much material it can be overwhelming.

My purposes are to help you navigate this rich landscape, and to prepare you for future studies. The presentation has been greatly influenced by Adrian Banner's successful *The Calculus Lifesaver*. Like that book, the goals are to teach you the material and mathematical thinking through numerous worked out problems in a relaxed and informal style. While you'll find the standard statements and proofs, you'll also find a wealth of worked out examples and lots of discussions on how to approach theorems. The best way to learn a subject is to do it. This doesn't just mean doing worked out problems, though that's a major part and one which sadly, due to limited class time, is often cut in courses. It also means *understanding* the proofs.

Why are proofs so important? We'll see several examples in the book of reasonable statements that turn out to be wrong; the language and formalism of proofs are the mathematician's defense against making these errors. Furthermore, even if your class isn't going to test you on proofs, it's worth having a feel as to why something is true. You're not expected on Day One to be able to create the proofs from scratch, but that's not a bad goal for the end of the course. To help you, we spend a lot of time discussing *why* we prove things the way we do and what is it in the problem that suggests we try one approach over another. The hope is that by highlighting these ideas, you'll get a better sense of why the theorems are true, and be better prepared to not only use them, but be able to prove results on your own in future classes.

Here are some common questions and answers on this book and how to use it.

- **What do I need to know before I start reading?** You should be familiar and comfortable with algebra and pre-calculus. Unlike the companion book *The Calculus Lifesaver*, the courses this book is meant to supplement (or be!) are far more varied. Some probability courses don't assume any calculus, while others build on real analysis and measure theory or are half probability and half statistics. As much as possible, we've tried to minimize the need for calculus, especially in the introductory chapters. This doesn't mean these chapters are easier—far from it! It's often a lot easier to do some integrals than find the "right" way to look at a combinatorial probability problem. Calculus becomes indispensable when we reach continuous distributions, as the Fundamental Theorem of Calculus allows us to use anti-derivatives to compute areas, and we'll learn that areas often correspond to probabilities. In fact, continuous probabilities are often easier to study than discrete ones *precisely* because of calculus, as integrals are "easier" than sums. For the most part, we avoid advanced real analysis, save for some introductory comments in the beginning on how to put the subject on a very secure foundation, and some advanced chapters towards the end.

- **Why is this book so long?** An instructor has one tremendous advantage over an author: they can interact with the students, slowing down when the class is having trouble and choosing supplemental topics to fit the interest of who's enrolled in a given year. The author has only one recourse: length! This means that we'll have more explanations than you need in some areas. It also means we'll repeat explanations throughout the book, as many people won't read it in order (more on that later). Hopefully, though, we'll have a good discussion of any concept that's causing you trouble, and plenty of fun supplemental topics to explore, both in the book and online.

- **The topics are out of order from my class! What do I do?** One of my professors, Serge Lang, once remarked that it's a shame that a book has to be ordered along the page axis. There are lots of ways to teach probability, and lots of topics to choose. What you might not realize is that in choosing to do one topic your instructor is often choosing to ignore many others, as there's only so much time in the semester. Thus, while there will be lots of common material from school to school, there's a lot of flexibility in what a professor adds, in what tools they use, and in when they cover certain topics. To aid the reader, we occasionally repeat material to keep the different chapters and sections as self-contained as possible (you may notice we said something to this effect in answering the previous question!). You should be able to jump in at any point, and refer to the earlier chapters and appendixes for the background material as needed.

- **Do I really need to know the proofs?** Short answer: yes. Proofs are important. One of the reasons I went into mathematics is because I hate the phrase "because I told you so." Your professor isn't right just because they're a professor (nor am I right just because this book was published). Everything has to follow from a sound, logical chain. Knowing these justifications will help you understand the material, see connections, and hopefully make sure you never use a result inappropriately, as you'll be a master at knowing the needed conditions. By giving complete, rigorous arguments we try to cut down on the danger of subtly assuming something which may not be true. Probability generates a large number of reasonable, intuitively clear statements that end up being false; rigor is our best defense at avoiding these mistakes. Sadly, it becomes harder and harder to prove our results as the semester progresses. Often courses have some advanced applications, and due to time constraints it's impossible to prove all the background material needed. The most common example of this in a probability course is in discussions of the proof of the Central Limit Theorem, where typically some results from complex analysis are just stated. We'll always state what we need and try to give a feeling of why it's true, either through informal discussions or an analysis of special cases, and end with references to the literature.

- **Why do you sometimes use "we," and other times use "I"?** Good observation. The convention in mathematics is to always use "we," but that makes it more formal and less friendly at times; those points argue for using "I." To complicate matters, parts of this book were written with various students of mine over the years. This was deliberate for many reasons, ranging from being a great experience for them to making sure the book truly is aimed at students. To continue the confusion, it's nice to use "we" as it gets you involved; we're in this together! I hope you can deal with the confusion our choice of language is causing us!

- **Could my school use this book as a textbook?** Absolutely! To assist we've provided many exercises at the end of each chapter that are perfect for homework; instructors can e-mail me at either Steven.Miller.MC.96 @aya.yale.edu or sjm1@williams.edu for additional problems, exams, and their solutions.

- **Some of the methods you use are different from the methods I learned. Who is right—my instructor or you?** Hopefully we're both right! If in doubt, ask your instructor or shoot me an e-mail.

- **Help! There's so much material—how should I use the book?** I remember running review sessions at Princeton where one of the students was amazed that a math book has an index; if you're having trouble with specific concepts this is a great way to zero in on which parts of the book will be most helpful. That said, the goal is to read this book throughout the semester so you won't be rushed. To help you with your reading, on the book's home page is a document which summarizes the key points, terms, and ideas of each section, and has a few quick problems of varying levels of difficulty. I'm a strong believer in preparing for class and reading the material beforehand. I find it very difficult to process new math in real-time; it's a lot easier if I'm at least aware of the definitions and the main ideas before the lecture. These points led to the online summary sheet, which highlights what's going on in each section. Its goal is to help prepare you for exploring each topic, and provide you with a quick assessment of how well you learned it; it's online at http://web.williams.edu/Mathematics/sjmiller/ public_html/probabilitylifesaver/problifesaver_comments.pdf.

To assist you, important formulas and theorems are boxed—that's a strong signal that the result is important and should be learned! Some schools allow you to have one or two pages of notes for exams; even if your school doesn't it's a good idea to prepare such a summary. I've found as a student that the art of writing things down helps me learn the material better.

Math isn't about memorization, but there are some important formulas and techniques that you should have at your fingertips. The act of making the summary is often enough to solidify your understanding. Take notes as you read each chapter; keep track of what you find important, and then check that against the summaries at the end of the chapter, and the document online highlighting the key points of each section.

Try to get your hands on similar exams—maybe your school makes previous years' finals available, for example—and take these exams under proper conditions. That means no breaks, no food, no books, no phone calls, no e-mails, no messaging, and so on. Then see if you can get a solution key and grade it, or ask someone (nicely!) to grade it for you. Another great technique is to write some practice exam problems, and trade lists with a friend. I often found that after an exam or two with a professor I had some sense of what they like, and frequently guessed some of the exam questions. Try some of the exercises at the end of each chapter, or get another book out of the library and try some problems where the solutions are given. The more practice you do the better. For theorems, remove a condition and see what happens. Normally it's no longer true, so find a counterexample (sometimes it is still true, and the proof is just harder). Every time you have a condition it should make an appearance somewhere in the proof—for each result try to know where that happens.

- **Are there any videos to help?** I've taught this class many times (at Brown, Mount Holyoke, and Williams). The last few times at Williams I recorded my lectures and posted them on YouTube with links on my home page, where there's also a lot of additional information from handouts to comments on the material. Please visit http://web.williams.edu/Mathematics/sjmiller/public_html/probabilitylifesaver/ for the course home pages, which include videos of all the lectures and additional comments from each day. As I've taught the course several times, there are a few years' worth of lectures; they're similar across the years but contain slight differences based in part on what my students were interested in. One advantage is that by recording the lectures I can have several special topics where some are presented as lectures during the semester, while others are given to the students to view at home.

- **Who are you, anyway?** I'm currently a math professor at Williams College. I earned my B.S. in math and physics from Yale, moved on and got a Ph.D. in math from Princeton, and have since been affiliated (in order) with Princeton, NYU, the American Institute of Mathematics, The Ohio State University, Boston University, Brown, Williams, Smith, and Mount Holyoke. My main research interests are in number theory and probability, though I do a lot of applied math projects in a variety of fields, especially sabermetrics (the art/science of applying math and stats to baseball). My wife is a professor of marketing; you can see a lot of her influence in what topics were included and how I chose to present them to you! We have two kids, Cam and Kayla, who help TA all my classes, from Probability to the Mathematics of Lego Bricks to Rubik's Cubes.

- **What's with those symbols in the margin?** Throughout the book, the following icons appear in the margin to allow you quickly to identify the thrust of the next few lines. This is the same notation as *The Calculus Lifesaver*.

– A worked out example begins on this line.

– Here's something really important.

– You should try this yourself.

– Beware: this part of the text is mostly for interest. If time is limited, skip to the next section.

I'm extremely grateful to Princeton University Press, especially to my editor Vickie Kearn and to the staff (especially Lauren Bucca, Dimitri Karetnikov, Lorraine Doneker, Meghan Kanabay, Glenda Krupa, and Debbie Tegarden) for all their help and aid. As remarked above, this book grew out of numerous classes taught at Brown, Mount Holyoke, and Williams. It's a pleasure to thank all the students there for their constructive feedback and help, especially Shaan Amin, John Bihn, David Burt, Heidi Chen, Emma Harrington, Intekhab Hossain, Victor Luo, Kelly Oh, Gabriel Ngwe, Byron Perpetua, Will Petrie, Reid Pryzant (who wrote the first draft of the coding chapter), and David Thompson. Much of this book was written while I was supported by NSF Grants DMS0970067, DMS1265673, and DMS1561945; it is a pleasure to thank the National Science Foundation for its assistance.

Steven J. Miller
Williams College
Williamstown, MA
June 2016
sjm1@williams.edu, Steven.Miller.MC.96@aya.yale.edu

PART I
GENERAL THEORY

CHAPTER 1 _____

Introduction

I suppose it is tempting, if the only tool you have is a hammer,
to treat everything as if it were a nail.
— ABRAHAM MASLOW, *The Psychology of Science* (1966)

Probability is a vast subject. There's a wealth of applications, from the purest parts of mathematics to the sometimes seedy world of professional gambling. It's impossible for any one book to cover all of these topics. That isn't the goal of any book, neither this one nor one you're using for a class. Usually textbooks are written to introduce the general theory, some of the techniques, and describe some of the many applications and further reading. They often have a lot of advanced chapters at the end to help the instructor fashion a course related to their interests.

This book is designed to both supplement any standard introductory text and serve as a primary text by explaining the subject through numerous worked out problems as well as discussions on the general theory. We'll analyze a few fantastic problems and extract from them some general techniques, perspectives, and methods. The goal is to get you past the point where you can write down a model and solve it yourself, to where you can figure out what questions are worth asking to start the ball rolling on research of your own.

First, similar to Adrian Banner's *The Calculus Lifesaver* [Ba], the material is motivated through a rich collection of worked out exercises. It's best to read the problem and spend some time trying them first *before* reading the solutions, but complete solutions are included in the text. Unlike many books, we don't leave the proofs or examples to the reader without providing details; I urge you to try to do the problems first, but if you have trouble the details are there.

Second, it shouldn't come as a surprise to you that there are a lot more proofs in a probability class than in a calculus class. Often students find this to be a theoretically challenging course; a major goal of this book is to help you through the transition. The entire first appendix is devoted to proof techniques, and is a great way to refresh, practice, and expand your proof expertise. Also, you'll find fairly complete proofs for most of the results you would typically see in a course in this book. If you (or your class) are not that concerned with proofs, you can skip many of the arguments, but you should still scan it at the very least. While proofs are often very hard, it's not nearly as bad following a proof as it is coming up with one. Further, just reading a proof is often enough to get a good sense of what the theorem is saying, or how to use it. My goal is not to give you the shortest proof of a result; it's deliberately wordy below to have a conversation with you on how to think about problems and how to go about proving results. Further, before proving a result we'll often spend a lot of time looking at special cases to build intuition; this is an incredibly valuable skill and will help you in many classes to come. Finally, we frequently discuss how to write and execute code to check our calculations or to get a sense of the answer. If you are going to be competitive in the twenty-first-century workforce, you need to be able to program and simulate. It's enormously useful to be able to write a simple program to simulate one million examples of a problem, and frequently the results will alert you to missing factors or other errors.

In this introductory chapter we describe three entertaining problems from various parts of probability. In addition to being fun, these examples are a wonderful springboard which we can use to introduce many of the key concepts of probability. For the rest of this chapter, we'll assume you're familiar with a lot of the basic notions of probability. Don't worry; we'll define everything in great detail later. The point is to chat a bit about some fun problems and get a sense of the subject. We won't worry about defining everything precisely; your everyday experiences are more than enough background. I just want to give you a general flavor for the subject, show you some nice math, and motivate spending the next few months of your life reading and working intently in your course and also with this book. There's plenty of time in the later chapters to dot every "i" and cross each "t".

So, without further fanfare, let's dive in and look at the first problem!

1.1 Birthday Problem

One of my favorite probability exercises is the **Birthday Problem**, which is a great way for professors of large classes to supplement their income by betting with students. We'll discuss several formulations of the problem below. There's a good reason for spending so much time trying to state the problem. In the real world, you often have to figure out what the problem is; you want to be the person guiding the work, not just a technician doing the algebra. By discussing (at great lengths!) the subtleties, you'll see how easy it is to accidentally assume something. Further, it's possible for different people to arrive at different answers without making a mistake, simply because they interpreted a question differently. It's thus very important to always be clear about what you are doing, and why. We'll thus spend a lot of time stating and refining the question, and then we'll solve the problem in order to highlight many of the key concepts in probability. Our first solution is correct, but it's computationally painful. We'll thus conclude with a short description of how we can very easily approximate the answer *if* we know a little calculus.

1.1.1 Stating the Problem

Birthday Problem (first formulation): How many people do we need to have in a room before there's at least a 50% chance that two share a birthday?

This seems like a perfectly fine problem. You should be picturing in your mind lots of different events with different numbers of people, ranging from say the end of the

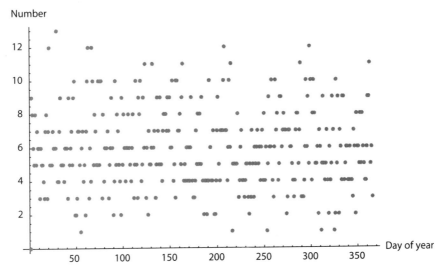

Figure 1.1. Distribution of birthdays of undergraduates at Williams College in Fall 2013.

year banquet for the chess team to a high school prom to a political fundraising dinner to a Thanksgiving celebration. For each event, we see how many people there are and see if there are two people who share a birthday. If we gather enough data, we should get a sense of how many people are needed.

While this may seem fine, it turns out there's a lot of hidden assumptions above. One of the goals of this book is to emphasize the importance of stating problems clearly and fully. This is very different from a calculus or linear algebra class. In those courses it's pretty straightforward: find this derivative, integrate that function, solve this system of equations. As worded above, this question isn't specific enough. I'm married to an identical twin. Thus, at gatherings for her side of the family, there are always two people with the same birthday!* To correct for this trivial solution, we want to talk about a *generic* group of people. We need some information about how the birthdays of our people are distributed among the days of the year. More specifically, we'll assume that birthdays are **independent**, which means that knowledge of one person's birthday gives no information about another person's birthday. Independence is one of the most central concepts in probability, and as a result, we'll explore it in great detail in Chapter 4.

This leads us to our second formulation.

Birthday Problem (second formulation): Assume each day of the year is as likely to be someone's birthday as any other day. How many people do we need to have in a room before there's at least a 50% chance that two share a birthday?

Although this formulation is better, the problem is *still* too vague for us to study. In order to attack the problem we still need more information on the distribution of

---

*This isn't the only familial issue. Often siblings are almost exactly *n* years apart, for reasons ranging from life situation to fertile periods. My children (Cam and Kayla) were both born in March, two years apart. Their oldest first cousins (Eli and Matthew) are both September, also two years apart. Think about the people in your family. Do you expect the days of birthdays to be uncorrelated in your family?

birthdays throughout the year. You should be a bit puzzled right now, for haven't we completely specified how birthdays are distributed? We've just said each day is equally likely to be someone's birthday. So, assuming no one is ever born on February 29, that means roughly 1 out of 365 people are born on January 1, another 1 out 365 on January 2, and so on. What more information could be needed?

It's subtle, but we are *still* assuming something. What's the error? We're assuming that we have a random group of people at our event! Maybe the nature of the event causes some days to be more likely for birthdays than others. This seems absurd. After all, surely being born on certain days of the year has nothing to do with being good enough to be on the chess team or football team. Right?

Wrong! Consider the example raised by Malcolm Gladwell in his popular book, *Outliers* [Gl]. In the first chapter, the author investigates the claim that date of birth is strongly linked to success in some sports. In Canadian youth hockey leagues, for instance, "the eligibility cutoff for age-class hockey programs is January 1st." From a young age, the best players are given special attention. But think about it: at the ages of six, seven, and eight, the best players (for the most part) are also the oldest. So, the players who just make the cutoff—those born in January and February—can compete against younger players in the same age division, distinguish themselves, and then enter into a self-fulfilling cycle of advantages. They get better training, stronger competition, even more state-of-the-art equipment. Consequently, these older players get better at a faster rate, leading to more and more success down the road.

On page 23, Gladwell substitutes the birthdays for the players' names: "It no longer sounds like the championship of Canadian junior hockey. It now sounds like a strange sporting ritual for teenage boys born under the astrological signs Capricorn, Aquarius, and Pisces. *March 11 starts around one side of the Tigers' net, leaving the puck for his teammate January 4, who passes it to January 22, who flips it back to March 12, who shoots point-blank at the Tigers' goalie, April 27. April 27 blocks the shot, but it's rebounded by Vancouver's March 6. He shoots! Medicine Hat defensemen February 9 and February 14 dive to block the puck while January 10 looks on helplessly. March 6 scores!*" So, if we attend a party for professional hockey players from Canada, we shouldn't assume that everyone is equally likely to be born on any day of the year.

To simplify our analysis, let's assume that everyone actually *is* equally likely to be born on any day of the year, even though we understand that this might not always be a valid assumption; there's a nice article by Hurley [Hu] that studies what happens when all birthdays are not equally likely. We'll also assume that there are only 365 days in the year. (Unfortunately, if you were born on February 29, you won't be invited to the party.) In other words, we're assuming that the distribution of birthdays follows a **uniform distribution**. We'll discuss uniform distributions in particular and distributions more generally in Chapter 13. Thus, we reach our final version of the problem.

Birthday Problem (third formulation): Assuming that the birthdays *of our guests* are independent and equally likely to fall on any day of the year (except February 29), how many people do we need to have in the room before there's at least a 50% chance that two share a birthday?

1.1.2 Solving the Problem

We now have a well-defined problem; how should we approach it? Frequently, it's useful to look at extreme cases and try to get a sense of what the solution should be. The worst-case scenario for us is when everyone has a different birthday. Since we're assuming there are only 365 days in the year, we *must* have at least two people sharing a birthday once there are 366 people at the party (remember we're assuming no one was born on February 29). This is **Dirichlet's famous Pigeon-Hole Principle**, which we describe in Appendix A.11. On the other end of the spectrum, it's clear that if only one person attends the party, there can't be a shared birthday. Therefore, the answer lies somewhere between 2 and 365. But where? Thinking more deeply about the problem, we see that there should be *at least* a 50% chance when there are 184 people. The intuition is that if no one in the first 183 people shares a birthday with anyone else, then there's at least a 50% chance that they will share a birthday with someone in the room when the 184th person enters the party. More than half of the days of the year are taken! It's often helpful to spend a few minutes thinking about problems like this to get a feel for the answer. In just a few short steps, we've narrowed our set of solutions considerably. We know that the answer is somewhere between 2 and 184. This is still a pretty sizable range, but we think the answer should be a lot closer to 2 than to 184 (just imagine what happens when we have 170 people).

Let's compute the answer by brute force. This gives us our first recipe for finding probabilities. Let's say there are n people at our party, and each is as likely to have one day as their birthday as another. We can look at all possible lists of birthday assignments for n people and see how often at least two share a birthday. Unfortunately, this is a computational nightmare for large n. Let's try some small cases and build a feel for the problem.

With just two people, there are $365^2 = 133,225$ ways to assign two birthdays across the group of people. Why? There's 365 choices for the first person's birthday and 365 choices for the second person's birthday. Since the two events are independent (one of our previous assumptions), the number of possible combinations is just the product. The pairs range from (January 1, January 1), (January 1, January 2), and so on until we reach (December 31, December 31).

Of these 133,225 pairs, only 365 have two people sharing a birthday. To see this, note that once we've chosen the first person's birthday, there's only one possible choice for the second person's birthday if there's to be a match. Thus, with two people, the probability that there's a shared birthday is $365/365^2$ or about .27%. We computed this probability by looking at the number of successes (two people in our group of two sharing a birthday) divided by the number of possibilities (the number of possible pairs of birthdays).

If there are three people, there are $365^3 = 48,627,125$ ways to assign the birthdays. There are $365 \cdot 1 \cdot 364 = 132,860$ ways that the first two people share a birthday and the third has a different birthday (the first can have any birthday, the second must have the same birthday as the first, and then the final person must have a different birthday). Similarly, there are 132,860 ways that just the first and third share a birthday, and another 132,860 ways for only the second and third to share a birthday. We must be very careful, however, and ensure that we consider *all* the cases. A final possibility is that all three people could share a birthday. There are 365 ways that that could happen. Thus, the probability that at least two of three share a birthday is 398,945 / 48,627,125, or about .82%. Here 398,945 is 132,860 + 132,860 + 132,860 + 365, the number of triples with at least two people sharing a birthday. One last note about the $n = 3$ case. It's always a good idea to check and see if an answer is reasonable. Do we expect there to be a greater chance of at least two people in a group of two sharing a birthday, or a group of two in a group of three? Clearly, the more people we have, the greater the chance of a shared birthday. Thus, our probability must be rising as we add more people, and we confirm that .82% is larger than .27%.

It's worth mentioning that we had to be *very careful* in our arguments above, as we didn't want to **double count** a triple. Double counting is one of the cardinal sins in probability, one which most of us have done a few times. For example, if all three people share a birthday this should only count as *one* success, not as three. Why might we mistakenly count it three times? Well, if the triple were (March 5, March 5, March 5) we could view it as the first two share a birthday, or the last two, or the first and last. We'll discuss double counting a lot when we do combinatorics and probability in Chapter 3.

For now, we'll leave it at the following (hopefully obvious) bit of advice: don't discriminate! Count each event once and only once! Of course, sometimes it's not clear what's being counted. One of my favorite scenes in *Superman II* is when Lex Luthor is at the White House, trying to ingratiate himself with the evil Kryptonians: General Zod, Ursa, and the slow-witted Non. He's trying to convince them that they can attack and destroy Superman. The dialogue below was taken from http://scifiscripts.com/scripts/superman_II_shoot.txt.

> General Zod: He has powers as we do.
> Lex Luthor: Certainly. But - Er. Oh Magnificent one, he's just one, but you are three (Non grunts disapprovingly), or four even, if you count him twice.

Here Non thought he wasn't being counted, that the "three" referred to General Zod, Ursa, and Lex Luthor. Be careful! Know what you're counting, and count carefully!

Okay. We shouldn't be surprised that the probability of a shared birthday increases as we increase the number of people, and we have to be careful in how we count. At this point, we could continue to attack this problem by brute force, computing how many ways at least two of four (and so on...) share a birthday. If you try doing four, you'll see we need a better way. Why? Here are the various possibilities we'd need to study. Not only could all four, exactly three of four, or exactly two of four share a birthday, but we could even have two pairs of distinct, shared birthdays (say the four birthdays are March 5, March 25, March 25, and March 5). This last case is a nice complement to our earlier concern. Before we worried about double counting an event; now we need to worry about forgetting to count an event! So, not only must we avoid double counting, we must be **exhaustive**, covering all possible cases.

Alright, the brute force approach isn't an efficient—or pleasant!—way to proceed. We need something better. In probability, it is often easier to calculate the **probability of the complementary event**—the probability that A doesn't happen—rather than determining the probability an event A happens. If we know that A doesn't happen with probability p, then A happens with probability $1 - p$. This is due to the fundamental relation that *something* must happen: A and not A are mutually exclusive events—either A happens or it doesn't. So, the sum of the probabilities must equal 1. These are intuitive notions on probabilities (probabilities are non-negative and sum to 1), which we'll deliberate when we formally define things in Chapter 2.

How does this help us? Let's calculate the probability that in a group of n people *no one* shares a birthday with anyone else. We imagine the people walking into the room one at a time. The first person can have any of the 365 days as her birthday since there's no one else in the room. Therefore, the probability that there are no shared birthdays when there's just one person in the room is 1. We'll rewrite this as 365/365; we'll see in a moment why it's good to write it like this. When the second person enters, someone is already there. In order for the second person not to share a birthday, his birthday must fall on one of the 364 remaining days. Thus, the probability that we don't have a shared birthday is just $\frac{365}{365} \cdot \frac{364}{365}$. Here, we're using the fact that probabilities of independent events are multiplicative. This means that if A happens with probability p and B happens with probability q, then if A and B are independent—which means that knowledge of A happening gives us no information about whether or not B happens, and vice versa—the probability that both A and B happen is $p \cdot q$.

Similarly, when the third person enters, if we want to have no shared birthday we find that her birthday can be any of $365 - 2 = 363$ days. Thus, the probability that she doesn't share a birthday with either of the previous two people is $\frac{363}{365}$, and hence the probability of no shared birthday among three people is just $\frac{365}{365} \cdot \frac{364}{365} \cdot \frac{363}{365}$. As a consistency check, this means the probability that there's a shared birthday among three people is $1 - \frac{365}{365} \cdot \frac{364}{365} \cdot \frac{363}{365} = \frac{365^3 - 365 \cdot 364 \cdot 363}{365^3}$, which is 398,945 / 48,627,125. This agrees with what we found before.

> Note the relative simplicity of this calculation. By calculating the **complementary probability** (i.e., the probability that our desired event doesn't happen) we have eliminated the need to worry about double counting or leaving out ways in which an event can happen.

Arguing along these lines, we find that the probability of no shared birthday among n people is just

$$\frac{365}{365} \cdot \frac{364}{365} \cdots \frac{365 - (n - 1)}{365}.$$

The tricky part in expressions like this is figuring out how far down to go. The first person has a numerator of 365, or $365 - 0$, the second has $364 = 365 - 1$. We see a pattern, and thus the nth person will have a numerator of $365 - (n - 1)$ (as we subtract one less than the person's number). We may rewrite this using the **product notation**:

$$\prod_{k=0}^{n-1} \frac{365 - k}{365}.$$

This is a generalization of the **summation notation**; just as $\sum_{k=0}^{m} a_k$ is shorthand for $a_0 + a_1 + \cdots + a_{m-1} + a_m$, we use $\prod_{k=0}^{m} a_k$ as a compact way of writing $a_0 \cdot a_1 \cdots a_{m-1} a_m$. You might remember in calculus that empty sums are defined to be zero; it turns out that the "right" convention to take is to set an empty product to be 1.

If we introduce or recall another bit of notation, we can write our expression in a very nice way. The **factorial** of a positive integer is the product of all positive integers up to it. We denote the factorial by an exclamation point, so if m is a positive integer then $m! = m \cdot (m - 1) \cdot (m - 2) \cdots 3 \cdot 2 \cdot 1$. So $3! = 3 \cdot 2 \cdot 1 = 6$, $5! = 120$, and it turns out to be *very* useful to set $0! = 1$ (which is consistent with our convention that an empty product is 1). Using factorials, we find that the probability that no one in our group of n

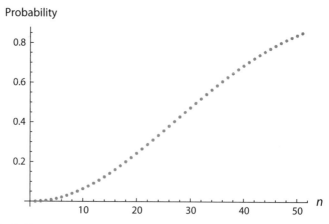

Probability

Figure 1.2. Probability that at least two of n people share a birthday (365 days in a year, all days equally likely to be a birthday, each birthday independent of the others).

shares a birthday is just

$$\prod_{k=0}^{n-1} \frac{365 - k}{365} = \frac{365 \cdot 364 \cdots (365 - (n-1))}{365^n}$$

$$= \frac{365 \cdot 364 \cdots (365 - (n-1))}{365^n} \frac{(365 - n)!}{(365 - n)!} = \frac{365!}{365^n \cdot (365 - n)!}. \quad (1.1)$$

It's worth explaining why we multiplied by $(365 - n)!/(365 - n)!$. This is a very important technique in mathematics, **multiplying by one**. Clearly, if we multiply an expression by 1 we don't change its value; the reason this is often beneficial is it gives us an opportunity to regroup the algebra and highlight different relations. We'll see throughout the book advantages from **rewriting algebra** in different ways; sometimes these highlight different aspects of the problem, sometimes they simplify the computations. In this case, multiplying by 1 allows us to rewrite the numerator very simply as 365!.

To solve our problem, we must find the smallest value of n such that the product is less than 1/2, as this is the probability that no two persons out of n people share a birthday. Consequently, if that probability is less than 1/2, it means that there'll be at least a 50% chance that two people do in fact share a birthday (remember: complementary probabilities!). Unfortunately, this isn't an easy calculation to do. We have to multiply additional terms until the product first drops below 1/2. This isn't terribly enlightening, and it doesn't generalize. For example, what would happen if we moved to Mars, where the year is almost twice as long—what would the answer be then?

We could use trial and error to evaluate the formula on the right-hand side of (1.1) for various values of n. The difficulty with this is that if we are using a calculator or Microsoft Excel, 365! or 365^n will overflow the memory (though more advanced programs such as Mathematica and Matlab can handle numbers this large and larger). So, we seem forced to evaluate the product term-by-term. We do this and plot the results in Figure 1.2.

Doing the multiplication or looking at the plot, we see the answer to our question is 23. In particular, when there are 23 people in the room, there's approximately a 50.7% chance that at least two people share a birthday. The probability rises to about 70.6% when there are 30 people, about 89.1% when there are 40 people, and a whopping 97% when there are 50 people. Often, in large lecture courses, the professor will bet someone in the class $5 that at least two people share a birthday. The analysis above shows that the professor is very safe when there are 40 or more people (at least safe from losing the bet; their college or university may frown on betting with students).

As one of our objectives in this book is to highlight coding, we give a simple Mathematica program that generated Figure 1.2.

```
(* Mathematica code to compute birthday probabilities *)
(* initialize list of probabilities of sharing and not *)
(* as using recursion need to store previous value *)
noshare = {{1, 1}}; (* at start 100% chance don't share a bday *)
share = {{1, 0}};   (* at start 0% chance share a bday *)
currentnoshare = 1; (* current probability don't share *)
For[n = 2, n <= 50, n++, (* will calculate first 50 *)
  {
    newfactor = (365 - (n-1))/365; (*next term in product*)
    (* update probability don't share *)
    currentnoshare = currentnoshare * newfactor;
    noshare = AppendTo[noshare, {n, 1.0 currentnoshare}];
    (* update probability share *)
    share = AppendTo[share, {n, 1.0 - currentnoshare}];
  }];
(* print probability share *)
Print[ListPlot[share, AxesLabel -> {"n", "Probability"}]]
```

1.1.3 Generalizing the Problem and Solution: Efficiencies

Though we've solved the original Birthday Problem, our answer is somewhat unsatisfying from a computational point of view. If we change the number of days in the year, we have to redo the calculation. So while we know the answer on Earth, we don't immediately know what the answer would be on Mars, where there are about 687 days in a year. Interestingly, the answer is just 31 people!

While it's unlikely that we'll ever find ourselves at a party at Marsport with native Martians, this generalization is very important. We can interpret it as asking the following: given that there are D events which are equally likely to occur, how long do we have to wait before we have a 50% chance of seeing some event twice? Here are two possible applications. Imagine we have cereal boxes and each is equally likely to contain one of n different toys. How many toys do we expect to get before we have our first repeat? For another, imagine something is knocking out connections (maybe it's acid rain eating away at a building, or lightening frying cables), and it takes two hits to completely destroy something. If at each moment all places are equally likely to be struck, this problem becomes finding out how long we have until a systems failure.

This is a common theme in modern mathematics: it's not enough to have an algorithm to compute a quantity. We want more. We want the algorithm to be *efficient* and easy to use, and preferably, we want a nice closed form answer so that we can see how the solution varies as we change the parameters. Our solution above fails miserably in this regard.

The rest of this section assumes some familiarity and comfort with calculus; we need some basic facts about the Taylor series of $\log x$, and we need the formula for the sum of the first m positive integers (which we can and do quickly derive). *Remember that* $\log x$ *means the logarithm of* x *base* e; *mathematicians don't use* $\ln x$ *as the derivatives of* $\log x$ *and* e^x *are "nice," while the derivatives of* $\log_b x$ *and* b^x *are "messy" (forcing us to remember where to put the natural logarithm of* b). If you haven't seen calculus, just skim the arguments below to get a flavor of how that subject can be useful. If you haven't seen Taylor series, we can get a similar approximation for $\log x$ by using the tangent line approximation.

We're going to show how some simple algebra yields the following remarkable formula: If everyone at our party is equally likely to be born on any of D days, then we need about $\sqrt{D \cdot 2 \log 2}$ people to have a 50% probability that two people share a birthday.

Here are the needed calculus facts.

- The Taylor series expansion of $\log(1 - x)$ is $-\sum_{\ell=1}^{\infty} x^\ell/\ell$ when $|x| < 1$. For x small, $\log(1 - x) \approx -x$ plus a very small error since x^2 is much smaller than x. Alternatively, the tangent line to the curve $y = f(x)$ at $x = a$ is $y - f(a) = f'(a)(x - a)$; this is because we want a line going through the point $(a, f(a))$ with slope $f'(a)$ (remember the interpretation of the derivative of f at a is the slope of the tangent to the curve at $x = a$). Thus, if x is close to a, then $f(a) + f'(a)(x - a)$ should be a good approximation to $f(x)$. For us, $f(x) = \log(1 - x)$ and $a = 0$. Thus $f(0) = \log 1 = 0$, $f'(x) = \frac{-1}{1-x}$ which implies $f'(0) = -1$, and therefore the tangent line is $y = 0 - 1 \cdot x$, or, in other words, $\log(1 - x)$ is approximately $-x$ when x is small. *We'll encounter this expansion later in the book when we turn to the proof of the Central Limit Theorem.*

- $\sum_{\ell=0}^{m} \ell = m(m + 1)/2$. This formula is typically proved by induction (see Appendix A.2.1), but it's possible to give a neat, direct proof. Write the original sequence, and then underneath it write the original sequence in reverse order. Now add column by column; the first column is $0 + m$, the next $1 + (m - 1)$, and so on until the last, which is $m + 0$. Note each pair sums to m and we have $m + 1$ terms. Thus the sum of *twice* our sequence is $m(m + 1)$, so our sum is $m(m + 1)/2$.

We use these facts to analyze the product on the left-hand side of (1.1). Though we do the computation for 365 days in the year, it's easy to generalize these calculations to an arbitrarily long year—or arbitrarily many events.

We first rewrite $\frac{365-k}{365}$ as $1 - \frac{k}{365}$ and find that p_n—the probability that no two people share a birthday—is

$$p_n = \prod_{k=0}^{n-1} \left(1 - \frac{k}{365}\right),$$

where n is the number of people in our group. A very common technique is to take the logarithm of a product. From now on, whenever you see a product you should have a **Pavlovian response and take its logarithm**. If you've taken calculus, you've seen sums. We have a big theory that converts many sums to integrals and vice versa. You may remember terms such as Riemann sum and Riemann integral. Note that we do not have similar terms for products. We're just trained to look at sums; they should be comfortable and familiar. We don't have as much experience with products, but as we'll see in a moment, the logarithm can be used to convert from products to sums to move us into a familiar landscape. If you've never seen why logarithms are useful, that's about to change. You weren't just taught the log laws because they're on standardized tests; they're actually a great way to attack many problems.

Again, the reason why taking logarithms is so powerful is that we have a considerable amount of experience with sums, but very little experience with products. Since $\log(xy) = \log x + \log y$, we see that taking a logarithm converts our product to a sum:

$$\log p_n = \sum_{k=0}^{n-1} \log \left(1 - \frac{k}{365} \right).$$

We now Taylor expand the logarithm, setting $u = k/365$. Because we expect n to be much smaller than 365, we drop all error terms and find

$$\log p_n \approx \sum_{k=0}^{n-1} -\frac{k}{365}.$$

Using our second fact, we can evaluate this sum and find

$$\log p_n \approx -\frac{(n-1)n}{365 \cdot 2}.$$

As we're looking for the probability to be 50%, we set p_n equal to 1/2 and find

$$\log(1/2) \approx -\frac{(n-1)n}{365 \cdot 2},$$

or

$$(n-1)n \approx 365 \cdot 2 \log 2$$

(as $\log(1/2) = -\log 2$). As $(n-1)n \approx n^2$, we find

$$n \approx \sqrt{365 \cdot 2 \log 2}.$$

This leads to $n \approx 22.49$. And since n has to be an integer, this formula predicts that n should be about 22 or 23, which is exactly what we saw from our exact calculation above. Arguing along the lines above, we would find that if there are D days in the year then the answer would be $\sqrt{D \cdot 2 \log 2}$.

Instead of using $n(n-1) \approx n^2$, in the Birthday Problem we could use the better estimate $n(n-1) \approx (n-1/2)^2$ and show that this leads to the prediction that we need $\frac{1}{2} + \sqrt{365 \cdot 2 \log 2}$. This turns out to be 22.9944, which is stupendously close to 23. It's amazing: with a few simple approximations, we can get pretty close to 23; with just a little more work, we're only .0056 away! We completely avoid having to do big products.

In Figure 1.3, we compare our prediction to the actual answer for years varying in length from 10 days to a million days and note the spectacular agreement—visually, we can't see a difference! It shouldn't be surprising that our predicted answer is so close for long years—the longer the year, the greater n is and hence the smaller the Taylor expansion error.

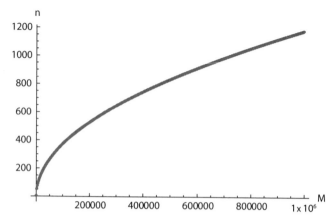

Figure 1.3. The first n so that there is a 50% probability that at least two people share a birthday when there are D days in a year (all days equally likely, all birthdays independent of each other). We plot the actual answer (black dots) versus the prediction $\sqrt{D \cdot 2\log 2}$ (red line). Note the phenomenal fit: we can't see the difference between our approximation and the true answer.

1.1.4 Numerical Test

After doing a theoretical calculation, it's good to run numerical simulations to check and see if your answer is reasonable. Below is some Mathematica code to calculate the probability that there is at least one shared birthday among n people.

```
birthdaycdf[num_, days_] := Module[{},
  (* num is the number of times we do it *)
  (* days is the number of days in the year *)
  For[d = 1, d <= days, d++, numpeople[d] = 0];
  (* initializes to having d people be where the share happens
  to zero *)

  For[n = 1, n <= num, n++,
  { (* begin n loop *)
   share = 0;
   bdaylist = {}; (* will store bdays of people in room here *)
   k = 0; (* initialize to zero people *)
   While[share == 0,
    {
     (* randomly choose a new birthday *)
     x = RandomInteger[{1, days}];
     (* see if new birthday in the set observed *)
     (* if no add, if yes won and done *)
     If[MemberQ[bdaylist, x] == False,
     bdaylist = AppendTo[bdaylist, x],
     share = 1];
     k = k + 1; (* increase number people by 1 *)
     (* if just shared a birthday add one from that person
     onward *)
     If[share == 1, For[d = k , d <= days, d++,
      numpeople[d] = numpeople[d] + 1];
     ]; (* records when had match *)
```

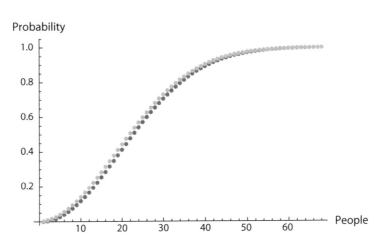

Figure 1.4. Comparison between experiment and theory: 100,000 trials with 365 days in a year.

```
    (* as doing cdf do from that point onward *)
    }]; (* end while loop *)
  }]; (* end n loop *);

bdaylistplot = {};
max = 3 * (.5 + Sqrt[days Log[4]]);
For[d = 1, d <= max, d++,
bdaylistplot =
    AppendTo[bdaylistplot, {d, numpeople[d] 1.0/num}]
    ]; (* end of d loop *)
 (* prints obs prob of shared birthday as a function of people*)
 Print[ListPlot[bdaylistplot, AxesLabel -> {People, Prob}]];
 Print[
   "Observed probability of success with 1/2 + Sqrt[D log(4)] people
   is ", numpeople[Floor[.5 + Sqrt[days Log[4.]]]]*100.0/num, "%."];
 (* this is our theoretical prediction *)
 f[x_] := 1 - Product[1 - k/days, {k, 0, Floor[x]}];
 (* this prints our obseerved data and our predicted at
 the same time using show *)
 Print[
  Show[Plot[f[x], {x, 1, max}],
   ListPlot[bdaylistplot, AxesLabel -> {People, Prob}]
   ]];
 theorybdaylistplot = {};
 For[d = 1, d <= max, d++,
  theorybdaylistplot = AppendTo[theorybdaylistplot, {d, f[d]}]];
 Print[
  ListPlot[{bdaylistplot, theorybdaylistplot},
   AxesLabel -> {People, Prob}]];
 ];
```

The above code looks at num groups, with days in the year, with various display options. It also computes our observed success rate at $1/2 + \sqrt{D \log 4}$. We record the results of one such simulation in Figure 1.4, where we took 100,000 trials in a 365-day year. Using the estimated point of $1/2 + \sqrt{D \log 4}$ led to a success rate of 47.8%, which is pretty good considering all the approximations we did. Further, a comparison of the cumulative probabilities of success between our experiment and our prediction is quite striking, which is highly suggestive of our not having made a mistake!

Figure 1.5. Larry Bird and Magic Johnson, game two of the 1985 NBA Finals (May 30) at the Boston Garden. Photo from Steve Lipofsky, Basketballphoto.com.

1.2 From Shooting Hoops to the Geometric Series

The purpose of this section is to introduce you to some important results in mathematics in general and probability in particular. While we'll motivate the material by considering a special basketball game, the results can be applied in many fields. It's thus good to have this material on your radar screen as you continue through the book. After discussing some generalizations we'll conclude with another interesting problem. Its solution is a bit involved and there's a nice paper with the solution, so we won't go through all the details here. Instead we'll concentrate on how to *attack* problems like this, which is a very important skill. It's easy to be frustrated upon encountering a difficult problem, and frequently it's unclear how to begin. We'll discuss some general problem solving techniques, which if you master you can then fruitfully apply to great effect again and again.

1.2.1 The Problem and Its Solution

The Great Shootout: Imagine that Larry Bird and Magic Johnson decide that instead of a rough game (see Figure 1.5), they'll just have a one-on-one shoot-out, winner takes all. (When I was growing up, these were two of the biggest superstars. If it would be easier to visualize, you may replace Larry Bird with Paul Pierce and Magic Johnson with Kobe Bryant, and muse on where I grew up and what year this section was written.) The two superstars take turns shooting, always releasing the ball from the same place. Suppose that Bird makes a basket with probability p (and

thus misses with probability 1 − p), while Magic makes a basket with probability q (and thus misses with probability 1 − q). If Bird shoots first, what is the probability that he wins the shootout?

Is this problem clear and concise? For the most part it is, but as we saw with the Birthday Problem it's worthwhile to take some time and think carefully about the problem, and make sure we're not making any hidden assumptions. There's one point worth highlighting: this is a mathematics problem, and not a real-world problem. We assume that Bird *always* makes a basket with probability p. He never tires, the crowd never gets to him (positively or negatively), and the same is true for Magic Johnson. Of course, in real life this would be absurd; if nothing else, after a year of doing nothing but shooting we'd expect our players to be tired, and thus shoot less effectively. However, we're in a math class, not a basketball arena, so we won't worry about endowing our players with superhuman stamina, and leave the generalization to "human" players to the reader.

While we chose to phrase this as a Basketball Problem, many games follow this general pattern. A common problem in probability involves finding the distribution of waiting times for the first successful iteration of some process. For example, imagine flipping a coin with probability p of heads and probability $1 - p$ of tails. Two (or more!) people take turns, and the first one to get something wins. There are many ways we can complicate the problem. We could have more people. We could also have the probabilities vary. We'll leave these generalizations for later, and stay with our simple game of hoops, for once we learn how to do this we'll be well-prepared for these other problems.

The standard way to attack this problem is to write down a number of probabilities and then evaluate their sum by using the Geometric Series Formula:

Geometric Series Formula: Let r be a real number less than 1 in absolute value. Then

$$\sum_{n=0}^{\infty} r^n = 1 + r + r^2 + r^3 + \cdots = \frac{1}{1-r}.$$

I'll review the proof of this useful formula at the end of this section. After first solving this problem by applying the geometric series formula, we'll discuss another approach that leads to a *proof* of the geometric series formula! We're going to use a powerful technique, which we'll call the **Bring It Over Method**. I'm indebted to Alex Cameron for coining this phrase in a Differential Equations class at Williams College. This strategy is important not just in probability, but also throughout much of mathematics, as we'll see shortly in some examples. It is precisely because this method is so important and useful that we've moved it to the beginning of the book. You should see from the beginning "good" math, which means math that's not only beautiful, but powerful and useful. There's a lot happening below, but you'll be in great shape if you can take the time and digest it.

First, we'll discuss the standard approach to solving this problem. For each positive integer n, we calculate the probability that Bird wins on his n^{th} shot. To get a sense of the answer, let's do some small n first. If $n = 1$, this means Bird wins on his first shot. In other words, he makes his first shot, which happens with probability p. If $n = 2$, then Bird wins on his second shot. In order for Bird to get a second shot, he and Magic must both miss their first shots. Since Bird misses his first shot with probability $1 - p$ and Magic misses his first shot with probability $1 - q$, we know that the probability that Bird misses his first, Magic misses his first, and Bird then makes his second shot is just $(1 - p)(1 - q)p = rp$, where we've set $r = (1 - p)(1 - q)$. Similarly, we see that if $n = 3$ then Bird must miss his first two shots, Magic must miss his first two shots, and Bird must make his third shot. The probability of this happening is $(1 - p)(1 - q)(1 - p)(1 - q)p = r^2 p$. In general, the probability Bird wins on his n^{th} shot is $r^{n-1} p$. Note the exponent of r is $n - 1$, as to win on his n^{th} shot he must miss his first $n - 1$ shots and then make his n^{th} shot.

We've thus broken the probability of Bird winning into summing (infinitely many!) simpler probabilities. We haven't counted anything twice, and we've taken care of all the different ways for Bird to win. If Bird wins, then he must make the first basket of the shoot-out at some n. In other words, his probability of winning is

$$\text{Prob(Bird wins)} \ = \ p + rp + r^2 p + r^3 p + \cdots \ = \ \sum_{n=0}^{\infty} r^n p \ = \ p \sum_{n=0}^{\infty} r^n,$$

where as before, $r = (1 - p)(1 - q)$. *Using the geometric series formula to evaluate the probability*, we see that

$$\text{Prob(Bird wins)} \ = \ \frac{p}{1 - r},$$

with $r = (1 - p)(1 - q)$.

Now we'll derive this probability *without* knowing the geometric series formula. In fact, we can use our probabilistic reasoning to derive an alternative proof of that formula. Let's denote the probability that Bird wins by x. We'll compute x in a different way than before. If Bird makes his first basket (which happens with probability p), then he wins. By definition, this happens with probability p. If Bird misses his first basket (which will happen with probability $1 - p$), then the only way he can win is if Magic misses his first shot, which happens with probability $1 - q$. But Magic missing isn't enough to ensure that Bird wins, though if Magic doesn't miss then Bird cannot win.

We've now reached a very interesting configuration. Both Bird and Magic have missed their first shots, and Bird is about to shoot his second shot. A little reflection reveals that if x is the probability that Bird wins the game *with Bird getting the first shot*, then x is also the probability that Bird wins after he and Magic miss their first shots. The reason for this is that it doesn't matter how we reach a point in this shootout. As long as Bird is shooting, his probability of winning in our model is the same regardless of how many times he and Magic have missed. This is an example of a **memoryless process**. The only thing that matters is what state we're in, not how we got there.

Amazingly, we can now find x, the probability that Bird wins! Recalling $r = (1 - p)(1 - q)$, we see that this probability is $p + (1 - p)(1 - q)x$, or

$$x \ = \ p + (1 - p)(1 - q)x$$

$$x - rx \ = \ p$$

$$x \ = \ \frac{p}{1 - r}.$$

We've now computed the probability Bird wins two different ways, the first using the geometric series and the second noting that we have a memoryless process. Our two expressions must be equal, so if we set these answers equal to one another we see that we've also proved the geometric series formula:

$$\text{Since } p \sum_{n=0}^{\infty} r^n = \frac{p}{1-r} \quad \text{we have} \quad \sum_{n=0}^{\infty} r^n = \frac{1}{1-r}$$

provided $p \neq 0$! In mathematical arguments you must *always* be careful about dividing by zero; for example, if $p = 0$ then $4p = 9p$, but this doesn't mean $4 = 9$. Of course, if $p = 0$ then Bird has no chance of winning and we shouldn't even be considering this calculation. By choosing appropriate values of p, q (see Exercise 1.5.30) we can prove the geometric series for all r with $0 \leq r < 1$.

This turns out to be one of the most important methods in probability, and in fact is one of the reasons this problem made it into the introduction. Frequently we'll have a very difficult calculation, but if we're clever we'll see it equals something that's easier to find. It's of course very hard to "see" the simpler approach, but it does get easier the more problems you do. We call this the **Proof by Comparison** or **Proof by Story** method, and give some more examples and explanation in Appendix A.6.

Our second approach to finding the probability of Bird winning worked because we have something of the form

$$\text{unknown} = \text{good} + c \cdot \text{unknown},$$

where we just need $c \neq 1$. We must avoid $c = 1$; otherwise, we'd have the unknown on both sides of the equation occurring equally, meaning we wouldn't be able to isolate it. If, however, $c \neq 1$, then we find unknown $= \text{good}/(1 - c)$.

Example 1.2.1 (Bring It Over for Integrals): *The **Bring It Over Method** might be familiar from calculus, where it's used to evaluate certain integrals. The basic idea is to manipulate the equation to get the unknown integral on both sides and then solve for it from there. For example, consider*

$$I = \int_0^{\pi} e^{cx} \cos x \, dx.$$

We integrate by parts twice. Let $u = e^{cx}$ and $dv = \cos x \, dx$, so $du = ce^{cx} dx$ and $v = \sin x \, dx$. Since $\int_0^{\pi} u \, dv = uv \big|_0^{\pi} - \int_0^{\pi} v \, du$, we have

$$I = e^{cx} \sin x \Big|_0^{\pi} - \int_0^{\pi} ce^{cx} \sin x \, dx = -c \int_0^{\pi} e^{cx} \sin x \, dx.$$

We integrate by parts a second time. Then, we again take $u = e^{cx}$ and set $dv = \sin x$, so $du = ce^{cx} dx$ and $v = -\cos x$. Thus,

$$I = -c \int_0^{\pi} e^{cx} \sin x \, dx$$
$$= -c \left[e^{cx} (-\cos x) \Big|_0^{\pi} - \int_0^{\pi} ce^{cx} (-\cos x) \, dx \right]$$

$$= -c \left[e^{\pi c} + 1 + c \int_0^\pi e^{cx} \cos x \, dx \right]$$

$$= -ce^{\pi c} - c - c^2 \int_0^\pi e^{cx} \cos x \, dx = -ce^{\pi x} - c - c^2 I,$$

because the last integral is just what we're calling I. *Rearranging yields*

$$I + c^2 I = -ce^{\pi c} - c, \tag{1.2}$$

or

$$I = \int_0^\pi e^{cx} \cos x \, dx = -\frac{ce^{\pi c} + c}{c^2 + 1}.$$

This is a truly powerful method—we're able to evaluate the integral not by computing it directly, but by showing it equals something known minus a multiple of itself.

Remark 1.2.2: *Whenever we have a complicated expression such as* (1.2), *it's worth checking the special cases of the parameter. This is a great way to see if we've made a mistake. Is it surprising, for example, that the final answer is negative for* $c > 0$? *Well, the cosine function is positive for* $x \leq \pi/2$ *and negative from* $\pi/2$ *to* π, *and the function* e^{cx} *is growing. Thus, the larger values of the exponential are hit with a negative term, and the resulting expression should be negative. (To be honest, I originally dropped a minus sign when writing this problem, and I noticed the error by doing this very test!) Another good check is to set* $c = 0$. *In this case we have* $\int_0^\pi \cos x \, dx$, *which is just 0. This is what we get in* (1.2) *upon setting* $c = 0$.

Remark 1.2.3 (Proof of the geometric series formula): *For completeness, let's do the standard proof of the geometric series formula. Consider* $S_n = 1 + r + r^2 + \cdots + r^n$. *Note* $rS_n = r + r^2 + r^3 + \cdots + r^{n+1}$; *thus* $S_n - rS_N = 1 - r^{n+1}$, *or*

$$S_n = \frac{1 - r^{n+1}}{1 - r}.$$

If $|r| < 1$, *we can let* $n \to \infty$, *and find that*

$$\lim_{n \to \infty} S_n = \sum_{n=0}^\infty r^n = \frac{1}{1 - r}.$$

The reason we multiplied through by r *above is that it allowed us to have almost the same terms in our two expressions, and thus when we did the subtraction almost everything canceled. With practice, it becomes easier to see what algebra to do to lead to great simplifications, but this is one of the hardest parts of the subject.*

Remark 1.2.4: *Technically, the probability proof we gave for the geometric series isn't quite as good as the standard proof. The reason is that for us,* $r = (1 - p)(1 - q)$, *which forces us to take* $r \geq 0$. *On the other hand, the standard proof allows us to take any* r *of absolute value at most 1. With some additional work, we can generalize our* *argument to handle negative r as well. Let $r = -s$ with $s \geq 0$. Then*

$$\sum_{n=0}^{\infty}(-s)^n = \sum_{n=0}^{\infty}s^{2n} - \sum_{n=0}^{\infty}s^{2n+1} = (1-s)\sum_{n=0}^{\infty}s^{2n}.$$

We now apply the geometric series formula to the sum of $s^{2n} = (s^2)^n$ and find that

$$\sum_{n=0}^{\infty}(-s)^n = (1-s)\cdot\frac{1}{1-s^2} = \frac{1-s}{(1-s)(1+s)} = \frac{1}{1+s} = \frac{1}{1-(-s)},$$

just as we claimed above. It may seem like all we've done is some clever algebra, but a lot of mathematics is learning how to **rewrite algebra** *to remove the clutter and see what's really going on. This example teaches us that we can often prove our result for a simpler case, and then with a little work get the more general case as well.*

 Remark 1.2.5: *As the math you do becomes more and more involved, you'll appreciate the power of* **good notation**. *Typically in probability we use q to denote $1 - p$, the complementary probability. In this problem, however, we use p and the next letter in the alphabet, q, for the two probabilities we care about most: the chance Bird has of making a basket, and the chance Magic has. We could use p_B for Bird's probability of getting a basket and p_M for Magic's; while the notation is now a bit more involved it has the advantage of being more descriptive: when we glance down, it's clear what item it describes. Along these lines, instead of writing x for the probability Bird wins we could write x_B. For this simple problem it wasn't worth it, but going forward this is something to consider.*

1.2.2 Related Problems

The techniques we developed for the Basketball Problem can be applied in many other cases; we give two nice examples below. The first is a great introduction to **generating functions**, which we explore in great detail in Chapter 19.

Example: Another fun example of the Bring It Over Method is the following problem: let F_n denote the nth Fibonacci number. Compute $\sum_{n=0}^{\infty}F_n/3^n$.

Recall that the **Fibonacci numbers** are defined by the recurrence relation $F_{n+2} = F_{n+1} + F_n$, with initial conditions $F_0 = 0$ and $F_1 = 1$. Once the first two terms in the sequence are specified, the rest of the terms are uniquely determined by the recurrence relation. We'll see recurrence relations again when we study betting strategies in roulette in §23.

We now apply our method to solve this problem. Let $x = \sum_{n=0}^{\infty}F_n/3^n$. In the argument below we'll re-index the summation in order to use the Fibonacci recurrence; it shouldn't be surprising that we use this relation, as it is *the* defining property of the Fibonacci numbers. We have

$$x = \sum_{n=0}^{\infty}\frac{F_n}{3^n}$$

$$= \frac{F_0}{1} + \frac{F_1}{3} + \sum_{n=2}^{\infty}\frac{F_n}{3^n}$$

$$= \frac{0}{1} + \frac{1}{3} + \sum_{m=0}^{\infty} \frac{F_{m+2}}{3^{m+2}}$$

$$= \frac{1}{3} + \sum_{m=0}^{\infty} \frac{F_{m+1} + F_m}{3^{m+2}}$$

$$= \frac{1}{3} + \sum_{m=0}^{\infty} \frac{F_{m+1}}{3^{m+1} \cdot 3} + \sum_{m=0}^{\infty} \frac{F_m}{3^m \cdot 9}$$

$$= \frac{1}{3} + \frac{1}{3} \sum_{n=1}^{\infty} \frac{F_n}{3^n} + \frac{1}{9} \sum_{n=0}^{\infty} \frac{F_n}{3^n}.$$

As $F_0 = 0$, we may extend the first sum in the last line over all n and find

$$x = \frac{1}{3} + \frac{x}{3} + \frac{x}{9},$$

which implies that $x = 3/5$.

It's annoying, but frequently in problems like the above you have to change the index of summation, moving it a bit. If you continue and take a course on differential equations, you'll do this non-stop when you reach the sections on series solutions. For another example along these lines, see the proof of the Binomial Theorem in Appendix A.2.3.

Example: We'll give one more example. Alice, Bob, and Charlie (whom you'll meet again if you take a cryptography course) are playing a game of cards. The first one to draw a diamond wins. They take turns drawing—Alice then Bob then Charlie then Alice and so on—until someone draws a diamond. After each person draws, if the card isn't a diamond it's put back in the deck and the deck is then thoroughly shuffled before the next person picks. What is the probability that each person wins?

WARNING: I hope the argument below seems plausible. I thought so at first, but it led to the wrong answer! After outlining it, we'll analyze what went wrong. As you read it below, see if you can find the mistake.

Let x denote the probability that Alice wins, y the probability that Bob wins, and z the probability that Charlie wins. Because there are 52 cards in a deck and 13 of these cards are diamonds, whomever is picking always has a $13/52 = 1/4$ chance of winning. The probability Alice wins is just

$$x = \frac{1}{4} + \frac{3}{4} \cdot \frac{3}{4} \cdot \frac{3}{4} x,$$

or $x = \frac{1}{4} + \frac{27}{64} x$, which implies that $\frac{37}{64} x = \frac{1}{4}$ or $x = \frac{16}{37}$. Why is this the answer? Either Alice wins on her first pick, which happens with probability $1/4$, or to win she, Bob, and Charlie all miss on their first pick, which happens with probability $(3/4)^3$. At this point, it's as if we just started the game. You should see the similarity to the Basketball Problem now.

Similarly, we find the probability that Bob wins is

$$y = \frac{3}{4} \cdot \frac{1}{4} + \frac{3}{4} \cdot \frac{3}{4} \cdot \frac{3}{4} \cdot \frac{3}{4} y.$$

That is to say, either Bob wins on his first pick or they all miss once, Alice misses, and then Bob gets to pick again. After cleaning up the algebra, we get $y = \frac{48}{175}$. If we argue analogously for Charlie, we find that $z = \frac{9}{37}$.

As always, it's extremely valuable to check our answer. We must have $x + y + z = 1$, since exactly one of them must win. While we could have computed z directly from our knowledge of x and y, we prefer this method because it gives us an opportunity to talk about testing answers. Whenever possible, you should try to find an answer two different ways as a check against algebra (or other more serious) errors. In our case, we have

$$x + y + z = \frac{16}{37} + \frac{48}{175} + \frac{9}{37} = \frac{6151}{6475} \neq 1.$$

So, what went wrong? These probabilities should sum to 1, but they don't; we're off by a little bit. The problem is that we didn't compute the probabilities correctly. We defined y to be the probability that Bob wins when *Alice* draws first. Thus, the equation for y isn't $y = \frac{3}{4} \cdot \frac{1}{4} + \left(\frac{3}{4}\right)^4 y$, but instead

$$y = \frac{3}{4} \cdot \frac{1}{4} + \left(\frac{3}{4}\right)^3 y.$$

Remember, y is the probability that Bob wins when *Alice* picks first. So, when we start the game over, it must be Alice picking, not Bob. More explicitly, let's look at the two terms above. The $\frac{3}{4} \cdot \frac{1}{4}$ comes from Alice picking and not getting a diamond, followed by Bob immediately picking a diamond. Since y is the probability Bob wins when Alice is picking, we need to get back to Alice picking. Thus, in the second term the factor $\left(\frac{3}{4}\right)^3$ represents Alice, then Bob, and finally Charlie picking non-diamonds. At this point, it is again Alice's turn to take a card, and thus from *here* the probability Bob wins is y.

Thus $y = \frac{3}{4} \cdot \frac{1}{4} + \left(\frac{3}{4}\right)^3 y$, as claimed. We can easily solve this for y, and find $y = \frac{12}{37}$. A similar argument gives $z = \frac{9}{37}$. Note that $x + y + z = \frac{16}{37} + \frac{12}{37} + \frac{9}{37} = 1$.

Alternatively, once we know x, we can immediately determine y by noting that $y = \frac{3}{4}x$. The intuition is simple: if we're calculating the probability that Bob wins, Alice must obviously not win on her first pick. After Alice fails on her first pick, it's Bob's turn. From this point forward, however, the probability that Bob wins is identical to the probability that Alice wins when Alice picks first, namely x. Therefore, $y = \frac{3}{4}x = \frac{12}{37}$. Similarly, we find that $z = \frac{3}{4} \cdot \frac{3}{4}x$, or $z = \frac{9}{37}$. It takes awhile to become comfortable looking at problems this way, but it is worth the effort. If you can correctly identify the memoryless components, you can frequently bypass infinite sums; it is far better to have a finite number of things on your "to-do" (or perhaps I should say "to-sum") list than infinitely many items!

We end this section with an appeal to you to learn how to write simple computer code. It is an incredibly useful, valuable skill to be able to numerically explore these problems as well as check your math. Let's revisit our incorrect logic, and let's write a simple program to see if our answer is reasonable. I often program in Mathematica because (1) it is freely available to me, (2) it has a lot of functions predefined that I like, (3) it's a fairly friendly environment with good display options, and (4) it's what I used when I was in college.

```
diamonddraw[num_] := Module[{},
  awin = 0; bwin = 0; cwin = 0; (* initialize win counts to 0 *)
  For[n = 1, n <= num, n++,
    { (* start of n loop *)
    diamond = 0;
    While[diamond == 0,
      { (* start of diamond loop, keep doing till get diamond *)
      (* randomly choose a card for each of three players,
      with replacement*)
      (* we'll order the deck so first 13 cards are the diamonds *)
      c1 = RandomInteger[{1, 52}];
      c2 = RandomInteger[{1, 52}];
      c3 = RandomInteger[{1, 52}];
      (* if one is a diamond we win and will stop *)
      If[c1 <= 13 || c2 <= 13 || c3 <= 13, diamond = 1];
      (* give credit to winner *)
      If[diamond == 1,
        If[c1 <= 13, awin = awin + 1,
          If[c2 <= 13, bwin = bwin + 1,
            If[c3 <= 13, cwin = cwin + 1]]]
        ]; (* end of if loop on diamond = 1 *)
      }]; (* end of while diamond loop *)
    }]; (* end of n loop *)
  Print["Here are the observed probabilities from ", num, " games."];
  Print["Percent Alice won (approx): ",   100.0 awin / num, "%."];
  Print["Percent Bob won (approx): ",   100.0 bwin / num, "%."];
  Print["Percent Charlie won (approx): ",   100.0 cwin / num, "%."];
  Print["Predictions (from our bad logic) were approx ", 1600.0/37,
   " ", 4800.0/175, " ", 900.0/37];
  ];
```

Playing one million games yielded:

- Percent Alice won (approx): 43.2202%.
- Percent Bob won (approx): 32.4069%.
- Percent Charlie won (approx): 24.3729%.
- Predictions (from our bad logic) were approx 43.2432%, 27.4286%, 24.3243%.

Thus while we're fairly confident about the probability for Alice, something looks fishy with our answer for Bob. One would hope with a million runs we would be close to the true answer; we'll return to figuring out how close we should be after we learn the Central Limit Theorem.

 Remark 1.2.6: *Remember how we said that $y = \frac{3}{4}x$ and $z = (\frac{3}{4})^2 x$? We can use this to solve for x. Since someone wins, the sum of the probabilities is 1:*

$$1 = x + y + z = x + \frac{3}{4}x + \frac{9}{16}x = \frac{37}{16}x,$$

*and thus x = 16/37! The reason we're able to so easily find x here is that there is a great deal of **symmetry**; all players have the same chance of winning when they pick. This would be true in the Basketball Problem only if p = q.*

1.2.3 General Problem Solving Tips

We end this section by discussing another Basketball Problem. I heard about this from a beautiful article by Yigal Gerchak and Mordechai Henig, "The basketball shootout: strategy and winning probabilities" (see [GH]). Our goal is not to go through all the mathematics to solve the problem; if you want the solution you can go to their paper. Instead, our purpose is to explain good ways to attack problems like this. The ability to analyze something new is a very valuable skill, but a hard one to master. The more problems you do, the more experience you gain and the more connections you can make. You'll start to see that a new problem has some features in common with something you've done before, which can give you a clue on how to start your analysis. Of course, the more problems you master, the better chance you have of seeing connections. Our goal below is to highlight some good strategies for investigating new problems outside your comfort zone. Here's the problem.

Problem: N people are in a basketball shootout. Each gets one shot, and they're told if they're shooting first, second, third, and so on. Whomever makes a basket from the furthest distance wins. If you are the kth person shooting, you know the outcome of the first $k - 1$ shots, and you know how many people will shoot after you. Where should you shoot from?

As with so many problems in this chapter, our first step is to make sure we understand the problem. We'll make several assumptions to simplify the problem. If after reading this section and their paper you're up for a challenge, try removing some of these assumptions and figuring out the new solutions.

- Let's assume all basketball players shoot from somewhere on the line connecting the two baskets. You might think this is an automatic assumption, since the players are shooting without any defenders pressuring them and thus all shots only depend on the distance. There is a flaw in that argument, however; the ball could bounce off the backboard, and thus perhaps the *angle* of the shot matters. If that's the case, we might need detailed information about how well people make different shots depending on both the distance and angle to the basket. Thus, let's make our lives simpler and assume everyone shoots from the same line.

- Next, we'll assume all players have the same ability. Of course this isn't true, but remember the great advice: *Walk before you run!* Always try to do simpler cases first. If we can't do the case when all players are the same, we have no chance of handling the general case.

- The description is vague as to what happens if two people make a basket from the same distance. We could say whoever made the shot first wins, in which case the other person would never shoot from the same place; however, they might shoot 10^{-10} centimeters further. To avoid such small ridiculous motions, let's just say if two people make a basket from the distance, whomever shot last gets the win. This avoids having to do a limiting argument, and really won't fundamentally change the solution.

- The probability of making a basket cannot increase as you move further away from the basket. While this should seem reasonable, it's important to realize we're making this assumption. Consider the following: extend your right hand to the sky. Try to touch your right shoulder with the thumb on your right hand. Now try to touch your right elbow with the same thumb. As the elbow is closer to the thumb when the arm is extended, it might seem reasonable to suppose it will be easier to reach, but this is clearly not the case.

- Related to the above, we'll assume the players can move so close to the basket that they can make a shot 100% of the time. This is a very useful assumption to include. Why? If the first $N - 1$ players miss, the last player automatically wins by moving really close to the basket. If this couldn't happen, then it would be possible for there to be no winners in the game.

Okay, it's now time to try to solve the problem. As our players are all identical, instead of measuring their distance to the basket in feet or meters, we can record where they shoot by the *probability* they make a shot from there. Thus if we're close to the basket our p should be close to 1, and it should be non-increasing as we move further back.

Before we can solve the problem, however, it's worthwhile to spend some time and think about notation. We need to encode the given information and our analysis in math equations. Notation is very important. We need a symbol to denote the probability of person 1 winning given that there are N people playing and that they shot at p *and all subsequent people shoot from their optimum locations!* Let's denote this by $x_{1;N}(p)$. Why is this **good notation**? We often use x to represent unknown quantities. It should be a function of how far away we shoot, and thus writing it as a function of p is reasonable. What about the subscripts? The first subscript refers to person one, while the second tells us how many people there are. As the two numbers play different roles, we separate them by a colon. It's not as clear what the notation should be for the second person as where they shoot depends on where the first person shoots. We'll return to this later.

Armed with our notation, we now turn to determining $x_{1;N}(p)$. Whenever you have a hard problem, a great way to start is to look at simpler cases and try to detect a pattern. If there's just one player it's clear what happens: they win! They just shoot from where they have a 100% chance of making it, and thus $x_{1;1}(1) = 1$. Note that we would never have them shoot from anywhere else if they're the only shooter.

What about two players? If you think about it, everything is determined by where the first person shoots. If they miss then the second player automatically wins, as we've said they can move close enough to the basket to be assured of making their shot. If however the first player makes a basket, then the second player shoots from the same spot (as we've declared that if two people make a basket from the same place, then whoever shot second wins).

Before we convert the above analysis to mathematical notation, let's try and get a feel for the solution. This is a very valuable step. If you have a rough sense of what the answer should be, you're much more likely to catch an algebra error. The first question to ask is: do we think the first player has a better than 50% chance or worse than a 50% chance of winning? Another way of putting this is: would you rather shoot first or second? For me, I'd rather shoot second. If the first person misses then I automatically win, while if they make a shot all I have to do is make the same shot they did. Thus, it seems reasonable to expect that $x_{1;2}(p) \leq 1/2$.

Let's assume the first player shoots at position p (remember this means their probability of making the shot is p). There are two possibilities.

1. Person one can make the basket (which happens with probability p), in which case the second person shoots. If this happens then the second person makes a basket with probability p, so in this case person one wins with probability $1 - p$.
2. Person one can miss the basket (which happens with probability $1 - p$), in which case the second person wins with probability 1 and the first person wins with probability 0.

Combining the two cases, we find

$$x_{1;2}(p) = p \cdot (1 - p) + (1 - p) \cdot 0 = p(1 - p).$$

We now want to find the value of p that maximizes the above expression; that will tell us where person one should shoot. If you know calculus you can take the derivative, set it equal to zero, and find that $p = 1/2$ gives the maximum. Alternatively, you can plot the function $x_{1;2}(p) = p(1 - p)$. This is a downward parabola with vertex at $p = 1/2$, and thus the maximum probability is 1/4 or 25%. Notice our answer is less than 50%, as expected.

We leave the rest of the analysis to the reader. I strongly encourage you to try the case of three shooters. For some problems the difficulty doesn't increase too much with increasing N, while for others new features emerge. Even figuring out good notation for 3 shooters is hard. For example, where the second person shoots will depend on whether or not the first person makes their shot. This observation does suggest one piece of good news: if the first person misses, the problem reduces to the two shooter case we just studied. Frequently we can make observations like this in our studies; you should always be on the lookout for simplifications, for reductions to earlier and simpler cases.

We end this section by explicitly culling out some useful observations on how to tackle new, hard problems.

General Problem Solving Strategies:

- Clearly define the problem. Be careful about hidden assumptions. Be explicit; if you need to assume something, do so but make note of the fact.
- Choose **good notation**. I've always been bothered by cosecant being the reciprocal of sine—shouldn't cosecant and cosine go together? In calculus we use F to denote the anti-derivative of f; by doing so, we make it easy to glance at the work and get a feel for what's happening.
- Do special cases first to build intuition. Walk before you run. Don't try to do the whole case at once; do some simpler cases first, and try to detect a pattern.

1.3 Gambling

No introduction to probability would be complete without at least a passing discussion of applications to gambling. This is both for historical reasons (a lot of the impetus for the development of the subject came from studying games of chance) and for current applications (consider how many billions of dollars are wagered, lost, and won in everything from football to poker to elections).

1.3.1 The 2008 Super Bowl Wager

I arrived at Williams in the summer of 2008. One of my favorite students relayed the story of a friend of his (let's call him Bob) who, in 2007, placed a $500 wager with Las Vegas that the Patriots would go undefeated in the regular season and continue on and win the Super Bowl. He received 1000 to 1 odds, so if he wins he walks away with $500,000, while if he loses he's down $500.

As a Patriots fan, that season is still a little hard to talk about (though easier after the win over the Seahawks in 2015), but I'll try. The Patriots *did* go undefeated in the regular season, becoming the first team to do so in a 16-game season. They won their two AFC play-off games, and advanced to the Super Bowl and faced the New York Giants. The Patriots beat the Giants in the last game of the regular season, but it was a close game.

In the middle of the third quarter, with the Patriots enjoying a small lead, Vegas calls Bob and offers to buy the bet back at 300 to 1 odds; this means that they'll give him $150,000 now to limit their exposure. Thus if Bob accepts, then Vegas immediately loses $150,000 but protects themselves from losing the larger $500,000; similarly it means Bob gets $150,000 but loses the opportunity to get $500,000.

Bob has faith in the Patriots and declines the offer, electing to go for the big payoff. I claim, and hope to convince you, that Bob made a *bad* choice; however, the reason Bob made a bad choice has nothing to do with the phenomenal catch by Giants wide receiver David Tyree on his helmet that kept the Giants' game-winning drive alive on their way to a huge upset win. Bob is living life on the edge: if the Patriots win he wins big, but if they lose he gets nothing. In the next subsection we'll look at a way for Bob to greatly minimize his risk; in fact, with a little bit of applied probability, Bob can ensure that he gets several hundred thousand dollars, *no matter who wins the game*!

1.3.2 Expected Returns

Right now Bob has bet $500 on the Patriots; he stands to receive $500,000 if the Patriots win but nothing if they lose. If the Patriots win with probability p, then p% of the time he makes $500,000 and $(1 - p)$% of the time he makes $0; also, no matter what, he loses the $500 he bet.

The problem for Bob is that he's in a very risky position, and depending on the outcome of the game he can have huge fluctuations in his personal fortune. He can protect himself by placing a secondary bet on the Giants. *If* he were to make protective bets at the start of the season he'd be in trouble due to how the payoffs are calculated, but Bob is in a fortunate position (which sadly he didn't realize). We're not at the start of the season—the Patriots *have* made it to the Super Bowl, and we know their opponent. He can now protect himself by betting on just the Giants. As a Patriots fan, I can understand the reluctance to do so; as a mathematician, however, it's the only sensible decision!

Imagine that for every $1 bet on the Giants you receive x if they win, and $0 if the Giants lose; as the Patriots were favored to win x must exceed 2. Why? Imagine the two teams were equal and each wins half the time. Then if $x = 2$ if we were to bet $1 then half the time we would get $2, half the time we would get $0, and thus on average we expect to get $1. Note this exactly equals the amount we wagered, so we should be indifferent to betting in this situation. As the Patriots were expected to win, however,

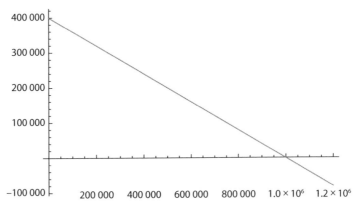

Figure 1.6. Plot of expected returns given an additional $\$B$ bet on the Giants, assuming that the Patriots have an 80% chance of winning and, if the Giants win, each dollar bet on them gives $3.

Vegas needs to give people an incentive to place money on the Giants. As the odds of the Giants winning was believed to be less than 50%, there had to be a bigger payoff if the Giants won to make the wager more fair, and hence $x > 2$.

For definiteness, let's assume that the probability the Patriots win is $p = .8$, that $x = 3$, and that we now bet $\$B$ on the Giants winning. *Let's also assume that the Super Bowl will continue until one team wins and thus there is no tie; if you don't like this we can always phrase things as the Patriots win or the Patriots don't win, and note that not winning may be different than losing.* How do our returns look? If the Pats win, which happens with probability p, we make $500,000; if the Giants win (which occurs with probability $1 - p$) we make $\$xB$; in both cases we have wagered $500 + \$B$.

Thus our expected return is

$$p \cdot \$500{,}000 + (1 - p)x \cdot \$B - \$500 - \$B;$$

we plot this in Figure 1.6.

Notice that the more we bet on the Giants, the lower our expected return is. This shouldn't be surprising, as we are assuming the Patriots win 80% of the time. In particular, if we bet a huge amount on the Giants we expect to lose a lot (the reason is that $(1 - p)x$ is less than 1).

At first, it appears that betting on the Giants is a bad idea—the more we bet on them, the lower our expected return. In the next subsection, however, we'll continue our analysis and show that this in fact *is* a good idea for most people.

1.3.3 The Value of Hedging

Figure 1.6 is misleading. Yes, the more we bet on the Giants the lower our expected return; however, this is not the right question to ask. Most people are risk averse. Which would you rather have: a guaranteed $10,000 or a .001% chance of winning a million dollars and a 99.999% chance of getting nothing? Most people would take the sure $10,000, especially when you calculated the expected return in the second situation: .001% of the time we get a million, while the rest we get nothing; thus we expect to make

$$.00001 \cdot \$1{,}000{,}000 + .99999 \cdot \$0 = \$100.$$

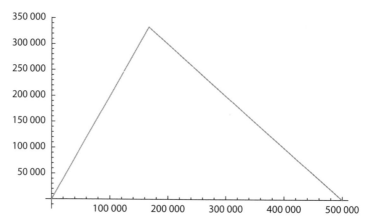

Figure 1.7. Plot of minimum guaranteed returns given an additional B bet on the Giants, assuming that the Patriots have an 80% chance of winning and, if the Giants win, each dollar bet on them gives $3.

While in the second situation when we win we win *big*, the chance is so low that the expected return is worse.

What if instead of a million dollars we now get a billion dollars in the second situation? In that case our expected return increases from $100 to $100,000. Now the situation isn't as clear. The expected value is greater in the second case, but most of the time we'll get nothing. Should we take the deal? The answer to that question is beyond the scope of this book, and falls to the realm of economics and psychology. It's worth briefly noting, though, what we are *not* being offered. We are *not* being offered the chance to play this game many times; we can only play once....

While the above problem is hard and involves personal choice, it's the wrong problem. What we'd rather do is have a situation where we can still win big, but no matter what we are still assured of getting something good. In general this is not possible; in the situation of Bob it fortunately is, and involves the beautiful concept of **hedging**. One of the hardest things to learn is to ask the right question. When we looked at the plot of the expected returns from a bet of B on the Giants, that was the wrong object to study. What we should be looking at is how much money are we guaranteed to make from a bet of B on the Giants.

Though the two questions sound similar, the answers are very different. If we have bets on both the Patriots and the Giants winning, then no matter what at least *one* of our bets must win. If the Patriots win we get $500,000; if the Giants win we get $x B$ (note that regardless of the outcome we lose our initial wager of $500 + B). Thus, no matter whether the Patriots win or the Giants win, we will get the minimum of $500,000 and $x B$. We display our minimum guaranteed winnings in Figure 1.7.

This plot is very different than Figure 1.6: our minimum return increases at first as we increase our bet on the Giants, and then decreases! Our minimum return is

$$\min(500000, x B) - 500 - B;$$

assuming $x = 3$ and $p = .8$ we find the critical bet is when $500000 = 3B$, or when B is about $166,667. At that special wager we're indifferent (from a financial point of view!) to whomever wins, and we are ensured of making approximately $332,833.

It's worth pausing and letting this sink in. By placing a large bet on the Giants ($166,667 is not small change for most of us!) we can make sure that we walk away with $332,833 *no matter who wins in the game*! At this point we're no longer gambling as there is no longer an element of chance!

1.3.4 Consequences

There is a lot more that could be said about this problem, but this is enough to highlight some key points. Most of the time in life you cannot eliminate all risk, but sometimes it is possible! Why was that an option here? The reason was that we had the chance of placing a second bet late in the season (either right before, or even during, the Super Bowl!).

Why did Bob fail to do this? Sadly, Bob never took a course in probability (an advantage you have over him and others). Psychologically, however, Bob was focused on the big payoff, on winning the huge bet. He was so focused on maximizing his return that he completely forgot about minimizing his losses, or, in other words, maximizing his minimum return. It's very easy in life to look at the wrong item (magicians are wonderful at misdirection); one of the goals of this book is to help you learn how to ask the right questions and look at the right quantities. A great example of this is the Method of Least Squares versus the Method of Absolute Values (see Chapter 24); depending on what matters most to you there are different "best" choices to what curve "best" fits the data.

In this betting case, we could use basic probability to calculate our expected return, and we saw that with a large chance of a Patriots win it made no sense, from the point of view of maximizing our expected winnings, to bet on the Giants. For most of us, however, that's the wrong problem. Most of us are risk averse, and we'd rather have a guaranteed $332,833 than a possible $500,000 (the expected value is $400,000, with 80% of the time us winning $500,000 and 20% of the time was walking away with nothing). It's very interesting who would choose which option for various probabilities; if the Patriots really will win 80% of the time then the expected value is better when we don't bet on the Giants, but it is a lot riskier. For me, it's worth a little smaller expected return to have no risk at all on a good payout.

Interestingly, when we look at the minimum return it's no longer a probability problem. If we change x then the minimum return plot in Figure 1.7 changes; however, the plot does not change if we change the probability p of the Patriots winning! Why? The reason is that we're not looking at our expected return now, we're just looking at the minimum return and thus it doesn't matter who wins as we always assume the outcome is whatever is the worst for us.

When looking at math, be it an equation or a figure, you want to get a *feel* for the behavior. Try playing around with some of the parameters and intuiting the resulting change. For example, we talked about what should happen if we change p; how do you think the shape changes if we increase x?

1.4 Summary

I hope you've enjoyed these problems. The Birthday Problem makes an appearance in almost every first course in probability (a quick Google search turns up hundreds of millions of hits), and for good reason. It's ideally suited to introduce the course. It involves so many of the most important issues, including some obvious ones, such as the notion of independence, when probabilities multiply, the dangers of double counting, and the perils of missing cases, as well as a few less obvious ones, such as the need to state a problem clearly, the advantages of introducing new functions (like the factorial function) to simplify expressions, the power of taking logarithms and using log laws, and ways to approximate the answers to difficult calculations.

The Basketball Shootout is a less clear choice. I almost gave a great problem connecting the Fibonacci numbers and gambling strategies for playing roulette in Las Vegas; don't worry—we'll hit that in Chapter §23 (or go to https://www.youtube.com/watch?v=Esa2TYwDmwA). The point is that a probability instructor has a great deal of freedom in designing a course and choosing examples. It's impossible for this book to perfectly align with any class, nor should it. What we can do is talk in great detail about how to attack a problem, emphasizing the techniques, discussing how to check your answer, and highlighting the dangers and pitfalls. These can be transferred to almost anything you'll see in your class. Further, by choosing a few less standard examples you get to see some things you wouldn't have otherwise. The Basketball Problem quickly introduces us to the concept of a memoryless game, which is crucial in much of game theory (as well as advanced topics in probability, such as Markov processes).

If you've seen calculus before, there's the added advantage of revisiting what seemed like a one-time trick, namely the "Bring It Over" Method where we got our unknown integral on both sides of an equation. A technique is a trick that can be used successfully again and again, and this is a great one. We'll say more about this in a moment.

There are many possible gambling problems to choose; I chose the one above because (1) I'm a Patriots fan (while the 2008 Super Bowl was a painful loss, this section was written shortly after Butler's great interception and the Pats 2015 Super Bowl triumph), (2) it illustrates applications of probability and issues of applying it in the real world, and (3) it provides a terrific opportunity to talk about asking the right question. In many previous classes you have been given the problem to solve, which is frequently a trivial modification of worked out examples you've seen; in the real world often the hardest part is figuring out what the problem is or what the metric for success will be. Are we concerned with maximizing our expected returns, or maximizing our minimum return and eliminating as much risk as possible?

It's now time to explore the subject in earnest. We're forced to order the chapters and topics; while our choice is defensible, be aware that it's not the only one. Your instructor and your book may choose to do things in another order, so if you're using this book to supplement your course text, just be aware that you may be hopping around a bit. To assist you, I've tried to make the chapters as self-contained as possible. This means that if you read this book cover to cover, you'll notice passages suspiciously like earlier ones. This isn't accidental; it's to make the book as easy to use as possible. If you're having trouble in your class on the Gamma distribution, you can jump in at that chapter.

The next chapter is pretty standard for all courses. We'll cover the basic concepts in probability and discuss the definitions. While most courses use calculus, not all do. This isn't a problem here; we can cover the building blocks without calculus. Where is calculus most useful? It's really needed in expanding our domain of discourse; more examples are available with calculus, as well as more methods for finding probabilities (in fact, the Fundamental Theorem of Calculus allows us to interpret probabilities as areas under curves, which can be computed through integration).

There's one significant issue, though, in the next chapter. How rigorously do we want to define everything? This is a very important question, and no answer is right for all. Typically a first course doesn't assume familiarity with real analysis, and things are a bit informal. I agree with this. I don't think it's the best idea for most students to have the advanced, rigorous formation hurled at them, being forced to digest it and to use it. That said, I think it's a mistake to be ignorant of these subtle, technical issues. A good compromise is to briefly discuss these issues so they're on your radar screen, then move them to the back of your mind until a future, more advanced, and more rigorous course. You should be aware of the dangers and how easy it is to build a theory on a shaky foundation (we'll see how this disastrously happened with Set Theory in Section 2.6).

Thus, be warned: parts of Chapter 2 are intense. We'll introduce the correct terminology to put the subject on solid ground, but then we won't do much else with it in this book. If your course isn't covering σ-algebras at all, you could safely skip that section, but you might as well spend a few minutes skimming it. Grab a nice drink or take a tasty treat to put yourself in a calm state of mind, and then give it a chance. For, after all, knowing why things are true is why you're taking a math course.

We end this introduction with one bit of advice. As you read this book, you'll notice me harping again and again on the power of this technique or the usefulness of another. The point of mathematics is not to solve one problem in isolation. You want to be able to solve similar ones too, and even better new ones as well. One of my hopes is that you'll become adept at identifying when each method can be profitably used. Here's an observation that's served me well. There are many variants of the **Principle (or Law) of the Hammer**. Abraham Maslow (1966) said: *"I suppose it is tempting, if the only tool you have is a hammer, to treat everything as if it were a nail."* Another nice phrasing, often attributed to Bernard Baruch, is: *"If all you have is a hammer, everything looks like a nail."*

There's lots of ways to look at this. One is that we each have our areas of expertise, and whenever we're given a problem we first see if we can somehow fit it in to a framework where we're comfortable and proficient. But there's another way to view this. If you're working on the same problems as everyone else with the same techniques at the same time, it's hard to distinguish yourself, it's hard to shine. Take your hammer, and move to the land of the screwdriver. You'll have a completely different perspective, and often what seems like an unsurmountable problem to them will seem easy to you. Richard Feynman was a great advocate of this method. He had a wonderful reputation of being able to solve integrals no one else could, and quickly too! Whenever he was given a pesky challenge, he knew his friend wasn't stupid and had tried all the standard methods. Feynman therefore didn't spend any time on that, trusting that if they could have solved it using their methods then they would have and the only way was to try something different. The quotation below is from his book *Surely You're Joking, Mr. Feynman.*

That book also showed how to differentiate parameters under the integral sign—it's a certain operation. It turns out that's not taught very much in the universities; they don't emphasize it. But I caught on how to use that method, and I used that one damn tool again and again. So because I was self-taught using that book, I had peculiar methods of doing integrals.

The result was, when guys at MIT or Princeton had trouble doing a certain integral, it was because they couldn't do it with the standard methods they had learned in school. If it was contour integration, they would have found it; if it was a simple series expansion, they would have found it. Then I come along and try differentiating under the integral sign, and often it worked. So I got a great reputation for doing integrals, only because my box of tools was different from everybody else's, and they had tried all their tools on it before giving the problem to me.

1.5 Exercises

Exercise 1.5.1 *Assume each person is equally likely to be born in any month. If two people's birth months are independent, what is the probability they were born in the same month? At most one month apart? Find the probability they were born at most k months apart for each $k \in \{0, 1, \ldots, 11, 12\}$. Are you surprised by how likely it is for two people to have birth months at most one apart?*

Exercise 1.5.2 *Keeping the same assumptions made in the Birthday Problem, assume the people enter the room one at a time. Which person is the most likely to be the first person to share a birthday with someone already in the room?*

Exercise 1.5.3 *We showed that 50% of the time at least two people will share a birthday when there are 23 people in the room. On average, how many people are needed before there's a pair with the same birthday?*

Exercise 1.5.4 *Prove the Taylor series of $\log(1 - u)$ is $-(u + u^2/2 + u^3/3 + \cdots)$.*

Exercise 1.5.5 *(Approximately) what is the average magnitude error between $\log u$ and its first order Taylor series approximation, $-u$, for $u \in [-1/10, 1/10]$? In other words, compute*

$$\int_{-1/10}^{1/10} |\log(1 - u) - (u)| \, du / (2/10).$$

Exercise 1.5.6 *Prove $\log_b(xy) = \log_b x + \log_b y$ (remember if $\log_b x = z$ then $x = b^z$).*

Exercise 1.5.7 *Keep track of the error terms in the Birthday Problem leading to the formulas $\sqrt{D \cdot 2 \log 2}$ and $1/2 + \sqrt{D \cdot 2 \log 2}$, and bound the error for a given D.*

Exercise 1.5.8 *How many people do you think are required before there's a 50% chance that at least three people share a birthday? Before there's a 50% chance that there are at least two pairs of people sharing a birthday? When the author taught probability at Mount Holyoke, there were no triple birthdays, but three pairs of shared birthdays among himself and the 31 students.*

Exercise 1.5.9 *Go back through the Birthday Problem and manipulate the parameters to find how many days in a year we would need to have a 75% probability of having a pair with the same birthday in a group of 23 people.*

Exercise 1.5.10 *Consider again our game where Larry Bird and Magic Johnson alternate shooting with Bird going first, and the first to make a basket wins. Assume Bird always makes a basket with probability p_B and Magic with probability p_M, where p_B and p_M are independent uniform random variables. (This means the probability each of them is in $[a, b] \subset [0, 1]$ is $b - a$, and knowledge of the value of p_B gives no information on the value of p_M.)*

1. *What is the probability Bird wins the game?*
2. *What is the probability that, when they play, Bird has as good or greater chance of winning than Magic?*

The hardest part of this exercise is interpreting just what is being asked.

Exercise 1.5.11 *Humans have 23 pairs of chromosomes; in each pair you receive one from the corresponding pair of your father and one from the corresponding pair of your mother. Assume that each parent randomly passes along a chromosome to their kids. Let's assume two parents always share no chromosomes (not a valid assumption, but it will simplify the analysis). How many chromosomes do you expect two siblings to share? Imagine now you have two married pairs; the two husbands are completely unrelated to each other and their wives, but the wives are identical twins and have identical chromosomes. Each wife has two kids; how many chromosomes do you expect are shared between kids of different wives here? What do you think is the probability that a wife's kid shares more chromosomes with their cousin than with their sibling? Write a simple program to try and compute this probability.*

Exercise 1.5.12 *Revisit the previous exercise, but now imagine the wives are triplets, each having one kid. What is the expected number of chromosomes shared by all three kids? Write a computer program to explore this exercise.*

Exercise 1.5.13 *Read up on "birthday attacks" to see an interesting application of the Birthday Problem to cryptography; the Wikipedia entry "birthday attack" and the links there are a good place to start.*

Exercise 1.5.14 *Here's a nice exercise to illustrate doing simpler cases first. There are 100 people waiting to board a plane. The first person's ticket says Seat 1; the second person in line has a ticket that says Seat 2, and so on until the 100th person, whose ticket says Seat 100. The first person ignores the fact that his ticket says Seat 1, and randomly chooses one of the hundred seats (note: he might randomly choose to sit in Seat 1). From this point on, the next 98 people will always sit in their assigned seats if possible; if their seat is taken, they will randomly choose one of the remaining seats (after the first person, the second person takes a seat; after the second person, the third person takes a seat, and so on). What is the probability the 100th person sits in Seat 100?*

Exercise 1.5.15 *Imagine Michael Jordan joins Bird and Magic so that there are now three shooters in the basketball game with probability of making a shot p_1, p_2, p_3. What is the probability of each shooter winning? What if there are n shooters with probabilities $p_1, p_2, p_3, \ldots, p_n$ of making each shot?*

Exercise 1.5.16 *Modify the basketball game so that there are 2013 players, numbered 1, 2, ..., 2013. Player i always gets a basket with probability $1/2i$. What is the probability the first player wins?*

Exercise 1.5.17 *Let's suppose Magic and Bird are playing a variation of the game in which for a player to win they must make their shot and the other player must miss their corresponding shot. For example, if Bird shot first and Bird missed his shot but Magic made his shot, Magic would win. However, if both Bird and Magic make their first shots, they would continue shooting. Denote the probability Bird makes a shot p and the probability Magic makes a shot q, and assume p and q are constants. What is the probability Bird wins? Does it matter who shoots first or second? Why? (Try to answer this last part before doing the math.)*

Exercise 1.5.18 *Is the answer for Example 1.2.1 consistent with what you would expect in the limit as c tends to minus infinity?*

Exercise 1.5.19 *Imagine we start with one bacterium. At every integer time t, all the bacteria present independently either split into two bacteria with probability p, or die* with probability $1 - p$. As a function of p, what is the probability that at some point all the bacteria have died (or, equivalently, what is the probability the bacteria persist forever)?

Exercise 1.5.20   Come up with generalizations to the previous exercise and try to solve them. For example, maybe the bacteria can split into different numbers of bacteria with different probabilities, or there are dependencies....

Exercise 1.5.21   The Sierpinski triangle is formed by starting with an equilateral triangle and then removing a triangle formed by connecting the midpoints of each leg of the original triangle. This process is then repeated with the three remaining triangles n times. Assuming that the probability of a dart landing in a given area is equal to the ratio of that area to the area of the original triangle, what is the likelihood that a dart thrown inside the original triangle hits a point that is still in the Sierpinski triangle?

Exercise 1.5.22   Alice, Bob, and Charlie are rolling a fair die in that order. They keep rolling until one of them rolls a 6. What is the probability each of them wins?

Exercise 1.5.23   Alice, Bob, and Charlie are rolling a fair die in that order. What is the probability Alice is the first person to roll a 6, Bob is the second, and Charlie is the third?

Exercise 1.5.24   Alice, Bob, and Charlie are rolling a fair die. What is the probability that the first 6 is rolled by Alice, the second 6 by Bob, and the third 6 by Charlie?

In the following three exercises, imagine that you are in a strange place where the probability a rabbit randomly drawn from the (very large) population is born in a given year is related to the year so that the probability they were born this year is $1/2$, last year $1/4$, and more generally n years ago $1/2^{n-1}$.

Exercise 1.5.25   If we have 20 rabbits, how many birth years do we expect to be shared?

Exercise 1.5.26   How many rabbits do we need for there to be at least a 50% chance that two share a birth year?

Exercise 1.5.27   What is the probability the first two rabbits to share a birth year were born this year?

Exercise 1.5.28   Write code that can simulate the Birthday Problem for a given year length. Find the average wait length for a pair of birthdays in a 365-day year over at least 10,000 trials.

Exercise 1.5.29   Write code that can simulate the Basketball Problem for given probabilities, p and q. Fix $q = .5$ and graph Bird's win percentage out of 1000 trials against the different values of p.

Exercise 1.5.30   Show that for any $r \in [0, 1)$ there is at least one choice of probabilities $p, q \in [0, 1]$ such that $p \neq 0$ and $(1 - p)(1 - q) = r$.

Exercise 1.5.31   Not surprisingly, there is a huge advantage in our basketball shootout to whomever goes first. To lessen this advantage, we now require the first person to get m baskets before the second gets n. If $p = q = 1/2$, what do you think would be a fair choice of m and n? Find the pair (m, n) with smallest sum such that the probability the first person wins is between 49% and 51%.

Exercise 1.5.32 *Try running the following Mathematica code for creating a list of Fibonacci numbers for n = 10 (or convert this to equivalent code in your preferred language). Try running it for n = 100 (you don't need to run it to completion to get the point). Explain why this code becomes infeasible to run for large n.*

```
n = 10;
Fibset = {};
F[1] := 1
F[2] := 1
F[i_] := F[i - 1] + F[i - 2]
For[i = 1, i <= n, i++,
  Fibset = AppendTo[Fibset, F[i]]];
Print[Fibset];
```

The following code works much better.

```
F[1] = 1; F[2] = 1;
Flist = {};
num = 100;
curr = F[2];
prev = F[1];
For[n = 3, n <= num, n++,
  {
   curr = curr + prev;
   prev =  curr - prev;
   Flist = AppendTo[Flist, curr];
   }];
Print[Flist]
```

Exercise 1.5.33 *Write code that can generate the first n Fibonacci numbers more efficiently, but still use the recurrence relation.*

Exercise 1.5.34 *Show that the Fibonacci numbers grow exponentially. (Hint: Find upper and lower bounds for the growth.)*

Exercise 1.5.35 *Here are the Mathematica commands to plot the various functions from the Super Bowl problem.*

```
f[p_, x_, B_] := 500000 p + (1 - p) B x - 500 - B
g[p_, x_, B_] := Min[500000, B x] - 500 - B
Plot[f[.8, 3, B], {B, 0, 1200000}]
Plot[g[.8, 3, B], {B, 0, 500000}]
Manipulate[Plot[g[p, x, B], {B, 0, 500000}],
    {p, 0, 1}, {x, 1, 10}]
```

Manipulate is a wonderful command, as it allows you to adjust several parameters and see how the figure changes. Explore what happens to our minimum return (given by a plot of g) as we vary x and p. Are you surprised by the observed relationship it has with x? Explicitly, investigate how the maximum minimum return and the location of the bet where we are indifferent (financially) on the outcome changes with x. Explain what you see.

Exercise 1.5.36 *Think of some examples in your life where you are exposed to risk, and what you can do to minimize your exposure.*

Exercise 1.5.37 *A major theme of the gambling section was the importance of asking the right question; we saw this also in discussing the Birthday Problem. Sadly, often in life problems are not clearly stated; this can be due to carelessness or due to the fact that* *it often isn't clear what should be studied! Recently I was reading an astronomy book to my kids. In one passage it said that if the sun was hollowed out then 1.3 million Earths could fit inside. Give two different interpretations for what this could mean! (Hint: You may assume the Earth and the sun are perfect spheres).*

Exercise 1.5.38   *Let's explore some more how people's risk preference affects decisions. There is an enormous difference in strategy depending on whether or not we are playing the game once or multiple times. Imagine two games: in the first you always earn $40, while in the second you get $100 half the time and you get $0 the other half. Which game would you rather play if you can only play once? What if you can play ten times? Or a hundred, or a thousand? Write a computer program to numerically explore the probability that it is better to play the second game repeatedly over the first game repeatedly.*

Exercise 1.5.39   *Fix real numbers $a, b,$ and c and consider the solutions to $ax^2 + bx + c = 0$. If we fix two of the quantities how do the location of the two roots change as we vary the third? Great questions to consider are whether or not we have real roots (and if so, how many) and how far apart the roots are (either on the real line or in the complex plane; recall the complex numbers are of the form $z = x + iy$, with $i = \sqrt{-1}$).*

Exercise 1.5.40   *Let's revisit the previous exercise. The Fundamental Theorem of Algebra asserts that any polynomial of degree n with complex coefficients has exactly n roots (we must count the roots with multiplicity; thus $x^4 + x^2 = 0$ has roots $0, 0, i,$ and $-i$). Can you find a bound for how far a root of a quadratic is from the origin as a function of its coefficients? In other words, if r is a root to $ax^2 + bx + c$ can you bound $|r|$ in terms of some function of $a, b,$ and c?*

Exercise 1.5.41   *Generalize the previous two exercises to polynomials of any fixed, finite degree.*

Exercise 1.5.42   *In the previous exercise you were asked to generalize to polynomials of fixed, finite degree. Do you think the answer is different if we consider infinite polynomials? For example, imagine we have a series $\sum_{n=0}^{\infty} a_n x^n$ which converges for all x. What can we say about the roots? Do you expect a generalization of the Fundamental Theorem of Algebra to hold? In other words, must it have roots (and if so, would you expect infinitely many)?*

Exercise 1.5.43   **(Differentiating under the integral sign)** *The following describes the method Feynman mentioned (see Chapter 11 on differentiating identities for additional examples); I found this exposition online at https://www3.nd.edu/~math/ restricted/CourseArchive/100Level/166/1662000S/Misc/DiffInt.pdf. The idea is to introduce a parameter α into an integral over x, call the resulting expression $f(\alpha)$, note $f'(\alpha)$ is the x-integral of the derivative of the integrand with respect to α, and hope that the x-integration yields an expression for $f'(\alpha)$ where we can find the anti-derivative, and then choose the value of α that reproduces the original integral. The example used is quite nice. Consider*

$$\int_0^1 \frac{x^5 - 1}{\log x} dx.$$

We replace 5 with α and define

$$f(\alpha) := \int_0^1 \frac{x^\alpha - 1}{\log x} dx.$$

Thus

$$f'(\alpha) \;=\; \frac{d}{d\alpha} \int_0^1 \frac{x^\alpha - 1}{\log x}\,dx \;=\; \int_0^1 \frac{d}{d\alpha} \frac{x^\alpha - 1}{\log x}\,dx \;=\; \int_0^1 \frac{x^\alpha \log x}{\log x}\,dx \;=\; \frac{1}{\alpha + 1}.$$

Since $f'(\alpha) = 1/(\alpha + 1)$ we must have

$$f(\alpha) \;=\; \log(\alpha + 1) + c$$

for some constant c. As $f(0) = 0$ we see $c = 0$ and thus $f(\alpha) = \log(\alpha + 1)$, and we obtain the beautiful result that our integral is just $f(5) = \log 6$.

Try some of the following from those notes. Good luck: it's sometimes non-trivial figuring out how to introduce the parameter in a way which will be useful. In the examples below we've already introduced a parameter; the notes remark that the $\cos(\alpha)$ in the third equality was a replacement of $1/2$.

$$\int_0^\pi \log(1 + \alpha \cos(x))\,dx \;=\; \pi \log \frac{1 + \sqrt{1 - \alpha^2}}{2}$$

$$\int_0^\pi \log(1 - 2u \cos(x) + \alpha^2)\,dx \;=\; \begin{cases} \pi \log \alpha^2 & \alpha^2 \geq 1 \\ 0 & \alpha^2 \leq 1 \end{cases}$$

$$\int_0^{\pi/2} \frac{\log(1 + \cos(\alpha) \cos(x))}{\cos(x)}\,dx \;=\; \frac{1}{2}\left(\frac{\pi^2}{4} - \alpha^2\right)$$

$$\int_0^1 x^\alpha (\log(x))^n\,dx \;=\; (-1)^n \frac{n!}{(\alpha + 1)^{n+1}}$$

$$\int_0^\infty \frac{dx}{(x^2 + \alpha^2)^{n+1}} \;=\; \frac{\pi}{2} \frac{1 \cdot 3 \cdot 5 \cdots (2n - 1)}{2 \cdot 4 \cdot 6 \cdots 2n \cdot \alpha^{2n+1}}.$$

CHAPTER 2 _____

Basic Probability Laws

A scrupulous writer, in every sentence that he writes, will ask himself at least four questions: What am I trying to say? 2. What words will express it? 3. What image or idiom will make it clearer? 4. Is this image fresh enough to have an effect?
—GEORGE ORWELL, *Politics and the English Language* (1946)

In the first chapter we discussed probabilities for three special problems, often emphasizing their conceptual foundations over rigorous mathematical precision. For the most part, there's no harm in this. Our overall intuition of the rules of probability is pretty solid, based on lots of experience in the real world. However, there are numerous examples in mathematics—and probability in particular—where our initial intuition failed spectacularly. As a result, mathematicians were forced to reexamine their fundamental premises. Though it's annoying to have to go back to square one, it's worthwhile in the end as this helps us to better understand the matters at hand. Much of this chapter deals with technical issues about sets and elements. This material isn't as exciting as learning how to break the bank in Vegas (see Chapter 23 for a warning on a popular method to win at roulette which we'll see is, sadly, fundamentally flawed, or view a short lecture by me on it at https://youtu.be/Esa2TYwDmwA); however, it is *very* important that we build our theory on a solid foundation. To show how easy it is to accidentally accept a false statement as true, we'll start with Russell's paradox. This is one of the most famous paradoxes in mathematics. It shows how statements that are "obviously" correct can turn out to be false, highlighting the need for careful proofs and justifying all the time you'll spend on the definitions below.

After Russell's paradox, we'll discuss some needed results from set theory and topology, and then move into the heart of this chapter—the foundations of probability. To do the subject justice and rigorously build the theory also requires some results from real analysis. While we'll mention these briefly for completeness in §2.6, building such a careful edifice is not our goal, nor is it the goal of most first classes. Instead, we want to emphasize the main ideas. For the most part, your everyday intuition and common sense is a great guide in understanding the axioms of probability and their consequences.

They're essentially a *perfect* guide if you remember just one warning: **be careful when dealing with infinities—strange things can, and do, happen!**

Even though it was decades ago, I still remember a passage from one of my high school physics books. It was giving a general overview of Einstein's Theory of Special Relativity. As we were clearly high school juniors or seniors, unprepared for the advanced mathematics of the subject, our book just described the results. One of the strangest is that if you're on a *very* fast train traveling at 75% of the speed of light and you're running inside the train at 50% of the speed of light, an observer on the ground doesn't see you moving at 125% of the speed of light; in other words, speed is not additive! How much are we off by? According to the theory, nothing travels faster than the speed of light. Denoting that by c, if the train's speed is v_{train} and you're running on the train at v_{run}, then an observer on the ground sees you travel at a speed of $(v_{\text{train}} + v_{\text{run}})/(1 + \frac{v_{\text{train}}}{c} \frac{v_{\text{run}}}{c})$. For our example, we get about 91% of the speed of light, not 125%. What about more real-world numbers? Assuming a very fast train going at 700 miles per hour and an enhanced Olympic sprinter going at 50 miles per hour, the speed isn't 750 mph but instead is approximately 749.9999999999416786522 mph!

The book then said something along the following lines: *While these results may seem counterintuitive, you have to remember that most people don't have experience traveling at three-fourths the speed of light.* I always found the passage funny, but over time I've also found it to be great advice. Be very careful not to pull too much out of an experience. For small speeds, velocity does seem additive, but that doesn't mean the pattern persists at enormously higher speeds, or that it's exactly additive. It's hard to blame centuries of physicists for not detecting the difference between 750 and 749.9999999999416786522; this was well within their measurement errors, and thus it's understandable that they thought velocities simply add.

We'll encounter similar issues in probability. However, as long as we're only dealing with finitely many sets having finitely many elements, everything is fine and our intuition is a wonderful guide; however, once infinities enter we must be careful. This shouldn't be too surprising, as we don't have any real experience with the infinite. It's dangerous to apply intuitions based on one set of experiences in another, different regime.

We end this alert with some wisdom from the great Mark Twain, who has a great quote summarizing our discussion: "We should be careful to get out of an experience only the wisdom that is in it—and stop there, lest we be like the cat that sits down on a hot stove-lid. She will never sit down on a hot stove-lid again—and that is well; but she will also never sit down on a cold one anymore."

2.1 Paradoxes

In this section we explain Russell's paradox from set theory, and discuss its implications in probability theory. You can safely skip this section if you want, as we don't need its results directly in what follows; however, if you can at least skim it, you'll get a sense of why mathematicians are so insistent about rigor and proving what seems to be a never-ending flood of "obviously true" statements. As you read on, think of how many words you use to describe either everyday items or mathematical concepts. Think about a few in particular, and try to hack away at the definition to the most basic level. Keep questioning, and see how far down you can go. For example, we all know what a continuous function is: we take our pen and draw a curve on the paper, never jumping, never creating holes, always moving left to right. If you say this and give a picture, your friend will have a good idea of what you mean, but this definition is an absolute nightmare for a mathematician. It's very imprecise, and doesn't lend itself to systematic attacks on a problem. To do things properly, we must first define what a function is, and only after that is clear may we move on to understanding its continuity. The goal below is to highlight how some words whose definition might seem "obvious" and clear to all are, in fact, not obvious at all. In fact, they are often quite hard to define precisely!

Hopefully you've seen sets in a previous course; but, just in case you haven't (or if it's been a long time), here's a quick refresher. If we have a set A, then $a \in A$ means a is an **element** of A, while $a \notin A$ means that a is not found in A. For example, 2004 and 2007 are elements of the set of all integers (we denote the set of all integers by $\mathbb{Z}$, where $\mathbb{Z}$ is from the German *zahl*, meaning number), but 2005.5 is not. Similarly, $3t^3 + t - 9$ is an element of the set of polynomials of degree 3, but $\cos(t)$ is not an element of that set as it isn't a degree 3 polynomial. Furthermore, in the definition of a set, read the colon ":" as "such that" (some authors use | instead of :). Thus $\{y : y = a + b\sqrt{5}, a, b \in \mathbb{Z}\}$ means the set of all elements y such that y can be written as $a + b\sqrt{5}$, with a and b as integers.

Enough preliminaries, it's time to meet the star witness. **Russell's Paradox** shows that relying on our intuition regarding what can be done with sets can get us into trouble. We present this paradox to demonstrate the subtleties involved in creating sets. In particular, not every collection we want to be a set actually is a set. You should have encountered many sets by now. Mathematical examples include the set of integers, the set of real numbers, or the set of primes; more entertaining examples include the set of people who have won a Super Bowl, the set of men who have landed on the moon, or the set of people who have won a Super Bowl and landed on the moon. This last set is, as of the writing of this book, the empty set (which is one of the most important sets of all)!

It's natural to assume that if P is any property, then the set of all objects with property P is a set. For example, $P(x)$ could mean x is an integer, x is a root of $1701x^2 + 1864x + 16309$, or x is a polynomial of degree at most 4 with integer coefficients. These all generate nice sets. The first is just the set of integers, $\mathbb{Z} = \{\ldots, -2, -1, 0, 1, 2, \ldots\}$. The second, by the quadratic formula, is $\{(-932 - i\sqrt{26872985})/1701, (-932 + i\sqrt{26872985})/1701\}$. The last is a bit harder to write down: it's just

$$\{p(t) = a_4 t^4 + a_3 t^3 + a_2 t^2 + a_1 t + a_0 : a_0, a_1, a_2, a_3, a_4 \in \mathbb{Z}\}.$$

For the last, we couldn't use x as the variable of the polynomial as we're using x to represent an arbitrary element of our set. This isn't a big deal—we just have to use another letter for our dummy variable; fortunately there are lots of good choices in the English alphabet (though most mathematicians would probably choose t given that x was taken).

Now let's look at a strange choice for $P(x)$, one you probably haven't seen before. Let's take $P(x)$ to mean that $x \notin x$. In other words, if $P(x)$ is true for an x then that x is not an element of itself, while if $P(x)$ is false then that x is an element of x. Most objects aren't elements of themselves. For example, $\mathbb{Z}$ is the set of all integers: $\{\ldots, -2, -1, 0, 1, 2, \ldots\}$. It's obvious that $\mathbb{Z} \notin \mathbb{Z}$, since the elements of $\mathbb{Z}$ are integers and not *sets* of integers. This property is at the heart of Russell's Paradox. If *any* collection of objects satisfying the given property were a set, we could form the set

$$\mathcal{R} = \{x : x \notin x\},$$

taking $P(x)$ to be $x \notin x$.

So is $\mathcal{R}$ a set? If so, what are its elements? If you're wondering whether $\mathcal{R} \in \mathcal{R}$, then you're on the right track. We somehow have to use the expression $x \notin x$, which means we need to make a choice for x. Taking $x = \mathcal{R}$ gives a natural candidate to investigate as it involves the collection we're trying to study (and it's nice to try to work that into the analysis!). There are two possibilities: either $\mathcal{R}$ is in $\mathcal{R}$ or $\mathcal{R}$ isn't in $\mathcal{R}$.

- First, let's assume that $\mathcal{R}$ is in $\mathcal{R}$. Because we've assumed that $\mathcal{R} \in \mathcal{R}$ and that $\mathcal{R}$ is the set of all objects that aren't elements of themselves, the definition of $\mathcal{R}$ tells us that $\mathcal{R} \notin \mathcal{R}$. But this is absurd. How can we have $\mathcal{R} \in \mathcal{R}$ and $\mathcal{R} \notin \mathcal{R}$? So, we see that our assumption that $\mathcal{R}$ is in $\mathcal{R}$ is false.

- The only possibility left is that $\mathcal{R}$ isn't in $\mathcal{R}$. Let's explore this case now. As we've already said, $\mathcal{R}$ is the set of all things that aren't elements of themselves. We're now assuming $\mathcal{R} \notin \mathcal{R}$, but by definition this is exactly what it means to be in $\mathcal{R}$! Again, we find the absurd result that $\mathcal{R} \in \mathcal{R}$ and $\mathcal{R} \notin \mathcal{R}$ *both* hold.

In other words, in either case we have the strange situation that $\mathcal{R} \in \mathcal{R}$ precisely when $\mathcal{R} \notin \mathcal{R}$. What does this mean? It means that our notion of what we can do with sets—and more specifically, of how we can form new sets from old sets—is fatally flawed. The solution of this paradox led to the foundations of modern set theory. One consequence of Russell's Paradox is that we can't form a set by simply collecting all objects with a given property. Fortunately for us, most of the sets we encounter in probability are nice, but it's important that we're aware of potential hazards, and that we get a healthy respect and appreciation for proofs.

Now that we're alerted to the dangers of loose definitions and informal arguments, let's revisit the foundations of set theory and build up the language we'll need to discuss probabilities.

2.2 Set Theory Review

Before we describe the common rules of probability, we must quickly state some facts from set theory and topology. These topics are essential for learning the language of probability, which is an important part of understanding the subject as a whole. If you aren't familiar with these terms, it will take some effort before you become fluent, but the time spent and energy expended will yield big dividends later. Imagine taking a biology class without knowing the names for the various organisms and compounds—it would be impossible! The same is true in mathematics. We can't do anything until we agree upon and learn a language.

If we want to talk about the probabilities of events, a natural framework involves taking some massive set Ω and assigning probabilities to various subsets of Ω. Surprisingly, it turns out that there's no general way to assign probabilities to every subset so that the probability function satisfies certain "natural" conditions. This is related to the Banach-Tarski paradox, which we briefly discuss in §2.6. To get around our inability to consistently assign probabilities to all possible subsets, we must be careful about which events we assign probabilities. For this reason, we require some standard facts from set theory and (point set) topology. Fortunately, these basic relations more than suffice for all of the objects encountered in a first probability course. For a more advanced class we need the notion of a σ-algebra, which we also briefly discuss in §2.6.

Let's start with some basic definitions and properties of sets. While different books use different notations, many of these properties should look familiar. Let $A, B, C, \ldots$ be sets of objects and let $a, b, c, \ldots$ be elements of these sets, respectively. We say $a \in A$ (read "a is an **element of** A," "a is **in** A," or "a **belongs to** A") if a is one of the objects in A, and $a \notin A$ if a isn't contained in A. For example, if A is the set of even numbers, then $24 \in A$ but 25 is not. If A is the set of teams that have won the World Series, then the Boston Red Sox are in A, but the Seattle Mariners are not—at least at the time of this book's writing. Unfortunately, it's often difficult to determine if an object is in a specific set. If we take A to be the set of all even numbers that are the sum of at most two primes, there's a simple way to check whether or not a given number is in A, but it's computationally expensive. We easily see 4, 100, and 1864 are in A because $4 = 2 + 2$, $100 = 47 + 53$ and $1864 = 3 + 1861$; what about $24601^{2013!1701!} + 2013!^{24601} \cdot 1701! + 3^{2012!}$? The famous Goldbach Conjecture says that *every* positive even number is in this set; however, we're a *long* way from proving this.

If every element of A is in B, we say that A is a **subset** of B, and we write $A \subset B$. We can also say that B is a **superset** of A and write $B \supset A$. It's worth noting that some books use slightly different notation. For example, for some authors $A \subset B$ means that A is not only contained in B, but also that B has some element not in A, implying that the inclusion is strict. Writing $A \subseteq B$ implies that $A \subset B$ and A equals B is allowed; this notation is much clearer but is often not used. Along similar lines, another notational convention writes $A \subsetneq B$ for A is a subset of B and A isn't equal to B. *For our purposes, when we write $A \subset B$ we mean that A is contained in B and* **may**, *in fact, equal B.* Why do we make this choice? The reason is that it isn't always clear whether equality is ruled out. Returning to our baseball example, if B is the set of all baseball teams and A is the set of all teams that have won a World Series, then $A \subset B$. This is a proper subset because there are some teams that haven't won a World Series, but unless you know some baseball history it's not clear that it's a proper subset. Or maybe the Mariners will pull it together by the second edition....

One of the most important sets is the **empty set** which is the unique set with no elements. We denote this set by $\emptyset$. The following are all examples of the empty set:

- Let A be the set of all even prime numbers exceeding 1000. Then $A = \emptyset$.
- Let A be the set of humans on *The Simpsons* with five fingers on a hand. Then $A = \emptyset$ since all characters have four fingers. For another TV example, let A be the set of all episodes of *Friends* not in the first or last season without the phrase "The One" in the title. Then $A = \emptyset$ since all episodes of *Friends* have the phrase "The One" in the title (except for the pilot and series finale).
- Let A be the set of all positive integers that cannot be written as the sum of at most four squares. Then $A = \emptyset$. It's not obvious that this is true, but it's a beautiful result from number theory. For instance, one of the many ways of writing 1729 (a very important number in mathematics) as a sum of four squares is $2^2 + 3^2 + 10^2 + 40^2$. In addition, there are 70 ways of writing 2013 as $d^2 + c^2 + b^2 + a^2$, where we've standardized the representation to have $d \leq c \leq b \leq a$; for example,

one representation is $2013 = 8^2 + 16^2 + 18^2 + 37^2$. For 2014 there are 72 such representations, 61 for 2015, and then only 8 for 2016 but 53 for 2017.

Hopefully it's clear that the empty set is unique, even though there are many different ways of referring to it. This brings up the important question of when two sets are the same. Here is a method that's often easy to use: **To prove $A = B$ just show $A \subset B$ and $B \subset A$.** If these two inclusions hold, then everything in A is in B and everything in B is in A, so they must be equal.

Given two sets A and B, we can form several new sets.

1. $A \cup B$: We read this as "A **union** B." It's the set of all objects in A *or* B, which allows for the possibility that an object is in both. We write this as

$$A \cup B = \{x : x \in A \text{ or } x \in B\}.$$

If we have several sets, we may take their union. We can write this in several ways: if it's a finite union, we write either $A_1 \cup A_2 \cup \cdots \cup A_n$ or $\cup_{i=1}^{n} A_i$; if it's an infinite union, we write $\cup_{i=1}^{\infty} A_i$. We can use a notation that works for each at the same time: $\cup_{i \in I} A_i$. Here, I might be a finite set, or it might be infinite. For an example, consider $A = \{1, 2, 3\}$ and $B = \{2, 3, 4\}$; then $A \cup B = \{1, 2, 3, 4\}$.

2. $A \cap B$: We read this as "A **intersection** B." It's the set of all objects in A *and* B. We write this as

$$A \cap B = \{x : x \in A \text{ and } x \in B\}.$$

If we have several sets, we may take their intersection. The notation is similar to what we did above: for a finite intersection, we write $A_1 \cap A_2 \cap \cdots \cap A_n$ or $\cap_{i=1}^{n} A_i$; for an infinite intersection, we write $\cap_{i=1}^{\infty} A_i$, or more generally $\cap_{i \in I} A_i$. We say two sets A and B are **disjoint** if $A \cap B = \emptyset$. We say a collection of sets $\{A_i\}_{i \in I}$ is **pairwise disjoint** if $A_j \cap A_k = \emptyset$ whenever $j, k \in I$ and $j \neq k$. Returning to $A = \{1, 2, 3\}$ and $B = \{2, 3, 4\}$, we see that $A \cap B = \{2, 3\}$.

3. A^c: We read this as the **complement** of A. It's the set of all elements that are not in A. Of all the set theory concepts, this is the hardest to understand because we need to know where A lives. Whenever there's a possibility of confusion, we try to make the notation more precise. If X is the space under discussion, so $A \subset X$, then we often write $X \setminus A$ for A^c. If we want the complement to be well-defined, we need to make clear what X is—with respect to what, in other words, are we taking the complement. For example, let's assume that we're looking at subsets of the integers $\mathbb{Z}$ and that A is the set of all even numbers. Then A^c is the set of all odd numbers. If, on the other hand, we're looking at subsets of the real numbers and A is the even numbers, then A^c is far more than just the odd numbers!

4. $A \times B$: We read this as the **Cartesian product** of A and B. It's the set of all pairs (a, b) with $a \in A$ and $b \in B$. If $A = B$, we often write A^2 for $A \times A$, or more generally A^n if we have n copies of A. The most common examples are $\mathbb{R}^n$ and $\mathbb{C}^n$, the set of all n-tuples of real numbers and all n-tuples of complex numbers. If $A = \{1, 2\}$ and $B = \{2, 3, 4\}$ then

$$A \times B = \{(1, 2), (1, 3), (1, 4), (2, 2), (2, 3), (2, 4)\}.$$

Here order matters; $\{1, 3\} \in A \times B$ but $\{3, 1\}$ is not. Giving a verbal interpretation, going one block east and three blocks north isn't the same as going three blocks east and one block north. Note the order we used for listing the elements of $A \times B$. We could've written $A \times B = \{(1, 3), (2, 3), (2, 2), (1, 2), (1, 4), (2, 4)\}$. Even though both of these listings are $A \times B$, we prefer the first. Why? It's a very nice, methodical way to enumerate the elements. The second seems willy-nilly, and going in no particular order leads to the dangerous possibility of missing an element.

5. $\mathcal{P}(A)$: We read this as the **power set** of A. It's the set of all subsets of A. If $A = \{x, y\}$, then $\mathcal{P}(A) = \{\emptyset, \{x\}, \{y\}, A\}$; note the elements of $\mathcal{P}(A)$ are themselves sets.

Let's do more examples. Let A_i be all integers whose remainder is i when we divide by 2010. So $A_{15} = \{\ldots, -1995, 15, 2025, \ldots\}$, which means $2025 \in A_{15}$, but 2024 is not. Some thought reveals that

$$A_0 \cup A_1 \cup \cdots \cup A_{2009} = \bigcup_{i=0}^{2009} A_i$$

is the set of all integers because any integer must have some remainder when we divide by 2010. It either has a remainder of 0, of 1, ..., or of 2009. Since we've exhausted all the possibilities, our union is the set of all integers.

What about the intersection

$$A_0 \cap A_1 \cap \cdots \cap A_{2009} = \bigcap_{i=0}^{2009} A_i?$$

This is the set of all integers whose remainder, when we divide by 2010, is 0 as well as 1 and 2 and so on. Since each number has a unique remainder, this intersection is empty. In fact, $A_0 \cap A_1 = \emptyset$, because there's no number with both a remainder of 0 and a remainder of 1 after it's divided by 2010. There isn't anything special about A_0 and A_1; if $j \neq k$ then $A_j \cap A_k = \emptyset$. In fact, the sets $\{A_i\}_{i=0}^{2009}$ are **pairwise disjoint**.

For the power set, if $A = \{x, y, z\}$ then $\mathcal{P}(A) = \{\emptyset, \{x\}, \{y\}, \{z\}, \{x, y\}, \{x, z\}, \{y, z\}, A\}$. We found these by first listing all subsets of A with 0 elements, then those with exactly one element, then those with exactly two elements, and then finally all of A. If A is finite, then the number of elements in $\mathcal{P}(A)$ is just $2^{\#A}$ (with $\#A$ denoting the number of elements in A). The easiest way to see this is that each element in A is either in a given subset, or it isn't. We thus have 2 choices for each element of A (take it or don't take it) when creating a subset, and hence there are $2^{\#A}$ possible subsets.

An interesting problem is to build up the integers (or, more generally, the reals) in a rigorous, set-theoretic way. Amazingly, this can be done assuming the existence of only one set, the empty set! We can form one set from the empty set $\emptyset$, namely $\{\emptyset\}$ (the set containing the empty set). From this, we can then form $\{\emptyset, \{\emptyset\}\}$, and continuing we would next get $\{\emptyset, \{\emptyset\}, \{\emptyset, \{\emptyset\}\}\}$. If we let $\emptyset$ correspond to 0, $\{\emptyset\}$ correspond to 1, $\{\emptyset, \{\emptyset\}\}$ correspond to 2 and so on, show that each "number" is a proper subset of all larger numbers (and thus set inclusion plays the role of "less than").

2.2.1 Coding Digression

As a major theme of this book is the need to be able to program well, let's take a short digression and explore one of the examples from a moment ago: writing a number as a sum of four squares. We'll give two short programs to compute the number of ways of representing an integer as a sum of four squares $d^2 + c^2 + b^2 + a^2$ with $d \leq c \leq b \leq a$.

There are several ways to code. We first give an efficient method: it takes longer to write than a simpler program as we have to think a bit about the upper bounds in some for statements, but it runs *much* faster. We then give a slower program that is faster to type. If you want to look at small numbers, say on the order of 2000, it doesn't matter which you use (and the second program is easier to write); for large numbers the difference is staggering: the efficient program takes 7.5 seconds to do 20,000, versus 48.7 seconds for the slower one; the discrepancy is even stronger for 40,000, as the first program takes 22.6 seconds while the second takes 194.2 seconds. This illustrates an important principle: if you are going to do something many times, or if the program will take awhile to run, it is well worth it to take some time to improve your code. It's amazing how rewriting a few for loops can save us so much.

First program (efficient but longer to code):

```
sumfoursquares[m_, print_] := Module[{},
  (* m is the number we are investigating *)
  (* will count how often m = a^2 + b^2 + c^2 + d^2 with *)
  (* a >= b >= c >= d *)
  (* if print \[Equal] 1 we print out all the quadruples that work *)
  count = 0; (*
  counts how often m is a sum of four appropriate squares *)
  list  = {}; (* stores the quadruples that work here *)
  (* next few lines are efficient looping *)
  (* as we want a >= b >= c >= d
     we can use that to restrict the for loops *)
  (* this saves us from computing items we don't need *)
  (* note b cannot be more than a, and also must be less than
     Sqrt(m-a^2) *)
  (* this allows us to restrict the for loops and run faster *)
  For[a = 0, a <= Sqrt[m], a++,
   For[b = 0, b <= Min[a, Sqrt[m - a^2]], b++,
    For[c = 0, c <= Min[b, Sqrt[m - a^2 - b^2]], c++,
     {
      (* we let d be what we need so that m = a^2+b^2+c^2+d^2 *)
      (* we then make sure d is an integer and at most c *)
      (* checking to be an integer is easy in Mathematica --
         use IntegerQ *)
      (* if conditions satisfied increment count, save values *)
      d = Sqrt[m - a^2 - b^2 - c^2];
      If[d >= 0 && d <= c && IntegerQ[d] == True,
       {
        count  = count + 1;
        list = AppendTo[list, {a, b, c, d}];
        }]; (* end of if loop *)
      }]; (* end of c loop *)
    ]]; (* end of b and end of a loops *)
  Print["The number of representations of ", m,
   " as a sum of four squares with a >= b >= c >= d is ", count, "."];
  If[print == 1, Print[list]];
  ]
```

Second program (slower but easier to code):

```
slowsumfoursquares[m_, print_] := Module[{},
  count = 0;
  For[a = 0, a <= Sqrt[m], a++,
   For[b = 0, b <= a, b++,
    For[c = 0, c <= b, c++,
     For[d = 0, d <= c, d++,
      If[a^2 + b^2 + c^2 + d^2 == m,
       {
        (* unlike other program fewer conditions to check *)
        (* note this doesn't use the IntegerQ function *)
        (* it has four for loops, not three *)
        count = count + 1;
        list = AppendTo[list, {a, b, c, d}];
        }];
       ]]]];
  Print["The number of representations of ", m,
   " as a sum of four squares with a >= b >= c >= d is ", count, "."];
  If[print == 1, Print[list]];
  ]
```

2.2.2 Sizes of Infinity and Probabilities

It's worth skimming this section, but for many classes there's no need to stress about learning all the definitions here. We include this material for completeness and general interest, as well as advanced courses.

One important result from set theory is that there are different magnitudes of infinity. At first, this seems strange; what do we even mean by the size of an infinite set? Shouldn't any two infinite sets be the same "size"? It turns out that the answer is a resounding "no." There is, in fact, a way to compare infinities, and counter-intuitively, some infinities turn out to be larger than others. In this subsection we'll talk a bit about orders of infinity and the implications for probability. If you want to see more, see Appendix C. Many introductory books only mention these issues in passing; our goal here is to provide enough detail to give you a sense of the subject, and help you understand why we study some sets but not others.

Before describing the different infinities, let's first establish some notation. A function $f : A \to B$ is a **one-to-one function** if distinct inputs are sent to distinct outputs. This means that the only way we could have $f(x) = f(y)$ is if $x = y$. If we let $f(x) = x^2$, then $f : [0, 2] \to [0, 4]$ is a one-to-one function, while $f : [-2, 2] \to [0, 4]$ is not (see Figure 2.1 (Top)). Note that the domain of definition plays a big role in whether or not a function is one-to-one. Some books use the word **injective** instead of one-to-one.

The other notion we need is that of an onto (or surjective) function. We say $f : A \to B$ is **onto** (or **surjective**) if given any $b \in B$ there's some $a \in A$ such that $f(a) = b$. For instance, our function $f : [0, 2] \to [0, 4]$ given by $f(x) = x^2$ is surjective (and so is $f : [-2, 2] \to [0, 4]$); to see this, just take $a = \sqrt{b}$. If instead we had $f : [0, 2] \to [-4, 4]$ given by $f(x) = x^2$, the situation is different. While we have the same rule, namely square the output, in this case our function isn't onto since no input can be sent to a negative number by squaring.

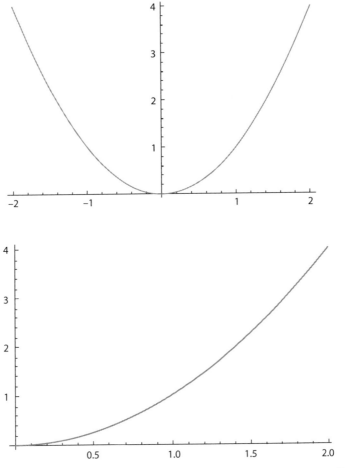

Figure 2.1. (Top) The surjective but not injective function $f : [-2, 2] \to [0, 4]$ given by $f(x) = x^2$. (Bottom) The bijection $f : [0, 2] \to [0, 4]$ given by $f(x) = x^2$.

Finally, if f is both one-to-one and onto, then we say f is a **bijection**. Our squaring function $f(x) = x^2$ with $A = [0, 2]$ and $B = [0, 4]$ is a bijection (see Figure 2.1 (Bottom)); if $B = [-4, 4]$ then it wouldn't be a bijection.

Armed with our notation, we can now discuss sizes of sets. The ordering of size goes from finite to countable to uncountable (we will not worry about the different levels of uncountable). Our purpose is not to write a Set Theory Lifesaver—at least for now!—so we'll only give a brief introduction; if you want a lengthier discussion, see Appendix C. We say a set A is finite of **size n** (or **cardinality n**) if there's a one-to-one correspondence between elements in A and the set $\{1, 2, \ldots, n\}$. If this is indeed the case, we can write $A = \{a_1, a_2, \ldots, a_n\}$. More formally, we have a function $f : \{1, 2, \ldots, n\} \to A$ such that no two distinct integers are sent to the same element in A, and, given any element $a \in A$, there's some $k \in \{1, 2, \ldots, n\}$ such that $f(k) = a$. In other words, there's a bijection between these two sets. We say a set A is **countable** if there's a one-to-one and onto function f from A to the positive integers. If A is neither finite nor countable, we say it's **uncountable**. We let $\#A$ or $|A|$ denote the **size** of A. Not only do countable sets have infinitely many elements, but their size is the smallest infinity.

Obviously, any countable set is larger than any finite set. What's surprising is that if A is a proper subset of B (so $A \subsetneq B$), it's possible for $|A| = |B|$. One illustrative example is the set of even positive integers E and the set of positive integers P. Clearly, $E \subsetneq P$, but there's a one-to-one function from E to P, namely $f(x) = x/2$. As a result, every element of E is matched with a unique element of P and vice versa. This is why it's natural to say that E and P have the same size.

In a set theory course, one proves that the positive integers, the integers $\mathbb{Z}$, the rationals $\mathbb{Q}$, and $\mathbb{Q}^n = \{(x_1, \ldots, x_n) : x_i \in \mathbb{Q}\}$—which is the set of all n-tuples of rational numbers—all have the same size. These sets are countable. On the other hand, the real numbers $\mathbb{R}$, the plane $\mathbb{R}^2$, and n-dimensional space $\mathbb{R}^n$ are all uncountable. This surprising proof uses a brilliant method from Cantor known as the diagonalization argument. For more details, please consult any comprehensive set theory text. See also Appendix C for more on countable sets.

After reading the paragraphs above, we hope you're wondering: *What does this have to do with probability?* As we'll see in §2.6, we can only talk about probabilities of countable unions of events. We cannot, that is, discuss probabilities of uncountable unions of events. Our discussion here is meant to prepare us for which sets we can use and which sets we cannot use in probability.

2.2.3 Open and Closed Sets

The last bit of terminology we need covers open and closed intervals—or more generally, sets. You might remember this notation from calculus; nevertheless, we'll just quickly go through the definitions below. Most first courses don't worry too much about the general cases, concentrating instead on intervals on the line. The higher dimension analogues are essential for studying functions of several variables, especially if we want to use results such as the Change of Variables Formula to convert a messy integration into a simpler one.

There are four types of **intervals** commonly studied:

- $[a, b] := \{x \in \mathbb{R} : a \leq x \leq b\}$
- $[a, b) := \{x \in \mathbb{R} : a \leq x < b\}$
- $(a, b] := \{x \in \mathbb{R} : a < x \leq b\}$
- $(a, b) := \{x \in \mathbb{R} : a < x < b\}$

(where $\mathbb{R}$ represents the real numbers). The first, $[a, b]$, is called a **closed interval** since it contains both endpoints. The last is called an **open interval** since it contains neither endpoint. The second and third are called half-open—or, naturally, half-closed—intervals. We'll frequently be assigning probabilities to subsets of the real line, and these intervals serve as the building blocks. In other words, if we understand the probabilities of these sets, we'll understand the probabilities of all the sets with which we'll be working.

Open and closed intervals are perfect for studying the real line $\mathbb{R}$, but what about $\mathbb{R}^2$ or even $\mathbb{R}^n$? What should our building blocks be? Or equivalently: How should we generalize intervals? A good choice is to use the **Cartesian product**, and study rectangles in two dimensions, boxes in three dimensions, and so on. For example, consider the set $[a, b] \times [c, d]$, which is $\{(x, y) : a \leq x \leq b, c \leq y \leq d\}$. To move up to three dimensions, we could add another interval $[e, f]$, but this is getting messy, and we'll have an alphabet soup if we go much higher. To keep the notation clean, consider the following:

$$[a_1, b_1] \times \cdots \times [a_n, b_n] = \{(x_1, \ldots, x_n) : a_i \leq x_i \leq b_i\}.$$

This is a nice, concise way to describe one of these sets in $\mathbb{R}^n$. I can't stress enough the importance of and need for **good notation**. Good notation allows you to glance down and get a sense of what exactly is happening; bad notation leaves you scratching your head and muttering under your breath (which isn't conducive to learning). It's worth spending some time thinking about the best way to present your ideas. You want people to be able to follow what you're saying. See how concisely we can write the box in $\mathbb{R}^n$ with this notation; we couldn't do that if we were writing it as $[a, b] \times [c, d] \times [e, f] \times \cdots$. In fact, we see that we should read $\mathbb{R}^2$ as $\mathbb{R} \times \mathbb{R}$, or pairs of all real numbers. Similarly $\mathbb{R}^3$ is $\mathbb{R} \times \mathbb{R} \times \mathbb{R}$, and so on.

While rectangles and boxes are convenient, they're not the only choice. Another possibility is to use circles in the plane and then spheres in three dimensions. Both approaches are useful; while the rectangles fit together better, the circles and spheres are more convenient for some theoretical calculations.

In practice, we might need to study probabilities of sets that have more than three dimensions. For example, we could have an economic model with 10 parameters and we want to know the measure or probability of the parameter values leading to a certain outcome. To study such situations, it's very convenient to have some general notation that works well for all dimensions. We need this notation to put probability on a really firm foundation; however, for many courses the emphasis is on a more informal approach, and thus these concepts may not be seen.

We define the **open ball** of radius r about a point $a = (a_1, \ldots, a_n) \in \mathbb{R}^n$ to be

$$B_a(r) := \{x = (x_1, \ldots, x_n) : (x_1 - a_1)^2 + \cdots + (x_n - a_n)^2 < r^2\}.$$

The **closed ball** $\overline{B}_a(r)$ is defined similarly, except with a $\leq$ instead of a $<$. Thus, the open ball is all points less than r units from a, which we can also think of as all points in an n-dimensional sphere centered at a with radius r. We say that a set $A \subset \mathbb{R}^n$ is **open** if, given any $a \in A$, there's some r—which may depend on a—such that $B_a(r) \subset A$. We also say that a set A is **closed** if its complement is open.

Consider the following example: let A be the set of all points $(x, y) \in \mathbb{R}^2$ such that $|xy| < 1$. Put differently, A refers to all of the points inside the region in Figure 2.2.

To see that this is an open set, consider any point inside. Let ρ be the shortest distance from our point to any of the four curves. Based on our calculus knowledge, we recognize that ρ both exists and is greater than zero. Therefore, the ball about our point with radius $\rho/2$ will be contained in A.

Here are two of the most common examples of open and closed sets.

- A circle or sphere without the boundary included is an open set; if we include the boundary we get a closed set.
- An interval, square, cube, and so on without the boundary included is an open set; similarly, an interval, square, cube, and so on with the boundary included is a closed set.

With that, we'll conclude this terminological section, which we admit is rather lengthy. There's just a lot of terms we need to introduce. It may not be immediately

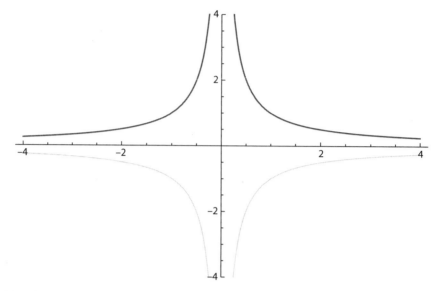

Figure 2.2. Set of points inside $|xy| \leq 1$. It's an infinite set: does it have finite or infinite area?

apparent that it was worth it, but trust me, it was. While learning definitions isn't the most enjoyable activity, our clear, efficient communication with others depends upon it. Now that we've taken the time to learn this language and to ensure that we're on the same page, we can reap the benefits!

 Try your hand at showing whether sets are open or not. If A is the set of all points (x, y) such that $(x/4)^2 + (y/3)^2 < 1$, show that A is open. If B is the set of all points (x, y) such that $(x/4)^2 + (y/3)^2 \leq 1$, show that B is closed.

2.3 Outcome Spaces, Events, and the Axioms of Probability

This section begins innocently enough, but as we progress to more delicate issues, the material becomes more technically difficult. The solutions of many of these delicate issues require courses in analysis and point set topology, which aren't necessary for introductory probability classes. We have plenty of everyday experience with probabilities, and even though there's some danger of being misled (as Russell's paradox showed us), for the most part your intuition will serve you well. As we discussed in the introduction to this chapter, the danger revolves around infinity—infinitely many sets, or sets with infinitely many elements. Whenever there's an infinity, we have to be very careful. If, however, we just have finitely many sets with finitely many elements, our intuition is usually trustworthy. We therefore present two sections on the axioms of probability, which should feel reasonable and right. This section and the next will give us an intuitive approach to the subject; readers interested in seeing the technical details should make sure to read §2.6.

Let's start by formalizing the intuitive notions we've been using. To talk about probabilities in a sensible manner, we need a few pieces of information. First, we must specify all possible outcomes and the probability that each outcome occurs.

We assume that all possible outcomes are subsets of some given set Ω. Consequently, Ω might be the set of positive integers and our events might be the number of times we need to toss a fair coin before it lands on "heads." For another example, imagine that Ω is the unit circle, and the events are subsets of the unit circle. In this case, consider a dart-throwing scenario. If we were to throw a dart at the unit circle, it would need to land somewhere. Each point on or inside the circle is a potential landing spot.

We call Ω the **sample space** or **outcome space** and the elements of Ω the **events**. This definition works in many cases. It's satisfactory, for example, if Ω is finite or countable. For probability spaces in general, though, it needs to be modified, as we discuss in §2.6.

Once we have our outcome space Ω, we want to assign probabilities to the different elements of Ω. To do this, we introduce the **probability function**, which we'll write as Prob. We denote the probability that event A happens by Prob(A), though often for brevity we write Pr (A).

 An example will help to clarify this. Suppose the first wheel of a slot machine has 20 symbols: 10 clubs, 5 hearts, 3 diamonds, and 2 spades. If each of the 20 objects are equally likely to appear when we play, what are our sample space and probability function?

While this problem may seem simple, there are a few subtleties lurking. There are two possible interpretations: are we able to distinguish among the different objects that have the same symbol? It's not unreasonable to assume that we can't tell the difference between the 10 clubs since they all look the same to us. If the wheel stops and we see a club, we can't tell if it's the first or the tenth club.

The sample space is the set of possible outcomes, so we have

$$\Omega = \{\spadesuit, \heartsuit, \diamondsuit, \clubsuit\}.$$

Since each of the twenty objects is as likely to appear as any other, the process is fairly straightforward. All we have to do in order to find the probabilities of observing each of these symbols is count how many objects equal that shape and then divide by the total number of objects, so

$$\Pr(\spadesuit) = \frac{2}{20}, \quad \Pr(\heartsuit) = \frac{5}{20}, \quad \Pr(\diamondsuit) = \frac{3}{20}, \quad \Pr(\clubsuit) = \frac{10}{20}.$$

There are two things to notice about this probability function. First of all, every probability is non-negative and at most 1. This is good—what would it mean for something to occur with probability -0.5 or 2? By multiplying by 100, you can convert probabilities to percentages. Thus a probability of .5 corresponds to 50%, or something happening half the time, while a probability of 1 corresponds to a probability of 100%, which means it must happen. Secondly, we can use the probability function to find the probability of other events, such as the probability that we get a club *or* a heart. Since 15 of the 20 objects are either clubs or hearts, and nothing is both a club and a heart, we should have

$$\Pr(\clubsuit \text{ or } \heartsuit) = \frac{15}{20} = \Pr(\clubsuit) + \Pr(\heartsuit).$$

These last two properties are actually general properties that we want our probability function to obey. More specifically, we **want** our probability function to satisfy the following wish list.

> **Wish list**:
>
> 1. For any event A, we have $0 \leq \Pr(A) \leq 1$, and if Ω is the outcome space, then $\Pr(\Omega) = 1$.
> 2. If $\{A_i\}$ is a pairwise disjoint collection of sets (which means $A_j \cap A_k$ is empty if $j \neq k$), then $\Pr(\cup_i A_i) = \sum_i \Pr(A_i)$.

The first condition states that no event can occur with probability greater than 1 or less than 0, and that *something* must happen. The second condition states that probability is additive in certain occasions; that is, if we have a collection of disjoint sets $\{A_i\}$, then the probability that one of them happens is the sum of their respective probabilities. While this held for our slot machine example when we looked at the probability of getting a club or a heart, *it turns out that this condition, while intuitively appealing, cannot be satisfied in general.* Not surprisingly, the problem has to do with an infinity. We need to be more careful about which events we assign probabilities in order to retain this property, which we'll cover in the next section. For the remainder of this section, we'll work through a few examples of different outcome spaces and probability functions.

Let's explore a consequence of these two conditions. We'll prove Prob($\emptyset$) = 0; this should be the case, as otherwise there would be a positive probability of nothing happening! From the first condition, we know Prob(Ω) = 1. We can write Ω as the disjoint union of Ω and $\emptyset$; while this may seem a bit strange, there can't be any element in common between Ω and $\emptyset$ as the empty set has no elements! Thus, by Property 2 we have Prob($\Omega \cup \emptyset$) = Prob(Ω) + Prob($\emptyset$); substituting gives $1 = 1 + $ Prob($\emptyset$), and we find Prob($\emptyset$) = 0 as was claimed.

Consider the following example. Suppose we want to keep track of the number of heads we flip over a sequence of two tosses of a fair coin. What's the sample space, and what's the probability function? Since we can only get 0, 1, or 2 heads, our sample space is the set $\{0, 1, 2\}$. For our probability function, notice that there are four possible outcomes of our coin tosses: we could flip heads-heads, heads-tails, tails-heads, or tails-tails. Since we're flipping a fair coin, each of these outcomes should be equally likely. Since only one of the outcomes involves us flipping two heads, we see that Prob(two heads) = 1/4. Using the same logic, we have Prob(one head) = 1/2 and Prob(no heads) = 1/4. To recap: in tracking the number of heads we flip in two tosses of a fair coin, we have

$$\Omega = \{0, 1, 2\}$$

$$\Pr(0) = \frac{1}{4}, \quad \Pr(1) = \frac{1}{2}, \quad \Pr(2) = \frac{1}{4}.$$

Now, let's tweak this example a bit. Suppose we're flipping a biased coin, which lands heads with probability 0.7. Again, we'll track the number of heads we flip over

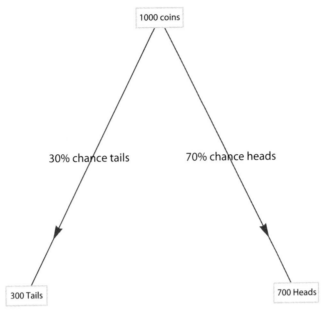

Figure 2.3. Expected results of flipping 1000 coins with each coin having a 70% chance of landing on heads and a 30% chance of landing on tails.

a sequence of two tosses. What are the sample space and probability function in this adjusted case? Since we can still only flip 0, 1, or 2 heads, the sample space is still {0, 1, 2}. Notice that even though we're looking at a different problem, the sample space hasn't changed at all. Therefore, the difference must lie in the probability function. But how can we find the probability that we flip two heads? Since the coin is no longer fair, we can't simply enumerate all the possible outcomes and take the ratios as we did before.

One way to work through problems like these is to use a **probability tree**, which is a nice way to visualize the possible outcomes and their relative frequencies. Suppose we were to repeat our coin-flipping experiment 1000 times. Since the probability of flipping a head is 0.7, we would expect the first flip to be a head 700 of the 1000 times, and a tail 300 out of 1000 times. We show this graphically in Figure 2.3.

What happens with the second flip of the coin? Of the 700 that we originally flipped heads, we expect 70% of those to flip heads on the second toss, for a total of 490 experiments ending in heads-heads. Proceeding similarly, we can fill out the next branch of our tree in Figure 2.4.

Remember, we're trying to find the probability function that goes along with our set $\Omega = \{0, 1, 2\}$, where we're counting the number of heads we get in our two-flip sequence. From our tree in Figure 2.4, we see that of our 1000 experiments, we expect to get two heads 490 times, one head 420 times (210 from heads-tails, and 210 from tails-heads), and zero heads 90 times. Thus, our probability function is

$$\Pr(0) = 0.09, \quad \Pr(1) = 0.42, \quad \Pr(2) = 0.49.$$

Our method in the last example is a common technique; calculating the expected results from a large number of trials and then dividing is often a good strategy. You may have also noticed that the probability of flipping two consecutive heads was the

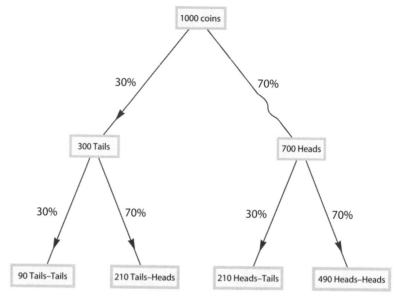

Figure 2.4. Expected results of flipping 1000 coins with each coin having a 70% chance of landing on heads and a 30% chance of landing on tails; each coin is flipped twice.

product of the probability of flipping a head on each toss. This is no coincidence. We can frequently find probabilities using the **multiplication rule**, which states: for certain A and B,

$$\Pr(A \cap B) \;=\; \Pr(A) \cdot \Pr(B).$$

So in our case, we would have Prob(head and head) = Prob(head)^2 = 0.49, which is indeed what we found. This formula **does not** hold in general. It assumes that the events A and B are *independent*. We'll give a rigorous definition and a lengthy discussion later in Chapter 4. For now, we'll argue informally. Two events A and B are **independent** if they don't affect each other in any way. This means that knowing one of the events happened (or didn't happen) doesn't affect our knowledge of the probability of the other happening. In our coin-tossing example, it's reasonable to assume that the outcome of one toss won't affect the outcome of the other toss. As a result, we feel justified in applying the multiplication rule here. Using it, we see the probability of flipping two heads is $.7 \cdot .7 = 0.49$, the probability of flipping no heads is $.3 \cdot .3 = 0.09$, and the probability of flipping one head is $2 \cdot .7 \cdot .3 = 0.42$. (Why a factor of two? Because we could either flip heads-tails or tails-heads.) Notice that these are the exact same probabilities we found using our tree method.

Remark 2.3.1: *When we created the probability tree for flipping a coin twice, we started with 1000 coins. Why did we choose 1000? It came from looking at the numbers in the problem and choosing the number of coins so that all elements in the tree would be integers. There's nothing wrong with choosing other numbers and potentially having fractions of coins. Of course, we don't need 1000; 100 coins would have sufficed and led to all integers along the tree. The reason I chose 1000 was to give myself some protection. I like doing this—just in case I made a small error in my calculations, I have a small safety factor built in.*

2.4 Axioms of Probability

We're now ready to list the basic rules that probability functions satisfy. Because discussing these in general requires some advanced analysis, specifically the σ-algebras of §2.6, we'll restrict our study to some important special cases. For finite or countable outcome spaces, the claims are considerably more straightforward. We'll state the axioms and then give examples and applications. As always, remember that your intuition and experiences can mislead you when infinities emerge!

The gist of §2.6 is that given an outcome space Ω, we can only define the probability function on a special set of subsets. We denote this set of special subsets by Σ, which is called a σ-**algebra**. This set has many nice properties: if A and B are in Σ, then so too are $A \cap B$ and $A \cup B$. Further, if Ω is finite or countable, we may choose Σ to be every possible subset. It's only when there are uncountably many elements in Ω that we must be careful. Uncountable is a higher infinity than countable; there are countably many integers but uncountably many reals; see Appendix C for more on sizes of infinities.

Fortunately, Σ contains more than enough subsets to keep us happy. For example, if Ω is the interval $[0, 1]$, then Σ contains any finite union or intersection of intervals. If the probability function satisfies a few nice properties (which we'll give in a minute) then we call the triple $(\Omega, \Sigma, \text{Prob})$ a **probability space**. We use the function Prob to assign probabilities to subsets of Σ, our special σ-algebra of sets where Prob is defined.

We end this introduction with Kolmogorov's Axioms of Probability, which we'll examine in greater detail and a bit more rigor in §2.6. Our purpose here is to gain some familiarity with them and see how reasonable they are. Right now we're being a bit vague (actually, we're being very vague!) on what's in the σ-algebra Σ; we'll talk about this more in §2.6. For now, think of everyday events like the outcome of rolling a die, or the toss of a coin, or the number of runs a team scores in baseball. Better yet, think of Ω as a finite set, and Σ is just the set of all possible subsets of Ω. Thus, if Ω has n elements, the 2^n subsets of Ω form Σ.

(Kolmogorov's) Axioms of Probability: $(\Omega, \Sigma, \text{Prob})$, where Ω is the outcome space and Σ, a σ-algebra, is a probability space if the probability function satisfies the following:

- If $A \in \Sigma$ then $\Pr(A)$ is defined and $0 \le \Pr(A) \le 1$.
- $\Pr(\emptyset) = 0$ and $\Pr(\Omega) = 1$.
- Let $\{A_i\}$ be a finite or countable collection of disjoint elements of Σ. Then $\Pr(\cup_i A_i) = \sum_i \Pr(A_i)$.

Before exploring their consequences, let's briefly look at these axioms. The first states that we can assign a probability to any event in the σ-algebra Σ, and that probability has to be between 0 and 1. These are clearly reasonable bounds; nothing should have a probability of happening below 0. No matter how bad a baseball team is, the team cannot score negative runs. Also, no matter what certain sports announcers say, no one can give 110% (or more!); no matter how strong a team is, it can't have a greater than 100% chance of winning.

The second axiom asserts that the probability that nothing happens is 0, while the probability that something happens is 1. Both these claims pass the smell test, and are in line with our experiences. Actually, as a nice exercise show that we don't need to assume $\Pr(\emptyset) = 0$; it turns out this follows from the other axioms (but it seems friendlier to include it, and doing so causes no harm).

The last is the most interesting. The key words are finite and countable. Notice this is the first occurrence of an infinity in our discussion. We'll see later that it's essential that we restrict ourselves to finite or countable sets; larger infinities will cause serious issues. However, if we restrict ourselves to a finite or countable collection, what it says is that probabilities are additive. The most important case is that if A and B are disjoint elements of Σ then $\Pr(A \cup B) = \Pr(A) + \Pr(B)$; in other words, if A and B cannot happen together then the probability one of them happens is the sum of the probabilities each happens.

Let's explore some examples of probability spaces, and then continue in the next section by isolating some important properties of these spaces.

 One of the best and most important examples of a probability space is where the σ-algebra is simply all subsets of a finite set Ω, which we'll write as $\Omega = \{\omega_1, \omega_2, \ldots, \omega_n\}$. We need to assign probabilities to the elements of Ω; the simplest way to do so is to have all items equally likely; thus $\mathrm{Prob}(a_k) = 1/n$ for $1 \le k \le n$. In this case, if A is any subset of Ω, then $\mathrm{Prob}(A)$ is just $\#A/\#\Omega$, where $\#S$ is the number of elements in S. This is often called the **counting model**. All the axioms hold with this definition of probability.

 For instance, imagine that we roll a fair die. In this case, we get either a 1, 2, 3, 4, 5, or 6, and each occurs with probability 1/6. If we take $A = \{1, 3, 5\}$, then $\mathrm{Prob}(A) = 3/6 = 1/2$, meaning that we have a 50% chance of rolling an odd number.

 Let's do another example along these lines. Let $\Omega = \{1, 2, 3, 4, 5, 6\}$ be the outcome of the roll of a fair die, let $A = \{2, 4, 6\}$ (the probability we roll an even number) and $B = \{3, 5\}$ (the probability we roll an odd prime). Then $\Pr(A) = 3/6$, as three of the six possible rolls are in A, $\Pr(B) = 2/6$, and $\Pr(A \cup B) = 5/6$. Notice $\Pr(A \cup B) = \Pr(A) + \Pr(B)$; this is true as A and B are disjoint. If instead we looked at $C = \{2, 3, 5\}$ (the probability we roll a prime), then $\Pr(A \cup C) = 5/6 \ne \Pr(A) + \Pr(C)$.

 For another example, let's consider the outcome space of 5 tosses of a fair coin. There are 32 possible sequences since each event, heads or tails, happen with probability 1/2 and the coin is tossed 5 times. As you look at the list below, think about how we chose to list them. It's very important that we don't miss anything, so we need a very good way to go through the possibilities. For example, notice each time we go down a level, we have one more total tail. Think also about how we chose to list the strings within a level.

- HHHHH,
- HHHHT, HHHTH, HHTHH, HTHHH, THHHH,
- HHHTT, HHTHT, HTHHT, THHHT, HHTTH, HTHTH, THHTH, HTTHH, THTHH, TTHHH,
- TTTHH, TTHTH, THTTH, HTTTH, TTHHT, THTHT, HTTHT, THHTT, HTHTT, HHTTT,
- TTTTH, TTTHT, TTHTT, THTTT, HTTTT,
- TTTTT.

Let's let A be the event that there are an even number of heads. What's $\Pr(A)$? Using the counting method, we look at our list of events and see that 16 of the 32 have an even number of heads, and thus $\Pr(A) = 16/32 = 1/2$.

 For the last problem, one possible approach, **which is wrong**, would be to say that there are three possible values for an even number of heads, namely 0, 2, and 4, and three possible values for an odd number of heads, namely 1, 3, and 5. Thus the probability of having an even number of heads is 3/6 or 1/2. While this is the right answer, it's a coincidence. The problem is that each possible number of heads isn't equally likely; we have zero heads with probability 1/32, two heads with probability 10/32, and four heads with probability 5/32. It's the individual strings of heads and tails that are equally likely, not the number of heads in the tosses.

 There's another way to attack the previous problem, and that's through the **method of symmetry**. There's nothing to distinguish heads and tails from each other; if we swap labels we haven't really changed anything. As we have five tosses, we either have an even number of heads or tails, but not both. Thus, the probability of having an even number of heads should be the same as the probability of having an even number of tails. As these are disjoint events and exactly one happens, the sum of their probabilities must be one. This forces each event to happen with probability 1/2.

 Our discussion above works for any odd number of tosses. As a nice exercise, try to tweak the above argument to the case when we have an *even* number of tosses. The problem is now that an even number of heads and an even number of tails aren't complementary events. It's still possible to show the probability is 1/2 by appealing to *some* symmetry of the problem.

2.5 Basic Probability Rules

Let's now record some basic rules about probability spaces—these are the consequences of our three axioms. I can't emphasize enough how important it is to remember that we're not assigning probabilities to all events, but only to *special* sets. Below we'll simply state a few facts that should feel right, and save a slightly more rigorous attack for §2.6.

We'll take a probability space $(\Omega, \Sigma, \text{Prob})$, which means Prob assigns probabilities to all elements of Σ, which is a collection of subsets of Ω. Specifically, $\Pr(\emptyset) = 0$, $\Pr(\Omega) = 1$, and if $A \in \Omega$ then $\Pr(A)$ is defined. Finally, the probability of a finite or countable disjoint union of elements of Σ is just the sum of their probabilities. The four properties below are very useful, and worth isolating and remembering.

Useful Rules for Probability Spaces: Let $(\Omega, \Sigma, \text{Prob})$ be a probability space. Then

1. **"Law of Total Probability"**: If $A \in \Sigma$, then $\Pr(A) + \Pr(A^c) = 1$. Equivalently, $\Pr(A) = 1 - \Pr(A^c)$.

---

2. $\Pr(A \cup B) = \Pr(A) + \Pr(B) - \Pr(A \cap B)$. This can be generalized. For example, if we have three events, then

$$\begin{aligned} \Pr(A_1 \cup A_2 \cup A_3) = {} & \Pr(A_1) + \Pr(A_2) + \Pr(A_3) \\ & - \Pr(A_1 \cap A_2) - \Pr(A_1 \cap A_3) \\ & - \Pr(A_2 \cap A_3) + \Pr(A_1 \cap A_2 \cap A_3). \end{aligned}$$

This is also known as **inclusion-exclusion**.

3. If $A \subset B$, then $\Pr(A) \le \Pr(B)$. If, however, A is a proper subset of B, we don't necessarily have $\Pr(A) < \Pr(B)$, but we do know for certain that $\Pr(B) = \Pr(A) + \Pr(B \cap A^c)$, where $B \cap A^c$ refers to all elements of B that aren't in A.

4. Let $A_i \subset B$ for all i. Then $\Pr(\cup_i A_i) \le \Pr(B)$.

Notice that the "Law of Total Probability" is written in quotes. This property is actually just a special case of a more general result known as the Law of Total Probability, which we'll discuss in §4.5.

In the next few subsections we'll prove the above properties and provide examples. We'll also end the chapter with an extensive section on how to guess what the answer could be in some cases (§2.7). In some sense that material can be ignored, as it is just a non-rigorous attempt to guess the answer; however, in another sense it is the most important part of the chapter! The reason is that as you go further in life, the formulas and expressions you need to work with become more and more involved. What we're doing is almost experimental mathematics, trying to sniff out the functional form of the answer before proving it. This is an incredibly useful skill to master, as it is much easier to prove something if you know what you're trying to prove! Thus, I strongly encourage you to carefully go through the analysis there, and try the additional problems.

2.5.1 Law of Total Probability

The first rule follows directly from the properties of the probability function. We know that A and A^c are disjoint sets whose union is Ω. Therefore, we have

$$1 = \Pr(\Omega) = \Pr(A \cup A^c) = \Pr(A) + \Pr(A^c),$$

with the last equality true because the probability of two disjoint events is the sum of the probabilities. $\qquad \square$

The Law of Total Probability is conceptually straightforward: the probability that A happens should be 1 minus the probability that A doesn't happen. This reformulation can be extremely useful, as it's often easier to calculate the probability that something doesn't happen than to find the probability that it does occur.

 For instance, imagine that Newman and Kramer (from *Seinfeld*) are playing a heated game of Risk. Believing that Ukraine is weak, Kramer has his forces attack. He rolls three dice and Newman rolls two. If you've never played Risk before, all that matters is that each die is fair, each roll is independent of the others, and higher numbers are better. It's not unreasonable to wonder how many sixes will be rolled. Thus, imagine we're trying to compute the probability that at least one six is rolled; let's call this event A. There are unfortunately *many* ways this could happen. We could roll exactly one six, exactly two sixes, and so on up to exactly five sixes. We'd have to compute the probability for each, and this begins to get messy quite quickly. Fortunately, it's easy to compute the probability of A^c, the event where we don't roll any sixes. In this case, as each die has a probability of 5/6 of landing on a non-six, the probability that all five aren't a six is just $(5/6)^5$. Thus $\Pr(A) = 1 - \Pr(A^c) = 1 - \left(\frac{5}{6}\right)^5 = \frac{4651}{7776} \approx 59.8\%$.

As a painful exercise, compute the probability of rolling exactly k sixes when we roll five fair die for each possible value of k. Doing so will help you appreciate the Law of Total Probability. It's not in the book because I've done this many times in my life, and I already have a healthy appreciation for the **Method of Complementary Probabilities**!

Finding the probability an event happens from the probability it doesn't happen is one of the greatest ways to solve problems. We've already seen a great example of this in Chapter 1 when we studied the Birthday Problem. It was a nightmare trying to keep track of all the different ways we could have at least two share a birthday, but it was incredibly simple to study the one way no one shares a birthday with another.

2.5.2 Probabilities of Unions

To prove the second rule, we can, for instance, use the fact that if we have a finite collection of disjoint sets C_i, then the probability of the union is the sum of the probabilities (our third probability axiom). We write $A \cup B$ as

$$A \cup B = (A \setminus (A \cap B)) \cup (B \setminus (A \cap B)) \cup (A \cap B), \qquad (2.1)$$

where $A \setminus (A \cap B)$ means all things in A that aren't also in B. It's helpful to read the above as splitting our set into things in A but not B, things in B but not A, and finally things in both A and B. We haven't double counted, and we have taken care of all possibilities. That said, the set-theoretic notation has completely overwhelmed the problem. It's easy to look up and get lost in the notation. So let's take a step back to readjust and look at the problem another way. We want to show that

$$\Pr(A \cup B) = \Pr(A) + \Pr(B) - \Pr(A \cap B).$$

Let's interpret the left- and right-hand sides of the proposed equality and show that they're referring to the same events.

The left-hand side is the probability of $A \cup B$. In other words, it's the probability of x being in A or B or both. But how should we interpret the right-hand side? To begin, $\Pr(A)$ and $\Pr(B)$ are the probabilities that an x is in A or B, respectively. At first, we might think that their sum is the probability an x is in either, but this assumption is incorrect because it overlooks the problems created by double counting. If an x is in both A and B, we've counted it twice—once as a part of A and again as a part of B. As a result, we must subtract the probability that x is in $A \cap B$. The claim then follows. $\qquad\square$

We give a second argument for this claim for two reasons. First, because this is such an important result, it helps to see multiple avenues that we can take to reach it. Secondly, this argument introduces a useful concept. We can also demonstrate this second rule graphically, using a **Venn diagram**. A Venn diagram is a way of depicting our outcome space Ω. We draw Ω as a rectangle, and then shade in or mark off subsets

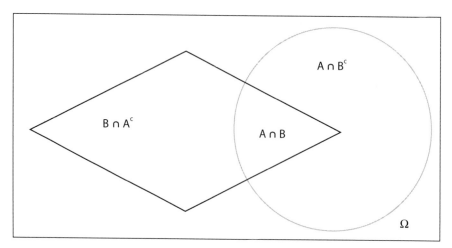

Figure 2.5. Venn diagram for two sets, A and B.

of Ω. Suppose that A and B are proper subsets of Ω that aren't disjoint. Then, we can draw the Venn diagram in Figure 2.5.

We see that we can split the region $A \cup B$ into the union of three disjoint pieces. Call C_1 the part of A that shares nothing with B, so that $C_1 = A \cap B^c$ (the points in A not in B); note $A \cap B^c$ is the same as $A \setminus (A \cap B)$, but is a more compact notation. Similarly, let C_2 be the part of B that shares nothing with A, so that $C_2 = B \cap A^c$. Finally, we'll set C_3 to be the part common to A and B, so that $C_3 = A \cap B$. We have $A \cup B = C_1 \cup C_2 \cup C_3$. This is exactly what that intimidating set equation, Equation (2.1), means. As we've said before, the probability of a countable union of disjoint sets is the sum of their respective probabilities. Therefore, we have

$$\Pr(A \cup B) = \Pr(C_1) + \Pr(C_2) + \Pr(C_3).$$

But we can rewrite this. Notice that $C_1 \cup C_3 = A$. Why? C_1 are points in A but not B, while C_3 are points in A and B. Thus the union is just points in A. Similarly $C_2 \cup C_3 = B$ and $C_3 = A \cap B$. Thus, we have

$$
\begin{aligned}
\Pr(A \cup B) & \\
&= \Pr(C_1) + \Pr(C_2) + \Pr(C_3) \\
&= (\Pr(C_1) + \Pr(C_3)) + (\Pr(C_2) + \Pr(C_3)) - \Pr(C_3) \\
&= \Pr(A) + \Pr(B) - \Pr(A \cap B).
\end{aligned}
$$

   In the algebra above, we did something clever: we did nothing! Doing nothing is one of the most important skills to master in mathematics. There are two good ways to not do anything: we can **add zero**, or we can **multiply by one**. These are powerful proof techniques. We added zero by adding $\Pr(C_3) - \Pr(C_3)$, as this allows us to get $\Pr(A)$ and $\Pr(B)$ terms. It takes a lot of practice to become comfortable with these arguments; see §A.12 for more on these techniques. The best advice I can give is to think about what you have and what you want to get, and what would be useful to help in that endeavor.

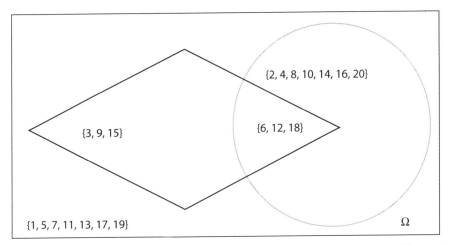

Figure 2.6. Venn diagram for $A = \{2, 4, 6, \ldots, 20\}$, $B = \{3, 6, 9, \ldots, 18\}$, and $\Omega = \{1, 2, 3, \ldots, 20\}$.

Let's illustrate this second method with an example (see Figure 2.6).

Assume $\Omega = \{1, 2, \ldots, 20\}$, and let each element have probability $1/20$—that way, if C is any subset of Ω, we have $\Pr(C) = \#C/20$. Let $A \subset \Omega$ be all the numbers divisible by 2, and let B be all the numbers divisible by 3. Then

$$
\begin{aligned}
A &= \{2, 4, 6, \ldots, 20\}, \quad \Pr(A) = 10/20 \\
B &= \{3, 6, 9, \ldots, 18\}, \quad \Pr(B) = 6/20 \\
A \cap B &= \{6, 12, 18\}, \quad \Pr(A \cap B) = 3/20 \\
A \cup B &= \{2, 3, 4, 6, 8, 9, 10, 12, 14, 15, 16, 18, 20\}, \quad \Pr(A \cup B) = 13/20;
\end{aligned}
$$

note that $13/20 = 10/20 + 6/20 - 3/20$.

The advantage of the above formula, $\Pr(A \cup B) = \Pr(A) + \Pr(B) - \Pr(A \cap B)$, is that it's often easier to compute the three probabilities on the right than the one probability on the left. On its surface, this contention seems absurd. Why would it be better to do three problems when we could get by with only doing one? The answer, of course, depends critically on the nature of that one problem in relation to the other three problems. Imagine instead that we changed Ω to $\{1, 2, \ldots, 10^{10}\}$. We can readily write down the three probabilities on the right, but we would have to resort to enumeration to find the probability on the left, $\Pr(A \cup B)$. *This formula allows us to figure out unions from our original set and intersections. For many problems, intersections are easier to work with than unions!* This is a very important observation. We'll use it numerous times over the course of this book.

As the union formula is so important, it's worth spending a bit more time thinking about it. Sometimes reading a math formula aloud can give a clue as to its answer; given our penchant for naming anything that could be considered a technique, let's call this one the **say it aloud technique**. For example, consider $\sin(\arcsin(x))$. In words, this is asking for the sine of the angle whose sine is x. This is clearly just x! If we want to find $\Pr(A \cup B)$, this is the probability that an element is in A or B or both. Writing it out like this, we can see that we're being asked to find the three probabilities $\Pr(A)$, $\Pr(B)$, and $\Pr(A \cap B)$.

 Let's do one more example to show how much easier intersections are than unions. Imagine we have a class with 27 students. Initially everyone is sitting somewhere. Let's say Valeri is in the first seat, and Charlotte is in the second seat. Everyone then leaves the class and returns; each person is equally likely to be sitting in any seat. What's the probability that either Valeri *or* Charlotte is sitting in their original seat?

Let A be the event that Valeri returns to her initial seat, and B be the event that Charlotte returns to her original seat. We want to find $\Pr(A \cup B)$. This isn't easy to compute directly, as we would need to go through all possible cases with at least one in the correct seat; however, $\Pr(A)$, $\Pr(B)$, and $\Pr(A \cap B)$ aren't too bad to find. There are 27 people, and there are thus 27! ways to arrange them among the 27 seats (remember $n! = n(n-1)(n-2)\cdots 3 \cdot 2 \cdot 1$, so $3! = 6$ and $4! = 24$; note also $(n+1)! = (n+1)n!$). To see this, we use the multiplication principle. There are 27 possible seats available to the first person who returns to the class, then 26 to the next person who returns, and so on. This leads to $\Pr(A) = \Pr(B) = (1 \cdot 26 \cdot 25 \cdots 3 \cdot 2 \cdot 1)/27! = 1/27$; this is because our special person must return to her seat (there's only one way this can happen), and then the remaining 26 people may sit anywhere.

What of $\Pr(A \cap B)$? This is just $(1 \cdot 1 \cdot 25 \cdot 24 \cdots 3 \cdot 2 \cdot 1)/27! = 1/(27 \cdot 26)$. Here the double 1's in the beginning are due to Valeri and Charlotte having to sit in specific seats (and it matters who sits where). Combining, we find

$$\Pr(A \cup B) = \frac{1}{27} + \frac{1}{27} - \frac{1}{27 \cdot 26} = \frac{17}{234},$$

or about 7.26%. This is almost 2/27 (which is about 7.41%). So the probabilities are almost additive, but it's a little less than the probability of A plus the probability of B due to potential double counting.

In our discussion above, we had to be very careful to answer the right question. If we observe some of the people entering the room, and see where they sit, then clearly this will affect the probabilities of where the next people sit. That is a *conditional probability*, as now the likelihood of someone taking a given seat depends on what other people are doing. We'll discuss conditional probabilities in Chapter 4.

 We'll see versions of this problem again. More generally, we can ask what is the probability that if we randomly shuffle n objects then at least one returns to where it started. As n grows, the answer converges to $(e-1)/e$.

2.5.3 Probabilities of Inclusions

The proof of the third claim is more straightforward. Suppose we have set A, which is a subset of B. What can we say about the respective probabilities of A and B? Intuitively, it seems that B should occur with at least the same probability as A, since every time A happens, B does as well. But how can we formalize this? Since A is a subset of B, we can write

$$B = A \cup (B \cap A^c).$$

Here $B \cap A^c$ means all elements of B not in A. We're just saying we can split B into two disjoint subsets: A and everything in B that isn't in A. Since A and $B \cap A^c$ are disjoint, we have

$$\Pr(B) = \Pr(A) + \Pr(B \cap A^c) \geq \Pr(A),$$

since $\Pr(B \cap A^c) \geq 0$. $\qquad\qquad\qquad\qquad\qquad\qquad\qquad\qquad\qquad\qquad\qquad\qquad$ □

Let's return to our die. Imagine we roll five fair die again. Let B be the event that all of the die land on an odd number, and let A be the event that all of the die land on a 1. Clearly, if A happens, then B happens; the event A is a proper subset of the event B. The probability of A is just $(1/6)^5$, while the probability of B is $(3/6)^5$. So, we see that $\Pr(A) \leq \Pr(B)$ — in fact, we have $\Pr(A) < \Pr(B)$ in this case.

Now that we know the third claim, the fourth follows in one short line: all we need to do is let $A = \cup_{i=1}^\infty A_i$. As each $A_i \subset B$, we have $\cup_{i=1}^\infty A_i \subset B$, and from here, we can use the third claim. $\qquad\qquad\qquad\qquad\qquad\qquad\qquad\qquad\qquad\qquad\qquad\qquad\qquad\qquad\qquad$ □

This leads us to a general principle: when we're given unions, we often want to choose new sets so that we have a disjoint union. One of the most common mistakes in probability is to double count. For example, maybe we're looking at how many consecutive heads we have when we flip a fair coin 100 times. If we let A_i be the event that we have at least i consecutive heads in the 100 tosses, then clearly these sets aren't disjoint (if we have 6 consecutive heads in a string, we also have 5 consecutive heads). For some problems, it makes sense to switch and study say B_i, the event where the longest run of consecutive heads in 100 tosses is *exactly* i; note the B_i's are disjoint.

2.6 Probability Spaces and σ-algebras

In the previous section, we laid out the axioms of probability. But in doing so, we omitted some technical details. In this section, we'll look at an example of the pitfalls we need to know about, and see how σ-algebras help us to formulate a rigorous treatment of probability.

We began this chapter by explaining Russell's paradox, which has profound consequences in studying sets. We now turn to another paradox, which has a very direct application to probability. *The material in this section is more advanced, and is often omitted in a first course. It's perfectly fine to just skim this section to get a sense of the challenges.*

The **Banach-Tarski paradox** has major ramifications when applied to probability. The following claim should seem reasonable: if we have a three-dimensional object, its volume doesn't change if we rotate it or translate it. In other words, spinning and sliding it shouldn't have any affect on volume. Consider a solid unit sphere. The Banach-Tarski paradox tells us that it's possible to divide the sphere into 5 disjoint pieces such that we can assemble 3 into a solid unit sphere and the other 2 into a disjoint solid unit sphere, by simply translating and rotating the 5 original pieces. Even though translating and rotating shouldn't change volumes, we've somehow managed to double the volume of our sphere! This construction depends entirely on the **(Uncountable) Axiom of Choice**, which states that given any collection of sets $(A_x)_{x \in J}$ indexed by some set J, there's a function f from J to the disjoint union of the A_x with $f(x) \in A_x$ for all x. Let's unwind that. It means we can form a new set by choosing an element a_x from each A_x. Our choice function is f. If we have a countable collection of sets this is reasonable: a countable set is in a one-to-one correspondence with $\mathbb{N}$. We know exactly when we'll reach a set to choose a representative by "walking through" the options. If we have an uncountable collection of sets, however, it isn't clear "when" we'd reach a given set to choose an element.

So what does all this mean for probability? It means that if we assume and accept the Axiom of Choice, which is useful in many branches of mathematics, then the concept of "volume" doesn't behave as we'd expect. Since we'll often use volumes to define probabilities, the difficulty is apparent. There are several possible resolutions to problems like this. One common solution is to only assign probabilities to particular sets. But to what sets would we restrict ourselves? The five cuts needed to double the sphere are highly non-constructible. We can't even describe them; we can only claim they exist by appealing to the Axiom of Choice. Alternatively, if we restrict ourselves to "nice" sets, then volume behaves well. This is done in advanced courses through sigma-algebras and measures, which we'll now briefly introduce.

We've seen that there are issues in trying to assign probabilities to all possible events in a consistent way. As an example of how things can go wrong if we're not careful, let's return to one of the scenarios we discussed briefly in the previous section. If we throw a dart randomly at the unit circle, it is reasonable to assume that it's probability of landing at any one point is the same as its probability of landing at any other point. Let's call this probability c. We then have the following situation.

- The outcome space is $\Omega = \{(x, y) : x^2 + y^2 \le 1\}$.
- Let $A_{x,y}$ be the event of the dart landing at the point (x, y). From what we've said above, we're assuming that $\Pr(A_{x,y}) = c$ if $x^2 + y^2 \le 1$ and 0 otherwise.

What should c be? It turns out that there's *no* value of c that works. We have the following disjoint union for our outcome space:

$$\Omega = \bigcup_{\substack{x,y \\ x^2+y^2 \le 1}} A_{x,y}.$$

Thus,

$$1 = \Pr(\Omega) = \sum_{\substack{x,y \\ x^2+y^2 \le 1}} \Pr(A_{x,y}) = \sum_{\substack{x,y \\ x^2+y^2 \le 1}} c.$$

If $c > 0$, then the sum over x and y is infinite, and if $c = 0$, that sum is zero. We don't get 1 in either case. Consequently, there's *no* way to assign probabilities so that the desired properties of a probability function, discussed in §2.4, hold.

So what went wrong? The problem lies in how we defined our events. Our definition works fine when there are only finitely many possibilities, but as we just saw, it can fail when there are infinitely many events. To get around this, we must be more careful about what we consider events. The technical solution involves studying the σ-**algebras** of sets. At long last, we'll rigorously define it! A σ-algebra Σ of Ω is a non-empty collection of subsets of Ω with the following properties.

Let Ω be a set, and Σ be a non-empty set of subsets of Ω. Then Σ is a σ-**algebra** if:

1. $A \in \Sigma$ implies $A^c \in \Sigma$, and
2. a countable union of subsets of Σ is in Σ: $\cup_{i=1}^{\infty} A_i \in \Sigma$ if each $A_i \in \Sigma$.

This means that we may take unions of sets in our σ-algebra and get back a set in our σ-algebra—so long as it's a countable union. Further, it means that the complement of anything in our σ-algebra is also in our σ-algebra. But this isn't all—we have a lot more that's true!

- $\emptyset$ and Ω are always in any σ-algebra Σ of Ω. To see this, let A be any element of Σ (there's at least one such A, as we're assuming Σ isn't empty). Since Σ is closed under complements, A^c is in Σ; since it's closed under unions, $A \cup A^c \in \Sigma$, but $A \cup A^c = \Omega$! Thus $\Omega \in \Sigma$, as is $\Omega^c = \emptyset$.
- The σ-algebra is closed under countable intersections. This means if $B_i \in \Sigma$, so is $\cap_{i=1}^{\infty} B_i$. To see this, let $A_i = B_i^c \in \Sigma$. We have $\cup_{i=1}^{\infty} A_i \in \Sigma$. We leave it to you to show that the complement of a union is the intersection of the complements. Thus, $(\cup_{i=1}^{\infty} A_i)^c = \cap_{i=1}^{\infty} A_i^c \in \Sigma$, but the latter is our intersection of the B_i's.

Given any set Ω, we can always find at least one σ-algebra: let $\Sigma = \{\emptyset, \Omega\}$. In other words, the only elements of Σ are the empty set and all of Ω. We leave it to the reader to check and see that the two properties hold in this case. This example is too trivial to really be useful, as it only allows us to assign probabilities to the events "something happens" and "nothing happens."

Let's look at a more interesting example. If $\Omega = \{1, 2, \ldots, n\}$ is a finite set, then $\Sigma = \mathcal{P}(\Omega)$ is a σ-algebra. Here $\mathcal{P}(\Omega)$ means the **power set** of Ω, which is just the set of all subsets of Ω. For example, if $\Omega = \{1, 2\}$ then

$$\mathcal{P}(\Omega) = \{\emptyset, \{1\}, \{2\}, \{1, 2\}\},$$

while if $\Omega = \{1, 2, 3\}$ then

$$\mathcal{P}(\Omega) = \{\emptyset, \{1\}, \{2\}, \{3\}, \{1, 2\}, \{1, 3\}, \{2, 3\}, \{1, 2, 3\}\}.$$

Again we leave it to the reader to check the two conditions hold; the reason this situation is easier than the general case is that we only have finitely many subsets, and thus we cannot have a countable union of distinct sets. It's also worth noting that if Ω has n elements then $\mathcal{P}(\Omega)$ has 2^n elements. Be careful, though; the elements of $\mathcal{P}(\Omega)$ are *subsets* of Ω, and not elements of Ω. There's a big difference between 1 and the set containing 1.

Let's return to developing our theory. Recall that our original problem is to define probabilities of events. So why are we talking about σ-algebras? The reason is that the solution to the dart problem involves lowering our sights. Instead of trying to define the probability of *any* subset of an outcome space Ω, we need only define the probability of elements of the σ-algebra of Σ. We call the elements of Σ our **events**.

Initially, we'd hoped to be able to talk about the probability of any subset of Ω. We've changed course; now, we're saying that certain subsets just won't have probabilities attached to them. With respect to real-world applications, this isn't a big problem because the σ-algebra sets contain all the "natural" sets we would like to study but *not* the pesky sets like those causing the Banach-Tarski paradox. Using this as our definition, we recover the properties we hoped would hold.

(Kolmogorov's) Axioms of Probability: Let Σ be a σ-algebra for some outcome space Ω. We can define a probability function Prob : $\Sigma \to [0, 1]$. In other words, we can assign a probability between 0 and 1 to each element in Σ that satisfies the following properties:

1. For any event $A \in \Sigma$, we have $0 \leq \Pr(A) \leq 1$. Some books call this the **First Axiom of Probability**.

2. If Ω is the outcome space, then $\Pr(\Omega) = 1$. This is sometimes called the **Second Axiom of Probability**.

3. If $\{A_i\}$ is a countable pairwise disjoint collection of elements of Σ, then $\Pr(\cup_i A_i) = \sum_i \Pr(A_i)$. As we hope you might expect, this is often called the **Third Axiom of Probability**. One of many immediate and important consequences of this is the **Law of Total Probability**, which we'll discuss later in greater detail: $\Pr(A) + \Pr(A^c) = 1$. Another is if $A \subset B$, then $\Pr(A) \leq \Pr(B)$.

We call a triple $(\Omega, \Sigma, \text{Prob})$ a **probability space**. A complete description of admissible σ-algebras is well beyond the scope of this book; it requires tools from analysis and point set topology. The interested reader should turn to [Fol]. In many cases, instead of specifying the σ-algebra, we're satisfied to say such things as, "and the σ-algebra generated from the following." What we mean will become clearer below during our discussion of various examples of common probability spaces.

- Let Ω be the positive integers, Σ all subsets of Ω, and Prob any non-negative function such that $\sum_{n=1}^{\infty} \Pr(\{n\}) = 1$, where $\Pr(\{n\})$ is the probability of getting n. This is one of the most important examples. It arises everywhere. For instance, waiting for the first head when tossing a fair coin, the number of points a team scores in a game, the leading digits of numbers on tax returns, stock prices (in cents), the number of days Homer Simpson goes without causing an accident at the nuclear plant, the number of people who want to see a movie at a given time, ... The list goes on and on.

- Let $\Omega = [0, 1]$ and take the σ-algebra generated by all open sets $(a, b) \subset [0, 1]$, and let $\Pr((a, b)) = b - a$. This is another important example. It's the one-dimensional equivalent of our dart-throwing problem. (Imagine some very sober people at a bar capable of throwing darts that always land on a given line!) As it turns out, this allows us to assign a probability to any point in $[0, 1]$. For simplicity, we assume $x \in (0, 1)$. We have

$$\{x\} = ([0, x) \cup (x, 1])^c.$$

The two intervals are both open in $[0, 1]$ and thus, in our σ-algebra as well. As a result, their complement is in our σ-algebra.

What's the probability of $\{x\}$? Well,

$$\Pr([0, x) \cup (x, 1]) \;=\; \Pr([0, x)) + \Pr((x, 1]) \;=\; x + (1 - x) \;=\; 1.$$

Therefore, $\Pr(\{x\}) = 0$. This seems to be the same problem we had before; the probability that the dart lands at x is 0, but surely it must land somewhere. Does this not imply that

$$1 \;=\; \Pr([0, 1])$$

$$= \; \Pr\left(\bigcup_{x \in [0,1]} \{x\}\right) \;=\; \sum_{x \in [0,1]} \Pr(\{x\})$$

$$= \; \sum_{x \in [0,1]} 0 \;=\; 0?$$

The answer is "No!" The explanation is as follows: we only know that the probability of a union is the sum of all the probabilities if we have a *countable* union of members of our σ-algebra. In this case, though, we have an uncountable union. If you're not familiar with countable and uncountable sets, see Appendix C.

We mentioned that trouble arises when we try to assign a probability to every event in an uncountable set. The example we looked at—throwing a dart at the unit circle—is actually misleading to a certain extent. It showed that we cannot have a *uniform* probability function on an uncountable set—that is, a probability function with the property that each element of the set occurs with the same probability. It turns out that we actually can't have a uniform probability function on a countably infinite set either. The real difference between countable and uncountable sets is the following: for an uncountable set A, there doesn't exist a probability function such that $\Pr(a) > 0$ for all $a \in A$ and $\sum_{a \in A} \Pr(a) = 1$. On the other hand, for a countable set B, we can find a probability function such that $\Pr(b) > 0$ for all $b \in B$ and $\sum_{b \in B} \Pr(b) = 1$.

We'll do the case of a countable infinity in full detail, and wave our hands a bit for the uncountable case. For the countable case, look at $\Pr(n) = 1/2^{n+1}$ if n is a non-negative integer, and 0 otherwise. Clearly the probabilities are non-negative. Using the geometric series expansion from Chapter 1, we find they sum to 1. Recall the geometric series formula says

$$\sum_{n=0}^{\infty} ar^n \;=\; \frac{a}{1 - r},$$

provided $|r| < 1$. In our case, $1/2^{n+1}$ corresponds to $1/2 \cdot 1/2^n$. Thus $a = 1/2, r = 1/2$, and the sum of the probabilities is $\frac{1/2}{1-1/2} = 1$, as desired.

For the uncountable case, we have to assign uncountably many events a positive probability. Let's look at the event A_n, which will be all elements of A whose probability is in $(\frac{1}{n+1}, \frac{1}{n}]$. There are countably many subsets A_n, and as each event in A has *some* positive probability, it must lie in *some* A_n. Thus

$$A \;=\; \bigcup_{n=1}^{\infty} A_n.$$

At least one of the A_n's must have infinitely many elements, as otherwise we would only have countably many elements in A (we need to use a few results from our set theory appendix, specifically a countable union of countable sets has countably many elements). So, there's some m such that A_m has infinitely many elements, and each element there has probability at least $\frac{1}{m+1}$. We've reached a contradiction — we've just shown A_m has infinite probability, which can't happen! □

 We end with one of the most troublesome sets in all mathematics, the standard non-measurable set. I'm including this example for completeness, but if ever there was a time to skim, this is it! Its existence depends on the (Uncountable) Axiom of Choice, and it's the cause of the Banach-Tarski paradox. In fact, it's the reason I don't like to assume the Axiom of Choice (but a lot of my good friends are algebraists and they need it, so for them...). Anyway, here's the set. We say two numbers x and y in $[0, 1]$ are equivalent if their difference is rational, and denote this by writing $x \sim y$. Let $[x] = \{y \in [0, 1] : y \sim x\}$. Using the Axiom of Choice, form a set $\mathcal{N}$ consisting of one element of each equivalence class of numbers in $[0, 1]$. Maybe $\mathcal{N}$ looks like $\{0, \sqrt{2}/2, \pi/4, \dots\}$. For a rational number r, let $\mathcal{N}_r = \{x + r : x \in \mathcal{N}\} \cap [0, 1]$.

A little algebra shows that if $r_1 \neq r_2$ then $\mathcal{N}_{r_1} \cap \mathcal{N}_{r_2}$ is empty. If not, say $x_1 + r_1 = x_2 + r_2$ and $x_1 \neq x_2$. Rearranging gives $x_1 = x_2 + r_2 - r_1$, implying $x_1 \sim x_2$, which contradicts the definition of $\mathcal{N}$ (which has exactly one representative of each equivalence class). We can therefore write $[0, 1]$ as a countable union of these $\mathcal{N}_r$'s: $[0, 1] = \cup_{r \in \mathbb{Q}} \mathcal{N}_r$.

As our sets live in $[0, 1]$, it's natural to associate probabilities to "lengths." We're going to add one more axiom to our list: if A is a set and $A + r$ is a translation, then A and $A + r$ have the same probability. This seems reasonable—simply sliding a set down shouldn't change its length, right? This is the one-dimensional analogue of our Banach-Tarski discussion, where we assumed that rotating and translating won't change volumes. Let's see what happens!

One of our probability axioms says that the probability of a countable disjoint union should be the sum of the probabilities. As each set is a translation of $\mathcal{N}_0$, and translating shouldn't effect the probability (remember we're using probability interchangeably with length), we run into trouble. The sum has to be 1, but there are infinitely many identical summands. If each is zero we get the total probability (or length) of $[0, 1]$ is 0, but if each summand is positive then the sum is infinity! There's *no* valid way to assign a probability to $\mathcal{N}$.

Though involved, the previous example has a great lesson for us. Our prior discussions involved having uncountably many sets. Here we only have countably many sets, but uncountably many elements arranged in a bad way (the structure of these sets are wildly different from the intervals $[a, b]$, which also have uncountably many elements). Again, you should view this example as planting a flag in the field warning you of the issues and the care needed.

2.7 Appendix: Experimentally Finding Formulas

Students often remark that while they can follow a proof where each detail is given, going from line to line, they don't feel as if they could have initiated the chain of logic. One of the hardest skills to learn in math is getting a sense of what is true, of what you should try to prove. The purpose of this section is to help you sniff out formulas. It's significantly easier to prove something if you know what you should try to prove!

Our main example from probability will be the formula for $\Pr(A \cup B)$. We'll start first with a result from calculus you should know well, and use that to introduce you to some important techniques. In a sense, what we're doing is almost experimental mathematics; we're looking at a few special cases and trying to use that to detect the general pattern. Note that the formulas we find are not rigorously proved, but at least we now have a target.

2.7.1 Product Rule for Derivatives

Earlier we talked about a wish list for probability rules. Of course, there's nothing to stop us from creating other lists. Let's revisit calculus. While the sum rule is very nice (the derivative of a sum is the sum of the derivatives), the product and chain rules are definitely *not* what we would wish for!

For example, if we have $h(x) = f(x)g(x)$, our dream result would be $h'(x) = f'(x)g'(x)$, or the derivative of a product is the product of the derivatives (and similarly we would hope that the derivative of a quotient was the quotient of the derivatives). Unfortunately, of course, this is not true and $h'(x) = f'(x)g(x) + f(x)g'(x)$. The first time students see this the formula appears quite strange and not at all intuitive; our goal below is to explain how one can guess this formula by trying special choices of f and g.

How should we choose f and g? We need to take two functions where not only do we know their derivatives but we also know the derivative of their product. This suggests we take f and g to be simple pure polynomials, such as $f(x) = x^m$ and $g(x) = x^n$ for two positive integers m and n. Note in this case we have $h(x) = x^{m+n}$ and $h'(x) = (m+n)x^{m+n-1}$.

Now, how do we sniff out the formula for the derivative of a product? It's reasonable to expect our answer to depend on the following inputs: $f(x)$, $f'(x)$, $g(x)$, $g'(x)$. This gives us the following:

$$f(x) = x^m, \quad f'(x) = mx^{m-1}$$
$$g(x) = x^n, \quad g'(x) = nx^{n-1}.$$

Notice that the power of x in $h'(x)$ is $m + n - 1$; thus it's natural to ask how we can use the information above to get terms with x to that power. There are two obvious ways: $f'(x)g(x) = mx^{m+n-1}$, and $f(x)g'(x) = nx^{m+n-1}$. Amazingly, if we add these two together we get $h'(x)$, which suggests that perhaps the derivative of $f(x)g(x)$ is $f'(x)g(x) + f(x)g'(x)$!

Again, it's important to note that this is *not* a proof, but it now gives us a target for our theoretical analysis. If you look at the standard proof of the product rule, one starts with the definition of the derivative and adds zero in a clever way so that these two terms pop out.

Finally, it's worth mentioning the value of looking at additional examples to support what we've seen. The more examples we have, the more confident we can feel. While it was natural to look at $f'(x)g(x)$ and $f(x)g'(x)$ (as those gave the correct power of x), there are lots of other expressions we could have had, ranging from $f(x)g(x)/x$ to $xf'(x)g'(x)$!

While there are infinitely many possibilities, some do seem more natural than others. Further, we can look at other examples to hopefully narrow down the list of possibilities, so long as we can find examples where we can compute the derivative of the product. Another great choice would be to take $f(x) = \sin(x)$ and $g(x) = \cos(x)$. This is a little harder to use, though, as we need to know some trig formulas to simplify the analysis (starting with $h(x) = \sin(x)\cos(x)$ can be rewritten as $h(x) = \frac{1}{2}\sin(2x)$); if we didn't recall that trig identity, we wouldn't have a product whose derivative was known (although even this requires us to know a simple result, a special case of the chain rule!).

2.7.2 Probability of a Union

We now come to our main objective, trying to explain how we can predict

$$\Pr(A \cup B) = \Pr(A) + \Pr(B) - \Pr(A \cap B).$$

It's reasonable to guess $\Pr(A \cup B)$ is a function of $\Pr(A)$, $\Pr(B)$, and $\Pr(A \cap B)$. Why? We start with sets A and B, and it's natural to also consider the intersection, $A \cap B$. The hope is that $\Pr(A \cup B)$ is some function of these. Let's start by assuming it's a polynomial, say

$$\begin{aligned}
\Pr(A \cup B) = {} & c_1 + c_2\Pr(A) + c_3\Pr(B) + c_4\Pr(A \cap B) \\
& + c_5\Pr(A)^2 + c_6\Pr(A)\Pr(B) \\
& + c_7\Pr(A)\Pr(A \cap B) + c_8\Pr(B)^2 + \cdots \, ;
\end{aligned}$$

note there are a lot of possible terms!

To prune the possibilities, we'll look at special choices of A and B and see what that implies about the coefficients c_i. Let's start simple, and see if we can find a good candidate if we just look at constant and linear terms (if that doesn't work we'll have to consider something more complicated). So, for now, our goal is to find c_1, c_2, c_3, c_4 such that

$$\Pr(A \cup B) = c_1 + c_2\Pr(A) + c_3\Pr(B) + c_4\Pr(A \cap B).$$

It's time to choose special values of A and B. The easiest sets to work with are the empty set $\emptyset$ and the entire space Ω, as we both know the probabilities of these sets and intersections and unions of them.

- If we take $A = B = \emptyset$, then $\Pr(A \cup B) = 0$ and $\Pr(A) = \Pr(B) = \Pr(A \cap B) = 0$. Thus we must have $c_1 = 0$.
- If we take $A = B = \Omega$, then $\Pr(A \cup B) = 1$ and $\Pr(A) = \Pr(B) = \Pr(A \cap B) = 1$. Thus we must have $c_2 + c_3 + c_4 = 1$.
- If we take $B = \emptyset$, then $\Pr(A \cup B) = \Pr(A)$ and $\Pr(B) = \Pr(A \cap B) = 0$, which implies $c_2 = 1$.
- We could repeat the above argument, now taking $A = \emptyset$, and we would find $c_3 = 1$; however, there is no need. By symmetry we should have $c_2 = c_3$, as we can't tell the difference between the sets A and B (note $\Pr(A \cup B) = \Pr(B \cup A)$).

We don't need to look at any more cases to get c_4, as we know $c_2 + c_3 + c_4 = 1$ and $c_2 = c_3 = 1$ and thus we must have $c_4 = -1$. Putting the pieces together, we are lead to conjecture that

$$\Pr(A \cup B) = \Pr(A) + \Pr(B) - \Pr(A \cap B).$$

Notice we're able to get to the right formula just by exploring some basic cases!

It's worth stepping back and looking at what we've done. In some sense, none of this was needed as we were given a proof of the formula for the probability of a union. In another sense, this is far more useful long term. Why? The purpose of this section is to help you figure out what the answer should be in new, unfamiliar situations, and hopefully gaining a familiarity with arguments like this will pay great dividends later.

2.8 Summary

In this chapter we paid our dues. In the next chapter, we earn the dividends. Now that you know the language of probability you can learn to speak the subject, but it's essential to first learn the terminology.

Language matters. A terrific example of this is George Orwell's *1984*, where the language Newspeak is designed to limit thought. Of course, our goal in mathematics is the opposite. It's more than just choosing good names, it's choosing *what* to name. We give names to the important concepts to call attention to them.

Here, we introduced the standard notions and notations of set theory, and some of point set topology. These terms were essential in helping us axiomatize probability, aiding us in avoiding paradoxes and pitfalls. We gave Kolmogorov's axiomatic foundation of probability, and explored the consequences both theoretically (in rules that hold for probabilities of events) and several examples.

We end with a quote from Orwell's *Politics and the English Language* (1946). Keep this in mind whenever you're starting a subject and mastering the definitions, or even better, when *you* get to choose notation. Orwell wrote:

A scrupulous writer, in every sentence that he writes, will ask himself at least four questions, thus: 1. What am I trying to say? 2. What words will express it? 3. What image or idiom will make it clearer? 4. Is this image fresh enough to have an effect?

2.9 Exercises

Exercise 2.9.1   *Consider the set $A = \{1, 2, \ldots, n\}$. If all subsets of A are equally likely to be chosen, what is the probability that a randomly selected subset of A contains 1? What is the probability that a randomly selected subset of A contains 1 and 2? What is the probability that a randomly selected subset of A contains 1 or 2?*

Exercise 2.9.2   *Consider the set $A = \{1, 2, \ldots, n\}$. If all subsets of A are equally likely to be chosen, what is the probability that a randomly selected subset of A has an even number of elements? Does the answer depend on n? If yes is it easier to handle the case of n even or n odd?*

Exercise 2.9.3   *Find sets A and B such that $|A| = |B|$, A is a subset of the real line and B is a subset of the plane (i.e., $\mathbb{R}^2$) but is not a subset of any line.*

Exercise 2.9.4   *Does $\mathcal{P}(A \cup B) = \mathcal{P}(A) \cup \mathcal{P}(B)$? Prove your answer.*

Exercise 2.9.5   *Does $\mathcal{P}(A \cap B) = \mathcal{P}(A) \cap \mathcal{P}(B)$? Prove your answer.*

Exercise 2.9.6   *Let $\mathcal{P}(X)$ denote the power set of a set X. Let A and B be two non-empty disjoint finite sets. Is there a relation between $\mathcal{P}(\mathcal{P}(A) \cup \mathcal{P}(B))$ and $\mathcal{P}(\mathcal{P}(A \cup B))$? If yes what is it?*

Exercise 2.9.7   *Let A be a finite set with $n > 0$ elements. Set $\mathcal{P}_1(A) = \mathcal{P}(A)$ and $\mathcal{P}_m(A) = \mathcal{P}_{m-1}(A)$ for $m \geq 2$. Is there a formula for the size of $\mathcal{P}_n(A)$? If yes what is it?*

Exercise 2.9.8   *We discussed how we can build up the integers from the empty set. Starting with the empty set, we could have instead created the sets $\{\emptyset\}$, $\{\{\emptyset\}\}$, $\{\{\{\emptyset\}\}\}$, $\{\{\{\{\emptyset\}\}\}\}$ and so on. Is this a better or worse choice than what we did earlier? Why?*

Exercise 2.9.9   *Two sets A and B are equal if every element in A is in B and every element in B is in A. Let n be any positive integer. Prove or disprove: there are only finitely many distinct sets with exactly n elements.*

Exercise 2.9.10   *Assume there is a one-to-one function f from a set A to the positive real numbers, but do not assume that f is onto. Must A have infinitely many elements?*

Exercise 2.9.11   *Write at most a paragraph on the Continuum Hypothesis.*

Exercise 2.9.12   *Suppose we have two events, A and B, with $\Pr(A) = .3$, $\Pr(B) = .6$, $\Pr(A \cap B^C) = .2$. What is $\Pr(A \cup B)$?*

Exercise 2.9.13   *Suppose there are 3 events A, B, and C, with $\Pr((A \cup B) \cap C) = .3$, $\Pr((A \cup C) \cap B) = .3$, $\Pr((B \cup C) \cap A) = .3$, and $\Pr(A \cap B \cap C) = .1$. What is $\Pr(((A \cup B) \cap C) \cup ((A \cup C) \cap B) \cup ((B \cup C) \cap A))$?*

Exercise 2.9.14   *When we proved $\Pr(A \cup B) = \Pr(A) + \Pr(B) - \Pr(A \cap B)$ we expressed the probability of a union in terms of probabilities of intersections; we'll see later that intersections are often easier to calculate (they correspond to each event that must happen, whereas unions are the trickier as at least one of the events happens). Guess a formula for $\Pr(A \cup B \cup C)$ in terms of sums and differences of intersections. What if you had four (or more) sets? The answer in the general case is the inclusion-exclusion formula (see Chapter 5).*

Exercise 2.9.15   *Give an example of an open set, a closed set, and a set that is neither open nor closed (you may not use the examples in the book); say a few words justifying your answer.*

Exercise 2.9.16   *Let $A = \{(x, y) : |x + y| < 1\}$, and let $B = \{(x, y) : |x + y| \leq 1\}$. Prove A is open and B is closed.*

Exercise 2.9.17   *Prove or disprove: a countable union of open sets is open.*

Exercise 2.9.18   *Prove or disprove: a countable intersection of open sets is open.*

Exercise 2.9.19   *In the probability tree with two independent tosses of a coin that lands on heads 70% of the time, we had four numbers in the final row. Is it a coincidence that the middle two numbers are equal? What would you expect if we had three tosses?*

Exercise 2.9.20   *Use the axioms of probability to prove that the method of counting works when there are finitely many elements in Ω and each element is equally likely to be chosen. That is, $\Omega = \{a_1, a_2, \ldots, a_n\}$ and $\Pr(a_i) = 1/n$ for all $i \in \{1, 2, \ldots, n\}$.*

Exercise 2.9.21   *Use the Axioms of Probability to prove that $\Pr(A) + \Pr(A^c) = 1$ and if $A \subset B$, then $\Pr(A) \leq \Pr(B)$.*

Exercise 2.9.22   *Let $\Omega = \{1, \ldots, 100\}$. Find a σ-algebra Σ that is neither $\{\emptyset, \Omega\}$ nor $\mathcal{P}(\Omega)$.*

Exercise 2.9.23 *Give another proof that the probability of the empty set is zero.*

Exercise 2.9.24 *Let $\mathcal{Z}_n = \{1, 2, \ldots, n\}$. How many subsets of $\mathcal{Z}_n$ are there such that we never choose two adjacent elements of $\mathcal{Z}_n$? Thus $\{1, 4, 6, 12, 15\}$ and $\{1, 4, 6, 12, 15, 20\}$ would both work for $\mathcal{Z}_{20}$, but $\{1, 4, 5, 12, 15\}$ would not. What if we additionally require n to be in the set: how many subsets are there now? In exercises like this it is often worthwhile to write a program or compute by hand the answer for small n and detect the pattern.*

Exercise 2.9.25 *Find the probability of rolling exactly k sixes when we roll five fair die for $k = 0, 1, \ldots, 5$. Compare the work needed here to the complement approach in the book.*

Exercise 2.9.26 *If f and g are differentiable functions, prove the derivative of the product $f(x)g(x)$ is $f'(x)g(x) = f(x)g'(x)$. Emphasize where you add zero.*

Exercise 2.9.27 *Let $\{A_n\}_{n=1}^{\infty}$ be a countable sequence of events such that for each n, $\mathrm{Prob}(A_n) = 1$. Prove the probability of the intersection of all the A_n's is 1.*

Exercise 2.9.28 *Imagine that fingerprint evidence is found at the scene of a crime. The probability that it matches a person by chance is 1 in 5,000. The evidence is compared to a database of 30,000 people's fingerprints. Find the probability it matches at least 1 person in the database by chance alone.*

Exercise 2.9.29 *Building on the previous exercise, if the guilty person is in the database, find the expected number of people in the database who match the fingerprint evidence.*

Exercise 2.9.30 *Assume the rate of transmission of HIV is between 1/100 and 1/1000 (this was the number reported in the Los Angeles Times, August 24, 1987). Taking the chance of transmission to be 1%, find the risk of becoming infected if you are exposed 100 times.*

Exercise 2.9.31 *In the previous exercise imagine instead that the rate of transmission is 1/1000; now what is the probability of becoming infected if you are exposed 1000 times? Find the risk of being infected if you are exposed to a virus that is transmitted with probability 1/n in each exposure if you are exposed n times, for large n.*

Exercise 2.9.32 *Imagine you are considering going on a hike today and want to know the probability it will rain before noon. Assume the probability of rain is equal to the percent chance of rain given by the news. Find the error in the following analysis: The hourly forecast on the news said that there is a 20% chance of rain from 8–9, a 30% chance from 9–10, a 30% chance it rains from 10–11, and a 10% chance it rains from 11–12. From this I conclude the probability of rain before noon is $(1 - .8)(1 - .7)(1 - .7)(1 - .9) \approx 35\%$. Is this a good conclusion? Explain.*

Exercise 2.9.33 *Imagine a team needs to win the last two games of the season to make the play-offs. The two teams it plays have complementary records (so if one wins p% of their games, the other wins $(1 - p)\%$). Assume if a team wins p% of their games you have a $(1 - p)\%$ chance of defeating them. What value of p maximizes your chance of making the play-offs?*

Exercise 2.9.34 *Write a program to approximate the average length of the longest run of heads or tails (whichever is longer) in 100 tosses of a coin that lands heads with fixed probability p. Try several values of p and report your results.*

Exercise 2.9.35 *Taking the union of sets is critical in probability. Mathematica has the built-in command "Union," which finds the union of the input sets and sorts them. Use more fundamental commands to write code that will take the union of two sets and delete duplicates.*

Exercise 2.9.36 *Mathematica has a built-in "Sort" function that will order the elements of a set from smallest to largest (or in other orderings if you give it additional input). Use more fundamental commands to sort a set with real-valued elements from smallest to largest.*

Exercise 2.9.37 *We saw that there are issues in assigning probabilities to an uncountable union of events; what if we have an uncountable intersection? Can we assign a probability there?*

Exercise 2.9.38 *Let Ω be an outcome space and A, B distinct events such that $\mathrm{Prob}(A) = \mathrm{Prob}(B) = 1$. What is true about the probability that A and B both happen? That neither of them happen? What is the probability of their **symmetric difference** (which is the set of all elements in one but not the other)?*

Exercise 2.9.39 *In §2.7 we discussed how to try and sniff out formulas by looking at special cases. Verify the product rule by looking at $f(x) = \sin(x)$, $g(x) = \cos(x)$, and $h(x) = f(x)g(x)$. You may use standard trig identities, such as $\sin(x)\cos(x) = \frac{1}{2}\sin(2x)$, and that the derivative of $\sin(2x)$ is $2\cos(2x)$.*

Exercise 2.9.40 *Imagine we try to guess the product rule for differentiation by taking the functions e^x, e^x, and e^{2x}. What goes wrong?*

Exercise 2.9.41 *In §2.7 we found a reasonable candidate for $\Pr(A \cup B)$; generalize that method to find a similar formula for $\Pr(A \cup B \cup C)$. Can you extend even further to $\Pr(A_1 \cup A_2 \cup \cdots \cup A_n)$?*

The next questions involve the **discriminant** Δ of a polynomial

$$f(x) = x^n + a_{n-1}x^{n-1} + \cdots + a_1 x + a_0,$$

which is defined as the square of the product of the differences of the roots. Thus if

$$f(x) = (x - r_1)(x - r_2)\cdots(x - r_n)$$

then

$$\Delta = \prod_{1 \le i < j \le n} (r_i - r_j)^2.$$

Note the discriminant vanishes if two of the roots are equal, and is non-zero otherwise. As the roots are functions (in general, complicated ones!) of the coefficients, the discriminant is a function of the coefficients of the polynomial, which for simplicity we have normalized to have leading coefficient 1. It would be too much of a detour to go into the importance and applications of the discriminant, though the fact that it detects whether or not there is a repeated root should be enough to suggest the value of studying it. The discriminant of the quadratic $f(x) = x^2 + bx + c$ is $b^2 - 4c$, while the discriminant of the cubic $f(x) = x^3 + Ax + B$ is $-4A^3 - 27B^2$ (note it suffices to study cubics of this form, as we can always replace x with $x - x_0$ to eliminate the quadratic term, and such a translation will not affect the difference of the roots). The following two exercises are amazing—using the techniques of §2.7 we are led to the correct form; unlike our computation of the probability of a union, however, in this case it is *not* immediately clear what the answer should be, and hence this technique is quite valuable here.

Exercise 2.9.42 *Consider the quadratic $f(x) = x^2 + bx + c$. In order to add three physical quantities, they must all have the same units. Thus if we pretend x is measured in units, then b must have units of meters and c must have units of meters-squared. As the discriminant is $(r_1 - r_2)^2$, it has units of meters-squared and it is natural to guess it must be of the form $\alpha_1 b^2 + \alpha_2 c$, as these are the simplest ways to get meters-squared (again, we could have $(b^4 + c^2)/(b^2 + 10c)$; our arguments are not proofs but merely experimental attempts to sniff out the correct functional form). Assuming a linear relation exists and $\Delta = \alpha_1 b^2 + \alpha_2 c$, determine α_1 and α_2 by looking at good choices of b and c.*

Exercise 2.9.43 *Let's generalize the previous exercise to the cubic $f(x) = x^3 + Ax + B$. Again viewing x as being in meters we find that the units of A should be meters-squared and of B meters-cubed in order for the three summands to be of the same dimension and amenable to addition. Now the discriminant $(r_1 - r_2)^2(r_1 - r_3)^2(r_2 - r_3)^2$ will have units meters^6, which suggests Δ should equal $\beta_1 A^3 + \beta_2 B^2$. Assuming such a form holds, find the values of β_1 and β_2 by making good choices for A and B where you can compute the roots (and hence the discriminant) easily.*

Exercise 2.9.44 *The ellipse $E_{a,b}$ is the set of points (x, y) such that $(x/a)^2 + (y/b)^2 \le 1$. Make a reasonable guess for the area of the ellipse as a function of a and b. (Hint: Your answer should be symmetric in a and b, i.e., you shouldn't be able to distinguish one from the other. Further, it should reduce to the formula for the area of a circle when $a = b$.) Unfortunately, even with those hints there are a lot of possibilities, such as $\pi \sqrt{\frac{a^2 + b^2}{2}}$; you'll need to explore special cases of a and b to eliminate candidates such as this. We will revisit this exercise when we discuss Monte Carlo integration.*

Exercise 2.9.45 *Generalize the previous exercise to guess the volume of an ellipsoid $(x/a)^2 + (y/b)^2 + (z/c)^2 \le 1$.*

CHAPTER 3 _____

Counting I: Cards

Every gambler knows
That the secret to survivin'
Is knowin' what to throw away
And knowin' what to keep.
— KENNY ROGERS, *The Gambler* (1978)

In Chapter 1 we tried, and hopefully succeeded, in motivating the study of probability. We talked about fun topics like birthdays and basketball. Then Chapter 2 happened. We came to a screeching stop, rolled up our sleeves, and were battered by definition after definition, axiom after axiom. While these are important and necessary, it's easy to be overwhelmed and frustrated. You probably don't want to sit and memorize definitions; you want to solve problems, you want to see how probability is used.

While this is impossible to do without the proper framework, at some point enough is enough. Instead of barreling forward and continuing with exploring the consequences of the axioms, let's put the theoretical development aside for a bit and look at some problems. So, for now, we won't discuss independence or conditional probabilities, even though these are two of the most important topics in the subject. We'll leave these for Chapter 4. Instead, let's apply what we've learned to a bunch of common problems, and hopefully reignite our interest in the subject.

What's nice is that we've already done enough to discuss a lot of interesting examples. We may informally appeal to some properties of independent events, but nothing beyond everyday experiences. Thus, not surprisingly, our examples are often drawn from everyday experiences in this chapter. It's not our place to write a history of the subject, but we'd be remiss not to mention its long connections with games of chance. Humanity has been rolling dice for over 5000 years, playing variants of what's now known as Backgammon. And, once games are being played, gambling isn't far behind. A lot of probability was developed to analyze games of chance, looking for optimal strategies. In this chapter we'll concentrate on some of these games, as well as more mundane examples.

We'll explore card games, such as poker, solitaire, and bridge. These are great opportunities to practice finding probabilities of discrete events, such as the odds of getting a specific hand. Before jumping into these problems, let me provide some motivation. At first, these may seem a bit far afield from probability, but remember: *probability and counting shouldn't be separated.* If we can count, we can find probabilities by dividing the number of events with a desired property by the number of possible events. So, when you read all the counting arguments in this chapter, keep this in mind: we need to master counting if we're to master probability. While you may not care about the games discussed, what matters is how we attack these problems. Even if you never play poker, or solitaire, or bridge, there are great lessons to be learned from counting problems in these games.

In fact, a lot of the issues that arise in this chapter involve the most fundamental notions in determining probabilities of events, such as independence and conditional probabilities. Thus, this chapter is a springboard to Chapter 4. In other words, we'll analyze these problems as much as we can using just the machinery of Chapter 2, and then see where we get stuck. To resolve these issues we'll need to flush out the basic theory more, which we'll do. After that, we'll return and do more of these basic probability problems, but armed with more powerful techniques.

Finally, some advice for those of you who are thinking about heading off to the casinos after reading this chapter: remember that the casinos employ mathematicians too, and they've read more than just a chapter! (For those who insist on heading out, read Chapter 23 or see https://www.youtube.com/watch?v=Esa2TYwDmwA.)

3.1 Factorials and Binomial Coefficients

Now that you're excited by thoughts of gambling success, I sadly have to ask you to hold on for a moment and master some basic combinatorics. We need a few functions to successfully attack these problems. We'll introduce these concepts as quickly as we can, and then move to the applications. So, the next two subsections won't have too many examples. There'll be a few to help highlight the concepts, but we're concentrating on motivating the definitions and their interpretations. Once these are solid, then we'll continue with the applications.

3.1.1 The Factorial Function

The first is the **factorial function**, which we've seen already. If n is a positive integer, we set $n! = n(n-1)(n-2)\cdots 3 \cdot 2 \cdot 1$; *it turns out to be convenient to extend this and let* $0! = 1$ (more on this later). This function grows extremely rapidly: $10! = 3,628,800$, $100!$ is approximately 10^{158}, while $1000!$ is greater than 10^{2500}. We can define this function **recursively**: $n! = n \cdot (n-1)!$; this means we obtain later values in terms of earlier values of the sequence.

There's a nice combinatorial interpretation: $n!$ is the number of ways of arranging n people when order matters. This is essentially the same idea as the multiplication rule we saw earlier. Imagine we have five people—Eli, Cam, Matt, Kayla, and Gabrielle— and we want to look at all the ways to order them. There are five ways to choose the first person. After that choice, there are four people left, and thus four choices for our second person. Maybe we chose Gabrielle first, and then our second choice will be one of Eli, Cam, Matt, or Kayla. Maybe it was Cam who was chosen, forcing our second choice to come from Eli, Matt, Kayla, or Gabrielle. Regardless, there are five choices for the first person, then four for the second, then three for the third, then two for the fourth, and finally just one person remains for the last position. In full glory, let's say

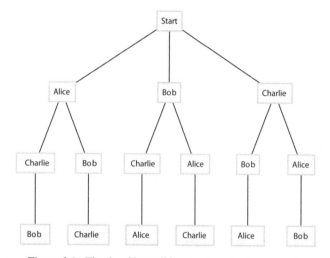

Figure 3.1. The $6 = 3!$ possible ways to order three people.

we choose Gabrielle first of the five options, then of the remaining four (Eli, Cam, Matt, and Kayla) we choose Cam, leaving us with three (Eli, Matt, and Kayla) and perhaps we then choose Matt. Now down to two (Eli and Kayla) let's say we take Kayla, and then finally we end with Eli.

It's a little hard to picture what's going on with five people, so we exhaustively look at three in Figure 3.1. You might also have noticed it's a bit confusing following the discussion because of the names. This wasn't accidental; it was to make a point (and to please my kids and my oldest nieces and nephews). **Good notation** is, well, good! This is why we typically choose names in alphabetical order, as it's much easier to follow. So, we'll name our three people Alice, Bob, and Charlie (though we could've used Ariel, Belle, and Cindy, or even Agamemnon, Brutus, and Caesar).

We said there are three possibilities for the first choice, and we see that in the tree. Each name appears one-third of the time as the first choice. While there are only two choices for the second person, we still have each name appearing one-third of the time as the second choice, and similarly for the third. This should seem reasonable. Each person has the same chance of being chosen first, second, or third, and therefore should appear equally often as a first, second, or third choice.

Of course, nothing says we have to choose all the people. If we want to choose 4 out of 9 people with order mattering, there are 9 choices for the first, 8 for the second, 7 for the third, and finally 6 for the fourth person. This is $9 \cdot 8 \cdot 7 \cdot 6$, but there's a far more illuminating way to write it. We get to use one of my favorite techniques, **multiplying by one**. How should we do it? If we choose *all* the people then the answer is 9!. Seeing the factorial emerge here, it's natural to try to find a way to bring in factorials here. Notice that $9 \cdot 8 \cdot 7 \cdot 6$ is the *start* of a factorial. It's missing $5 \cdot 4 \cdot 3 \cdot 2 \cdot 1$ or 5!. This suggests that a great way of writing the answer is

$$9 \cdot 8 \cdot 7 \cdot 6 = \frac{9 \cdot 8 \cdot 7 \cdot 6 \cdot 5!}{5!} = \frac{9!}{5!}.$$

More generally, if we want to choose k of n people, with order mattering, it's just

$$n(n-1)(n-2)\cdots(n-(k-1))$$
$$= \frac{n(n-1)(n-2)\cdots(n-(k-1))\cdot(n-k)!}{(n-k)!} = \frac{n!}{(n-k)!}.$$

The above is another great example of a powerful technique: **multiplying by one**. This allows us to **rewrite algebra** in a simpler, more illuminating way. It is well worth the time needed to become familiar with this powerful method.

For me, the hardest part in calculations like this is getting the factors right, namely, how far down do we go? The final factor is $n-(k-1)$, not $n-k$, when we're choosing just k from n with order important. There are a few tricks to make sure you don't make a mistake here. One is to do a special case first. We did choosing 4 from 9, and we saw the last factor in the product was 6. We need to write 6 in terms of 9 and 4, and we see $6 = 9-(4-1)$, not $9-4$. You could also note that the first factor is 9 or $9-0$, and see that we're subtracting one less than the person we're at, and so the final subtraction must involve $k-1$ and not k (remind you of the Birthday Problem?). It's always a good idea to test your formula on a case you know.

Let's do another example. Say there are 70 senior math majors at Williams College. How many ways are there to choose 30 of them to take Probability, where the order in which they're chosen matters? The answer's "just" $70!/(70-30)! = 70!/40!$. Notice how easily we can write down the answer—this is the advantage of the factorial function!

If you were looking at the text carefully, you should've noticed I put the word just in quotes above. Why? I can easily compute 5! or 6!, but 70!? (I know, it looks strange ending a sentence with an exclamation point followed by a question mark.) While it's easy in principle to compute 70! or 40!, it's a lot of multiplications. Can we say anything about how big these numbers are? These were carefully chosen—most calculators can handle these, but a little more (like 100!) might lead to an overflow error. Stirling's formula provides a very good estimate for $n!$ for n large, which we'll discuss in Chapter 18. Nonetheless, if you're interested, the number of ways is approximately $1.46 \cdot 10^{52}$; the exact answer is

14, 681, 146, 334, 564, 331, 088, 939, 671, 869, 953, 268, 066, 486, 845, 440, 000, 000.

Let's end this section with one more example to drive home just how enormous these factorials are. You might think the above problem is a bit unnatural; do we really think 30 out of the 70 senior math majors at Williams want to take probability with me, and if so, who cares what order they sign up?

Here's a more natural example. Consider a **standard deck of cards**. We're going to draw examples from cards throughout the book, so let's make sure everyone is familiar with them. There are 52 cards in four suits (spades ♠, hearts ♡, diamonds ◇, and clubs ♣). Each suit has 13 cards, numbered 2, 3, 4, 5, 6, 7, 8, 9, 10 and then the four special cards J (jack), Q (queen), K (king), and A (ace). How many ways are there to order a deck of cards?

Well, there are 52 choices for the first card. Maybe we get the ace of spaces (A♠) or the 5 of diamonds (5♢). Once we choose our first card, we have 51 choices for our next card, and so on and so on. Thus there are 52! ways to order the deck of cards, or

$$80, 658, 175, 170, 943, 878, 571, 660, 636, 856, 403, 766, 975,$$
$$289, 505, 440, 883, 277, 824, 000, 000, 000, 000$$

possibilities! How big is this? It's roughly $8.01 \cdot 10^{67}$, or 15 orders of magnitude greater than our made up example.

It's easy to write a number like 10^{67} on the blackboard; it's harder to get a sense of just how colossal it is. So, let's try to put it in perspective. Rounding up, say there are 10 billion (10^{10}) people on the Earth. Imagine each person can do one shuffle a second. It would take about $8.01 \cdot 10^{57}$ seconds to go through all the possibilities. There are about $3.2 \cdot 10^7$ seconds in a year, so it would require almost $2.6 \cdot 10^{50}$ years! And remember, this is in the special case when the entire population of the world devotes all of its time to shuffling, and somehow shuffles really, really fast!

It's worth pausing and reflecting on the factorial function and how fast it grows. We plot the factorial function $f(n) = n!$ against the squaring function $g(n) = n^2$ in Figure 3.2. Even though initially the squaring function is larger, by the time $n = 4$, the factorial function is greater, and around $n = 7$ the factorial is so much larger that we can just barely see the two of them. Continuing to $n = 10$ gives $10! = 3, 628, 800$, which is over 36,000 times the size of 10^2.

 If you looked at Figure 3.2 *very* carefully, you might have noticed something strange. We've only defined the factorial function for *integer* inputs, but we plotted it for all real values from 1 to 10. What gives? By now hopefully you're convinced that the factorial function is useful. It's so useful, in fact, that it's a shame and a crime to limit it to just integer inputs. Mathematicians (and statisticians, physicists, engineers, ...) have found it useful to extend the factorial function to *all* real numbers (and even complex ones too!). We call this extension the Gamma function, where $\Gamma(n) = (n-1)!$ for n a non-negative integer. We'll devote a lot of time to this function in Chapter 15, and even explain why it's "natural" to have that pesky shift (if you were going to generalize the factorial function, wouldn't you want $f(n) = n!$ and not $(n-1)!$?). Okay, while there's a right parenthesis in the mix, we did end the last sentence in a grammatically correct way with an exclamation point, a question mark, *and* a period—one of the many bonuses of studying factorials.

We conclude this section by isolating the definition and interpretation of the factorial function.

Factorial function: If n is a positive integer, then $n! = n \cdot (n-1) \cdots 1$, and $0! = 1$. We may interpret $n!$ as the number of ways to order n people; if n equals 0 we should interpret this as there is only one way to do nothing.

3.1.2 Binomial Coefficients

The factorial function arises in problems where order matters; however, there are plenty of times when order is unimportant. For example, imagine we're back on the elementary school playground and we're picking kids for kickball teams. After a while everyone's

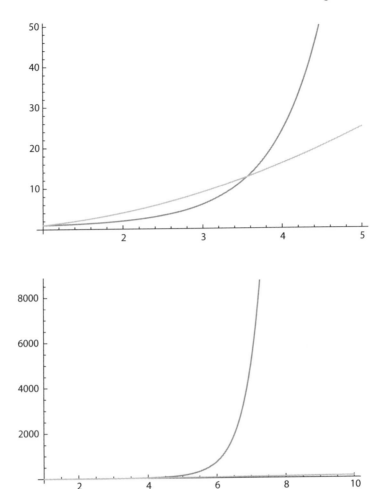

Figure 3.2. Plots of the factorial function versus the squaring function. Notice how much faster the factorial function grows. For n less than about 3.56, the squaring function is larger; however, once n exceeds 3.56 the factorial function is significantly greater than the squaring function. When $n = 6$ we have $6! = 720$ versus $6^2 = 36$; the difference is even more pronounced at $n = 7$, where $7! = 5040$ as compared to $7^2 = 49$.

on one of two teams. It doesn't matter what order you were assigned to a team, all that matters is which team you're on. Or consider a game of poker. You're dealt five cards; it doesn't matter what order you received them, all that matters is what five cards did you get.

Let's look at the last problem a bit more closely. If we want to know how many ways we can get 5 cards from a deck of 52, the answer is just

$$52!/(52 - 5)! \ = \ 52!/47! \ = \ 52 \cdot 51 \cdot 50 \cdot 49 \cdot 48,$$

or 311,875,200. Imagine we're dealt (in this order) 5♣, 6♢, 7♡, 7♠, J♣. While this is a different deal then getting the cards in the order J♣, 5♣, 7♠, 6♢, 7♡, at the end of the day we have the same five cards in each. We need a way to take into account the fact that we can reach the same final hand of five cards in many different ways.

Whenever you have a problem, try to get a feel for the answer. We know there are 52!/47! ways to choose 5 cards from 52 when order matters. We now want to forget about the ordering. Thus, whatever the answer is, it must be *less than* 52!/47!. How much less than? That's the question. Well, in the end all that matters is which 5 cards we have, not the order in which they arrive. How many ways are there to order 5 cards? That's easy—there are 5! ways to arrange 5 cards. Another way of viewing this is that there are 5! ordered sets of 5 cards that contain the same five cards. *All* of these orderings generate hands with the same five cards, just in a different order. Thus, we must *divide* 52!/47! by 5!, giving the number of ways of choosing 5 cards from 52, when order doesn't matter, is just

$$\frac{52!/47!}{5!} = \frac{52!}{5!47!} = 2,598,960.$$

We've gone from about 311 million possibilities to a tad over two and a half million. As expected, there are far fewer possibilities when order doesn't matter.

Let's do another example to showcase what's happening. We'll use good notation and say we have five people: Alice, Bob, Charlie, Dan, and Eve. How many ways are there to choose 3 people when order doesn't matter?

There are 5! or 120 ways to order the five people. In full gory, here they are:

{Alice, Bob, Charlie, Dan, Eve} {Alice, Bob, Charlie, Eve, Dan} {Alice, Bob, Dan, Charlie, Eve}
{Alice, Bob, Dan, Eve, Charlie} {Alice, Bob, Eve, Charlie, Dan} {Alice, Bob, Eve, Dan, Charlie}
{Alice, Charlie, Bob, Dan, Eve} {Alice, Charlie, Bob, Eve, Dan} {Alice, Charlie, Dan, Bob, Eve}
{Alice, Charlie, Dan, Eve, Bob} {Alice, Charlie, Eve, Bob, Dan} {Alice, Charlie, Eve, Dan, Bob}
{Alice, Dan, Bob, Charlie, Eve} {Alice, Dan, Bob, Eve, Charlie} {Alice, Dan, Charlie, Bob, Eve}
{Alice, Dan, Charlie, Eve, Bob} {Alice, Dan, Eve, Bob, Charlie} {Alice, Dan, Eve, Charlie, Bob}
{Alice, Eve, Bob, Charlie, Dan} {Alice, Eve, Bob, Dan, Charlie} {Alice, Eve, Charlie, Bob, Dan}
{Alice, Eve, Charlie, Dan, Bob} {Alice, Eve, Dan, Bob, Charlie} {Alice, Eve, Dan, Charlie, Bob}
{Bob, Alice, Charlie, Dan, Eve} {Bob, Alice, Charlie, Eve, Dan} {Bob, Alice, Dan, Charlie, Eve}
{Bob, Alice, Dan, Eve, Charlie} {Bob, Alice, Eve, Charlie, Dan} {Bob, Alice, Eve, Dan, Charlie}
{Bob, Charlie, Alice, Dan, Eve} {Bob, Charlie, Alice, Eve, Dan} {Bob, Charlie, Dan, Alice, Eve}
{Bob, Charlie, Dan, Eve, Alice} {Bob, Charlie, Eve, Alice, Dan} {Bob, Charlie, Eve, Dan, Alice}
{Bob, Dan, Alice, Charlie, Eve} {Bob, Dan, Alice, Eve, Charlie} {Bob, Dan, Charlie, Alice, Eve}
{Bob, Dan, Charlie, Eve, Alice} {Bob, Dan, Eve, Alice, Charlie} {Bob, Dan, Eve, Charlie, Alice}
{Bob, Eve, Alice, Charlie, Dan} {Bob, Eve, Alice, Dan, Charlie} {Bob, Eve, Charlie, Alice, Dan}
{Bob, Eve, Charlie, Dan, Alice} {Bob, Eve, Dan, Alice, Charlie} {Bob, Eve, Dan, Charlie, Alice}
{Charlie, Alice, Bob, Dan, Eve} {Charlie, Alice, Bob, Eve, Dan} {Charlie, Alice, Dan, Bob, Eve}
{Charlie, Alice, Dan, Eve, Bob} {Charlie, Alice, Eve, Bob, Dan} {Charlie, Alice, Eve, Dan, Bob}
{Charlie, Bob, Alice, Dan, Eve} {Charlie, Bob, Alice, Eve, Dan} {Charlie, Bob, Dan, Alice, Eve}
{Charlie, Bob, Dan, Eve, Alice} {Charlie, Bob, Eve, Alice, Dan} {Charlie, Bob, Eve, Dan, Alice}
{Charlie, Dan, Alice, Bob, Eve} {Charlie, Dan, Alice, Eve, Bob} {Charlie, Dan, Bob, Alice, Eve}
{Charlie, Dan, Bob, Eve, Alice} {Charlie, Dan, Eve, Alice, Bob} {Charlie, Dan, Eve, Bob, Alice}
{Charlie, Eve, Alice, Bob, Dan} {Charlie, Eve, Alice, Dan, Bob} {Charlie, Eve, Bob, Alice, Dan}
{Charlie, Eve, Bob, Dan, Alice} {Charlie, Eve, Dan, Alice, Bob} {Charlie, Eve, Dan, Bob, Alice}
{Dan, Alice, Bob, Charlie, Eve} {Dan, Alice, Bob, Eve, Charlie} {Dan, Alice, Charlie, Bob, Eve}
{Dan, Alice, Charlie, Eve, Bob} {Dan, Alice, Eve, Bob, Charlie} {Dan, Alice, Eve, Charlie, Bob}
{Dan, Bob, Alice, Charlie, Eve} {Dan, Bob, Alice, Eve, Charlie} {Dan, Bob, Charlie, Alice, Eve}
{Dan, Bob, Charlie, Eve, Alice} {Dan, Bob, Eve, Alice, Charlie} {Dan, Bob, Eve, Charlie, Alice}
{Dan, Charlie, Alice, Bob, Eve} {Dan, Charlie, Alice, Eve, Bob} {Dan, Charlie, Bob, Alice, Eve}
{Dan, Charlie, Bob, Eve, Alice} {Dan, Charlie, Eve, Alice, Bob} {Dan, Charlie, Eve, Bob, Alice}
{Dan, Eve, Alice, Bob, Charlie} {Dan, Eve, Alice, Charlie, Bob} {Dan, Eve, Bob, Alice, Charlie}
{Dan, Eve, Bob, Charlie, Alice} {Dan, Eve, Charlie, Alice, Bob} {Dan, Eve, Charlie, Bob, Alice}
{Eve, Alice, Bob, Charlie, Dan} {Eve, Alice, Bob, Dan, Charlie} {Eve, Alice, Charlie, Bob, Dan}

{Eve, Alice, Charlie, Dan, Bob} {Eve, Alice, Dan, Bob, Charlie} {Eve, Alice, Dan, Charlie, Bob}
{Eve, Bob, Alice, Charlie, Dan} {Eve, Bob, Alice, Dan, Charlie} {Eve, Bob, Charlie, Alice, Dan}
{Eve, Bob, Charlie, Dan, Alice} {Eve, Bob, Dan, Alice, Charlie} {Eve, Bob, Dan, Charlie, Alice}
{Eve, Charlie, Alice, Bob, Dan} {Eve, Charlie, Alice, Dan, Bob} {Eve, Charlie, Bob, Alice, Dan}
{Eve, Charlie, Bob, Dan, Alice} {Eve, Charlie, Dan, Alice, Bob} {Eve, Charlie, Dan, Bob, Alice}
{Eve, Dan, Alice, Bob, Charlie} {Eve, Dan, Alice, Charlie, Bob} {Eve, Dan, Bob, Alice, Charlie}
{Eve, Dan, Bob, Charlie, Alice} {Eve, Dan, Charlie, Alice, Bob} {Eve, Dan, Charlie, Bob, Alice}

Remember to think about how we choose to list them. It's very important that we don't miss anything, so we need a good, systematic way of going through the possibilities.

For this problem, we don't want to choose five people, but rather just three. That's fine—all we have to do is forget about the last two people in each choice above. We're left with the slightly more manageable looking set of 60 ways to choose 3 people from 5 with order mattering (this is $5!/(5-3)! = 5!/2! = 60$). Notice 60 equals $120/2!$. It turns out there's a lot of meaning to this. You can view this as saying that once we choose our first three people, there are $2!$ ways to complete our list to a list of 5 as there are $2!$ ways to choose the last two people (with order counting). We'll revisit this perspective again in the next stage.

{Alice, Bob, Charlie}	{Alice, Bob, Dan}	{Alice, Bob, Eve}	{Alice, Charlie, Bob}
{Alice, Charlie, Dan}	{Alice, Charlie, Eve}	{Alice, Dan, Bob}	{Alice, Dan, Charlie}
{Alice, Dan, Eve}	{Alice, Eve, Bob}	{Alice, Eve, Charlie}	{Alice, Eve, Dan}
{Bob, Alice, Charlie}	{Bob, Alice, Dan}	{Bob, Alice, Eve}	{Bob, Charlie, Alice}
{Bob, Charlie, Dan}	{Bob, Charlie, Eve}	{Bob, Dan, Alice}	{Bob, Dan, Charlie}
{Bob, Dan, Eve}	{Bob, Eve, Alice}	{Bob, Eve, Charlie}	{Bob, Eve, Dan}
{Charlie, Alice, Bob}	{Charlie, Alice, Dan}	{Charlie, Alice, Eve}	{Charlie, Bob, Alice}
{Charlie, Bob, Dan}	{Charlie, Bob, Eve}	{Charlie, Dan, Alice}	{Charlie, Dan, Bob}
{Charlie, Dan, Eve}	{Charlie, Eve, Alice}	{Charlie, Eve, Bob}	{Charlie, Eve, Dan}
{Dan, Alice, Bob}	{Dan, Alice, Charlie}	{Dan, Alice, Eve}	{Dan, Bob, Alice}
{Dan, Bob, Charlie}	{Dan, Bob, Eve}	{Dan, Charlie, Alice}	{Dan, Charlie, Bob}
{Dan, Charlie, Eve}	{Dan, Eve, Alice}	{Dan, Eve, Bob}	{Dan, Eve, Charlie}
{Eve, Alice, Bob}	{Eve, Alice, Charlie}	{Eve, Alice, Dan}	{Eve, Bob, Alice}
{Eve, Bob, Charlie}	{Eve, Bob, Dan}	{Eve, Charlie, Alice}	{Eve, Charlie, Bob}
{Eve, Charlie, Dan}	{Eve, Dan, Alice}	{Eve, Dan, Bob}	{Eve, Dan, Charlie}

Let's put in bold all the ones that have Alice, Bob, and Charlie.

{Alice, Bob, Charlie}	{Alice, Bob, Dan}	{Alice, Bob, Eve}	**{Alice, Charlie, Bob}**
{Alice, Charlie, Dan}	{Alice, Charlie, Eve}	{Alice, Dan, Bob}	{Alice, Dan, Charlie}
{Alice, Dan, Eve}	{Alice, Eve, Bob}	{Alice, Eve, Charlie}	{Alice, Eve, Dan}
{Bob, Alice, Charlie}	{Bob, Alice, Dan}	{Bob, Alice, Eve}	**{Bob, Charlie, Alice}**
{Bob, Charlie, Dan}	{Bob, Charlie, Eve}	{Bob, Dan, Alice}	{Bob, Dan, Charlie}
{Bob, Dan, Eve}	{Bob, Eve, Alice}	{Bob, Eve, Charlie}	{Bob, Eve, Dan}
{Charlie, Alice, Bob}	{Charlie, Alice, Dan}	{Charlie, Alice, Eve}	**{Charlie, Bob, Alice}**
{Charlie, Bob, Dan}	{Charlie, Bob, Eve}	{Charlie, Dan, Alice}	{Charlie, Dan, Bob}
{Charlie, Dan, Eve}	{Charlie, Eve, Alice}	{Charlie, Eve, Bob}	{Charlie, Eve, Dan}
{Dan, Alice, Bob}	{Dan, Alice, Charlie}	{Dan, Alice, Eve}	{Dan, Bob, Alice}
{Dan, Bob, Charlie}	{Dan, Bob, Eve}	{Dan, Charlie, Alice}	{Dan, Charlie, Bob}
{Dan, Charlie, Eve}	{Dan, Eve, Alice}	{Dan, Eve, Bob}	{Dan, Eve, Charlie}
{Eve, Alice, Bob}	{Eve, Alice, Charlie}	{Eve, Alice, Dan}	{Eve, Bob, Alice}
{Eve, Bob, Charlie}	{Eve, Bob, Dan}	{Eve, Charlie, Alice}	{Eve, Charlie, Bob}
{Eve, Charlie, Dan}	{Eve, Dan, Alice}	{Eve, Dan, Bob}	{Eve, Dan, Charlie}

Notice that 6 out of the 60 sets above have the same three people (Alice, Bob, and Charlie). The only difference among those three sets is the order of the three people; however, if we don't care about order, then these 6 sets all give rise to the same *unordered* set of three people. We may write that set {Alice, Bob, Charlie}, but of course *any* of the six arrangements is fine.

The key fact is that the number 6 above isn't just any old number. You should view it as 3!, the number of ways of ordering 3 people. What we're doing is *removing* the order. Essentially, we're collapsing things down, and saying we count all of these six orderings the same. While we highlighted the occurrences of the set Alice, Bob, and Charlie, there's nothing special here about their names. The same logic applies to the set of Alice, Dan, and Eve, or more generally to any set of three names.

We can now isolate the final answer. We want to count how many ways there are to choose 3 people from 5 when the order in which we choose doesn't matter. If order mattered, it would be $5!/(5-3)!$. We then "remove" the ordering by dividing by 3!, and thus the answer is $5!/3!(5-3)!$ or $5!/3!2!$, which is just 10. Here's our final list, the 10 distinct sets of sets of 3 people from 5:

{Alice, Bob, Charlie}	{Alice, Bob, Dan}	{Alice, Bob, Eve}	{Alice, Charlie, Dan}
{Alice, Charlie, Eve}	{Alice, Dan, Eve}	{Bob, Charlie, Dan}	{Bob, Charlie, Eve}
{Bob, Dan, Eve}	{Charlie, Dan, Eve}		

It's worth thinking about our choice of ordering. There's lots of ways to do it, but we want to make sure we don't forget anything. One easy way is to go through all the ones that have Alice and Bob. After that (which gives us 3), we look at ones with Alice and Charlie that we don't already have (this gives us 2 more). We then do Alice and Dan, gaining another. There aren't any with Alice and Eve that we haven't got, so we move on and look at Bob and Charlie. We already had one (involving Alice), but we get two new ones now (involving Dan and involving Eve). We then move on to look at triples with Bob and Dan and get another new one, and finally conclude with Charlie and Dan (there's no need to continue to Charlie and Eve or Dan and Eve, as we've already hit all of those).

As dedicated as I am to writing a good, useful book, I'm not patient enough to type out all the lists in this section! Nor should I be—this is an ideal task for a computer. There are lots of programs you could use. I'm a big fan of Mathematica and WolframAlpha, and I got these lists by typing

```
Permutations[{Alice, Bob, Charlie, Dan, Eve}]
Permutations[{Alice, Bob, Charlie, Dan, Eve}, {3}]
```

(I then had to clean up the spacing and format things a little better, but search and replace made fast work of that). I encourage you to use a computer to explore the various topics. Work out some problems, get a sense of what things look like.

Now that we've analyzed our problem in great detail, it's time to generalize. The number of ways to choose k people from n is $n!/k!(n-k)!$. As we'll keep meeting this expression throughout the course (and beyond), it deserves its own name (just like the factorial function got a name). We call this a **binomial coefficient**, and write it as $\binom{n}{k}$. We're assuming that n and k are non-negative integers and $k \leq n$; if $k > n$ we set $\binom{n}{k} = 0$. This makes sense; we needed a way to define 0!, and we had good arguments to make that 1 (either viewing it as an empty product, or saying there's only one way to do nothing). What's the story for $\binom{4}{5}$? Well, that's asking us how many ways are there to choose 5 objects from 4 when order doesn't matter. There's *no* way to do this, and thus setting it to zero is reasonable.

Above we discussed how $\binom{4}{5}$ should be zero; what do you think $\binom{4}{-1}$, or more generally $\binom{n}{k}$ where k is a negative integer, should be? There is a generalization of the factorial function to the **Gamma function**, defined by

$$\Gamma(s) := \int_0^\infty e^{-x} x^{s-1} dx$$

when the real part of s is greater than -1. Integrating by parts gives $\Gamma(n+1) = n!$ whenever n is a non-negative integer. More is true, as $\Gamma(s+1) = s\Gamma(s)$, and this suggests the answer to how to extend the binomial coefficients to negative values. We'll see the Gamma function in Chapter 15.

The method above is extremely powerful; I often call it **Proof by Story**. For another example, see Appendix A.6, where we prove one of the most important binomial identities: $\binom{n}{k} + \binom{n}{k+1} = \binom{n+1}{k+1}$. This is the basis of Pascal's triangle, which gives us the Binomial Theorem (see Appendixes A.2.3 and A.6 for a statement and proof). We'll need this very important fact later on.

3.1.3 Summary

All we're going to do here is quickly recap what we've done. We've just introduced two of the most important functions in combinatorics and probability, so let's look at them one last time. It's often a good idea to revisit earlier concepts after you've done more advanced ones, as you can put things in perspective and get a better lay of the land.

It turns out that if we have a set of objects, there are two very natural questions we can ask.

- How many ways are there to order the objects?
- How many ways are there to choose some of the objects if order does not matter?

The first is called **permutations** and its solution involves the factorial function, while the second is called **combinations** and leads to binomial coefficients. If we have n objects, there are $n!$ ways to order them; equivalently, we say there are $n!$ **permutations** or $n!$ ways to **permute** them. Permutations play a big role in many fields of mathematics, and are especially important in group theory.

In any list of most common mistakes on a probability assignment, high on that list has to be confusing permutations and combinations. Always ask yourself: *does order matter here?* If it does, think permutations and factorials; if it doesn't, think combinations and binomial coefficients.

It's worth ending with one very important fact about combinations: $\binom{n}{k} = \binom{n}{n-k}$. We give a proof of this in §A.6, but as it has a wonderful proof based on a supremely important observation, we repeat it here as well. Of course, there's an obvious way to try to prove it: write out the two expressions and see that they're equal. We have

$$\binom{n}{k} = \frac{n!}{k!(n-k)!}, \quad \binom{n}{n-k} = \frac{n!}{(n-k)!(n-(n-k))!} = \frac{n!}{(n-k)!k!},$$

and these are clearly the same since we can reorder the multiplication in the denominator. The key idea for the algebra here is that $k = n - (n-k)$. It's natural to try something like this, as we have an $(n-k)!$. Of course, another way to approach this problem is to just write out what $\binom{n}{k}$ and $\binom{n}{n-k}$ are; if we do this, for the latter we would get $n - (n - k)$, which we then note is k and then see the two terms are the same.

So, we've proved $\binom{n}{k} = \binom{n}{n-k}$, but I don't find it particularly enlightening. Here, in my view, is a better approach. What does $\binom{n}{k}$ *mean*? It's the number of ways to choose k people from n people when order doesn't matter. Let's say we have k tickets to Opening Day at Fenway Park, and we have n people who want to go. There's $\binom{n}{k}$ ways to choose a set of k friends to come; however, if we're choosing k to go, we could look at this as choosing $n - k$ *not* to go. How many ways are there to choose $n - k$ people? That's just $\binom{n}{n-k}$. Voilà—the two must be the same! Choosing k from n to go is the same as choosing $n - k$ from n to exclude! This is another example of **Proof by Story**.

3.2 Poker

It's hard to imagine a probability class going an entire semester without hitting games of chance. In fact, one time I taught probability I had a very motivated student who wanted to take the class even though he didn't have any of the prerequisites. He was an avid poker player, and wanted to get better at placing bets. So, in his honor, we'll lead off our applications with **poker**. We go through a lot of examples. There's no need to do them all; I encourage you to read a few and try others on your own, and then compare with the book. You should also write some code to numerically explore the probabilities; for some of the problems there are wonderful coding tricks that can greatly simplify your life, and we'll discuss those at length.

3.2.1 Rules

There are lots of variants of the game. We'll give a quick outline of the major rules; see Wikipedia's page for more. Most versions use a standard deck of 52 cards. There are 13 cards in four suits (spades, hearts, diamonds, and clubs), numbered 2 through 10 and then J (jack), Q (queen), K (king), and A (ace). Often players are dealt five cards. They may or may not be allowed to replace some cards with others, and whoever has the best hand wins (though some variants give some of the money to whomever has the worst hand). Sometimes certain cards are declared "wild" and can be used as anything, which gives you the possibility of getting five of a kind. We display the standard poker hands in Figure 3.3.

Hopefully the pictures in Figure 3.3 make the various types of hands clear. We actually haven't displayed the lowest hand, which is simply nothing special! The first interesting hand is having a **pair**, or exactly two of one number. Here, it's a pair of fives. If two people happen to have the same pair, you look at the highest card outside the pair to determine who wins. As this is a math book and not a guide to playing poker, we'll stop here and let you consult other sources to determine what happens when two or more people have the same type of hand (i.e., what are the tiebreakers).

The next highest hand is **two pairs** of different numbers (tens and twos in our picture), then **three of a kind** (nines here). After this, we get a fundamentally new feature: the **straight**. A straight is five cards in increasing order. They don't have to be in the same suit, and an ace may be taken as low or high. Thus possible straights include A♣, 2♡, 3♡, 4♢, 5♣, as well as 10♠, J♠, Q♢, K♡, A♣.

Immediately above a straight is a **flush**, five cards of the same suit. Continuing up, next is a **full house**, which is one three of a kind and one pair. Above that is **four of a kind**, and then finally at the very top is a **straight flush**.

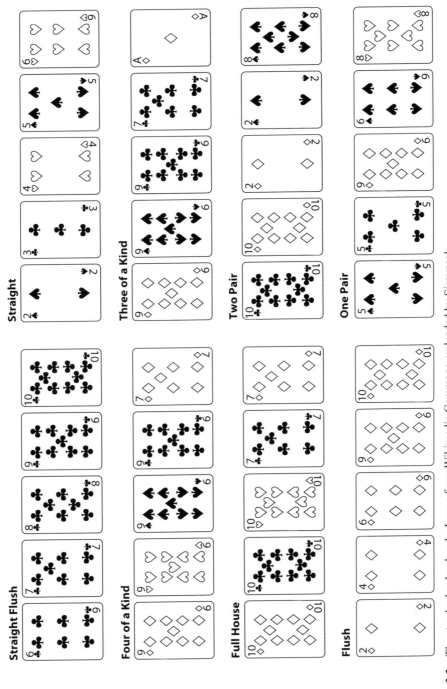

Figure 3.3. The standard poker hands. Image from Wikimedia Commons, uploaded by Sissyneck.

Alright. If you've never seen poker before, you at least have a rough idea of what it is. Again, the order of the hands (from lowest to highest) is pair, two pair, three of a kind, straight, flush, full house, four of a kind, straight flush. This leads naturally to two questions.

The first is: why these? Simple answer: why not! Okay, that's a bit of a cop-out, and my little kids would correctly continue asking. People have played for years and found these to be fun, there's advantages to standardization, yada yada yada. This question is outside the province of this book. Go to Wikipedia and look up non-standard hands if you want to see more options (growing up, we played with kangaroo straights all the time). As mathematicians, we'll accept this order as given.

The second question *is* within our bailiwick: given these hands, why this order? Ah, we can be useful here! If the hands are ordered properly, as you climb the chain there should be fewer and fewer of each. We can check this. It'll take some time, but let's compute the probability of getting each of the ten possibilities *if we're randomly given 5 of 52 cards*. Notice that this might not be the same as the probability of getting one of these hands in a game (there could be wild cards, you might have 7 cards available, you might have the ability to replace some of your cards). That's fine, the point is to get a sense of how to do these calculations. Even if you're not interested in poker, these are great problems to help you master probabilities.

In all of these problems, we'll be dividing by the number of ways to choose 5 cards from 52 with order not mattering, or $\binom{52}{5} = 52!/5!47!$, which is 2,598,960. That's right, there's a tad over two and a half million possible hands! We'll try whenever possible to compute things multiple ways. This is in part because some people prefer one approach to another, so it's good to show each one; however, the main reason is to gain confidence in the answer. If we get the same answer from two different techniques, we're more confident about the answer. Conversely, if we get different answers we know we made a mistake.

Finally, I'll occasionally give incorrect approaches to finding these probabilities. I'm not giving incorrect answers because I'm careless—hopefully my students caught and removed all of those errors! No, these are deliberate mistakes. Don't worry, I'll quickly tell you they're wrong and point out the error. I think it's useful to see a bunch of problems worked out with reasonable sounding methods, albeit false ones. These are some of the most common errors people make in the subject, and I want you to be aware of these pitfalls.

Needless to say, all of this leads to a very long section. Feel free to skip around and look at the calculations and problems you find interesting. Now, on to the probabilities!

3.2.2 Nothing

We should first compute the probability that we don't have any of the special hands. This means we never have two cards with the same number, nor do we have all 5 cards in the same suit, nor do we have 5 numbers in a row. Ack!

Let's lower our sights. It's not too bad to figure out the probability we don't have two cards with the same number, and ignore the possibility of getting a straight or a flush. Don't be incensed—we'll calculate those probabilities later, and we can then just subtract them off. Also, we'll see later that these two events are rare. They don't contribute significantly, so our error is small.

There are 13 numbers (2 through 10 and then J, Q, K, and A). We have to pick 5 of them so there aren't any repeats; we can do that $\binom{13}{5}$ ways. Why? This is just the definition of a binomial coefficient—we need to choose 5 of 13 with order immaterial. For each of these five numbers, our card could be either a spade, heart, diamond, or club; we thus have 4 choices for each.

Using the multiplication rule, the total number of possibilities is $\binom{13}{5} \cdot 4^5$; however, there's a much better way of viewing this. Look at it as $\binom{13}{5}\binom{4}{1}\binom{4}{1}\binom{4}{1}\binom{4}{1}\binom{4}{1}$. The five factors of $\binom{4}{1}$ arise from having to choose a suit (from four options) a total of five times. We get an answer of 1,317,888, and thus the probability is $1317888/2598960 = 2112/4165$, or approximately 0.507083 or 50.7083% (remember the denominator is $\binom{52}{5}$). That's right, there's about a 50% chance you'll have nothing special.

As always, let's see if we can check this by computing the answer another way. We have 52 cards. For our first card, we can choose any of the 52 cards. For our second card, we can't have the same number, so we can choose any of 48. For the third, it can't be the same number as either of the first two cards. That excludes 8 of the 52 cards, leaving us with 44 choices for the third card. Similarly we see there are 40 choices for the fourth card and 36 for the final, giving us a grand total of $52 \cdot 48 \cdot 44 \cdot 40 \cdot 36$.

Before we multiply it out, stop and think. Do you believe this is the same as before, namely 1,317,888? If you're reading this, at least a minute or two should have passed since you read the previous sentence. I strongly urge you to stop reading. Stop! Please! Think for a few minutes as to whether or not these numbers are the same. Don't multiply things out. Once you have your guess, then read on.

Okay, it's time to multiply. We find $52 \cdot 48 \cdot 44 \cdot 40 \cdot 36$ equals 158,146,560. Hmm. That's quite a bit larger than 1,317,888. Alright, they're not the same. How much larger is it? Well, if we subtract we get 156,828,672. It turns out that for problems like this, subtracting is *not* the way to go. We should **take ratios**. Doing that, we find $158, 146, 560/1, 317, 888 = 120$. That's interesting—the ratio of these two numbers turns out to be an integer. (This is why we take ratios: we're often off by multiplicative factors because of adding or removing order, and ratios help us find these mistakes.)

Actually, more is true. It's not just any integer, it's a really *nice* integer. We're doing a combinatorial problem. Factorials are flying all over the place. Eventually, we realize that 120 is 5!. Ah. We're off by a factor of 5!, and we have 5 cards. This can't be a coincidence.

It isn't. When we used our second method, we accidentally ordered the cards. Let's look at the start of our argument again: *For our **first card**, we can choose any of the 52 cards.* Do you see it now? We added order! We talked about first card and second card and so on. We don't care about the order in which we get our cards, all we care about is which cards we have. We need to divide by 5! to kill the ordering we accidentally did. Thus, there are $52 \cdot 48 \cdot 44 \cdot 40 \cdot 36/5!$ possibilities.

We end this section with one more way of viewing the problem. It's perfectly fine to include order, so long as we divide by the number of *ordered* hands. That's $52!/(52 - 5)! = 52!/47! = 311, 875, 200$. Attacking the problem this way, we find that the probability of no repeated numbers is $158, 146, 560/311, 875, 200$. In lowest terms, that's just $2112/4165$, our old friend from before (i.e., about 50.7083%).

How else could we view this? I like writing our product as $\binom{52}{1}\binom{48}{1}\binom{44}{1}\binom{40}{1}\binom{36}{1}$. For me, this highlights what we're doing: we're choosing one card from 52, then one from 48, and so on.

We end with some Mathematica code to simulate the probability that we do not have the same number twice. There are lots of ways to code something like this; the hard part is determining how many distinct numbers are drawn. We can also do this with a MemberQ function, and as we create our hand form a subset of values not previously seen.

```
nothing[numdo_] := Module[{},
  count = 0; (* counts number of successes *)
  deck = {}; (* initializes deck to empty *)
  (* creates deck; don't care about suit here *)
  (* have 1,1,1,1, 2,2,2,2, ... *)
  For[i = 1, i <= 13, i++,
    For[j = 1, j <= 4, j++, deck = AppendTo[deck, i]]];

  For[n = 1, n <= numdo, n++,
    {
      (* many ways to see if have 5 distinct cards *)
      (* will keep track of card values *)
      (* initialize to empty each time *)
      For[i = 1, i <= 13, i++, card[i] = 0];
      hand = RandomSample[deck, 5]; (* chooses 5 card hand *)
      (* the following records that we drew the value card[[i]] *)
      (* for each of the 5 cards. We only care if we have a *)
      (* value or not, so not worrying about multiple times *)
      For[k = 1, k <= 5, k++, card[hand[[k]]] = 1];
      (* only way sum below is 5 if 5 different numbers *)
      If[Sum[card[i], {i, 1, 13}] == 5, count = count + 1];
    }]; (* end of n loop *)
  Print["Theory says probability 5 distinct numbers is ", 2112/41.65,
    "%."];
  Print["Observed probability is ", 100. count/numdo, "%."];
  ] (* end of module *)
```

Running the code ten million times yields:

```
Theory says probability 5 distinct numbers is 50.7083%.
Observed probability is 50.7099%.
```

3.2.3 Pair

We're moving on up. We're going from having no repeated numbers to having exactly one repeated number, a pair. Though the problem may sound easy, it turns out there's some subtleties involved. We'll consider a succession of similar problems of increasing difficulty.

Problem 3.2.1: *What's the probability that in a five card hand, we have exactly two kings?*

 This means that two of the five cards must be kings, and the remaining three are chosen from the 48 non-king cards. There are $\binom{4}{2}$ ways to choose two kings from four kings, and $\binom{48}{3}$ ways to choose three cards from 48. Thus, with order not mattering, there are $\binom{4}{2}\binom{48}{3} = 103{,}776$ ways to choose exactly two kings and three non-kings; there are $\binom{52}{5} = 2{,}598{,}960$ ways to choose five cards from 52 (when order doesn't matter); thus the probability that a hand of five has exactly two kings is just $103776/2598960 = 2162/54145 \approx 0.0399298$, or about 4%. $\square$

Here is some simple code to calculate the probability of exactly two kings.

```
twokings[numdo_] := Module[{},
  count = 0; (* counts number of successes *)
  deck = {1, 1, 1, 1}; (* initializes deck to four kings *)
  (* creates rest of deck; write 0 for non-king *)
  For[i = 5, i <= 52, i++,  deck = AppendTo[deck, 0]];
  For[n = 1, n <= numdo, n++,
  {
    hand = RandomSample[deck, 5]; (* chooses 5 card hand *)
    (* if sum is 2 then have exactly two kings! *)
    If[Sum[hand[[i]], {i, 1, 5}] == 2, count = count + 1];
  }]; (* end of n loop *)
  Print["Probability of exactly 2 kings is ", 2162/541.45, "%."];
  Print["Observed probability is ", 100. count/numdo, "%."];
  ] (* end of module *)
```

Running 10,000,000 trials yields:

```
Probability of exactly 2 kings is 3.99298%.
Observed probability is 3.99252%.
```

Problem 3.2.2: *What's the probability that in a five card hand, we have exactly one pair?*

Building on the previous problem, it's tempting to say that since there's a 4% chance of having a pair of kings, and there are 13 possible pairs, then there should be about a 52% chance of having exactly one pair. This analysis, unfortunately, is fundamentally flawed for two reasons. The first is that we could have two pairs, say kings and jacks; if we just multiply by 13 we've double counted this hand. The second is we could have a full house (a pair of kings and three jacks!). We must be very careful neither to double count nor to allow the three cards from our non-pair to be the same number.

What's the best way to count, making sure we don't forget any possibility and making sure we don't double count? Well, there are 13 different card values; we could first choose the value we want for our pair, and then make sure we choose our three remaining cards from three different values of the remaining 12 values. For example, we might choose to have a pair of jacks, and then we must make sure that the remaining three cards are different values from each other and are not jacks.

There are $\binom{13}{1} = 13$ ways to choose a value, and then $\binom{4}{2}$ ways to choose a pair from that value. We have 48 cards remaining; we must choose three, and these three must all have different values. There are several ways to proceed. We have to choose three values from 12; there are $\binom{12}{3} = 220$ ways to do this. For each choice, we have to then choose one of four cards (since there are four suits); there are $\binom{4}{1} = 4$ ways this may be done. Thus, we find there are $\binom{12}{3}\binom{4}{1}^3 = 14{,}080$ ways to choose three cards with no value repeated and none of these values equaling our first choice. This implies that there are $\binom{13}{1}\binom{4}{2} \cdot \binom{12}{3}\binom{4}{1}^3 = 1{,}098{,}240$ ways to choose five cards so that we have one and only one pair. As we're doing our choice with order not mattering, to find the probability we divide by the number of ways of choosing 5 cards from 52, which is $\binom{52}{5} = 2{,}598{,}960$. This implies that the probability of getting exactly one pair is $1{,}098{,}240/2{,}598{,}960 = 352/833 \approx 0.422569$. Note that this is indeed *close* to our guess of 52%, and not surprisingly it's lower (as it excludes the possibility of having two pairs or a pair and three of a kind). $\qquad\square$

It's instructive to look at other ways to compute the probability of just a pair. Do you think the following arguments are correct, or are there mistakes? If there are mistakes, where are they? Here we go.

- We want to have exactly one pair. There are 52 possibilities for the first card, which will start our pair. Once we've chosen that, there are 3 possibilities for the second card, which completes the pair. The next card must be a different number, so there are 48 choices for it, the following must be another new number, so there are 44 possibilities for it, and finally the last card must be a different number than the first four, so it has 40 options. We get $52 \cdot 3 \cdot 48 \cdot 44 \cdot 40$.
- Alternatively, there's $\binom{13}{1}$ ways to choose the number for our pair, and then $\binom{4}{2}$ to choose two of the four cards with that number. We need three more cards which can't have this number, and which can't share a number. We have 48 choices for the next card, then 44 for the one after that, and finally 40 for the last card.

So, are either of these right? The first method gives 13,178,880 and the second produces 6,589,440. Neither agrees with the correct answer of 1,098,240. If we look at the ratios of our two new possible answers and the correct answer, we get 12 for the first ratio and 6 for the second. These are both nice answers, so perhaps we've just made a very small mistake. We have—the point of doing these problems is to drive home some of the dangers.

Let's start with the second approach. When you see 6 in a combinatorial problem, one of the first thoughts you should have is: aha—maybe this is really a 3!. So, what could give us a 3!? In the second approach we started off as before, but then we chose our last three cards and accidentally gave them an order! We went from unordered (binomial coefficient) to ordered (with phrases like 48 cards, 44 cards, 40 cards). We need to remove the ordering from the last three cards. There's 3! or 6 ways to order 3 cards, so we should divide 6,589,440 by 6; doing that, we regain the correct answer!

Okay, that explains where we went wrong in the second method; what's happening in the first? The only difference between these two is that the first method starts with $52 \cdot 3 = 156$, while the second starts with $\binom{13}{1}\binom{4}{2} = 78$. This mistake is more serious, and harder to see and explain. We chose our first card from 52; already that's bad as that puts in order. We then have 3 choices giving the same number to the next card; however, why is it the second card? Why couldn't we have chosen these two in the opposite order? In other words, instead of choosing 8♡ first and 8♠ second, we could've started with 8♠ and then received 8♡. Our mistake was putting an order among the two cards with the same number. There are $2! = 2$ ways to order two numbers. We've overcounted by a factor of 2. Note that when we divide by 2, the first answer is 6,589,440, exactly what we got in the second approach (and which we've already seen how to correct).

Whenever there are multiple ways which seem reasonable, it's worthwhile to write some code and gather data. We modify our code to compute exactly two kings. There are lots of ways to keep track of how often we have the same number twice; a particularly straightforward approach is to have an array and save the number of each value there, but of course there are other approaches (my favorite involves using primes for the cards and unique factorization!). One simple one, used below, is to just keep track of whether or not we have a 1, a 2, a 3, ..., a 13 (using those numbers for the 13 different values). We then add the number of distinct values, and we have exactly one pair if and only if that sum is four.

```
onepair[numdo_] := Module[{},
  count = 0; (* counts number of successes *)
  deck = {}; (* initializes deck to empty *)
  (* creates deck; don't care about suit here *)
  (* have 1,1,1,1, 2,2,2,2, ... *)
  For[i = 1, i <= 13, i++,
   For[j = 1, j <= 4, j++, deck = AppendTo[deck, i]]];
  For[n = 1, n <= numdo, n++,
   {
    (* many ways to see if have only one pair *)
    (* will keep track of card values *)
    (* initialize to empty each time *)
    For[i = 1, i <= 13, i++, card[i] = 0];
    hand = RandomSample[deck, 5]; (* chooses 5 card hand *)
    (* the following records that we drew the value card[[i]] *)
    (* for each of the 5 cards. We only care if we have a *)
    (* value or not, so not worrying about multiple times *)
    For[k = 1, k <= 5, k++, card[hand[[k]]] = 1];
    (* only way sum below is 4 is if have exactly one pair! *)
    If[Sum[card[i], {i, 1, 13}] == 4, count = count + 1];
   }]; (* end of n loop *)
  Print["Theory says probability one pair is ", 352/8.33, "%."];
  Print["Observed probability is ", 100. count/numdo, "%."];
  ] (* end of module *)
```

Doing 10,000,000 trials yields strong support for our first approach.

```
Theory says probability one pair is 42.2569%.
Observed probability is 42.2502%.
```

> **Ratio Method**: It's worth highlighting the **ratio method** to compare answers. What's really nice is that often these ratios are integers, and give a clue as to where we went wrong. For example, if you see a ratio of 2, 6, 24, 120, 720 et cetera, you should be thinking factorials! Also, be ever vigilant about introducing order where it shouldn't be.

3.2.4 Two Pair

As we dwelled so long on one pair, we'll calculate the probability of having exactly two pair somewhat quickly. We need to choose two numbers, each to be repeated twice, and then a third distinct number. How many ways are there to choose two numbers from 13? That's just $\binom{13}{2}$, and for each of those numbers we must choose 2 of the 4 suits, which can be done in $\binom{4}{2}$ ways each time. Finally, the last card must be a different number. We have $\binom{11}{1}$ ways to choose one of the remaining 11 numbers, and then $\binom{4}{1}$ ways to choose the suit. Combining, we get $\binom{13}{2}\binom{4}{2}^2\binom{11}{1}\binom{4}{1} = 123,552$. Thus the probability is $123,552/2,598,960$, or about 4.7539%.

3.2.5 Three of a Kind

Three of a kind isn't horrible; the real issue is making sure the other two cards are different numbers. There's $\binom{13}{1}$ ways to choose the number we'll get three times; once we've chosen that number, we need to choose 3 of the 4 suits, and there's $\binom{4}{3}$ ways for this. Okay, time for the remaining two cards. We need to choose two different numbers, which happens in $\binom{12}{2}$ ways. For each number we get to choose one of four suits, which we can do $\binom{4}{1}$ ways. Thus, the number of possibilities is $\binom{13}{1}\binom{4}{3}\binom{12}{2}\binom{4}{1}^2$, or 54,912. Thus the probability of exactly three of a kind in a five card hand is $54{,}912/2{,}598{,}960$, which is $88/4165 \approx 0.0211285$ (or about 2.11285%).

How else might we compute this? We need to choose three distinct numbers from 13; that's just $\binom{13}{3}$. One of those numbers is repeated three times, and there's $\binom{3}{1}$ ways to choose one of three numbers. For the number repeated three times, we need to take 3 of the 4 suits, which is $\binom{4}{3}$. For the other two numbers, we have four choices for each, giving us two factors of $\binom{4}{1}$. Putting this all together, the number of five card hands with exactly three of a kind is $\binom{13}{3}\binom{3}{1}\binom{4}{3}\binom{4}{1}^2$. So, once again, what do you think? Is this right? Have we double counted anything? Have we forgotten anything? Did we accidentally add order?

Drum roll please.... Multiplying everything out gives 54,912, the same as before! This is the first time one of our "alternate" approaches actually gave the same answer! Note that this *does not* mean that our answer is correct. If we have two different answers then clearly one is wrong; however, just because two different methods give the same answer doesn't mean that answer is right. Of course, the more complicated the problem is, the harder it'll be for two approaches to agree unless they happen to be right.

The logic here was very good. We neither double counted, nor did we accidentally introduce order. Notice how careful we were. We kept using binomial coefficients and not multiplying by numbers; that's a good sign that we're respecting order. We started with $\binom{13}{3}$ and then hit that with $\binom{3}{1}$, being oh so careful not to distinguish the location of our triply repeated number.

3.2.6 Straights, Flushes, and Straight Flushes

It's fitting to do straights and flushes together. The reason is we have to be careful as a straight flush counts as both. We'll first find how many straights there are, allowing the straights to also be flushes. We'll then find how many flushes there are (allowing them to be straights as well). Finally, we'll calculate how many straight flushes there are. Subtracting this from the first two will yield our answers.

How many straights are there? We need five consecutive numbers, though we can get them in any order. Using T for 10, it could be A2345, 23456, 34567, 45678, 56789, 6789T, 789TJ, 89TJQ, 9TJQK, TJQKA. So, there are ten possibilities. For each of the ten possibilities, there are four choices for each position, as each number can be any suit. Thus, there are $10 \cdot 4^5 = 10{,}240$ possible straights.

What about flushes? Remember these are hands where all five cards are the same suit. There are $\binom{4}{1}$ to choose this special suit, and then all that's left is to choose 5 numbers, which we can do in $\binom{13}{5}$ ways. Thus, there are $\binom{4}{1}\binom{13}{5} = 5148$ flushes.

Finally, how many straight flushes are there? All that changes from our straight argument is that the factor of 4^5 is replaced with $\binom{4}{1}1^5$ (there are four ways to choose the suit, and once the suit is chosen then the suit of the five cards is forced upon us). This gives $10 \cdot 4 = 40$ straight flushes.

We can now collect the pieces. The number of straights but not straight flushes is $10,240 - 40 = 10,200$, giving a probability of $10240/2598960 = 128/32487 \approx 0.00394004$ or about .394004%. Similarly the number of flushes but not straight flushes is $5148 - 40 = 5108$, giving a probability of $5108/2598960$ or about .19654%. Finally, the number of straight flushes is 40, for a percentage of about 0.00153908%.

3.2.7 Full House and Four of a Kind

We're almost done—are you still loving poker? These calculations may seem long and endless, but remember these only need to be done once. After we know these values, we know how to rank the hands, and can assess the relative probabilities of winning and losing.

Alright, let's do a full house, which is three of one number and two of another. There are $\binom{13}{1}$ ways to choose the number we'll have three times, and then $\binom{4}{3}$ ways to choose 3 of the 4 suits (i.e., to choose three of that number). For the remaining two cards, they must have the same number. There's $\binom{12}{1}$ ways to choose one of the remaining 12 numbers, and then $\binom{4}{2}$ ways to choose two different suits. Combining, we get $\binom{13}{1}\binom{4}{3}\binom{12}{1}\binom{4}{2} = 3744$. Thus the probability of a full house is $3744/2598960$, or about 0.144058%.

Could we have made the following argument: We have to choose two numbers, which happens $\binom{13}{2}$ ways. There's then $\binom{4}{3}$ ways to choose three suits for the first number, and $\binom{4}{2}$ for the second, giving us $\binom{13}{2}\binom{4}{3}\binom{4}{2} = 1872$.

This is a first for us—instead of being too large, this time we're too small! Taking ratios, we see $3744/1872 = 2$. How should we interpret this? We're off by a factor of 2. It's probably not a double counting issue, more likely we forgot something. But what? It's subtle, but the troublesome part is the phrase *first number*. Maybe the triple is the second number, not the first! We ignored that case! We need to include the number of ways of choosing which of the two numbers is chosen to get three; that's done with a $\binom{2}{1}$. Thus, the answer should be (and is) $\binom{13}{2}\binom{2}{1}\binom{4}{3}\binom{4}{2}$, which is indeed 3744. We can view this as saying there are $\binom{13}{2}$ ways to choose the two numbers, and then $\binom{2}{1}$ ways to choose which of the numbers occurs three times and which occurs twice. What's left is $\binom{4}{3}\binom{4}{2}$, which is the number of ways to have three of one number and two of another *where we have fixed which number occurs thrice and which occurs twice*. In other words, we can view the product as arising from two pieces: $\binom{13}{2}\binom{2}{1} = 13 \cdot 12$ is the number of ways to choose the two numbers that occur, complete with the specification of which happens thrice and which twice, and $\binom{4}{3}\binom{4}{2}$, the number of ways to choose 3 of the first number and 2 of the second.

In situations like these, it's very useful to be able to write some simple code to numerically investigate probabilities. This is especially important if we're unclear as to whether or not we accidentally added some extra order, or forgot a factor. Here is a simple program to explore the probability a randomly chosen hand is a full house. Note that if we label the cards well, there's a nice trick which allows us to easily tell if we have a full house.

```
fullhousesearch[num_] := Module[{},
   (* creates a deck of cards, as only care about numbers *)
   (* we ignore suits and write each number four times *)
   cards = {};
   fullhouse = 0;
   For[d = 1, d <= 13, d++,
     For[i = 1, i <= 4, i++, cards = AppendTo[cards, d]]];
   (* main code here, do num times *)
   For[n = 1, n <= num, n++,
     {
       (* code randomly chooses 5 from 52 cards and sorts. *)
       (* for the analysis below it's easy to check to see *)
       (* if we have a full house if the hand is sorted *)
       (* we have a full house if the first three are the same *)
       (* and the last two are the same, or the first two are *)
       (* the same and the last three are *)
       hand = Sort[RandomSample[cards, 5]];
       If[hand[[1]] == hand[[2]] && hand[[4]] == hand[[5]],
         If[hand[[3]] == hand[[2]] || hand[[3]] == hand[[4]],
           fullhouse = fullhouse + 1]];
     }];
   Print["Percent of time got full house is ",
     100.0 fullhouse/num, "."];
   (* we now print out the predictions, and see that the *)
   (* first is very close to our numerics *)
   Print["The [predictions were 0.144058% and 0.072029%."];
 ];
```

Typing `fullhousesearch[1000000]`, which means we are randomly choosing a million hands, yielded about 0.1447% as full houses; this is almost equal to our first proposed answer (which is approximately 0.144058%), and differs from our second proposed answer (of about 0.072029%) by essentially a factor of two!

 We've reached the end. Our last task is to find the probability of four of a kind. We have $\binom{13}{1}$ ways to choose this lucky number, and then must take all four suits (which happens only 1 way, which is indeed $\binom{4}{4}$). Finally, we have to take our final card. It can be any of the remaining 48, giving us $\binom{13}{1}\binom{4}{4}\binom{48}{1} = 624$ possibilities, or the very small probability of about 0.0240096%.

3.2.8 Practice Poker Hand: I

We end our studies of poker hands with a few final questions. It's a nice way to review what we've done, and make sure everything is solid. In this section we'll use slightly different notation, emphasizing different aspects of the calculation. It's useful to have several perspectives for attacking problems, as what clicks for one person might not for another. Here we use distinct letters for distinct numbers, emphasizing the different patterns.

Problem 3.2.3: *What's the probability that, in a five card hand, we have at least two cards with the same value?*

We have to be careful to make sure we enumerate all possibilities. Let A, B, C and so on denote distinct values—for now, we don't care about the suit. We could have exactly one pair ($AABCD$), three of a kind but not a full house ($AAABC$), four of a kind ($AAAAB$), exactly two pairs ($AABBC$), or a full house, namely a pair and three of a kind ($AABBB$). Note these events are mutually disjoint, and exhaust all possibilities. We need only count how many ways each can happen.

We calculated the first, $AABCD$, in §3.2.3, and got 1,098,240 ways. We turn to the rest. We could just quote our earlier results for these, but instead let's quickly do them again from scratch. It's a good way to review what we've done, and of course some readers may have skipped some of the previous sections! We'll have less comments this time through, as you can always go back to the earlier sections for more detail.

For three of a kind but no full house, we see that the number of ways is $\binom{13}{1}\binom{4}{3} \cdot \binom{12}{2}\binom{4}{1}^2 = 54912$ (there are $\binom{13}{1}$ ways to choose the value we have three times, then $\binom{4}{3}$ ways to choose three of those four cards, then $\binom{12}{2}$ ways to choose two of the remaining 12 values, and then for each of those $\binom{4}{1}$ way of choosing one card).

For four of a kind, $AAAAB$, it's just $\binom{13}{1}\binom{4}{4} \cdot \binom{12}{1}\binom{4}{1} = 624$. The factor of $\binom{12}{1}\binom{4}{1} = 48$ arises as we have 12 remaining values, and once we specify a value we have to choose one of four cards. Another way to find this number is to note that we can choose any of the remaining 48 cards, and there are $\binom{48}{1} = 48$ ways of doing this.

For exactly two pairs, $AABBC$, we first have to specify the two values that we take twice. There are $\binom{13}{2}$ ways to choose the two values. Once we have chosen a value, we must choose two of four cards; there are $\binom{4}{2}$ ways of doing this. We are left with $52 - 8 = 44$ cards that are neither of these values, and we must choose one of them; there are $\binom{44}{1}$ ways of doing this. Thus the number of ways is $\binom{13}{2}\binom{4}{2}^2 \cdot \binom{44}{1} = 123552$.

Finally, we must compute the number of ways of getting a full house, $AABBB$. We have to be a little careful, as it matters which value we get twice and which we get three times. There are $\binom{13}{2}$ ways to choose the two values, and then $\binom{2}{1}$ ways to choose which one occurs twice and which occurs thrice. This leads to $\binom{13}{2}\binom{2}{1}\binom{4}{2}\binom{4}{3} = 3744$. Alternatively, we have to choose two values from 13. Let the first value be the one that will be the pair, and the second value the one that will be the triple. There are $\binom{13}{1}\binom{4}{2}$ ways to get the pair, and then as there are 12 values left, $\binom{12}{1}\binom{4}{3}$ ways to get the triple. This leads to a count of $\binom{13}{1}\binom{4}{2} \cdot \binom{12}{1}\binom{4}{3}$, which also equals 3744. We see again that there can be many ways to count the same event.

We now sum the above, and find the number of ways of getting at least two cards with the same value in a hand of five cards is

$$1098240 + 54912 + 624 + 134768 + 3744 = 1281072,$$

which implies that the probability of getting at least two cards with the same value is $1281072/2598960 = 2053/4165 \approx 0.492917$. Thus there's almost a 50% chance of having at least two cards with the same value in a hand with five cards! □

There is another, *significantly simpler*, way to compute the probability we have at least two cards with the same value. Earlier we showed the probability that we have *no* repeated value is 2112/4165, and thus using the Law of Complementary Probability the answer to our question is

$$1 - \frac{2112}{4165} = \frac{2053}{4165},$$

exactly what we computed earlier but with far less work! Why did we spend so much time on such a long, tedious computation? **To highlight that it is a long, tedious calculation and that you should always be asking yourself: Is this the best way to attack the problem?** It is very dangerous to enumerate all cases; the more cases there are, the more likely you are to either miss a case or do one incorrectly. If a calculation looks painful, see if there is a better approach.

Exercise 3.2.4: *Suppose we are playing a very simple game of poker in which both players have just two cards. Each player can look at one of the cards, but not the other. One of your cards is a 10. Your opponent is not very good at the game and accidentally reveals to you that he has an ace. What is the probability you have a winning hand? Assume in the case in which both players have the same high card you lose.*

3.2.9 Practice Poker Hand: II

Consider a five card poker hand, where the first two cards are the ace of spades and the 8 of diamonds, which we denote $\{A\spadesuit, 8\diamondsuit\}$. We must add three more cards to have a hand. As the two cards are in different suits and more than 4 apart in our ordering, it's impossible to get a straight or a flush.

Example 3.2.5: *Starting with $\{A\spadesuit, 8\diamondsuit\}$, what is the probability that the next three cards will give a hand having exactly one pair, where that pair is neither two aces nor two 8s?*

There are 50 cards remaining; however, as we're told that we don't have a pair of aces or 8s, we're drawing three cards from $50 - 6 = 44$ cards. Why 44? We clearly can't pick either $A\spadesuit$ or $8\diamondsuit$, as these are already in our hand. We're told we don't end up with a pair of aces or 8s, so we can't pick any of $\{A\heartsuit, A\diamondsuit, A\clubsuit, 8\spadesuit, 8\heartsuit, 8\clubsuit\}$. This leaves 44 cards. We are told that we have exactly one pair, so we must pick either two 2s and a card that isn't a 2, 8, or ace, or we must pick two 3s and a card that isn't a 3, 8, or ace, et cetera. How many possibilities are there? We have 11 numbers that are neither an 8 nor an ace: $\{2, 3, 4, 5, 6, 7, 9, 10, J, Q, K\}$. We must pick one of these: there are $\binom{11}{1} = 11$ ways to choose one number. Once we've picked that number, there are four cards of that denomination; for example, if we decided we wanted to have a pair of 6s, we must now choose two from $\{6\spadesuit, 6\heartsuit, 6\diamondsuit, 6\clubsuit\}$. There are $\binom{4}{2} = 6$ ways to do this. We then need to choose the last card of our hand. It can't be an 8, ace, or the same number we just chose. There are $52 - 12 = 40$ possibilities, and thus $\binom{40}{1} = 40$ ways to choose one of them.

We see that the number of ways of choosing three cards to get exactly a pair that isn't 8s or aces, given that we start with an ace and an 8, is $\binom{11}{1}\binom{4}{2}\binom{40}{1} = 11 \cdot 6 \cdot 40 = 2640$. To find the probability, we need to divide by the number of ways of choosing 3 cards from the remaining 50. That's just $\binom{50}{3} = 50 \cdot 49 \cdot 48/3! = 19,600$, so the probability is $2640/19600 = 33/245$ or approximately 13.4694%. $\qquad\square$

Hopefully I've harped enough already that if you *can* solve a problem multiple ways, then you *should* solve the problem multiple ways. It's a great way to check and catch errors. Can we find another valid way to view this problem?

Let's attack the problem by using an approach where order matters. We want exactly one pair. Similar to the problem in §3.2.8, letting B, C, and D denote distinct numbers (we can't use A as we're already using that letter for an ace), our hand must be

$A8BBC$, $A8BCB$, or $A8BCC$. For the first of these (the $A8BBC$ pattern), we have 44 possibilities for the third card (it can be anything that isn't an ace or an eight), then 3 possibilities for the fourth card (it has to be the same number but a different suit than the third card), then 40 possibilities for the final card (it can't be an ace, an eight, or whatever number the third and fourth cards are), for a total of $44 \cdot 3 \cdot 40 = 5280$ possibilities.

If instead we have $A8BCB$, we see there are again 44 choices for the third card, 40 choices for the fourth card (it has to be a different number than the first three; as the first three all have distinct numbers, 12 of the 52 cards can't be chosen and thus only 40 remain), and just 3 for the last (it has to be the same number as the third card). We find there are $44 \cdot 40 \cdot 3 = 5280$ choices that work.

Finally, we turn our attention to $A8BCC$. There are still 44 choices for the third card, 40 choices for the fourth card, and since the fifth card must be the same number but a different suit as the fourth, there are 3 choices for the final card. We again get $44 \cdot 40 \cdot 3 = 5280$ possibilities.

Adding everything up, we have $5280 + 5280 + 5280 = 15,840$ *ordered* choices for the remaining three cards that give a hand with exactly one pair that isn't a pair of 8s or a pair of aces. There are $50 \cdot 49 \cdot 48 = 117,600$ ways to choose our three remaining cards *when order matters*. Dividing, we find the probability of our desired hand is $15840/117600 = 33/245$ or about 13.4694%.

Hooray! We got the same answer as before. The moral of this example is that there's no mandatory way to approach these problems. If you prefer to live in a nice, ordered world: go ahead! If instead you want to rebel against authority and live a life without order, that works too! Just remember, though, to be consistent. Don't mix and match. For example, in the second method it would've been disastrous to calculate the number of ordered ways to finish our hand and then divide by $\binom{50}{3}$, the number of ways of choosing 3 cards when order doesn't matter. If you start with order then you must end with order, and vice versa.

3.3 Solitaire

Poker is one of the biggest social card games, appearing everywhere from tournaments televised on ESPN to after hours on the Enterprise D in *Star Trek: The Next Generation*. Let's go to the opposite extreme, and look at some of the many variants of solitaire. There are hundreds of popular (and not so popular) ways to play. Typically we use one deck, with a goal related to aces (or building on aces).

Below we'll look at undoubtedly the most important problem in the theory of solitaire: What's my chance of winning? It turns out this is an exceptionally difficult question, but we can at least start the analysis. Most people play solitaire to pass the time, so perhaps winning isn't the most important thing. Perhaps a better question would be: What's the probability of having an entertaining, enjoyable game? As you can imagine the nightmares of quantifying that question, we'll stick to winning!

A few minutes on the internet will give you more information than you could possibly want about how to play these games (as well as their history, societies devoted to them, programs to automatically solve them, and so on). We'll *briefly* describe the games below, but only briefly. Hopefully you've got some experience playing these games. If not, the actual rules and goals aren't too important—a broad overview is enough. Thus, we won't describe all the moves, just enough at times to get partial answers to our questions; you can easily go online for more.

3.3.1 Klondike

One of the most common and popular versions of **solitaire** (at least before Microsoft expanded the reach of FreeCell) is called **Klondike** (or **patience**). There's a plethora of ways to play. They all start from the same board setup. Cards are placed face-down in 7 piles. The first pile has 1 card, the second 2, and so on until we have 7 in the seventh. We then turn over the top card in each pile. That takes care of 28 cards. The remaining 24 cards are kept together, and we'll discuss their use later. See Figure 3.4 for an illustration.

There are four open foundations at the top. Whenever an ace is exposed, it's moved and starts a foundation. After you have an ace, you can put the two of that suit on top of it. After that, you can then place the three of that suit on top of it, and so on. The goal is to get all 52 cards into the foundations.

You can always move an exposed card from one of the seven piles to another *if* the exposed card in the other pile is an opposite color and one number higher. In the game depicted in Figure 3.4, we could move the queen of spades to the king of diamonds, or the seven of spades to the eight of diamonds. After a card is moved, you turn over the next card on that pile. If one of the seven piles is ever empty, you can move any exposed king to it (or any chain starting with a king).

There are lots of variants for what to do with the remaining 24 cards. A popular choice is to cycle through them, exposing every third card as you go through. Whenever you show a card, you have the choice of moving that exposed card to a foundation (if you can), an open pile (if there's one), or building on an exposed number in a pile.

Our main question is what's the chance of winning, but there are others as well. How many possible games are there? One solution is to say there are 52!, but if we're willing to consider equivalence classes, it's less. For example, if we switch all hearts with diamonds, there's essentially no change....

Let's concentrate on winning below. We can't give a complete analysis due to the complexity of the problem. Computer searches and computer aided proofs have demonstrated that over 80% of these games are winnable and at least 8% cannot be won. What of the rest? Well, there's still work to be done. There's a large class of games whose status is open.

Some games are so bad that to lump them in with unwinnable is a grave injustice; .25% of games are so bad that there aren't any valid moves. These games are called unplayable, and form a particularly nasty subset of unwinnable games. Thinking back to our initial discussion, these unplayable games are a perfect candidate to analyze. Not only are they related to winning, but they're also related to fun: who wants to play a game where you can't do anything, ever?

 Let's try to get a lower bound for the number of unplayable games. We're not going to go for the best bound. Instead, let's go for some low fruit. Let's try to think of a configuration where we can't do anything. What if only black cards are exposed on the pile, and every third card in the set we cycle through is black. Would that be unplayable? It seems so, as we can't move a card because only red cards can be placed on black cards.

So, would exposing just black cards lead to an unplayable game? *Almost!* We need to make sure we don't expose a black ace, as we can move either of the two black aces to the foundation. These little caveats are common in problems like this; all too frequently there are small, pesky cases to exclude.

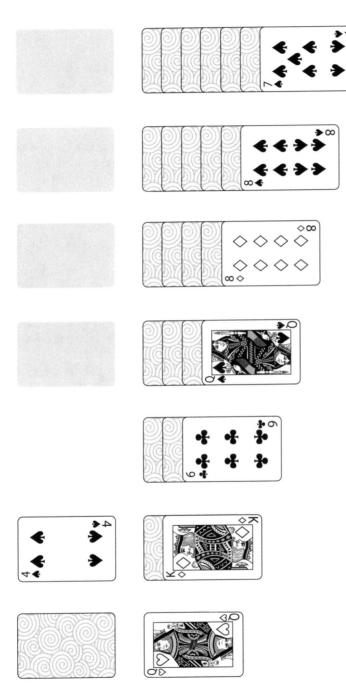

Figure 3.4. Starting position of a Klondike game. Image from Wikimedia Commons, posted by user Andreas Rosdal.

Okay, how do we determine the probability of such an unplayable game? Well, we'll have to choose 7 black, non-aces to expose on our 7 piles. For the remaining 24 cards, we cycle through and expose every third. Thus 8 cards are exposed here, and these need to be black non-aces as well. Thus there are 15 spaces that must be assigned black non-aces. There are 26 black cards, 24 of which aren't aces. The answer? There's $\binom{24}{15}$ ways to choose 15 black non-aces from the 24 possibilities.

Now what? For me, this is where the problem gets tricky, and why I wanted to talk about it. Until you've done mountains of problems, it's very easy to mix up when things should be ordered, and when they shouldn't. Our word choices suggest binomial coefficients, which is unordered. But we've seen that order matters for a game. We've got to be very careful.

Probably the easiest way to solve this problem is to tackle it using order. All 52 cards must be assigned somewhere, so let's give each position a number. We'll number in a non-standard way. The first number, position 1, is the card exposed on the first pile. Position 2 is the card exposed on the second pile. We continue to position 7, the exposed card on the seventh pile. Position 8 is the first location among the remaining 24 cards that will be exposed, followed by position 9 (the second to be exposed), continuing to position 15 (the last exposed). It doesn't matter how we number the remaining $52 - 15 = 37$ locations.

We have 24 (which is $24 - 0$) possibilities for the first position, 23 (which is $24 - 1$) for the second, 22 (which is $24 - 2$) for the third, and so on down to 10 (which is $24 - 14$) for the fifteenth position. Multiplying these together gives $24 \cdot 23 \cdots 10 = 24!/9!$ (which we may view as $24!/(24 - 15)!$). After we fill these 15 spots, we have 37 cards remaining, and there are 37! ways to arrange them among the remaining positions, so the number of unplayable configurations due to only seeing black non-aces is $(24!/9!)37!$. As there are 52! configurations, the probability of getting one of these unplayable setups is $(24!/9!)37!/52!$. This is a lot of factorials of decently sized numbers, and you need to be careful not to get an overflow error. Mathematica fortunately can whip out the answer. It's $11/37701755$, or approximately $2.9 \cdot 10^{-5}$ percent!

We can reach this answer another way, which is worth highlighting. This time we'll build on the binomial approach. We saw there are $\binom{24}{15}$ ways to choose 15 black non-aces with order not mattering. Unlike previous problems, now we *add* order. There are 15! ways to order these 15 cards for the 15 positions. Thus the number of ordered ways to place 15 black non-aces in these 15 slots is $\binom{24}{15}15!$ (which is just $24!/9!$, as before). How many ordered ways are there to choose 15 cards from 52? There are $\binom{52}{15}$ ways to choose 15 cards with order not mattering, and then 15! ways to order them. Thus the probability of a configuration of 15 exposed black non-aces is just $\left(\binom{24}{15}15!\right) / \left(\binom{52}{15}15!\right)$. Notice the 15! cancel, giving $\binom{24}{15}/\binom{52}{15}$, which is $11/37701755$ as before.

That's interesting—we're back where we started, but instead of a mess involving four factorials and two divisions we have a ratio of two binomial coefficients. This seems like a cleaner, nicer way of writing it. It should be possible to get here by staring at $(24!/9!)37!/52!$. I want to show you how to do this, as these algebra manipulations are very useful in showcasing what's going on.

Look closely at $(24!/9!)37!/52!$. Think about binomial coefficients. We have a 24! and a 9!; they're screaming for a 15!. Let's oblige them by **multiplying by one** (one of my favorite techniques, see §A.12 for more on this). Multiplying by $1 = 15!/15!$ gives $(24!/9!15!)15!37!/52!$. Here the miracle happens—notice 15 plus 37 is 52, so we really have a second binomial coefficient; multiplying by one allows us to **rewrite the algebra** in a way that *greatly* clarifies the answer. We move those factors to the denominator, using $a/b = 1/(b/a)$, and get $(24!/9!15!)/(52!/15!37!)$. We have choices in how to write these binomial coefficients. The first is either $\binom{24}{9}$ or $\binom{24}{15}$, while the second is $\binom{52}{15}$ or $\binom{52}{37}$. Hopefully, it's clear which are the better choices. We should take the versions with 15, as that makes the two similar. So, we could've also reached $\binom{24}{15}/\binom{52}{15}$ from our first approach, and this has a nice combinatorial interpretation for what we've done.

This is quite small, much less than the claimed .25%. Can we do better? Sure—we can double it. How? There's nothing special about black. Instead, let's look at using just reds.

Okay, we've moved up to $5.8 \cdot 10^{-5}$ percent. That's still a long way from .25%. Sadly, it becomes more involved if we go further. The real killer is having no mix of black and red cards. We could allow both so long as there aren't any adjacent numbers. As a nice exercise, try to bump this number up as much as you can as simply as you can. For more information, see the links on Wikipedia's Klondike page.

3.3.2 Aces Up

One of my favorite solitaire games is **Aces Up**. It goes as follows: shuffle the deck and then deal four cards face up. You now have four piles. If the top card on a pile has the same suit but has a lower number than the top card of another pile, that card can be moved into the discard pile. For example, if there are four piles and the top cards are 4♦, 3♥, 6♠, and 2♦, then we can move the 2♦ into the discard pile. If there was a card below the 2♦, that's now the top card of the pile; if there was no card below it, we now have a free pile and we can move *any* top card on to that pile. The goal of the game is to end with just the four aces showing (as obviously we can't do better than that!).

Growing up, I was told that having 10 or fewer cards at the end was a very good game, and that if you end up with the four aces you win. I've passed many hours playing this game, but one thing always bothered me growing up. I knew any game with cards has a luck component, but this game seemed to have a very high luck component; in other words, no matter how skillfully I played (I thought I was good!), I frequently lost! While sometimes I would lose big, there were many times when I was doing quite well until the very end, when I ran into what I now know are the laws of probability.

How? It's not hard to show that while you can easily have just three piles in the beginning, by the fortieth card you are guaranteed to see at least one card from each suit, and thus from this point onward you will always have at least four piles and thus at least four cards; this is just the **Pigeon-Hole Principle** (if you haven't heard of this wonderful observation and its far-reaching consequences, see §A.11 for a brief discussion).

The worst thing that could happen to you is to have your last four cards be in different suits. Why? If this happens, there's no way any card can be moved to the discard pile. If you were lucky enough to be down to just the four aces, you end up with 8 cards (the 8 cards are the four aces which were showing, and then the 4 cards in the 4 different suits which were then dealt); if you had more, it's even worse!

Similar to our study of Klondike, we won't compute the true probability of losing. We'll concentrate on this special case, as again it's a quite annoying type of losing. We called our pesky configurations in Klondike unplayable; let's call these unmovable, as it's not our fault the last four cards can't be moved.

 What's the probability of an unmovable game? In other words, how often are the last four cards in different suits? Is the probability of this happening large enough that we should look for a different solitaire game to play (fortunately there are many!), or is it small enough so that we shouldn't really worry about it? Obviously the answers to *these* questions depend on your personal preferences; all we can do here is compute the probability of choosing four cards and having them all be in different suits.

There are $\binom{52}{4} = 270,725$ ways to choose four cards from 52 where order doesn't matter, and $52 \cdot 51 \cdot 50 \cdot 49 = 6,497,400$ ways to choose four cards from 52 where order does matter; obviously we don't want to look at each possibility!

We have to choose one card from each suit: there are $4! = 24$ ways to choose the order of the four suits (♠♡◇♣ to ♣◇♡♠); in each suit, there are 13 ways to choose a card. Thus, with order counting, there are $4! \cdot 13^4 = 685,464$ ways to choose four cards from four different suits. In other words, there's about a 10% chance ($685464/6497400 = 2197/20825 \approx 0.105498$) of getting an unmovable game. This is much worse than the .25% from Klondike. Ten percent is pretty significant, and remember, this is just a lower bound on losing.

We can count this another way. Let's look at it without specifying the order of suits. For our first card, we may choose *any* of the 52 cards; there are $\binom{52}{1} = 52$ ways of doing this. Our second card can be anything so long as it's in a different suit; thus, there are $\binom{39}{1} = 39$ ways of choosing a second card. Similarly our third card cannot be in either of the first two suits, so there are $\binom{26}{1} = 26$ ways to choose the third card, and finally there are $\binom{13}{1} = 13$ ways to choose the last card. Note that $\binom{52}{1}\binom{39}{1}\binom{26}{1}\binom{13}{1} = 685,464$, exactly as before! We divide by the number of ways to choose 4 cards with order mattering, and get the same answer as before.

There's even another way to calculate this quantity! Our problem is equivalent to the following: we have four boxes; each box has 13 cards of a suit, and we need to choose one card from each box. There are 13 ways to choose a card from each box, and thus $13^4 = 28,561$ ways of choosing four cards, one from each suit. It's essential to note that in this method, order doesn't matter. In other words, we haven't said whether or not spades was the first or fourth suit chosen, only that we have chosen one spade. There are $\binom{52}{4} = 270,725$ ways to choose four cards from 52 when order doesn't matter. Thus, the probability that we have one card in each suit is just $28561/270725 = 2197/20825$, exactly as before!

 Takeaways: While it can be fun to analyze the odds of winning a game, the reason I enjoy this problem so much is that it shows you there can be more than one right way to the correct answer; in fact, in this case there are at least three! Two of the ways involved counting ordered sequences of four cards, while the third considered unordered. As long as you're careful, you can frequently attack a problem in several ways. The important thing to remember, of course, is not to mix and match: if the numerator is the number of ordered sets of four cards satisfying our condition (one in each suit), the denominator should be the number of ordered ways of having four cards.

 Additional questions: As an aside, a little more inspection leads to additional fun problems (as well as the knowledge that this is a hard game to win!). If either the last four cards or the previous four cards are all in the same suit, it will be very hard to win. What's the probability that this happens? It should be around 20%. Another difficulty is that, even if we don't have four cards in the same suit, it's possible that the highest spade is placed on the spade pile, and so on. What are the probabilities of these events?

3.3.3 FreeCell

No discussion of solitaire would be complete without at least a brief mention of **FreeCell**. Variants go back to the 1920s. While a computer version was available in 1978 (by Paul Alfille for the PLATO system), the really important milestone was when Microsoft included it in Win32s in the early 1990s. Since then FreeCell has been installed and readily available on millions of machines. Countless users have spent (or is it misspent?) hours playing the game.

We'll briefly describe how the game is setup and played; see Wikipedia (or Microsoft's FreeCell help) for more. All of the cards are dealt face up, with four columns of 7 cards and four columns of 6 cards. The goal is similar to Klondike in that you want to build up four foundations. There are of course rules as to which cards can be moved and when. If you know them, great; if not, no worries. You can go online and read more if you're interested. Our goal is not a complete analysis of the game, but rather to use the game to highlight how easy it is to accidentally double count.

Continuing our theme, we're going to concentrate on whether or not every game is solvable. Microsoft's program has initial game configurations saved, so different people can share the fun of looking at the same boards. Of the original 32,000, all have been solved except game number 11,982. In the later version with one million games (the first 32,000 are the same as before), the only ones that haven't been solved are 11,982, 146,692, 186,216, 455,889, 495,505, 512,118, 517,776, and 781,948. While this isn't a proof, with a million data points it sure looks like almost all games are solvable; at the very least, this is enormously better odds than what we see in Aces Up or Klondike.

Obviously if you don't know the rules of the game the following discussion may be a bit hard to follow. So, you'll just have to trust me that the following board leads to an unsolvable game. I grabbed this position from a Web page of Hans Bodlaender, posted Fri Jul 12 16:18:16 MDT 1996. It's online at http://www.staff.science.uu.nl/~bodla101/d.freecell/node2.html#SECTION0002.

2♠	2♣	2♦	2♡	A♠	A♣	A♦	A♡
8♠	8♣	8♦	8♡	7♠	7♣	7♦	7♡
K♠	K♣	K♦	K♡	6♦	6♡	6♠	6♣
Q♦	Q♡	Q♠	Q♣	5♠	5♣	5♦	5♡
J♠	J♣	J♦	J♡	4♦	4♡	4♠	4♣
10♦	10♡	10♠	10♣	3♠	3♣	3♦	3♡
9♠	9♣	9♦	9♡				

Again, we're not going to go into the details of how the game is played. Like Klondike, it involves moving red cards (the hearts ♡ and diamonds ♦) on to the black cards (the spades ♠ and the clubs ♣), or vice versa. If we were to switch all the clubs with spades and all the hearts with diamonds, the above would still be unsolvable. It's also not an all or nothing; we could switch just some of the pairs (say J♠ with J♣, and 5♦ with 5♡).

This leads to the natural question: how many unwinnable configurations can we get from trivially modifying the above? There are a lot of ways to think about this; the danger is trying to combine them without double counting.

First, there are 26 pairs of each number with matching colors. Each of these can be in either of two configurations. This gives 2^{26} configuration from swapping pairs.

Second, we could switch the ordering of the columns. There are $4! \cdot 4!$ possible column orderings, since we can order the first 4 columns in 4! ways and the second 4 columns in 4! ways. However, some of the column switches are the same as row switches. For example, if we were to switch all the pairs in the first two columns and then switch the columns, we would be back to the original arrangement. This raises concerns over double counting.

It turns out that when we account for all the possible switches of pairs, columns with the same numbers and colors are indistinguishable. If we have 2 pairs of indistinguishable elements to order there are 6 ways to do this (AABB, ABAB, ABBA, BBAA, BABA, BAAB). Similarly, there are 6 ways to arrange the second set of 4 columns. This gives $6 \cdot 6$ ways to rearrange the columns.

There are a total of $2^{26} \cdot 6 \cdot 6 = 2,415,919,104$ ways to rearrange columns and flip rows. While this seems large, it is tiny compared to the 52! total starting configurations, which is more than $8 \cdot 10^{67}$.

As you can see, it's not the case that all the column moves can be combined with all the row moves. Since there're dependencies here, we can't just use our multiplication rule to multiply the number of possibilities each generates. While we'll say far more about such issues when we reach independence in Chapter 4, it's good to have these issues on your radar. The number of unwinnable boards arising from trivial tweaks of this configuration is minuscule; however, the issues we've encountered in our search are some of the most important in all of probability. It's really easy for dependencies to sneak in. You must be ever vigilant to avoid double counting.

3.4 Bridge

I come from a long line of card players, and am continuing that tradition with my kids. One of my favorites is **bridge**, which represents my only successful foray into the world of intramural sports. Interestingly, when I was in grad school at Princeton the intramural league was comprised of the undergraduate eating clubs and the Graduate College. At the end of the season the two graduate teams were tied in first, and never bothered playing one last game to determine the winner. If we had, though, I'm sure my team would've been soundly defeated by the other, which interestingly was captained by Adrian Banner, author of *The Calculus Lifesaver*!

So, with 100% of the current Lifesavers involved in bridge, how can we not have at least a few problems? We'll quickly summarize enough of the rules to get a feel for what happens. We'll completely ignore how to bid, and instead concentrate on determining probabilities of various deals.

For our purposes, all we need to know about bridge is that there are four players in two teams of two, each of whom is randomly given 13 cards from a standard deck. The players bid to determine whether or not there will be a special suit which is designated **trump** (the bidding is quite involved; there are lots of conventions, and there's no way to do it justice in a few paragraphs, so we won't even try). The goal is to win as many tricks as possible. A trick is a play of four cards, with people playing cards in order with whomever won the previous trick playing first, with the order of the remaining players proceeding clockwise from the first player. The highest card played wins the trick. The advantage of having a trump is that, if you are out of the suit being played, you may play a trump card and win the trick. Thus, if there's a trump suit, it becomes extremely important to know the distribution of cards in the opponents' hands. For example, if one opponent has a lot of trump, then they have fewer cards in the remaining three suits, and thus there's a greater chance they can trump in earlier. It's thus of primary importance to have estimates for the probability an opponent has a given number of trump.

We'll first calculate how many different bridge games there are, then investigate an important problem on the distribution of trumps among the opposition. We'll give two different ways to compute the answer, and try to reconcile why they violently disagree!

3.4.1 Tic-tac-toe

It may seem strange to begin an analysis of bridge by looking at **tic-tac-toe** (also called **wick wack woe** or **noughts and crosses**), but there's a very good reason for this. Bridge is complicated. Even ignoring the bidding, it involves a 52 card deck being split into four hands. Tic-tac-toe is far simpler, and is played and enjoyed by kids in playgrounds all over the world (though it's not clear if it's enjoyed by people beyond elementary school). Let's first build some intuition by looking at this game, and then move on to bridge.

In tic-tac-toe, players alternate putting X's and O's on a 3 × 3 grid; whomever gets three in a row first wins (horizontally, vertically, or diagonally); if no one gets three in a row then the game is a tie. Below is an example of a tic-tac-toe game, with X going first.

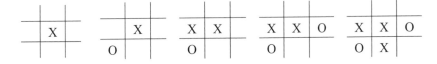

So, how many possible tic-tac-toe games are there? This turns out to be a lot harder than you might expect. It's easy to get an upper bound on the number of games. There are 9 places for the first move, then 8 for the second, 7 for the third, and so on, leading to 9! or 362,880 possible games. That's a lot of games for little kids, and should keep them busy for a long time.

However, this isn't the right answer for two reasons. The first is that if we rotate the board 90 degrees, or if we flip the board horizontally or vertically, things really haven't changed. Consider the following sequence, obtained by rotating the board 90 degrees clockwise (the opening move looks the same as it was in the center, but the second move is now in the upper left corner and not the lower left).

	O	
X		

	O	X
	X	

	O	X
	X	
	O	

	O	X
X	X	
	O	

O	X	
X	X	O
	O	

O	X	X
X	X	O
	O	

O	X	X
X	X	O
	O	O

O	X	X
X	X	O
O	O	X

Should this really be counted as a different game? Rotating or flipping the board doesn't really change anything, so these really shouldn't be counted as distinct. Thus, 9! greatly overestimates.

How many first moves are there? There are really just 3: middle, side, corner. Using this observation, we've pruned the number of games from 9! to $3 \cdot 8!$, saving a factor of 3. Can we make more progress? Sure! How many responses are there to an opening move? Let's break it into the three cases.

- If someone goes in the center, there are really only two responses: a corner or a side. From this point onward, there are 7! ways to fill in the board, for a total of $2 \cdot 7! = 10,080$ games.

- If someone goes in the side, there are five essentially distinct responses: the two corners that border that side are the same after flipping the board, and the two other corners are also the same. That gives two moves. There's the middle, bringing us up to three. Then there's the two sides diagonally from our initially chosen side; these are equivalent and add one more move, and now we're at four moves. Finally, there's the side opposite our initial side, so we end at 5 moves. Thus there are 5 different second moves in this situation, and then 7! ways for the remaining 7 moves, for a total of $5 \cdot 7! = 25,200$.

- Finally, let's consider an opening move in the corner. A similar analysis as the last case also gives there are five moves (center, opposite corner, adjacent side, non-adjacent side, non-opposite corner) for the response, and then 7! ways to finish the board, so we again get $5 \cdot 7! = 25,200$.

Adding up the three cases, the total number of games is now at most $10,800 + 25,200 + 25,200 = 61,200$. It's still a lot, but it's much better than the $9! = 362,880$ we started with; we're saving almost a factor of 6 (improving on our earlier savings of a factor of 3).

I encourage you to continue this analysis. We're still overcounting, as symmetries will allow us to continue pruning. However, there's something else we've completely ignored. We said there were two reasons 9! isn't the right answer. We've discussed one issue, namely rotating and flipping boards. What's the other? Not all games go the distance! If someone wins, they often win in less than 9 moves. For example, if you start in the center and your opponent is foolish enough to choose a side, you can ensure a win in 7 moves.

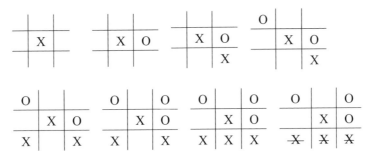

At this point, the game is done—there's no need to figure out where to put the final O and X. The purpose of this book is not to set you up as the king of tic-tac-toe on the playgrounds, so we're going to stop our counting here and just say there are 765 different games, far less than our initial bound of 362,880. That said, we've learned an important fact. There are often lots of ways to view a situation, and sometimes things that look different are really equivalent, and should be counted as the same.

 Here are some tic-tac-toe problems to play with: (1) prove there are 765 distinct configurations for tic-tac-toe games (if not, please fix the Wikipedia page!); (2) figure out an optimal strategy for going first, and an optimal strategy for going second. For example, if you go first and your opponent foolishly takes a side, you can win. Come up with moves for each possibility, and resolve never to lose a game of tic-tac-toe again.

For some nice, additional reading, see

- http://www.btinternet.com/~se16/hgb/tictactoe.htm
- http://www.mathrec.org/old/2002jan/solutions.html.

 The following fun problem is related to our tic-tac-toe counting. Imagine that, due to budgetary cuts, we have a 5 × 5 chess board. It is possible to put 5 queens on the board so that 3 pawns can safely be placed. Find that configuration! (Hint: Instead of trying all the possible ways to put down 5 queens, you can use the ideas from our tic-tac-toe analysis to cut down on the possibilities. While this works, if you're clever you can save even more time by noting a powerful equivalence. See http://www.youtube.com/watch?v=aMorr1h4Egs.)

3.4.2 Number of Bridge Deals

In the previous subsection we counted the number of distinct games of tic-tac-toe. We started with an upper bound of 9! or 362,880, and then showed that if we consider games that are the same after rotating or flipping that number falls to at most 61,200; if we had worked harder and longer, we could've gotten that down to 765 (though getting to that number requires taking into account additional symmetries).

What about bridge? Well, we have four players, and each are dealt 13 cards. There are 52! ways to order the cards in a deck, so (assuming the four players are distinct), there are 52! or about $8 \cdot 10^{67}$ distinct deals.

However, just like tic-tac-toe, this isn't the number we want. Once the dealing is done, all that matters are which 13 cards each player gets, not the order the cards arrived. So 52! is way off. What's the right answer? Well, there are $\binom{52}{13}$ ways to choose 13 cards for the first player, then $\binom{39}{13}$ ways to choose 13 of the remaining $52 - 13 = 39$ cards for the next player, then $\binom{26}{13}$ choices for the third (remember only 26 cards are free at this point, and we need to grab 13). There's no choice left for the final person—they have to get the remaining thirteen cards. This fits our patterns, as $\binom{13}{13} = 1$, confirming that there's but one way to assign the remaining cards.

Multiplying everything out, we find the number of distinct hands that can be dealt (i.e., the four people are distinguishable, but it doesn't matter the order in which the cards are received) is

$$\binom{52}{13}\binom{39}{13}\binom{26}{13}\binom{13}{13} = 53,644,737,765,488,792,839,237,440,000,$$

or about $5 \cdot 10^{28}$. It's still a gigantic number, but it's more than 39 orders of magnitude smaller than our original upper bound!

Armed with the above calculation, let's find the probabilities of some interesting events. I remember a great passage in the probability book I used as an undergraduate (*Introduction to Probability* by Douglas G. Kelly) about **perfect deals**. In a perfect deal each of the four players receives all 13 cards in a suit. As there are occasional reports of perfect deals, it's worth investigating how likely they are. Assuming the deck is well mixed, there are four choices for the first player (they may get all the spades, hearts, diamonds, or clubs), then three choices for the next person, two for the third person, and finally the last person's hand is forced. There are thus $4! = 24$ perfect deals. Dividing this by the number of distinguishable hands we get a probability of less than 10^{-25} percent! It's thus spectacular that this could ever be reported or observed (assuming the cards were truly mixed). To get a sense of how unlikely this is, go through the analysis we did in §3.1.1 to get a feel for the factorial function. Imagine 10 billion people dealing a hand a second, and see how many *trillions* of years it would take!

What I remember enjoying is that while many perfect deals were reported, no **semi-perfect deals** made the news. A semi-perfect deal has only two of the four people receiving all cards in a suit. The implication, of course, is that there's something fishy about all those perfect deals.

Let's go one step further, and consider just the number of ways that one of the people at the table could get all 13 cards in a suit. Remember in Chapter 1 how it took us three tries to ask the Birthday Problem correctly? We'll encounter a similar problem here, which is another reason for the choice of this problem. There's no shortage of things to count; what I like about this problem is we get another example of the need to be very careful and precise with our language.

Warning: what follows is some of the peskiest counting you'll ever see! It's very easy to accidentally assume something and add some structure. Read what follows carefully. Be aware. Be alert. As you're reading, ask if the counting is believable, or if you think some order was accidentally added. After we go through the arguments, we'll return and look at them again. Of course, since I'm warning you to read the following carefully, it does seem like there's a mistake waiting to be found. The problem is that it's so easy to add small amounts of order without even knowing it. So, eyes open and read on!

Without further ado, let's *try* to calculate the probability that a lucky (or should we say unbelievably lucky) person gets all thirteen cards in a suit. There are $\binom{4}{1}$ ways to choose that lucky person, and there are exactly 4 hands they can get. Why? We have to choose one of the four suits, which happens $\binom{4}{1} = 4$ ways, though it's better to think of it as happening 4 ways *without* a binomial coefficient, as what's really going on is we're making an ordered choice of one suit from 4 (more on this later). What about the remaining three people? The number of ways to give them the remaining cards is $\binom{39}{13}$, then $\binom{26}{13}$, and finally $\binom{13}{13}$ for the last. Thus, the total number of deals with someone getting a one-suited hand is just $\binom{4}{1}\binom{4}{1}\binom{39}{13}\binom{26}{13}\binom{13}{13}$, which is 1,351,649,569,165,862,400 or about $1.3 \cdot 10^{18}$. Dividing by the number of hands (approximately $5 \cdot 10^{28}$), we see the probability that a person gets all their cards in one suit is one in $1/39,688,347,475$, or about $2.5 \cdot 10^{-11}$ percent.

Before going further, it's worth noting a good way to find the probability. We shouldn't compute the number of hands and divide by the number of possible hands. Why? A lot of the factors cancel, and if we remove them first *before* doing our algebra, we make our lives easier and minimize the danger of computational overflow errors. The ratio we need is

$$\frac{\binom{4}{1}\binom{4}{1}\binom{39}{13}\binom{26}{13}\binom{13}{13}}{\binom{52}{13}\binom{39}{13}\binom{26}{13}\binom{13}{13}} = \frac{4 \cdot 4}{\binom{52}{13}} = \frac{1}{39688347475}.$$

We've answered a question (we'll see later if our answer is even correct!), but is this the question we wanted to answer? What we hopefully did was to find the probability that *at least* one person at the table has all their cards in one suit. Is that what we wanted, or did we want instead the probability that *exactly* one person has all their cards in the same suit? If we want exactly one, then we need to remove the hands where two, three, or four people have one-suited hands.

I like this feature of the problem. From a practical point of view, there's really no difference as it's so much rarer to have two or more one-suited hands than just one. It's not a dangerous error for this problem, but it can be for others and thus you should make sure you're well aware of these issues.

Okay, time to find how many hands have exactly one person being one-suited. What's nice is that it's impossible to have exactly three people one-suited, as that forces the final person to be one-suited as well. We already know how many hands have everyone one-suited. It's just 24.

Time to determine how many hands have exactly two people one-suited. There are $\binom{4}{2} = 6$ ways to choose the two special people to get one-suited hands. Clearly the order in which they're chosen doesn't matter; all that matters is which two people are chosen. You might think that the next item is to say that there are $\binom{4}{2}$ ways to assign two suits to each of our two special people, but this is emphatically *not correct!* The problem is we've been assuming the four people are distinguishable; thus it matters who gets which suit, and it's not $\binom{4}{2} = 6$ but instead $4 \cdot 3 = 12$ (four choices for the suit for the first person, three choices of suit for the second). Remember back to when we did everyone being one-suited. We said there were $4! = 24$ ways to assign the suits, coming from $4 \cdot 3 \cdot 2 \cdot 1$. So, there are $\binom{4}{2} = 6$ ways to choose the two special people, and then $4 \cdot 3 = 12$ hands for them. A great way to view $4 \cdot 3$ is that it equals $\binom{4}{2} \cdot 2!$; there are $\binom{4}{2}$ ways to choose two suits from four suits, and then $2!$ ways to order the two suits to the two players.

We're left with assigning the remaining 26 cards to the last two people. There are $\binom{26}{13}$ ways to give 13 cards to the third player, and then the fourth person has their hand forced (there's only 1 or $\binom{13}{13}$ ways to choose 13 cards from 13). So, we have $\binom{26}{13}\binom{13}{13} = 10,400,600$ ways to assign these remaining cards. The problem, of course, is that some of these assignments may lead to four people having one-suited hands. How do we take care of that? Well, after we've assigned the first 26 cards we have 26 remaining. We have to give 13 to the third person; there are $\binom{26}{13}$ ways to do this. How many of these ways lead to them getting a one-suited hand? Exactly two. We've already given out all the cards in two suits to the first two people. These 26 cards are all the cards in two suits. We thus need to modify our calculation. We should have $\binom{26}{13} - 2 = 10,400,598$ ways of assigning 13 cards to the third person, and then $\binom{13}{13} = 1$ way to finish by giving the fourth person their cards. A small change, but an important one. Combining, the number of ways to have exactly two people with one-suited hands is

$$\binom{4}{2} \cdot 4 \cdot 3 \cdot \left(\binom{26}{13} - 2 \right) \binom{13}{13} = 748,843,056.$$

If we haven't made any mistakes, we can now find the number of hands with exactly one person one-suited. We just take the number of hands with at least one person one-suited and subtract the number of hands where two or more people are one-suited. We said there were $4 \cdot 4 \cdot \binom{39}{13}\binom{26}{13}\binom{13}{13}$ or 1,351,649,569,165,862,400 with at least one person one-suited. We subtract the number of hands with exactly two people one-suited, which is 748,843,056, and then subtract the number of hands with everyone one-suited, which is 24. Thus, we get a final answer of 1,351,649,568,417,019,320. It's smaller than our original answer, but just barely.

Of course, the big question is: are we right? How confident are you that we haven't made a mistake? Is there another way to find the answer to serve as a check? Fortunately, there is, so we must try it. We can modify the argument we used to find exactly two people are one-suited.

Here's another approach to finding the number of hands with exactly one person one-suited. There are $\binom{4}{1}$ ways to choose our lucky person, and then 4 hands they can have (corresponding to the four suits). For the next person, there are $\binom{39}{13}$ possible hands. Of these, exactly 3 are one-suited and must be discarded, so there are $\binom{39}{13} - 3$ possible hands for the second person. What about the third? This is where things get tricky. *If* the second person has one card in each suit, then it's impossible for the third or fourth person to be one-suited; however, if the second person only has two suits (they can't have just one as we removed that possibility), then there's a danger.

Arg. This is becoming a combinatorial nightmare! We keep splitting into cases and subcases of cases. No wonder so many people throw their hands up in disgust over these counting problems. Alright. We'll break things into cases.

- Case 1: The second person is missing exactly one suit (if they're missing two suits then they have a one-suited hand, and we don't want to allow that). There are $\binom{3}{2}$ ways to choose two suits. That gives us 26 cards, we need to take 13. We can do that $\binom{26}{13}$ ways; however, two of those ways result in a one-suited hand, which isn't good. So there are $\binom{3}{2}(\binom{26}{13} - 2) = 31,201,794$ hands for the second person that have exactly two suits. At this point, the third person has to get 13 cards, but there's a danger that they might get all 13 cards in one suit. Fortunately, there's only one way that can happen, as three of the suits are already represented in the first two people's hands and one suit is in no one's hand so far. Thus, the number of valid assignments to the third person is $\binom{26}{13} - 1$, and then the fourth person's hand is determined. Collecting everything, we find the number of hands in this case is

$$\binom{3}{2}\left(\binom{26}{13} - 2\right)\left(\binom{26}{13} - 1\right) = 324,517,347,474,606.$$

- Case 2: What if the second person isn't missing a suit? There are $\binom{39}{13} - 3$ ways to give them 13 cards, making sure they don't have a one-suited hand (that's why we have a minus 3). We have $\binom{3}{2}\binom{26}{13} - 2 = 31,201,798$ hands for the second person that have exactly two suits. Thus there are $\binom{39}{13} - 3 - \left(\binom{3}{2}\binom{26}{13} - 2\right) = $ 8,091,223,643 hands for the second person that have at least one card in the three remaining suits. From this point on, there's no danger of either the third or fourth person getting a one-suited hand, as the first two people have at least one card in each suit between them. There are thus $\binom{26}{13}$ ways to give the third person a hand now, and $\binom{13}{13}$ possibility for the fourth. Combining, we find the number of hands in this case is

$$\left(\binom{39}{13} - 3 - \binom{3}{2}\left(\binom{26}{13} - 2\right)\right)\binom{26}{13}\binom{13}{13}$$
$$= 84,153,580,662,988,200.$$

We can now collect and find our answer. There are 16 possibilities for choosing a special person to have a one-suited hand *and* choosing the suit. We thus multiply the number of hands from Case 1 and Case 2 by 16 and add, and get (drum roll please)

$$1,351,649,568,167,404,896.$$

Drat (though please feel free to use a stronger expletive here). It's not the same as our earlier answer of 1,351,649,568,417,019,320. It doesn't differ by much, but it is a little lower. Our first answer is larger by 249,614,424. Which one is right? Is either? I've gone through several different numbers in writing up this problem!

 I'm not going to tell you which answer is right for several reasons. First, you should have the fun of going through the calculations yourself! Second, and more important, regardless of which (if either) is the right answer, it should be clear that *neither* of these methods are good. There's too much adding and subtracting in what feels like an ad hoc manner. We have to keep remembering to count carefully, and since there's no methodical way to plow through the calculations, it's very likely that we'll make a small error at some point. We need a better way, and we'll find one in the **Method of Inclusion-Exclusion**. This is one of the most important techniques for counting, and it's earned an entire section, Chapter 5. We'll revisit this problem there, and see how much simpler this approach makes life. In brief, the Method of Inclusion-Exclusion is a very spelled out procedure to compute the probabilities of difficult events in terms of basic ones. View these last few pages as motivation to learn advanced theory!

If you've been observant, you're probably wondering why I haven't talked about writing code and numerically investigating the probability. Well, here's some code! The interesting part is finding a clean, efficient way to check if at least one of the hands is one-suited. We give an okay method below, and describe a *much* better way in Exercise 3.7.34; see also Remark 3.4.1.

```
onesuit[numdo_] := Module[{},
  count = 0; (* counts the success *)
  deck = {};
  For[i = 1, i <= 4, i++,
   For[j = 1, j <= 13, j++, deck = AppendTo[deck, i]]];
  For[n = 1, n <= numdo, n++,
   {
    (* low probability of success, print out every 10% *)
    (* so know how much we've done *)
    If[Mod[n, numdo/10] == 0, Print["Have done ", 100. n/numdo, "%."]];
    (* ranomly shuffle deck *)
    mix = RandomSample[deck];
    (* partition deck into four hands of 13 *)
    hands = Partition[mix, 13];
    (* set onesuited to 0, if stays 0 then no onesuited hand *)
    (* if at least 1 then at least one hand is onesuited *)
    onesuited = 0;
    (* check each of four hands to see if onesuited *)
    (* if cards 2 through 13 are the same as first card then
       first hand is onesuited, if cards 14 through 26 are the
       same then second hand is one suited, and so on. *)
    For[i = 1, i <= 4, i++,
     {
      (* load in the current hand *)
      possibleonesuited = 1; (* initialize to 1 *)
      (* lower possibleonesuited to 0 once two cards differ *)
      currenthand = hands[[i]];
      (* checks to see if cards 2 through 13 match first card *)
      (* if don't match decrease possibleonesuited to 0 *)
      (* if possibleonesuited ends at 1 must be onesuited *)
      For[j = 2, j <= 13, j++,
       If[currenthand[[j]] != currenthand[[1]],
         possibleonesuited = 0];
       ]; (* end of j loop *)
      If[possibleonesuited > 0, onesuited = onesuited+1];
      }]; (* end of i loop *)
    (* if onesuited > 0 then at least one hand is one suited! *)
    (* note if two hands onesuited only count once *)
    If[onesuited > 0, count = count + 1];
    }]; (* end of n loop *)
  Print["Observed probability is ", 100. count/numdo, "%."];
  ]
```

Normally I'd report the theoretical probability (but that's in some dispute) and the observed probability; unfortunately, in ten million trials *no* one-suited hand was found! This is due to the fact that our probability is so small; we need better ways to investigate the enormous number of deals in order to see the true probability. While there are enormous gains that can be made by shifting to a faster environment than Mathematica, a lot of work is still needed. We leave that as an exercise to the interested reader.

Remark 3.4.1: *It's worth ending with a short remark on coding. Note we handled it by creating a holding variable,* `possibleonesuited`, *and adjusting it by doing 12 comparisons. We could instead do*

```
If[Product[currenthand[[j]], {j,1,13}] == currenthand[[1]]^13,
  onesuited = onesuited+1];
```

which is a nice, concise way to code the problem, and since 4^{13} is only $67,108,864$, the numbers are not overwhelmingly large. That said, we leave it for those interested in programming to find the most efficient, least intensive memory approach; one possibility is to break some of the counting loops as soon as two suits are seen in a hand.

3.4.3 Trump Splits

One of the biggest problems in **bridge** (or the simpler variant I grew up playing, **whist**) is trying to estimate the probabilities of various trump splits in the opposition. Frequently these probabilities are miscalculated. Below we'll explain the problem and the correct solution, and examine the mistake people often make.

The way bridge and whist are played is that if you bid highest, your partner's cards are turned over for all to see. Thus, you know exactly how many trump are in your team's hands and how many are in your opponents'; however, you don't know how many trump are in an individual opponent's hand.

The natural question becomes: if your team has n of the 13 trump, what is the probability one opponent has k and the other opponent has $13 - n - k$ trump? For example, if you have 8 trump that means 5 are missing; what are the probabilities they are split 5–0, 4–1, or 3–2?

 Let's first consider a 5–0 split: this means that of the 13 cards one player has, exactly 5 of them are from the missing 5 trump, and the other 8 are from the remaining 21 cards that aren't trump (there are 26 cards to be assigned to these two players, 5 are trump and thus the other 21 are non-trump). There are $\binom{2}{1}$ ways to choose which opponent gets the five trump cards, $\binom{5}{5}$ ways to give that special person all five remaining trump, and then $\binom{21}{8}$ ways to choose 8 of the remaining 21 cards to fill out their hand to 13 cards. Thus, there are $\binom{2}{1}\binom{5}{5}\binom{21}{8} = 406{,}980$ ways this could happen. Note that once we have specified the cards one opponent has, the other opponent's hand is determined. As there are $\binom{26}{13} = 10{,}400{,}600$ ways of choosing 13 of 26 cards for our opponent (and choosing all of one opponent's cards means we completely specify the other opponent's hand as well), we see the probability of a 5–0 split is $406980/10400600 = 9/230$, or about 3.91%.

What about a 4–1 split? We again have $\binom{2}{1}$ ways to choose the player who gets 4 trump, then $\binom{5}{4}$ ways to give them 4 trump, and then $\binom{21}{9}$ ways to give them 9 more cards to round out their hand. Again, at this point the other person's hand is completely determined, as it's just the remaining cards. We find there are $\binom{2}{1}\binom{5}{4}\binom{21}{9} = 2{,}939{,}300$ such hands, for a probability of $2939300/10400600 = 13/46$ or about 28.26%.

Finally, let's do a 3–2 split. We begin with $\binom{2}{1}$ ways to choose the person getting 3 trumps. There are $\binom{5}{3}$ ways to give them 3 of the 5 trump, then $\binom{21}{10}$ ways to choose 10 non-trump cards, for a total of $\binom{2}{1}\binom{5}{3}\binom{21}{10} = 7{,}054{,}320$. The probability of this is therefore $7054320/10400600 = 78/115$, or around 67.83%.

There's a natural consistency check available: let's add the three probabilities and see if they sum to 1. Equivalently, we can add the number of possibilities for each of the three cases, and see if that sums to $\binom{26}{13} = 10,400,600$. We find $406,980 + 2,939,300 + 7,054,320 = 10,400,600$. While this isn't a proof that we haven't made a mistake, it's a pretty good sign all is well!

Let's consider another approach to this problem. *While this approach* **should** *seem reasonable, it has a very subtle error; it's worth reading this very carefully and slowly to master this fundamental issue.* We need to assign 26 cards (5 trump, 21 non-trump) to two people. Each gets 13 cards. For each trump card, why not look to see whether or not it goes to the first person or the second person. Clearly each trump card is equally likely to be in either person's hand. Thus, each card has a 50% chance of landing with person 1, and a 50% chance of landing with person 2.

Arguing along these lines, the probability that all the trumps end in the first person's hand is 1/32. Similarly there's a 1/32 chance that all end with the second person, giving the probability of a 5–0 split as 2/32 or about 6.25%. Hmm. This is very different than what we had before, where our first calculation gave the probability was 9/230, or about 3.91%.

Let's move on to the next case, a 4–1 split. We first send four of the five cards to the first person. There are $\binom{5}{4}$ ways to choose four of five trump to give person 1, and each chosen card has a 50% chance of going to person 1. The remaining card has a 50% chance of going to person 2. Thus, the probability that the first person gets exactly four trump would be $\binom{5}{4}(1/2)^4(1/2) = 5/32$, or 15.625%. To get a 4–1 split we have to double, as either person 1 or 2 can get the four trump, giving the probability of a 4–1 split as 10/32 or 31.25%. This is close to, but more than, the 28.26% we found earlier.

The final case, a 3–2 split, is handled similarly. If the first person gets three trump, the odds of that happening (assuming we assign the five trump cards one at a time) is $\binom{5}{3}(1/2)^3(1/2)^2 = 5/16$, or 31.25%. Doubling (as either person could get the three trump) gives 62.5%. It's close, but it's not 67.83%.

Clearly we've made a mistake, but where? In situations like this, it's a great idea to try a smaller case first, build up some intuition, and then return to the original, larger problem. Given my penchant for naming things, we'll call this the **Method of the Simpler Example**. We'll go through a simpler case, and then revisit our discussion above. We'll see why the second method (using the plausible 50% arguments) is wrong.

Imagine we have four cards, two trump and two non-trump, and we want to divide these up among two players, Alice and Bob. So each person gets two cards, and all hands are equally likely. How many hands are there? There's $\binom{4}{2}\binom{2}{2} = 6$ (we have $\binom{4}{2}$ ways to assign two cards to Alice, and then Bob's is forced). How many hands have all the trump in one hand? There's only two: either all the trump go to Alice, or all go to Bob. Thus, there's a 2/6 or about a 33% chance of a 2–0 trump split.

Let's repeat our second argument here. We look at the first trump card; it has a 50% chance of going to Alice and a 50% chance of going to Bob, as it's equally likely to be in both cases. The same is true for the second trump card. Before, we said the probability they both end up in Alice's hand is $1/2 \cdot 1/2$ (as each card has a 50% chance of going to the first person), and similarly the probability they're both with Bob is $1/2 \cdot 1/2$. Adding these probabilities we get 1/2 or 50%, much larger than 1/3 or 33%. Where did we go wrong?

The problem is we're mixing up events. Yes, it's true that *each* card is equally likely to be in either hand. That's correct. What's not correct is that *given* the first trump goes to Alice *then* the second trump also goes to Alice with probability 1/2. In other words,

TABLE 3.1.
The six ways to have the four cards split evenly between the two players.

	Hand 1	Hand 2	Hand 3	Hand 4	Hand 5	Hand 6
Person 1	{1♠, 2♠}	{1♠, 3♡}	{1♠, 4♡}	{2♠, 3♡}	{2♠, 4♡}	{3♡, 4♡}
Person 2	{3♡, 4♡}	{2♠, 4♡}	{2♠, 3♡}	{1♠, 4♡}	{1♠, 3♡}	{1♠, 2♠}

if we know that Alice got one trump, she's now *less* likely to get the second trump! (For another example of this phenomenon, see the ABBA problem from §5.1.1.)

There's lots of good ways to think about this. The first trump card has to go to someone, so let's say for definiteness it goes to Alice. Remember, we only care whether or not the trump split is 2–0; we don't care *which* person gets both. So, Alice got the first trump. Needing two cards, Alice needs one more and three cards remain. *From this point on*, each of the three cards is as likely to go to Alice as any other card. Thus, the odds she gets the second trump *is not* 1 in 2 but instead 1 in 3.

Here's another way to look at this. The first trump went to Alice. There are three cards left; Alice needs one more card while Bob needs two more cards. Thus each of the remaining cards is twice as likely to go to Bob as Alice. Putting it another way, the three remaining slots are all equally likely for the three remaining cards, so the second trump goes to Bob a total of two out of three times, thus it goes to Alice a total of one out of three times.

Yes, each card has a 50% chance of being in either hand, but that's before *any* cards are dealt. Try this out: while each card has a 50% chance of being in either hand, after Alice gets two cards then there's *zero* chance that she gets either remaining card. We can't just multiply the probabilities.

It might help to write things out. Let's say spades is trump, and our four cards are 1♠, 2♠, 3♡, and 4♡ (I deliberately chose the numbers to be increasing and the other two to be the same suit to make this example as visually simple as possible). Table 3.1 shows all the ways to divide the cards into two hands of two. We'll always write spades before hearts, and we'll write a smaller number first. The reason is the hand {1♠, 2♠} could be written {2♠, 1♠}; either is fine as both represent the same hand. Anyway, here are the six possible hands.

Spend a few minutes staring at this table. We see that 1♠ is in half the hands (hands 1, 2, and 3), and 2♠ is also in half the hands (hands 1, 4, and 5). Okay, each is occurring in 50% of the hands. How often are they together? Only twice (hand 1 and 6) out of six, which means the probability of having a 2–0 split is 2/6 or 1/3 and *not* 1/2.

 Whenever you're unsure, draw a picture, do a special case, and try to build intuition. Here we can *see* what's going on. Pictures are far more powerful than words. Now, for many problems (such as trying to figure out the probability of exactly one person having a one-suited hand of 13 cards), it's just not practical to write everything out. That doesn't mean, though, that we should give up on pictures. Instead, concentrate on a smaller case, and try to see the key features.

 The multiplication rule has failed! What went wrong? Why does it fail? The problem is that information about one event imparts information on the other event. This leads to the notions of independence and dependence. We'll discuss these in great detail in Chapter 4, as we're now hopefully thoroughly motivated to delve back into theory.

3.5 Appendix: Coding to Compute Probabilities

3.5.1 Trump Split and Code

In §3.4.3 we computed the probability of having a 5–0 trump split in bridge (this means you and your partner have 8 trump and the two opponents share the remaining 5 trump among themselves) several different ways, and we got different answers. Whenever you can think of multiple approaches and they give different answers, it's a *great* idea to write some code to numerically investigate. Here is a simple Mathematica program to simulate a large number of deals of 26 cards to two people, with exactly 5 trump among the 26 cards.

Before giving the code (which is extensively commented), we describe some of the choices and approaches we took to solving the issue. This is not the worst way, but it is not the best and afterwards we'll give another approach.

The first choice we have is how to represent the cards. Remember that our goal is to calculate a given probability; we can throw away extraneous information. At the end of the day, all we care about is whether or not five trump are all in one hand. Thus let's just call the 26 outstanding cards 1, 2, ..., 26 and declare that cards 1, 2, 3, 4, and 5 are the trump. By representing the cards as numbers we can use simple functions to check if they're all in one hand. For example, instead of checking to see if 1, 2, 3, 4, and 5 are all in one hand (which requires five member identity checks) we just sort the cards in each hand and sum the five lowest numbers; that sum is 15 if and only if the hand contains our five trump!

```
fiveohsplit[num_] := Module[{},
   (* calculating probability get a 5-0 trump split in bridge *)
   (* assuming first team pair has 8 trump, other has 5 *)
   (* thus only care about 26 cards in deck *)
   (* we'll call the cards 1 thru 26, and 1 to 5 are the trump *)
   deck = {};
   For[i  = 1, i <= 26, i++, deck = AppendTo[deck, i]];
   count = 0; (* keeps track of number of times get 5-0 *)
   For[n = 1, n <= num, n++, (* n loop, does num deals *)
    {
     (* creates the two hands. chooses 13 cards randomly fromm 26 *)
     (* sorts the 13 cards so if the first five cards are 1 to 5 *)
     (* then the first player got all 5 missing trump *)
     (* we can easily check by seeing if the sum of first five is 15 *)
     handone = Sort[RandomSample[deck, 13]];
     (* now make the second hand *)
     (* we have 26 cards,
     put cards into second hand if not in first *)
     (* use the MemberQ command for this *)
     handtwo  = {};
     For[i = 1, i <= 26, i++, If[MemberQ[handone, i] == False,
       handtwo =  AppendTo[handtwo, i]]];
     handtwo = Sort[handtwo];
     (* use an or statement as just care if one of two hands has *)
     (* all the missing trump; again nice that can check by sum! *)
     If[Sum[handone[[i]], {i, 1, 5}] == 15
        || Sum[handtwo[[i]], {i, 1, 5}] == 15, count = count + 1];
    }]; (* end of n loop *)
```

```
(* Prints obsesrved percentage and then the different theories *)
Print["Observed percentage 5-0 is ", 100. count/num, "%."];
Print["Theory from Binomials: ", 100.0 2 Binomial[5, 5]
 Binomial[26 - 5, 13 - 5]/Binomial[26, 13], "%."];
Print["Theory from Cond Prob Prod: ",
 100.0 2 (13/26) (12/25) (11/24) (10/23) (9/22), "%."];
Print["Theory from each card 1/2 chance: ", 100.0 2 (1/2)^5, "%."];
];
```

Dealing a million hands we found:

```
Observed percentage 5-0 is   3.91590%.
Theory from Binomials:       3.91304%.
Theory from Cond Prob Prod: 3.91304%.
Theory from 2 * (1/2)^5:     6.25000%.
```

We now present another approach. As we only care if a card is trump or not, let's write a 1 for the five trump and a 0 for the 21 non-trump. This gives us a simpler deck. We then assign 13 cards randomly to one person; if the sum of their cards is 0 or 5 then we have a 5–0 trump split, otherwise we don't. Note how much easier this approach is. We don't have to look at the other person's hand, we aren't using member identity checks. It is frequently a lot easier to write the code if you strip away the non-essentials. Thus: ***Spend time thinking about your code.*** This becomes very important when you need efficiency (such as our search for the probability of having a one-suited hand); it's less important in cases like this, but it's not bad to get into good habits.

```
trumpsplit[numdo_] := Module[{},
   count = 0;
   deck = {}; (* initialize deck to empty *)
   For[n = 1, n <= 5, n++, deck = AppendTo[deck, 1]];
   For[n = 6, n <= 26, n++, deck = AppendTo[deck, 0]];
   For[n = 1, n <= numdo, n++,  (* main loop of code *)
   {
    hand = RandomSample[deck, 13]; (* randomlly choose 13 cards *)
    numtrump = Sum[hand[[k]], {k, 1, 13}];
    (* note numtrump is 0 or 5 if we have a 5-0 split *)
    If[numtrump == 0 || numtrump == 5, count = count + 1];
    (* count is our counter, counts how often have 5-0 *)
    (* we use || for or; would use &&
    for and use two equal signs for comparison*)
    }]; (* end of n loop *)
   Print["Two theories: binomial gave ", 6.25,
    "%, cond prob gave 3.913%."];
   Print["We observe ", 100. count/numdo, "%."];
   ];
```

3.5.2 Poker Hand Codes

We now explore how to simulate the probability of a few poker hands. Our first task is to find the probability of exactly two kings (if we wanted at least two kings all we would change is "numkings == 2" to "numkings >= 2" below). This time our deck has four 1s, representing the four kings, and forty-eight 0s, representing the non-kings. We then randomly choose 5 cards, and count the sum of the values. If the sum is two then we have exactly two kings! This illustrates the power of the method—no checking to see if certain cards are in our hand; by representing the cards in the deck with well-chosen numbers, determining if a hand satisfies our conditions becomes trivial.

```
twokings[numdo_] := Module[{},
  deck = {}; (* initialize deck to empty *)
  (* 1 is a king, 0 non-king *)
  For[n = 1, n <= 4, n++, deck = AppendTo[deck, 1]];
  For[n = 5, n <= 52, n++, deck = AppendTo[deck, 0]];
  count = 0; (* initialize num of successes to 0 *)
  For[n  = 1, n <= numdo, n++,
   {
    hand = RandomSample[deck, 5]; (* 5 card hand *)
    numkings = Sum[hand[[k]], {k, 1, 5}];
    If[numkings == 2, count = count + 1];
    }]; (* end of n loop *)
  Print["Theory predicts prob exactly two kings is ",
   100.0 Binomial[4, 2] Binomial[48, 3]/ Binomial[52, 5], "%."];
  Print["Observed probability is ", 100.0 count/numdo, "%."];
  ];
```

Dealing a million hands we found:

```
Theory predicts prob exactly two kings is 3.99298%.
Observed probability is 3.9965%.
```

Let's make the problem a bit harder and ask now for a full house of kings and queens; thus we have either two kings and three queens, or three kings and two queens. This time we make kings 1, queens 10, and everything else 0. If we have a full house the sum must be either 32 or 23, and thus a simple OR comparison (coded in Mathematica by ||) is all we need. See Exercise 5.1.1 for more on this method.

```
fullkingqueens[numdo_] := Module[{},
  deck = {}; (* initialize deck to empty *)
  (* 10 is a queen, 1 is a king, 0 non-king *)
  For[n = 1, n <= 4, n++, deck = AppendTo[deck, 1]];
  For[n = 5, n <= 8, n++, deck = AppendTo[deck, 10]];
  For[n = 9, n <= 52, n++, deck = AppendTo[deck, 0]];
  count = 0; (* initialize num of successes to 0 *)
  For[n  = 1, n <= numdo, n++,
   {
    hand = RandomSample[deck, 5]; (* 5 card hand *)
    numkings = Sum[hand[[k]], {k, 1, 5}];
    (* want full house of Qs and Ks *)
    (* sum is either 23 or 32! *)
    If[numkings == 32 || numkings == 23, count = count + 1];
    }]; (* end of n loop *)
  Print["Theory predicts prob full house (Qs and Ks) is ",
   100.0 Binomial[2, 1] Binomial[4, 3] Binomial[4, 2]/
    Binomial[52, 5], "%."];
  Print["Observed probability is ", 100.0 count/numdo, "%."];
  ];
```

Dealing ten million hands (we dealt ten times as many as before because the probability is so low we need more data to be confident of the true value) we found:

```
Theory predicts prob full house (Qs and Ks) is 0.00184689%.
Observed probability is 0.00168%.
```

We end with the more general problem of counting a full house, where now we don't care how we form it. It was a lot easier when we specified which two suits, but fortunately we can tweak the above idea. What we'll do is represent the thirteen numbers by $1, 10, 100, \ldots, 10^{12}$, forming a deck by taking four of each. We then create a hand of five and sum the values. We have a full house if and only if all the digits of the sum are 0, 2, and 3 (and 2 and 3 occur). We can easily check if 2 and 3 occur by using the IntegerDigits command, which gives us a list of the decimal digits of our sum, and then using the MemberQ function twice (with an or statement) to make sure 2 and 3 occur.

```
fullhouse[numdo_] := Module[{},
    count = 0; (* this is our count variable, records successes *)
    deck = {}; (* initializing deck; cards are 1, 10, 100, 1000, ... *)
    For[n = 1, n <= 13, n++,
     For[j = 1, j <= 4, j++,
      deck = AppendTo[deck, 10^(n - 1)]
      ]]; (* end of for loops, deck created *)
    For[n = 1, n <= numdo, n++,
     {
      hand = RandomSample[deck, 5]; (* randomly chooses 5 cards *)
      value = Sum[hand[[k]], {k, 1, 5}]; (*sums the value of cards *)
      (* next lines is a nice trick. if our hand is 10002011 this means *)
      (* we have one number twice, the 1000; and have one 1, one 10, *)
      (* and one 10000000. It's to check! the only way to have *)
      (* a full house is to have one 2 and one 3 among digits,
      or the product of non-zero digits is a 6. However, it is easier
      to do MemberQ to check on 2, 3; we show both so you can see each*)
      valuelist = IntegerDigits[value];
      If[MemberQ[valuelist, 2] == True &&
        MemberQ[valuelist, 3] == True, count = count + 1];

      (* this is how to do as a product of digits *)
      (* ----------------- *)(*
      trimlist = {};
      For[j = 1, j <= Length[valuelist], j++,
      If[valuelist[[j]] > 1, trimlist = AppendTo[trimlist,
      valuelist[[j]]]
      ] ];
      If[Product[trimlist[[j]],{j,1,Length[trimlist]}] \[Equal] 6,
      count = count+1];
      --------------- *)

      }]; (* end of n loop *)
    Print["Prob of a full house: ",
     100.0 Binomial[13, 2] Binomial[2, 1] Binomial[4,
       3] Binomial[4, 2] / Binomial[52, 5], "%."];
    Print["Observed prob: ", 100.0 count/numdo, "%."];
    ];
```

Dealing two million hands we found:

```
Prob of a full house: 0.144058%.
Observed prob: 0.14615%.
```

3.6 Summary

While we've covered three of the most important card games (as well as the ever popular tic-tac-toe), we've only scratched the surface of a few great questions. That said, you're armed with great methods that can handle a lot of these problems, but not always easily or cleanly. We've seen how quickly the number of cases can grow, and how a simple problem can have us sitting in fear of either double counting something, or forgetting to include a possibility. While we should be able to solve a lot of these problems by sticking it out and hacking away, these brute force approaches aren't pretty. We need a better attack.

One of my favorite songs (especially the Muppet version) is Kenny Rogers's "The Gambler". One of his couplets is: "Every gambler knows that the secret to survivin', Is knowin' what to throw away and knowin" what to keep. This is great advice for probability! You need to learn how to count. In our discussions we were constantly throwing away terms and adding in terms. It works, but it's a mess. In Chapter 4 we'll learn the secrets to surviving these calculations.

3.7 Exercises

Exercise 3.7.1 *Many functions and sequences are defined recursively. In the Fibonacci sequence $\{F_n\}$ subsequent terms are defined as the sum of the previous two terms; the first two terms are usually taken to be either 0 and 1 or 1 and 2 (depending on the problem, each definition has advantages and disadvantages). Write down a recursive formula for the Fibonacci sequence. Use that formula to show that $\sqrt{2}^n \leq F_n \leq 2^n$ for all $n \geq 6$.*

Exercise 3.7.2 *Find the probability of exactly ten heads in 60 tosses of a fair die.*

Exercise 3.7.3 *What is the probability that two cards from a well-shuffled deck sum to 21? Assume a 10, jack, queen, and king are worth 10 and an ace is worth 11.*

Exercise 3.7.4 *Suppose we have a 40 card deck (a normal deck of 52 with the 7, 8, and 9 cards removed). What is the probability of drawing a straight in a five card hand? A flush? A royal flush (a straight starting with 10, where each card is of the same suit)? Are these probabilities different from a 52 card hand?*

Exercise 3.7.5 *How many ways are there to group 6 people into groups of 3 with order not mattering?*

Exercise 3.7.6 *How many ways can we group 6 people into pairs with order not mattering?*

Exercise 3.7.7 *The **double factorial** of n, denoted n!!, is the product of every other number down to 2 (if we start with an even number) or 1 (if we start with an odd number); thus $7!! = 7 \cdot 5 \cdot 3 \cdot 1 = 105$, and not $(7!)! = 5040!$ (note that 5040! is about 10^{16473}). The reason this is the definition of the double factorial is that this expression occurs far more frequently than the factorial of a factorial. Show that the number of ways to group 2n objects into n pairs of 2 is $(2n - 1)!!$; here the pairs are unordered and unlabeled; thus a pairing $\{\{2, 4\}, \{1, 3\}\}$ is the same as $\{\{2, 4\}, \{3, 1\}\}$ which is the same as $\{\{1, 3\}, \{2, 4\}\}$.*

Exercise 3.7.8 *More generally, in how many ways can kn people be grouped into n groups of k without order mattering?*

Exercise 3.7.9 *For each $m \in \{100, 1000, 10000\}$, determine how many zeros there are at the end of $m!$. Is there a nice formula for the number of concluding zeros for $10^k!$?*

Exercise 3.7.10 *Imagine we have a deck with s suits and N cards in each suit. We play the game **Aces Up**, except now we put down s cards on each turn. What is the probability that the final s cards are all in different suits?*

Exercise 3.7.11 *Consider all generalized games of **Aces Up** with C cards in s suits with N cards in a suit; thus $C = sN$. What values of s and N give us the greatest chance of all the cards being in different suits? Of being in the same suit?*

Exercise 3.7.12 *Consider an $n \times n$ tic-tac-toe board, with n being an even integer. Accounting for all symmetries, how many distinct opening moves are there? What if n is odd?*

Exercise 3.7.13 *Find and explain where double counting has occurred in the following enumeration of the number of arrangements of FreeCell hands using column and pair swaps. There are 52 cards. Each card can be in either of 2 positions. This gives 2^{52} arrangements. We then have two sets of 4 columns. Each set can be ordered in 4! ways. This gives a total of $2^5 \cdot 2 \cdot 4! \cdot 4! \approx 2.6 \times 10^{18}$ arrangements.*

Exercise 3.7.14 *We can also create unwinnable games in FreeCell by reordering the numbers in each pair of columns, so long as none of the 2's, 8's, or aces are shown. How many new orderings does this give? (If you know the rules of FreeCell, you can see how far it is possible to push this argument. For example, you could have aces showing as long as there are no 2s and 8s and as soon as you pulled the aces off, the game would end.)*

Exercise 3.7.15 *Suppose you and your partner have $13 - k$ trump cards between the two of you in bridge, so that there are k trump cards split between your two opponents. What is the probability of a $(k - n) - n$ split? This means one opponent has $k - n$ and the other has n.*

Exercise 3.7.16 *A regular straight is five cards (not necessarily in the same suit) of five consecutive numbers; aces may be high or low, but we are not allowed to wrap around. A kangaroo straight differs in that the cards now differ by 2 (for example, 4 6 8 10 Q). What is the probability someone is dealt a kangaroo straight in a hand of five cards?*

Exercise 3.7.17 *For a fixed n, for what value of k does the binomial coefficient $\binom{n}{k}$ achieve its maximum value?*

Exercise 3.7.18 *Suppose we start at the point 0. We move one right with probability p and one left with probability q. After n steps, what's the probability we are at location k, where k is how far we are to the right of 0? (Hint: Consider even and odd n separately.)*

Exercise 3.7.19 *The double factorial is defined as the product of every other integer down to 1 or 2; thus $6!! = 6 \cdot 4 \cdot 2$ while $7!! = 7 \cdot 5 \cdot 3 \cdot 1$. One can write $(2n - 1)!!$ as $a!/(b^c d!)$ where a, b, c, and d depend on n; find this elegant formula! (Hint: b turns out to be a constant, taking the same value for all n.)*

Exercise 3.7.20   *In horse racing bets are placed in terms of odds, so that if you bet on a horse that has odds of 5:1 it pays out five dollars for every day you bet. What must the probability be that this horse wins for the bet to be fair?*

Exercise 3.7.21   *In the Kentucky Derby there are 20 horses. Imagine that regardless of which horse you choose each bet is equally fair (or unfair!). One horse is given odds at 2:1, one horse is given odds at 5:1, three are at 7:1, six are at 20:1, and the other nine are at 50:1. How much does the average bet of 1 dollar return?*

Exercise 3.7.22   *One of the options in betting is to bet on a trifecta, in which you pick the first three horses in order. How many different trifectas are possible in a race with 20 horses?*

Exercise 3.7.23   *A common betting strategy is to box the bet, which means you bet all possible trifectas over the horses you chose. Find the number of bets this requires.*

Exercise 3.7.24   *There are twenty students in a probability class with ten boys and ten girls. There is one row of twenty seats. What is the probability (assuming randomly assigned seating) that girls and boys alternate?*

Exercise 3.7.25   *There are twenty students in a probability class with ten boys and ten girls. There is one row of twenty seats. What is the probability (assuming randomly assigned seating) that we never have three consecutive boys or three consecutive girls? What if there were 2n students?*

Exercise 3.7.26   *Imagine you are interviewing for jobs with two firms. At Firm A, there are three interviewers. Two interviewers vote in favor of you getting the job with probability p. Their votes are independent. The third just flips a fair coin to decide which way he votes. The next day you interview with Firm B. There is only one interviewer, who selects you for the job with probability p, independently of the votes of any of yesterday's interviewers. Find which firm is more likely to select you for the job. (This may depend on p.)*

Exercise 3.7.27   *Redo the previous exercise, but assume all three interviewers with Firm A vote for you independently with probability p (i.e., there is no coin flip).*

Exercise 3.7.28   *This is a famous problem: you are on a game show and are to select one of three doors. Behind one door is the prize of your dreams. Behind the other two lay magnificent piles of dung. You pick a door, and the game show host purposely opens one of the doors you did not choose such that it always reveals dung. Do you switch your choice?*

Exercise 3.7.29   *Craps is a game with two dice. If in the first turn the two die sum to 7 or 11, you win. If they sum to 2, 3, or 12, you lose. If they sum to anything else, you get what is called a "point." If this occurs, you continue rolling until you either get the same "point" as you got on the first roll or a 7; if you roll a 7 before rolling your "point," you lose, and you win if you roll the "point" before a 7. What is your probability of winning?*

Exercise 3.7.30   *Cameron and Kayla are running against each other for president of College Council. There are N voters (N is an arbitrary, fixed positive integer) and a ballot reader tallies votes one by one. What is the probability that, during the tallying process, there is no point in time where Cameron and Kayla have an equal number of votes?*

Exercise 3.7.31   *Consider two Lego blocks each with six bumps and six places the bumps can fit. If the two blocks are different colors, find how many unique ways there are to put them together. (Keep in mind the various symmetries!) Assume blocks must connect at right angles to each other.*

Exercise 3.7.32   *Repeat the previous exercise but with indistinguishable blocks.*

Exercise 3.7.33   *Write a program that can be used to experimentally verify the probabilities of different poker hands calculated in this section via simulation.*

Exercise 3.7.34   *In computing the probability of a one-suited hand, the challenge was efficiently checking whether or not one of the four players has all their cards in the same suit. While we've been denoting the suits by 1, 2, 3, and 4 there is a better way: let's make the suits 1, 100, 10000, and 1000000. Show that someone is one-suited if and only if the sum of the cards in their hand is 13, 1300, 130000, or 13000000. Would it have also worked using suits of 1, 10, 100, and 1000? Could we use 0 as a suit?*

CHAPTER 4 _____

Conditional Probability, Independence, and Bayes' Theorem

C-3P0: Sir, the possibility of successfully navigating an asteroid field is
  approximately 3,720 to 1.
Han Solo: Never tell me the odds.
— *The Empire Strikes Back* (1980)

Earlier we talked about how to find probabilities of events. In Chapter 2 we concentrated on the foundations of the subject—the axioms and some of the most important consequences. We then continued in Chapter 3 by applying these results to some standard problems. We chose to take our problems mostly from card games as card games are both fun and have historically played an important role in the development of probability, but we could have drawn examples from many other fields too.

As we studied these problems, we ran into difficulties. My favorite was the trump split from §3.4.3. We gave two different methods, and discussed in great length why the second method failed and promised to return to that. Now it's time to honor that pledge. Why did the second argument fail? It seemed reasonable, it looked like it was based on a sound foundation, namely that each card has an equal chance of being in each hand. What went wrong?

Events don't happen in isolation, and knowing one happens can influence the chance of another. When doing our calculations, we often have additional information, and we want to know what's the probability something happens, *given that something else happens*. This is called conditional probability. There are many good reasons for spending so much time on finding simple formulas for conditional probabilities in terms of more basic results. The first, obviously, is that we need to be able to find these probabilities! The second isn't as immediately clear, but is more important. We're going to discuss at great lengths what a reasonable guess is for the answer and how to test it, and only after that turn to the proof. As you continue your studies, either in classes or in a job, eventually you'll reach the boundaries of knowledge. The next step will be up to *you*! It's to prepare you for that fun (but somewhat frightening) day that we're going to talk and talk and talk about how to view these formulas. The thought process is very important; this is a skill that is practiced too infrequently. Most of the time we're just presented with formulas or equations, as if Moses went back to Mount Sinai and returned with Tablets for Math and the Sciences. Our goal here is to help prepare you for the discovery ahead.

The following problem is a great warm-up for conditional probabilities. Three students enter a room and a white or black hat is placed on each person's head. The color of each hat is determined by a fair coin toss, with the outcome of one coin toss having no effect on the others; thus each person is equally likely to have either hat, independent of the other students. Each person can see the other players' hats but not his own. No communication of any sort is allowed, except for an initial strategy session before the game begins. Once they have had a chance to look at the other hats, the players must simultaneously guess the color of their own hats or pass. The group shares a $3 million prize if at least one player guesses correctly and no players guess incorrectly. If just one person speaks and always says white then they win half the time. Amazingly, it's possible to do better than 50%; this should seem quite surprising, as each person cannot see their hat and thus should only be right half the time! We'll revisit this problem later in §4.2.3, but of course feel free to pause and think about this now.

4.1 Conditional Probabilities

In probability class and in life, we frequently want to know the probability a certain event happens. Suppose, for instance, that you're planning an outdoor wedding in Philadelphia in April; you might want to know the probability that it's going to rain (okay, this example might be more meaningful in a few years; if you want, imagine you want to head there to see the Phillies play baseball). You could take the total number of rainy days in Philadelphia in the previous year and divide by 365, but is that the best we can do? Philadelphia typically gets much more rain in April than it does during the rest of the year, so using the year-round average wouldn't be too helpful. We want to know (or estimate) what's going to happen in April; more specifically, we want to know the probability that it's going to rain *given* that it's April. This type of probability, in which we use information about an event to restrict our focus to a subset of the sample space, is known as a **conditional probability**. We write $\Pr(A|B)$, which reads "the probability of A given B." Put differently, it's the probability of event A occurring conditional on event B having already occurred.

Conditional probabilities show up everywhere. Consider this short list of possible applications.

- In sports: What are the odds that Everton wins a "football" (or, as we say in America, a soccer) game, given that they're down 1-0 in the 75th minute? What's the probability of a football team getting a first down, given that it's fourth and one? What's the likelihood a hockey team will score given that they've pulled their goalie? What are the chances a baseball team with J. D. Drew hits four back-to-back-to-back-to-back home runs? What if they don't have J. D. Drew?*

- In health care: What are the odds this patient suffers a heart attack given that he's sixty-five and doesn't smoke? What's the probability that this drug cures the

---

*As of 2017, this has happened only seven times; twice J. D. Drew has been one of the four batters (interestingly, he was the second home run each time). So, if you want to have a credible threat of back-to-back-to-back-to back home runs....

Figure 4.1. The big wheel from *The Price is Right*. Image is copyright © the Pittsburgh Cultural Trust.

patient, given that the previous two drugs did not? What's the likelihood that I have a rare disease, given that I've just tested positive for it?

- In politics: What are the odds that she'll win the election given that she's polling at 53%? While there are many others, here's a famous one from the 1960s: Given that the Soviet Union is deploying nuclear missiles to Cuba, what's the probability they'll go to war if the U.S. blockades Cuba?

The common thread among these problems is that we use information at hand to move from a general probability question—what are the odds she'll win the election?—to a more specific question—given this polling data, what are the odds she'll win the election? This is the world of conditional probabilities. If at this point you're thinking to yourself, "Don't we almost *always* have some kind of additional data points available?", you're entirely right. In many ways, conditional probability problems are the most common, most applicable forms of probability. Therefore, it's very important that we know how to work with them.

 Let's do one more example. In my childhood I often watched *The Price Is Right* with my Grandma (Great Grammy to my kids). While many of their games heavily involve probability, one part of the show beautifully illustrates conditional probability, the Showcase Showdown. Three people spin the big wheel (see Figure 4.1), which has twenty-one numbers from zero to one dollar in increments of five cents. Each player has up to two spins; if you spin twice your score is the sum of your spins. Whomever is closest to a dollar without going over wins. As we're in a probability class and not jumping up and down with the crowd chanting on stage, we don't need to worry about what happens in a tie. Let's pretend we spin first. What's the distribution of values for our first spin? For our second spin? The probability distributions for the two spins are *not* the same. If we have a low first spin, say a 10, we'll try to spin near the 90 to get a dollar; if we have a high first spin, we might even choose not to spin again! If we imagine now that we're the third person, what's the distribution of our sum? That's clearly going to be conditional on the first two players. If the first two players go over, there's no chance the third person will spin twice, and thus the third person won't go over; however, if the high score of the first two players is a 95, then there's a very high chance the third person will have to spin twice and will go over.

4.1.1 Guessing the Conditional Probability Formula

At this point, two things should be clear. First, conditional probabilities are useful and pop up all over the place, and second, we need a way to find them! Actually, we're a bit greedy and want more than just *a* way to find conditional probabilities; we want a *good* way to find them. What should *good* mean when referring to the conditional probability $\Pr(A|B)$? It should mean a nice simple formula relating it to simpler events.

 Before we give the formula, let's see if we can figure out some of its features. We discussed in the introduction to this chapter why it's a good use of our time to try and predict the shape of the answer. You should always try to sniff out a formula before starting an investigation. If you can get a sense of its features, not only do you often catch mistakes, but you might, if you're really lucky, get a feeling for how the proof will go. A first guess is that $\Pr(A|B)$ should hopefully depend on the two basic events, $\Pr(A)$ and $\Pr(B)$.

Is it possible that there's some nice function F such that

$$\Pr(A|B) = F(\Pr(A), \Pr(B))?$$

No! Why not? *Always look at **extreme cases**.* In one case let's take A to be B, and in the other, let's take A to be B^c (read the complement of B, or the event that B doesn't happen). These are natural cases to consider, as we're looking at the situation when A and B are as close to each other as possible or as far apart. Further, we need choices of A and B where we can easily compute the conditional probabilities, and (as we'll see in a minute) these work beautifully there. In many other problems it's a good idea to try the entire space Ω or the empty set $\emptyset$; unfortunately here those choices won't help us.

We need to assign a value to all the probabilities. The simplest case is when they are all equal: $\Pr(A) = \Pr(B) = \Pr(B^c)$; as $\Pr(B) + \Pr(B^c) = 1$, this means each of these three events has probability $1/2$. We then have

$$1 = \Pr(B|B) = F(1/2, 1/2) = \Pr(B^c|B) = 0.$$

To see this, just read the statements aloud. We read $\Pr(B|B)$ as the probability that B happens, given that B happens; this is just 1. Similarly $\Pr(B^c|B)$ is the probability that B doesn't happen, given that B happens; this is simply 0. If the answer only depended on $\Pr(A)$ and $\Pr(B)$ *and not on the nature of the events A and B*, then $\Pr(B|B)$ would have to equal $\Pr(B^c|B)$, an absurdity. We've just proven that the conditional probability *must* depend on more than just $\Pr(A)$ and $\Pr(B)$. That's a pretty good return for looking at just one extreme case.

Again, this is incredible! We may not have the formula yet, but we know what it isn't. Let's now see what it *is*. We know that the answer has to be more than just a function of $\Pr(A)$ and $\Pr(B)$. What else could play a roll? Let's try reading the expression again: we say the probability of A given B. This means that both A and B should happen, so maybe we should add $\Pr(A \cap B)$ to our bag, and look for a formula like

$$\Pr(A|B) = G(\Pr(A), \Pr(B), \Pr(A \cap B))$$

(remember intersections mean both happen). We also know that

- $\Pr(B|B) = 1$,
- $\Pr(B^c|B) = 0$, and
- $0 \leq \Pr(A|B) \leq 1$.

TABLE 4.1.
These are the possible outcomes for events A and B. If we know that event B has happened, we
need only worry about the events in B's row.

	A	A^c
B	$A \cap B$	$A^c \cap B$
B^c	$A \cap B^c$	$A^c \cap B^c$

There's a simple expression using our three building blocks that has these three properties: $\Pr(A \cap B)/\Pr(B)$ (while we assume $\Pr(B) > 0$, notice that if it were zero then
$\Pr(A \cap B)$ would also be zero and our expression would become the indeterminate 0/0).
This is 1 if $A = B$, 0 if $A = B^c$, and since $A \cap B \subset B$, this ratio is always between 0 and
1. We'll explain more in the exercises how to arrive at this guess (see also §2.7 for more
on sniffing out formulas), but for now let's do an example and see how our guess fares.

4.1.2 Expected Counts Approach

We're going to use a method called the **Expected Counts Approach**. Suppose that
you go out fishing one day, and you have the following set of rules: you stop fishing
once you catch a fish, or after you've been on the water for four hours (whichever
comes first). Let's also imagine that there's a 40% chance that you catch a trout, a
25% chance you catch a bass, and a 35% chance you don't catch anything. Notice
that the percentages sum to 100%, and that you never catch more than one fish in a
day. Now, if we know that you caught a fish one day, what are the odds that fish was a
trout? Suppose that you went fishing 1000 times (we'll see soon why 1000 is a good
number). Then we would expect that 400 times you would catch a trout, 250 times
you would catch a bass, and 350 times you would leave empty-handed. So given that
you caught a fish, what's the probability that it's a trout? We are now **restricting** our
scope to the 650 times you caught a fish. Exactly 400 of these times the caught fish
was a trout. Therefore, the probability of getting a trout *given that a fish was caught*
is 400/650, or approximately 61.5%. To check our formula, if A is the probability
of catching a trout, and B the probability we catch something, then $\Pr(A) = .40$,
$\Pr(B) = .40 + .25 = .65$, $\Pr(A \cap B) = .40$ (if A happens then we've caught a fish,
in which case B trivially happens). These imply $\Pr(A \cap B)/\Pr(B) = .40/.65$, which
could be written as 400/650, exactly matching our earlier answer!

The Expected Counts Approach works because all of our probabilities are simply
being scaled up by a common factor. Instead of writing 400 and 250, we could
have written 1000·Pr(trout) and 1000·Pr(bass). So why bother with this approach?
Mostly, it's because using expected counts is often a more intuitive way to solve
conditional probability problems than using a memorized formula. Instead of dealing
with probability ratios, it's easier to think about the expected number of times each
event happens, and then take the appropriate ratio. In the fish example, the choice of
1000 total fishing excursions was completely arbitrary; we could've picked any number.
We took 1000 as it made all of our expected counts integers. In general, choose a large
number divisible by all the numbers occurring in the problem.

Another approach we could use is to deal with probability ratios directly. Suppose
that we have two events, A and B, and we want to know $\Pr(A|B)$. It can be helpful to
frame a scenario like this in table form, as shown in Table 4.1.

If we restrict our universe to the possible outcomes of A and B, then there are four outcomes we might reach. (Note that A^c and B^c mean "not A" and "not B.") If, however, we know event B has happened, then our universe is further restricted to B's row of Table 4.1, and the only possible outcomes are $A \cap B$ and $A^c \cap B$. The conditional probability $\Pr(A|B)$ is therefore given by

$$\Pr(A|B) \;=\; \frac{\Pr(A \cap B)}{\Pr(A \cap B) + \Pr(A^c \cap B)} \;=\; \frac{\Pr(A \cap B)}{\Pr(B)},$$

where the final step follows from the fact that $(A \cap B) \cup (A^c \cap B) = B$, and the probability of a finite, disjoint union is the sum of the two probabilities (this is one of our axioms of probability).

Let's record what we've found.

Conditional Probability: Let B be an event such that $\Pr(B) > 0$. Then the conditional probability of A given B is

$$\Pr(A|B) \;=\; \Pr(A \cap B)/\Pr(B). \tag{4.1.2}$$

If you were really reading carefully, you might've noticed a new condition snuck into the box above: $\Pr(B) > 0$. If $\Pr(B) = 0$, then B cannot happen. If B cannot happen, it doesn't make sense to talk about the probability A happens given B happens! Fortunately if $\Pr(B) = 0$ then $\Pr(A \cap B)$ is also 0, and we have the indeterminate ratio 0/0, which warns us that we are in dangerous waters.

Even though we've already had a lengthy discussion about why this formula is reasonable, there's one more point worth making. Our formula is actually a generalization of regular probabilities! We can view $\Pr(A)$ as $\Pr(A|\Omega)$, because Ω always happens, and thus if we're told Ω occurs, this knowledge doesn't change our probability calculation. Explicitly, $\Pr(\Omega) = 1$ as Ω is just our outcome space. It's always legal in math to multiply by 1; learning how to do this in a good way is one of the most important skills to master, and one of the hardest. **Multiplying by 1** is a powerful technique, often allowing us to rewrite the algebra in a more manageable, or a more enlightening, way. For conditional probabilities, since $A = A \cap \Omega$, we find

$$\Pr(A) \;=\; \frac{\Pr(A \cap \Omega)}{1} \;=\; \frac{\Pr(A \cap \Omega)}{\Pr(\Omega)};$$

note the final expression is just our formula for conditional probability! The point of all of this is to help see (again!) where our conditional probability formula comes from, and to gain experience with the thought process so that you can find similar formulas in the future.

4.1.3 Venn Diagram Approach

Let's look at some conditional probability examples.

Example: In a given town, the probability that an adult owns a car or a house is 90%. If 70% of the adults in the town own cars and 50% own houses, what's the probability that a homeowning adult also owns a car?

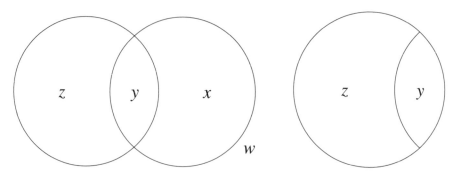

Figure 4.2. (Left) Venn diagram for the homeowner and car problem. (Right) Restricting to just homeowners.

Solution: Let's first think about what probability we're actually trying to find. Put simply, we want the probability that a homeowning adult owns a car. We can reword this as the probability an adult owns a car given that he or she is a homeowner. If we let A be the event of owning a car and B be the event of owning a house, then we want to find $\Pr(A|B)$. From Equation (4.1.2), we know that $\Pr(A|B) = \Pr(A \cap B)/\Pr(B)$. We're given that $\Pr(B) = 0.5$, so all we need is $\Pr(A \cap B)$. Unfortunately, we don't know $\Pr(A \cap B)$, the probability that someone owns a house and a car. Since we don't know this probability, we can't substitute into our formula.

All is not lost, though, as we do know $\Pr(A \cup B) = .90$. Since we know the probability someone owns either a house or a car (or both), we can figure out $\Pr(A \cap B)$ since

$$\Pr(A \cup B) = \Pr(A) + \Pr(B) - \Pr(A \cap B) = .7 + .5 - \Pr(A \cap B).$$

Since $\Pr(A \cup B) = .9$, we see $\Pr(A \cap B) = .3$. *Now* we can use Equation (4.1.2) to find the conditional probability. We get that the probability a homeowner also owns a car is $\Pr(A \cap B)/\Pr(B) = 0.3/0.5 = 0.6$.

Let's draw a **Venn diagram** to help visualize the question (Figure 4.2). In almost any math class, at some point you should hear your professor say, "Draw a picture." This advice might be given in passing, or your instructor may be jumping up and down and throwing chalk to call attention to the message. However it's delivered, listen to this advice; you ignore it at your own peril. Often just drawing a picture is enough to highlight the different relationships. What we need to do is pass from talking about the probabilities of events A and B to all the different parts in the Venn diagram. To do this, we'll introduce new variables (w, x, y, z) for the probabilities of the four regions in Figure 4.2.

Here x denotes the percentage of people who own cars only and z denotes the percentage of people who own houses only, while y denotes the percentage of people who own both houses and cars, and finally w refers to the percentage of people without houses and cars; thus we know

$$w + x + y + z = 1.$$

In terms of the original problem, $\Pr(A) = x + y$, $\Pr(B) = y + z$, and $\Pr(A \cap B) = y$.

While this is a good start, obviously it's just a start as we want to find out what w, x, y, and z are. We'll write down some relations involving these variables using information from the question. For example, we're told 90% of the people own a house or a car, which translates to $x + y + z = .9$. Arguing similarly for the rest gives the following equations:

$$x + y + z = .9 \qquad (4.1)$$

$$x + y = .7 \qquad (4.2)$$

$$y + z = .5. \qquad (4.3)$$

There are lots of ways to solve this system of equations and get the values of x, y, and z. We can subtract (4.2) from (4.1), giving $z = 0.2$. Then subtract (4.3) from (4.1) to find $x = 0.4$. We can now pick off y; it's just .3, and thus the solution is

$$x = .4$$
$$y = .3$$
$$z = .2.$$

Thus the probability of a homeowner also owning a car is just .3. Now, if we want the conditional probability of owning a car given that you own a home, *that* is represented by $y/(y + z)$ (as $\Pr(B) = y + z$). We find $\Pr(A|B) = .3/(.3 + .2) = .6$.

Was the problem easy or hard to follow? While it is standard to label variables in order $(x, y, \dots)$ or sets in order $(A, B, \dots)$, it makes it hard to follow the arguments as we have to keep remembering what each stands for. One solution is to have descriptive names. We could use x_{car}, but this could be confusing: does this represent the number of people who own a car, or just a car? We could do $x_{only\ car}$, but now the labels are getting long! We could choose letters that correspond to the item under discussion; thus instead of A we could use C (for *car*) and instead of B we could use H (for *house*). It's worth spending some time thinking about names for variables instead of just going in order, as **good notation** can make the arguments significantly easier to follow.

4.1.4 The Monty Hall Problem

The following problem is a famous one and has generated a lot of discussion. It's a great example of how probabilities surface in popular culture.

Example—The Monty Hall Problem: Say you are participating in a game show in which you are given the choice between three doors, knowing that you will get to keep what is behind it. Behind one of the doors is a new car, but behind the other two there are goats. After you initially pick your door, let's say door 1, the host of the shows opens door 2, and there is a goat behind it. After revealing the goat, the host allows you to keep your initial guess or switch your guess to door 3. What do you do? (The name comes from the game show *Let's Make A Deal*, where Monty Hall was the original host.)

Solution: Unless you prefer goats to cars, you should switch your guess to the door you did not initially pick. This less than intuitive answer can be better understood through conditional probabilities. It's important to make explicit an assumption lurking in the background: if you didn't choose the car, Monty won't open the door revealing the car (if he did then there would be no more suspense and no more decisions to be made; this would lead to unexciting tv and hence won't happen). Thus *if* the car is still available *then* Monty's choice is force while *if* you have the car *then* he can choose either. Before reading further, think about why this suggests you should switch.

For your initial guess, the probability you pick the door that contains the car is 1/3. If you do not choose to switch doors, this remains the probability you will win the car. Now consider the probability at door 3 contains a car, *given* that the host just revealed at door 2 contains a goat. Using our conditional probability formula in which A is the event that door 3 contains a car and B is the event that the host opens door 2, we get

$$\Pr(A|B) = \frac{\Pr(A \cap B)}{\Pr(B)} = \frac{1/3}{1/2} = \frac{2}{3}.$$

So, if you apply your knowledge of conditional probability and switch doors, your probability of winning a car improves from 1/3 to 2/3!

4.2 The General Multiplication Rule

4.2.1 Statement

Another consequence of Equation (4.1.2) is the General Multiplication Rule.

General Multiplication Rule: We have

$$\Pr(A \cap B) = \Pr(A|B) \cdot \Pr(B).$$

The above rule follows by multiplying the definition of the conditional probability of A given B by the probability of B. While at first it suggests the above formula only holds if the probability of B is non-zero, direct inspection shows it is still true in that case. The General Multiplication Rule is a handy way to find the probability that two events A and B both occur *if* we can easily calculate the conditional probability and the probability of B. (Of course, by symmetry this should also be $\Pr(A \cap B) = \Pr(B|A)\Pr(A)$; more on this later.)

4.2.2 Poker Example

Let's do another example. Watch carefully to see if this is a good way to tackle the problem.

Exactly Three Example: Suppose you're playing poker with a standard deck of cards. What's the probability of getting *exactly* three of a kind if your first three cards are two jacks and a four? (To be clear: if we have three of a kind and a pair, that counts as only having one number occurring three times, but four of a kind doesn't count; we want to have just three of one number: no more, no less.)

Solution: The very first thing we need to do is realize that this *is* a conditional probability problem. The key phrase is "if your first three cards..."; as soon as we see an "if" or a "given that," we should be bracing for the possibility that we're going to need to use conditional probabilities. This is indeed a conditional probability question: what's the probability of getting a certain hand *given* a certain initial setup. We could take the approach we took in the previous example and find Pr(hand and setup) and Pr(setup), and then take the ratio. Unfortunately in this case that approach is a computational nightmare. There is a more efficient way to solve this problem. Let's instead think about how we can possibly get three of a kind. Since there are only two cards left to deal, the only way to get three of a kind is to get exactly one more jack or exactly two more fours (since we only have 5 cards, we can't have 3 jacks and 3 fours). Let's examine each possibility. Since there are 49 cards left and three of these are fours, the odds of getting a four on the first draw is 3/49, and then the odds of getting another four on the second draw (given that we got a four on the first draw) is 2/48. Therefore, the odds of getting a hand with exactly three fours is $\frac{3}{49} \cdot \frac{2}{48} = \frac{1}{392}$, or approximately 0.26 percent.

As a brief aside, notice that the odds of drawing two fours is also a conditional probability. The General Multiplication Rule tells us that $\Pr(A \cap B) = \Pr(A \mid B) \cdot \Pr(B)$. So, Pr(four on second draw AND four on first draw) = Pr(four on first draw)·Pr(four on second draw | four on first draw) = $\frac{3}{49} \cdot \frac{2}{48}$, just as we just discovered above. We're fortunate here—it's very easy to calculate the conditional probability of getting a four on the second draw given a four on the first draw, and in fact this is why conditional probabilities can be so powerfully applied and used.

Now we turn to the other case, getting exactly one jack. There are 49 cards remaining: 2 are jacks and 47 are non-jacks. We can either get a jack followed by a non-jack, which happens with probability $\frac{2}{49} \cdot \frac{47}{48}$, or we can get a non-jack followed by a jack, which happens with probability $\frac{47}{49} \cdot \frac{2}{48}$. These two probabilities are the same, and their sum is $2 \cdot \frac{2 \cdot 47}{48 \cdot 49} = \frac{47}{588} \approx 0.08$, or about 8 percent. Also observe that we are interpreting the problem as it is okay to end with three jacks and two fours; if you want to disallow this possibility that's fine, and the analysis would change slightly here as now there would only be 46 cards available (not 49).

Now, given the probabilities of getting three fours and three jacks, what's the probability that we get three of a kind? Since it's impossible to have three jacks AND three fours as we only get five cards, we see that the probability of three of a kind is just the probability of three jacks plus the probability of three fours, or

$$\frac{1}{392} + \frac{47}{588} = \frac{97}{1176} \approx 8.25\%.$$

 Wait a minute! Something seems strange about our solution to the last problem. We realized it was a conditional probability problem, but we didn't use $\Pr(A|B) = \Pr(A \cap B)/\Pr(B)$! What gives? Don't worry—you were right to see the problem as a conditional probability one. We did use conditional probabilities, just in a different form. They surfaced when we talked about the probability of the second card taking certain values *given* the values of the first card. While we could have solved it with the traditional formula, it turns out to be easy to attack it another way. What we just did was to use the given information to directly find the probability of the event A given B; let's now do the problem the more "traditional" way.

 The first thing we need to do is figure out what A and B are. Event B is the first three cards in a five card hand are two jacks and a four, while event A is we have exactly three jacks or exactly three fours in our five cards. Let's find $\Pr(B)$ first, as that's easier. It doesn't matter that we draw five cards; we can have anything for the last two. What matters is that the first three cards contain two jacks and a four. A nice way to find this

TABLE 4.2.
An analysis of the eight case.

Outcome	1 says	2 says	3 says	1 is	2 is	3 is	Outcome
WWW	B	B	B	×	×	×	Lose
WWB			B			✓	Win
WBW		B			✓		Win
BWW	B			✓			Win
WBB	W			✓			Win
BWB		W			✓		Win
BBW			W			✓	Win
BBB	W	W	W	×	×	×	Lose

is to see it equals $\binom{4}{2}\binom{4}{1}/\binom{52}{3} = 6/5525$. Why? We have to choose two of the four jacks and we have to choose one of the fours, and then divide by the number of ways to choose 3 cards from 52. This is a little faster than looking at the three different ways to have two jacks and a four: JJ4, J4J, 4JJ.

We're left with the hard event. How do we find the probability of $A \cap B$? This means we have either 3 jacks or 3 fours *and* the first two cards have two jacks and a four. Well, one way to proceed is to choose the first three cards and then look at the remaining two cards. Notice, however, that as soon as we start to do this we're attacking the problem the way we did before!

The lesson to be learned here is that sometimes it's easier to use the conditional probability formula, while other times it's easier to incorporate the given information and compute directly.

4.2.3 Hat Problem and Error Correcting Codes

We return to the riddle from earlier in the chapter, where three people have three hats placed on them. The idea is that each person speaks *only if* (do you hear the conditional probability) they see two hats of the same color, in which case they say the opposite color. Why does this work? There are 8 possible hat assignments: WWW, WWB, WBW, BWW, WBB, BWB, BBW, BBB. Notice that two of these have all hats of the same color. In this case everyone speaks and says the opposite color. Everyone is wrong. There are six cases where two hats are one color and the third is the other. In these cases, only one person sees two hats of the same color. She speaks and is correct, and the other two are silent. Thus we win in 6 of the 8 cases, or 75% of the time! Notice that each person is still wrong half the time they speak; what we are able to do is align the wrong answers (so when we are wrong, we are *really* wrong!); we illustrate this Table 4.2.

This is not just a make-work example of conditional probabilities; it plays a key role in the creation of error correcting codes (methods of transmitting information so that the receiver cannot only detect an error was made, but can also fix it without needing additional information!). For more information, see [CM] (or go online and look up the **Hamming (3,1) code**).

4.2.4 Advanced Remark: Definition of Conditional Probability

We end with an advanced remark; if you're not interested you can safely skip this section as it's independent of the rest of the book. Why is the conditional probability simply "defined" as $\Pr(A|B) = \Pr(A \cap B)/\Pr(B)$? Can we derive it from the Axioms of Probability we laid out earlier? Almost; we need to assume just a bit more, which we'll now do. Suppose we're working in a probability space (Ω, Σ, P), and an event B occurs that changes our probability distribution to P_B. What's the probability of an event A happening now? We need to make two assumptions: first, since B has happened, we want $P_B(A) = 0$ for all $A \subset B^c$; and second, we don't want the relative magnitudes of our probabilities to change. That is, for A, $C \subset \Omega$ not disjoint from B, we want

$$\frac{P_B(A)}{P_B(C)} = \frac{P(A \cap B)}{P(C \cap B)}.$$

Notice that this gives $P_B(A) = P(A|B) = \alpha P(A \cap B)$ for some constant α. To see this, we do our standard trick of making good choices for our sets. The two most likely candidates are to either take $C = B$ or to take $A = B$. As we want A to be a general set, let's try the first. If now $C = B$ then $P_B(C) = 1$ (as $C = B$ and the probability B happens, given B happens, is 1!) and $P(C \cap B) = P(B)$ (since $C \cap B$ is now just B). We find $P_B(A)$ equals $P(A \cap B)/P(B)$. Thus $\alpha = P(B)$.

 Let's do an example. Let A_s be the event that the sum of the rolls of two independent, fair die is s, and let B be the event that the second roll is a three. Let's look at $P_B(A_s)$, the probability the sum of the two rolls is an s given the second roll is a 3. Since the first roll is in $\{1, 2, \ldots, 6\}$, the sum must be between 4 and 9. Thus $P_B(A_s) = 0$ if $s \le 3$ or $s \ge 10$, so we concentrate on $4 \le s \le 9$.

First, note $P(B) = 1/6$, as one-sixth of the time we roll a three. What is $P(A_s \cap B)$ for $4 \le s \le 9$? For each such s there is one and only one roll of the first die such that it sums to s when added to three. Therefore, of the 36 pairs of die rolls, each such A_s consists of just one pair; for example, A_4 is the pair of rolls $(1, 3)$, A_5 is $(2, 3)$, and so on. Thus $P(A_s \cap B) = 1/36$ for these s. Putting the pieces together, our formulas above tell us the conditional probability $P_B(A_s) = P(A_s|B)$ is $P(A \cap B)/P(B) = (1/36)/(1/6) = 1/6$ for $4 \le s \le 9$, and 0 otherwise.

As a quick check, this result is very reasonable. Given that B happens, we cannot have a sum less than 4 or greater than 9, and the remaining six candidates should all be equally likely, which is exactly what we found. Note that the conditional probability of A_s given B is very different than the probability of A_s, as $P(A_2) = P(A_{12}) = 1/36$, $P(A_3) = P(A_{11}) = 2/36, \ldots, P(A_7) = 1/6$.

4.3 Independence

One of the most important concepts in probability is that of **independence**. Informally, we say that two events are independent if they don't affect one another. This means that knowledge of one event happening gives no information about the chance of the other happening. We've already given this cursory definition in the previous pages. For example, if you're flipping a coin, the odds that it lands heads shouldn't be affected by whether it landed heads or tails on the previous flip. It follows then, that unless something weird is going on, we can think of successive flips of a coin as independent events. While independence may seem like a rather intuitive concept, it can be tricky at times. So, we need a clear, unambiguous definition; thanks to the language of conditional probability, we have one. Note the definition below is marked as "tentative." There's a reason for that, as it subtly assumes an additional fact that is not needed. See if you can figure out what was accidentally assumed before reading on.

Tentative definition of independence: Consider the events A and B. We tentatively say that A and B are **independent** if

$$\Pr(A|B) \;=\; \Pr(A).$$

That is, A and B are **independent** if the occurrence of event B doesn't change the probability that event A also occurs.

If we think about this definition for a moment, we realize that it gets at the heart of independence. If $\Pr(A|B) = \Pr(A)$, then learning that event B has happened gives us no additional information about event A. In our coin-flipping example, knowing that the coin landed heads on the last flip tells us nothing about the probability that it will land heads on the next flip.

As you might imagine, it's *wonderful* to work with independent events. In §4.2 we developed the General Multiplication Rule:

$$\Pr(A \cap B) \;=\; \Pr(A|B) \cdot \Pr(B).$$

Suppose, however, that A and B are independent events. This implies $\Pr(A|B) = \Pr(A)$, and thus

$$\Pr(A \cap B) \;=\; \Pr(A) \cdot \Pr(B). \tag{4.4}$$

This is an incredibly nice result, since we can now find the probability of two events occurring without having to deal with conditional probability calculations. We'll see many more instances of the wonders of independence when we talk about random variables in later chapters.

Why did we call the above a "tentative" definition of independence? Remember that when dealing with the conditional probability of A given B that we require the probability of B to be non-zero. Thus we cannot talk about independence if B never happens. *Further, there should be a symmetry!* If A and B are independent, then B and A are independent and our definition should reflect that.

We see that our original definition is lacking. Fortunately we've already seen the solution in (4.4). Notice the symmetry there, as well as the fact that this equation makes sense if any of the probabilities are zero. Thus, the correct definition of independence of two events is

---

Independence (two events): Two events A and B are independent if

$$\Pr(A \cap B) \;=\; \Pr(A) \cdot \Pr(B).$$

What about independence of three or more events? Our definition of independence of two events can be extended to describe three or more events.

Independence (three events): Events A, B, and C are **mutually independent** if

1. $\Pr(A \cap B \cap C) = \Pr(A) \cdot \Pr(B) \cdot \Pr(C)$, and

2. *any* two of the three events are independent.

Note the second condition means that $\Pr(A \cap B) = \Pr(A) \cdot \Pr(B)$ and similarly for $\Pr(A \cap C)$ and $\Pr(B \cap C)$. It seems reasonable that if we have a set of n events which are independent, then any subset of them should be independent; that's the meat of condition (2) above. That said, there should also be some statement relating all the events; that's condition (1). More generally, events $A_1, \ldots, A_n$ are called independent if similar formulas hold for *any* combination of these events.

Independence (general case): Events $A_1, \ldots, A_n$ are **mutually independent** if

1. $\Pr(A_1 \cap \cdots \cap A_n) = \Pr(A_1) \cdots \Pr(A_n)$, and

2. *any* non-empty subset of $\{A_1, \ldots, A_n\}$ are mutually independent.

A big caveat for independence of three or more events is that any combination of two of those events may be independent of each other, but three or more might be dependent. For example, roll a die twice. Let

- A denote the event that the first time the die shows an even number,
- B the second time the die shows an even number, and
- C the sum of the first two numbers is even.

We can see that

$$\Pr(A \cap B) \ = \ \Pr(A) \cdot \Pr(B)$$

$$\Pr(A \cap C) \ = \ \Pr(A) \cdot \Pr(C)$$

$$\Pr(B \cap C) \ = \ \Pr(B) \cdot \Pr(C).$$

However, in this case

$$\Pr(A \cap B \cap C) \ \neq \ \Pr(A) \cdot \Pr(B) \cdot \Pr(C),$$

as $\Pr(A \cap B \cap C)$ is the probability of getting an even number the first time and an even number again the second time (if the first two rolls are even, then the sum *must* be even—this is what causes the problem). We thus have $\Pr(A \cap B \cap C) = \frac{1}{4}$, but if the three events were independent then, according to the formula,

$$\Pr(A \cap B \cap C) \ = \ \Pr(A) \cdot \Pr(B) \cdot \Pr(C) \ = \ \frac{1}{2} \cdot \frac{1}{2} \cdot \frac{1}{2} \ = \ \frac{1}{8}.$$

What's going on here? The problem is that while any two of the three events are independent, if we know A and B happen then C *must* happen; the clue that the three events are dependent is the presence of the word "must." If something "must" happen, we have information!

For fun, try to come up with a set of events A, B, C, D such that any three are independent but all four are not! Additionally, find events $A, B,$ and C such that $\Pr(A \cap B \cap C) = \Pr(A)\Pr(B)\Pr(C)$ but the events are not independent.

Note for students with statistics experience: The notion of independence is highly important in statistics. For students with some prior experience in the subject, this section likely called the idea of correlation to mind. For those without a background in statistics, two random variables X and Y are said to be correlated if they have a linear relationship. The correlation between two variables is stated as a correlation coefficient between -1 to 1; X and Y are said to be uncorrelated if they have a correlation coefficient of 0. The notions of independence and being uncorrelated are similar, but it is important to note that they are *not* interchangeable. The independence of two random variables implies they are uncorrelated, but two variables being uncorrelated does not imply that they are independent. Only one direction of implication exists because independence requires the two variables to be fully unrelated while two variables need only have no linear relationship to be uncorrelated. Variables that depend quadratically on one another could be uncorrelated but would not be independent.

4.4 Bayes' Theorem

At this point, it seems appropriate to circle back, just in case you've been wondering about a certain feature of the General Multiplication Rule. This rule tells us that

$$\Pr(A \cap B) = \Pr(A|B) \cdot \Pr(B).$$

Let's flip this around! Since $B \cap A = A \cap B$, don't we also have

$$\Pr(A \cap B) = \Pr(B|A) \cdot \Pr(A)?$$

Can we do this? Indeed we can! Remember $A \cap B$ and $B \cap A$ are the same events; they both represent A and B both happening. We must therefore have $\Pr(A \cap B) = \Pr(B \cap A)$. This implies the following result, known as Bayes' Theorem.

Bayes' Theorem: The General Multiplication Rule implies

$$\Pr(B|A) \cdot \Pr(A) = \Pr(A|B) \cdot \Pr(B)$$

for events A and B. Therefore, so long as $\Pr(B) \neq 0$, we have

$$\Pr(A|B) = \Pr(B|A) \cdot \frac{\Pr(A)}{\Pr(B)}.$$

The key idea above is **commutativity**: $A \cap B = B \cap A$. We'll see consequences of commutativity again and again; a great one involves convolutions and moment generating functions of sums of independent random variables (see the proof of commutativity of convolution in §19.5).

TABLE 4.3.
This is the breakdown of expected dog owners and flu cases in our population. Since 8 of the 15 sick people are dog owners, the probability that the person is a dog owner given that he or she is sick is simply 8/15 = 0.53.

	Sick	Not Sick	Total
Dog owner	8	32	40
Non dog owner	7	53	60
Total	15	85	100

In Section 4.6, we'll explore a more general version of Bayes' Theorem. While it's an extremely useful result, many students have trouble using it at first. I'd say it's at the top of the list of sources of greatest confusion in an introductory probability class (also high on that list are the discrete combinatorics problems from Chapter 3—so take heart, as we're hitting the worst material first!).

To help, we'll solve many of these problems using the expected counts approach too. Bayes' Theorem is used, misused, and debated in many settings. One of my favorite places to go to hear discussions on Bayes' Theorem is a courtroom. That's right, a courtroom! If you think about this for a moment, it's not so unreasonable. The prosecution is trying to show, *given all this information*, that there's a very high probability (their lingo for this high probability is "beyond a reasonable doubt") that the defendant is the guilty party; meanwhile, the defense is trying to argue the opposite. It all comes down to calculating probabilities given information. So, as hard as a probability class or assignment may seem at the time, and as worried as you are that your life depends on a good grade, take comfort in the fact that, for you, it doesn't! Here's a nice story about using Bayes' rule in court, taken from http://www.guardian.co.uk/law/2011/oct/02/formula-justice-bayes-theorem-miscarriage. Sadly, a judge ruled Bayes' rule can no longer be used. Of the many posts, my favorite was from pseudosp1n (posted 3 October 2011 12:03AM): "Aren't you supposed to be tried by a jury of your peers? If I'm ever on trial and the jury is incapable of understanding Bayes' theorem then I wouldn't consider them remotely my peers and certainly not fit to pass judgement upon me."

We'll do one of the standard Bayes' problems now. It's boring, and it probably doesn't seem that important. Fair enough, these are valid criticisms, but as problems like this are so common in courses, it's worth doing an example. As a reward, afterwards we'll do an outstanding problem with enormous applications!

Example: Suppose 40% of all adults own dogs and that 20% of all dog-owning adults get the flu every year. If 15% of the entire adult population gets the flu every year, what's the probability that someone owns a dog given that they got the flu?

Solution: We'll call D the event of owning a dog and F the event of getting the flu; *while we could denote these events respectively by A and B, this notation is far more suggestive.* We want to find the probability that someone owns a dog given that they got the flu—or mathematically, $\Pr(D|F)$. We're told that the probability of someone owning a dog is 0.4, $\Pr(D) = 0.4$, and we're told that the probability of getting the flu is 0.15, $\Pr(F) = 0.15$. Finally, we're given that 20% of all dog-owning adults get the flu every year, so $\Pr(F|D) = 0.2$. Therefore, Bayes' Theorem tells us

$$\Pr(D|F) = \Pr(F|D) \cdot \frac{\Pr(D)}{\Pr(F)} = 0.2 \cdot \frac{0.4}{0.15} \approx 0.53.$$

If the Bayes' Theorem approach doesn't help you understand that the probability is 0.53, you could think of this using expected counts, as in Table 4.3. Imagine that there are 100 people in our population. (We choose 100 people because it allows us to get integer numbers of adults when we solve the questions. You can choose any number you want.) We would expect 40 of them to own dogs, and 8 of those 40 people to get the flu. Since 15% of the population gets sick every year, we would expect 15 people to get sick. Therefore, of the 15 sick people, 8 of them were dog owners. Thus, the probability of an adult being a dog owner, given that he or she got the flu, is $8/15 \approx 0.53$. The expected counts approach is a friendly way to do these problems without explicitly using Bayes' formula.

We now move on to the promised, important problem. In our study we'll need to introduce the notion of a partition. We'll cover it briefly here, and go over it in detail in §4.5.

Example: Nationwide, tuberculosis (TB) affects about 1 in every 15,000 people. Suppose that there's a TB scare in your town, and for simplicity assume that the rate of incidence of TB in your town is the same as the national average. Just to be safe, you go to the doctor to get tested for the disease. The doctor tells you that the test has a 1% false positive rate—which is to say that for every 100 healthy people, one will test positive. The doctor also reveals that the test has a 0.1% false negative rate—similarly, for every 1000 sick people, only one will test negative. Suppose that you test positive. What's the probability that you have TB?

Solution: Before reading through this solution, take a guess about the probability. Many people guess the probability is high, frequently somewhere in the 90+ percent range. After all, the test came back positive! In reality, however, it turns out to be pretty small. This is a great problem because it demonstrates how non-intuitive probabilistic reasoning can be.

Once again, we're dealing with conditional probabilities. Now, we've been given Pr(positive|healthy) and Pr(negative|sick), and we want to find Pr(sick|positive). Notice that we are again using good, descriptive labels for our events; I cannot encourage you enough to do this. These problems are complicated enough as is; take the time and choose good notation so you can easily glance down and see what everything represents.

Bayes' Theorem tell us that

$$\Pr(\text{sick}|\text{positive}) = \frac{\Pr(\text{sick})}{\Pr(\text{positive})} \cdot \Pr(\text{positive}|\text{sick}).$$

If we're given that the probability of being sick is 1 in 15,000 and that the probability of testing positive given that you are sick is 0.999, then what's the probability of getting a positive result? It's very important to realize that we aren't directly given this information; we need to figure out Pr(positive) from the given information.

We can do this by using a **partition**. We'll discuss these in greater detail in §4.5; we just need the case of a partition into two possibilities (for us, it'll be "sick" and "not sick").

A partition splits an event into disjoint events. Let A and B be events, then

$$A = (A \cap B) \cup (A \cap B^c)$$

writes A as a union of two disjoint events. By our axioms of probability,

$$\Pr(A) = \Pr(A \cap B) + \Pr(A \cap B^c).$$

We can simplify further. Our conditional probability results tell us

$$\Pr(A \cap B) = \Pr(A|B)\Pr(B) \quad \text{and} \quad \Pr(A \cap B^c) = \Pr(A|B^c)\Pr(B^c).$$

This is why we call these partitions; we've broken the probability that A happens into two terms, one involving B occurring and one involving B^c occurring. Since it's impossible for both B and B^c to happen, we've partitioned the probability into two disjoint cases.

We use the partition "sick" and "not sick." So B is the event sick, B^c the event healthy (i.e., "not sick"), and A is the event of testing positive. We find

$$\Pr(\text{positive}) = \Pr(\text{positive}|\text{sick})\Pr(\text{sick}) + \Pr(\text{positive}|\text{healthy})\Pr(\text{healthy})$$

$$= 0.999 \cdot \frac{1}{15000} + 0.01 \cdot \frac{14999}{15000} \approx 0.01.$$

This gives

$$\Pr(\text{sick}|\text{positive}) = \frac{1/15000}{0.01} \cdot 0.999 \approx 0.0066.$$

Were you expecting the probability to be that low?

Let's attempt our tried-and-true expected counts approach here. Suppose that we had 15 million people. (Again, 15 million is just a convenient number for us to obtain integer number of people; it's coming from the 15000 we saw in the problem, with several padding zeros to hopefully ensure all derived quantities are integral.) Then we would expect 1000 of them to be infected (from 15 million times the infection rate of 1/15000) and the rest to not be infected. Of those infected, we would expect 999 of them to test positive, and of the 14,999,000 people not infected, we would expect 149,990 of them to test positive. So there are $149{,}990 + 999 = 150{,}989$ positive tests in this population and only 999 of them belong to sick people. Therefore, the odds of being sick given that you tested positive are $999/(149{,}990 + 999) \approx 0.0066$. Intuitively, the reason that the result is so low is that the initial prevalence of the disease is so small.

The expected counts approach can be seen graphically in the tree below.

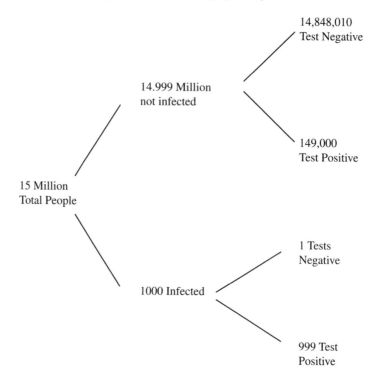

You can also interpret the test results in another light. Before you knew the results of the test, you would expect that you had about a 1 in 15,000 chance of having tuberculosis. After having gotten the positive result, your chances are now on the order of 1 in 150, or 100 times more likely.

On another note, what would happen to the probability that one has TB if there's a 50% false positive rate? Bayes' Theorem tell us that

$$\text{Pr(positive)} = \text{Pr(positive|sick)Pr(sick)} + \text{Pr(positive|healthy)Pr(healthy)}$$

$$= 0.999 \cdot \frac{1}{15000} + 0.50 \cdot \frac{14999}{15000} \approx 0.50.$$

Now we find

$$\text{Pr(sick|positive)} = \frac{1/15000}{0.50} \cdot 0.999 \approx 0.00013.$$

This result is 5 times smaller than the case where the false positive rate was 0.01.

It's worth isolating our partition fact. This is useful for a variety of problems, and we'll expand on it in the next section.

Partitions in two: Let B be an event so that $0 < \text{Pr}(B) < 1$ (and thus the same is true for the event B^c, or the event that B does not happen). Then we may partition the event A into two disjoint events, $A \cap B$ and $A \cap B^c$, and

$$\Pr(A) = \Pr(A|B)\Pr(B) + \Pr(A|B^c)\Pr(B^c).$$

This formula is quite reasonable. Reading it aloud, we see the probability that A happens is (the probability A happens given B) times (the probability B happens), plus (the probability A happens given B does not happen) times (the probability B doesn't happen). We're breaking into two cases: what happens if B happens, plus what happens if B doesn't.

Example: There are two urns with red and blue balls in them. The first one contains 40% red balls, and the second contains 20% red balls. Your friend picks randomly from one of the urns and draws a red ball. What are the odds it came from the first urn?

Solution: We want to find Pr(ball came from first urn|the ball is red). Using Bayes' Theorem, the chance of this happening is

$$\text{Pr(first urn|red)} = \frac{\text{Pr(first urn)}}{\text{Pr(red)}} \cdot \text{Pr(red|first urn)}.$$

Since the urns were picked at random, we know the probability that the ball came from the first urn is 0.5. Also, we're given that the probability the ball is red, given that it came from the first urn, is 0.4. So what's the probability that the ball actually is red? Since the events A (drawing from the first urn) and B (drawing from the second urn) form a partition, we see that

$$\text{Pr(red)} = \Pr(A) \cdot \text{Pr(red|}A) + \Pr(B) \cdot \text{Pr(red|}B)$$

$$= 0.5 \cdot 0.4 + 0.5 \cdot 0.2 = 0.3.$$

Thus, the probability that the ball came from the first urn, given that it's red, is clearly $0.5 \cdot 0.4/0.3 = 0.67$.

Did you follow all of that? Let's return to the expected counts argument to provide a more intuitive explanation. Imagine that your friend repeated this process 1000 times, replacing the ball after every draw. Then we would expect him to pick from the first urn 500 times, and 200 of those times (500 times 40%) to pick a red ball. Similarly, we would expect him to pick from the second urn 500 times and 100 of those times to pick a red ball. So, of the 300 times he picked a red ball, in 200 of those trials he picked from the first urn, meaning the probability of coming from the first urn given that the ball is red is $200/300 \approx 0.67$.

4.5 Partitions and the Law of Total Probability

In the last section we explored Bayes' Theorem, which is a monumental result that allows us to switch between two related conditional probabilities—namely $\Pr(A|B)$ and $\Pr(B|A)$. If you want, ***think of this as the analogue to interchanging orders of integration, or interchanging a sum and a derivative.*** We frequently want to change the order of operations. Often we can't, as order matters. Typically the square-root of a sum is not the sum of the square-roots. That's true here, as it's rare for $\Pr(A|B) = \Pr(B|A)$; however, all is not lost. We don't care whether or not these are equal, but only that we can easily find one given another. Why? Often one of the two expressions is simpler to find, and this provides the bridge to get to the other.

Unfortunately, Bayes' Theorem can be hard to use in practice. We need to know both $\Pr(A)$ and $\Pr(B)$ explicitly. Often, we can't be told precisely what $\Pr(B)$ or $\Pr(A)$ are, but we'll instead be given some conditional probabilities from which we can calculate $\Pr(B)$ or $\Pr(A)$. Taking this approach, we can formulate a more general version of Bayes' Theorem. But in order to do this, we must first sharpen our knowledge of the Theory of Partitions.

Before we start, here's a bit of terminology. Remember two sets A and B are disjoint if $A \cap B = \emptyset$. That is, A and B are disjoint if they have no elements in common.

A **partition** of a sample space S is a countable collection of sets $\{A_1, A_2, \dots\}$ with the following properties.

1. If $i \neq j$, then A_i and A_j are disjoint. We often write $A_i \cap A_j = \emptyset$ to indicate that their intersection is empty.
2. The union of all the A_i's is the entire sample space: $\bigcup_i A_i = S$.

Essentially, a partition is a collection of sets which together contain every possible outcome exactly once. We've seen this several times, partitioning our space into B and not B (or B and B^c). This is just the more general situation.

For example, imagine that we're looking at the possible outcomes from rolling a six-sided die. Our sample space in this case is $S = \{1, 2, 3, 4, 5, 6\}$. Suppose that we then formulate the two following sets:

$$A_1 = \{s \in S | s \text{ is even}\} = \{2, 4, 6\}$$
$$A_2 = \{s \in S | s \text{ is odd}\} = \{1, 3, 5\}.$$

Do A_1 and A_2 form a partition of S? They have no elements in common—because a positive integer cannot be both even and odd—and together they contain every possible value for S since every positive integer is either even or odd. So, A_1 and A_2 do indeed form a partition of S.

But what if we had picked the following two sets?

$$A_1 = \{s \in S | s < 4 \text{ or } s \text{ is even}\} = \{1, 2, 3, 4, 6\}$$
$$A_2 = \{s \in S | s \text{ is odd}\} = \{1, 3, 5\}.$$

Do these form a partition of S? We notice that $A_1 \cup A_2 = S$, so every possible outcome is represented. However, $A_1 \cap A_2 = \{1, 3\}$, so these are not disjoint sets, and therefore we don't have a partition.

Now what, you may be asking, is so great about partitions? We promise you that it's not just a new term for you to use to impress your friends. It's often the case that we're interested in learning something about the probability of an event A happening, but we're only given information about the probability of A *and* some other events B_i *simultaneously* happening as well.

For instance, suppose we know that the probability of owning a dog and getting sick with the flu in a given year is 0.08 and that the probability of getting sick and not owning a dog is 0.07. What's the probability that a person randomly drawn from this population gets sick?

As always, we could attack this through the expected counts method. Imagine that we had 100 people. Then eight of them would get sick and own a dog, and seven of them would get sick and not own a dog. Since you either own a dog or you don't own a dog, we've accounted for all the sick people. Hence, the probability of getting sick is 0.15. Notice that the events of owning a dog and not owning a dog form a partition of our probability space. As such, we can write the probability that a given individual gets sick as

$$\Pr(\text{Sick}) = \Pr(\text{Sick and dog}) + \Pr(\text{Sick and no dog}).$$

Before stating the general result, we first review some notation. Given a set $\{B_1, B_2, \dots\}$ by $\sum_n$ we mean the sum over all indices. If we want to deal with both finite and infinite partitions at the same time and if we want to be very formal, we can write $\{B_n\}_{n \in I}$ for the elements of our partition, and $\sum_{n \in I} \Pr(A|B_n) \cdot \Pr(B_n)$ for the sum. Frequently we just write $\sum_n$ as it is understood.

Law of Total Probability: If $\{B_1, B_2, \dots\}$ form a partition for the sample space S (into at most countably many pieces), then for any $A \subset S$ we have

$$\Pr(A) = \sum_n \Pr(A|B_n) \cdot \Pr(B_n).$$

We should have $0 < \Pr(B_n) < 1$ for all n as the conditional probabilities aren't defined otherwise (note if a B_n has probability zero then it isn't needed, as that piece is hit by the factor $\Pr(B_n) = 0$, while if it is 1 then all the other factors are unnecessary).

Proof: Consider the set $G = \bigcup_n A \cap B_n$. Any element of G is in both A and at least one of the B_n's, so we can rewrite $G = A \cap \left(\bigcup_n B_n \right)$. However, we know that the B_n's form a partition for S, so $\bigcup_n B_n = S$, which means that $G = A \cap S = A$. Therefore, we have

$$\Pr(A) = \Pr(G) = \Pr\left(\bigcup_n (A \cap B_n) \right).$$

To make things cleaner, write $A_n = A \cap B_n$. This simplifies our expression to

$$\Pr(A) = \Pr\left(\bigcup_n A_n \right).$$

Since the B_n's form a partition of S, we know that the A_n's are pairwise disjoint. Consequently, by the additivity of disjoint sets,

$$\Pr(A) = \Pr\left(\bigcup_n A_n \right) = \sum_n \Pr(A_n) = \sum_n \Pr(A \cap B_n). \qquad \square$$

You may have noticed something strange with respect to our previous example about dogs and the flu: *Who is going to keep statistics on the total percent of the population that owns a dog and gets the flu?* It's far more likely that you would have information about the percent of dog owners who get the flu and the percent of non-owners who get the flu. But these are just conditional probabilities! The percent of dog owners who get the flu is the same as the probability of getting the flu given that you own a dog. So while the above theorem is true, it's not always useful because we'll rarely have probabilities in the form $\Pr(A \cap B_i)$. How can we tailor the theorem to our conditional probability needs? We saw earlier that the probability of the events A and B happening is given by

$$\Pr(A \cap B) = \Pr(A|B) \cdot \Pr(B),$$

so we can rewrite the Law of Total Probability like this:

$$\Pr(A) = \sum_{i=1}^{n} \Pr(A|B_i) \cdot \Pr(B_i).$$

We could also write this as $\Pr(A) = \sum_{i=1}^{n} \Pr(B_i|A) \cdot \Pr(A)$. But this (for $\Pr(A) \neq 0$) just simplifies to $\sum_{i=1}^{n} \Pr(B_i|A) = 1$, which isn't too interesting.

The next problem might mean more to you in a few years....

According to some studies, Caesarean section is used in about 30% of the births in the United States and 0.2% of the babies don't survive. Of the births in which Caesarean section isn't used, 99.1% of the babies survive. Given these data, what percentage of the newborn babies will survive?

Solution: We want to find Pr(babies survive). Let A denote the event that a baby survives after birth, B the type of birth. Specifically, let B_1 represent Caesarean section, B_2 non-Caesarean section. Then

$$
\begin{aligned}
\Pr(A) &= \sum_{i=1}^{n} \Pr(A|B_i) \cdot \Pr(B_i) \\
&= \Pr(A|B_1) \cdot \Pr(B_1) + \Pr(A|B_2) \cdot \Pr(B_2) \\
&= (1 - 0.002) \cdot 30 + (.991) \cdot (1 - .30) \\
&= .9931 = 99.31\%.
\end{aligned}
$$

This method works because we're able to create two disjoint partitions which are Caesarean section and non-Caesarean section births. By applying the Law of Total Probability, we find that 99.31% of the babies will survive.

4.6 Bayes' Theorem Revisited

Now that we've developed the Theory of Partitions and the Law of Total Probability, we can formulate a more general version of Bayes' Theorem. Before we found $\Pr(A|B)$ by setting

$$\Pr(A|B) = \frac{\Pr(B|A) \cdot \Pr(A)}{\Pr(B)},$$

but now suppose that we have some partition $\{A_1, A_2, \ldots, A_n\}$ of our sample space S. So, we can write $\Pr(B) = \sum_{i=1}^{n} \Pr(B|A_i) \cdot \Pr(A_i)$. Thus, Bayes' Theorem becomes as follows.

Bayes' Theorem: Let $\{A_i\}_{i=1}^{n}$ denote a partition of the sample space. Then

$$\Pr(A|B) = \frac{\Pr(B|A) \cdot \Pr(A)}{\sum_{i=1}^{n} \Pr(B|A_i) \cdot \Pr(A_i)}.$$

Frequently one takes A to be one of the sets A_i.

The Boston Red Sox are about to have a home game at Fenway Park. In recent years, it's rained on average 30% of the game days. On days when it rains, 85% of the time the forecast was for rain; on days when it doesn't rain, only 20% of the time was rain predicted. What's the probability that it rains on the day of a Red Sox game given that the weather forecast predicts rain on that day?

Solution: We want to find the probability that it rains on the day of the Red Sox game given that the weather forecast predicts rain. Let A_1 denote the event that it rains on game day, A_2 the event that it won't rain on game day, and B the event that the forecast predicts rain.

$$\begin{aligned}
\Pr(A_1|B) &= \frac{\Pr(B|A_1) \cdot \Pr(A_1)}{\sum_{i=1}^{n} \Pr(B|A_i) \cdot \Pr(A_i)} \\[2mm]
&= \frac{\Pr(B|A_1) \cdot \Pr(A_1)}{\Pr(B|A_1) \cdot \Pr(A_1) + \Pr(B|A_2) \cdot \Pr(A_2)} \\[2mm]
&= \frac{.85 \cdot 0.30}{0.85 \cdot 0.30 + 0.20 \cdot (1 - 0.30)} \\[2mm]
&\approx 0.708333.
\end{aligned}$$

Thus we find that the probability it rains on the day of the Red Sox game, given that the weather forecast predicts rain on that day, is about 0.64557 or around 65%.

4.7 Summary

We devoted this chapter to expanding the general theory. We saw earlier that while our basic rules of probability suffice to solve problems, they're not always easy to use. There's a great need to isolate out useful results to simplify these calculations. The two big concepts in this chapter are independence and conditional probabilities.

Our dream is to find independent events, as the probability of both happening is simply related to the individual probabilities, as they don't interact. Most of the time, however, we'll have dependent events. Knowledge of one happening affects the odds of the other. One of my favorite quotes along these lines is from *The Empire Strikes Back*. The Millennium Falcon is being pursued by Star Destroyers and TIE fighters. Han Solo, captaining the Falcon, decides to take the ship into an asteroid field in an attempt to shake the pursuers.

> C-3PO: Sir, the possibility of successfully navigating an asteroid field is approximately 3,720 to 1.
>
> Han Solo: Never tell me the odds.

Even in the future (or is it in the past as *Star Wars* supposedly took place a long time ago) we find conditional probabilities. C-3PO isn't worried about Han Solo's piloting skills in general; he's worried about him piloting the Falcon *given* that they're in an asteroid field.

We developed several formulas involving conditional probabilities and independent events. The culmination was Bayes' Theorem. The idea is that we can often partition our space into simpler events, where frequently it's easy to find the probabilities of these smaller events. We then need a way to piece things together; Bayes' Theorem provides this framework.

I view a lot of this like algebra: often one way of looking at an expression is simpler and more illuminating than others. The point of Bayes' Theorem is to allow us to pass from one set of probabilities (which are either given or hopefully easy to find) to others (which frequently are either not given to us or are harder to find). It's a lot like changing orders of integration in multivariable calculus: sometimes one perspective leads to easier algebra than another.

In the next chapter we'll continue our study of probabilities of discrete events, but now armed with these additional tools (as well as some which we'll develop).

4.8 Exercises

Exercise 4.8.1 *Returning to our guess for* $\Pr(A|B)$, *we were trying to write it as a function of* $\Pr(A), \Pr(B)$, *and* $\Pr(A \cap B)$. *As we are looking at conditional probabilities, it's reasonable to look at a ratio where* $\Pr(B)$ *is the denominator (we're trying to figure out the probability something happens given that B has occurred; we can view this as adjusting our space so now the probability B happens is 1 and other probabilities are adjusted accordingly). The simplest formula we could have would be*

$$\Pr(A|B) = \frac{\alpha \Pr(A) + \beta \Pr(A \cap B)}{\Pr(B)};$$

by looking at extreme cases show the only option that makes sense is $\alpha = 0$ *and* $\beta = 1$. *Note of course that this is not a proof that this is the right formula, just evidence in favor of it. For more along these lines see §2.7, especially §2.7.2.*

Exercise 4.8.2 *Can an event be dependent on another event that has 0 probability of happening? Justify your answer.*

Exercise 4.8.3 *Find 3 events such that* $Pr(A \cap B \cap C) = \Pr(A) \cdot \Pr(B) \cdot \Pr(C)$, *but at least two of the events are dependent on each other.*

Exercise 4.8.4 *If* A, B, *and* C *are independent then* $\Pr(A \cap B \cap C) = \Pr(A) \times \Pr(B)\Pr(C)$, $\Pr(A \cap B) = \Pr(A \cap B)$, $\Pr(A \cap C) = \Pr(A)\Pr(C)$, *and* $\Pr(B \cap C) = \Pr(B)\Pr(C)$; *thus there are four conditions to check. If* $A_1, \ldots, A_n$ *are independent how many conditions are there to check?*

Exercise 4.8.5 *Consider the variation of the Basketball exercise explained in Exercise 17 in Chapter 1. Suppose Bird shoots second and the probability that Bird* *makes his shot depends on whether or not Magic made his shot this round (imagine Bird shoots either better or worse under pressure). Let the probability Bird makes his shot equal $a \cdot p$ given that Magic makes his shot and $\frac{p}{a}$ given that Magic missed his shot $(0 \le a \cdot p \le 1, 0 \le \frac{p}{a} \le 1)$. Find the probability that Bird wins.*

Exercise 4.8.6 *Imagine you are walking in the woods. So far you have encountered a cardinal, three deer, and two chipmunks. You want to predict the next animal you will see. Your first impulse is to set the probability of seeing a cardinal at 1/6, of seeing a deer at 3/6, and of seeing a chipmunk at 2/6. Something, however, seems troubling. In your calculation the probability of seeing a squirrel is 0, which is downright nutty. One approach would be to account for unseen elements by pretending to have seen the unseen and have the count of unseen animals be one; thus the probability of an animal other than a cardinal, deer, or chipmunk is 1/7. Laplace proposed a different fix for the unseen-elements exercise. He added one to the count of every species. If we have s species of animals, and if we see s_k of species k, Laplace suggests that the probability of observing something from species k should be $(s_k + 1)(1 + \sum_{k=1}^{s}(s_k + 1))$, and therefore the probability of observing something from an unseen species should be $1/(1 + \sum_{k=1}^{s}(s_k + 1))$. In our example, Laplace's approach would set the probability of seeing a cardinal at 2/9, a deer at 4/9, a chipmunk at 3/9, and a new species at 1/9. Why does Laplace's approach beat ours? In particular, what parameter is Laplace's equation sensitive to that ours is not?*

Exercise 4.8.7 *Suppose the probability the Red Sox have an OPS over .800 given that they won the World Series in a given year is 70%. (If you don't know what OPS is, it's a hitting statistic in baseball and a higher number is better. All of these numbers are made up, but the Red Sox did have an OPS above .800 in both 2004 and 2007, but not in 2013). The probability that the Red Sox have an OPS over .800 is 30% and the probability the Red Sox win the World Series is 10%. What is the probability the Red Sox win the World Series in a given year if the team OPS is above .800?*

Exercise 4.8.8 *If three dice are rolled and sum to 7, what is the probability at least one of the dice is a 1?*

Exercise 4.8.9 *If two dice are rolled and one of the dice is odd, what is the probability the product of the two dice has exactly two prime factors (not necessarily distinct)?*

Exercise 4.8.10 *If we are dealt a hand of five cards and at least two of them are the same number, what is the probability that we have at least 3 of the same number?*

Exercise 4.8.11 *A prisoner is given an interesting chance for parole. He's blindfolded and told to choose one of two bags; once he does, he is to reach in and pull out a marble. Each bag has 25 red and 25 blue marbles, and the marbles all feel the same. If he pulls out a red marble he is set free; if it's a black, his parole is denied. What is his chance of winning parole?*

Exercise 4.8.12 *The setup is similar to the previous exercise, except now the prisoner is free to distribute the marbles among the two bags however he wishes, so long as all the marbles are distributed. He's blindfolded again, chooses a bag at random again, and then a marble. What is the best probability he can achieve for being set free? Prove your answer is optimal.*

Exercise 4.8.13 *Consider we have 11 squares numbered from left to right 0–10. If we are at square k, the probability we move left is $.1 \cdot k$ and the probability we move right* is $1 - .1 \cdot k$. *If we start at square 5, given that we end up on or past square 7 what is the probability our first move was to the right?*

Exercise 4.8.14 *Assume $0 < \text{Prob}(X), \text{Prob}(Y) < 1$ and X and Y are independent. Are X^c and Y^c independent? (Note X^c is "not X," or $\Omega \setminus X$.) Prove your answer.*

Exercise 4.8.15 *An insurance company insures 10,000 homes for $5,000 dollars each per year. If a home is destroyed, the owner is payed $100,000 dollars. Assume any given house being destroyed is independent of damage to the other houses. Each house is destroyed with probability p in a given year. Find the probability the insurance company makes money in a given year (in terms of p).*

Exercise 4.8.16 *Are the assumptions in the previous exercise reasonable? Explain why or why not.*

Exercise 4.8.17 *Would you expect the probability that a car is in a given spot on a road to be independent of the probability that other cars are in given spots on the road? Explain.*

Exercise 4.8.18 *The letter "H" appears about 6% of the time in the English language. The letter "T" appears around 9% of the time. The bigram "TH" appears about 1.5% of the time. Given that the previous letter is a "T," find the probability the next letter is an "H."*

A person's blood type is determined by a single gene. Assume that the probability a person inherits one allele is completely independent of the other allele they have inherited. There are three possible alleles for this gene, A, B, and O. If a person is AA, or AO, they have blood type A. If they are BB or BO they have bloodtype B. If they are OO they have blood type O and if they are AB they have phenotype AB. Assume that 45% of alleles are O, 40% are A, and the other 15% are B.

Exercise 4.8.19 *Given that Justine has type O blood, find the probabilities for each blood type for both of her parents.*

Exercise 4.8.20 *Find the probability Justine's brother also has type O blood.*

Exercise 4.8.21 *People with type A blood can get blood transfusions from people with either type A or type O blood. Find the percentage of people from which those with type A blood can get blood.*

Exercise 4.8.22 ***Texas Hold'em*** *is a common variant of poker where each player is dealt two cards and then must bet. The hands are the same as in other types of poker. Five more cards are revealed and shared by all the players. The players make the best hand possible out of these five cards and the two in their hands. It is incredibly useful in Texas Hold'em to know the probability of winning given the two cards you are dealt. Write code that approximates via simulation the probability of different hands given an initial two cards. (You may assume only three more cards are flipped, to avoid having to deal with maximizing the hands rank out of seven cards.)*

Exercise 4.8.23 *Adapt your code from the previous exercise to simulate the probability you win if you start with 3♠ and 3♡ and your opponent has A♡ and 8◇.*

Exercise 4.8.24 *Let Ω have n elements. How many partitions are there of Ω into exactly k sets, for each $k \in \{1, \ldots, n\}$?*

Exercise 4.8.25   *Let's revisit the previous exercise. Assume all possible partitions of Ω are equally likely. If $n = 4$ what is the probability that all sets in the partition have the same number of elements? Can you answer this for general n?*

Exercise 4.8.26   *We flip a coin that is heads with probability .2 and tails with probability .8. We then roll two fair (and independent) die if the coin comes up heads, or roll three fair die if it comes up tails. What is the probability the sum of the dice is 3?*

Exercise 4.8.27   *Prove that if $A \subset B$, then $\Pr(A) \leq \Pr(B)$ using the Law of Total Probability.*

Exercise 4.8.28   *Consider the intervals $[1, 102]$ and $[1001, 1102]$. We pick one of these randomly, with each being chosen with equal likelihood. After picking either interval, pick a random integer m inside that interval, with all integers inside the range equally likely to be chosen. What is the probability m is a prime number in the interval $[1001, 1102]$? What is the probability m is a prime in either range? Notice both intervals have the same number of integers; is the probability a number is prime dependent only on the number of integers in the range?*

CHAPTER 5 _____

Counting II: Inclusion-Exclusion

*Vizzini: But it's so simple. All I have to do is divine from what I know
of you: are you the sort of man who would put the poison into his
own goblet or his enemy's? Now, a clever man would put the poison
into his own goblet, because he would know that only a great fool
would reach for what he was given. I am not a great fool, so I can
clearly not choose the wine in front of you. But you must have known
I was not a great fool, you would have counted on it, so I can clearly
not choose the wine in front of me.*
Man in Black: You've made your decision then?
Vizzini: Not remotely.
—THE BATTLE OF WITS IN *The Princess Bride* (1987)

We introduced the factorial function and binomial coefficients in Chapter 3, and made productive use of them in successfully attacking many counting problems. Many, but not all. The number of cases and subcases multiplied so quickly in some straightforward questions that we quickly became frustrated (I know I was—I spent several hours plowing through the bridge deals!) We saw how hard it was to keep track of all the different possibilities, and the dangers of both double counting some configurations, and forgetting others. A great example of this was the perfect deal problem from §3.4.2: what is the probability that exactly one person is dealt all cards in one suit? Clearly, we need better ways to navigate the counting.

This chapter is structurally similar to Chapter 3. We'll quickly review factorials and binomial coefficients in §5.1, with two new twists. The first involves circular orderings, which is between the extremes of order matters and order doesn't. The second comes from counting how many clothing ensembles have certain properties. We'll do it by brute force and by a clever perspective. This clever idea isn't a one hit wonder, but can be used time and time again to simplify the counting in a variety of situations. We'll thus use it as motivation to segue into our new technique, the **Method of Inclusion-Exclusion**, and then explore several new questions and revisit some old.

We'll continue and spend some time on some famous combinatorics problems. At first it might not seem like this is a proper topic for a probability class. Sure, some people find combinatorics fun, but shouldn't that be left to a discrete math class? While I agree that the deeper theory belongs in another book, the basics *must* be covered here as, after all, much of probability comes from counting. We look at how many successes there are relative to how many possibilities. A great example of this are the lotteries, especially those where you have to pick k numbers from N in any order. While it's not too bad to figure out your odds of losing if a number can't repeat, things get quite interesting if you're allowed to reuse numbers. I'll show you a wonderfully clever idea to handle that twist. Finally, we'll do some standard problems, such as the famous Mississippi question (someone gives you the word Mississippi and wants to know the probability a six letter—possibly nonsense—word has exactly three s's). This isn't just a make-work problem, though in a rushed lecture it may seem so. The solution to this can be expressed in terms of **multinomial coefficients**, which generalize the binomial coefficients.

5.1 Factorial and Binomial Problems

Everyday life provides a wealth of opportunities to apply our results on the factorial function and binomial coefficients. Below is a small, but representative, sample of the types of problems you can solve. Sometimes we'll phrase it as a counting problem (how many ways can we do X), while other times we'll give it as a probability exercise (what're the odds that Y happens). It's easy to pass from one to another; for example, to go from a counting to a probability problem all we have to do is divide the number of outcomes we desire by the number of possible outcomes.

5.1.1 "How many" versus "What's the probability"

We're going to state the problem first and then discuss a *serious* issue with it, and only after a long (but important) discussion will we turn to solving it.

Question: *Imagine there are 20 people applying for positions at U. S. Robots and Mechanical Men. Assume 10 of them have no experience, 7 of them have exactly one year of experience, and 3 of them have exactly two years' experience. Assume everyone is equally likely to be chosen. (1) How many ways are there to choose 6 people to work for the company? (2) What's the probability a specific group of 6 people is chosen? (3) What's the probability that none of the 6 hires has any experience?*

Before solving the problem, it's worth carefully looking at the phrasing. We saw how long it took us in the Birthday Problem (Chapter 1) to properly formulate the question. This problem *seems* to be well-posed, but interestingly *part of it is not!* There's a small subtlety, an assumption that people implicitly make. In fact, you're almost surely *supposed* to make this assumption. Many teachers and books don't even realize the assumption was made.

What's the issue? There's a big difference between the first part (how many ways are there) and the second part (what's the probability). We defined a probability space in §2.4. It's a triple involving the outcome space Ω and a function which assigns probabilities to any subset of Ω in a distinguished collection of sets (the σ-algebra). If we ask a *how many* question, we're only asking for a list of the possibilities.

If we want to compute probabilities, we need to know something about the probability function. Somehow, we need to get information about how it assigns probabilities to sets. The phrase that's supposed to help us is: *Assume everyone is equally likely to be chosen.* This phrase is supposed to specify the probability function, but unfortunately it's ambiguous. The most natural assumption, the one you're expected to make, is that each *group* of k people is as likely as any other group of k people. This immediately implies that each person is as likely to be chosen as any other; however, we can assign unequal probabilities to groups yet still have each person equally likely.

Let's do a simpler example to highlight this. Imagine we have four people: Agnetha, Benny, Björn, and Anni-Frid, and we want to form a group of two. Yes, we're not using the standard Alice, Bob, Charlie, and Dan—patience, there's a good reason for these names! Let's answer the *how many* question first. There are $\binom{4}{2} = 6$ ways to choose two people from four. The groups are {Agnetha, Benny}, {Agnetha, Björn}, {Agnetha, Anni-Frid}, {Benny, Björn}, {Benny, Anni-Frid}, and finally {Björn, Anni-Frid}. There are six groups, and each person is in exactly three groups. What about the *what's the probability* question? If all the groups are equally likely to be chosen, then the probability of getting any group of two is just 1/6. Each person is also equally likely to be chosen (as they're in exactly three groups), and is chosen with probability 1/2.

Now it's time to explain our choice of names, why we wanted groups of two, and another perfectly valid way to interpret the phrasing. These are four very famous people: Agnetha Fältskog, Benny Andersson, Björn Ulvaeus, and Anni-Frid Lyngstad. Together they formed ABBA, one of the most popular and (in my uncool opinion) one of the greatest music groups of all time. Agnetha and Björn were married, as were Benny and Anni-Frid. Both couples divorced, and shall we say each person would prefer *not* to be grouped with their former spouse. There are *still* six ways to choose two people from four, but let's assign a probability of 0 to the groups {Agnetha, Björn} and {Benny, Anni-Frid}, and a probability of 1/4 to {Agnetha, Benny}, {Agnetha, Anni-Frid}, {Benny, Björn}, and {Björn, Anni-Frid}. The probability of any person being chosen is still 1/2. For example, Benny is in two groups, and each group is chosen with probability 1/4, so the probability he's picked is 1/4 + 1/4, or 1/2. Everyone is still equally likely to be chosen, but it's not the case that all groups are equally likely.

Which interpretation is correct? *Both* are valid interpretations and legitimate assignments of a probability function. While in problems like this it's typically the first view that's meant, whoever is giving you the problem really should use careful, precise language. Thus, usually the proposer means that each *group* is equally likely to be chosen. *If* each group is equally likely to be chosen, *then* each person is also equally likely to be chosen and has the same chance of being with any person or subset of people as any other equally sized group. The converse is not true. Just because each person is equally likely to be chosen doesn't mean that each group is equally likely to be chosen. In our ABBA example, two of the six groups are assigned zero probabilities. They're still counted if we're asked to solve a *how many* question, but they don't add to the probability in a *what's the probability* question.

This was a long spiel, and it won't be the last time you hear it in this book because it deserves repeating. Be precise. Yes, it often leads to what look like long-winded statements, but you want to make sure everyone's on the same page. Imagine someone asked you to choose a collection of two-element people from Agnetha Fältskog, Benny Andersson, Björn Ulvaeus, and Anni-Frid Lyngstad such that each person is equally likely to be chosen. As the feuds are well-known in many circles (the band passed on a quarter of a billion dollars to reunite!), they might assume it was *obvious* that Agnetha and Björn can't be together, nor Benny and Anni-Frid. Be clear when you write. Don't be the source of misunderstandings.

5.1.2 Choosing Groups

After this long discussion, the solution is relatively straightforward. As our digression was long, let's repeat the problem. We'll use the first interpretation; namely, each group is equally likely to be chosen.

Question: *Imagine there are 20 people applying for positions at U. S. Robots and Mechanical Men. Assume 10 of them have no experience, 7 of them have exactly one year of experience, and 3 of them have exactly two years' experience. Assume everyone is equally likely to be chosen. (1) How many ways are there to choose 6 people to work for the company? (2) What's the probability a specific group of 6 people is chosen? (3) What's the probability that none of the 6 hires has any experience?*

(1) In this problem, the first thing to notice is that order isn't important. A worker could be the company's first choice, or he could barely make the cut; but in either scenario, he is still hired, regardless of where in line he was ranked. We've seen how to count the number of ways people can be picked when order doesn't matter; it's just the binomial coefficients. We want to choose 6 from 20, so the answer is $\binom{20}{6} = \frac{20!}{6!14!} = 38,760$.

(2) We'll use the first interpretation, namely that each group of 6 people is as likely to be chosen as any other. As there are $\binom{20}{6} = 38,760$ distinct groups, the probability any specific set of 6 is hired is just $1/38760$.

(3) To find out the probability that none of the six hires has any work experience, we calculate how many different ways only people without work experience can be hired. This is simply a matter of selecting our six workers from only the group of workers without experience. There are ten such workers, so we see there are $\binom{10}{6} = \frac{10!}{6!4!} = 210$ ways six inexperienced workers may be hired. In our example every group of six is equally likely to be chosen, so the probability of none of the workers having any experience is simply the number of ways workers with no experience can be chosen divided by the number of total possible selections. Since we now know both values, this is a quick calculation: $\frac{210}{38,760} = \frac{7}{1292}$, or about a .54% chance that none of the hires will have any experience.

We end with one last problem. Our goal here is to emphasize when order matters, and when it doesn't.

Problem 5.1.1: *Assume there are 12 people running for student government.*

- *How many ways are there to choose 4 of them?*
- *How many ways are there to choose one of them to be president, one to be vice president, one to be treasurer, and one to be secretary, given that no one can hold two positions?*
- *How many ways are there to choose one of them to be president, one to be vice president, one to be treasurer, and one to be secretary, allowing people to hold any number of offices.*

Solution: The first problem is $\binom{12}{4} = 495$; all positions are equal, and we just have to choose 4 out of 12 with order not mattering. The second is $12 \cdot 11 \cdot 10 \cdot 9 = 12!/8! = 11,880$, as now order matters (there are 12 ways to choose the president, then 11 for the vice president, and so on; remember, it matters who is president, who is vice president, and so on). For the final problem, the answer is $12 \cdot 12 \cdot 12 \cdot 12 = 12^4 = 20{,}736$. Each person can hold any number of positions; unlike the previous problem, choosing one person for the presidency doesn't diminish the number of candidates for the other positions. There's always 12 people available. We see that, depending on what we mean by choosing 4 people from 12, there are three possible answers! This further emphasizes the point that it is important to clearly state a problem. It is not enough to ask how many ways to choose 4 people from 12 candidates, you need to specify whether order matters and also whether a person can be chosen more than once.

5.1.3 Circular Orderings

We've done lots of problems where we've had to worry about ordered choices or unordered choices. The next problem is a compromise. Why must we go to such extremes? Are there cases where some of the ordering matters and some where it doesn't? Yes, and we've already seen one example: the tic-tac-toe problem from §3.4.1. If we consider the board ordered, so each of the nine squares is distinguishable from the others, then there are 9 opening moves. If, however, we allow ourselves to rotate or flip, there are only three distinct first moves. This is a powerful aid in analyzing the game. Let's do another example along these lines.

 Example 5.1.2: *Consider the following related problems.*

(1) How many ways are there for 5 people to line up, where order matters?

*(2) How many ways are there for 5 people to sit at a **circular table**, where all that matters is the relative ordering? In other words, imagine the table has unit radius, we list the person sitting at $(1, 0)$ first and then list everyone counter-clockwise. We say the ordering Alice, Bob, Eve, Charlie, Dan is the same as Eve, Charlie, Dan, Alice, Bob.*

(1) Since order matters in the first problem, each person's position in the line is important. We can place any one of the five people in the first position in line. For the second position in line, there are only four people available, and so there are only four people that can fill this spot. As we consider the third, fourth, and fifth positions in line, we see that by this logic there are three, two, and one person(s) available to fill each respective spot in line. We multiply these numbers together, and get a product representing the total number of unique ways these five people can order themselves in line: $5 \times 4 \times 3 \times 2 \times 1 = 5! = 120$.

(2) Now in the second problem suppose these same five people sit at a circular table. There's no "first" and there's no "last" place in line. If we rotate the table and the chairs, we don't change the **relative ordering**. We might as well select one person to be special, and rotate until they're at the $(1, 0)$ position. For obvious reasons, this is called a **circular ordering** problem.

With our fixed person, we can now describe this circular problem in terms of a linear list; this list is a clever way of rewriting the problem into something simpler. Now to find all the orderings around the table, we need only to find the number of orderings of a group of four people, where there's a first and a last. It doesn't matter if we view our special, fixed person as first in line, or last, or any position. What matters is that for any order we rotate until this special person is in the special place, leaving us with one fewer person to sit but now having order matter. As we learned from the first part of this example, there are $4! = 24$ ways these four remaining people can arrange themselves in a line in which order matters. Thus we see that there will be 24 ways these five people can arrange themselves around a circular table, in which order matters.

In general, there are $(n-1)!$ different ways that n people can arrange themselves around a circular table when order matters. Note that it makes sense that there are less ways for people to arrange themselves around a table than in a line: moving the first person in line to the last position, and pushing everyone else forward one position would yield a new ordering in a linear setting; but around a table, this would be equivalent to rotating everyone one spot down the table, which doesn't yield a new ordering.

A nice problem related to sitting at a table is Conway's napkin problem. We quote from the entertaining paper by Claesson and Petersen (available online at http://arxiv.org/abs/math/0505080). "At a particular table, n men are to be seated around a circular table. There are n napkins, exactly one between each of the place settings. Being doubly cursed as both men and mathematicians, they are all assumed to be ignorant of table etiquette. The men come to sit at the table one at a time and in random order. When a guest sits down, he will prefer the left napkin with probability p and the right napkin with probability $q = 1 - p$. If there are napkins on both sides of the place setting, he will choose the napkin he prefers. If he finds only one napkin available, he will take that napkin (though it may not be the napkin he wants). The third possibility is that no napkin is available, and the unfortunate guest is faced with the prospect of going through dinner without any napkin!" There are lots of questions to be asked (and answered), such as how many people are expected to get a napkin, and what the probability is that everyone gets a napkin.

Similar to the napkin problem is Dijkstra's dining philosophers problem from computer science. The basic idea is that there are n people around a circular table and between each pair of people is a chopstick. A person can pick up the chopstick on their left or their right as long as it is available, but they can only eat after they pick up both chopsticks. When someone finishes eating, they put down their chopsticks in any order. This is an important problem because the chopsticks represent critical resources in the computer and the philosophers around the table represent the different processes running. Some obvious questions to ask are how many people do we expect to be able to eat, what is the probability that nobody gets to eat, and how does altering the constraints of the problem alter the results. Clearly, if everybody always picks up the chopstick on their right first, nobody will get to eat, which translates to your computer stopping. We could alter the problem and explore how changing the probability of choosing a given chopstick first could affect the probability that nobody gets to eat. This problem is related to sitting at a table, but it also is related to the idea that changing the interpretation of a problem can change the outcome. It is also an example of a real-life problem to which we can apply these concepts in order to better understand what is going on.

Another example of the sitting at a table problem is King Arthur and the round table. You could figure out how many ways there are to arrange Arthur's knights and calculate the probability that a given knight is seated next to Arthur, something that would be considered desirable. You could also imagine that Arthur might not want his wife, Guinevere, seated next to Lancelot, so you could calculate the number of ways to arrange the knights, given that Lancelot cannot be seated next to Guinevere. You could further complicate this by adding that Lancelot will want to be seated across from Guinevere so that they can exchange secretive glances, reducing the number of possible arrangements. You could also imagine that more important people are usually seated next to Arthur, so a given knight is seated next to Arthur with probability p, and you could figure out the probability of any given arrangement. This may seem like a ridiculous example, but it is not necessarily about calculating the actual values, but rather seeing how these concepts appear in the real world and how learning these approaches can help you reason about different problems. As we can see, there are a lot of different problems involving sitting around a table.

5.1.4 Choosing Ensembles

Even if you don't care about fashion (and, since this is a book for mathematicians, that's a real possibility!), the next problem is important. Don't skip it—it provides a great motivation for not just one but *two* of the most important viewpoints in probability!

 Problem 5.1.3: *You're getting dressed, and have 5 different pairs of socks, 3 different pairs of pants, and 4 different pairs of shirts to choose from. Assume exactly two pairs of socks are red, none of the pants are red, and only one shirt is red. You must choose exactly one pair of socks, one pair of pants, and a shirt. (1) How many different ensembles can you make? (2) It's school pride day and your school's color is red; how many different ensembles can you make where at least one item is red?*

(1) Two outfits are different if they have a different pair of socks, pants, or a different shirt. To find the number of different outfits, we want to figure out how many unique ways we can choose a pair of socks, pants, and a shirt to make an outfit. Since there are 5 pairs of socks, 3 pairs of pants, and 4 shirts, we simply multiply $5 \times 3 \times 4 = 60$ to see that there are 60 different outfits we can make.

(2) There are two ways an outfit can contain a red component: the outfit can have either a pair of red socks or red shirt (or both). To find how many outfits contain a red component, we first count how many outfits contain just a red pair of socks. To make an outfit with a red pair of socks, we must choose one of the two red pairs of socks, but any shirt and pair of pants will do. This would give us a wardrobe of two pairs of socks, 3 pants, and 4 shirts to choose from, or $2 \times 3 \times 4 = 24$ total outfits with red socks. Similarly, since there's only one red shirt, we have a wardrobe of 5 pairs of socks, 3 pants, and 1 shirt that make an outfit with a red shirt. This gives us $5 \times 3 \times 1 = 15$ total outfits with a red shirt.

It might be easy to think all that remains is to sum up the total number of outfits including a red shirt and a red pair of socks, but this would be premature. Consider an outfit with both a red shirt and a red pair of socks. We would have counted this outfit twice: once when counting outfits with red socks (we did not say the shirt could not be red) and once when counting outfits with red shirts (likewise, we did not say the socks could not be red). Thus we must count the number of total outfits with *both* red socks and a red shirt, and subtract that number from our total in order to only count them once.

There are two pairs of red socks and only one red shirt to choose from, so this gives us a wardrobe of 2 pairs of socks, 3 pairs of pants, and 1 shirt from which any outfit will have both a red shirt and a red pair of pants. This gives us a total of $2 \times 3 \times 1 = 6$ outfits with both red socks and a red shirt.

We now are ready to see how many outfits have *at least* one red item: we add the number of outfits with a red pair of socks to the number of outfits with a red shirt, and subtract off the number of outfits with both red socks and a red shirt. Doing so, we get $24 + 15 - 6 = 33$ total outfits with at least one red component. Since we know from our previous work that there are 60 possible outfits, assuming each outfit is equally likely to be picked we see there's a $33/60 = 11/20$ or a 55% probability that a random outfit has at least one red component.

The method used of counting the number of outfits with one red component and then subtracting outfits with two red components uses what is called the **Inclusion-Exclusion Principle**. It's a great technique, and frequently simplifies pesky counting. We'll discuss it in detail in §5.2. What it does is to frequently break one very hard calculation into a lot of simpler ones, but then requires us to do some combinatorics to combine those answers to get the solution to the original problem. As a rule of thumb, it's often better to do a lot of simpler problems than one hard problem.

 There's another good way to approach this problem. Every outfit will either contain a red element or not. If we know the total number of possible outfits *and* the number of outfits without red, we can find the number of outfits *with* red through subtraction. From our work above we already know there are 60 different outfits. We need only find how many outfits contain no red elements. To do this, we must select outfits from only part of the wardrobe. There are only three pairs of non-red socks, three pairs of pants, and three non-red shirts. This gives us $3 \times 3 \times 3 = 27$ possible outfits containing no red elements. Since there are 60 possible outfits, there then must be $60 - 27 = 33$ outfits with at least one red element. Happily this is the same result we came to with our first method as well.

We can extract a general technique from this too.

Complementary Events: For many problems, the easiest way to find the probability of A is to find the probability that A does not happen, as $\Pr(A) = 1 - \Pr(A^c)$. This is extremely useful in problems with the phrase *at least one*, as the complementary event is just none occurring.

An application of this problem that will appeal to college students is, given a wardrobe of n shirts, m pairs of pants, and k pairs of socks, how many days before you have to do laundry if you cannot wear the same outfit more than once. To make it more realistic, you could allow for the same pair of pants to be worn multiple times and not allow certain color combinations. This is one of the most important problems that college students attempt to solve. In real life it is a bit more complicated because some shirts don't match certain pants and sometimes you don't need to wear socks. Also, there are more articles of clothing to consider and weather considerations, but we can imagine applying the same techniques and at least getting an estimate.

In conclusion, looking at this simple fashion question led us to two important and powerful techniques in probability: the Method of Inclusion-Exclusion and the Method of Complementary Events; that's a pretty good return of investment for getting dressed!

5.2 The Method of Inclusion-Exclusion

In Chapter 3 we developed some of the properties of the factorial function and binomial coefficients, and saw how useful these were in solving probability problems. Unfortunately, for some complicated problems it was painful to find a good way to order the counting so we got all the cases, and never counted something twice. The **Method** of **Inclusion-Exclusion** (or the **Inclusion-Exclusion Principle**) is designed to order the counting in such a way that we avoid these issues, and are left with easier algebra. This is a non-trivial goal, and an important one. If you can simplify the algebra you can sometimes see patterns you might not have noticed.

The following is one of my favorite examples of the power of simplifying algebra. Imagine someone asked you to add

$$
\begin{array}{rl}
  & 51 - 47 \\
+ & 132 - 51 \\
+ & 611 - 132 \\
+ & 891 - 611 \\
+ & 1234 - 891 \\
+ & 2013 - 1234.
\end{array}
$$

One way, of course, is to say the first difference is 4, the next is 81, the following 479, and so on, and then add these. A better way is to notice that this is a **telescoping sum**. We have 51 twice (once with a positive sign, once with a negative sign), so these two occurrences cancel. Similarly the two instances of 132 cancel each other out, and so on until we're left with $2013 - 47$, or 1966. Notice how much easier this second approach is—by arranging the algebra in a convenient way, we can make our lives much easier.

We'll first state the Inclusion-Exclusion Principle and then describe why it works and do many examples.

5.2.1 Special Cases of the Inclusion-Exclusion Principle

Before stating the Inclusion-Exclusion Principle in general, let's look at some special cases and get a feeling for what it is, and why it can be so useful. Given a collection of sets there are two natural operations: unions and intersections. We use unions when we want to talk about the probability that *at least one* of the events happens while we use intersections when we want the probability that *all* happen. Frequently it is significantly easier to compute the probability that all the events happen, rather than the probability that at least one does. For example, think back to our work on perfect deals. It isn't too bad to find the probability a fixed number of people have one-suited hands, but we saw it was very involved to find the probability that at least one does.

 Remark 5.2.1: *In many inclusion-exclusion problems the probability of a set is just the number of elements in the set divided by the number of elements in the space. To emphasize this special case we'll sometimes write $|A|$ for the number of elements of a set, rather than $\Pr(A)$; however, the more general case allows different elements to have different probabilities, and if you want to have the most general formula possible you should use probabilities of sets and not cardinalities of sets.*

The way inclusion-exclusion works is that we express the probability of unions in terms of probabilities of intersections. Lots of intersections. This is another example of the technique of **doing a lot of simple problems rather than one hard problem**. The idea is we express the probability of a union in terms of sums and differences of probabilities of intersections.

We've already seen an example of this in our formula for the probability of a union of two sets:

$$
\Pr(A \cup B) = \Pr(A) + \Pr(B) - \Pr(A \cap B).
$$

Let's rewrite this in a way that will generalize nicely. Instead of A and B we'll call the two sets A_1 and A_2, and we find

$$\Pr(A_1 \cup A_2) = \Pr(A_1) + \Pr(A_2) - \Pr(A_1 \cap A_2)$$

$$= \sum_{i=1}^{2} \Pr(A_i) - \sum_{1 \le i < j \le 2} \Pr(A_i \cap A_j).$$

While the final expression on the right seems like an enormous overkill (especially as the final sum is just one term!), this suggests how to generalize the formula. As we have already explained why this formula is true, let's move on to the formula for three sets and talk about why that holds:

$$\Pr(A_1 \cup A_2 \cup A_3) = \Pr(A_1) + \Pr(A_2) + \Pr(A_3) - \Pr(A_1 \cap A_2)$$

$$-\Pr(A_1 \cap A_3) - \Pr(A_2 \cap A_2) + \Pr(A_1 \cap A_2 \cap A_3),$$

which we can write as

$$\Pr(A_1 \cup A_2 \cup A_3) = \sum_{i=1}^{3} \Pr(A_i) - \sum_{1 \le i < j \le 3} \Pr(A_i \cap A_j)$$

$$+ \sum_{1 \le i < j < k \le 3} \Pr(A_i \cap A_j \cap A_k)$$

$$= \sum_{i_1=1}^{3} \Pr(A_{i_1}) - \sum_{1 \le i_1 < i_2 \le 3} \Pr(A_{i_1} \cap A_{i_2})$$

$$+ \sum_{1 \le i_1 < i_2 < i_3 \le 3} \Pr(A_{i_1} \cap A_{i_2} \cap A_{i_3}).$$

Although the notation is beginning to look nightmarish in the final line, it's actually a good choice and will make things easier later. What we're doing is very similar to the notation in multivariable calculus. If we're just working in 2 or 3 dimensions we often use x, y, z for variables and $\vec{i}, \vec{j}, \vec{k}$ for directions; however, as we move on to an arbitrary number of directions, using different letters of the alphabet is not feasible. In that case, we switch to $x_1, \ldots, x_n$ and $\vec{e}_1, \ldots, \vec{e}_n$. We are taking a similar approach here by using i's with subscripts.

We illustrate the case of three sets in Figure 5.1. What we want to do is count how many elements are in the union $A_1 \cup A_2 \cup A_3$, which we denote $|A_1 \cup A_2 \cup A_3|$. If the sets were disjoint, we would just add the number of elements in each set: $|A_1| + |A_2| + |A_3|$. Unfortunately, if there are common elements we have double counted; remember double counting is one of the greatest dangers in the subject. Note any element in two sets has been counted twice. For example, any $x \in A_1 \cap A_2$ was counted once in $|A_1|$ and once in $|A_2|$. Thus we need to subtract $|A_1 \cap A_2|$, and similarly $|A_1 \cap A_3|$ and $|A_2 \cap A_3|$. Now we have no longer double counted elements in *exactly two of the sets*; each of these elements is now counted once.

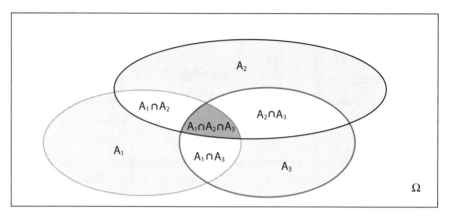

Figure 5.1. Illustration of inclusion-exclusion with three sets.

Unfortunately our counting is slightly off. We have correctly dealt with elements that are in exactly one or exactly two of the three sets, but if an element is in all three sets it is no longer counted. The reason is we counted it three times through $|A_1|, |A_2|, |A_3|$, but then removed it three times from the three intersections $|A_1 \cap A_2|, |A_1 \cap A_3|, |A_2 \cap A_3|$. The solution is clear: we need to add it back once, and must add in $|A_1 \cap A_2 \cap A_3|$. Now every element is handled correctly; the net number of times it is counted is 1 if it is in the union, and 0 otherwise. We thus find

$$|A_1 \cup A_2 \cup A_3|$$
$$= \sum_{i_1=1}^{3} |A_{i_1}| - \sum_{1 \le i_1 < i_2 \le 3} |A_{i_1} \cap A_{i_2}| + \sum_{1 \le i_1 < i_2 < i_3 \le 3} |A_{i_1} \cap A_{i_2} \cap A_{i_3}|,$$

or more generally if each element doesn't necessarily have the same probability:

$$\Pr(A_1 \cup A_2 \cup A_3) = \sum_{i_1=1}^{3} \Pr(A_{i_1}) - \sum_{1 \le i_1 < i_2 \le 3} \Pr(A_{i_1} \cap A_{i_2})$$
$$+ \sum_{1 \le i_1 < i_2 < i_3 \le 3} \Pr(A_{i_1} \cap A_{i_2} \cap A_{i_3}).$$

 Remark 5.2.2: *In our analysis above it's hard to miss all the conditions on the subscripts for the sets. For the intersection of two sets, for example, we don't have i_1 and i_2 each ranging from 1 to 3, but rather $1 \le i_1 < i_2 \le 3$. The reason is that we must avoid double counting sets, as $A_1 \cap A_3$ is the same as $A_3 \cap A_1$. Listing all the valid pairs we find*

$$(i_1, i_2) \in \{(1, 2), (1, 3), (2, 3)\}.$$

As the pattern may not be clear, let's look at $1 \le i_1 < i_2 \le 4$. In that case we have

$$(i_1, i_2) \in \{(1, 2), (1, 3), (1, 4), (2, 3), (2, 4), (3, 4)\},$$

while if $1 \le i_1 < i_2 < 5$ we have

$$(i_1, i_2) \in \{(1, 2), (1, 3), (1, 4), (1, 5), (2, 3), (2, 4), (2, 5), (3, 4), (3, 5), (4, 5)\}.$$

*Going up to 3 we had 3 pairs, going up to 4 we had 6 pairs, and going up to 5 we had 10 pairs. Gathering data is a great way to build intuition; if you don't see the pattern yet do a few more, but what we have is consistent and the number of pairs with $1 \le i_1 < i_2 \le n$ is $\binom{n}{2}$. Recalling the definition of the binomial coefficients, this is quite reasonable: we need to choose two distinct indices from $\{1, 2, \ldots, n\}$. While order does not matter, for **definiteness** in writing down the expression we denote the smaller one by i_1 and the larger one by i_2. The pattern continues, and the number of triples with $1 \le i_1 < i_2 < i_3 \le n$ is $\binom{n}{3}$. This observation will be extremely valuable later, as in many cases all the events with the same number of intersections occur the same number of times; thus we can often compute just one and multiply by an appropriate binomial coefficient.*

5.2.2 Statement of the Inclusion-Exclusion Principle

We now state the general inclusion-exclusion formula, and leave its justification to §5.2.3; note it's the natural extension of our formula from the union of three sets.

Inclusion-Exclusion Principle: Consider sets $A_1, A_2, \ldots, A_n$. Denote the number of elements of a set S by $|S|$ and the probability of a set S by $\Pr(S)$. Then

$$\left| \bigcup_{i=1}^{n} A_i \right| = \sum_{i=1}^{n} |A_i| - \sum_{1 \le i < j \le n} |A_i \cap A_j| + \sum_{1 \le i < j < k \le n} |A_i \cap A_j \cap A_k|$$

$$- \cdots + (-1)^{n-2} \sum_{1 < \ell_1 < \ell_2 < \cdots < \ell_{n-1} \le n} |A_{\ell_1} \cap A_{\ell_2} \cap \cdots \cap A_{\ell_{n-1}}|$$

$$+ (-1)^{n-1} |A_1 \cap A_2 \cap \cdots \cap A_n|;$$

this also holds if we replace the size of all the sets above with their probabilities.

We may write this more concisely. Let $A_{\ell_1 \ell_2 \ldots \ell_k} = A_{\ell_1} \cap A_{\ell_2} \cap \cdots \cap A_{\ell_k}$ (so $A_{12} = A_1 \cap A_2$ and $A_{489} = A_4 \cap A_8 \cap A_9$). Then

$$\left| \bigcup_{i=1}^{n} A_i \right| = \sum_{i=1}^{n} |A_i| - \sum_{1 \le i < j \le n} |A_{ij}| + \sum_{1 \le i < j < k \le n} |A_{ijk}| - \cdots$$

$$+ (-1)^{n-2} \sum_{1 < \ell_1 < \ell_2 < \cdots < \ell_{n-1} \le n} |A_{\ell_1 \ell_2 \cdots \ell_{n-1}}| + (-1)^{n-1} |A_{12 \cdots n}|.$$

If the A_i's live in a finite set and we use the counting measure where each element of our outcome space is equally likely, we may replace all $|S|$ above with $\Pr(S)$.

As this is a long formula, let's take a few minutes and parse it. The left-hand side has just one term, $\cup_{i=1}^{n} A_i$. This is the event that *at least one of the A_i's happens*. What about the events on the right-hand side? $A_i \cap A_j = A_{ij}$ is the event that *both A_i and A_j happen*. Similarly $A_i \cap A_j \cap A_k = A_{ijk}$ is the event that all three of A_i, A_j, and A_k

occur. Thus the point of the Inclusion-Exclusion Principle is to reduce the determination of a union of events to a combination of intersections. While there are a lot more terms on the right-hand side, typically intersections are easier to count than unions.

We've seen a proof of the inclusion-exclusion formula in the simple case of just two events: $\Pr(A \cup B) = \Pr(A) + \Pr(B) - \Pr(A \cap B)$. If $A \cap B$ is empty then the last term on the right doesn't happen, and the probability at least one happens is the sum of the probabilities; however, if the intersection is non-empty then excluding the last term on the right will mean that we've double counted.

Let's do a quick example. Imagine we roll two die (each having the numbers 1, 2, 3, 4, 5, and 6), and we want to know how many rolls have a 1. Let A_1 be the event the first die is a 1 and A_2 the event that the second die is a 1. We want $|A_1 \cup A_2|$. Let's find the various quantities. First, $|A_1| = 6$ as $A_1 = \{(1, 1), (1, 2), (1, 3), (1, 4), (1, 5), (1, 6)\}$. Similarly $|A_2| = 6$ as $A_2 = \{(1, 1), (2, 1), (3, 1), (4, 1), (5, 1), (6, 1)\}$ and $|A_1 \cap A_2| = |A_{12}| = 1$ as $A_{12} = \{(1, 1)\}$. Putting these together, we have

$$|A_1 \cup A_2| = |A_1| + |A_2| - |A_{12}| = 6 + 6 - 1 = 11,$$

and there are exactly 11 rolls of the dice that have a 1. If we wanted to do probabilities, we would find $\Pr(A_1) = \Pr(A_2) = 1/6$, $\Pr(A_{12}) = 1/36$, and

$$\Pr(A_1 \cup A_2) = \Pr(A_1) + \Pr(A_2) - \Pr(A_{12}) = \frac{1}{6} + \frac{1}{6} - \frac{1}{36} = \frac{11}{36}.$$

Inclusion-Exclusion with equally likely sets: In many inclusion-exclusion problems, all the sets A_i have the same size, all the sets $A_i \cap A_j = A_{ij}$ have the same size, all the sets $A_i \cap A_j \cap A_k = A_{ijk}$ have the same size, and so on. This makes the counting much easier, reducing the formula to

$$\left| \bigcup_{i=1}^{n} A_i \right| = n|A_1| - \binom{n}{2}|A_{12}| + \binom{n}{3}|A_{123}| - \cdots + (-1)^{n-1}|A_{12\cdots n}|.$$

In the above note, the binomial coefficients are coming from the number of ways to choose the sets $A_{i_1, i_2, \ldots, i_\ell}$. As this is such an important part of the formula we quickly review the reason (see Remark 5.2.2 for more details). For example, if we look at A_{ij} we need to consider all indices where $1 \leq i < j \leq n$. This is the same as asking how many ways are there to choose two distinct numbers from 1 through n when order doesn't matter (we just make the larger of these two j and the smaller of these two i). The answer is $\binom{n}{2}$, and a similar argument explains the presence of the other binomial coefficients.

5.2.3 Justification of the Inclusion-Exclusion Formula

Our proof of the Inclusion-Exclusion Principle is based on the **Binomial Theorem**, which says

$$(x + y)^n = \sum_{k=0}^{n} \binom{n}{k} x^k y^{n-k};$$

see §A.2.3 for a proof. One of the great things about formulas such as this is that they hold for all x and y; choosing different values leads to different useful expressions.

One great choice is to take $x = y = 1$, which gives $2^n = \sum_{k=0}^{n} \binom{n}{k}$. Notice that if we have n elements then there are exactly $\binom{n}{k}$ subsets of size k. As every subset has some size, if we add up how many there are of each size we have to get the total number of subsets, which is just 2^n (either an element is taken or it isn't for each possible subset).

For the purposes of this section, though, we want $x = -1$ and $y = 1$. Then we get

$$0 = (-1 + 1)^n = \sum_{k=0}^{n} \binom{n}{k}(-1)^k 1^{n-k} = \sum_{k=0}^{n}(-1)^k\binom{n}{k}.$$

We're now ready to prove the claim. We have

$$\left|\bigcup_{i=1}^{n} A_i\right| = \sum_{i=1}^{n}|A_i| - \sum_{1 \le i < j \le n}|A_i \cap A_j| + \sum_{1 \le i < j < k \le n}|A_i \cap A_j \cap A_k|$$

$$+ (-1)^{n-2}\sum_{1 < \ell_1 < \ell_2 < \cdots < \ell_{n-1} \le n}|A_{\ell_1} \cap A_{\ell_2} \cap \cdots \cap A_{\ell_{n-1}}|$$

$$+ (-1)^{n-1}|A_1 \cap A_2 \cap \cdots \cap A_n|.$$

We'll do this by counting how often an element x is counted by each side of the equation. If you want, you can safely skim the argument below to get a flavor of why it's true. We'll do more cases than we need to and in greater detail than required to help drive home the calculation; thus feel free to just read the first case or two and then skip to the general argument.

- *Case 0: The element x is in none of the A_i's.* Since x is in none of the A_i's, it isn't counted in the union, and doesn't contribute to the sum on the left-hand side. Further, it can't be in any of the intersections as it isn't in any of the A_i's, so it doesn't contribute to any of the sums on the right-hand side.

- *Case 1: The element x is in exactly one of the A_i's.* Such elements contribute 1 to the sum on the left-hand side, as it's in the union. How about the right-hand side? Since it's in only one A_i, it isn't in any of the intersections and is only counted in one of the A_i's. For definiteness, imagine x is only in A_3. Then it's counted in the union on the left, and is only counted in the set A_3 on the right.

- *Case 2: The element x is in exactly two of the A_i's.* Now things get a bit more interesting. Our x is counted in the union on the left. What about the right? Let's say $x \in A_3$ and $x \in A_7$, though our argument would hold for any two indices. We then count x once from the set A_3, once from the set A_7, and once from the set $A_3 \cap A_7$. What's the net effect of this? In the end, the element x is counted $1 + 1 - 1 = 1$ time on the right, exactly how often it's counted on the left. Note that x isn't in any other set. For example, it can't be in $A_4 \cap A_7$ as $x \notin A_4$, nor can it be in $A_3 \cap A_7 \cap A_8$ as it isn't in A_8. The only sets it can be in must arise from intersections of A_3 and A_7. There are three such sets we can form: A_3, A_7, and $A_3 \cap A_7$.

- *Case 3: The element x is in exactly three of the A_i's.* As always, x is in the union and therefore is counted on the left-hand side. For definiteness, let's say x is in A_3, A_7, and A_8. Which sets on the right-hand side have x? It's the seven possibilities coming from intersections of $\{A_3, A_7, A_8\}$: A_3, A_7, A_8, $A_3 \cap A_7$, $A_3 \cap A_8$, $A_7 \cap A_8$, and $A_3 \cap A_7 \cap A_8$ (there are seven possibilities and not eight as we must choose at least one set, which eliminates the empty set). The sets A_3, A_7, A_8 are all counted positively, the sets $A_3 \cap A_7$, $A_3 \cap A_8$, $A_7 \cap A_8$ are all counted negatively (they have a minus sign in front), and finally $A_3 \cap A_7 \cap A_8$ is counted positively. Thus the number of times x is counted on the right is $3 - 3 + 1 = 1$.

It's important to realize that there's a better way of writing $3 - 3 + 1$; we can write this as $\binom{3}{1} - \binom{3}{2} + \binom{3}{3}$, where the first summand is the number of ways to choose 1 index from three, the second is the number of ways to choose two indices from 3, and finally the last is the number of ways to choose 3 indices from three. Here's our seven sets, broken into those involving just one of our special indices, then just two, and finally all three.

In fact, there's an even better way to write $3 + 3 - 1$; it equals $(-1)^{1-1}\binom{3}{1} + (-1)^{2-1}\binom{3}{2} + (-1)^{3-1}\binom{3}{3}$.

- *Case k: The element x appears in exactly $k \geq 3$ of the A_i's.* The argument is similar to our previous cases, with just a bit more algebra at the end. Clearly x is counted in the union on the left. What's happening on the right? We might as well assume x occurs in $A_1, A_2, \ldots, A_k$; this is just for convenience–the argument is the same for any set of k indices. There are $\binom{k}{1}$ ways to choose just one index from k indices. Thus the contribution on the right from $\sum_{i=1}^{n} |A_i|$ is $\binom{k}{1} = (-1)^{1-1}\binom{k}{1}$ (we put in the factor $(-1)^{1-1}$ to write this the same way as the later expressions).

There are $\binom{k}{2}$ ways to choose exactly two indices from k. Thus the contribution from the right from $\sum_{1 \leq i < j \leq n} |A_i \cap A_j|$ is $\binom{k}{2}$. Remember, though, that there's a negative sign in front of these summands, so the contribution is really $-\binom{k}{2} = (-1)^{2-1}\binom{k}{2}$.

Continuing, there are $\binom{k}{j}$ ways to choose exactly j indices from k indices, so the contribution from intersections involving exactly j sets is $(-1)^{j-1}\binom{k}{j}$ (remember, we have to keep track of the negative signs on the right-hand side).

We do this for all j from 0 to k, and find the number of times x is counted on the right-hand side is

$$(-1)^{1-1}\binom{k}{1} + (-1)^{2-1}\binom{k}{2} + (-1)^{3-1}\binom{k}{3} + \cdots + (-1)^{k-1}\binom{k}{k}.$$

Since we believe this should equal 1, let's add $-1 + 1$. We're just **adding zero**, which doesn't change the sum but allows us to **rewrite the algebra**; this allows us to see a useful relationship and apply the Binomial Theorem. This is a powerful technique, and one of my favorites (see §A.12 for more on it). We find

$$(-1)^{1-1}\binom{k}{1} + (-1)^{2-1}\binom{k}{2} + (-1)^{3-1}\binom{k}{3} + \cdots + (-1)^{k-1}\binom{k}{k}$$

$$= -1 + (-1)^{1-1}\binom{k}{1} + (-1)^{2-1}\binom{k}{2} + (-1)^{3-1}\binom{k}{3} + \cdots$$

$$\cdots + (-1)^{k-1}\binom{k}{k} + 1$$

$$= -\left[(-1)^0\binom{k}{0} + (-1)^1\binom{k}{1} + (-1)^2\binom{k}{2} + (-1)^3\binom{k}{3} + \cdots\right.$$

$$\left.\cdots + (-1)^k\binom{k}{k}\right] + 1$$

$$= -(-1+1)^k + 1 = 1,$$

where the last line follows from the Binomial Theorem (we used $1 = (-1)^0\binom{k}{0}$ to set up the application of the Binomial Theorem).

As the above argument is long, it's worth going back and seeing what we did. The left-hand side is simple–either an element x is in the union, or it isn't. The right-hand side is more involved. If x is in just one A_i then it's counted in just one set, but if it's in only two A_i's then it's counted in 3 sets, if it's in exactly three A_i's then it's counted in 7 sets, and in general if it's in exactly j of the A_i's then it's counted $2^j - 1$ times. However, sometimes we count it positively and sometimes we count it negatively. When we look at the *net* number of times x is counted, we get the same number on the left- and right-hand sides, proving the formula. The key step in the algebra was to use the Binomial Theorem, noting that $(-1+1)^k = 0$.

5.2.4 Using Inclusion-Exclusion: Suited Hand

 It's time for examples! Let's return to the one-suited hand problem from §3.4.2. We first find the probability of getting a hand where at least one person has a one-suited hand (so we deal 13 cards to each of 4 people, and want at least one of them to have all their cards in the same suit). The main idea of inclusion-exclusion is the following: *Frequently it is easier to do a lot of simple calculations than one complicated calculation.* We'll see it's a lot easier to count the number of ways that a subset of people are one-suited, regardless of the others, than to calculate the number of ways that just a single person is one-suited; we use the former and inclusion-exclusion to compute the latter.

The first step is to choose our events. We'll let A_1 be the event that the first person has a one-suited hand, regardless of what the other three people have. We define A_2, A_3, and A_4 similarly. Note we're not saying *what* suit someone has when they're one-suited, just that it's a one-suited hand. The event $A_1 \cap A_2 = A_{12}$ is then person 1 and person 2 *both* being one-suited, with nothing claimed about the hands of the third and fourth persons. Here's where things get interesting. Technically, the event $A_1 \cap A_2 \cap A_3$ means the first three people are each one-suited, but of course that forces the fourth person to be one-suited as well!

Okay, let's now find the number of hands with at least one person one-suited. What's nice, and this is often the case in inclusion-exclusion problems, is that $|A_1| = |A_2| = |A_3| = |A_4|$, and $|A_{12}| = |A_{13}| = \cdots = |A_{34}|$ and so on. Claims like this follow from symmetry. The number of ways for players 1 and 3 to have something special and then 2 and 4 being arbitrary is the same as 2 and 4 being special and 1 and 3 being arbitrary.

What is $|A_1|$? There's $\binom{4}{1}$ ways to choose the suit for the first person's hand, and then $\binom{13}{13} = 1$ way to give them the 13 cards in that suit. We now have to assign the remaining cards to the other people. We have $\binom{39}{13}$ ways to give the second person

13 cards from the remaining 39, $\binom{26}{13}$ ways to give the third person their 13 cards from the 26 that are left, and finally $\binom{13}{13} = 1$ way to give the fourth person their hand. Thus,

$$|A_1| = \binom{4}{1}\binom{13}{13}\binom{39}{13}\binom{26}{13}\binom{13}{13} = 337{,}912{,}392{,}291{,}465{,}600.$$

Note that the above allows other people to be one-suited as well! This is okay, and in fact is what makes inclusion-exclusion work so easily. The event A_1 is simply the first person is one-suited, regardless of what happens with the other three people. This is a significantly easier calculation than determining the number of ways that only the first person is one-suited, and in fact we use this simpler calculation of just person one's hand to figure out the number of ways to have just the first person one-suited.

What about $|A_i \cap A_j| = |A_{ij}|$? We have two special people who have to receive all their cards in just one suit. There's $4 \cdot 3$ ways to assign the two suits to these two people: we have 4 choices for the first and then 3 for the second. Another way to look at this is we have $\binom{4}{2}$ ways to choose two suits and then 2! ways to order it among the two people, and $\binom{4}{2}2! = 4 \cdot 3$. Once we've specified who gets which suit, there's $\binom{13}{13} = 1$ way to give the first person their hand and $\binom{13}{13} = 1$ way to give the second person their cards. We now have 26 cards left; there's $\binom{26}{13}$ ways to give 13 cards to the third player, and then only $\binom{13}{13} = 1$ way to give the fourth their hand. Thus,

$$|A_{12}| = 124{,}807{,}200.$$

We now move on to triples, $A_i \cap A_j \cap A_k = A_{ijk}$; however, once three people are one-suited then the fourth must be one-suited too. We could go through the same choosing calculation as before, or we could just note that there's 4! hands that work (four possibilities for the first person, then 3 for the second, 2 for the third, and finally 1 for the fourth, giving $4 \cdot 3 \cdot 2 \cdot 1 = 4! = 24$). Thus

$$|A_{123}| = |A_{124}| = |A_{134}| = |A_{234}| = |A_{1234}| = 24.$$

We now use the inclusion-exclusion formula to find the number of hands with at least one person one-suited. Remember, all the probabilities below only depend on the *number* of the indices, not which ones we have; this is a great boon in calculating the answer. It's just

$$\left| \bigcup_{i=1}^{4} A_i \right| = \sum_{i=1}^{4} |A_i| - \sum_{1 \le i < j \le 4} |A_{ij}| + \sum_{1 \le i < j < k \le 4} |A_{ijk}| - |A_{1234}|$$

$$= |A_1| \sum_{i=1}^{4} 1 - |A_{12}| \sum_{1 \le i < j \le 4} + |A_{123}| \sum_{1 \le i < j < k \le 4} - |A_{1234}|$$

$$= |A_1| \cdot 4 - |A_{12}| \cdot \binom{4}{2} + |A_{123}| \cdot \binom{4}{3} - |A_{1234}| \cdot 1$$

$$= |A_1|4 - |A_{12}|6 + |A_{123}|4 - |A_{1234}|$$

$$= 1{,}351{,}649{,}568{,}417{,}019{,}272.$$

There are some other interesting examples of the Inclusion-Exclusion Principle. Given some equation like $x + y = 12$, you could use the Inclusion-Exclusion Principle to determine how many different ways to distribute 12 among the two variables if x and y both have to be positive integers. You could restrict the values to make them be greater than or less than certain values, or you could say that they cannot be equal. For example, suppose we say $x \leq 4$ and $y \geq 6$, then we could figure out how many solutions there are by first calculating the number of possible solutions, then subtracting the ones with $x \geq 5$ and the ones with $y \leq 5$. Since there are some solutions where both $x \geq 5$ and $y \leq 5$, we have subtracted those solutions twice, and we need to add them back in. There are probably better ways to solve a problem like this, and usually you are more concerned with what the solution is rather than how many there are, but there are times when you just want to know how many solutions there are to a given equation. While the Inclusion-Exclusion Principle may not always be pretty, it always works.

An important component of the Inclusion-Exclusion Principle is properly choosing your sets. Choosing your A_i's wisely will make a problem much easier to solve. In the above example, if we had not seen that we could subtract all of the solutions that violated our constraints and instead tried to calculate all of the solutions that fit, the cases would be more complicated. Let's do one final example to illustrate this.

Consider a group of students, Alice, Bob, Charlie, and David, who will be giving presentations. Alice does not want to be the first person to present, and David does not want to be the last person to present. How many different ways can the presentations be ordered? Now, without restrictions, there are $4! = 24$ ways to arrange the students. We let A_i be the occurrence of putting person i into a position that they are not allowed in. So, A_1 is the number of ways that Alice can be put into position 1, which is $3! = 6$. $A_2 = 0$ and $A_3 = 0$ because there are no restrictions on where Bob and Charlie can be ordered. Finally, A_4 is the number of orderings that will result in David going last, which is $3! = 6$. Now, some orders will result in Alice going first and David going last, and we have subtracted these twice, so we need to add them back in. The number of ways for Alice to go first and David to go last is $2! = 2$. Since there are no restrictions on Bob and Charlie, we do not need to consider any other overlapping sets. So, we combine these values with the Inclusion-Exclusion Principle to see that $24 - (6 + 0 + 0 + 6) + 2 = 14$ and we have 14 possible orders for the presentations. Our choices of sets made this problem fairly straightforward.

5.2.5 The At Least to Exactly Method

We found the number of ways at least one person is one-suited; as the number of ways to assign the cards is $\binom{52}{13}\binom{39}{13}\binom{26}{13}\binom{13}{13}$, we just have to divide our number by this to get the probability of at least one person being one-suited.

Imagine now that we want the probability of *exactly* one person being one-suited. For all practical purposes, it's the same as at least one person being one-suited, as it's so unlikely to have two or more people one-suited; however, as mathematicians we want *exact* answers.

We can find this by a combination of the inclusion-exclusion formula and a powerful principle. As everything should have a name so we can refer to it easily, let's call it the **At Least to Exactly Method**.

> **At Least to Exactly Method**: Let $N(k)$ be the number of ways for *at least* k things to happen, and let $E(k)$ be the number of ways for *exactly* k things to happen. Then $E(k) = N(k) - N(k+1)$. Equivalently,
>
> $$\text{Prob(exactly } k \text{ happen)}$$
> $$= \text{Prob(at least } k \text{ happen)} - \text{Prob(at least } k+1 \text{ happen)}.$$

Don't forget this simple principle—it allows us to painlessly pass from knowledge of *at least* events to knowledge of *exactly* events. We'll see it again as it's the basis of the powerful cumulative distribution method (see Chapter 7 for the next occurrence).

We now apply this to find the number of deals with exactly one person being one-suited. We'll do this by first finding the number of ways to have at least two people one-suited, and then subtract this from the number of ways to have at least one person one-suited.

Let $A_{12}, A_{13}, A_{14}, A_{23}, A_{24}, A_{34}$ be the six different events of two people each being one-suited. We're using slightly different notation than before as this notation is more suggestive. By looking at A_{14}, we know this means persons 1 and 4 are one-suited; if instead we labeled these six events as $B_1, B_2, \ldots$ then it wouldn't be clear what each event indicated (and this will become very important in a moment). The notation can easily become messy in problems like this; instead of explicitly writing the union of these events we'll just say the union of the events.

If we want to use the inclusion-exclusion formula to find the probability of the union of these six events (in other words, the number of ways to have at least two people be one-suited), we need to figure out the counts of certain events.

- The first are the six events A_{ij} (with $i \neq j$). Fortunately all six of these events occur the same number of times (as we computed previously), which is

$$|A_{12}| = |A_{13}| = |A_{14}| = |A_{23}| = |A_{24}| = |A_{34}| = 124,807,200.$$

The contribution from these six sets is therefore

$$6 \cdot 124,807,200 = 748,843,200.$$

- We now come to the hard calculation. What's $|A_{ij} \cap A_{\ell m}|$? The answer depends on whether or not i, j, ℓ, m are four distinct indices, three distinct indices, or two distinct indices (remember $i \neq j$ and $\ell \neq m$, but ℓ or m may be either i or j). If we have four distinct indices, then this is just the event A_{1234}, and we know there are only 24 such hands. What if there are three distinct indices? This means that three people were dealt one-suited hands, which forces the first person to be one-suited too. Thus if there are three distinct indices then again it's the event A_{1234}, and again there are only 24 such possibilities. What if there are only two distinct indices, say $A_{12} \cap A_{12}$? This is the simplest intersection one can ask for: it's just itself, and we know $|A_{12}| = 124,807,200$.

Okay, we know the size of the possible intersections; now we need to figure out how many of each we have. There are $\binom{6}{2} = 15$ ways to choose two of the six events $A_{12}, A_{13}, A_{14}, A_{23}, A_{24}, A_{34}$. Of these 15 ways, 3 have all four indices distinct: choose A_{12}, A_{34} or A_{13}, A_{24} or A_{14}, A_{23}. It would be a mistake to argue that there are six ways since we can choose the first set any way and then there is exactly one choice for the second. Why is this wrong? It creates an *order* among the sets. The best way to see there are only three ways is to choose an *ordered* set, and then remove the order by dividing by 2! (the number of ways to order two elements). The number of ordered pairs with all indices distinct is $3 \cdot 2 = 6$ (we may choose any of the six for the first set, and then the second set is forced); thus the number of unordered pairs is $6/2! = 3$. As we have to choose two distinct sets from the six, there's no way to have exactly two distinct indices. Thus the remaining 12 must all have exactly three distinct indices, and again this then forces all the indices to be distinct. This is great—all 15 intersections have the same size, $4! = 24$. Therefore, the contribution from the intersection of pairs is just $-15 \cdot 4! = -360$.

- We now look at intersections of three distinct sets from $\{A_{12}, A_{13}, A_{14}, A_{23}, A_{24}, A_{34}\}$. There are $\binom{6}{3} = 20$ ways to choose three of these sets. No matter which three sets we choose, we must have at least three distinct indices (the easiest way to see this is we already had at least three distinct indices when we only chose two sets!). Thus all 20 of these sets have three people being one-suited, which immediately implies the fourth person is one-suited. Thus, each of the 20 sets has 24 elements, for a contribution of $20 \cdot 24 = 480$.

- We now look at intersections of four distinct sets. There are $\binom{6}{4} = 15$ ways to choose 4 sets from six, and each intersection is just A_{1234} (we have all indices), so the contribution from this case is $-15 \cdot 24 = -360$.

- Next are intersections of five sets. There are $\binom{6}{5} = 6$ ways to do this. We again get A_{1234} each time, for a total contribution of $6 \cdot 24 = 144$.

- Finally, we have the intersection of all six sets. There's $\binom{6}{6} = 1$ way to do this, and again we get A_{1234}, for a contribution of $-1 \cdot 24 = -24$.

Putting the pieces together, the inclusion-exclusion formula gives us the probability of at least two people being one-suited (remember this is the union of the six events) is

$$748, 843, 200 - 360 + 480 - 360 + 144 - 24 = 748, 843, 080.$$

We can now find the number of hands where exactly one person is one-suited. In §5.2.4 we showed the number of hands with at least one person being one-suited is 1,351,649,568,417,019,272; we've just showed that the number of hands with at least two people being one-suited is 748,843,080. Thus the number of hands where exactly one person is one-suited is 1,351,649,567,668,176,192.

Remember, if it's possible to check an answer, check the answer. In this case, there's a nice check we can do for $748, 843, 200 - 360 + 480 - 360 + 144 - 24$. The first number is 748,843,200, which is $|A_{12}| + |A_{13}| + |A_{14}| + |A_{23}| + |A_{24}| + |A_{34}|$. If a deal has three people being one-suited then in fact all four are one-suited. There are 24 such deals, and each deal is included in the sum of the sizes of these six sets. Thus, this sum is the total number of hands with at least two people being one-suited *except* that the 24 hands that have everyone one-suited are counted six times instead of once. We must therefore subtract $5 \cdot 24 = 120$. Looking back to our original expression, we notice that $-360 + 480 - 360 + 144 - 24 = -120 = -5 \cdot 24$. It's terrific that we are able to interpret the algebra in a slightly different way and see this agreement.

5.3 Derangements

If we have n objects, there are $n!$ ways to arrange them with order mattering. There are lots of problems where we care about arrangements with special properties. For example, imagine that we have n people. We address n envelopes and n letters. Unfortunately, the person who's supposed to put the letters in the envelopes is having a bad day, and just puts the letters in willy-nilly. How many ways can they put in a letter so that *no* letter is placed in the correct envelope? This is called a **derangement** and is typically considered a bad event! If we can find this then we can answer the related question of how many arrangements have at least one letter in the correct envelope (it's just $n!$ minus the number of derangements).

Of course, we can do more than count; we can find probabilities. If each of the $n!$ arrangements is equally likely, then the probability of a derangement is just the number of derangements divided by $n!$. We'll see later in this section that this has a beautiful limit as $n \to \infty$, and then discuss some applications.

5.3.1 Counting Derangements

So, how many of the $n!$ orderings have no element returned to where it starts? This means the 1st element cannot be in the first spot, nor the 2nd element in the second spot, and so on. For example, $\{2, 3, 4, 1\}$ is a derangement as each number is moved, while $\{3, 2, 4, 1\}$ is not a derangement as 2 is in the second position.

It turns out to be much easier to look at the related problem, where we count how many ways there are for at least one element to return to its starting point. Why is this easier? Remember the statement of the Inclusion-Exclusion Principle (see §5.2.2). We show how to write an *at least* event in terms of intersections of events, and intersections are often easy to compute. To get the number of derangements, we just subtract the number of non-derangements from $n!$.

Whenever using inclusion-exclusion, one of the earliest decisions is what to make the events. We're trying to find the number of ways at least one element is not moved. Let's make A_i be the event that i isn't moved. Then $A_1 \cup A_2 \cup \cdots \cup A_n$ is the event that at least one item occurs in the same place in the rearrangement. This is precisely what we want to compute, and by the inclusion-exclusion formula its cardinality equals

$$\left| \bigcup_{i=1}^{n} A_i \right| = \sum_{i=1}^{n} |A_i| - \sum_{1 \le i < j \le n} |A_i \cap A_j| + \cdots + (-1)^{n-1} |A_1 \cap \cdots \cap A_n|.$$

Fortunately, this expression simplifies. The n sets A_i all have the same size, $|A_1|$. The reason is symmetry—we can always relabel and assume it's the first element that's not moved. Similarly the $\binom{n}{2}$ sets $A_i \cap A_j$ all have the same size, $|A_i \cap A_j| = |A_{12}|$. Continuing, we find the $\binom{n}{3}$ sets $A_i \cap A_j \cap A_k$ have the same size, $|A_i \cap A_j \cap A_k| = |A_{ijk}|$, and of course similar results hold for the rest. Thus

$$\left| \bigcup_{i=1}^{n} A_i \right| = n|A_i| - \binom{n}{2}|A_{ij}| + \cdots + (-1)^{n-1}|A_{12\cdots n}|.$$

We now have to compute the various quantities on the right-hand side. Fortunately, they're not too bad. When we find these numbers below, we spend a little time writing the result in an algebraically useful way. The pattern shouldn't be clear at first, but by the end we'll have an "a-ha" moment.

- $|A_1| = (n-1)!$. Why? The first element is fixed, and thus in our new arrangement it must be in the first position. We now have $n-1$ objects that must be permuted, and there are $(n-1)!$ ways to order these $n-1$ elements. There are $n = \binom{n}{1}$ such sets. Note $n \cdot (n-1)! = n!$.

- $|A_{12}| = (n-2)!$. The reasoning is similar to above. The first two elements are fixed. We have $n-2$ objects left, and the number of ways to order $n-2$ elements is $(n-2)!$. There are $\binom{n}{2}$ such sets. Note

$$\binom{n}{2} \cdot (n-2)! = \frac{n(n-1)}{2} \cdot (n-2)! = \frac{n!}{2}.$$

- $|A_{12\cdots j}| = (n-j)!$. To see this, observe that the first j are taken care of. We're left with $n-j$ elements, and so the answer is $(n-j)!$. There are $\binom{n}{j}$ such sets. Note

$$\binom{n}{j} \cdot (n-j)! = \frac{n(n-1)\cdots(n-(j-1))}{j!} \cdot (n-j)! = \frac{n!}{j!}.$$

It sometimes takes awhile to see the pattern. The first gave us $n!$ and the second $n!/2$; it's only after doing the general case (where we got $n!/j!$) do we see that we should write the first as $n!/1!$ and the second as $n!/2!$. Don't worry—you're not expected to see the first as $n!/1!$ on the first pass; however, once you see $n!/j!$ *then* you should go back and try to see if there's a general pattern. This also suggests that we write the final term, arising from $|A_{12\cdots n}|$, as $n!/n!$ instead of 1.

Collecting, the inclusion-exclusion formula gives

$$\left| \bigcup_{i=1}^{n} A_i \right| = \binom{n}{1}|A_1| - \binom{n}{2}|A_{12}| + \cdots + (-1)^{n-1}\binom{n}{n}|A_{12\cdots n}|$$

$$= n(n-1)! - \binom{n}{2}(n-2)! + \cdots + (-1)^{j-1}\binom{n}{j}(n-j)! + \cdots$$

$$\cdots + (-1)^{n-1}\binom{n}{n}(n-n)!$$

$$= \frac{n!}{1!} - \frac{n!}{2!} + \frac{n!}{3!} - \frac{n!}{4!} + \cdots + (-1)^{n-1}\frac{n!}{n!}.$$

So, we've found our formula counting the number of non-derangements. The number of derangements is thus $n!$ minus this. Letting D_n denote the number of derangements of $\{1, 2, \ldots, n\}$, we find

$$D_n = n! - \frac{n!}{1!} + \frac{n!}{2!} - \frac{n!}{3!} + \frac{n!}{4!} + \cdots + (-1)^n\frac{n!}{n!}.$$

In the next subsection we'll see that this formula can be rewritten in a more illuminating manner.

5.3.2 The Probability of a Derangement

In §5.3.1 we used inclusion-exclusion to count how many derangements there are of the set $\{1, 2, \ldots, n\}$; remember a derangement is a reordering of these numbers so that each number moves to a position that it wasn't originally in. Thus $\{3, 4, 2, 1, 5, 7, 6\}$ is not a derangement of $\{1, 2, 3, 4, 5, 6, 7\}$ as 5 is still the fifth element; however, $\{2, 3, 1, 7, 6, 5, 4\}$ is. We saw the number of derangements of n objects, which we'll denote by D_n, is

$$D_n = n! - \frac{n!}{1!} + \frac{n!}{2!} - \frac{n!}{3!} + \frac{n!}{4!} - \cdots + (-1)^{n-1}\frac{n!}{n!}.$$

We're now going to spend some time and see a nicer way of viewing this formula.

The first thing to note is that each term has an $n!$, so we should pull that out. We get

$$D_n = n! \left(1 - \frac{1}{1!} + \frac{1}{2!} - \frac{1}{3!} + \frac{1}{4!} - \cdots + (-1)^n \frac{1}{n!}\right).$$

We're making real progress here, as $n!$ is not any old number but rather a *very* relevant quantity: it's the number of permutations of n objects! Thus, if we divide both sides by $n!$, we get the fraction of all permutations that are derangements:

$$\frac{D_n}{n!} = 1 - \frac{1}{1!} + \frac{1}{2!} - \frac{1}{3!} + \frac{1}{4!} - \cdots + (-1)^n \frac{1}{n!}.$$

If you've seen calculus, the right-hand side should look a lot like an old friend, the series expansion for e^x. The **exponential function** is one of the most important in all of mathematics (see §B.5 for a quick review). One definition of it is

$$e^x = \sum_{k=0}^{\infty} \frac{x^k}{k!} = 1 + x + \frac{x^2}{2!} + \frac{x^3}{3!} + \cdots.$$

What we have is *almost* e^{-1}; the difference is that we only have finitely many terms.

This isn't a huge deal. The series expansion for e^x converges very rapidly. For instance, if we take $n = 10$ then e^{-1} is approximately 0.3678794412, while $D_n/n!$ is about 0.367879464, a difference of around .000000023. In other words, when n is 10 we're already very close to e^{-1}.

What does all this mean? It means that if we take n elements and consider a random permutation, the probability that *no* element is left unchanged is quite high: it's about $1/e$, or approximately 36.79%. More than one out of every three permutations have every number moving! Later in the book we'll return to this problem and tackle related questions, such as what's the expected number of elements that don't move in a random permutation.

5.3.3 Coding Derangement Experiments

As always, it's worth spending a few moments to write some code to check our theory. It's a bit more complicated here as the probability of a derangement is a function of n, so we have to remember that it's only $1/e$ in the limit.

```
derangement[n_, numdo_] := Module[{},
  count = 0; (* set counter of successes to 0 *)
  (* prediction for finite n and n --> oo *)
  theory = Sum[(-1)^k/k!, {k, 0, n}];
  limit = 1/E;
  people = {}; (* this will be our list of people *)
  For[i = 1, i <= n, i++, people = AppendTo[people, i]];
  For[m = 1, m <= numdo, m++, (* main loop *)
   {
    mix = RandomSample[people]; (* randomly mix people *)
    found = 0; (* set found to 0, if becomes 1 someone fixed *)
    For[i = 1, i <= n, i++,
     {
      If[mix[[i]] == i,
       {
        found = 1;
        i = n + 1; (* exit loop: why keep computing! *)
        }];
      }];
    If[found == 1, count = count + 1]; (* if found=
    1 increase count *)
    }];
  Print["Theory is ", 100.  theory, "%."];
  Print["Limit is ", 100.  limit, "%."];
  (* we want prob derangement so it's 1 - prob someone fixed *)
  Print["Observe ", 100. - 100. count/numdo, "%."];
  ]
```

For example, if we run ten million trials with $n = 5$ we find:

```
Theory is 36.6667%.
Limit is 36.7879%.
Observe 36.6783%.
```

The reason we had to run so many trials is that the prediction for finite n and the $n \to \infty$ limit differ by a small amount when $n = 5$; later using the Central Limit Theorem you'll learn how to construct confidence intervals around simulated values and see if they support the prediction. The situation is very different when $n = 20$, as there is negligible difference between the finite n and the limit:

```
Theory is 36.7879%.
Limit is 36.7879%.
Observe 36.8023%.
```

5.3.4 Applications of Derangements

We've counted how many derangements of $\{1, \ldots, n\}$ there are, and we've found a nice formula for the probability of getting a derangement (the formula is so nice that we can deduce the limiting behavior as n tends to infinity). What we haven't done yet is explain why we care about them.

There's a lot of places in advanced math where derangements pop up, but that feels like cheating. Instead, let's discuss an example from communication theory. If you've never done **graph theory** before, here's a very quick crash course! A **graph** is a collection of points, called the **vertices** (singular **vertex**) and a set of **edges** (each edge connects exactly two of the vertices). If the graph is **simple** then we have at most one edge between any pair of vertices; a repeated edge is called a **compound edge**. Frequently we disallow **self-loops**, which means each edge must connect two distinct vertices. We say a graph is **bipartite** if we can split its vertices into two sets, A and B, such that each edge connects a vertex in A with one in B (so we have no edges between vertices in A, or between vertices in B).

Many problems can be reformulated to involve graphs. A terrific example is in communication theory. Think of each vertex as a computer, and the edges are connections between the computers (alternatively, you could think of the vertices as cities, and the edges roads). We want a well-connected system: we want any computer to be able to quickly reach any other computer in the network. One way to do this is to connect each computer with every other; we call the resulting graph the **complete graph**, as it contains all possible edges. Unfortunately, this is also very expensive. If we have n computers, we need $\binom{n}{2} = n(n-1)/2$ edges to get all the connections. While this does a great job for connectivity, often it's prohibitively expensive.

This has led people to search for networks with far fewer edges, but still great connectivity. For example, in the complete graph it takes only 1 step to get from any vertex to any other, but the cost is having $n(n-1)/2$ edges. Typically, a great trade-off is to have the number of edges grow linearly in n and the number of steps to get from any vertex to any other to grow like $\log n$.

Again, our aim here is *not* to develop the theory (see for example [HJ] for more reading), but to alert you to what's out there. Amazingly, while there are some deep, explicit constructions of these desirable graphs using properties from number theory, it turns out almost all graphs do a great job.

There's some nice algorithms to create random bipartite graphs. Let

$$A = \{1, 2, \ldots, n\}$$

be one set of n vertices, and $B = \{1, 2, \ldots, n\}$ another. If we choose a random permutation of $B = \{1, 2, \ldots, n\}$, we then connect vertex i in A with the vertex in B corresponding to what's in the ith position of our permutation. For example, if $n = 5$ and the permutation is $(4, 2, 3, 5, 1)$, then we connect 1 in A with 4 in B, 2 in A with 2 in B, and so on.

If we take a few random permutations, we actually get a pretty good graph. It's cheap (not too many edges), but is typically well-connected. The problem is we want our graph to be simple—we don't want to have multiple edges between the same two vertices. The reason is that once two vertices are connected then they're connected, and the edges are better used elsewhere. In computer science, if you have an algorithm you're frequently concerned with how efficient it is or, in other words, how long it takes to run.

In our situation, every time we take a new permutation we can keep it *only* if it doesn't introduce a compound edge. Thus, derangements are *good* for us, and we want to know how likely we are to get one. Of course, it's more complicated than just looking for derangements. We need to make sure the permutations are pairwise derangements. If we only need two permutations this isn't too bad; however, as each permutation must shuffle elements relative to not just the initial ordering but from all the other permutations as well the problem quickly becomes challenging.

For those who are still struggling to see how useful derangements are, consider the following examples. Say there are n people at a restaurant and they all happen to be wearing the same coat. Suppose that they all take their coats off and store them in the closet until they are finished eating. An obvious, and important, question to ask is what is the probability that nobody leaves the restaurant with their own coat if they just grab a coat randomly as they go. Similarly, you could ask what is the probability that at least one person leaves with their own coat. This is a perfect example of derangements. It does not happen too often, and usually not with large numbers of people, but there are occasions where people take the wrong coat. To make this a more common example, you could replace the coat with a hat or an umbrella (if you're a fan of the TV show *How I Met Your Mother*, make it a yellow umbrella), items that are more likely to look alike. What is the probability that you will end up with someone else's luggage after a flight? (This is why a lot of people put some colorful bling on their suitcases!) These are all rather basic and insignificant examples, but they show how these concepts can be applied in your daily life.

5.4 Summary

As you continue with your education, you'll have more and more facts thrown at you. While it's tempting to try to memorize all of them for exams, that's a bad policy for life. I'm obviously not recommending you forget everything, but it's so easy to look up facts now. A few minutes browsing the Web can yield treasure troves of facts and relations. What you want to master is how to apply these facts to build theories.

This is why I'm emphasizing the techniques. If you remember the methods, you can re-derive the facts quickly if needed. Thus, you shouldn't memorize how we determined the number of poker hands or derangements, but instead you should master the statement of inclusion-exclusion, so you can use it when needed.

Of course, this is easier said than done. There are usually three major stages to solving problems.

- *Understand the question.* Sometimes the problem is clearly stated, but often there are several different interpretations, and you have to think and decide which is meant. Make sure you understand the definitions of all the terms.

- *Determine an attack.* Frequently the terms in the problem provide clues on how to proceed. For example, we've seen how different phrases suggest different methods. We've remarked you need to know the definitions of all terms. Additionally, you should know what theorems use these terms, as this gives you a starting point for your attack.

- *Execute the attack.* Once you've decided on a method, you now have to do it. If you're going to proceed by inclusion-exclusion, you have to decide what your sets are, and determine their probabilities.

Looking at a problem and a blank piece of paper can be very intimidating. I've known many students who don't want to try anything until they're sure it'll work, and end up just staring and staring. Don't be afraid to try something. If it works, great; if not, try something else. Give a method a chance, but if it's not working don't be afraid to set it aside for awhile and try another approach.

One last piece of advice: make sure you consider all possibilities. Probability is famous for having some pesky problems with subtle counting issues; it's easy to double count some items and ignore others. Make sure you take everything into account and cover all the cases. To drive the point home, we'll end with an amusing scene from *The Princess Bride*. Vizzini has captured Buttercup, and is going to kill her; the Man in Black is pursuing Vizzini and will stop at nothing to save her. Vizzini, with a dagger at Buttercup's throat, warns him that if he takes another step then she dies. The Man in Black takes two wine goblets. Hiding them from Vizzini, he places the poison in one and then places a glass in front of each of them.

> *Man in Black:* All right. Where is the poison? The battle of wits has begun. It ends when you decide and we both drink, and find out who is right . . . and who is dead.
>
> *Vizzini:* But it's so simple. All I have to do is divine from what I know of you: are you the sort of man who would put the poison into his own goblet or his enemy's? Now, a clever man would put the poison into his own goblet, because he would know that only a great fool would reach for what he was given. I am not a great fool, so I can clearly not choose the wine in front of you. But you must have known I was not a great fool, you would have counted on it, so I can clearly not choose the wine in front of me.
>
> *Man in Black:* You've made your decision then?
>
> *Vizzini:* Not remotely.

The scene goes on (and on and on) as Vizzini gives a long chain of reasoning in his attempt to determine which glass has the poison. He keeps looking for all possible items that could be relevant. In the end, he distracts the Man in Black by saying he sees something. When the Man in Black turns, Vizzini quickly switches the glasses, and then drinks from the one now in front of him. As the Man in Black doesn't make a move to avoid drinking, Vizzini is convinced that he's tricked him and has the safe glass. Unfortunately for him, Vizzini makes one of the classic blunders in probability. He forgets one possibility, and dies.

> *Buttercup:* And to think, all that time it was your cup that was poisoned.
>
> *Man in Black:* They were both poisoned. I spent the last few years building up an immunity to iocane powder.

Ah, the dangers of missing a case!

5.5 Exercises

Exercise 5.5.1   *If two teams are equally likely to win each game, what is the probability a series of 7 games takes at least 5 games to be decided (that is, before 1 team wins 4 games)? What is the probability the series takes exactly 5 games to be decided?*

Exercise 5.5.2   *How many positive integers less than or equal to 100 are not divisible by 2, 3, or 11?*

Exercise 5.5.3   *Prove that a randomly selected positive integer chosen from $\{1, 2, 3, \ldots, n\}$ is a perfect square with vanishing probability as $n \to \infty$. For the brave: what if you replace "perfect square" with "prime number"?*

Exercise 5.5.4   *How many ways can we arrange ten distinct objects numbered 1–10 so that 5 is not in either of the first two spots and 10 is not in either of the last two spots?*

Exercise 5.5.5 *Using the Method of Inclusion-Exclusion, count how many hands of 5 cards have at least one ace. You need to determine what the events A_i should be. Do not find the answer by using the Law of Total Probability and complements (though you should use this to check your answer).*

Exercise 5.5.6 *One of the greatest applications of the Method of Inclusion-Exclusion is the computation of the sum of reciprocals of the twin primes; discrepancies between results on computers with different processors led to the discovery of the Intel Pentium Bug, and huge financial costs to the company. Read up about this ([Ni1, Ni2] are good starts) and write a short review.*

Exercise 5.5.7 *Imagine that you are in class with 20 students. On the first day the professor decides that each student should shake hands and introduce him or herself to every other student, how many hand shakes must happen for this to occur?*

Exercise 5.5.8 *In the previous exercise, it is tempting to say that the number of introductions is equal to the number of ways to group the 20 students into ten pairs, since we could imagine that we pair the students and then have them introduce themselves to their partners and we look at all possible pairings. This gives $\frac{20!}{10! \cdot 2^{10}} = 654,729,075$ introductions, which is far above the actual number. Where is the mistake in this argument?*

    For the next two exercises, consider a combination lock with 4 circular dials, all of which must be turned to the unique pass code for the lock to open. Each dial has the number 0 through 9 on it.

Exercise 5.5.9 *Find the number of possible lock combinations.*

Exercise 5.5.10 *If we define distance to be the minimum of the sum of the number of units we must turn each of the four dials to reach the code, find the average distance between a randomly set lock combination and the pass code.*

Exercise 5.5.11 *Imagine we have beads, each of which has a letter on it. We have a different letters, and a lot of each bead (more precisely, more than p of each bead). We want to make them into bracelets, but since we enjoy math so much, we want to make sure each bracelet has p different beads on it, where p is a prime number. Find the number of distinct bracelets that can be formed.*

Exercise 5.5.12 *Use the result from the previous exercise to prove Fermat's Little Theorem, $a^p \equiv a \pmod{p}$ where a is an integer and p is prime.*

Exercise 5.5.13 *How many distinct, simple graphs can be drawn with n vertices given that each edge is undirected (that is, given vertices a and b, the edge (a, b) is equivalent to (b, a))?*

Exercise 5.5.14 *How many distinct graphs can be formed from n vertices if the edges are directed (that is, given vertices a and b, the edge (a, b) is distinct from (b, a))?*

Exercise 5.5.15 *The degree of vertex a, written $d(a)$, is equal to the number of edges that connect to that vertex. Loops count twice. Prove that for a graph with vertices $a_1, a_2, \ldots, a_n$ $\sum_{i=1}^{n} d(a_i) = 2E$, where E is the number of edges in the graph.*

Exercise 5.5.16 *One of the best known problems in graph theory is the Traveling Salesman problem, which asks given a set of cities what is the shortest distance a* *traveling salesman could take so that he visits every city at least once? This optimization problem turns out to be incredibly difficult to find answers to in practice. How many possible routes could the traveling salesman take between n cities so that he visits each city exactly once?*

Exercise 5.5.17 *The inclusion-exclusion formula gives an exact expression for the probability of a union in terms of sums and differences of probabilities of intersections. Prove that if we truncate the sum-difference of the intersections after a positive term (i.e., after we have just added terms) that we have an upper bound for the probability of the union; similarly show that if we truncate after a negative term (i.e., after we have just subtracted terms) that we have a lower bound for the probability of the union. This is a very important result, as it allows us to get upper and lower bounds for probabilities, and frequently in practice the later terms in the formula have negligible contributions.*

Exercise 5.5.18 *As the number of objects to be rearranged goes to infinity, what is the probability that at least one object is returned to its starting spot? What is the probability half the objects return to their starting places?*

Exercise 5.5.19 *How many ways are there to order 8 people (four wives and their four husbands) along a circle, assuming that all that matters is the relative ordering and that no wife sits next to a wife and no husband sits next to a husband?*

Exercise 5.5.20 *How many ways are there to order 8 people (four wives and their four husbands) along a circle, assuming that all that matters is the relative ordering and that no person sits next to their spouse?*

Exercise 5.5.21 *Consider a table that is a regular n-gon, with a seat at each vertex and then two more on each side, giving us a total of 3n seats. How many ways are there to sit 3n people at this table, if all that matters is the relative ordering?*

Exercise 5.5.22 *Consider a circular table with n seats. How many ways are there to sit $m \le n$ people at the table, if all that matters is the relative ordering of the people? If you don't see the answer immediately, try special cases of n and m and see if you can find a pattern.*

Exercise 5.5.23 *How many arrangements of $\{1, 2, 3, 4\}$ have two numbers in their initial position? Have exactly two numbers in their initial position? Have at most one number in its initial position?*

Exercise 5.5.24 *The number of derangements of n objects can be written n!. Prove the recurrence relation n!=(n-1)(!(n-1))(!(n-2)). (Hint: Proof by Story is a really good choice!)*

Exercise 5.5.25 *Write code to verify the probability of a derangement for large sets via simulation.*

Exercise 5.5.26 *Plot the proportion of permutations of n objects that are derangements for different n and comment on how this number converges to its limiting behavior.*

Exercise 5.5.27 *Write code that can be used to find an approximate solution to Conway's napkin problem for a given probability of preferring the left napkin p and number of people n.*

Exercise 5.5.28   *If King Arthur, his wife Guinevere, and his 4 knights are seated around a table, how many ways can they be arranged if one of the knights, Lancelot, is always seated across from Guinevere?*

Exercise 5.5.29   *There are 5 people who all brought the same umbrella to a restaurant and left it in the closet. If everybody randomly grabs an umbrella when they leave, what is the probability that nobody leaves with their own umbrella?*

Exercise 5.5.30   *How many different outfits can you make with 4 shirts, 2 sweaters, 3 pairs of shorts, 2 pairs of pants, and 5 pairs of socks (using exactly one of each item)? What is the probability that if you randomly select an outfit, you will be appropriately dressed for an extremely hot day (meaning you are wearing shorts and a shirt)? If 2 of the shirts and 1 of the sweaters are blue and 2 pairs of shorts and 1 pair of pants are red, what is the probability that a randomly chosen outfit either includes a blue top or red bottoms, but not both?*

Exercise 5.5.31   *Given the numbers 1 through 10, how many ways can they be arranged so that each even number is always placed in an even position?*

CHAPTER 6 _____

Counting III: Advanced Combinatorics

Dr. Ray Stantz: What do you mean choose? We don't understand.
Gozer: Choose. Choose the form of the destructor.
— Ghostbusters (1984)

We've already introduced a lot of powerful functions and techniques to attack counting problems, ranging from the factorial function and binomial coefficients to the Method of Inclusion-Exclusion. In this chapter we'll apply these to more problems. There's a virtually inexhaustible reserve of counting problems available for a probability class. Long term, this is good, as the more situations you see, the better you'll be at recognizing what to do in either later classes or the real world.

Short term, however, this is a real challenge. Sometimes it's not clear that there are a few common themes running through all the exercises. Our goal in this chapter is to call attention to these common themes. We've seen how to solve lots of different problems; here, we'll introduce some other popular ones, as always highlighting what these have in common with what we've done before, and explaining what new twists these bring.

Obviously no class has time to cover all these topics; that's the advantage of a book. You should read the areas you find particularly interesting (or the topics that you're covering in your course!). The material here is independent of the rest of the book, but the perspectives can be invaluable going forward. On a personal note, the math riddles page I run (located at http://mathriddles.williams.edu/, typically one of the top hits when googling "math riddles") and several of my research papers are due to the ideas behind the elegant solution to the cookie problem from §6.3.

Here's a quick summary of the problems in this chapter.

- §6.1: Basic Counting: The problems from §6.1.1 and §6.1.2 involve standard counting problems, specifically, when does order matter, and how to deal with certain options being excluded. We end in §6.1.3 by considering sampling with and without replacement, which lead to very different results.
- §6.2: Word Orderings: It might seem like busywork figuring out how many different words we can make from the letters in Mississippi, but the solution to this problem is related to one of the most important scientific concerns of all time! In

the course of studying these problems we'll encounter the multinomial coefficients, which as their name suggest do indeed generalize binomial coefficients.

- §6.3: Partitions: We end this chapter by discussing partitions. We've seen that even though we could do the needed counting for many problems, we don't want to as it quickly becomes too involved. These problems all have *elegant* solutions. If you ask a mathematician what makes a solution good, you'll often hear elegance. Similar to inclusion-exclusion, we'll see how the right way of looking at problems leads to very pleasant solutions, completely bypassing the annoyance of case upon case upon case.

In terms of big ideas, the third section is the most important. Many classes don't do counting at this level, which is a shame as there are some really great ideas here.

6.1 Basic Counting

In this section we return to some basic counting problems. The point is to give you some more practice in going through the cases, making sure you exhaust all the possibilities without forgetting any. The problems in §6.1.1 and §6.1.2 are similar to many we've done before. Those in §6.1.3 introduce a new wrinkle, as there we have sampling with and without replacement.

6.1.1 Enumerating Cases: I

There are so many problems we could choose to emphasize the need to be careful when counting. The following is inspired by many years of travel in cars before iPods.

 Example 6.1.1: *Imagine we have a CD player in our car. This player can hold 6 CDs. When we put it on randomize, after it plays a song from a CD it randomly chooses another CD, with each CD equally likely to be chosen (including the CD currently being played). We play 10 songs with the randomizer on and all 6 CD slots filled. How many possibilities are there for which CDs are chosen? How many of these possibilities never have CD 2, 3 or 6 played, or, equivalently, what is the probability that we never hear a song from CD 2, 3, or 6? Imagine now we have an improved randomizer, which doesn't play two songs from the same CD in a row. Now what is the probability we never hear a song from CD 2, 3, or 6?*

Remember, when doing problems like this, make sure you understand the problem before choosing a method to solve it. As you read the problem, think about whether or not order matters.

Solution: When we first put the CD player on randomize, there are 6 possible CDs it can choose. When picking the next song, the CD player will again randomize the CDs and pick again from the 6. Thus, there are $6 \cdot 6 = 36$ possibilities for the first two songs. Each time the randomizer chooses a new song, it chooses a CD, multiplying the possibilities by 6. Thus after 10 songs there are $6^{10} = 60,466,176$ possibilities. Note order matters in this problem: if only two songs were selected by the randomizer, picking CD 4 followed by CD 3 is a distinct way to select these two CDs from CD 3 followed by CD 4.

We can count directly to find out how many possibilities never involve CDs 2, 3, or 6. When the randomizer selects the first song, there are only three choices it can make that don't involve CDs 2, 3, or 6: namely CDs 1, 4, or 5. Each time it chooses a CD, it must choose from this pool of only three CDs, and so after ten songs there will be only $3^{10} = 59,049$ outcomes that never choose CDs 2, 3, or 6. This gives us a probability of $\frac{59,049}{60,466,176} = \frac{1}{1,024}$, or a little under 0.1% that these CDs are never selected.

For our improved randomizer, we have a different number of possible outcomes. The first song will again be chosen from one of the six CDs; but when the randomizer selects subsequent songs, as it never chooses from the CD that just played, there are only five CDs to choose from. So there are 6 choices for the first song, and then 5 choices for each of the remaining 9 songs, giving us a total of $6 \cdot 5^9 = 11,718,750$ possible outcomes.

To find the total number of possibilities that never play songs from CDs 2, 3, or 6 now, we have three possible CD choices for our first song: CDs 1, 4, or 5. After the first song, however, whichever CD was just selected will not be chosen next, so there are only two remaining CDs to choose from that aren't CDs 2, 3, or 6. This is true each time after the first song is chosen. Thus we have $3 \cdot 2^9 = 1,536$ possible outcomes that never play a song from these CDs, and so the probability is $\frac{1,536}{11,718,750} = \frac{256}{1,953,125}$ that songs from those CDs never play, just over 0.01%.

6.1.2 Enumerating Cases: II

For our second example, we again consider a case with restrictions on what can come next, and go through the careful bookkeeping analysis.

Example 6.1.2: *Imagine we live in a state where license plates for cars are constructed according to the following rules: all license plates start with three letters (and each letter is equally likely to be chosen) followed by three numbers (and all numbers are equally likely to be chosen). (1) How many different license plates are there? (2) Assume now that we're not allowed to have any vowels (A, E, I, O, or U) or any even numbers. How many license plates are there? (3) What's the probability two people have exactly four of their six digits equal?*

We'll solve this problem in the standard way, by a careful enumeration of all possibilities. As you're reading the solution, though, think about how pleasant or unpleasant it is, and whether or not there's a simpler way.

Solution to (1): Figuring out how many different possible license plates there are is relatively straightforward. Since the first three characters are letters, we have 26 possible choices (A through Z) for each. The next three characters are numbers, each having ten choices (0 through 9). There are, then, $26 \cdot 26 \cdot 26 \cdot 10 \cdot 10 \cdot 10 = 17,576,000$ total license plates possible under these restrictions.

Solution to (2): If vowels and even numbers aren't allowed, we simply recalculate our totals with modified pools. That is, instead of a total of 26 letters to choose from, there are now only 21, and instead of 10 numbers, we have only 5. This gives us a total of $21 \cdot 21 \cdot 21 \cdot 5 \cdot 5 \cdot 5 = 1,157,625$ possible license plates.

Figure 6.1. One possible license plate that starts with three letters followed by three numbers. Image from War (Wikipedia user), https://commons.wikimedia.org/wiki/File:Alaska_License_Plate_10432.jpg.

Solution to (3): To have exactly four of six digits (a digit is one of the symbols on the license plate, i.e., either one of the three numbers or letters) equal is to say that exactly two of six digits are different. There are $\binom{6}{2} = 15$ ways exactly two digits can be different. Why? We have to pick two of the six digits to differ; this is the definition of the binomial coefficient. Because some digits are numbers and others are letters, and numbers have a smaller pool to pick from than letters, we evaluate this on a case-by-case basis. Adding up each case will give us the total number of ways in which two license plates can have exactly four digits equal.

We list the fifteen possible ways for two digits from six to be chosen: (1) two letters: first and second, first and third, second and third; (2) two numbers: fourth and fifth, fourth and sixth, fifth and six; (3) a letter and a number: first and fourth, first and fifth, first and sixth, second and fourth, second and fifth, second and sixth, third and fourth, third and fifth, third and sixth. We'll see below that all pairs in each class give the same contribution. *Keep the following in mind as you read the analysis below: if we are trying to find how often two license plates agree in exactly four places, this is the same as fixing one license plate and asking how often a randomly chosen second plate matches it in exactly four places.* After our first approach we'll see how this insight speeds up the calculation.

- First suppose every digit but the **first and second** are equal. Without loss of generality, pick one of the license plates to look at first. There's a letter for the first digit and a letter for the second, giving us $26 \cdot 26$ choices for those two letters. We must ensure that the second license plate has different values in these two places than the first license plate. Thus we have 25 possible letters for the first spot (as they must be different than the first letter on the first license plate) and, likewise, 25 possible letters for the second. We are left with choosing the remaining four places, which are the same for the two license plates. There are 26 possibilities for the third digit (as we must choose a letter), and then 10 choices for each of the remaining digits (as they can be any number). Multiplying we find that the number of pairs of license plates differing only in the first and second positions equals

$$26^2 \cdot 25^2 \cdot 26 \cdot 10^3 \ = \ 10{,}985{,}000{,}000.$$

The only thing special about looking at the first two digits is that they're both letters. We get the same number of pairs if only the **first and third** or only the **second and third** differ, and thus the number of pairs of license plates where all but two letters agree is three times the above, or 32,955,000,000.

- Now suppose every digit but the **fifth and sixth** were equal; remember these are digits 0 through 9. There are $10 \cdot 10$ choices for the digits on the first plate in these two positions, and then $9 \cdot 9$ for the second to differ in these spots. There are 26^3 possibilities for the first three digits (which are the same for both), and 10 options for the fourth digit, giving

$$10^2 \cdot 9^2 \cdot 26^3 \cdot 10 \ = \ 1,423,656,000.$$

The same logic holds for the case when the **fourth and fifth** or the **fourth and sixth** digits are chosen as well, so we again multiply by three and find there are 4,270,968,000 possibilities when the two plates agree save for two numbers.

- Finally, suppose the **third and fourth** digits were chosen to be different. The third digit is a letter; so there are 26 choices for that value on the first plate, and then 25 choices for the value on the second plate. Similarly there are 10 choices for the first plate's value at the fourth spot, and then 9 for the second place there. The remaining spots must be the same; there are 26^2 ways to choose the two remaining letters and 10^2 ways to choose the remaining two numbers. Once more, this isn't unique to the third and fourth digits. Any time one letter and one number are selected to be different, there will be 15,210,000 ways that can happen. This will be the case when the **first and fourth, first and fifth, first and sixth, second and fourth, second and fifth, second and sixth, third and fifth**, and **third and sixth** digits are chosen, so we multiply by 9 and find there are

$$9 \cdot 26 \cdot 25 \cdot 10 \cdot 9 \cdot 26^2 \cdot 10^2 \ = \ 35,591,400,000$$

possibilities.

Adding the different possibilities gives 72,817,368,000 pairs of license plates where exactly four digits are the same. The number of pairs is $17,576,000^2$, thus the probability is about 0.0235719%.

After such a long analysis as we had for part (3), it's worth seeing if there's a better way. What else could we have tried? In some cases, it's easier to calculate the probability of the *complementary event* (or count the number of ways the complementary event can happen). Sadly, the complementary event is having none, one, two, three, five, or six digits agreeing. Sometimes there's no way around it—you just have to roll up your sleeves and persevere.

Thus, while we are stuck with enumerating all the cases above, maybe there is a better way to do this calculation. We already had some savings by noticing symmetries. For example, the number of ways just the **first and second** digits of the two pairs differ is the same as the number of ways just the **first and third** or just the **second and third** differ. *Rather than looking at all pairs of plates, we can fix one license plate and look at the probability the next license plate agrees in exactly four locations.* For definiteness, we could even assume the plate was AAA000. Observations like this can help streamline coding or simulations. Below is some code to numerically explore.

We first give a program that generates two plates and counts the spots that agree, and then comment on the changes needed if we assume the first plate is AAA000, as well as the computational savings that gives.

```
licensecheck[numdo_] := Module[{},
  count = 0;
  Print["Theory predicts ", 100. (72817368000) / (26^6 10^6), "%."];
  For[n = 1, n <= numdo, n++,
   {
    numagree = 0;
    For[j = 1, j <= 3, j++,
     {
       (* representing letters by 1 through 26 *)
       (* choose two, one for each plate, if agree increment *)
       (* agree counter by 1. Then do similar for three numbers *)
       x = RandomInteger[{1, 26}];
       y = RandomInteger[{1, 26}];
       If[x == y, numagree = numagree + 1];
       u = RandomInteger[{0, 9}];
       v = RandomInteger[{0, 9}];
       If[u == v, numagree = numagree + 1];
       }]; (* end of j loop *)
    If[numagree == 4, count = count + 1];
    (* the following few lines outputs the results every time *)
    (* we do another 10% of the sampling. *)
    If[Mod[n, numdo/10] == 0,
     {
       Print["At ", 100. n/numdo, "percent, observe ",
         SetAccuracy[100. count/n, 5], "%."];
       }];
    }]; (* end of n loop *)
  Print["Observe percentage ", SetAccuracy[100. count/numdo, 5], "%."];
  ]
```

Running 10,000,000 trials gives the following:

```
Theory predicts 0.0235719%.
Observe percentage 0.0236%.
```

We end with a few remarks on efficiency. As the probability was so small we needed to run a large number of simulations to be reasonably confident in our answer; in other words, since the answer was so close to zero we needed many trials to be comfortable on how slightly it is above zero. Fortunately it's very cheap to randomly choose numbers, and even running on low battery power on a flight to Denver I was able to do 10,000,000 trials in 169.9, 238.8, and 170.1 seconds (there are fluctuations in the run-time due, among other items, to other processes running on my machine, though I shut most down).

If we use the observation that without loss of generality we can fix the first plate to be AAA000 (we represent A by 1), we can change the code.

```
x = RandomInteger[{1, 26}];
If[x == 1, numagree = numagree + 1];
u = RandomInteger[{0, 9}];
If[u == 0, numagree = numagree + 1];
```

Note that we have decreased the number of random variables we must draw by 50%, from four to two. There are still many calculations to be done, but the run-times for 10,000,000 trials were now 136.8, 141.3, and 136.4 seconds. If we throw away the longer run-time of 238.8, the code generating both plates took about 170 seconds while generating just one plate took about 139 seconds, which means it's running in about 5/6ths the time, a sizeable savings!

6.1.3 Sampling With and Without Replacement

Let's consider the following **jar problem**. As you read it, ask yourself if it's well-formulated, or if there are multiple ways of interpreting it.

Example 6.1.3: *Imagine we have four jars. In each jar there are 100 marbles, and each marble is either purple or gold. The first jar has exactly 10 purple marbles, the second exactly 30 purple marbles, the third exactly 60 purple marbles, and the last exactly 90 purple marbles. (1) Assume we draw 5 marbles from the first jar: what is the probability that we have at least 4 purple marbles? (2), (3), and (4) Repeat this for the other three jars. (5) If we were to choose a jar at random, what is the probability we have at least 4 purple marbles when we draw 5?*

As stated, the problem is too vague to solve. What does it mean to draw 5 marbles from a jar? Are we pulling out 5 marbles one at at time, or are we pulling out a marble, recording its color, replacing it, and then pulling a marble out again? Not surprisingly, the two methods lead to two different answers. The first is called **sampling without replacement**, while the second is called **sampling with replacement**. We first describe the solution when we sample without replacement, and then with. For many problems, sampling without replacement is more natural; a great example are lotteries, where numbers cannot be repeated. However, there are times when we do have replacement (such as sampling points for Monte Carlo integration), so it's important to be able to do both.

I almost included one more source of confusion. How would you interpret the problem if instead of saying, "The first jar has exactly 10 purple marbles," it said, "The first jar has 10 purple marbles"? While the second is a bit sloppy, it's almost always correct to assume the writer meant *exactly*, even if it isn't explicitly stated. It does make the problem more wordy to keep writing exactly, but it removes all danger of confusion. We'll first solve the problem assuming no replacement.

Solution to (1) without replacement: Let's consider the first jar, with 10 purple marbles and 90 gold ones. There are two ways we can draw at least 4 purple marbles: by drawing exactly 4 purple marbles and 1 gold one, or by drawing 5 purple marbles. We add up the probability of each case to find the probability of either case, and thus the probability of at least 4 purple marbles.

- First consider the case of drawing 5 purple marbles. When drawing the first, there are 10 purple marbles to pick out of a total 100, so the probability is 10/100. Once we've picked out the first purple marble, though, there are only 9 purple ones out of a total 99 left. The probability of having a purple marble on the second pick (given a purple marble on the first pick) is 9/99. The probabilities of the remaining 3 marbles being purple are 8/98, 7/97, and 6/96, respectively. Multiplying all these probabilities together, we find the probability of drawing 5 purple marbles is

$$\frac{10}{100} \cdot \frac{9}{99} \cdot \frac{8}{98} \cdot \frac{7}{97} \cdot \frac{6}{96} = \frac{30240}{9034502400} = \frac{1}{298760}.$$

There's another way to view this: $\binom{10}{5} / \binom{100}{5}$ (there are $\binom{100}{5}$ ways to choose 5 marbles without replacement, and $\binom{10}{5}$ ways to choose all 5 marbles from the set of 10 purple.

• Next we calculate the probability of drawing 4 purple marbles and 1 gold one. Note that there are five possible ways this can happen: if the gold marble is picked first and the remaining 4 marbles are purple; if the gold marble is picked second and the other 4 are purple; and so on. We must add up all these probabilities. First suppose the gold marble was picked first. The probability of picking out a gold marble is 90/100. With 10 purple marbles still in the bag but only 99 total marbles remaining, the probability of the second marble being purple is 10/99. With similar reasoning from above, we see that the probabilities of picking the remaining purple marbles are 9/98, 8/97, and 7/96. Thus we can calculate the probability of picking out 1 gold marble followed by 4 purple marbles:

$$\frac{90}{100} \cdot \frac{10}{99} \cdot \frac{9}{98} \cdot \frac{8}{97} \cdot \frac{7}{96} = \frac{453600}{9034502400} = \frac{3}{59752}.$$

Now suppose the gold marble was picked last. The probability of picking out the first 4 purple marbles is the same as when we were picking out 5 purple marbles: 10/100, 9/99, 8/98, and 7/97. But instead of drawing 1 last purple marble, we would like to draw 1 of the 90 gold marbles. Since there are only 96 marbles remaining in the bag, this gives us a final probability of 90/96. Multiplying these together, we get

$$\frac{10}{100} \cdot \frac{9}{99} \cdot \frac{8}{98} \cdot \frac{7}{97} \cdot \frac{90}{96} = \frac{453600}{9034502400} = \frac{3}{59752}.$$

Notice that this is the same exact probability as drawing the gold marble first. This is no coincidence. Consider the denominators of both products: they are identical, because no matter which marble we select, we're always reducing the number of marbles in the jar by one. In the numerator, because we're always picking out 1 gold marble and 4 purple ones without replacement, we see the same numbers only in different orders. Because multiplication is **commutative**, we always arrive at the same result: a probability of 3 in 59,752. Since there are $\binom{5}{1} = 5$ ways to draw 1 gold marble and 4 purple ones (the gold marble first, second, third, fourth, or fifth), we can multiply this probability by five to find the probability of selecting exactly 4 purple marbles: P(4 purple) $= 5 \cdot \frac{3}{59752} = \frac{15}{59752}$.

Another way to view the calculation is that it equals $\binom{10}{4}\binom{90}{1} / \binom{100}{5}$, which is again $\frac{15}{59752}$. To see this, there are $\binom{10}{4}$ ways to choose 4 purple marbles, $\binom{90}{1}$ way to take a gold one, and $\binom{100}{5}$ ways to choose 5 marbles from the 100. In problems like this it is very easy to forget structure, or to accidentally add it. It's thus very good to be able to do it multiple ways. The first method is longer, but to me feels less controversial. We're looking at the five different sequences; if we write P for drawing a purple marble and G for getting a gold, they are PPPPG, PPPGP, PPGPP, PGPPP, and GPPPP. We calculate the probability of each of these five and add. For the second approach, there is no factor of $\binom{5}{1}$ to account for these options. That's taken care of in the product $\binom{10}{4}\binom{90}{1}$; we don't care about the *order* in which the 5 marbles are drawn, only that we take 4 purple and 1 gold.

As we initially discussed, the probability of selecting at least 4 purple marbles is equal to the probability of selecting exactly 4 or exactly 5 purple marbles. Thus the probability of selecting at least 4 purple marbles is $\frac{1}{298760} + \frac{15}{59752} = \frac{19}{74690}$, or about a .025% probability.

Solution to (2) without replacement: Next we calculate the probability for the **second jar**, which contains 30 purple marbles and 70 gold ones. Calculating the probability of drawing all 5 purple marbles follows the same analysis as for the first jar; we need only to incorporate the new numbers of this second jar into our work. The probability of drawing all 5 purple marbles is

$$\frac{30}{100} \cdot \frac{29}{99} \cdot \frac{28}{98} \cdot \frac{27}{97} \cdot \frac{26}{96} = \frac{17,100,720}{9,034,502,400} = \frac{1131}{597520}.$$

Using binomial coefficients, we would get $\binom{30}{5}/\binom{100}{5}$, which is the same number.

Similarly, to calculate the probability of drawing exactly 4 marbles, we start by calculating the probability of drawing 4 purple marbles followed by 1 gold marble:

$$\frac{30}{100} \cdot \frac{29}{99} \cdot \frac{28}{98} \cdot \frac{27}{97} \cdot \frac{70}{96} = \frac{46,040,400}{9,034,502,400}.$$

Multiplying this by five to account for the five different orderings of 4 purple marbles and 1 gold being selected, we get $5 \cdot \frac{46,040,400}{9,034,502,400} = \frac{230,202,000}{9,034,502,400}$, which is $\frac{435}{17072}$. If we were to use binomial coefficients, we would get $\binom{30}{4}\binom{70}{1}/\binom{100}{5}$; similar to the previous problem, we do not include a factor of $\binom{5}{1} = 5$ here, as the product of the two binomial coefficients already gives us all the possibilities.

Finally, we get the probability of drawing at least 4 purple marbles by summing the probabilities of drawing exactly 4 and exactly 5 purple marbles:

$$P = \frac{17,100,720}{9,034,502,400} + \frac{230,202,000}{9,034,502,400} = \frac{247,302,720}{9,034,502,400} = \frac{4,089}{149,380},$$

or about 2.74%.

Solution to (3) without replacement: For the **third jar**, we now have 60 purple marbles and only 40 gold ones. The probability of drawing 5 purple marbles is

$$\frac{60}{100} \cdot \frac{59}{99} \cdot \frac{58}{98} \cdot \frac{57}{97} \cdot \frac{56}{96} = \frac{655,381,440}{9,034,502,400}.$$

The probability of drawing exactly 4 purple marbles and 1 gold one is

$$5 \cdot \left(\frac{60}{100} \cdot \frac{59}{99} \cdot \frac{58}{98} \cdot \frac{57}{97} \cdot \frac{40}{96} \right) = 5 \cdot \frac{468,129,600}{9,034,502,400} = \frac{2,340,648,000}{9,034,502,400}.$$

Thus the probability of drawing at least 4 purple marbles from this jar is

$$P = \frac{655{,}381{,}440}{9{,}034{,}502{,}400} + \frac{2{,}340{,}648{,}000}{9{,}034{,}502{,}400} = \frac{2{,}996{,}029{,}440}{9{,}034{,}502{,}400} = \frac{260{,}072}{784{,}245},$$

or about 33.16%. The binomial approach gives the same answer, written as $\binom{60}{5}/\binom{100}{5}$ $+ \binom{60}{4}\binom{40}{1}/\binom{100}{5}$.

Solution to (4) without replacement: Finally, for the **fourth jar**, we have 90 purple marbles and only 10 gold ones. The probability of drawing 5 purple marbles is

$$\frac{90}{100} \cdot \frac{89}{99} \cdot \frac{88}{98} \cdot \frac{87}{97} \cdot \frac{86}{96} = \frac{5{,}273{,}912{,}160}{9{,}034{,}502{,}400}.$$

The probability of drawing 4 purple marbles and 1 gold one is

$$5 \cdot \left(\frac{90}{100} \cdot \frac{89}{99} \cdot \frac{88}{98} \cdot \frac{87}{97} \cdot \frac{10}{96} \right) = 5 \cdot \frac{613{,}245{,}600}{9{,}034{,}502{,}400} = \frac{3{,}066{,}228{,}000}{9{,}034{,}502{,}400}.$$

Thus the probability of drawing at least 4 purple marbles equals

$$P = \frac{5{,}273{,}912{,}160}{9{,}034{,}502{,}400} + \frac{3{,}066{,}228{,}000}{9{,}034{,}502{,}400} = \frac{8{,}340{,}140{,}160}{9{,}034{,}502{,}400} = \frac{43{,}877}{47{,}530},$$

or about 92.3%. Using binomial coefficients, we get the same value, only written as $\binom{90}{5}/\binom{100}{5} + \binom{90}{4}\binom{10}{1}/\binom{100}{5}$.

Solution to (5) without replacement: To find the probability of drawing at least 4 marbles when one jar is picked at random, we simply multiply the probability of picking a given jar by the probability of picking at least 4 purple marbles from that jar. Since each jar is equally likely to be picked, each one has a 1/4 probability of being picked. We sum all these probabilities together to find the probability when a jar is picked at random:

$$P = \left(\frac{1}{4} \cdot \frac{779{,}919}{3{,}065{,}902{,}643} \right) + \left(\frac{1}{4} \cdot \frac{4{,}089}{149{,}380} \right) + \left(\frac{1}{4} \cdot \frac{260{,}072}{784{,}245} \right) + \left(\frac{1}{4} \cdot \frac{43{,}877}{47{,}530} \right)$$

$$\approx 0.0000635961975 + 0.00684328558 + 0.0829052146 + 0.230785819$$

$$\approx 0.320597915,$$

or just over 32.0597%.

We now solve the problem with replacement. This means that marbles are picked one at a time, and each time a marble is selected, the choice is noted and the marble is placed back in the jar before selecting the next marble. We'll see that this somewhat simplifies our analysis.

Solution to (1) with replacement: The **first jar** has 10 purple marbles and 90 gold ones. As when drawing marbles without replacement, there are two ways we can draw at least 4 purple marbles: by drawing exactly 4 purple marbles (and 1 gold one) or exactly 5 purple marbles.

- We now calculate the probability of drawing 5 purple marbles. Since there are 10 purple marbles out of a possible 100, the probability of drawing the first marble is 10/100. Indeed, since each marble is replaced, the probability of drawing each purple marble is 10/100. Thus the probability of drawing all 5 purple marbles equals

$$\frac{10}{100} \cdot \frac{10}{100} \cdot \frac{10}{100} \cdot \frac{10}{100} \cdot \frac{10}{100} = \left(\frac{1}{10}\right)^5 = \frac{1}{100,000}.$$

- We now need to find the probability of drawing 4 purple marbles and 1 gold one. Again, the same analysis we used on drawing marbles without replacement follows here: we need to take into account that the gold marble may be picked first, second, third, fourth, or fifth; but the probability of each of these events is equal, so we can calculate the probability of all of these events by multiplying the probability of one of them by five.

  Without loss of generality, suppose we select the gold marble first. There are 90 gold marbles, so the probability will be 90/100. Next we select a purple marble. Since the first marble was replaced, there are 10 purple out of a total of 100 marbles, so the probability of the marble being purple is 10/100. Since this second marble is also replaced, the probability of drawing each of the 3 remaining purple marbles will also be 10/100. The probability of drawing a gold marble followed by 4 purple marbles, then, is

$$\frac{90}{100} \cdot \frac{10}{100} \cdot \frac{10}{100} \cdot \frac{10}{100} \cdot \frac{10}{100} = \frac{9}{10} \cdot \left(\frac{1}{10}\right)^4 = \frac{9}{100,000}.$$

Recalling that we must multiply this value by five (as $\binom{5}{1} = 5$, i.e., there are 5 ways to choose one position to be gold) to find the probability of drawing 4 purple marbles and 1 gold marble, we get $5 \cdot \frac{9}{100,000} = \frac{45}{100,000} = \frac{9}{20,000}$.

Finally, the probability of drawing at least 4 purple marbles is the probability of drawing exactly 4 or exactly 5 purple marbles:

$$P = \frac{1}{100,000} + \frac{9}{20,000} = \frac{46}{100,000} = \frac{23}{50,000} = 0.00046,$$

or 0.046%.

Solution to (2) with replacement: Next consider the **second jar**, with 30 purple marbles and 70 gold ones. Since the analysis is exactly the same as with the first jar, we jump to the calculations. The probability of drawing exactly 5 purple marbles here is

$$\frac{30}{100} \cdot \frac{30}{100} \cdot \frac{30}{100} \cdot \frac{30}{100} \cdot \frac{30}{100} = \left(\frac{3}{10}\right)^5 = \frac{243}{100,000}.$$

The probability of drawing 4 purple marbles and 1 gold is:

$$5 \cdot \left(\frac{70}{100} \cdot \frac{30}{100} \cdot \frac{30}{100} \cdot \frac{30}{100} \cdot \frac{30}{100} \right) = 5 \cdot \frac{567}{100,000} = \frac{567}{20,000}.$$

Thus the probability of drawing at least 4 purple marbles equals

$$P = \frac{243}{100,000} + \frac{567}{20,000} = \frac{243}{100,000} + \frac{2,835}{100,000} = \frac{3,078}{100,000} = 0.03078,$$

or 3.078%.

Solution to (3) with replacement: Moving on to the **third jar**, we now have 60 purple marbles and 40 gold ones. The probability of drawing 5 purple marbles will be

$$\frac{60}{100} \cdot \frac{60}{100} \cdot \frac{60}{100} \cdot \frac{60}{100} \cdot \frac{60}{100} = \left(\frac{3}{5} \right)^5 = \frac{243}{3125}.$$

The probability of drawing 1 gold and 4 purple marbles in this jar would be

$$5 \cdot \left(\frac{40}{100} \cdot \frac{60}{100} \cdot \frac{60}{100} \cdot \frac{60}{100} \cdot \frac{60}{100} \right) = 5 \cdot \frac{162}{3125} = \frac{162}{625}.$$

Thus the probability of drawing at least 4 purple marbles would be

$$P = \frac{243}{3125} + \frac{162}{625} = \frac{1053}{3125} = 0.33696,$$

or 33.696%.

Solution to (4) with replacement: Finally, consider the **fourth jar** containing 90 purple marbles and only 10 gold ones. The probability of drawing 5 purple marbles is

$$\frac{90}{100} \cdot \frac{90}{100} \cdot \frac{90}{100} \cdot \frac{90}{100} \cdot \frac{90}{100} = \left(\frac{9}{10} \right)^5 = \frac{59,049}{100,000}.$$

The probability of drawing 1 gold and 4 purple marbles in this jar would be

$$5 \cdot \left(\frac{10}{100} \cdot \frac{90}{100} \cdot \frac{90}{100} \cdot \frac{90}{100} \cdot \frac{90}{100} \right) = 5 \cdot \frac{6,561}{100,000} = \frac{6,561}{20,000}.$$

Thus the probability of drawing at least 4 purple marbles from the fourth jar equals

$$P = \frac{59,049}{100,000} + \frac{6,561}{20,000} = \frac{91,854}{100,000} = 0.91854,$$

or 91.854%.

Solution to (5) with replacement: We find the probability of drawing at least 4 marbles when one jar is picked at random the same way we do when the marbles were selected without replacement. Each of the four jars is chosen with probability 1/4, and we thus multiply the probabilities of each jar having at least 4 marbles by 1/4 and add.

$$P = \left(\frac{1}{4} \cdot 0.00046\right) + \left(\frac{1}{4} \cdot 0.03078\right) + \left(\frac{1}{4} \cdot 0.33696\right) + \left(\frac{1}{4} \cdot 0.91854\right)$$
$$\approx 0.000115 + 0.007695 + 0.08424 + 0.229635$$
$$\approx 0.321685,$$

or 32.1685%.

After surviving such a long calculation, let's see if it's reasonable. Our goal in the discussion below is to highlight how computer simulations can help us be confident about the correctness of our answers. Notice for this problem that the two interpretations are fairly close to each other. For example, the probability of getting at least 4 purples is about 32.0597% when we sample without replacement, and 32.1685% when we sample with replacement. It's a small difference, but it's a reasonable one. Why? If we sample without replacement, it gets harder and harder to draw purples, and thus the probability of at least 4 purples should be a little less. Imagine the ridiculous extreme where we wanted at least 91 purples in 100 draws; it's impossible to do this without replacement, but possible (albeit extremely unlikely) with replacement.

To get a feel for this problem, I wrote a short Mathematica code and investigated the probabilities of getting at least 4 purples in the two situations. Computers are a *great* tool. We'll devote an entire chapter, Chapter 25, to discussing how to write code to investigate problems such as this. I did each 100,000 times. The first time I did it, I observed a probability of getting at least 4 purple (without replacement) of 32.289%, slightly higher than the observed probability of 31.907% (with replacement). The next time I did 100,000 simulations the results were reversed, with 32.008% (with replacement) slightly less than 32.114% (without replacement). There's an important lesson here—we need to be able to figure out how large our simulations should be to obtain a given confidence about our answer. We'll talk more about this when we cover standard deviations.

```
marblecheck[num_] := Module[{},
    countwith = 0; (* count successes with replacement *)
    countwithout = 0;  (* count successes without replacement *)
    list = {}; (*  creating a list of 1, 2, ..., 100 *)
    For[m = 1, m <= 100, m++, list = AppendTo[list, m]];
    p[1] = .1; p[2] = .3; p[3] = .6; p[4] = .9; (* the probabilities *)
    For[n = 1, n <= num, n++ (* main loop *)
      {
        (* first do with replacement, then without *)
        (* randomly chooses 1,2,3,4 given uniform random generator *)
        (* could have used x = RandomInteger[{1, 4}]; *)
        x = Floor[4*Random[]] + 1;
        numgold = 0; (* counts the number of gold *)
        For[i = 1, i <= 5, i++,
         If[Random[] > p[x], numgold = numgold + 1]];
        If[numgold <= 1, countwith = countwith + 1];

        (* without replacement, a bit harder *)
        y = Floor[4*Random[]] + 1; (* randomly choose 1,2,3,4 *)
        numgold = 0; (* set count of gold to zero *)
```

```
templist = RandomSample[list, 5]; (* choose 5 from our list *)
cutoff = Floor[p[y]*100]; (* cutoff value for comparisons *)
numgold = 0;
For[m = 1, m <= 5, m++,
  If[templist[[m]] > cutoff, numgold = numgold + 1]];
If[numgold <=  1, countwithout = countwithout + 1];

}]; (* end of n loop *)

Print["Observed probability at least four purple (without replacement)
    is ", 100 countwithout/num 1.0, "% (32.0597 predicted)."];
Print["Observed probability of getting at least four purple (with
    replacement) is ", 100 countwith/num 1.0, "% (32.1685 predicted)."];
Print["Did ", num, " iterations."];
]; (* end of module *)

For[nn = 1, nn <= 5, nn++, (* code to run for various amounts *)
{
marblecheck[10^nn];
Print[" "];
}]
```

6.2 Word Orderings

We've discussed in great detail probabilities of various hands in games of cards. Here's a nice, related question. Instead of a deck of cards we have a word (often chosen to be Mississippi, though my home state of Massachusetts would work well), and instead of dealing cards to people we give letters. Here's a bunch of natural questions to ask.

- How many k letter words can you form from an n letter word? Here the order of the letters matters, but we can't distinguish say the first s in Mississippi from the second, third or fourth s.
- Tweak the above question, and require that the new word have at least one letter repeated.
- Go one step further, and insist that we must have at least one letter repeated back to back. For example, imagine we start with the word "baboon" and want a four letter word with a double letter in its spelling. Then "boon" is acceptable as the two o's are next to each other, but "bono" would fail as now the o's aren't adjacent.

We'll show how to answer questions like these. In the course of our studies, we'll meet multinomial coefficients, which generalize binomial coefficients.

Finally, we promised in the introduction to this chapter that we'd give an application. Hundreds of years ago, research was disseminated differently than it is today. People often wanted to stake their claims to a result before all the work was done and checked. One way to do that was to publish a Latin anagram, an ordering of the letters of the Latin phrase describing their discovery! (Remember, in the 1600s most scientists understood Latin, so the difficulty was not in the language.)

For example, Hooke's Law states that the force of a string is proportional to the displacement. He established his priority by publishing `ceiiinosssttuv`. This is the alphabetical ordering of the Latin phrase `Ut tensio sic vis`, which means "As the extension, so the force."

Imagine you're given `ceiiinosssttuv`. How many phrases can you form using all these letters? Suddenly, the subject isn't as silly as it seemed.

6.2.1 Counting Orderings

We first consider the easier case when all the letters are distinct, and then move to the more general case.

Consider the word MAINE. There are five letters, and no letter is repeated. How many five letter words can we form? It's just 5!: there are 5 choices for the first letter, then 4 for the second, 3 for the third, 2 for the fourth, and finally the last letter is forced. So we have 120 possible words.

Now consider MISSISSIPPI. The same reasoning we used on MAINE won't work here because if we were to switch two of the S characters, we would still have the same original word, whereas MAINE has no repeated characters and thus any switch would always be a new word.

Let's start with an easier word, where there's only one repeated letter: ALABAMA. For us, this is a great state as we have four A's, and no other letter repeated. We want to find how many different seven letter words we can make. If instead we started with a word with no repeated letters, we'd have $7! = 5,040$ possibilities. While this doesn't solve the problem, it provides some valuable guidance: we know the answer is less than 5,040.

One of the greatest attributes of a mathematician is **laziness**. It may seem strange, but it's actually *good* to be lazy, at least if you're lazy in a good way. How are mathematicians lazy? We love reducing new problems to previously solved ones. Sometimes we can do it perfectly, though more likely we need to tweak things a bit.

Imagine instead of the word ALABAMA we had the word $A_1LA_2BA_3MA_4$. What we've done is place subscripts on the A's. We've added structure—we can now distinguish the different A's. Therefore, there are 7! ways to rearrange the letters of this word; however, for our original problem, all that matters is the *relative* location of the A's, and not which A occurs where. Thus, we don't want to count $A_3MLA_2BA_1A_4$ differently than $A_4MLA_1BA_2A_3$, as we are interested in **distinguishable words**.

What matters is which spots get A's, not which A's go into each spot. As we have 4 A's, there are 4! ways to order them for insertion into the four spots chosen. We've thus overcounted by a factor of 4!, and hence the number of distinct seven letter words we can make from Alabama is $7!/4! = 210$, much smaller than 5,040.

It's worth studying our solution for a bit. We reached it by making the A's distinguishable, as this reduced us to a situation we knew how to do. Then, once we had that solution, we removed the distinguishing marks. There are 4! ways to distinguish the A's, so we had to divide by 4! (as each of these markings lead to an indistinguishable word once the markings are gone).

There's another way to view the calculation. We have 7 letters: 4 are A's, 3 are non-A's (and we don't have any repeated non-A's). There are $\binom{7}{4}$ ways to choose 4 spots of the 7 for A's. Of the remaining 3 spots, there are 3! ways to insert our 3 remaining letters. Thus, the number of **distinguishable arrangements** of ALABAMA is $\binom{7}{4} \cdot 3!$ $= \frac{7!}{4!3!} \cdot 3! = 7!/4!$, exactly what we got before.

Let's look at KANSAS. We start by changing it to $KA_1NS_1A_2S_2$. Now all letters are distinct, and we have $6! = 720$ distinct words. We now remove our markings on the A's and the S's. There are 2! ways to order the A's, and 2! ways to order the S's. Thus, the number of distinguishable six letter words is $\frac{6!}{2!2!}$, or 180.

We get the same answer with our binomial approach, but it's a bit more involved now. There are $\binom{6}{2}$ ways to choose 2 of the six spots for an A. We now have 4 spots left, and there are $\binom{4}{2}$ ways to choose 2 of the remaining 4 spots for our two S's. Finally, there are 2 spots left and two distinct letters (neither of which is an A or an S), and there are thus 2! ways to place our remaining two letters. Combining, we find the number of distinguishable words is

$$\binom{6}{2}\binom{4}{2}2! = \frac{6!}{4!2!} \cdot \frac{4!}{2!2!} \cdot 2! = \frac{6!}{2!2!} = 180,$$

exactly as before.

KANSAS is a small enough word that we can see the 2!2! possibilities by a relatively painless enumeration. Let's take the ordered word $A_1NKS_1A_2S_2$. We can switch the order of the A's or the order of the S's without changing the word's appearance (if we drop the subscripts). So, we obtain $A_1NKS_2A_2S_1$, $A_2NKS_1A_1S_2$, and $A_2NKS_2A_1S_1$. These all give the same unlabeled word, ANKSAS, and thus we must divide our labeled count of 6! by 2!2!.

Now, let's look at a "better" state (at least for such problems!).

It's time to look at MISSISSIPPI, one of the greatest states for developing the theory. We have three different repeated letters; there are two P's, four I's, and four S's, as well as a singleton M. As always, let's add markings to the letters and write $MI_1S_1S_2I_2S_3S_4I_3P_1P_2I_4$. There are 11! orderings, as all the letters are distinct. We now remove the markings.

Let's first look at the four S's. Once we choose four out of the eleven places to put the S's, there are 4! ways to put down these labeled S's. As all of these are indistinguishable once we remove the labels, we've overcounted by a factor of 4!, and thus must divide 11! by 4! to remove the S labels.

What about the I's? It's the same story. Again, once we've chosen the four positions for the I's there are 4! ways to order our marked I's. We now remove these markings, as we don't care which I is where. We see that, in our list of 11!/4! words, we've again overcounted by a factor of 4!. Thus the number of words, once we remove the markings for S and I, is 11!/(4!4!).

We're almost there. It's time to deal with the P's. The same argument tells us we've overcounted in our list of 11!/(4!4!) words by a factor of 2!, and thus we must divide by 2!. We reach our final answer: there are $11!/(4!4!2!) = 34,650$ distinguishable words that can be formed from MISSISSIPPI.

Let's revisit the above calculation, from the point of view of binomial coefficients. We have $\binom{11}{4}$ ways to choose the four places to put our S's. Next, we have $\binom{7}{4}$ ways to choose four of the remaining seven positions for our four I's. Continuing, we have $\binom{3}{2}$ ways to choose two of the three spots left for our two P's, which leaves us with $\binom{1}{1}$ way to put the M in the remaining space. Multiplying, we get

$$\binom{11}{4}\binom{7}{4}\binom{3}{2}1! = \frac{11!}{4!7!}\frac{7!}{4!3!}\frac{3!}{2!1!}\frac{1!}{1!1!} = \frac{11!}{4!4!2!1!} = 34,650,$$

agreeing with our previous answer!

While factors which are 1 do not change a multiplication, it's worth including them. The reason is they act as important placeholders, and can help make sure we don't forget something. In the problem above note the sum of the numbers in the denominator factorials (4, 4, 2, and 1) add up to the number in the numerator's factorial (11); this is not an accident.

We can isolate a formula for these types of problems.

Distinguishable reorderings: Consider a word with k distinct letters, and N total letters. Let the first letter occur n_1 times, the second n_2 times, ..., and the kth letter n_k times (so $n_1 + \cdots + n_k = N$). Then the number of distinguishable words that can be formed is $N!/(n_1! \cdots n_k!)$.

Using the above formula, we see the number of distinguishable words that can be formed from the thirteen letter word MASSACHUSETTS is $13!/4!2!2! = 64,864,800$. Here the distinct letters are A, C, E, H, M, S, T, and U, with multiplicities 2, 1, 1, 1, 1, 4, 2, and 1; we usually don't bother writing 1! in the division as 1! equals 1, but perhaps we should as a placeholder: $13!/(2!1!1!1!1!4!2!1!)$. Note that again the sum of the numbers in the denominator factorials (2, 1, 1, 1, 1, 4, 2, and 1) add up to the number in the numerator (13).

6.2.2 Multinomial Coefficients

Building on our successful study counting words, we introduce a nice generalization of binomial coefficients.

Multinomial coefficients: Let N be a positive integer, and let $n_1, n_2, \ldots, n_k$ be non-negative integers that sum to N ($n_1 + \cdots + n_k = N$). The associated multinomial coefficient is

$$\binom{N}{n_1, n_2, \ldots, n_k} = \frac{N!}{n_1! n_2! \cdots n_k!}.$$

As always, the first thing you should do when encountering a generalization of a previous concept or definition is see if it reduces to the old one in special cases.

Let's consider the case $k = 2$. Then $n_1 + n_2 = N$, implying $n_2 = N - n_1$, and we find

$$\binom{N}{n_1, n_2} = \frac{N!}{n_1! n_2!} = \frac{N!}{n_1!(N - n_1)!} = \binom{N}{n_1}.$$

Thus, when $k = 2$ the multinomial coefficients are just the binomial coefficients.

Alright, the multinomial coefficients generalize the binomial coefficients. How can that help us? Remember the binomial coefficients arise from expanding polynomials:

$$(x + y)^N = \sum_{n=0}^{N} \binom{N}{n} x^n y^{N-n}.$$

We can look at this as saying that the number of times $x^n y^{N-n}$ appears in the expansion of $(x + y)^N$ is $\binom{N}{n}$. A great interpretation of this is that we have to choose exactly n of the N factors to be x, which forces the remaining $N - n$ to be y.

What if we consider $(x + y + z)^N$? If we expand this out, we'll get expressions of the form $x^{n_1} y^{n_2} z^{n_3}$. We must have $0 \leq n_1, n_2, n_3 \leq N$, as we can't get a higher power than N from expanding (which happens if we take that variable each time), nor a lower power than 0 (which happens if we never take it). More, however, is true: $n_1 + n_2 + n_3 = N$. Why does this hold? For each of the N factors $x + y + z$ we either take x, y, or z, and thus the total number chosen must be N. Therefore, there are integers a_{n_1, n_2, n_3} such that

$$(x + y + z)^N = \sum_{\substack{0 \leq n_1, n_2, n_3 \leq N \\ n_1 + n_2 + n_3 = N}} a_{n_1, n_2, n_3} x^{n_1} y^{n_2} z^{n_3}.$$

We'll sketch the proof that a_{n_1, n_2, n_3} equals the multinomial coefficient $\binom{N}{n_1, n_2, n_3}$. One way to see this is to use our **laziness principle**. Remember, mathematicians are the good type of lazy. We want to reduce new problems to ones previously solved. We know about binomial coefficients and the Binomial Theorem, so let's reduce this more general problem to repeated applications of these.

The key insight is to write $x + y + z$ as $x + (y + z)$; we're now back into having two choices: x or not x (with not x being $(y + z)$ of course). **Proof by grouping** is another great method for attacking a variety of problems. We used this in §A.3 to prove the derivative of a sum of three functions is the sum of the three derivatives, just using the fact that the derivative of a sum of two functions is the sum of the two derivatives. We'll also see it again in some counting problems in §6.3.3.

From the Binomial Theorem, we have

$$(x + (y + z))^N = \sum_{n_1=0}^{N} \binom{N}{n_1} x^{n_1} (y + z)^{N-n_1},$$

where we used the suggestive notation of n_1 for the number of x's chosen. We now use the Binomial Theorem again, this time expanding $(y + z)^{N-n_1}$:

$$(y + z)^{N-n_1} = \sum_{n_2=0}^{N-n_1} \binom{N - n_1}{n_2} y^{n_2} z^{N-n_1-n_2}.$$

Putting the pieces together, we find

$$
\begin{aligned}
(x + y + z)^N &= \sum_{n_1=0}^{N} \binom{N}{n_1} x^{n_1} \sum_{n_2=0}^{N-n_1} \binom{N-n_1}{n_2} y^{n_2} z^{N-n_1-n_2} \\
&= \sum_{n_1=0}^{N} \sum_{n_2=0}^{N-n_1} \binom{N}{n_1} \binom{N-n_1}{n_2} x^{n_1} y^{n_2} z^{N-n_1-n_2} \\
&= \sum_{\substack{0 \le n_1, n_2, n_3 \le N \\ n_1 + n_2 + n_3 = N}} \binom{N}{n_1} \binom{N-n_1}{n_2} x^{n_1} y^{n_2} z^{n_3},
\end{aligned}
$$

where the last line is a great, concise way of rewriting what we have.

All that remains is to show $\binom{N}{n_1} \binom{N-n_1}{n_2}$ is the multinomial coefficient $\binom{N}{n_1,n_2,n_3}$ with $n_3 = N - n_1 - n_2$. Well,

$$
\begin{aligned}
\binom{N}{n_1} \binom{N-n_1}{n_2} &= \frac{N!}{n_1!(N-n_1)!} \frac{(N-n_1)!}{n_2!(N-n_1-n_2)!} \\
&= \frac{N!}{n_1! n_2! (N-n_1-n_2)!} = \frac{N!}{n_1! n_2! n_3!},
\end{aligned}
$$

exactly as we claimed!

The proof of the general case proceeds similarly. Imagine we have

$$
(x_1 + x_2 + \cdots + x_k)^N.
$$

Let's look at the term $x_1^{n_1} x_2^{n_2} \cdots x_k^{n_k}$. We must have $0 \le n_1, \ldots, n_k \le N$ and $n_1 + n_2 + \cdots + n_k = N$. We get this term by choosing x_1 from exactly n_1 of the N terms (there are $\binom{N}{n_1}$ ways to do this), then by choosing x_2 from exactly n_2 of the remaining $N - n_1$ factors (there are $\binom{N-n_1}{n_2}$ ways to do this), and so on, up to choosing x_k from exactly n_k of the remaining n_k factors (there's $\binom{n_k}{n_k} = 1$ way to do this). Multiplying, we see the number of these terms is

$$
\begin{aligned}
&\binom{N}{n_1} \binom{N-n_1}{n_2} \binom{N-n_1-n_2}{n_3} \cdots \binom{N-(n_1+\cdots+n_{k-1})}{n_k} \\
&= \frac{N!}{n_1!(N-n_1)!} \frac{(N-n_1)!}{n_2!(N-n_1-n_2)!} \frac{(N-n_1-n_2)!}{n_3!(N-n_1-n_2-n_3)!} \cdots \frac{n_k!}{n_k!0!} \\
&= \frac{N!}{n_1! n_2! \cdots n_k! 0!} = \frac{N!}{n_1! n_2! \cdots n_k!} = \binom{N}{n_1, n_2, \ldots, n_k},
\end{aligned}
$$

where $n_k = N - (n_1 + \cdots + n_{k-1})$.

This argument shows us where the multinomial coefficients come from, and lets us see their combinatorial interpretation. The final formula, with the numerator $N!$ and the denominator k non-negative integers summing to N, is just our formula for the number of distinguishable words.

Figure 6.2. The author discussing the cookie problem. Photo courtesy of Susmita Paul.

6.3 Partitions

In this section we introduce a marvelous perspective that eliminates a lot of tedious case analysis for combinatorial problems. We'll meet it when we study the cookie problem, and conclude by showing its applicability to understanding lotteries.

6.3.1 The Cookie Problem

Below we describe a combinatorial problem which contains many common features of the subject. As our statement involve cookies, we'll call it the **cookie problem**. Other texts will have different names. Though the **stars and bars problem** is frequently used, I prefer calling it the cookie problem as I'm a huge fan of Cookie Monster, pictured in Figure 6.2.

 Without further commentary, here's the problem. Assume we have 10 identical cookies and five distinct people. How many different ways can we divide the cookies among the people, such that all 10 cookies are distributed?

Since the cookies are identical, we can't tell which cookies a person receives; we can only tell how many. We could enumerate all possibilities. This is a good strategy for problems that seem intractable. Unfortunately, we have to be very careful. One of the most common wrong answers is to say the solution is $5^{10} = 9,765,625$. People reach this by looking at what happens to each cookie: as each cookie can go to any of the five people, there are 5 options for the first cookie, then 5 for the second cookie, and so on. The problem is that this argument assumes the cookies are distinguishable. Having the first person get the first 9 cookies and the second person get the tenth is indistinguishable from the second person getting the first cookie and the first person getting the last 9 cookies.

Let's count. We'll go by the highest number of cookies received by a person. There are 5 ways to have one person receive 10 cookies. There are $5 \cdot 4 = 20$ ways to have one person receive 9 and another receive 1. Instead of viewing it as $5 \cdot 4$ we could view it as

$\binom{5}{2}2!$; there are $\binom{5}{2}$ ways to choose two people to get cookies, and then 2! ways to order them (i.e., one gets 9 and one gets 1).

So far the counting isn't too bad, but that changes when the maximum is 8. There are $5 \cdot 4$ ways to choose one person to get 8 and one to get 2; however, we also have the case of one person getting 8 and two people getting 1. For that case, there are 5 ways to choose the person who gets 8, and then $\binom{4}{2} = 6$ ways to choose two of the remaining four people to get 1; thus in this case there are a total of 30 different ways to have 8, 1, and 1. We can also view this through binomial coefficients. We have to choose three of five people to get cookies; there are $\binom{5}{3}$ ways to do this. We now order the three people. Naively we might expect the answer to be 3!, but remember we have 8, 1, and 1 cookies; thus the two people getting one cookie are indistinguishable (i.e., they both have 1 cookie, not 8 and not 0). Thus we have to divide by 2!, and the answer is $\binom{5}{3}3!/2! = 30$. If we forget to divide by 2! we're off by a factor of 2 (though it is better to view it as being off by a factor of 2!). And so on.

The killer here is the phrase "and so on." As we hit more and more cases where the number of cookies given to the most fortunate continues to decline, there are more and more cases and subcases to remember. With enough time and care we could make it work, but this approach isn't illuminating. It's not **elegant**.

While in principle we can solve the problem, in practice this computation becomes intractable, especially as the numbers of cookies and people increase. In other words, I don't want to write this out, and even if I did I doubt you would want to read it! Fortunately, there's a better way. An elegant way.

Cookie Problem: The number of distinct ways to divide C identical cookies among P different people is $\binom{C+P-1}{P-1}$.

Let's prove the claimed formula. Consider $C + P - 1$ cookies in a line, and number them 1 to $C + P - 1$. Choose $P - 1$ cookies. There are $\binom{C+P-1}{P-1}$ ways to do this—this is essentially the definition of the binomial coefficients. Cookie Monster (the original, non-politically correct Cookie Monster, who's always up for eating cookies) helps out by eating these $P - 1$ cookies. This divides the remaining cookies into P sets: all the cookies up to the first chosen (which gives the number of cookies the first person receives), all the cookies between the first chosen and the second chosen (which gives the number of cookies the second person receives), and so on. This divides C cookies among P people. Note different sets of $P - 1$ cookies correspond to different partitions of C cookies among P people, and every such partition can be associated to choosing $P - 1$ cookies as above. Intuitively, we have C cookies and add $P - 1$ "cookies" to act as boundaries for cookie separation. We then choose the $P - 1$ which will separate the cookies into P distinct stacks. It was necessary to add the $P - 1$ initially because these "cookies" will not be distributed, and the total number of cookies distributed must sum to C.

For example, if we have 10 cookies and five people, say we choose cookies 3, 4, 7, and 13 of the $10 + 5 - 1$ cookies to be our boundaries:

This corresponds to person 1 receiving two cookies, person 2 receiving zero, person 3 receiving two, person 4 receiving five, and person 5 receiving one cookie.

To drive home the point, if we have 8 cookies and four people, say we choose cookies 4, 7, and 11 of the $8 + 4 - 1$ cookies to be our boundaries:

This corresponds to person 1 receiving three cookies, person 2 receiving two, person 3 receiving three, and person 4 receiving zero.

We can now answer the original question: the number of distinct ways to divide 10 identical cookies among five people is $\binom{10+5-1}{5-1} = \binom{14}{4} = 1001$. It's almost unbelievable how simply this solves the problem. It's so much better than the brute force enumeration we started!

There's a number theory interpretation to what we've done. There are two related ways to view the word "partition." The first is that we're adding **partitions**, and the second is that we're partitioning a number N. For the original problem of dividing 10 cookies among five students, this is the same as counting the number of solutions to $x_1 + x_2 + x_3 + x_4 + x_5 = 10$, where x_i is the number of cookies person i receives (and is thus a non-negative integer). Our cookie principle tells us that the number of valid divisions is $\binom{14}{4} = 1001$, and so there are 1001 non-negative integer tuples solving $x_1 + \cdots + x_5 = 10$.

I like the number theory interpretation, as it suggests a related problem and solution. What if we have some notion of fairness, and will only consider assignments where each of our five people gets at least one of the 10 cookies? Before we had $x_i \geq 0$; now we have $x_i \geq 1$. We can incorporate this known information by writing $x_i = y_i + 1$, where $y_i \geq 0$. This allows us to rewrite our problem of $x_1 + \cdots + x_5 = 10$, each $x_i \geq 1$, as $(y_1 + 1) + \cdots + (y_5 + 1) = 10$ and $y_i \geq 0$. But this last expression is just $y_1 + \cdots + y_5 = 10 - 5 = 5$ and $y_i \geq 0$. In other words, the only effect of the constraints is to convert this to another, related cookie problem (just with fewer cookies to divide). The answer is therefore $\binom{5+5-1}{5-1} = \binom{9}{4} = 126$.

We can also explain this using our original understanding of the problem. Because each person gets 1 cookie, and there are five people, we already know where 5 of the cookies will be. Because the cookies are indistinguishable, the fact that each person already has one cookie means that there are now only $10 - 5 = 5$ cookies left to distribute. We therefore have $\binom{5+5-1}{5-1} = \binom{9}{4} = 126$ as before. It is important to have multiple interpretations of a problem as sometimes different interpretations can hold profound insights into an issue.

There's a lot of fun, related problems that we could study, but doing so will really take us far afield, so let's stop here. If you want to see some more problems, I've collected a few in §6.3.3. The more problems you do, the more you're able to internalize these arguments and the better prepared you'll be for seeing how to use these methods in new problems in the future. So, if you're interested and have the time, look at these problems and the methods used to solve them. It's also fine to move on to §6.3.2, where we apply this vantage point to attack the lottery.

6.3.2 Lotteries

Lotteries are a huge industry. How huge? Easily over \$50 billion a year just in state lotteries! In this section we'll concentrate on calculating your odds of getting all the numbers right in various lotteries. Of course, while you can still win some money if you have some numbers right, our goal here is not to help you figure out whether or not you should play (hint: you shouldn't!); we'll talk more about that when we do expected values in §9.2.

We start with a standard type of lottery.

Example 6.3.1: *The Probability Lottery Jackpot is now $25 million. The way the lottery works is as follows: there are 50 balls numbered from 1 to 50. Six of these balls are randomly chosen, which means that no number can be chosen twice. If you happen to choose all the numbers correctly, you win the $25 million; if even one of your numbers wasn't chosen, you lose. What are your odds of winning? Assume it costs $1 to play a choice of six numbers, and you win only if you have all the numbers right. Is it worthwhile to buy all possible choices (assuming, somehow, that you have enough money to make all these bets!)?*

Solution: The first thing to realize in this problem is that since a numbered ball can't be chosen twice, each ball must be chosen without replacement. Also, since all that is required is that a player selects the correct numbers, order isn't important. Thus the answer is $\binom{50}{6}$, the number of ways to choose 6 objects from 50 when order doesn't matter:

$$\binom{50}{6} = \frac{50!}{6!44!} = \frac{50 \cdot 49 \cdot 48 \cdot 47 \cdot 46 \cdot 45}{6 \cdot 5 \cdot 4 \cdot 3 \cdot 2 \cdot 1} = 15{,}890{,}700.$$

For a player to win, he must correctly guess all six numbers, so his guess would have to match the exact outcome. That is, the probability of winning would be precisely $1/15{,}890{,}700$. Not great odds.

However, if a player had $15{,}890{,}700$ to spend, he could theoretically buy every possible outcome, guaranteeing that he would win the jackpot; and with a prize of $25 million, this would definitely be a worthwhile bet.

Our lottery commissioner has just heard that the jackpot is so high that there's a danger of a group of rich people just buying up all the tickets. I can easily see some rich people deciding to do this, as there's no risk and it's almost $10 million in profit (or close to a doubling of the investment). Of course, there are difficulties and expenses in purchasing every possible ticket, but people are clever and can often find ways to resolve such issues. Thus our lottery commissioner is worried, and she decides to institute some changes.

Example 6.3.2: *The lottery now changes its rule; each number from 1 to 50 is equally likely to be chosen each time; thus the winning six numbers can now include repeats. How many ways are there for six numbers to be chosen? What's your probability of winning? If it still costs $1 to play, is it worth betting on all possible combinations?*

Once the lottery changes its rule and allows repeats, however, we must rethink our strategy. As was the case before, order doesn't matter, but this time repetition is allowed. There's an equation which will tell us how many different outcomes we have in this situation, but it appears confusing at first glance. *To see how difficult and delicate these problems are, we first give reasonable sounding approaches which give the* wrong *answer. We'll then analyze why they failed, and conclude with a correct analysis.* I know it can be confusing to look at wrong answers, but I strongly recommend doing so. It's worth seeing what errors people make so you can be aware of these tendencies and on your guard.

Faulty analysis I: There are 50 choices for each position, so there are 50^6 possibilities; however, this has order mattering. As we have 6 positions, we should divide by 6!, and find the number of possible winning tickets is

$$\frac{50^6}{6!} = 21,701,388.88888\ldots.$$

Analysis: It's pretty easy to see that this can't be right, as we don't end up with an integer! Where did we go wrong? The problem was dividing by 6!. If all of our numbers are distinct then, yes, we have overcounted by a factor of 6!. What if, however, there are two sets of two numbers repeated? Then we've overcounted by 2!2!. If one number is repeated three times and another twice, our factor is now 3!2!. We sadly can't remove the order just by dividing by 6!.

Faulty analysis II: As repeats are allowed, there are many more possibilities. Since a number may be selected up to six times, we can imagine there are six copies of each number to select from. With fifty numbers, this gives us a pool of 300 total balls, and we need to choose 6. Order still doesn't matter, so we use combinations again to find the total number of outcomes:

$$\binom{300}{6} = \frac{300!}{6!294!} = 962,822,846,700.$$

Analysis: We'll see later that the correct answer is 28,989,675; this is a little more than the 15,890,700 from the no-repeat lottery, and significantly less than the suggested answer of 962,822,846,700. Why is the suggested answer off by so much? The reason comes from creating six copies of a number. The copies are now distinguishable, as we're assuming we have 300 distinct numbers in our bag. If you want, think of a number such as 17 as coming in 6 different colors (17 red, 17 blue, 17 black, ...), or think of it as coming with six different subscripts $(17_1, 17_2, \ldots, 17_6)$. The problem is we don't care *which* 17's we get, just how many. Thus this method adds some structure that doesn't really belong, and that's why it counts too many. We don't want to distinguish between any of the 17's, all we want to do is keep track of how many 17's are chosen.

First solution: One way that isn't too bad is to write it as

$$\sum_{i_1=1}^{50} \sum_{i_2=i_1}^{50} \sum_{i_3=i_2}^{50} \sum_{i_4=i_3}^{50} \sum_{i_5=i_4}^{50} \sum_{i_6=i_5}^{50} 1.$$

Why is this a solution? Given any winning lottery number, let's reorder it and write the tuple from smallest to largest. Let's say that, in increasing order, it's $(i_1, i_2, \ldots, i_6)$. Note i_1 can take on any value from 1 to 50. What about i_2? It has to be at least as large as i_1, but no larger than 50. Similarly $i_3 \in \{i_2, \ldots, 50\}$, and so on. Evaluating the sum, we get 28,989,675.

While the sum may look unwieldly, there are ways to evaluate sums like this. If we consider the more general case of having N distinct numbers and choosing 6 (order doesn't matter, repeats allowed), then

$$\sum_{i_1=1}^{N}\sum_{i_2=i_1}^{N}\sum_{i_3=i_2}^{N}\sum_{i_4=i_3}^{N}\sum_{i_5=i_4}^{N}\sum_{i_6=i_5}^{N} 1 = \frac{N^6 + 15N^5 + 85N^4 + 225N^3 + 274N^2 + 120N}{720}.$$

Second solution: It's possible to bypass all these sums and just write the answer down! The key idea is to adopt the viewpoint we learned in §6.3.1 when studying the cookie problem. Let's modify that analysis to apply to our lottery situation. What was the key idea in the cookie problem? It was a one-to-one correspondence between partitions and the valid sets we wished to count.

Rather than just doing 50 balls, we'll do the more general case of N balls. So we have N numbers (one number on each ball), and we must choose k balls, where repeat ball selections are allowed. All that matters is how many of each number we choose, not the order. Suppose we have $N + k - 1$ balls in a row, and we choose k of them. There are $\binom{N+k-1}{k}$ ways to do that. Taking $N = 50$ and $k = 6$, we get

$$\binom{50 + 6 - 1}{6} = \frac{55!}{6!49!} = 28{,}989{,}675,$$

which is precisely what we got with the six nested sums!

At this point, it shouldn't be clear why this works; however, spend a few minutes thinking about this before reading on. You have a hint that partitions somehow matter— see if you can find the connecting argument. This is a very hard problem, one of the harder probability problems I know.

Okay, so why does this work? For ease of exposition, we'll use our numbers of $N = 50$ and $k = 6$ for the rest of the problem, though the argument holds in general. We've chosen 6 of 55 numbered balls. Let's say we chose balls $1 \le j_1 < j_2 < j_3 < j_4 < j_5 < j_6 \le 55$. We need to convert these choices to a six digit lottery number. We'll explain the correspondence through a specific choice of 6 balls.

Suppose we selected the following 6 balls: 1, 5, 8, 9, 22, 30. We now remove these balls from the line. We now have six spaces in the line, and these spaces correspond to the 6 lottery balls on our ticket through the following recipe: simply count up the balls remaining from the start until you reach a space and then add one to find the value of that lottery ball.

For example, the first number picked was 1, so the first ball was removed from the line. Starting at the left, there are no balls between the beginning of the line and the first space—the first space occurred at the first possible place. Adding one, we get $0 + 1 = 1$ and so our first lottery ball is the number 1. The next empty space is the one at the place of the fifth ball of the 55 balls—since the first ball was also removed, there are only 3 balls remaining from the start of the line until this second space (namely, balls 2, 3, and 4). Thus our second lottery ball is $3 + 1 = 4$.

The next chosen ball is 8; there are 5 unchosen balls before it (balls 2, 3, 4, 6, and 7), so it represents a lottery ball with the number 6 $(5 + 1)$. Since the next chosen ball is 9, which immediately follows 8, there are still only 5 unchosen balls before it as well; thus this also represents a lottery ball with the number 6. Continuing we see there are

17 balls remaining before 22, so the fifth lottery ball is 18 (17+1). Finally, there are 24 balls remaining before 30, so the last lottery ball is 25 (24+1). Thus, the six lottery balls are 1, 4, 6, 6, 18, and 25.

Through this process, each of the 6 chosen balls out of our original 55 yields a lottery ball. Just as six balls of the same value can be chosen in the lottery, 6 dots can be chosen in a row to yield that result; and because 6 dots are removed, this process never yields a number larger than 50 (55 − 6 + 1). Each choice of our 6 removed balls corresponds to an outcome of lottery balls picked (ordered from least to greatest); thus the number of ways we can choose 6 of the 55 balls to remove gives the number of possible outcomes from the lottery if repeated balls are allowed. As we discussed above, we know exactly how many ways there are to choose these 6 dots: $\binom{55}{6} = 28{,}989{,}675$.

Therefore if it still costs \$1 to play with a jackpot of \$25 million, it's no longer profitable to bet on every possible outcome. Our commissioner did well!

As always, let's look at some extreme cases to get a better feel for the argument above. If we remove the first 6 balls, then we get the ticket 1, 1, 1, 1, 1, 1. If instead we remove the last six balls (50, 51, 52, 53, 54, and 55), then our ticket is just 50, 50, 50, 50, 50, 50. It's very important to make sure we never choose a lottery number greater than 50. We'll show we can't get 51; the other numbers are handled similarly. If we want 51, then we have to select either 52, 53, 54, or 55. The more balls removed in the beginning, the lower our number. Thus if we're trying to get a large number like 51, our best chance is to take all of 52, 53, 54, and 55; if we can't do it in this case, we can't do it at all. Remember, though, we still need to choose 6 balls. We've only accounted for 4. Thus we must choose at least 2 balls before (and possibly including) 51, and that's enough to make sure our lottery ball is at most 50 (the largest it can be is $51 - 2 + 1 = 50$).

One last remark on this problem. If you can write some computer code and get some answers, you have a chance of detecting a pattern. Don't avoid experimental mathematics–data is good! It took less than 10 seconds for Mathematica to evaluate the sum on my computer. If you type

```
Sum[Sum[Sum[Sum[Sum[Sum[1, {i6, i5, 50}], {i5, i4, 50}],
{i4, i3, 50}], {i3,i2, 50}], {i2, i1, 50}], {i1, 1, 50}]
```

or even better, type

```
Sum[1, {i1,1,50}, {i2,i1,50}, {i3,i2,50}, {i4,i3,50},
{i5,i4,50}, {i6,i5,50}]
```

a few moments later you'll get 28,989,675. Instead of doing 50 numbers we could of course do 49 or 48 or 51, and gather some more data points. If you think the answer might look like a binomial coefficient, it's natural to look for something of the form $\binom{f(N,k)}{g(N,k)}$ for some nice functions f and g of the number of balls N and the number k that we must choose. If we play around and look at various binomial coefficients, after a while we'll hopefully stumble upon $\binom{55}{6} = 28{,}989{,}675$. If we've looked at not just $N = 50$ but also some other N, we'd get a few more data points and might guess that $f(N, k) = N + k - 1$ and $g(N, k) = k$. Keep k fixed and vary N, and then fix N and vary k; this will help give a good sense of the true behavior.

If you've seen Mathematical Induction (if not, it's discussed in §A.2), you know that it's much easier to do an inductive proof when you know what the answer should be. For example, it's not horrible to prove by induction that the sum of the first N integers is $N(N+1)/2$, or the sum of the first N squares is $N(N+1)(2N+1)/6$. There are similar formulas for sums of cubes and sums of kth powers; however, if you don't know the formula, the induction is much harder! For more, see Exercise 6.5.11.

So, what's the big takeaway? Do smaller problems. Build intuition. Detect a pattern. Instead of doing 50 balls and selecting 6, try maybe 8 balls and selecting 3. By looking at these simpler cases you can often sniff out what's happening, and then use that to guide you in attacking the general case.

6.3.3 Additional Partitions

While the following questions aren't always directly related to probability problems we'll study in this book, they're all natural generalizations of the cookie problem. I like them as they all have nice solutions *if* you can find the right perspective. As a major goal of this book is to help you develop your problem solving skills, I've included a few more problems and solutions here for fun.

 In solving equations in integers, often slight changes in the coefficients can lead to wildly different behavior and very different sets of solutions. Determine the number of non-negative integer solutions to (1) $x_1 + x_2 = 1996$; (2) $2x_1 + 2x_2 = 1996$; (3) $2x_1 + 2x_2 = 1997$; (4) $2x_1 + 3x_2 = 1996$; (5) $2x_1 + 2x_2 + 2x_3 + 2x_4 = 1996$; and (6) $2x_1 + 2x_2 + 3x_3 + 3x_4 = 1996$.

Solution: (1) There are 1997 solutions. This is because we can choose x_1 to be any value in $\{0, 1, \ldots, 1996\}$, and then once x_1 is chosen x_2 is uniquely determined. This is equivalent to the cookie problem with 1996 cookies and two people $\binom{1996+2-1}{2-1} = \binom{1997}{1} = 1997$. (2) Dividing both sides by two, we see this equation is equivalent to $x_1 + x_2 = 998$. Arguing as in (1), there are 999 solutions. (3) There are no solutions as the left-hand side is always even and the right-hand side is always odd. (4) Clearly once we choose either x_1 or x_2 then there's at most one choice for the other. Let us choose x_2 first. As $x_1 = (1996 - 3x_2)/2$, we see $1996 - 3x_2$ must be even. Thus we must have x_2 is even. This means that x_2 can be 0, 2, 4, all the way up to 664 (as $666 \cdot 3 = 1998$). For each of the 333 choices of x_2 there's exactly one choice of x_1 that works. Thus, there are 333 solutions.

(5) is similar to (2); if we divide by 2 we get $x_1 + x_2 + x_3 + x_4 = 998$. This is just the cookie problem with 998 cookies and four people, so the answer is $\binom{998+4-1}{4-1} = \binom{1001}{3}$, or 166,666,500. Notice this is a lot more than the others. The reason is that we have four variables, where before we had just two. With two variables, once we specified one value then the other was determined; with four variables, we have a lot more freedom.

Of all these problems, however, (6) is my favorite. It's another example of the power of **grouping**, which we used so effectively in §6.2.2 to understand the multinomial coefficients (see §A.3 for another example). We may rewrite $2x_1 + 2x_2 + 3x_3 + 3x_4 = 1996$ as $2(x_1 + x_2) + 3(x_3 + x_4) = 1996$. This now looks very similar to (4). Let $y_1 = x_1 + x_2$ and $y_2 = x_3 + x_4$. If y_2 is odd there are no solutions, as $2y_1 = 1996 - 3y_2$, which is impossible as the left-hand side is even and the right-hand side is odd. If, however, y_2 is even then there is a unique choice for y_1, namely $(1996 - 3y_2)/2$. We are left with solving $x_3 + x_4 = y_2$, $y_1 = x_1 + x_2 = (1996 - 3y_2)/2$, where $y_1 \geq 0$, $y_2 \geq 0$ and are even. We have two independent cookie problems. There are $\binom{y_2+2-1}{2-1} = \binom{y_2+1}{1} = y_2 + 1$ solutions to $x_3 + x_4 = y_2$. Similarly there are $\binom{\frac{1}{2}(1996-3y_2)+2-1}{2-1} = \frac{1}{2}(1996 - 3y_2) + 1 = 999 - \frac{3}{2}y_2$ solutions to the second, so the total number of solutions is

$$\sum_{\substack{y_2=0 \\ y_2 \text{ even}}}^{\lfloor 1996/3 \rfloor} (y_2 + 1)\left(999 - \frac{3}{2}y_2\right) = \sum_{y=0}^{\lfloor 1996/6 \rfloor} (2y + 1)(999 - 3y) = 37,092,537,$$

where $\lfloor n \rfloor$ is the floor function, the largest integer that is less than or equal to n.

It's worth harping on our solution to (6). What did we do? We were given *one* hard problem, and replaced it with *two* easier problems. Often this is a good exchange: I'd rather do lots of easy computations and then combine them, rather than one really hard one.

Our next problem has a nice combinatorial feel, and is useful for a variety of counting problems.

Let $\mathcal{M}$ be a set with $m > 0$ elements, $\mathcal{W}$ a set with $w > 0$ elements, and $\mathcal{P}$ a set with $m + w$ elements. For $\ell \in \{0, \ldots, m + w\}$, prove

$$\sum_{k=\max(0,\ell-w)}^{\min(m,\ell)} \binom{m}{k}\binom{w}{\ell - k} = \binom{m + w}{\ell}.$$

Solution: The proof is actually one line, provided we look at the problem correctly. We're going to use the **story method**, and **count our quantity two different ways**. Let's think of $\mathcal{M}$ as a set of m men, $\mathcal{W}$ as a set of w women, and $\mathcal{P}$ as a set of $m + w$ people (with m of them men and w of them women). How many ways can we form a set of ℓ people from $\mathcal{P}$? So long as $0 \leq \ell \leq m + w$, this is just the definition of the binomial coefficient $\binom{m+w}{\ell}$—we choose ℓ of the $m + w$ people.

Okay, this gives us the right-hand side, now for the left. If we have a group of ℓ people, there must be *some* number of men chosen. Let's denote the number of men chosen by k. Since we want to choose ℓ people from the $m + w$ people, k cannot be too small, nor can it be too large. Clearly $0 \leq k \leq m$, but we can do better. If $k < \ell - w$, then there's no way we can get ℓ people with exactly k of the men, so we must also have $k \geq \ell - w$. Similarly we can't have $k > \ell$ as we only want ℓ people. Thus $\max(0, \ell - w) \leq k \leq \min(m, \ell)$. For such k, there are $\binom{m}{k}$ ways to choose k men from m, and then $\binom{w}{\ell-k}$ ways to choose $\ell - k$ women from w women. Thus the number of groups with exactly k men is $\binom{m}{k}\binom{w}{\ell-k}$. As each group has to have *some* number of men, summing over all valid k (which is the left-hand side) must count all the groups, which is $\binom{m+w}{\ell}$.

This problem illustrates a general phenomenon. Often there's a really nice, elegant solution for sums of products of binomial coefficients. The difficulty is *finding* the interpretation. If you want to really appreciate the power of the right perspective, try to solve this problem directly from the definition of the binomial coefficients, expanding them out and trying to do the algebra—good luck!

We end with one more problem. The original cookie problem assumed every cookie was distributed. What if some cookies are held back? Specifically, how many ways are there to divide N cookies among k people, where we *do not* have to give out all the cookies? In the course of solving this problem we'll discover yet another nice identity for sums of binomial coefficients.

Solution: One of the hardest steps in solving a problem is trying to figure out what method to use. The more experience you get, the easier things become. A great guide is to look for similar problems which we know how to solve. We know there are $\binom{C+P-1}{P-1}$ ways to distribute C cookies among P people. We can't immediately use that here, some of the cookies might not be distributed; however, with a little bit of work we can use this.

This is similar to other instances of the **laziness principle**. We reduce our problem to a bunch of simpler ones and then combine. Assume we distribute n cookies and keep $N - n$ cookies. The number of ways to do this is $\binom{n+k-1}{k-1}$ (we have n cookies and k people). As n may be any integer from 0 to N, the total number of ways to distribute the cookies is $\sum_{n=0}^{N} \binom{n+k-1}{k-1}$.

While this is a solution, it's not a particularly illuminating one. What does this binomial sum equal? Fortunately, if we look at what we're doing the right way, we can immediately write down the answer to this sum. We're trying to count how many ways we can distribute N identical cookies among k distinct students, where some cookies might be kept. Let's imagine there's one more person, either a special student or perhaps the kind author of this book. Let's just give any cookies not sent to the k students to this lucky person! All we've done is just say the leftover cookies go to a person. We're now doing a *new* cookie problem. We're distributing N cookies among $k + 1$ people! We know how to solve this problem; there's just $\binom{N+k+1-1}{k+1-1} = \binom{N+k}{k}$ ways to do this.

So, not only did we solve the problem in an elegant manner with a beautiful solution, but we got a formula for sums of binomial coefficients too:

$$\sum_{n=0}^{N} \binom{n+k-1}{k-1} = \binom{N+k}{k}.$$

6.4 Summary

The following couplet from *Ghostbusters* does a great job summarizing the issues we've faced in this chapter. It's the end of the movie, and the four Ghostbusters are trying to stop Gozer from destroying New York. Gozer tells them to choose.

> *Dr. Ray Stantz:* What do you mean choose? We don't understand.
> *Gozer:* Choose. Choose the form of the destructor.

This chapter is all about choices. You get to choose the method to attack a problem. It's okay if you don't understand at first how things will work out. Make a choice, pray for the best, and move on; if things don't work out, try again.

Here, we saw some great ways to bypass long, tedious calculations. The more of these problems you do, the easier it'll be to choose the elegant path through the computations. We saw in particular the power of partitions, where often we could convert a painful counting problem into a simple choice problem (how many ways are there to choose some number of elements from a larger set that's related to the initial problem). We also saw the advantages of regrouping expressions, and reducing to simpler problems.

6.5 Exercises

Exercise 6.5.1   *In §6.1.1 we looked at several problems with a randomizer and six CDs. We saw that if we're playing 10 songs then there are almost six times as many possibilities if we use all six CDs as there are in the case where we can never play the same CD twice in a row. What happens if we play n songs as $n \to \infty$?*

Exercise 6.5.2   *Continuing the CD exercise from above, if we play 10 songs from six CDs, where each CD is equally likely to be chosen for each song, what is the probability that we hear at least 2 consecutive songs from the same CD?*

Exercise 6.5.3   *In a best of seven series, two teams play each other and the first to win four games wins the series. How many different orderings of wins are there in such a series if no team ever wins two games in a row?*

Exercise 6.5.4   *Assume in a best of seven series (see the previous exercise) each team has a 50% chance of winning each game. What is the probability the series ends after exactly four, five, six, and seven games? Are you surprised by which is the most likely?*

Exercise 6.5.5   *Let n be a positive integer, and consider n special decks where deck d consists of cards numbered $1, 2, \ldots, d$. We choose a number uniformly at random from 1 to n, say m_1. We then go to deck m_1 and choose a number uniformly at random from that deck, say m_2. We now go to deck m_2 and choose a number uniformly at random from m_2, say m_3. What happens as we continue this process? Be as specific as you can (in terms of what must happen by a certain time, and what probably happens by a certain time). Write a computer program to numerically simulate this game for various n to help formulate some conjectures.*

Exercise 6.5.6   *Generalize the previous exercise to the best of a $2n + 1$ series. What is the probability the series ends after exactly k games, for $k \in \{n + 1, n + 2, \ldots, 2n + 1\}$. Are you surprised by which k is the most probable?*

Exercise 6.5.7   *Consider the jar problems from §6.1.3. What if we dump each jar into a giant bucket, so we now have 400 marbles, with 190 purple and 210 gold. What is the probability of drawing at least 4 purple marbles if we pick 5? How does this answer compare to what we found when we chose one of the four draws (with each jar equally likely to be chosen)? Are the answers different? Are you surprised?*

Exercise 6.5.8   *Consider a standard deck of cards. We keep picking a card uniformly from the remaining cards until we get a king. What is the probability we see at least one queen before we pick a king? What is the probability we see all four queens before we pick a king?*

Exercise 6.5.9   *How many words can we make from MAINE, where now we allow the word to have any length (but each letter may still be used at most once)?*

Exercise 6.5.10   *Prove the distinguishable reorderings formula from §6.2.1.*

Exercise 6.5.11 *Try to find closed form expressions for the following:*

$$\sum_{i_1=1}^{N} 1, \quad \sum_{i_1=1}^{N}\sum_{i_2=i_1}^{N} 1, \quad \sum_{i_1=1}^{N}\sum_{i_2=i_1}^{N}\sum_{i_3=i_2}^{N} 1.$$

Make a conjecture—do you believe the answer is always a polynomial in N of degree equal to the number of summands? What do you think the coefficient will be of the leading term?

Exercise 6.5.12 *In the lottery problem we used*

$$\sum_{i_1=1}^{N}\sum_{i_2=i_1}^{N}\sum_{i_3=i_2}^{N}\sum_{i_4=i_3}^{N}\sum_{i_5=i_4}^{N}\sum_{i_6=i_5}^{N} 1 = \frac{N^6 + 15N^5 + 85N^4 + 225N^3 + 274N^2 + 120N}{720}.$$

*Prove this formula. (Hint: These are **Stirling numbers**.)*

Exercise 6.5.13 *Similar to the previous two exercises, find formulas for $\sum_{n=1}^{N} n^k$ for $k \in \{3, 4, 5\}$. (Hint: If we approximate the sum by an integral, we see that the sum is approximately $N^{k+1}/(k+1)$. It turns out more is true; these sums are always polynomials in N of degree $k+1$ with leading coefficient $1/(k+1)$.)*

Exercise 6.5.14 *How many ways are there to assign 15 people into 5 distinct groups, each of which contains 3 people? What if the groups are not distinct?*

Exercise 6.5.15 *Generally, how many ways are there to assign x people into n groups of $x_1, x_2, \ldots, x_n$ groups where $x_1 + x_2 + \cdots + x_n = x$.*

Exercise 6.5.16 *Consider $\binom{x}{x_1, x_2, \ldots, x_n}$ where $x_1 + x_2 + \cdots + x_n = x$. For a fixed n, what values of $x_1, x_2, \ldots, x_n$ maximize $\binom{x}{x_1, x_2, \ldots, x_n}$? You may assume x is divisible by n, since the argument applies to other cases, but the notation is a bit easier in this case. Justify your answer.*

Exercise 6.5.17 *Suppose we are distributing exactly 10 cookies among five people. But we want to distribute the cookies more fairly, so that no person has more than twice as many cookies as the person with the least. How many ways are there to do this? What if we are distributing exactly 15 cookies among five people?*

Exercise 6.5.18 *We are going to divide 15 identical cookies among four people. How many ways are there to divide the cookies if all that matters is how many cookies a person receives? Redo this exercise but now only consider divisions of the cookies where person i gets at least i cookies (thus person 1 must get at least 1 cookie, and so on).*

Exercise 6.5.19 *Redo the previous exercise (15 identical cookies and four people), but with the following constraints: each person gets at most 10 cookies (it's thus possible some people get no cookies).*

Exercise 6.5.20 *Find the probability that I get exactly 10 1's and 10 2's in 60 tosses of a fair die.*

Exercise 6.5.21 *Find the error in the following argument: First, we find the number of distinct combinations of rolls: since we have twenty fixed objects (1s and 2s), how many ways can we split 40 objects (tosses) among 4 objects (numbers 3, 4, 5, and 6)? We do not care about the order; we only care about how many of each number we toss.*

Exercise 6.5.22 *If a TV show that is only broadcast in a small town has 200 viewers, 50 of whom are between 11 and 20, is this sufficient evidence that this population is over-represented, given that the total population of the town contains 300 people aged 0 to 10, 200 people from 11 to 20 years, 150 people from 21 to 30, 150 from 31 to 45, 150 people from 46 to 60, and finally 150 people over 61?*

Exercise 6.5.23 *Find the probability of k successes being drawn without replacement in n draws given that the total population has N objects, K of which are successes. (Hint: If you are stuck, generalize the previous exercise.)*

Exercise 6.5.24 *Imagine a housing lottery with 100 students in which each student just writes down which building they would like to live in. There are 5 buildings, A, B, C, D, and E. Buildings A,B, and C house 20 people each, building D houses 30, and building E houses 10. If each student is equally likely to want to live in any of the 5 buildings and all of their choices are independent, what is the probability that each student is able to live in their first choice building?*

Exercise 6.5.25 *Find the probability that there exists a housing assignment in which each student is able to live in one of their first three choices.*

Exercise 6.5.26 *Consider a lottery in which 10 numbers are chosen from the integers between 1 and 50 with replacement and equal probability of any number appearing on each draw. You just need to pick the 10 numbers, in any order. Are all tickets equally likely to win? If not, which ticket should you pick and what are the odds you win with this ticket?*

Exercise 6.5.27 *Consider the cookie problem, except now we have only two people and instead of having cookies, we have indivisible piles of cookies. We want to partition them as fairly as possible between the two people. Randomly generate a set of piles of cookies. (You can choose the maximum pile size and number of piles, but don't make it so small that it is trivial or so large that it will take a long time to analyze.) Write Mathematica code that randomly partitions your cookie set many times. What is the fairest partition you find?*

Exercise 6.5.28 *Instead of randomly partitioning the piles many times, write code that starts with a random partition, then searches all possible switches of a single pile until it finds the switch that comes closest to equalizing the partitions. Have the computer repeat this process until no more switches can be made that increase the equality of the partitions. Compare this partition to the one in the previous exercise. This is known as a "greedy algorithm" and can be used to find approximate solutions for many types of optimization problems.*

Exercise 6.5.29 *Consider the cookie problem with 20 cookies and 5 people, but allow all cookies to be broken into halves and fourths. How many ways are there to distribute cookies? Now allow all cookies to be entirely continuous such that each person can have any rational or irrational number (ex. $\pi/6$) of cookies. How many ways can these cookies be distributed?*

Exercise 6.5.30 *Amy, Ben, and Cathy are playing with a standard deck of 52 playing cards. Each one draws a card in order. When one person gets a spade, the game ends and that person wins. Calculate the probability of each person winning with and without replacement. (Hint: think back to the basketball problems we studied.)*

PART II
INTRODUCTION TO RANDOM VARIABLES

CHAPTER 7 _____

Introduction to Discrete Random Variables

Lorraine: Hey, don't I know you from somewhere?
George: Yes, yes, I'm George, George McFly, and I'm your density.
— Back to the Future (1985)

In the previous chapters we stated the axioms of probability and learned how to calculate the probabilities of certain discrete events, such as hands in card games or lotteries. These are, of course, only a small subset of what we wish to study. The goal of this chapter is to introduce the concept of a **random variable** and study a few special cases.

Informally, a random variable is a map from our outcome space to the real numbers. We'll first talk about discrete random variables, and then see the changes that occur in the continuous case. Random variables arise everywhere, from looking at the speeds of molecules in a box to how many runs a team scores in baseball to the number of people desiring to fly between two cities to how well students do on probability exams. You hopefully get the point. They're everywhere and are a key ingredient in describing and modeling the real world.

7.1 Discrete Random Variables: Definition

In this section we define discrete random variables. We'll get to the definition by first considering an enlightening example, extracting the definition from our study.

Imagine we toss a fair coin three times. Each toss has a 50% chance of landing on heads, and a 50% chance of landing on tails. We thus have eight possible outcomes in our outcome space Ω:

$$\Omega = \{TTT, TTH, THT, HTT, THH, HTH, HHT, \text{ and } HHH\}.$$

We have to assign a probability to each element of Ω. As each coin is heads with probability 1/2, and the three tosses are independent, each element of Ω happens with probability $1/2 \cdot 1/2 \cdot 1/2 = 1/8$. Therefore, we have our outcome space and our probability function (the σ-algebra is just all possible subsets, as we have a finite outcome space).

There are lots of questions we could ask. The most natural is whether or not we get a certain sequence of tosses. We've already answered this above: each sequence occurs one-eighth of the time. There are, however, other good questions to ask. A great one is to ask how many heads we get. Perhaps every time we get a head we earn a million dollars. If this is so, we don't care what *order* the heads appear; only the number of heads matters. In fact, I chose to order the eight outcomes as I did precisely to make answering this question easier. We see we get zero heads once, one head three times, two heads three times, and three heads once.

We can represent this with a function. We define a function X from Ω to the real numbers $\mathbb{R}$ by setting $X(\omega)$ equal to the number of heads in $\omega \in \Omega$. Thus $X(HHT) = 2$, $X(TTT) = 0$, and so on.

There are other functions we could define. Maybe we're interested in the number of tails. If so, we can set $Y(\omega)$ equal to the number of tails. Note $X(\omega) + Y(\omega) = 3$. This shouldn't be too surprising, as we have three tosses and whatever tosses aren't heads must be tails. This suggests another good random variable to study, the excess of heads to tails. Going back to the monetary motivation, perhaps now we get one million dollars for each head, but have to pay out a million dollars for each tail. This random variable is the same as $1,000,000 \cdot (X(\omega) - Y(\omega))$.

We end with three more functions. We define $X_i : \Omega \to \mathbb{R}$ by setting $X_i(\omega) = 1$ if the ith toss is a head, and 0 if the ith toss is a tail. Thus $X_1(HHT) = 1$, $X_2(HHT) = 1$, and $X_3(HHT) = 0$. We have the following important relationship:

$$X(\omega) \; = \; X_1(\omega) + X_2(\omega) + X_3(\omega).$$

In words, the total number of heads is equal to the number of heads on the first toss plus the number on the second plus the number on the third.

> There's a very important point lurking here which is easy to miss, and in fact is often sadly glossed over. We are *not* adding heads; what we are doing is adding the number of heads. It's a fine distinction, but an important one. What's the sum of a head and a tail? They're two different objects, and we can't add them. We can, however, assign a 1 to each head and a 0 to each tail and then add those numbers. (We'll see later that for some problems it's more convenient to assign -1 to a tail; it all depends on whether or not we want the sum to denote the number of heads, or the excess of heads over tails.)

While we haven't defined random variables yet, our discussion above shows why we want them to be real-valued functions on elements in the outcome space. We want to be able to build more complicated functions out of simpler ones. Armed with the above motivation, we now introduce the definition of a discrete random variable.

> **Discrete random variables**: A discrete random variable X is a real-valued function from a discrete outcome space Ω (this means Ω is finite or at most countable). Specifically, to each element $\omega \in \Omega$ we assign the real number $X(\omega)$.

As another example, imagine we roll two fair, independent die, and let R be the sum of the two rolls. As each die has six sides (with the numbers 1 through 6 on the sides), there are 36 possible rolls, going from (1,1) to (6,6), and 11 possible sums (the smallest sum is 2, the largest is 12). Thus $R((1, 1)) = 2$ and $R((3, 5)) = 8$. We should say a bit about the notation. We apply our random variable to elements of the outcome space. For us, the elements of the outcome space are pairs, such as $\omega = (3, 5)$. Thus $R(\omega) = R((3, 5))$. It may seem a bit absurd to have that extra set of parentheses, but it helps us keep track of everything. We're applying the function R to the pair $(3, 5)$ and getting 8 as our output. There's a great temptation to just write $R(3, 5)$ instead, but try to fight it!

7.2 Discrete Random Variables: PDFs

So far we've defined random variables and seen we can add two (or more) random variables to make a new random variable. There's another very important property. Remember a probability space is a triple. It's more than just a set of outcomes Ω; we also have a probability function defined on a set of subsets of Ω (these allowable sets are the σ-algebra). It therefore makes sense to talk about the probability our random variable takes on different values. Two natural questions to ask are what is $\mathrm{Prob}(X = x)$ (the probability our random variable is exactly x) or $\mathrm{Prob}(X \leq x)$ (the probability our random variable is at most x).

For our example of flipping a fair coin three times from §7.1, we have $\mathrm{Prob}(X_i = 1) = 1/2$, $\mathrm{Prob}(X_i = 0) = 1/2$, and all other values have zero probability of happening. Things are more interesting for X, the number of heads in three tosses of a fair coin. As each of the eight outcomes are equally likely, all we have to do is count how often X equals x and divide by 8. Equivalently, we just multiply by $1/8$ for each outcome with x heads. Thus

$$\mathrm{Prob}(X = 0) = 1/8, \quad \mathrm{Prob}(X = 1) = 3/8,$$
$$\mathrm{Prob}(X = 2) = 3/8, \quad \mathrm{Prob}(X = 3) = 1/8;$$

all other values have zero probability. Not surprisingly, each value is non-negative and the sum of all these values is 1.

Let's do another example. We'll still flip our coin three times, but this time let's assume that it lands on a head with probability p and a tail with probability $1 - p$. To emphasize the dependence on p, we'll write B_p for the number of heads from three tosses of this coin. We're using the letter B for our random variable to emphasize that the coin may be *biased*, and we're adding a subscript p to highlight the fact that the probability of a head is now p.

Our outcome space hasn't changed; it's still the eight events

TTT, TTH, THT, HTT, THH, HTH, HHT, and HHH.

What *has* changed is our probability function. If $p \neq 1/2$ then these eight events are no longer equally probable.

- The probability of TTT is $(1 - p)^3$,
- the probability of TTH, THT, and HTT is $p(1 - p)^2$,

- the probability of THH, HTH, and HHT is $p^2(1 - p)$,
- and finally the probability of HHH is p^3.

To find the probability that B_p equals 2 we just add the probabilities of THH, HTH, and HHT, and find $\text{Prob}(B_p = 2) = 3p^2(1 - p)$. For definiteness, let's take $p = 4/5$ (so we're four times as likely to get a head as a tail). We find

$$\text{Prob}(B_{4/5} = 0) = 1/125, \quad \text{Prob}(B_{4/5} = 1) = 12/125,$$
$$\text{Prob}(B_{4/5} = 2) = 48/125, \quad \text{Prob}(B_{4/5} = 3) = 64/125;$$

all other values have zero probability. Just like our previous example, the sum of these values is 1, indicating that something does happen.

Even though we have the same possible outcomes as before, the probability of getting two heads is different in the two cases. This is due to the fact that we have different probability functions. These lead to the next important definition associated to random variables, that of a **probability density function (pdf)**.

Probability density function (pdf) for discrete random variables: Let X be a random variable on a discrete outcome space Ω (so Ω is finite or at most countable). The **probability density function** of X, often denoted f_X, is the probability that X takes on a certain value:

$$f_X(x) = \text{Prob}(\omega \in \Omega : X(\omega) = x).$$

Note that some books use the phrase **probability mass function** instead of probability density function. **The pdf always takes on a value greater than or equal to zero, and always sums to one.**

As always, we should say a bit about the notation, as different books have different styles. I like having the subscript X on the probability density function. This reminds us that this density is associated to the random variable X. This is especially helpful when we have a plethora of random variables.

Returning to our fair coin example, there's a nice, closed formula for the probability density function. Recall the non-zero probabilities are

$$\text{Prob}(X = 0) = 1/8, \quad \text{Prob}(X = 1) = 3/8,$$
$$\text{Prob}(X = 2) = 3/8, \quad \text{Prob}(X = 3) = 1/8.$$

We may write this concisely as

$$\text{Prob}(X = k) = \begin{cases} \binom{3}{k} \frac{1}{8} & \text{if } k \in \{0, 1, 2, 3\} \\ 0 & \text{otherwise}, \end{cases}$$

where $\binom{n}{k} = \frac{n!}{k!(n-k)!}$ is the number of ways of choosing k objects from n when order does not matter.

It's easy to check this works: $\binom{3}{0} = \binom{3}{3} = 1$ and $\binom{3}{1} = \binom{3}{2} = 3$; however, just checking isn't particularly enlightening. Why are there binomial coefficients? Would

TABLE 7.1.
The probabilities of the sum of rolls of two independent, fair die. The shape of the output suggests the probabilities of the different events.

- Prob($R = 2$) = 1/36, from the pair (1,1).
- Prob($R = 3$) = 2/36, from the pairs (1,2), (2,1).
- Prob($R = 4$) = 3/36, from the pairs (1,3), (2,2), (3,1).
- Prob($R = 5$) = 4/36, from the pairs (1,4), (2,3), (3,2), (4,1).
- Prob($R = 6$) = 5/36, from the pairs (1,5), (2,4), (3,3), (4,2), (5,1).
- Prob($R = 7$) = 6/36, from the pairs (1,6), (2,5), (3,4), (4,3), (5,2), (6,1).
- Prob($R = 8$) = 5/36, from the pairs (2,6), (3,5), (4,4), (5,3), (6,2).
- Prob($R = 9$) = 4/36, from the pairs (3,6), (4,5), (5,4), (6,3).
- Prob($R = 10$) = 3/36, from the pairs (4,6), (5,5), (6,4).
- Prob($R = 11$) = 2/36, from the pairs (5,6), (6,5).
- Prob($R = 12$) = 1/36, from the pair (6,6).

we get a similar answer if we used the weighted coin? The binomial coefficients arise because we need to choose some number of tosses to be heads, and thus a similar formula holds for any coin.

Let's consider a weighted coin with probability p of a head and $1 - p$ of a tail.

- If we want to get no heads, then all tosses must be tails. There are $\binom{3}{0} = 1$ way to choose none of the tosses to be heads; once we've chosen which tosses are heads the remaining are of course tails. As the probability of a tail is $1 - p$, the probability of no heads is $\binom{3}{0}(1 - p)^3$.
- To get exactly one head, we must choose one of the three tosses to be a head. We have $\binom{3}{1} = 3$ ways to do this. Each head happens with probability p and each tail with probability $1 - p$. Thus the probability of exactly one head is $\binom{3}{1}p(1 - p)^2 = 3p(1 - p)^2$. What's happening here is that each of the three strings TTH, THT, and HTT are equally likely, occurring with probability $p(1 - p)^2$, and there are $\binom{3}{1}$ such triples.
- The case of exactly two heads is almost identical, and we find that happens with probability $\binom{3}{2}p^2(1 - p)$.
- The case of three heads is handled similarly to that of no heads, and we find its probability is $\binom{3}{3}p^3$.

If we take $p = 4/5$, we regain the formulas we found earlier. This is called the **Binomial Distribution** (with parameters 3 and 4/5), and while we'll discuss it in much greater depth in §12.2, we'll say a few quick words about it now. It's very nice to have a closed form expression, as we can then find any probability by simple substitution. Later we'll show that if we toss n coins where each is heads with probability p, then the probability of getting exactly k heads is $\binom{n}{k}p^k(1 - p)^{n-k}$ for $k \in \{0, 1, \ldots, n\}$ and 0 otherwise. Thus, this distribution is a pleasure to work with, as we have a clean formula for each probability.

Let's do another example and revisit rolling two fair die. It turns out there's a nice formula for the probability density function. Each of the 36 pairs happens with probability 1/36. Letting R denote the sum of the two rolls, all we need to do to find the probability R equals r is to count up the number of pairs that sum to r. We give the answer in Table 7.1; note again everything is non-negative and sums to 1.

Notice the pattern in the numbers: 1/36, 2/36, 3/36, rising up to 6/36 and then falling down to 1/36. You can also see this pattern by looking at the text itself! We can write this concisely as

$$\text{Prob}(R = r) = \begin{cases} \frac{6-|r-7|}{36} & \text{if } r \in \{2, 3, \ldots, 12\} \\ 0 & \text{otherwise.} \end{cases}$$

The absolute value starts at its maximum when $r = 2$, rises linearly to its minimum at $r = 7$, and then decreases linearly back to its maximum when $r = 12$. This gives a maximum probability of 6/36 when $r = 7$ (as we're not subtracting anything), and minimum probabilities of 1/36 at $r = 2$ or 12.

Don't worry if you didn't see this concise way of writing the answer. It's not easy sniffing out patterns like this. The more problems you do, though, the easier it becomes and the more patterns you can recognize. For example, let's look at the numbers a bit more and try to see what clues there are to help us find this answer. Note the range of r is from 2 to 12, and the 7 in $|r - 7|$ is the average of 2 and 12. If we had two fair die with each having n sides, we'd expect the probability of rolling a sum of r to be $\frac{n-|r-(n+1)|}{n^2}$ if $r \in \{2, 3, \ldots, 2n\}$ and 0 otherwise. One way to view this is that every time the sum increases, so long as we haven't reached the midpoint, we have one more possibility than before; when we increase beyond the midpoint, we have one fewer possibility each time.

We've made a lot of progress. We've done two examples, and even seen how to generalize them. Unfortunately, we're only generalizing in certain ways. For example, in the die problem we increased the number of sides of our die. Another modification would be to increase the number of die we roll. The formulas become far more involved as the number of die increase. Fortunately, for many problems we don't need to know the exact probability of each outcome, but only the probability of getting outcomes in certain ranges. This leads to the Central Limit Theorem, the topic of Chapter 20.

 As a nice exercise, try and find the probability density function for the sum of three rolls of a fair die, and then for four. (Hint: Four die are actually easier than three die, as we can group the four die into two sets of two.)

7.3 Discrete Random Variables: CDFs

Let's recap what we've done in the previous sections. We've defined discrete random variables and their probability density functions (pdfs). Our two big examples of discrete random variables are the number of heads of three tosses of a fair coin, and the sum of the numbers on a pair of fair die. We calculated the probability density functions for each.

There's a related concept which is very important, that of the **cumulative distribution function (cdf)**. Though this concept is more useful for continuous random variables than discrete random variables, it still has its uses here, and is thus worth studying. Before diving into our discussion of the cdf, let's quickly preview what it gives us. The cdf turns out to be a very useful tool to help us pass from knowledge of one random variable to knowledge of a later random variable. This is called the **Cumulative Distribution Function Method**, and we'll describe it in §13.2.4.

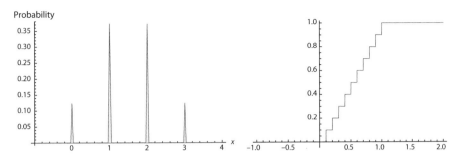

Figure 7.1. The probability distribution function (left) and the cumulative distribution function (right) for the random variable of the number of heads in three tosses of a fair coin.

> **Cumulative distribution function (cdf) for discrete random variables**: Let X be a random variable on a finite or at most countable discrete outcome space Ω. Recall that the **probability density function** of X, often denoted f_X, is the probability that X takes on a certain value. The **cumulative distribution function**, often denoted F_X, is the probability that X is at most a certain value. We write these as
>
> $$f_X(x) = \text{Prob}(\omega \in \Omega : X(\omega) = x)$$
> $$F_X(x) = \text{Prob}(\omega \in \Omega : X(\omega) \leq x).$$

We talked about the **good notation** for the pdf earlier, and explained that the point of the subscript is to help us remember what random variable is associated to what density. Okay, that explains the subscript. Why do we use a capital letter for the cumulative distribution function? The reason is an analogy from calculus (and, as we'll see later, this is a hint on why the cdf is a more useful concept for continuous than discrete random variables). There, we used f to denote a function and F to denote its anti-derivative, with the anti-derivative related to the area under the curve. We have a similar situation here (we'll see the connection is even more pronounced when we study continuous random variables). The cumulative distribution function is adding all the probabilities up to a given point, which is quite similar to integration and thus why we transfer that notation to here. *As always, it's worth spending time thinking about notation, and choosing something so that if you quickly glance down you know what each quantity is.*

Let's look at tossing the fair coin three times. Let F_X denote the cumulative distribution function, so $F_X(x) = \text{Prob}(X \leq x)$. We have

$$F_X(x) = \begin{cases} 0 & \text{if } x < 0 \\ 1/8 & \text{if } 0 \leq x < 1 \\ 4/8 & \text{if } 1 \leq x < 2 \\ 7/8 & \text{if } 2 \leq x < 3 \\ 1 & \text{if } x \geq 3. \end{cases}$$

Notice the cumulative distribution function is discontinuous, with jumps at 0, 1, 2, and 3. See Figure 7.1, where we plot both its pdf and cdf.

The situation is even worse if we have more heads. There is a "nice" formula for the cumulative distribution function of the number of heads in n tosses of a fair coin. The probability of getting exactly k heads is $f_X(k) = \binom{n}{k}(1/2)^k(1 - 1/2)^{n-k} = \binom{n}{k}1/2^n$ for $k \in \{0, 1, \ldots, n\}$. The cumulative distribution function F_X just adds all of these up to a given point. Thus

$$F_X(m) = \begin{cases} 0 & \text{if } m < 0 \\ \sum_{k=0}^{m} \binom{n}{k}1/2^n & \text{if } 0 \leq m \leq n \\ 1 & \text{if } m \geq n. \end{cases}$$

Unfortunately, there are no nice, simple, closed form expressions for partial sums of binomial coefficients. There are of course nice expressions if we add none or all of them.

Let's do a more general case than just three coins, which will suggest a general fact about cumulative distribution functions. Consider the case where we have n coins, and each coin is a head with probability p and a tail with probability $1 - p$. Let X be the random variable which equals the number of heads. We've already discussed its probability density function; it's just

$$f_X(k) = \begin{cases} \binom{n}{k}p^k(1 - p)^{n-k} & \text{if } k \in \{0, 1, \ldots, n\} \\ 0 & \text{otherwise.} \end{cases}$$

In writing down the pdf, we looked at $X = k$ instead of $X = x$. It doesn't matter what letter we use for the values our random variable takes on. The standard convention is to use letters like i, j, k, ℓ, m, and n for integers, and letters like x, y, and z for real numbers. It doesn't matter, but it helps to use a k instead of an x as this alerts us to the values being integers.

To find the cumulative distribution function F_X at m, we just sum over all integers $k \leq m$. If $m < 0$ then our sum is empty, and thus $F_X(m) = 0$ for such m. If $m \geq n$, then the sum is 1 by the Binomial Theorem (see Theorem 2.7 for a review and proof), since

$$\sum_{k=0}^{n} \binom{n}{k}p^k(1 - p)^{n-k} = (p + (1 - p))^n = 1.$$

There's an important fact worth isolating here. The cumulative distribution function at x is the probability our random variable takes on a value at most x. If there are only finitely many values it can take on (such as the sum of two die, or the number of heads tossed in n coins, or perhaps the number of heads minus the number of tails in n tosses), then if x is sufficiently large and negative the cumulative distribution function is zero from that point and before; similarly, if x is sufficiently large and positive the cumulative distribution function is 1 from that point and onward. For example, if we're looking at the number of heads minus the number of tails in n tosses, once x is less than $-n$ then the cumulative distribution function is zero, as there is no probability of having a value less than $-n$; similarly, its value is 1 once we look past n, as we can't have more than n heads more than tails. Let's highlight this observation.

Limiting behavior of cumulative distribution functions: Let F_X be the cumulative distribution function of a discrete random variable X. Then

$$\lim_{x \to -\infty} F_X(x) = 0, \qquad \lim_{x \to \infty} F_X(x) = 1,$$

and if $y > x$ then $F_X(y) \geq F_X(x)$.

We have to use limits above even though our random variable is defined on a discrete set. We'll give an example that shows the limits are necessary; in other words, there is a random variable such that for any finite x we have $0 < F_X(x) < 1$. Our example needs the geometric series formula (see §1.2 for more on it), which we quickly review. There are two versions, a finite and an infinite version. The finite version is

$$\sum_{n=k}^{j} ar^n = \frac{ar^k - ar^{j+1}}{1 - r}.$$

If $|r| < 1$ then we can let j tend to infinity, obtaining

$$\sum_{n=k}^{\infty} ar^n = \frac{ar^k}{1 - r}.$$

 Imagine our outcome space Ω is the set of all integers and the probability of n is zero if n is zero, and $1/2^{|n|+1}$ if $n \neq 0$. Using the geometric series formula, we see that this is indeed a probability function. It's clearly non-negative; all that's left is to check it sums to 1. By symmetry, the sum of the positive terms equals the sum of the negative terms, so we're reduced to showing the sum over $n \geq 1$ is $1/2$. Using the infinite version of the geometric series formula, we find that

$$\sum_{n=1}^{\infty} \frac{1}{2^{n+1}} = \frac{1}{2} \sum_{n=1}^{\infty} \frac{1}{2^n} = \frac{1}{2} \frac{1/2}{1 - 1/2} = \frac{1}{2},$$

so it is a probability function.

Let's find the cumulative distribution $F_X(m)$. We'll break it up into three cases: m is zero, positive, or negative. From symmetry, we know $F_X(0) = 1/2$. If m is positive, all we need to do is sum from 0 to m and then add $1/2$, as the sum of the negative values is $1/2$. For m a positive integer, we get

$$F_X(m) = \frac{1}{2} + \sum_{k=1}^{m} \frac{1}{2^{k+1}} = \frac{1}{2} + \frac{1/4 - 1/2^{m+2}}{1 - 1/2} = 1 - \frac{1}{2^{m+1}}.$$

What if $m = -|m|$ is negative? We find

$$F_X(-|m|) = \sum_{k=-\infty}^{-|m|} \frac{1}{2^{|k|+1}} = \frac{1}{2} \sum_{k=|m|}^{\infty} \frac{1}{2^k} = \frac{1}{2} \frac{1/2^{|m|}}{1 - 1/2} = \frac{1}{2^{|m|}}.$$

As a quick check, this makes sense: if $m = -1$ we get $1/2$, which corresponds to the fact that half the time we have a negative value. Also, if $\lfloor r \rfloor$ represents the **floor** of the real number r (i.e., the largest integer at most r), then $F_X(r) = F_X(\lfloor r \rfloor)$ and it is enough to find the cumulative distribution function at the integers. We find

$$F_X(m) = \begin{cases} 1/2^{|m|} & \text{if } m \text{ is a negative integer} \\ 1/2 & \text{if } m = 0 \\ 1 - 1/2^{m+1} & \text{if } m \text{ is a positive integer.} \end{cases}$$

We do see that

$$\lim_{m \to -\infty} F_X(m) = 0, \qquad \lim_{m \to \infty} F_X(m) = 1;$$

however, for any finite m we have $0 < F_X(m) < 1$. Thus, it's only as we head down to negative infinity that we get 0, or up to positive infinity that we get 1. This is why we need to take limits.

Before we turn to our final example, we recall a needed fact about sums of integers: if n is a non-negative integer, then

$$\sum_{k=1}^{n} k = \frac{n(n + 1)}{2}.$$

A proof is given in §A.2.1, but we can give a short proof here as well. Take the numbers $1, 2, \ldots, n$ and add the numbers $n, n - 1, \ldots, 1$. We now have n pairs, with each pair summing to $n + 1$. This equals $n(n + 1)$ and is twice our desired sum, so the sum is $n(n + 1)/2$. Symbolically, if S is the sum then

$$2S = (1 + \cdots + n) + (n + \cdots + 1) = (1 + n) + \cdots + (n + 1) = n(n + 1),$$

so $S = n(n + 1)/2$. We can use this to find the sum of a consecutive block of numbers:

$$m + (m + 1) + \cdots + n = \sum_{k=1}^{n} k - \sum_{\ell=1}^{m-1} \ell = \frac{n(n + 1)}{2} - \frac{(m - 1)m}{2}.$$

The argument above is another example of the **At Least to Exactly Method**. We used that to find the probability of having a value of exactly k by taking the probability of at least k and subtracting the probability of at least $k + 1$. The methods that we're seeing throughout this book are not isolated tricks, but powerful techniques that you will use again and again as you continue your studies.

We now do one last example, returning to the die problem. In §7.2 we found the probability density function for the random variable that's the sum of two rolls of a fair die. It was

$$\text{Prob}(R = r) = \begin{cases} \frac{6 - |r - 7|}{36} & \text{if } r \in \{2, 3, \ldots, 12\} \\ 0 & \text{otherwise.} \end{cases}$$

One of the themes of this book is that there's often many different ways to write something, and sometimes one way is more illuminating than another. It's hard to add absolute values, so we prefer to write the probabilities as

$$\text{Prob}(R = r) = \begin{cases} \frac{r-1}{36} & \text{if } r \in \{2, 3, \ldots, 7\} \\ \frac{13-r}{36} & \text{if } r \in \{7, 8, \ldots, 12\} \end{cases}$$

and zero otherwise (note both definitions agree when $r = 7$, so we may use either).

We can now find the cumulative distribution function F_R by summing consecutive integers. Clearly $F_R(m) = 0$ if m is less than 2. In the sum we'll encounter in a moment, it starts as a sum over r from 2 to m; we change variables and replace it with a sum over ℓ from 1 to $m - 1$; doing so forces us to replace each occurrence of r with $\ell + 1$ (in other words, we write $r = \ell + 1$, so $\ell = r - 1$). For $2 \leq m \leq 7$ we have

$$F_R(m) = \sum_{r=2}^{m} \frac{r-1}{36} = \frac{1}{36} \sum_{\ell=1}^{m-1} \ell = \frac{1}{36} \frac{(m-1)m}{2} = \frac{(m-1)m}{72}.$$

We replaced r with $\ell = r - 1$. We didn't have to use the letter ℓ; we could use any letter. The standard convention, unfortunately, is to use the same letter (in this case, r). While many books will just replace r with $r - 1$, I've found it confuses many students learning the subject. There's really no harm with this, as r is just a dummy variable, but there is a danger that r in one line isn't quite the same as r in another.

What about $F_X(m)$ for $m \geq 8$? We know the contribution from $r \leq 7$ is $\frac{(7-1)7}{72} = \frac{7}{12}$, so all we need to do is add to that the contribution from $8 \leq r \leq m$. For $m \in \{8, 9, \ldots, 12\}$ we find

$$F_R(m) = \frac{7}{12} + \sum_{r=8}^{m} \frac{13-r}{36}$$

$$= \frac{7}{12} + \frac{1}{36} \sum_{\ell=13-m}^{5} \ell$$

$$= \frac{7}{12} + \frac{1}{36} \left(\frac{5(5+1)}{2} - \frac{(13-m-1)(13-m)}{2} \right)$$

$$= 1 - \frac{(13-m-1)(13-m)}{72}.$$

Collecting all the pieces, we see

$$F_R(m) = \begin{cases} 0 & \text{if } m < 2 \\ \frac{(m-1)m}{72} & \text{if } m \in \{2, 3, \ldots, 7\} \\ 1 - \frac{(13-m-1)(13-m)}{72} & \text{if } m \in \{7, 8, \ldots, 12\} \\ 1 & \text{if } m \geq 12. \end{cases}$$

Whenever possible, *check your answer!* Note that the two expressions agree when $m = 7$. Further, when $m = 12$ we get $F_R(12) = 1$, which is good as we can't roll a sum greater than 12. Thus, it's quite likely that our algebra is correct.

As we want to motivate coding to get a sense of an answer, this is an important point to keep in mind. We often want to figure out how to turn a real-life scenario into numbers so that we can code it. Depending on what we want to figure out, we may want to choose tails as 0 and heads as 1, or tails as -1 and heads as 1. As another example, consider a deck of cards. We may want to construct a set with four iterations of each number 1 through 13, or we may prefer the set of integers from 1 to 52, where we consider each interval of 13 as one suit.

Here is some Mathematica code to create a histogram of the sum of two dice rolls. By doing a large number of simulations we should have very accurate estimates of the true values; we set up our program to compare our numerics to the predicted values.

```
diceroll[num_] := Module[{},
  allsums = {}; (* store sums here *)
  For[n = 1, n <= num, n++, (* main loop *)
    {
    (* next lines print where we are every 10\% *)
    (* this is a good trick for long code so know how much done *)
    If[Mod[n, num/10] == 0,
      Print["Have done ", 100. n/num, "%."]];
    die1 = RandomInteger[{1, 6}]; (* chooses die 1 value *)
    die2 = RandomInteger[{1, 6}]; (* chooses die 2 value *)
    AppendTo[allsums, die1 + die2]; (* adds sum to list *)
    }];
  For[i = 2, i <= 12, i++,
  (* counts how often we roll an i in our list *)
  (* Mathematica has a nice Count function for this *)
  (* If not can go through list and record  how often each *)
  (* Alternatively could update counts when generate die1+die2 *)
  (* We do it this way as need the list for the histogram plot *)
  Print["Percent of time rolled ", i, " is ",
    100.0 Count[allsums, i]/num, "%, and theory
      predicts ", 100.0 (6 - Abs[7 - i])/36, "%."];
  ];
  (* Prints a histogram and scales to area under curve is 1 *)
  Print[Histogram[allsums, Automatic, "Probability"]];
  ];
Timing[diceroll[100000]] (* runs and times how long program takes *)
```

While it is good practice when coding to have a list of what you are doing and test using trials to make sure the code is giving you the answer you want, be careful. Lists are *very* expensive to work with and store in memory. This program took .31 seconds to do 10,000 rolls, 19.69 seconds to do 100,000, and 79.64 seconds to do 200,000. The cost is most definitely *not* linear in expanding, and as I'm impatient I gave up on having it do one million rolls. What went wrong with such a simple problem? It's expensive to store large lists and it slows down the computer enormously. For this problem, we really don't need to save all these values; we just want to keep track of how often we get each sum, we don't care about the *order* in which we get the sums. Thus, rather than saving our results in a list we should have an array of the 11 possible outcomes (from 2 to 12) and increment, as done in the following code.

```
betterdiceroll[num_] := Module[{},
  For[i = 2, i <= 12, i++, number[i] = 0]; (* initialize counts to 0 *)
```

```
For[n = 1, n <= num, n++, (* main loop *)
 {
  (* Prints out every time do 10\%, good habits for longer runs *))
  If[Mod[n, num/10] == 0, Print["Have done ", 100. n/num, "%."]];
  die1 = RandomInteger[{1, 6}]; (* chooses die 1 value *)
  die2 = RandomInteger[{1, 6}]; (* chooses die 2 value *)
  roll = die1 + die2; (* calculates sum, increments right counter *)
  number[roll] = number[roll] + 1;
  }];
 list = {}; (* stores in list the probabilities of each sum *)
 For[i = 2, i <= 12, i++,
 {
  list = AppendTo[list, {i, 100.0 number[i]/num}];
  Print["Percent of time rolled ", i, " is ", 100.0 number[i]/num,
   "%, and theory predicts ", 100.0 (6 - Abs[7 - i])/36, "%."];
  }];
 Print[ListPlot[list]]; (* prints list *)
 ];
Timing[betterdiceroll[100000]] (* runs and times how long program takes *)
```

It took .72 seconds to do 100,000 rolls, and 6.8 seconds to do one million rolls. Notice that here we do have essentially a linear increase in run-time with the number of rolls.

7.4 Summary

In this chapter we met one of the central objects of study, discrete random variables; we'll see continuous random variables in the next chapter. Discrete random variables are useful in modeling a variety of problems. They allow us to assign a number to events in our outcome space; this can range from the temperature at a weather station to the pressure in a box to the number of bacteria in a culture. In order to be useful, however, we need ways to extract information about these random variables. That's where the probability density function enters.

George McFly's slip in the movie *Back to the Future* is a nice way of highlighting the true nature of a random variable.

> *Loraine:* Hey, don't I know you from somewhere?
> *George:* Yes, yes, I'm George, George McFly, and I'm your density. I mean, I'm your destiny.

We use destiny to denote something that must happen, a preordained sequence of events. In the movie, George has a crush on Loraine. He's finally prodded to talk to her, but he's so nervous that he misspeaks. He meant to say that they were meant for each other, that they were destined for each other. Instead, he uses the word density. We use densities to talk about random variables, to describe the probabilities that one of a multitude of possibilities may happen.

While the words are far apart, it turns out a synthesis of the two concepts is often very powerful. One of my favorite examples is the $3x + 1$ **problem**. We start with a positive integer, call it a_0, as the seed. We then define a sequence by setting

$$a_{n+1} = \begin{cases} 3a_n + 1 & \text{if } a_n \text{ is odd} \\ a_n/2 & \text{if } a_n \text{ is even.} \end{cases}$$

For example, if we start with $a_0 = 7$ we get the sequence

$$7 \to 22 \to 11 \to 34 \to 17 \to 52 \to 26 \to 13 \to 40 \to 20 \to 10$$
$$\to 5 \to 16 \to 8 \to 4 \to 2 \to 1 \to 4 \to 2 \to 1 \to \cdots.$$

It's not hard to show that once we hit 1, we eternally cycle from 1 to 4 to 2 to 1. It's conjectured that no matter what positive integer we choose as our seed, we eventually reach this cycle. Given any such integer, it's predetermined what happens (if it terminates, we can even discover this after enough iterations); each seed has a destiny. However, we can gain a lot of insight as to what happens by studying a related system.

 Remark: It's technically more convenient to study the related sequence

$$b_{n+1} = \frac{3b_n + 1}{2^k}, \quad \text{where } 2^k | 3b_n + 1 \text{ but } 2^{k+1} \nmid 3b_n + 1;$$

in other words, we start with an odd number, multiply by 3, and add 1 (which gives an even number), and then remove as many powers of 2 as we can. Note that if b_n is odd then $3b_n + 1$ is even, and thus there will be a positive number of powers of 2 to remove. We expect that half the time we can only remove one power of 2, one-fourth of the time we can remove exactly two powers of 2, one-eighth of the time we can remove exactly three powers of 2, and so on.

We use this to model the deterministic system with a random one. We have a deterministic map on the odd integers, defined by $T(2m + 1) = (6m + 4)/2^{n(6m+4)}$, where $n(6m + 4)$ is the number of powers of 2 we can remove from $6m + 4 = 3(2m + 1) + 1$. We consider the related process that sends an input x to $3x/2$ half the time, to $3x/4$ one-fourth of the time, $3x/8$ one-eighth of the time, et cetera. Notice this is very similar to the deterministic map T. There are two differences. The first is that this map is probabilistic. The second is we have dropped the $+1$ in T; the hope is that for large inputs there will be negligible changes by this (if x is of size 10^{100}, for the probabilistic process there is essentially no difference with or without the $+1$; however, for the deterministic process there can be a great change in the number of powers of 2 removed). While this process is random, it turns out to do a wonderful job describing and predicting properties of our deterministic system, and highlights the great power and utility of random variables. For more on this fascinating problem, see the survey articles [Lag1, Lag2], as well as the research collection [Lag3].

We end this chapter by taking a moment to reflect on some of the examples we've analyzed. Why were we able to get a nice closed form expression for some examples and not others? The reason is that we were able to use the geometric series formula for one, and a formula for sums of integers for another. In our two "successful" cases, we got a nice, closed form expression for the cdf because we had a good summation result. Unfortunately, most sums don't have nice, closed form expressions. It's not a coincidence we studied the problems we did, as the geometric series formula and the sum of integers are the two best known good summation formulas. In general, the cumulative distribution function of a discrete random variable can't be simplified to a nice, closed form expression. The situation is markedly different for continuous random variables. What's the difference? In continuous random variables, we'll have integrals rather than sums. We'll have the Fundamental Theorem of Calculus at our disposal, which allows us to get nice, closed form answers.

7.5 Exercises

Exercise 7.5.1   *Describe three real-world applications of discrete random variables. Do not use the ones already described in the chapter.*

Exercise 7.5.2   *The alternating harmonic series is*

$$\sum_{n=1}^{\infty} \frac{(-1)^{n+1}}{n} \; = \; 1 - \frac{1}{2} + \frac{1}{3} - \frac{1}{4} \cdots .$$

It is known to sum to $\log(2)$. *Is*

$$\Pr(X = n) \; = \; \frac{1}{\log(2)} \frac{(-1)^{n+1}}{n}$$

a probability density function? Explain.

Exercise 7.5.3   *Fix positive integers* k *and* n *with* $k \le n$. *Show that* $\Pr(M = m) = \binom{m-1}{k-1} / \binom{n}{k}$ *for* $k \le m \le n$ *is a probability density function.*

Exercise 7.5.4   *For what value of* C *is* $\Pr(X = n) = C/n!$ *for all* n *in the non-negative integers a probability density function?*

Exercise 7.5.5   *We saw in the example with adding the outcome of two fair, independent die that the sum of two identical random variables tends toward the middle. More precisely, if we add two independent identically distributed random variables with outcome spaces of arithmetic sequences, the sum tends to be near twice the average of the sequences. Why is this?*

Exercise 7.5.6   *What do we expect* $X - X$ *to be around if* X *is a discrete uniform random variable on* $\{m, m + 1, \ldots, n\}$? *Why?*

Exercise 7.5.7   *What is the probability distribution for the number of rolls it takes with a six sided die before we get a six?*

Exercise 7.5.8   *Let* $f(n) = \frac{1}{n}$ *for* $n \in \{1, 2, 3, \ldots \}$; *is* f *a probability mass function? Why or why not? If it isn't, is there a constant* C *such that* $g(n) = Cf(n)$ *is?*

Exercise 7.5.9   *Let* $f(n) = \frac{1}{n(n+1)}$ *for* $n \in \{1, 2, 3, \ldots \}$; *is* f *a probability mass function? Why or why not? If it isn't, is there a constant* C *such that* $g(n) = Cf(n)$ *is?*

Exercise 7.5.10   *Let* $f(n) = \frac{1}{n(n+1)(n+2)}$ *for* $n \in \{1, 2, 3, \ldots \}$; *is* f *a probability mass function? Why or why not? If it isn't, is there a constant* C *such that* $g(n) = Cf(n)$ *is?*

Exercise 7.5.11   *Imagine we have four independent die, where the first two equally take on values in* $\{1, 2, 3, 4, 5, 6\}$ *and the second two equally take on values in* $\{100, 200, 300, 400, 500, 600\}$. *What is the distribution of the sum of the four rolls?*

Exercise 7.5.12   *If we have a coin with probability* p *of being heads, what is the probability distribution for how long we need to wait for a head? Verify that this is a probability distribution.*

Exercise 7.5.13   *Consider a coin with probability* p *of heads. Find the probability density function for* X_2, *where* X_2 *is how long we must wait before we get our second head. What about waiting for* k *heads? What about getting two consecutive heads?*

Exercise 7.5.14   *We toss n fair coins. Every coin that lands on heads is tossed again. What is the probability density function for the number of heads after the second set of tosses (i.e., after we have retossed all the coins that landed on heads)?*

Exercise 7.5.15   *What is the probability a fair coin will land heads up exactly 5 times in 8 flips? What is the probability of getting at most 5 heads in 8 flips?*

Exercise 7.5.16   *Show that*

$$\text{Prob}(X = k) = \begin{cases} \binom{n}{k} p^k (1-p)^{n-k} & \text{if } k \in \{0, 1, \ldots, n\} \\ 0 & \text{otherwise} \end{cases}$$

is a probability distribution for $0 \le p \le 1$. (Hint: Use the Binomial Theorem.)

Exercise 7.5.17   *Consider random variables X and Y, where X is the maximum of three rolls of a fair die and Y is equal to the roll of a single fair die. Find $P(X > Y)$.*

Exercise 7.5.18   *Let X be a distinct random variable. Must $\Pr(X + X = 2x) \ge \Pr(X = x) + \Pr(X = x)$? Prove or disprove.*

Exercise 7.5.19   *If we flip 10 fair, independent coins and then re-flip all the heads, what is the probability distribution for the total number of heads?*

Exercise 7.5.20   *You have $10 dollars and are offered the following wager. You will flip a fair coin 5 times. For each head you double your money, for each tail you lose half your money. Find the probability density function for your amount of money after 5 flips. Is it a good wager assuming that each dollar you win is equally valuable to each dollar you lose? Would it matter if you started with $32 instead of $10?*

Exercise 7.5.21   *Consider the following two investments. You can buy Stock A, which has a value that increases by a factor of 1.1 each year with probability 1/2 and keeps the same value with probability 1/2. Alternatively, you can purchase Bond B which has a guaranteed increase in value of 1.05 each year. Assuming that you can only purchase one of these investments and you are going to hold your position over a 5 year period, which investment gives a better expected return.*

Exercise 7.5.22   *Building on the previous exercise, now either you buy Stock A, which stays the same with probability 1/3, increases by a factor of 1.1 with probability 1/3, and increases by a factor of 1.2 with probability 1/3 each year, or you can buy Bond B, which has a guaranteed increase in value of 1.1 each year. Comment on which investment you would choose.*

Exercise 7.5.23   *The following exercise relates directly to the popular board game Risk. We avoid going into details about the rules of the game, and will instead just focus on a probability related aspect. Consider 5 fair dice, a_1, a_2, a_3, b_1, b_2, which are independent uniform random variables on $\{1, 2, \ldots, 6\}$. Find the probability both $\max(a_1, a_2, a_3) > \max(b_1, b_2)$ and the second largest of the $a_i > \min(b_1, b_2)$ (in other words, the highest roll of the a's exceeds the highest roll of the b's, and similarly for the second highest).*

Exercise 7.5.24   *With notation the same as the previous exercise, find the probability that $\max(b_1, b_2) > \max(a_1, a_2, a_3)$ and $\min(b_1, b_2)$ is larger than the second largest of the a's.*

Exercise 7.5.25 *Consider the random variable X, where X is equal to the number of typos you make in a given page. Describe properties that you would expect its pdf to have. One thing to consider is whether you believe that each typo is independent of other typos or not. Justify your model.*

Exercise 7.5.26 *A bag contains 4 red chips, 3 white chips, and 5 black chips. You pick three chips from the bag blindly. Let X denote the number of red chips you choose. Find the pdf for the distribution of X.*

Exercise 7.5.27 *Imagine you are traveling and have 5 flights to catch. Unfortunately, you didn't plan well and have a 0 minute layover, that is if a flight is late and the next flight is on time you will miss your flight, ruining all your travel plans. Make the unreasonable assumptions that all planes that are late are equally late, say 30 minutes, and that airline delays are independent, so that if your first flight is delayed it does not have any effect on the probability any of your later flights are delayed. Also, make the more reasonable assumption that airplanes are delayed 25% of the time. What is the probability you do not miss any of your 5 flights?*

Exercise 7.5.28 *Bull Durham once said: "18 strikeouts, new league record; 18 walks, another new league record." In baseball, a batter strikes out if they get three strikes before four balls, and a walk occurs when the fourth ball comes before the third strike. Assume that a pitcher is un-hittable, but so wild that he is expected to walk half of the batters and strike out the other half. What must this pitcher's probability of throwing a strike be for this to be true?*

Exercise 7.5.29 *Assume a hitter has the following possibilities each at bat: Home run, triple, double, single, fly out, walk, ground out, and strike out. Let's say each possibility but the last has a 1/10 chance of occurring, and there is a 3/10 chance of a strike out. If this hitter has four plate appearances each game, what are the chances of the following:*

1. *Wearing a golden sombrero (4 strikeouts)?*
2. *Hitting for the cycle (home run, triple, double, and single in any order)?*
3. *Hitting at least 2 home runs?*
4. *Getting on base each time?*
5. *Getting on base at least twice?*
6. *Getting on base at most twice?*
7. *Hitting .500 or over (number of hits greater than or equal to number of outs)?*

Exercise 7.5.30 *Make a histogram showing the probability $X = n$ if X is a random variable whose value corresponds to the number of heads in 3 tosses of a fair coin. Redo with 6 tosses, then 10, then 20, then 40.*

Exercise 7.5.31 *Redo the previous exercise but now instead of plotting the number of heads plot the number of heads minus the number of tails.*

Exercise 7.5.32 *Plot the probability density function for the number of heads in 3 tosses of a fair coin. Redo with 6 tosses, then 10, then 20, then 40.*

Exercise 7.5.33 *Plot the cumulative density function for the number of heads minus tails in 3 tosses of a fair coin. Redo with 6 tosses, then 10, then 20, then 40.*

CHAPTER 8 _____

Introduction to Continuous Random Variables

I don't agree with mathematics; the sum total of zeros is a
frightening figure.
— STANISLAW J. LEC, *More Unkempt Thoughts* (1968)

Not surprisingly, there are many similarities between the theory of continuous and discrete random variables, as well as a very important difference. The difference is that for continuous random variables we have integrals, while for discrete we have sums. How profound is this difference? Well, anyone who has taken calculus knows of the Fundamental Theorem of Calculus. We'll review this theorem in a little bit; briefly, it allows us to find areas under curves by doing integrals. For many functions, these integrals can be done without too much difficulty, and we have exact answers for our probabilities. There is, sadly, no corresponding theory for sums. In general it's much harder to evaluate a sum and get a nice, simple, closed form expression (as we saw in §7.3, though of course there are many discrete distributions where we can execute the summation). This is why continuous random variables are often easier to handle, and more desirable as a closed form expression allows us to see how the answer changes as we vary the parameters.

This chapter is preparatory for our discussion of many of the common continuous random variables in later chapters. We'll describe some of their uses and many of their properties. To make those chapters self-contained, we'll often go through identical arguments as we did in the discrete random variables chapters. It's not bad seeing these arguments multiple times. The general framework is the same; the only difference is that we'll have integrals to evaluate and not sums.

Before getting to these continuous probability distributions however, we'll first quickly review some results from calculus and other needed material in this chapter. Without fail, in almost every math class the part that gives students the greatest heartache is the material assumed known from earlier classes. In probability this is especially dangerous, as sometimes it has been a year (or years!) since a student has seen derivatives and integrals. Fortunately a brief refresher course is usually enough. If you would like a more detailed review of these concepts, I urge you to read Adrian Banner's *The Calculus Lifesaver*, where all this material is carefully worked out. To gauge how well you remember your techniques of differentiation and integration, I've written and solved over 50 calculus problems; these (first the statements and then detailed solutions) and other supplementary material (such as a review of the Change of Variables Theorem and some calculus review lectures) are available in the online supplements at http://web.williams.edu/Mathematics/sjmiller/public_html/probabilitylifesaver/.

The plan of attack for this chapter is to

1. review the Fundamental Theorem of Calculus and its applications to probability, and then

2. discuss the issues that arise in determining the probability of singleton events for continuous random variables.

8.1 Fundamental Theorem of Calculus

The Fundamental Theorem of Calculus is one of the most important tools in studying continuous random variables. In fact, it's the reason we can get such nice, closed form expressions time after time. We'll quickly review the meaning of the different terms in its expression, and then show you why this is such an important theorem. Sadly, most calculus teachers do not emphasize the connections between integration and probability. Because of this, many students are often unimpressed and unmotivated when they see these arguments in calculus. We'll see in just a little bit that the Fundamental Theorem of Calculus allows us to compute areas under curves, and these areas correspond to probabilities of events!

Let's quickly review some terminology from calculus. If f is a function, then another function F is said to be an **anti-derivative** or an **(indefinite) integral** of f if $F'(x) = f(x)$. Note that the anti-derivative isn't unique: if $F'(x) = f(x)$ and $G(x) = F(x) + C$ for some fixed constant C, then $G'(x) = f(x)$ as well; this is why we say *an* anti-derivative and not *the* anti-derivative. It turns out that this is the only obstruction; specifically, if F and G are two anti-derivatives of f then they must differ by a constant.

Whenever you see functions and statements, try to assign values and make a story. For example, let's have $f(x)$ represent how fast we're traveling at time x, and $F(x)$ represent where we are at time x. Note that this is a consistent story, as the rate of change of where we are is just our speed. We can now interpret the above spiel on anti-derivatives. Imagine we have two friends, say Floyd and Grover, with $F(x)$ and $G(x)$ giving their respective locations at time x. Assume now that their speeds are the same; thus $F'(x) = G'(x)$. Well, if they're always traveling at the same speed then the distance between them must be constant, so there must be a C with $G(x) = F(x) + C$. (We chose the somewhat unusual names of Floyd and Grover so that the letters of our functions corresponded to our people, hopefully making it easier for you to keep track of which function is associated to which person. I always recommend taking a few moments to come up with good notation.)

The next item we need is the notion of a **piecewise continuous function**, which is a function which is continuous except at finitely many points. For example, consider the function in Figure 8.1 on the interval [0, 4]. It has three points where it's discontinuous, namely $x = 1, 2$, and 3, and is continuous everywhere else. This is thus a piecewise continuous function.

Note that the definition requires there to be just finitely many discontinuities. If there were infinitely many, we might not be able to use the standard tools from differential and

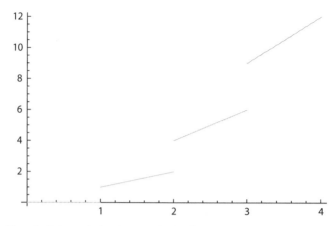

Figure 8.1. Plot of $f(x) = x\lfloor x \rfloor$, where $\lfloor t \rfloor$ is the floor function, returning the greatest integer at most t.

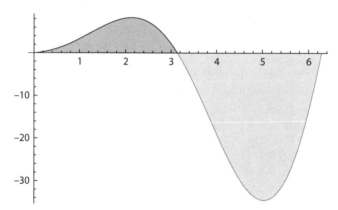

Figure 8.2. The area under $f(x) = (x + 1)^2 \sin x$ for $0 \le x \le 2\pi$. The area above the x-axis is counted positively and the area below the x-axis is counted negatively.

integral calculus. If you take a course in real analysis, many of these assumptions can be weakened, as the standard integral (also known as the Riemann integral) is replaced with the more powerful Lebesgue integral.

We have just one more definition. The area between the curve $y = f(x)$ and the x-axis from $x = a$ to $x = b$ is denoted $\int_a^b f(x)dx$; this is called the **definite integral**. This is a signed quantity, which often confuses students as how can area be negative? It's because we count area *above* the x-axis and below the curve as positive, and area *below* the x-axis and above the curve as negative. For example, we consider $f(x) = (x + 1)^2 \sin x$ in Figure 8.2. The area under the curve up to $x = \pi$ is counted positively, while the area from π to 2π is counted negatively.

If you want, think of the area above the x-axis as money you have (your assets) and the area below the x-axis as money you owe (your debits). We all agree the minus sign is quite important here—your net worth is the difference of how much you have and how much you owe. We can't just add the two magnitudes together, we must include the sign.

If you know a little bit about where we're going in this chapter, you might be tempted to think that while a calculus student needs to worry about issues such as this, we're safe in a probability class as all probability densities are non-negative. Hence, if we're finding areas under curves, we'll never have to worry about this! Unfortunately, probability densities are not the only functions we integrate. When we compute means (the average value of a random variable) we'll encounter functions that can be negative, and thus these concerns matter.

Okay, enough preliminaries. We can now state the **Fundamental Theorem of Calculus**.

Fundamental Theorem of Calculus: Let f be a piecewise continuous function, and let F be any anti-derivative of f. Then

$$\int_a^b f(x)dx = F(b) - F(a).$$

In words: the (signed) area under the curve $y = f(x)$ from $x = a$ to $x = b$ is the anti-derivative of f at b minus the anti-derivative of f at a.

It's easy to miss what the Fundamental Theorem of Calculus is stating. Many people mistakenly believe that the symbol $\int_a^b f(x)dx$ means $F(b) - F(a)$; nothing could be further from the truth! The symbol *means* the area under the curve $y = f(x)$ from $x = a$ to $x = b$; it's a deep *theorem* that this area can be computed by subtracting an anti-derivative at two points.

In the next section we'll see how this is used in probability.

8.2 PDFs and CDFs: Definitions

We want to build on our success with discrete random variables and construct a similar theory for continuous random variables. Before seeing how the definition changes from the discrete case to the continuous case, we'll review some of the issues and complications of continuous random variables. We sadly *can't* resolve these issues fully; to do so requires advanced courses in analysis, well beyond the scope and goal of this book. Fortunately, there's no need. There's a wealth of problems where these advanced techniques aren't needed; not surprisingly, those are the ones we'll study. So, as you read the next few paragraphs on your way to the definition of a continuous random variable, remember *why* you're reading them! The point is to alert you to the dangers that lie ahead, and how careful we must be.

Our goal is to define a continuous random variable on an outcome space. It turns out this is a bit harder than the case when the outcome space is finite or countable. Remember a probability space has three components: an outcome space Ω, a probability function Prob, and a σ-algebra of subsets where Prob is defined (and Prob is not defined elsewhere). This means we can't find the probability of an arbitrary subset of Ω, but only certain **subsets**, and this causes some difficulty.

It's essential that we have such a restriction. The standard example is choosing a number randomly in [0, 1] such that each number is equally likely to be chosen. If we try to do this, we run into trouble. There are only two possibilities: the probability is positive, or it's zero. If it's positive, then since we have infinitely many points in [0, 1] the total assigned probability exceeds 1, which is impossible. What if the probability of any number is zero? We want to say this implies nothing happens, as the sum of zero is zero. We can't quite do this, as it's an *uncountable* sum, and our rule on additivity (the probability of a disjoint union is the sum of the probabilities) only holds for countable unions. Instead, what we do is talk about the probability of getting a number in an interval. In this case, we want all intervals of the same length to have the same probability. The only natural choice is to say the probability of getting a value in $[a, b] \subset [0, 1]$ is $b - a$.

What does this have to do with continuous random variables? Remember that, after defining discrete random variables, we defined the probability density function and the cumulative distribution function. We wanted to find $\text{Prob}(X = x)$ and $\text{Prob}(X \leq x)$. To make sure both of these are computable in the continuous case, the following two sets must be in our σ-algebra: $\{\omega \in \Omega : X(\omega) = x\}$ and $\{\omega \in \Omega : X(\omega) \leq x\}$. In other words, the set of all elements that evaluate to x or to at most x under X must be in our σ-algebra.

We thus have to think very carefully about the probability spaces we study. In this book, we stick to the standard examples for continuous random variables. Our outcome spaces will be intervals, the real numbers $\mathbb{R}$, "nice" subsets of the plane $\mathbb{R}^2$, the plane $\mathbb{R}^2$, nice subsets of $\mathbb{R}^3$, and so on. You should be imagining intervals, or circles or rectangles or boxes. Fortunately, there's a wealth of great examples on these nice regions. The most important examples come from the half-line $[0, \infty)$ or the real line $\mathbb{R} = (-\infty, \infty)$. Remember, though, a probability space is more than just an outcome space Ω; we also have a probability function defined on a σ-algebra (i.e., on certain subsets of Ω). Thus if our outcome space were $[0, \infty)$ we might as well extend it to be $(-\infty, \infty)$ and just assign zero probability to anything negative. The advantage of this is that it allows us to use one notation for all random variables defined on a subset of the real numbers.

Continuous random variable, probability density function (pdf), and cumulative distribution function (cdf): We say X is a **continuous random variable** if there is a real-valued function f_X, called the **probability density function (pdf)** of X, which satisfies

1. f_X is piecewise continuous;
2. $f_X(x) \geq 0$;
3. $\int_{-\infty}^{\infty} f_X(t)dt = 1$.

Sometimes in an abuse of notation we call f_X the **density function**.

The **cumulative distribution function (cdf)** $F_X(x)$ of X is the probability of X being at most x:

$$F_X(x) = \text{Prob}(X \leq x) = \int_{-\infty}^{x} f_X(t)dt.$$

More generally, we can consider a continuous random variable on $\mathbb{R}^n$. The general case requires us to understand what it means for a function of several variables to be integrable (in other words, what is the multidimensional analogue of a piecewise continuous integrable function; see Exercise 8.6.4); fortunately in many cases the density function is continuous, non-negative, and of course integrates to 1.

It should be clear why we spent so much time reviewing integration. Probabilities can often be expressed as areas under curves, and calculus **(specifically, the Fundamental Theorem of Calculus)** tells us how to compute these areas and hence the desired probabilities! Similar to our discussion in §7.3, we put a subscript of X on each function to remind ourselves that these are associated to the random variable X. This is a very useful convention when we have several random variables floating around in the problem; if, however, there's just one random variable we often suppress the subscript in the interest of space. Finally, note the choice of lowercase f for the probability density function and uppercase F for the cumulative distribution function. This is *not* accidental and is meant to make you think about the interplay between a function and its anti-derivative in calculus.

We briefly remark on the conditions f_X must satisfy to be a probability density function. The first is that it's piecewise continuous; the purpose of this is to have the tools of integration (specifically, the Fundamental Theorem of Calculus) available. Instead of a piecewise continuous function we could consider a Lebesgue integrable function. Don't worry if you haven't seen this—the purpose of advanced analysis classes is to see just how far we may weaken these conditions and still have integrable functions! The second condition is that f_X is non-negative; this is because probabilities can never be negative. The final is that f_X must integrate to 1 over the entire space; this just means "something happens," i.e., X takes on some value!

We are thus reduced to looking at functions that satisfy these three conditions; any time we find such a function is a cause for celebration, as we've just found a probability density function for a continuous random variable. The big question, obviously, is how easy is it to find such functions? We'll start the next section with an example of this, and when we finish, we'll see that we've discovered a general principle to crank out valid probability densities.

8.3 PDFs and CDFs: Examples

In this section we'll take a cookbook problem and explore it. Cookbook means that this is a random variable that, almost surely, no one cares about! I chose it to illustrate the key features of the theory. We'll study the important random variables in great detail in later chapters.

 Let's look at an example. Consider the function

$$f_X(x) = \begin{cases} 2 + 3x - 5x^2 & \text{if } 0 \le x \le 1 \\ 0 & \text{otherwise;} \end{cases}$$

is there a random variable X which has this as its density? To answer this question, we must check whether or not f_X satisfies the three conditions from §8.2: *It must be piecewise continuous, non-negative, and integrate to 1.*

Clearly the density is piecewise continuous (in fact, it's continuous if we restrict the space to $x \ge 0$). It's also non-negative. Seeing this requires a little bit of work, but there are several ways to attack the algebra. The easiest is to note that

$$f_X(x) = 2 + 3x - 5x^2 = (1 - x)(2 + 5x)$$

when $0 \leq x \leq 1$. The factor $1 - x$ is always non-negative in this region, as is $2 + 5x$; thus their product is also non-negative, which is what we wanted to show. We could also see that $f_X(x)$ is non-negative by noting that if $0 \leq x \leq 1$, then $x^2 \leq x$ and so

$$2 + 3x - 5x^2 \geq 2 + 3x - 5x = 2 - 2x = 2(1 - x) \geq 0.$$

All that is left is to make sure that it integrates to 1. We have

$$
\begin{aligned}
\int_{-\infty}^{\infty} f_X(x)dx &= \int_0^1 \left(2 + 3x - 5x^2\right) \\
&= 2x\Big|_0^1 + \frac{3x^2}{2}\Big|_0^1 - \frac{5x^3}{3}\Big|_0^1 \\
&= 2 + \frac{3}{2} - \frac{5}{3} = \frac{11}{6}.
\end{aligned}
$$

Arg. Everything looked so good and so promising, but sadly this is *not* a probability distribution. It's not enough that it satisfies two out of the three conditions—all *three* conditions must be met!

If we had to choose one of the three conditions to fail, the third is the one to pick. Why? It's the easiest to remedy. It's hard to change the general shape of the function; however, we can rescale it by a constant and make it integrate to 1. Consider

$$g_X(x) = \frac{6}{11} f_X(x).$$

There isn't that much difference between these two functions. As f_X is piecewise continuous, so too is g_X. Similarly, as f_X is non-negative so too is g_X. What about the third condition? *Drawing on our calculus knowledge, we know we can pull constants out of integrals.* So then the integral of g_X is just $6/11$ths of the integral of f_X; as the integral of f_X is $11/6$ this means the integral of g_X is $\frac{6}{11} \cdot \frac{11}{6} = 1$, and thus g_X is a probability density! We can also see this by doing the integration directly:

$$
\begin{aligned}
\int_{-\infty}^{\infty} g_X(t)dt &= \int_0^1 \frac{6}{11} \left(2 + 3t - 5t^2\right) dt \\
&= \frac{6}{11} \int_0^1 \left(2 + 3t - 5t^2\right) dt \\
&= \frac{6}{11} \left[2t\Big|_0^1 + \frac{3t^2}{2}\Big|_0^1 - \frac{5t^3}{3}\Big|_0^1\right] \\
&= \frac{6}{11} \left[2 + \frac{3}{2} - \frac{5}{3}\right] \\
&= \frac{6}{11} \cdot \frac{11}{6} = 1.
\end{aligned}
$$

It shouldn't be surprising how similar this integration is to before, as the only change is that we've multiplied everything by $6/11$.

While we have shown that

$$g_X(x) = \begin{cases} \frac{6}{11}(2 + 3x - 5x^2) & \text{if } 0 \le x \le 1 \\ 0 & \text{otherwise} \end{cases} \tag{8.1}$$

is a probability density, we've in fact shown *a lot* more. We've actually found a general procedure to construct probability densities!

Normalizing potential densities: If f_X is a piecewise continuous function that is never negative and has finite integral, then

$$g_X(x) = \frac{f_X(x)}{\int_{-\infty}^{\infty} f_X(t)dt}$$

is a probability density. Another way of phrasing this is as follows: there is a c such that

$$g_X(x) = cf_X(x)$$

is a probability density, and

$$c = \frac{1}{\int_{-\infty}^{\infty} f_X(t)dt}.$$

Now that we know g_X is a probability density, it's useful to compute its cumulative distribution function, G_X. We have

$$G_X(x) = \int_{-\infty}^{x} g_X(t)dt.$$

It's worth briefly remarking on the notation. We need a dummy variable for integration. **If we want to find the probability of taking on a value of at least x, then x can't be the dummy variable of integration.** This is why we're using a t now, even though earlier we used an x.

Unfortunately, since g_X is defined piecewise, we have to be a bit careful. Fortunately two of the three cases are readily handled. If $x \le 0$ then $G_X(x) = 0$; this is because there is no probability before 0, or alternatively $g_X(t) = 0$ for $t \le 0$. We can similarly handle $x \ge 1$; for such x we have $G_X(x) = 1$. Why? The function $g_X(x) = 0$ for $x \ge 1$; thus no new probability is found as we continue to increase x past 1. We've already accounted for all the probability by the time we hit $x = 1$, and thus $G_X(x) = 1$ for $x \ge 1$.

We are left with the interesting part, $0 \le x \le 1$. For such x, we have

$$G_X(x) = \int_{-\infty}^{x} \frac{6}{11}\left(2 + 3t - 5t^2\right) dt$$

$$= \frac{6}{11} \int_{0}^{x} \left(2 + 3t - 5t^2\right) dt$$

$$= \frac{6}{11} \left[2t \Big|_0^x + \frac{3t^2}{2} \Big|_0^x - \frac{5t^3}{3} \Big|_0^x \right]$$

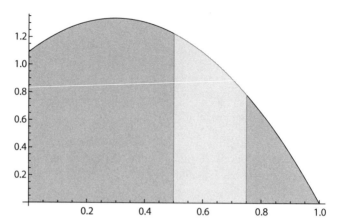

Figure 8.3. The density $g_X(x) = \frac{6}{11}(2 + 3x - 5x^2)$ for $0 \le x \le 1$ and 0 otherwise. The area under the curve is 1, and the area from 1/2 to 3/4 (the dark shaded region) is 91/352.

$$= \frac{6}{11}\left(2x + \frac{3x^2}{2} - \frac{5x^3}{3}\right)$$

$$= \frac{6(12x + 9x^2 - 10x^3)}{11}.$$

We can use the cumulative distribution function to quickly find probabilities of various events. For example, let's find the probability that X takes on a value in $(1/2, 3/4]$, where X is a random variable with the density g_X defined in (8.1). We plot the probability in Figure 8.3.

The answer is

$$\text{Prob}(X \in (1/2, 3/4]) = \int_{1/2}^{3/4} g_X(t)dt$$

$$= G_X(3/4) - G_X(1/2)$$

$$= \left.\frac{6(12x + 9x^2 - 10x^3)}{11}\right|_{x=3/4} - \left.\frac{6(12x + 9x^2 - 10x^3)}{11}\right|_{x=1/2}$$

$$= \frac{315}{352} - \frac{224}{352} = \frac{91}{352}.$$

There's a very important lesson to learn from our attempt to find the probability that X (with density g_X) took a value in $(1/2, 3/4]$. We studied an event which is an open interval on the left (we don't include 1/2) and a closed interval on the right (we do include 3/4). We are asking for the probability that X is *greater than* 1/2 as well as *at most* 3/4. There is a bit of asymmetry here in how we treat the two endpoints. However, we don't have to worry too much about this because the probability X takes on any given value is zero. The fancy way of saying this is that, for continuous random variables, the probability of a singleton event is zero.

If X is a continuous random variable with probability density function f_X, then the following four probabilities are all equal:

1. the probability that X is in $[a, b]$;
2. the probability that X is in $(a, b]$;
3. the probability that X is in $[a, b)$;
4. the probability that X is in (a, b).

The reason is that, for continuous random variables, the probability that X takes on any specific value is zero. For example, from the definition of the cumulative distribution function we have

$$F_X(b) - F_X(a) + \text{Prob}(X = a) = \int_a^b f_X(t)dt.$$

As the probability of singletons is zero, our expression simplifies to

$$F_X(b) - F_X(a) = \int_a^b f_X(t)dt.$$

Going back to the density g_X from (8.1), we saw that it was easy to find the cumulative distribution function G_X for $x \le 0$ or $x \ge 1$. While we're not always fortunate enough to have a nice, closed form expression for the cdf, we do know a little about its behavior. Specifically, we have the following result, which is analogous to the similar result from discrete random variables.

Let X be a continuous random variable with cumulative distribution function F_X. Then

$$\lim_{x \to -\infty} F_X(x) = 0 \quad \text{and} \quad \lim_{x \to \infty} F_X(x) = 1.$$

Further, F_X is a non-decreasing function: if $y > x$ then $F_X(y) \ge F_X(x)$.

This is clear if we look at it the right way. As X is a random variable, it must take on some value. If $x \to \infty$ then we're allowing that value to be anything, and thus the probability should tend to 1; *this is also clear from the fact that we require our probability density functions to integrate to 1.* Alternatively, if we send $x \to -\infty$ we restrict further and further what that value can be, and thus the probability tends to zero.

Finally, while we have seen there is a nice relation between the probability density function f_X and the cumulative distribution function F_X, so far we've only seen it in one way. Namely, given the density we can find the cumulative distribution function. Is it possible to go the other way? In other words, can we recover f_X from F_X? Of course we can! We just have to use the opposite process from integration: differentiation!

If X is a continuous random variable with cumulative distribution function F_X, then it has a probability density function f_X such that $F'_X(x) = f_X(x)$ for all but finitely many x.

Sadly we have to be a little careful above. Remember our densities need only be piecewise continuous. While the cumulative distribution function is continuous, it need not be differentiable everywhere; at the finitely many points where it's not differentiable we don't expect $F_X'(x)$ to equal $f_X(x)$ (as F_X' doesn't even exist at these points!). Fortunately the probability of a singleton event is zero, and thus this does not really affect anything.

Here's an example of how we can recover the density function. Assume someone tells us our random variable X has cumulative distribution function $F_X(x)$ which vanishes for $x \le 0$ and is $1 - e^{-x}$ for $x \ge 0$. The density has to be zero for $x \le 0$, so we concentrate on $x \ge 0$. Differentiating $F_X(x) = 1 - e^{-x}$ gives

$$F_X'(x) \ = \ -e^{-x}(-1) \ = \ e^{-x};$$

thus the density is

$$f_X(x) \ = \ \begin{cases} e^{-x} & \text{if } x \ge 0 \\ 0 & \text{otherwise.} \end{cases}$$

8.4 Probabilities of Singleton Events

This section is a bit more advanced, and can safely be skimmed or skipped. Its purpose is to elaborate on some issues that arise when we discuss the probabilities assigned by continuous random variables.

In the last section, we argued that for a continuous random variable X the probability that X takes on any specific value is zero. The fancy way of saying this is that the probability of a singleton event is zero, or

$$\text{Prob}(X = x) = 0.$$

This is true for any value of x. Taking unions, we find

$$\text{Prob}(X \in \{x_1, x_2, \dots, x_n\}) \ = \ \sum_{i=1}^{n} \text{Prob}(X = x_i) \ = \ \sum_{i=1}^{n} 0 \ = \ 0.$$

We can write this more compactly as

$$\text{Prob}\left(X \in \bigcup_{i=1}^{n} \{x_i\} \right) \ = \ 0.$$

We now write

$$[a, b] \ = \ \bigcup_{x=a}^{b} \{x\},$$

and thus shouldn't

$$\text{Prob}(X \in [a, b]) = \text{Prob}\left(X \in \bigcup_{x=a}^{b} \{x\}\right) = 0?$$

This is absurd—we've just shown that, no matter what continuous random variable we take and no matter what interval we consider, the probability of X taking on a value in that interval is zero! We could take our interval to be $(-\infty, \infty)$, and we would still find zero probability!

The above seems to be saying that nothing ever happens. Something is clearly wrong! It's actually good when we have something like this—the conclusion is so absurd that we revolt; the danger is when we have wrong conclusions that seem plausible, as they might slip past our defenses and not arouse any suspicion.

Anyway, returning to our problem, what gives? How can every continuous random variable assign zero probability to every event? The problem stems from what seems to be a very reasonable assumption but which, in fact, is false. Our argument crucially used that the probability of a sum of disjoint events is the sum of the probabilities of the events. This is true if it's a finite sum, or even a **countably** infinite sum, but not necessarily true if it's an uncountable sum.

It all comes back to the beginning of the book (see Chapter 2, especially §2.6), when we were very careful to describe what events we could and could not assign probabilities to. We can assign probabilities to singletons; that's no problem. The problem is that we cannot assert that the probability of an uncountable union of disjoint events is the sum of the probabilities.

We review countable and uncountable sets in Appendix C. We've seen many finite and countable sums in our lives. For example, if we wanted to do the Riemann sum for the area under the curve $y = x^2$ from 0 to 1 with n partitions, we would have

$$\sum_{k=0}^{n-1} \frac{k^2}{n^2} \cdot \frac{1}{n},$$

and we obtain the area by taking the limit as $n \to \infty$ (giving us a countable sum). If we wanted to compute the probability that a fair coin eventually lands on heads, it would be

$$\sum_{k=1}^{\infty} \frac{1}{2^k}$$

because the probability that the first toss to be a head occurs on the nth toss is $1/2^n$. This is because we need the first $n - 1$ tosses to be tails (which happens with probability $1/2^{n-1}$) and the nth toss to be a head (which happens with probability $1/2$).

We have lots of experience and familiarity with finite and countable sums. What would an uncountable sum look like? Let's image that S is some uncountable set, and we want to evaluate

$$\sum_{x \in S} a_x.$$

For ease of exposition, let's assume that $a_x \geq 0$ for all x. There is only one way for this sum to be finite, and that is for a_x to equal zero for all but countably many x. If this happens, of course, we no longer have a legitimately uncountable sum, but rather we have a countable sum!

Why must all but countably many of the a_x's vanish? Let

$$S_n = \left\{ x : x \in S \text{ and } \frac{1}{n+1} < a_x \le \frac{1}{n} \right\}.$$

Note

$$S = \bigcup_{n=0}^{\infty} S_n$$

(where by S_0 we mean all $x \in S$ with $a_x > 1$). Remember the a_x's are non-negative, so

$$\sum_{x \in S} a_x \ge \sum_{x \in S_n} a_x.$$

If there were infinitely many elements in an S_n, then the sum over $x \in S_n$ would be infinite, implying the sum over all of S is infinite, which contradicts our assumption that the sum is finite. Thus, each S_n is finite, and a countable union of finite sets is countable, so we conclude that all but countably many of the a_x's must vanish for $\sum_{x \in S} a_x$ to be finite.

8.5 Summary

This chapter is a natural sequel to the previous one, as we redo everything but now for continuous random variables. There are issues in translating our results and notation from the land of discrete random variables, though. Stanislaw J. Lec (*More Unkempt Thoughts*, 1968) is perhaps on the right path, writing: "I don't agree with mathematics; the sum total of zeros is a frightening figure."

I think "frightening" is a bit much, probably chosen more for humor than mathematical appropriateness, but this quote does illustrate the key difference between the continuous and discrete cases. In probability, an uncountable sum of zero can be any finite number from 0 to 1! This requires us to be very careful in developing the theory. In particular, we can't find the probabilities of arbitrary events, but are restricted to special collections. **Namely, we are often restricted to countable events.**

Just like the discrete case, we have probability density functions and cumulative distribution functions. Though we've only briefly touched upon the cdf here, we'll see in Chapter 10 how useful it is. In particular, it'll allow us to easily pass from knowledge of one random variable to a related one.

8.6 Exercises

Exercise 8.6.1   *Let f be a continuous function and assume $F'(x) = G'(x) = f(x)$. Prove $F(x) - G(x)$ is constant.*

Exercise 8.6.2   *Let f be a continuous function and assume $F''(x) = G''(x) = f(x)$ and $F(x) = G(x)$ for at least k different choices of x; what is the smallest k which ensures F and G are always equal?*

Exercise 8.6.3 *Prove that a function which is continuous except at finitely many points on an interval of finite length is Riemann integrable. What if it is discontinuous at infinitely many points?*

Exercise 8.6.4 *A continuous, non-negative function $f : \mathbb{R} \to \mathbb{R}$ is integrable if*

$$\lim_{A,B \to \infty} \int_{-A}^{B} f(x)dx$$

exists and is independent of how A and B tend to infinity. Show $f(x) = x\exp(-|x|)$ is integrable, while $g(x) = x/(1 + x^2)$ is not. What about $\sin(x)/x$?

Exercise 8.6.5 *Could the following function f_X be the probability density function for some random variable X:*

$$f(x) = \begin{cases} 4x^2 + 5x + 2 & \text{if } 0 \le x \le 2 \\ 0 & \text{otherwise?} \end{cases}$$

If it isn't, find a C if possible such that the function $g(x) = Cf(x)$ is a probability density function, or prove no such C exists.

Exercise 8.6.6 *Describe three real-world applications of continuous random variables. Do not use the ones already described in the chapter.*

Exercise 8.6.7 *Label each random variable as either discrete or continuous; for some of these you might be able to make either work depending on your interpretation!*

(a) *T, the time at which you get to class.*
(b) *T', the time at which you get to class truncated to the minute.*
(c) *N, the number of kids who sign up to take a probability class.*
(d) *G, the grade you receive on a test scored out of 100 points.*
(e) *H, the average height of the **students** in your probability class.*
(f) *S, the number of bricks it will take to build the new science building at Williams College.*
(g) *D, the distance between our star Sol and the star Wolf 359.*

Exercise 8.6.8 *Find a discrete random variable, or prove none exists, with probability density function f_X such that $f_X(x) = 2$ for some x between 17 and 17.01.*

Exercise 8.6.9 *Find a continuous random variable, or prove none exists, with probability density function f_X such that $f_X(x) = 2$ for all x between 17 and 17.01.*

Exercise 8.6.10 *Let X be a continuous random variable with pdf f_X satisfying $f_X(x) = f_X(-x)$. What can you deduce about F_X, the cdf?*

Exercise 8.6.11 *Verify that $f_X(x) = 2x \cdot e^{-x^2}$ for $0 \le x < \infty$ and 0 for $x < 0$ is a probability density function. What is the corresponding cumulative distribution function, $\mathcal{F}_X$?*

Exercise 8.6.12 *Is $F_X(x) = e^{-x}$ for all x a cumulative distribution function? If so, what is the corresponding probability density function?*

Exercise 8.6.13 *Is $F_X(x) = 1 - \frac{x^2}{1+x^2}$ a cumulative distribution function? If so, what is the corresponding probability density function?*

Exercise 8.6.14 *Is $F_X(x) = \frac{1}{2} + \frac{x}{2\sqrt{1+x^2}}$ a cumulative distribution function? If so, what is the corresponding probability density function?*

Exercise 8.6.15 *Let X be a continuous random variable. (a) Prove F_X is a non-decreasing function; this means $F_X(x) \le F_X(y)$ if $x < y$. (b) Let U be a random variable with cdf $F_U(x) = 0$ if $u < 0$, $F_U(x) = x$ if $0 < x < 1$, and $F_U(x) = 1$ if $1 < x$. Let F be any continuous function such that F is strictly increasing and the limit as x approaches negative infinity of $F(x)$ is 0 and the limit as x approaches positive infinity is 1. Prove $Y = F^{-1}(U)$ is a random variable with cdf F. (This is an extremely important exercise; as it allows us to generate many random variables if we can generate a uniform random variable.)*

Exercise 8.6.16 *A mixed random variable is a random variable whose probability density function takes on positive probability at a finite or countably infinite number of points and behaves like a continuous random variable elsewhere. Consider the mixed variable X, where the $\text{Prob}(X = n) = \frac{1}{3^n}$ where n is an integer and the continuous part of X has probability density function f defined on the non-integer real numbers, where $f(x) = \frac{1}{2x^2}$ for $x > 1$ and 0 elsewhere. Show that this is a valid probability density function.*

Exercise 8.6.17 *Describe two real-world applications for mixed random variables.*

Exercise 8.6.18 *Imagine we have a dartboard with unit radius (centered at the origin) in which the likelihood a dart lands in a given area is proportional to that area. What is the pdf for the distance the dart lands from the origin?*

Exercise 8.6.19 *Consider the dartboard from the previous exercise. What is the pdf for the distance in the x-direction the dart lands from the origin?*

Exercise 8.6.20 *If the dart lands a distance x from the origin with uniform probability 1, what does that say about the likelihood a dart lands in any given area in terms of x?*

Exercise 8.6.21 *Find a continuous probability density function that is positive (non-zero) for all x.*

Exercise 8.6.22 *If f and g are both probability density functions, for what a and b is $af + bg$ necessarily a probability density function?*

Exercise 8.6.23 *Let X and Y be two continuous random variables with densities f_X and f_Y. (a) For what c is $cf_X(x) + (1 - c)f_Y(x)$ a density? (b) Can there be a random variable with pdf equal to $f_X(x)f_Y(x)$?*

Exercise 8.6.24 *Is there a C such that $f(x) = C \exp(-x - \exp(-x))$ is a probability density function? Here $-\infty < x < \infty$.*

Exercise 8.6.25 *Imagine you are traveling and have 2 flights to catch. Unfortunately, you didn't plan well and have a 0 minute layover, that is if your first flight is late and the next flight is on time then you will miss your second flight, ruining all your travel plans. Make the unreasonable assumptions that airline delays are independent, so that if your first flight is delayed it does not have any effect on the probability that your later flight is delayed. Assume also that 75% of the time flights are on-time, and if the flight is late then the probability it is exactly t minutes late is $1/t^2$ for any real $t \ge 0$. What is the probability you do not miss your second flight?*

Exercise 8.6.26 *Write some code to model and simulate the previous plane question. Extend the simulation to more than 2 flights and find the probability you make all of these flights.*

Exercise 8.6.27 *Plot the cumulative distribution function for the mixed random variable X, where $\Pr(X = n) = 1/3^n$ where n is an integer and the continuous part of X has probability density function f defined on the non-integer real numbers, where $f(x) = 1/2x^2$ for $x > 1$ and 0 elsewhere.*

Exercise 8.6.28 *Let X be chosen uniformly on $[0, 1]$, so its density is $f_X(x) = 1$ for $0 \leq x \leq 1$ and 0 otherwise. What is the probability the first digit of X is an even number? What is the probability that every digit of X is even?*

CHAPTER 9 _____

Tools: Expectation

Winsto n Zeddemore: We had the tools! We had the talent!
Peter Venkman: It's Miller Time!
— GHOSTBUSTERS (1984)

Very soon we'll delve into example after example after example of random variables. Yes, there are many, and they're all worth knowing. The more random variables and distributions you know, the more likely you are to find an appropriate one to model a problem you care about. That's why we'll have a a bunch of chapters, each devoted to one or two random variables.

The purpose of this chapter and the next few chapters is to concentrate on some *similarities* among all the different random variables. In particular, there are some tools and techniques which can be fruitfully applied to understand all of them. If you can master these methods, you can analyze almost any random variable. We'll concentrate on five items: expectation and moments (which lead to concepts such as the mean, standard deviation, and variance) in this chapter, convolutions (which allow us to combine independent random variables) and changing variables (which allow us to pass from knowledge of one random variable to another) in Chapter 10, and differentiating identities (which often facilitate finding the mean, variance, and other moments) in later chapters.

The later chapters on specific random variables all follow the same pattern: we'll choose a probability density function and then study the associated random variable. The calculations are similar from chapter to chapter; the biggest change is the difficulty in doing the integrals or sums. This ranges from very easy (in the case of the uniform distribution) to impossible (in the case of normal distributions). Sadly, the later situation is more common. It's very rare to be able to evaluate integrals in a nice, closed form expression, and sums are typically worse! In practice we must resort to numerical approximations or series expansions. While we can get whatever accuracy we need in general, this is a major problem; we'll talk more about this at great length later.

Before delving into these special distributions, we're going to invest some time in learning some general tools to study continuous probability distributions. Some of these we've already seen, others we'll see in much greater detail later. Our first tool is that of expected values and moments. We'll study continuous and discrete random variables at the same time. Essentially the only difference in the *definitions* is replacing sums with integrals. In fact, in order to give a unified presentation, often advanced analysis books have a more general notion of the integral to handle both cases. We won't do that here; we'll write our results in both sum and integral form.

The presentation here is a bit different than you'll find in most textbooks. The reason is that this book is designed to supplement any book. Thus, what I've done in this chapter (and the next two) is collect a bunch of the most important techniques in one place, and describe the theory. While there are a few examples of how to use these results, most of the real applications are saved for the later chapters. If you're using this book as a supplement, you've probably seen a lot of these examples and applications already, and this chapter and the next few are meant to serve as a central depository, storing all the facts you'll need. If you're just reading this book as your text, don't worry: we'll get to the examples soon. One of my math professors, Serge Lang, wrote in one of his books that it's a shame books must be ordered along the page axis. This is one of many situations where there are multiple good, valid choices on how to order, and some choice just has to be made.

9.1 Calculus Motivation

The main concept in this chapter is that of a moment. Before we give the definition in §9.2, we first revisit a concept from calculus that, while at first may seem unrelated, actually provides an excellent motivation. If you haven't seen calculus before, continuous random variables will be a bit of a challenge, as most of our techniques to study them involve calculus. If that's the case, you can skim or skip this section and move on to the next, concentrating on the discrete random variables.

Taylor series: If f is differentiable n times (with $f^{(k)}(x)$ denoting the kth derivative at x), its nth order Taylor series at the point a is

$$T_n(x) := f(a) + f'(a)(x-a) + \frac{f''(a)}{2!}(x-a)^2 + \cdots + \frac{f^{(n)}(a)}{n!}(x-a)^n$$

$$= \sum_{k=0}^{n} \frac{f^{(k)}(a)}{k!}(x-a)^k.$$

We call $f^{(k)}(a)/k!$ the kth **Taylor coefficient** of f about a. In many applications we want the Taylor series about the origin, so $a = 0$ (some books call this the **Maclaurin series**). The Taylor series takes knowledge of the behavior of a function and its derivatives at a point, and uses that to estimate its values elsewhere.

In Figure 9.1 we compare the sine function to its third order Taylor series, which is $x - x^3/3!$, and its seventh order Taylor series, which is $x - x^3/3! + x^5/5! - x^7/7!$. While both approximations do well, it's not surprising that the seventh order does better. What we're doing is taking into account more and more information about our function.

Look at it this way. Imagine you want to understand a function f over a large interval centered at the origin. If you could only have one piece of information, you should ask for $f(0)$. Knowing this, what should you predict for $f(.1)$ or $f(-.2)$? Our best guess is

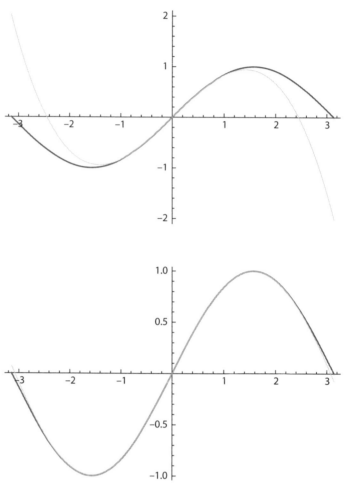

Figure 9.1. Plot of $\sin x$ against its third order Taylor series (top) and its seventh order Taylor series (bottom).

that they too are $f(0)$; we have no idea if f is increasing or decreasing, so we hedge our bets and approximate f by a constant function. Without any additional information, this is the best we can do, and is the zeroth order Taylor series expansion: $T_0(x) = f(0)$.

But what if we *do* have more information? Now, in addition to knowing $f(0)$, let's assume we have one more fact about our function. A great choice is to find out the value of $f'(0)$. Why is this such a good choice? The derivative gives us the instantaneous rate of change. This means that locally (in other words, for x near 0) our function is well approximated by $f(0) + f'(0)x$, which is $T_1(x)$. Think of it this way: $f(0)$ is your starting location at time 0, $f'(0)$ is your initial speed, so if your speed is constant then at time x you've moved $f'(0) \cdot x$, so you're now at $f(0) + f'(0)x$.

It's a little more complicated incorporating the next bit of information, $f''(0)$ (note the physical interpretation of this is acceleration, or the rate of change of the velocity). You might guess that the second order Taylor series would be $f(0) + f'(0)x + f''(0)x^2$, but you'd be slightly off; the final factor should be $f''(0)x^2/2!$. Why do we divide by 2!? The way to view the nth order Taylor series expansion of f at the origin is that

TABLE 9.1.
The general third order Taylor series of $f(x)$ at the origin.

$T_3(x)$	$=$	$f(0) + f'(0)x + \frac{f''(0)}{2!}x^2 + \frac{f'''(0)}{3!}x^3$	$T_3(0)$	$= \quad f(0)$
$T_3'(x)$	$=$	$f'(0) + f''(0)x + \frac{f'''(0)}{2!}x^2$	$T_3'(0)$	$= \quad f'(0)$
$T_3''(x)$	$=$	$f''(0) + f'''(0)x$	$T_3''(0)$	$= \quad f''(0)$
$T_3'''(x)$	$=$	$f'''(0)$	$T_3'''(0)$	$= \quad f'''(0)$

$T_n(0) = f(0)$, $T_n'(0) = f'(0)$, and so on up to $T_n^{(n)}(0) = f^{(n)}(0)$. In other words, we match the value of our approximation to the given function, and then make sure the first n derivatives agree at the origin. If we didn't divide by 2!, the second derivative of $T_n(x)$ wouldn't agree with the second derivative of $f(x)$ at the origin. We illustrate this in Table 9.1, where we consider a general third order Taylor series at the origin, and show that it and its first three derivatives at the origin agree with f.

The moral of the story is that the more information we include, the better the approximation. Figure 9.1 provides powerful evidence supporting this claim. It's interesting to see how far this can be pushed. The hope is that if we know all the Taylor coefficients then we know the function. Equivalently, the sequence of Taylor coefficients uniquely specifies the function. Unfortunately, that's not the case (see (19.3) of §19.6 for an example), and this is responsible for many pesky problems in probability.

Though there are issues, there are still many situations where knowing the Taylor coefficients convey a lot of useful information about the function. We'll now generalize this idea to help us understand densities of random variables.

9.2 Expected Values and Moments

In §9.1 we discussed how the Taylor coefficients, and the function built out of them, the Taylor series, provide a wealth of information about the function. We now find similar quantities for probability density functions.

Remember that a random variable X is defined on a probability space; that means we have a set of outcomes Ω, and a probability function defined on a σ-algebra of subsets of Ω. Our random variable has a probability density function (pdf) attached to it, which tells us the probability it takes on certain values. We'll define the moments of a pdf below. We have a similar situation as in §9.1; there the Taylor coefficients provided information about the function, while here the moments will convey information about the pdf.

Expected values, moments: Let X be a random variable on $\mathbb{R}$ with probability density function f_X. The **expected value** of a function $g(X)$ is

$$\mathbb{E}[g(X)] = \begin{cases} \int_{-\infty}^{\infty} g(x) \cdot f_X(x)dx & \text{if } X \text{ is continuous} \\ \sum_n g(x_n) \cdot f_X(x_n) & \text{if } X \text{ is discrete.} \end{cases}$$

> The most important cases are when $g(x) = x^r$. We call $\mathbb{E}[X^r]$ the rth **moment** of X, and $\mathbb{E}[(X - \mathbb{E}[X])^r]$ the rth **centered moment** of X.

As long as we can do the sum or the integral, we can calculate the expected values and moments; of course, there are choices of g where the expectation can fail to exist. The question, of course, is *why* we would want to. What information do these give? To really appreciate how much information these give, we have to wait until we cover generating functions (starting in Chapter 19), but we can say a few words here. Briefly, the centered moments are like the Taylor coefficients. Just as knowing more and more Taylor coefficients gave us a better and better approximation to the function, knowing more and more moments gives us a better understanding of the shape and properties of the probability density function.

In §9.3 we'll concentrate on the two most important moments, the mean (which is the first moment) and the variance (which is the second centered moment). We end this section with a few examples of how to find expectations.

In §8.3 we saw that

$$f_X(x) \;=\; \begin{cases} \frac{6}{11}(2 + 3x - 5x^2) & \text{if } 0 \le x \le 1 \\ 0 & \text{otherwise} \end{cases}$$

is the density of a continuous random variable. We'll calculate (1) the rth moment for all $r \ge 0$, (2) the expected value of $g(X) = e^X$, and (3) the expected value of $g(X) = 1/X$. Note that since our density $f_X(x)$ is zero if x is outside $[0, 1]$, we can restrict all the integrals below to be from 0 to 1. This is a typical cookbook problem. It's been deliberately cooked up—no one cares about this random variable or its density; however, it's been carefully created so that it highlights some of the important issues. We'll see that we need to be comfortable with many of our calculus techniques if we're going to have success with moments.

(1) To find the rth moment for $r \ge 0$, we must find $\mathbb{E}[X^r]$, which is

$$\mathbb{E}[X^r] \;=\; \int_0^1 x^r \cdot \frac{6}{11}(2 + 3x - 5x^2)\,dx$$

$$=\; \frac{6}{11}\int_0^1 (2x^r + 3x^{r+1} - 5x^{r+2})\,dx$$

$$=\; \frac{6}{11}\left(\frac{2x^{r+1}}{r+1}\Big|_0^1 + \frac{3x^{r+2}}{r+2}\Big|_0^1 - \frac{5x^{r+3}}{r+3}\Big|_0^1 \right)$$

$$=\; \frac{6}{11}\frac{7r + 11}{(r+1)(r+2)(r+3)}.$$

If we take $r = 1$ we get the first moment, which in this case is 9/22. The integrals would make sense even if r were negative, so long as $r > -1$ (if $r \le -1$ the integral of x^r times the constant term diverges).

(2) To find the expected value of e^X, we compute

$$\mathbb{E}[e^X] = \int_0^1 e^x \cdot \frac{6}{11}(2 + 3x - 5x^2)dx$$

$$= \frac{6}{11}\left(2\int_0^1 e^x dx + 3\int_0^1 xe^x dx - 5\int_0^1 x^2 e^x dx\right).$$

To finish, we need to recall some techniques from calculus, specifically integration by parts for the second and third integrals (the first integral is just $e^x\big|_0^1 = e - 1$).
We have

$$\int_0^1 u\,dv = u(x)v(x)\Big|_0^1 - \int_0^1 v\,du.$$

For the second integral, a great choice is to take $u = x$ and $dv = e^x dx$. Doing so gives us $du = dx$ and $v = e^x$, so the new integrand (the hard part) is just e^x. This is much friendlier than the initial integrand of xe^x. If instead we had tried $u = e^x$ and $dv = xdx$, then we would've found $du = e^x dx$, $v = x^2/2$, resulting in the nastier integrand $\frac{1}{2}x^2 e^x$. In other words, we started with an integral which was hard to evaluate because our exponential was hit with a polynomial, and our choice of u and v actually *increased* the degree of the polynomial! This is why the other choice is better, as it lowers the degree. We find

$$\int_0^1 xe^x dx = xe^x\Big|_0^1 - \int_0^1 e^x dx = e - (e - 1) = 1$$

(to save time, we recalled our earlier result that $\int_0^1 e^x dx = e - 1$).

We're left with $\int_0^1 x^2 e^x dx$. We integrate by parts. Let $u = x^2$ and $dv = e^x dx$ (this is the "right" choice, as the degree of the polynomial decreases by 1). We find $du = 2xdx$, $v = e^x$, and

$$\int_0^1 x^2 e^x dx = x^2 e^x\Big|_0^1 - 2\int_0^1 xe^x dx = e - 2$$

(remember we've just shown $\int_0^1 xe^x dx = 1$, which we use here).
Combining these three integrals, we get

$$\mathbb{E}[e^X] = \frac{6}{11}(2 \cdot (e - 1) + 3 \cdot 1 - 5 \cdot (e - 2)) = 6 - \frac{18e}{11}.$$

(3) Finally, we turn to finding the expected value of $g(X) = 1/X$. This leads us to study the integral

$$\mathbb{E}\left[\frac{1}{X}\right] = \int_0^1 \frac{1}{x} \cdot \frac{6}{11}(2 + 3x - 5x^2)dx$$

$$= \frac{6}{11}\left(2\int_0^1 \frac{dx}{x} + 3\int_0^1 dx - 5\int_0^1 xdx\right).$$

While the last two integrals exist, the first does not. The anti-derivative of $1/x$ is $\log x$, but we evaluate this at 1 (where we get 0) and at 0 (where we get $-\infty$). Thus the expectation here does not exist. The problem is the function $1/x$ blows up too rapidly near the origin to be integrable (see Exercise 9.10.3).

 We now do a discrete example. Let X be the number of heads in three tosses of a fair coin. We studied this random variable in great detail in §7.1 and §7.2, and showed $f_X(x) = 0$ unless $x \in \{0, 1, 2, 3\}$, and

$$f_X(0) = \frac{1}{8}, \quad f_X(1) = \frac{3}{8}, \quad f_X(2) = \frac{3}{8}, \quad f_X(3) = \frac{1}{8}.$$

Let's find the expected value of X^2:

$$\mathbb{E}[X^2] = \sum_{k=0}^{3} k^2 \cdot f_X(k) = 0^2 \cdot \frac{1}{8} + 1^2 \cdot \frac{3}{8} + 2^2 \cdot \frac{3}{8} + 3^2 \cdot \frac{1}{8} = 3.$$

This answer should look interesting. We flipped three fair coins, and the expected value of X^2 (i.e., the square of the number of heads) is 3. Is this a coincidence, or would we get 4 if we did four coins? Let's see!

Let Y_n denote the number of heads in n tosses of a fair coin (we've added a subscript to our random variable so we can quickly see how many coins are tossed). We discussed this random variable in §7.2; it's a binomial random variable. We saw its density was

$$f_{Y_n}(k) = \begin{cases} \binom{n}{k} 2^{-n} & \text{if } k \in \{0, 1, \ldots, n\} \\ 0 & \text{otherwise.} \end{cases}$$

Thus

$$\mathbb{E}[Y_n^2] = \sum_{k=0}^{n} k^2 \binom{n}{k} 2^{-n} = \frac{1}{2^n} \sum_{k=0}^{n} k^2 \binom{n}{k}.$$

Unfortunately, this sum doesn't look too promising. Oh well. Rolling up our sleeves and plunging into the computations, we find

$$E[Y_1^2] = \frac{1}{2}, \quad E[Y_2^2] = \frac{3}{2}, \quad \mathbb{E}[Y_3^2] = 3, \quad \mathbb{E}[Y_4^2] = 5,$$

$$\mathbb{E}[Y_5^2] = \frac{15}{2}, \quad \mathbb{E}[Y_6^2] = \frac{21}{2}, \ldots.$$

Our pattern didn't last too long! Okay, it's nothing as simple as n, but amazingly our guess isn't too far from the truth. If we look at all polynomials of degree 2, we find $f(x) = \frac{1}{4}n^2 + \frac{1}{4}n = n(n+1)/4$ happens to agree with all the values above! Why did we take polynomials of degree 2? Well, we're squaring things, so it seems reasonable to try and fit a quadratic. We'll see in Chapter 11 how to prove formulas like the one we're claiming through the technique of differentiating identities.

These two examples are illustrative of what often happens. Usually the expected values from a continuous random variable yield tractable integrals, while those for discrete random variables frequently lead to sums involving binomial coefficients. It's much harder to evaluate these sums in closed form than the integrals, but there are advanced techniques (differentiating identities is one of them) that work well in some cases.

9.3 Mean and Variance

It's time to continue our story. We started with Taylor coefficients, and then defined moments for a probability density function. Our goal here is to try to flesh out a bit more of this analogy with calculus. In particular, I want to show you how these moments provide information about our densities. The two most important moments are the first and the second centered moments. They're so important they get their own names: mean and variance.

Mean and variance: Let X be either a continuous or a discrete random variable with density f_X.

1. The **mean** (or **average value** or **expected value**) of X is the first moment. We denote it by $\mathbb{E}[X]$ or μ_X (if the random variable is clear, we often suppress the subscript X and write μ). Explicitly,

$$\mu = \begin{cases} \int_{-\infty}^{\infty} x \cdot f_X(x)dx & \text{if } X \text{ is continuous} \\ \sum_n x_n \cdot f_X(x_n) & \text{if } X \text{ is discrete.} \end{cases}$$

2. The **variance** of X, denoted σ_X^2 or $\text{Var}(X)$, is the second centered moment, or equivalently the expected value of $g(X) = (X - \mu_X)^2$; again, we often suppress the subscript X if the random variable is clear, and write σ^2. Writing it out in full,

$$\sigma_X^2 = \begin{cases} \int_{-\infty}^{\infty} (x - \mu_X)^2 f_X(x)dx & \text{if } X \text{ is continuous} \\ \sum_n (x_n - \mu_X)^2 f_X(x_n) & \text{if } X \text{ is discrete.} \end{cases}$$

As $\mu_X = \mathbb{E}[X]$, after some algebra (see Lemma 9.5.3) one finds

$$\sigma^2 = \mathbb{E}[(X - \mathbb{E}[X])^2] = \mathbb{E}[X^2] - \mathbb{E}[X]^2.$$

This relates the variance to the first two moments of X, and is useful for many calculations. The **standard deviation** is the square-root of the variance, or $\sigma_X = \sqrt{\sigma_X^2}$.

3. **Technical caveat**: In order for the mean to exist, we want $\int_{-\infty}^{\infty} |x| f_X(x)dx$ (in the continuous case) or $\sum_n |x_n| f_X(x_n)$ (in the discrete case) to be finite.

Why are these so important? The mean is the expected or average value. If we were to choose many values from this distribution again and again and average our outcomes, our average should be very close to μ_X. The standard deviation tells us what scale to expect fluctuations about the mean. The smaller the standard deviation, the more tightly the mass of our distribution is concentrated about the mean.

Let's do an example. Imagine we have two sections of a probability class. Each section has 5 people (Alice, Bob, Cam, Danie, and Eli in the first section, who respectively score 40, 45, 50, 55, and 60, and Fred, Gabrielle, Henry, Igor, and Justine in the second, who respectively score 0, 25, 50, 75, and 100). It pains me to write things rigorously, as in practice everyone is always informal in problems like this; however, it's good to see the gory details at least once. We want all students to count equally, so in each section, each student is assigned a probability of 1/5 or 20%.

We'll let X denote the score of someone in the first section (so $X(\text{Eli}) = 60$) and Y the score of someone in the second section (so $Y(\text{Gabrielle}) = 25$). The means are readily found:

$$\mu_X = 40 \cdot \frac{1}{5} + 45 \cdot \frac{1}{5} + 50 \cdot \frac{1}{5} + 55 \cdot \frac{1}{5} + 60 \cdot \frac{1}{5} = 50$$

$$\mu_Y = 0 \cdot \frac{1}{5} + 25 \cdot \frac{1}{5} + 50 \cdot \frac{1}{5} + 75 \cdot \frac{1}{5} + 100 \cdot \frac{1}{5} = 50.$$

So, both sections have the same mean. If all we know is the mean, then we can't distinguish between the two sections.

We now find the variance:

$$\sigma_X^2 = (40 - 50)^2\frac{1}{5} + (45 - 50)^2\frac{1}{5} + (50 - 50)^2\frac{1}{5} + (55 - 50)^2\frac{1}{5}$$
$$+ (60 - 50)^2\frac{1}{5} = 50$$

$$\sigma_Y^2 = (0 - 50)^2\frac{1}{5} + (25 - 50)^2\frac{1}{5} + (50 - 50)^2\frac{1}{5} + (75 - 50)^2\frac{1}{5}$$
$$+ (100 - 50)^2\frac{1}{5} = 1250.$$

There's a lot we can learn from this example. The first is that even though the two random variables have the same mean, they have strikingly different variances. This shouldn't be too surprising. Variances measure how spread out the values are about the mean. It's either the integral of $(x - \mu_X)^2$, or the sum of $(x_n - \mu_X)^2$. The higher the probability of X being close to μ_X, the smaller the variance. In our example, the first section had their scores tightly bound near the mean, ranging from 40 to 60, while the second class had a great spread, going from 0 to 100.

What else can we learn? The second takeaway is the *size* of the variance. Notice how large it is in the second section: 1250. As this is significantly higher than all the numbers in the problem, we should really stop and think about the physical significance of the variance. A great way to answer this is to assign units to our numbers; thus instead of a grade of 50 imagine we earn 50 dollars, or are 50 meters from home. Both sections would have an average distance of 50 meters, and the variances would be 50 meters-squared in the first section, and 1250 meters-squared in the second. Why is the unit meters-squared? Probabilities are unitless, and since we've put a unit of meters on our random variable, when we square it we get meters-squared. This suggests that the variance is the wrong quantity to study; we should look at its square-root, the standard deviation. Why? That has the same units as the mean (in this case meters). We get a standard deviation of $\sqrt{50} \approx 7.07$ meters in the first section, and $\sqrt{1250} \approx 36.4$ meters in the second. Not only are these values now measured in the same units as the mean, but the numbers are on a similar scale. It's worth recording and emphasizing this fact.

> **Variance versus the standard deviation**: The advantage the standard deviation has over the variance is that it has the same units as the mean. Thus, it's the standard deviation that provides the natural scale to view fluctuations about the mean.

To try and get a good sense as to what the moments tell you, think back to the Taylor series of §9.1. First, the more Taylor coefficients we know, the more we know about the shape and the behavior of the function. We can begin to see a similar statement holding for probability density functions. Even though we've only looked at the mean and the variance, already we can see how knowing more moments allows us to distinguish between different pdfs. Second, we viewed the Taylor coefficients as adding bits of information in a natural progression. If we could know only one fact about our function in order to extrapolate its behavior on an interval, we would choose to know its value at the center. If we could know two values, then it's logical to next choose the first derivative at the center, and then the second derivative at the center if we can know three, and so on. For a random variable, we know the sum of the probabilities must equal 1, so the zeroth moment is always the same. Knowing the mean is then like knowing the $f(0)$, knowing the variance (how spread out we are about the mean) is like knowing $f'(0)$ (how fast we are changing), et cetera.

To put it another way, if you can only find out one piece of information about a random variable, a great choice is its average value. If you get another wish and can find out one more tidbit, you then ask how spread out it is. We'll answer the obvious question of 'what comes next' in §9.7.

 Time for some examples. Let's revisit rolling two fair die, with R the random variable denoting the sum of rolled numbers. We found its pdf in §7.2:

$$\text{Prob}(R = r) = \begin{cases} \dfrac{6 - |r - 7|}{36} & \text{if } r \in \{2, 3, \ldots, 12\} \\ 0 & \text{otherwise.} \end{cases}$$

To find the mean, we just plug away:

$$\mu_R = \sum_{r=2}^{12} r \frac{6 - |r - 7|}{36} = 2 \cdot \frac{1}{36} + 3 \cdot \frac{2}{36} + \cdots + 12 \cdot \frac{1}{36} = 7.$$

While we could find a nice expression for the answer by splitting the sum into two parts ($r \leq 7$ and $r \geq 8$) and using formulas for sums of integers and sums of squares, it's just not worth it. We only have 11 terms—we can evaluate this without too much trouble; it would be a very different story if we had a million sided die! Our answer also passes the plausibility test. Our pdf is symmetric about 7, which suggests that 7 should be the mean.

What about the variance? That's

$$\sigma_R^2 = \sum_{r=2}^{12} (r-7)^2 \cdot \frac{6 - |r-7|}{36}$$

$$= (-5)^2 \cdot \frac{1}{36} + (-4)^2 \cdot \frac{2}{36} + \cdots + 5^2 \cdot \frac{1}{36} = \frac{35}{6},$$

which means the standard deviation is $\sqrt{35/6} \approx 2.42$.

 We end with a remark about the technical condition in the definition of the mean. It's not enough for the integral defining the mean to exist; we want the corresponding integral involving the absolute value of the integrand to be finite as well. While this may not seem too important now, it's needed in many later arguments (see for example those in §9.6). This finiteness assumption allows us to exchange orders of integration and summation (i.e., we can use Fubini's theorem), and that's a great assistance. Essentially, this means we don't have to worry too much about convergence.

 For example, consider the following probability density function:

$$f_X(x) = \frac{1}{\pi} \frac{1}{1 + x^2}.$$

To see this is a density, it helps to recall that $\frac{d}{dx} \arctan(x) = 1/(1 + x^2)$; this is why we have the normalization constant of $1/\pi$. If you haven't seen this or don't recall this, don't worry. The value of the normalization constant isn't that important. What really matters is that there is *some* normalization constant (we'll talk a bit more about finding this constant in §15.9). As our function never blows up and decays rapidly at infinity, by the integral test we see the integral is finite. Thus after some rescaling it will be a probability distribution. (It's a very important one, the **Cauchy distribution**.)

What is the mean? To find it we must evaluate

$$\int_{-\infty}^{\infty} x \cdot \frac{1}{\pi} \frac{1}{1 + x^2} dx = \frac{1}{\pi} \int_{-\infty}^{\infty} \frac{x}{1 + x^2} dx.$$

It's tempting to say this integral is zero, as we have an odd function and we're integrating over a symmetric region. Unfortunately, this is an improper integral, and we have to be a bit more careful. For the integral to exist, we should get the same answer no matter how we go to infinity. In other words, we need

$$\lim_{\substack{A \to \infty \\ B \to \infty}} \int_{-A}^{B} \frac{1}{\pi} \frac{x}{1 + x^2} dx$$

to converge to the same value, no matter how A and B tend to infinity. If we take $A = B$ we always get zero, while if we take $B = 2A$ we get

$$\lim_{A \to \infty} \int_{-A}^{2A} \frac{1}{\pi} \frac{x}{1 + x^2} dx = \lim_{A \to \infty} \left(\int_{-A}^{A} \frac{1}{\pi} \frac{x}{1 + x^2} dx + \int_{A}^{2A} \frac{1}{\pi} \frac{x}{1 + x^2} dx \right).$$

Note that the first integral equals zero because we are integrating an odd function over a symmetric region. Thus, we see

$$\lim_{A \to \infty} \int_{-A}^{2A} \frac{1}{\pi} \frac{x}{1+x^2} dx = \lim_{A \to \infty} \int_{A}^{2A} \frac{1}{\pi} \frac{x}{1+x^2} dx;$$

however, the last integrand is essentially $x/x^2 = 1/x$ for x large. If we want to be rigorous, the integrand is larger than $1/2x$. Integrating that from A to $2A$ gives $\frac{1}{2}(\log(2A) - \log(A)) = \frac{1}{2} \log 2$. This isn't zero, so the value of the integral *does* depend on how we tend to infinity!

In order to avoid these technicalities, this is why we require a bit more in the definition of the mean. It's not enough for the integral to exist—we have to spell it out a bit more as it is an improper integral.

We end with two last comments. First, as probability densities are non-negative, we never have to put them inside the absolute value. Second, we don't have to worry about any of this for the variance, as the variance is the integral of $(x - \mu_X)^2 f_X(x)$, which is always non-negative. It's only when the integrand is sometimes positive and sometimes negative that we have to worry about some delicate convergence issues.

9.4 Joint Distributions

In many problems, it's not enough to be able to compute moments (primarily the mean and variance) of a random variable. We often need to work with the sum of random variables. We've seen a few examples of this. Our first encounter was writing the number of heads in three tosses of a fair coin, X, as a sum of the number of heads on the ith toss: $X = X_1 + X_2 + X_3$. Fortunately, much of our theory generalizes to this case. We won't consider the most general situation below, contenting ourselves with a few comments afterwards.

Joint probability density function: Let $X_1, X_2, \ldots, X_n$ be continuous random variables with densities $f_{X_1}, f_{X_2}, \ldots, f_{X_n}$. Assume each X_i is defined on a subset of $\mathbb{R}$ (the real numbers). The joint density function of the tuple $(X_1, \ldots, X_n)$ is a non-negative, integrable function $f_{X_1, \ldots, X_n}$ such that for every nice set $S \subset \mathbb{R}^n$ we have

$$\text{Prob}\,((X_1, \ldots, X_n) \in S) = \int \cdots \int_S f_{X_1, \ldots, X_n}(x_1, \ldots, x_n) dx_1 \cdots dx_n,$$

and

$$f_{X_i}(x_i) = \int_{x_1 = -\infty}^{\infty} \cdots \int_{x_{i-1} = -\infty}^{\infty} \int_{x_{i+1} = -\infty}^{\infty} \cdots \int_{x_n = -\infty}^{\infty}$$
$$f_{X_1, \ldots, X_{i-1}, X_{i+1}, \ldots, X_n}(x_1, \ldots, x_{i-1}, x_{i+1}, \ldots, x_n) \prod_{\substack{j=1 \\ j \neq i}}^{n} dx_j.$$

We call f_{X_i} the **marginal density** of X_i, and obtain it by integrating out the other $n - 1$ variables.

TABLE 9.2.

The joint density of (X, Y), where X is the number of heads in the first 3 tosses and Y is the number of heads in the last 2 tosses of 5 independent tosses of fair coins.

	$Prob(Y = 0)$	$Prob(Y = 1)$	$Prob(Y = 2)$	
Prob($X = 0$)	1/32	2/32	1/32	1/8
Prob($X = 1$)	3/32	6/32	3/32	3/8
Prob($X = 2$)	3/32	6/32	3/32	3/8
Prob($X = 3$)	1/32	2/32	1/32	1/8
	1/4	2/4	1/4	

The n random variables $X_1, \ldots, X_n$ are independent if and only if

$$f_{X_1, \ldots, X_n}(x_1, \ldots, x_n) \ = \ f_{X_1}(x_1) \cdots f_{X_n}(x_n).$$

For discrete random variables, replace the integrals with sums.

While we frequently wish that each random variable is independent of the others, this is not always the case. The joint density function is a way to codify how the variables depend on each other, as it gives the probability of observing a given n-tuple.

In our definition above we assumed each random variable was defined on the real numbers for convenience, but we could easily allow the different variables to live on different spaces. All that matters is we have some measure, some way to integrate over the n-tuples.

 A great way to visualize the joint density function, especially if we have discrete variables, is through a table. Let's consider the case where we toss five fair coins, and all the tosses are independent of each other. Let X be the number of heads in the first three tosses, and Y the number of tails in the last two tosses. We want to find the joint density of the pair (X, Y). Note the possible values are (m, n), where $m \in \{0, 1, 2, 3\}$ and $n \in \{0, 1, 2\}$. As everything is independent, the probability that $X = m$ **and** $Y = n$ (for m and n in these ranges) is just the product of the probability $X = m$ **and** the probability $Y = y$. We find

$$\Pr((X, Y) \ = \ (m, n)) \ = \ \begin{cases} \binom{3}{m} 2^{-3} \cdot \binom{2}{n} 2^{-2} & \text{if } m \in \{0, \ldots, 3\} \text{ and } n \in \{0, 1, 2\} \\ 0 & \text{otherwise.} \end{cases}$$

We display this in Table 9.2. The entries are the probabilities of (X, Y) equaling (m, n). Note the first entry is the value of X and the second is the value of Y. If we sum over a row, that means we're looking at all pairs (X, Y) where Y can take on any value. But this is just the definition of the marginal density of X! So, to find the marginal density of X, we just sum each row. Similarly, if we sum over a column that means we consider all possible values of X, and we get the marginal density of Y.

We can check and see if the random variables are independent (they better be, as we chose X and Y independently!). We must show that $f_{X,Y}(m, n) = f_X(m) f_Y(n)$

TABLE 9.3.
The joint density of (U, V), where U is the number of heads in the first 3 tosses and V is the number of heads in the last 2 tosses of 5 tosses of fair coins.

	$Prob(V = 0)$	$Prob(V = 1)$	$Prob(V = 2)$	
$Prob(U = 0)$	1/16	1/16	0/16	1/8
$Prob(U = 1)$	2/16	3/16	2/16	3/8
$Prob(U = 2)$	1/16	3/16	2/16	3/8
$Prob(U = 3)$	0/16	1/16	1/16	1/8
	1/4	2/4	1/4	

for all m, n; we have to check this for all pairs. Clearly it's enough to check this for $m \in \{0, 1, 2, 3\}$ and $n \in \{0, 1, 2\}$. For example, $f_{X,Y}(1, 2) = 6/32$, and $f_X(1)f_Y(2) = 3/8 \cdot 2/4 = 6/32$. If you check the other 11 possibilities you'll find they all work too, and thus X and Y are independent.

In Table 9.3 we give a different joint density. There's a lot of interesting features here, and a lot of similarities with the example from Table 9.2.

First, note each entry is non-negative and the sum of all the entries is 1. This immediately implies that the sum of the column sums is 1, as is the sum of the row sums. Why? If we sum the three column sums we've just summed all the entries, which we know is 1; the same is true for the sum of the row sums. Thus, we have a joint probability distribution.

Notice that the marginal density of U is the same as the marginal density of X in the previous example, and the marginal density of V is the same as Y's; however, the joint density is *very* different. Are U and V independent? To show they're independent we must show $f_{U,V}(m, n) = f_U(m)f_V(n)$ for all (m, n); however, to show they're dependent we only need to find one pair that doesn't work. There are two good pairs to study: $(m, n) = (0, 2)$ or $(3, 0)$. Why are these good pairs to test? The joint probabilities for these pairs is zero, but f_U and f_V aren't zero for any of these m or n! If you prefer the actual calculation, it's

$$f_{U,V}(0, 2) = 0 \neq \frac{1}{8} \cdot \frac{2}{4} = f_U(0)f_V(2).$$

Though they aren't independent, there's still a nice story for these two random variables. While U is the number of heads in the first three tosses and V is the number of heads in the last two, *I glued the first and fourth coins together*. Thus those are always both heads or both tails, and there's a clear dependence between U and V! Instead of having 32 possibilities (8 options for the first three tosses times 4 options for the last two) we now only have 16 possibilities (8 options for the first three tosses, but after tossing the first three coins there's only two options for the last two, since the fourth coin is now determined and only the fifth coin is free).

We end with an example of a continuous joint density. We set

$$f_{X,Y}(x, y) = \begin{cases} 1/\pi & \text{if } x^2 + y^2 \leq 1 \\ 0 & \text{otherwise.} \end{cases}$$

The factor of $1/\pi$ is due to the fact that the area of the unit disk is π. Remember, we need

$$\int_{x=-\infty}^{\infty}\int_{y=-\infty}^{\infty} f_{X,Y}(x,y)dxdy = 1;$$

this is why we need to include the factor of $1/\pi$. There's a really nice interpretation for this: we're choosing a point randomly in the unit disk, and the probability we're in a given subregion of the disk is just the ratio of the area of the subregion to the area of the disk.

There are two good questions to ask: what are the marginals, and are X and Y independent? Let's find the marginal of X (the marginal of Y follows similarly). We might as well assume $x \in [-1, 1]$, as otherwise $f_X(x) = 0$. We now integrate over all y. The only non-zero contribution is when $x^2 + y^2 \le 1$. Since x is fixed, this means we only need to worry about y from $-\sqrt{1-x^2}$ to $\sqrt{1-x^2}$, and we find

$$f_X(x) = \int_{y=-\sqrt{1-x^2}}^{\sqrt{1-x^2}} \frac{1}{\pi}dy = \frac{2}{\pi}\sqrt{1-x^2}$$

for $x \in [-1, 1]$, and 0 otherwise.

As a check, let's see that this integrates to 1. We have

$$\int_{-\infty}^{\infty} f_X(x)dx = \int_{-1}^{1} \frac{2}{\pi}\sqrt{1-x^2} = \frac{4}{\pi}\int_0^1 \sqrt{1-x^2}dx.$$

We use trigonometric substitution (you were warned about the need to remember integration techniques!). Setting $x = \sin\theta$, we find $dx = \cos\theta d\theta$, $\sqrt{1-x^2} = \sqrt{1-\sin^2\theta} = \cos\theta$ (we don't need to worry about a minus sign—this is why we rewrote the integration earlier to be from 0 to 1), and x ranging from 0 to 1 becomes θ runs from 0 to $\pi/2$. We get

$$\int_{-\infty}^{\infty} f_X(x)dx = \frac{4}{\pi}\int_0^{\pi/2} \cos\theta \cdot \cos\theta d\theta = \frac{4}{\pi}\int_0^{\pi/2} \cos^2\theta d\theta.$$

We now have to integrate the square of cosine. We could use its anti-derivative is $\frac{1}{2}\theta + \frac{1}{4}\sin(2\theta)$, or we could note that by symmetry this integral is one-fourth the integral from 0 to 2π. I'll leave the first approach to you, and do the second as it illustrates a nice symmetry.

We first recall a useful fact: $\cos(\theta) = \sin(\theta + \frac{\pi}{2})$. Thus if we integrate $\cos^2\theta$ over a full period (i.e., from 0 to 2π), we get the same answer if instead we integrate $\sin^2\theta$ over this region. We'll use this fact below to give a nice, quick proof that f_X integrates to 1. Frequently it is easier to evaluate an integral over the entire region than over a subset; this is due to our ability to exploit symmetries. We'll see more examples of this when we compute the normalization constant of the Gaussian density in Chapter 14.

By symmetry, we have

$$\int_0^{\pi/2} \cos^2\theta d\theta = \frac{1}{4}\int_0^{2\pi} \cos^2\theta d\theta = \frac{1}{4}\int_0^{2\pi} \sin^2\theta d\theta = \int_0^{\pi/2} \sin^2\theta d\theta.$$

After a little bit of algebra, we find

$$\int_0^{\pi/2} \cos^2 \theta d\theta = \frac{1}{8} \int_0^{2\pi} (\cos^2 \theta + \sin^2 \theta) d\theta = \frac{1}{8} \int_0^{2\pi} 1 d\theta = \frac{\pi}{4}.$$

Substituting, we finally obtain

$$\int_{-\infty}^{\infty} f_X(x) dx = \frac{4}{\pi} \cdot \frac{\pi}{4} = 1.$$

Though this isn't a proof, we're now very confident that we haven't made an algebra mistake in computing f_X as its integral is 1.

Similarly, we can find the marginal density of Y:

$$f_Y(y) = \begin{cases} (2/\pi)\sqrt{1 - y^2} & \text{if } y \in [-1, 1] \\ 0 & \text{otherwise.} \end{cases}$$

We now turn to independence. Does $f_{X,Y}(x, y) = f_X(x) f_Y(y)$? No! We can easily see this by looking at $(4/5, 4/5)$, as $f_{X,Y}(4/5, 4/5) = 0$ (since this point is outside the unit circle) but $f_X(4/5) = f_Y(4/5) = 3/5$.

Actually, we didn't need to go to all this trouble. If x is close to 1, then the possible values of y are severely limited, and we can "see" the dependencies.

9.5 Linearity of Expectation

The purpose of this section is to prove one of the most important and useful facts about expectation: it's linear! The consequences are far-reaching; we'll see time and time again that many seemingly hopeless problems fall easily to linearity. Our main result is the following theorem.

Theorem 9.5.1 (Linearity of Expectation): *Let $X_1, \ldots, X_n$ be random variables, let $g_1, \ldots, g_n$ be functions such that $\mathbb{E}[|g_i(X_i)|]$ exists and is finite, and let $a_1, \ldots, a_n$ be any real numbers. Then*

$$\mathbb{E}[a_1 g_1(X_1) + \cdots + a_n g_n(X_n)] = a_1 \mathbb{E}[g_1(X_1)] + \cdots + a_n \mathbb{E}[g_n(X_n)].$$

Note the random variables are not assumed to be independent. Also, if $g_i(X_i) = c$ (where c is a fixed number) then $\mathbb{E}[g_i(X_i)] = c$.

In words, the above says the expected value of a sum is the sum of the expected values, and, as promised, in later chapters we'll see far-reaching consequences. Before sketching the proof, we briefly remark on the technical assumption that the expectation of the absolute values are finite. There are a few quantities that are undefined in mathematics, such as $\infty \cdot 0, 0/0, \infty/\infty$, and $\infty - \infty$; our assumption is to avoid the latter. We want to be able to freely interchange orders of integration, and not worry about integrals existing. This assumption allows us to use Fubini's theorem whenever needed to switch the orders of integration, which we'll see is very helpful.

 The proof below is a bit long, and slightly technical in places. It's fine to just skim it on a first pass, but it's worth mulling over. Remember, it's a one-time cost. Once we prove the theorem, we can then just use linearity of expectation. We never have to open up the nuts and bolts of this argument again! In a sense, it's a lot like all those results we proved in calculus. We found the derivatives of functions like x^n or $\sin x$ once, and then freely used the derivatives afterwards.

Sketch of the proof of Theorem 9.5.1: We'll only look at the $n = 2$ case, as a similar argument holds in general. We'll also assume our random variables are continuous and use integral notation, though of course similar arguments hold in the discrete case.

As a warm-up, let's look at the case where $X_2 = X_1$; as we only have one random variable, let's call it X. We want to evaluate $\mathbb{E}[a_1 g_1(X) + a_2 g_2(X)]$. We just use the definition, and see

$$
\begin{aligned}
\mathbb{E}[a_1 g_1(X) + a_2 g_2(X)] &= \int_{-\infty}^{\infty} (a_1 g_1(x) + a_2 g_2(x)) \, f_X(x) dx \\
&= \int_{-\infty}^{\infty} a_1 g_1(x) f_X(x) dx + \int_{-\infty}^{\infty} a_2 g_2(x) f_X(x) dx \\
&= a_1 \int_{-\infty}^{\infty} g_1(x) f_X(x) dx + a_2 \int_{-\infty}^{\infty} g_2(x) f_X(x) dx \\
&= a_1 \mathbb{E}[g_1(X)] + a_2 \mathbb{E}[g_2(X)].
\end{aligned}
$$

This wasn't too bad. Further, the proof suggests why the result is true: *sums and integrals are linear (the integral of a sum is the sum of the integrals), and expectation seems to inherit this linearity.*

The situation is a bit more complicated if X_1 and X_2 are different random variables. The problem is that we now have a *joint* density function. We have a two-dimensional integral. As each $\mathbb{E}[g_i(X_i)]$ involves just one variable, each is a one-dimensional integral. This observation suggests the proof: we need to integrate out one of the variables and reduce a two-dimensional integral down to the sum of two different one-dimensional integrals (if we had n variables, we would have to reduce from n-dimensional integrals down to one-dimensional ones).

Again, we'll just do the case when we have two variables; the general case follows similarly (and it's a good idea for you to fill in the details!). To simplify the calculation below, we take $a_1 = a_2 = 1$ and we let $g_i(X_i) = X_i$. You can go through the proof and tweak it to handle the more general case; I prefer doing the simpler case so the idea isn't buried in the notation.

So, let f_{X_1, X_2} be the joint density function of X_1 and X_2 (see §9.4 for a review of joint densities). Set $X = X_1 + X_2$; our goal is to show $\mathbb{E}[X] = \mathbb{E}[X_1] + \mathbb{E}[X_2]$. Our starting point is

$$
\mathbb{E}[X] = \int_{x=-\infty}^{\infty} x f_X(x) dx.
$$

We *have* to start this way. *Always, always, always* pay heed to definitions. This is how we define the expected value, so this is how we should start; this is very similar to how the various differentiation rules in calculus are proved (they all begin by starting with the definition of the derivative). We *must* begin with an integral (or a sum) of the random variable X. In order to make progress, of course, we *want* to replace this integral with integrals involving x_1 and x_2. These will initially be two-dimensional integrals, which we then replace with one-dimensional integrals by integrating out a variable. As we're going to have a lot of integrals over different variables, we include the variables in the bounds of integration to help keep track of what is being integrated.

We now replace the integral over x with an integral over x_1 and x_2. How do we do this? Well, to find the probability that $X = x$ we integrate over all x_1, with $x_2 = x - x_1$. We get

$$
\begin{aligned}
\mathbb{E}[X] &= \int_{x=-\infty}^{\infty} x \cdot \Pr(X = x)dx \\[2mm]
&= \int_{x=-\infty}^{\infty} x \left[\int_{x_1=-\infty}^{\infty} f_{X_1,X_2}(x_1, x - x_1)dx_1 \right] dx \\[2mm]
&= \int_{x_1=-\infty}^{\infty} \int_{x=-\infty}^{\infty} x f_{X_1,X_2}(x_1, x - x_1)dx\,dx_1.
\end{aligned}
$$

We now change variables, and let $x_2 = x - x_1$ (so $x = x_1 + x_2$). In the calculation below our goal is to integrate out one of the random variables, converting the joint density to a marginal density. Why is this useful? The resulting integral is the expectation of *one* of our random variables, which we'll know. We get

$$
\begin{aligned}
\mathbb{E}[X] &= \int_{x_1=-\infty}^{\infty} \int_{x_2=-\infty}^{\infty} (x_1 + x_2) f_{X_1,X_2}(x_1, x_2)dx_2\,dx_1 \\[2mm]
&= \int_{x_1=-\infty}^{\infty} \int_{x_2=-\infty}^{\infty} x_1 f_{X_1,X_2}(x_1, x_2)dx_2\,dx_1 \\[2mm]
&\quad + \int_{x_1=-\infty}^{\infty} \int_{x_2=-\infty}^{\infty} x_2 f_{X_1,X_2}(x_1, x_2)dx_2\,dx_1 \\[2mm]
&= \int_{x_1=-\infty}^{\infty} x_1 \left[\int_{x_2=-\infty}^{\infty} f_{X_1,X_2}(x_1, x_2)dx_2 \right] dx_1 \\[2mm]
&\quad + \int_{x_2=-\infty}^{\infty} x_2 \left[\int_{x_1=-\infty}^{\infty} f_{X_1,X_2}(x_1, x_2)dx_1 \right] dx_1 \\[2mm]
&= \int_{x_1=-\infty}^{\infty} x_1 f_{X_1}(x_1)dx_1 + \int_{x_2=-\infty}^{\infty} x_2 f_{X_2}(x_2)dx_1 \\[2mm]
&= \mathbb{E}[X_1] + \mathbb{E}[X_2].
\end{aligned}
$$

This completes the proof in the special case when $n = 2$, but a similar argument holds in general.

We're left with showing that if $g_i(X_i) = c$ then $\mathbb{E}[g_i(X_i)] = c$. This follows immediately from the definition of a probability density:

$$\mathbb{E}[g_i(X_i)] \ = \ \mathbb{E}[c] \ = \ \int_{-\infty}^{\infty} c f_{X_i}(x_i) dx_i \ = \ c \int_{-\infty}^{\infty} f_{X_i}(x_i) dx_i \ = \ c \cdot 1 \ = \ c,$$

completing the proof of the theorem.   □

In the proof above, it's worth looking at what we *didn't* use in the proof. I know typically we're concerned with what we *use* in the proof, but here what isn't used is just as important. *Nowhere* in the proof do we assume that the random variables are independent. This is wonderful, as it means the result holds *even if* the random variables are dependent! Finally, we don't have to go through the hassle of writing down the joint density anymore; we can attack problems with linearity of expectation. Let's isolate a few key consequences.

Lemma 9.5.2: *Let X be a random variable with mean μ_X and variance σ_X^2. If a and b are any fixed constants, then for the random variable $Y = aX + b$ we have*

$$\mu_Y \ = \ a\mu_X + b \quad \text{and} \quad \sigma_Y^2 \ = \ a^2 \sigma_X^2.$$

Proof: We could prove the above by going back to the definition, but it's easier (and a lot more fun!) to use Theorem 9.5.1. Let's do the mean first. As $\mu_X = \mathbb{E}[X]$ and $\mu_Y = \mathbb{E}[Y]$, we have

$$\mathbb{E}[Y] \ = \ \mathbb{E}[aX + b] \ = \ a\mathbb{E}[X] + b\mathbb{E}[1] \ = \ a\mu_X + b,$$

as claimed. We're now seeing the dividends for sloshing through the proof of Theorem 9.5.1; we no longer have to write the densities down explicitly!

We now turn to the variance. As always, the starting point is the definition. Fortunately for us, the definition of the variance involves expectation, which means we can use Theorem 9.5.1 again!

$$\begin{aligned}
\sigma_Y^2 \ &= \ \mathbb{E}[(Y - \mu_Y)^2] \\
&= \ \mathbb{E}\left[((aX + b) - (a\mu_X + b))^2\right] \\
&= \ \mathbb{E}\left[(aX - a\mu_X)^2\right] \\
&= \ \mathbb{E}[a^2(X - \mu_X)^2] \ = \ a^2 \mathbb{E}[(X - \mu_X)^2] \ = \ a^2 \sigma_X^2. \qquad \square
\end{aligned}$$

The claims in Lemma 9.5.2 are quite reasonable. If we rescale our random variable by a, it makes sense that the mean is rescaled by a. Think of this as changing from measuring in feet to inches; our new values are 12 times that of our old. This should also affect the standard deviation by rescaling that by $|a|$ (we need an absolute value as a might be negative), which changes the variance by a factor of a^2 (which is always non-negative). If now we add b, that increases the mean by b but has no effect on the variance (if everything is shifted by the same amount, the fluctuations about the mean aren't changed).

We end with one last consequence of linearity of expectation, a much simpler formula to find variances.

Lemma 9.5.3: *Let X be a random variable. Then*

$$\mathrm{Var}(X) = \mathbb{E}[X^2] - \mathbb{E}[X]^2.$$

This lemma tells us that order really matters. Variances are non-negative; in fact, unless *all* the probability is concentrated at one point the variance is positive. The right-hand side above is the expected value of X^2 minus the square of the expected value of X. This is thus typically positive, and illustrates that squaring *before* taking the expectation is not the same as squaring *after*.

Proof: The proof is an almost pleasurable calculation, as we know expectation is linear. As always, we start with the definition of our term (in this case, the variance) and start manipulating the algebra. In the analysis below we use $\mathbb{E}[aY] = a\mathbb{E}[Y]$ and $\mathbb{E}[b] = b$ for any random variable Y and any constants a and b. We have

$$
\begin{aligned}
\mathrm{Var}(X) &= \mathbb{E}[(X - \mu_X)^2] \\
&= \mathbb{E}[X^2 - 2\mu_X X + \mu_X^2] \\
&= \mathbb{E}[X^2] - \mathbb{E}[2\mu_X X] + \mathbb{E}[\mu_X^2] \\
&= \mathbb{E}[X^2] - 2\mu_X \mathbb{E}[X] + \mu_X^2 \\
&= \mathbb{E}[X^2] - 2\mu_X \cdot \mu_X + \mu_X^2 \\
&= \mathbb{E}[X^2] - \mu_X^2 \;=\; \mathbb{E}[X^2] - \mathbb{E}[X]^2,
\end{aligned}
$$

as claimed. □

Whenever you have a choice in how to compute something, it's worth reflecting a bit on the merits of each. Is this a better formula than the definition? What are its advantages and disadvantages?

It happens that this *is* a great formula. It allows us to find the second centered moment (the variance) if we know the first and second moments. It's often both theoretically easier and computationally easier to find $\mathbb{E}[X^2]$ or $\mathbb{E}[X]$ when given data than to first find the mean, then subtract the mean from all the values, then square these new values, and so on. Notice there's a lot more algebra with the definition—we have the same number of squarings, but a lot more subtraction. This suggests that our new formula has merit.

Of course, with the power of modern computers subtraction is no longer a tedious chore, but the more operations you do, the more chance of rounding errors and other mistakes. This new formula is simpler, and useful.

Let's use it to find the variance in the number rolled on a fair die. If X is the roll, then X takes on the values 1, 2, 3, 4, 5, or 6 with probability 1/6, and has zero probability of any other value. The mean is just

$$\mu_X = \mathbb{E}[X] = 1 \cdot \frac{1}{6} + 2 \cdot \frac{1}{6} + 3 \cdot \frac{1}{6} + 4 \cdot \frac{1}{6} + 5 \cdot \frac{1}{6} + 6 \cdot \frac{1}{6} = \frac{21}{6} = 3.5.$$

The second moment is

$$\mathbb{E}[X^2] = 1^2 \cdot \frac{1}{6} + 2^2 \cdot \frac{1}{6} + 3^2 \cdot \frac{1}{6} + 4^2 \cdot \frac{1}{6} + 5^2 \cdot \frac{1}{6} + 6^2 \cdot \frac{1}{6} = \frac{91}{6}.$$

Thus the variance is

$$\text{Var}(X) = \frac{91}{6} - 3.5^2 = \frac{35}{12} \approx 2.92.$$

Compare this to

$$\text{Var}(X) = (1 - 3.5)^2 \cdot \frac{1}{6} + (2 - 3.5)^2 \cdot \frac{1}{6} + \cdots + (6 - 3.5)^2 \cdot \frac{1}{6};$$

the algebra *is* nicer with the new formula.

This is an example of a general phenomenon in mathematics. Often we have a definition for a concept, but to actually find it in practice the definition is a bit unwieldy, and a different formula is used. If you've taken multivariable calculus, there's a good example there: compare the definition of the directional derivative, involving a limit, with the dot product involving the gradient, which is used in practice.

9.6 Properties of the Mean and the Variance

Before continuing our study of moments, we first state and prove a very useful fact.

Theorem 9.6.1: *If X and Y are independent random variables, then*

$$\mathbb{E}[XY] = \mathbb{E}[X]\mathbb{E}[Y].$$

A particularly important case is

$$\mathbb{E}[(X - \mu_X)(Y - \mu_Y)] = \mathbb{E}[X - \mu_X]\mathbb{E}[Y - \mu_Y] = 0.$$

Proof: Sadly, the proof doesn't follow immediately from Theorem 9.5.1. The problem is that theorem deals with sums, and we have a product. This means we *must* return to the definition of the expectation. For definiteness, we'll assume our random variables are continuous, with joint density function $f_{X,Y}$. We have

$$\mathbb{E}[XY] = \int_{-\infty}^{\infty} \int_{-\infty}^{\infty} xy \cdot f_{X,Y}(x, y) dx dy.$$

To be truly rigorous, perhaps we should go through the notational horror of first defining a random variable $Z = XY$, and then saying

$$\mathbb{E}[Z] = \int_{-\infty}^{\infty} z f_Z(z) dz,$$

and then going from this to our double integral. This argument is standard, and as we've done it several times already, let's just jump to the double integral and move on.

We need to use the hypothesis that X and Y are independent. Imagine that X has mean zero, and $Y = X$. In that case, $\mathbb{E}[X]\mathbb{E}[Y] = 0$, but $\mathbb{E}[X^2]$ is non-negative (and almost surely positive). *Whenever you have given information, you should always ask yourself what facts, theorems, and results it gives you.* One item that comes to mind is that the joint density function of two independent random variables is the product of the marginal density functions. Or, for this problem,

$$f_{X,Y}(x, y) = f_X(x) f_Y(y).$$

Using this in the double integral for $\mathbb{E}[XY]$ yields

$$\mathbb{E}[XY] = \int_{x=-\infty}^{\infty} \int_{y=-\infty}^{\infty} xy f_X(x) f_Y(y) dy dx$$

$$= \int_{x=-\infty}^{\infty} x f_X(x) dx \int_{y=-\infty}^{\infty} y f_Y(y) dy = \mathbb{E}[X]\mathbb{E}[Y].$$

If X and Y are independent, so too are $X - \mu_X$ and $Y - \mu_Y$; these are just simple shifts. Note that the expected value of each is zero. Remembering that the expected value of a constant is that constant, we see that

$$\mathbb{E}[X - \mu_X] = \mathbb{E}[X] - \mathbb{E}[\mu_X] = \mu_X - \mu_X = 0.$$

Substituting now gives the important special case. □

While all the moments are important, some are more important than others. The first two (the mean and the variance) are by far the key ones to know for any distribution. They tell us what to expect, and what size fluctuations we should see. They have many nice properties.

Theorem 9.6.2 (Means and Variances of Sums of Random Variables):
Let $X_1, \ldots, X_n$ be random variables with means $\mu_{X_1}, \ldots, \mu_{X_n}$ and variances $\sigma_{X_1}^2, \ldots, \sigma_{X_n}^2$. If $X = X_1 + \cdots + X_n$, then

$$\mu_X = \mu_{X_1} + \cdots + \mu_{X_n}.$$

If the random variables are independent, then we also have

$$\sigma_X^2 = \sigma_{X_1}^2 + \cdots + \sigma_{X_n}^2 \quad \text{or} \quad \text{Var}(X) = \text{Var}(X_1) + \cdots + \text{Var}(X_n).$$

> *In the special case when the random variables are independent and identically distributed (so all the means equal μ and all the variances equal σ^2), then*
>
> $$\mu_X = n\mu \quad \text{and} \quad \sigma_X^2 = n\sigma^2.$$

We've seen time and time again that, whenever we meet a theorem, the first thing we should do is see how crucial the hypotheses are. Do we really need the random variables to be independent for the variance formula to hold? Absolutely yes! Let's think for a minute about which two random variables are the most dependent. After a little thought, you might come to X and X or X and $-X$. Let's look at the latter. It's not too hard to study $X + (-X)$; this random variable is always zero, and thus it has zero variance. Our theorem won't hold in this case (assuming of course that X has positive variance).

We now turn to the proof. It's a bit technical, so don't worry if you just skim it on a first pass. The main ingredient is that the expected value of a sum is the sum of the expected values.

Proof sketch of Theorem 9.6.2: We'll just do the case when we have two variables; the general case follows similarly (and it's a good idea for you to fill in the details!). We'll also give the proof when X_1, X_2 are continuous; just replace the integrals with sums for the discrete case.

The first part of the theorem is just the special case of Theorem 9.5.1 when each $a_i = 1$ and each $g_i(X_i) = X_i$.

Okay, this shows the expected value of a sum is the sum of the expected values. What about the variance? We'll see things are more involved. We'll do the proof two ways. The first is the "slick" proof, using the power of linearity of expectation (see Theorem 9.5.1) so we don't have to write down the integrals explicitly. We have

$$\sigma_X^2 = \mathbb{E}[(X - \mu_X)^2]$$

$$= \mathbb{E}\left[\left((X_1 + X_2) - (\mu_{X_1} + \mu_{X_2})\right)^2\right]$$

$$= \mathbb{E}\left[\left((X_1 - \mu_{X_1}) + (X_2 - \mu_{X_2})\right)^2\right]$$

$$= \mathbb{E}\left[(X_1 - \mu_{X_1})^2 + 2(X_1 - \mu_{X_1})(X_2 - \mu_{X_2}) + (X_2 - \mu_{X_2})^2\right].$$

We now use linearity of expectation to write the expected value of the sum as the sum of the expected values, giving

$$\sigma_X^2 = \mathbb{E}\left[(X_1 - \mu_{X_1})^2\right] + \mathbb{E}\left[2(X_1 - \mu_{X_1})(X_2 - \mu_{X_2})\right] + \mathbb{E}\left[(X_2 - \mu_{X_2})^2\right]$$

$$= \sigma_{X_1}^2 + 2\mathbb{E}\left[(X_1 - \mu_{X_1})(X_2 - \mu_{X_2})\right] + \sigma_{X_2}^2.$$

In general, we're stuck at this point as the expected value of a product need not be the product of the expected values; however, it *is if our factors are independent*! Now we can see why we assumed the random variables are independent. This is precisely what we need to handle the cross term, which Lemma 9.5.2 now tells us is zero. Therefore

$$\sigma_X^2 = \sigma_{X_1}^2 + \sigma_{X_2}^2,$$

completing the proof (the special case when all the means and all the variances are the same follows immediately).　□

Again, let's explore the consequences of our new statement. We know variances must be non-negative, thus the variance of X above must be non-negative. Note that the formula for σ_X^2 involves only the *squares* of the a_i's; this is good and will make sure it's non-negative. Further, it makes sense that if we double all the a_i's then the variance quadruples (and the standard deviation doubles); think of this as just changing the scale of our measurements. We see this beautifully reflected in our formula; the variance for X does indeed quadruple (so its standard deviation does double) if we double all the a_i's. Obviously this isn't a proof, but it's nice that we can see that the statement is reasonable.

Important results should be proved, if possible, multiple times. We'll now give another proof of Theorem 9.6.2. It's essentially the same proof as before; the only difference is how we write the algebra. It's worth seeing both arguments, as some people prefer one way of viewing the algebra over the other. Personally, I like the "slickness" of the expectation heavy proof, but I do feel more comfortable at times seeing the integrals explicitly written out.

Proof of Theorem 9.6.2 (again): Starting from the definition of the variance, we see

$$\sigma_X^2 = \int_{x_1=-\infty}^{\infty} \int_{x_2=-\infty}^{\infty} \left((x_1+x_2) - (\mu_{X_1} + \mu_{X_2})\right)^2 f_{X_1,X_2}(x_1,x_2)dx_2dx_1.$$

We rewrite part of the integrand as

$$\left((x_1+x_2) - (\mu_{X_1} + \mu_{X_2})\right)^2 = \left((x_1 - \mu_{X_1}) + (x_2 - \mu_{X_2})\right)^2$$

$$= (x_1 - \mu_{X_1})^2 + 2(x_1 - \mu_{X_1})(x_2 - \mu_{X_2}) + (x_2 - \mu_{X_2})^2.$$

Substituting, this gives us three integrals for σ_X^2. Two are easily evaluated, but the third is quite difficult. We get

$$\sigma_X^2 = \int_{x_1=-\infty}^{\infty} \int_{x_2=-\infty}^{\infty} (x_1 - \mu_{X_1})^2 f_{X_1,X_2}(x_1,x_2)dx_2dx_1$$

$$+ \int_{x_1=-\infty}^{\infty} \int_{x_2=-\infty}^{\infty} 2(x_1 - \mu_{X_1})(x_2 - \mu_{X_2}) f_{X_1,X_2}(x_1,x_2)dx_2dx_1$$

$$+ \int_{x_1=-\infty}^{\infty} \int_{x_2=-\infty}^{\infty} (x_2 - \mu_{X_2})^2 f_{X_1,X_2}(x_1,x_2)dx_2dx_1.$$

For the first, we do the x_2 integration first. This integrates out x_2, and we're left with the marginal of X_1. Similarly, for the third we switch orders and do the x_1 integration first, getting the marginal of X_2. The resulting integrals are then just the variance of X_1 and X_2:

$$\sigma_X^2 = \int_{x_1=-\infty}^{\infty} (x_1 - \mu_{X_1})^2 f_{X_1}(x_1)dx_1$$

$$+ \int_{x_1=-\infty}^{\infty} \int_{x_2=-\infty}^{\infty} 2(x_1 - \mu_{X_1})(x_2 - \mu_{X_2})f_{X_1,X_2}(x_1, x_2)dx_2dx_1$$

$$+ \int_{x_2=-\infty}^{\infty} (x_2 - \mu_{X_2})^2 f_{X_2}(x_2)dx_2$$

$$= \sigma_1^2 + \sigma_2^2 + 2\int_{x_1=-\infty}^{\infty} \int_{x_2=-\infty}^{\infty} (x_1 - \mu_{X_1})(x_2 - \mu_{X_2})f_{X_1,X_2}(x_1, x_2)dx_2dx_1.$$

All that remains is showing the last integral is zero. As X_1 and X_2 are independent, $f_{X_1,X_2}(x_1, x_2) = f_{X_1}(x_1)f_{X_2}(x_2)$. The integral becomes

$$\int_{x_1=-\infty}^{\infty} (x_1 - \mu_{X_1})f_{X_1}(x_1)dx_1 \int_{x_2=-\infty}^{\infty} (x_2 - \mu_{X_2})f_{X_2}(x_2)dx_2$$

$$= \mathbb{E}[X_1 - \mu_{X_1}]\mathbb{E}[X_2 - \mu_{X_2}] = 0,$$

which finishes the proof. □

Application: Amazingly, Theorem 9.6.2 plays a big role in designing optimal **investment portfolios**! Here's a brief sketch of how it's used. Imagine we have two stocks with variable returns. Let X_1 denote our return from the first and X_2 from the second. For simplicity, let's assume they both cost \$1 per share, both have an average return of \$3, and both have a variance of \$2. Our goal is to build a portfolio that makes as much money as possible, with as little risk as possible. *If* the two stocks are independent, we can minimize risk by investing in each!

The math is simplified by the two having the same price. Imagine we have \$1 to invest. If we buy w parts of a share of the first stock, we can buy $1 - w$ shares of the second. We denote this by writing $S = wX_1 + (1 - w)X_2$. We first calculate the expected value:

$$\mathbb{E}[S] = \mathbb{E}[wX_1 + (1 - w)X_2] = w\mathbb{E}[X_1] + (1 - w)\mathbb{E}[X_2]$$

$$= w \cdot \$3 + (1 - w) \cdot \$3 = \$3.$$

Note that it doesn't matter what w is; we always expect to make \$3.

What about the variance? That's more interesting. As the two random variables are independent and each has a variance of 2,

$$
\begin{aligned}
\mathrm{Var}(S) \;&=\; \mathrm{Var}(wX_1 + (1-w)X_2) \\[2mm]
&=\; w^2\mathrm{Var}(X_1) + (1-w)^2\mathrm{Var}(X_2) \\[2mm]
&=\; \left(w^2 + (1-w)^2\right)\cdot 2.
\end{aligned}
$$

Note the variance of our investment depends on w. What w minimizes the variance? Using calculus (or even plotting), we see it's $w = 1/2$. The resulting variance is $(1/4 + 1/4)\cdot 2 = 1$.

Ahh. The variance of the combined investment is 1, and this is *half* the variance of our original investment. What's happened is we are able to keep the same level of returns, but with significantly less risk (since the variance is smaller, the chance of large fluctuation is smaller). Of course, this also means we have less of a chance of being far above \$3, but most people are quite happy to make this trade (giving up a chance for higher returns to remove the chance of losing money). This is just one of many examples illustrating the power and utility of these concepts.

We end with one last bit of notation.

Covariance: Let X and Y be random variables. The covariance of X and Y, denoted by σ_{XY} or $\mathrm{Cov}(X, Y)$, is

$$
\sigma_{XY} \;=\; \mathbb{E}\left[(X - \mu_X)(Y - \mu_Y)\right].
$$

Note $\mathrm{Cov}(X, X)$ equals the variance of X. Also, if $X_1, \ldots, X_n$ are random variables and $X = X_1 + \cdots + X_n$, then

$$
\mathrm{Var}(X) \;=\; \sum_{i=1}^{n} \mathrm{Var}(X_i) + 2 \sum_{1 \le i < j \le n} \mathrm{Cov}(X_i, X_j).
$$

We've essentially prove this in the proof of Theorem 9.6.2; the only change is that instead of using independence to show the cross terms vanish, now we allow them to remain and note that they are precisely what we're calling the covariance.

So far, it looks like all we've done is give a name to a term. It turns out covariances frequently arise (as most random variables are not independent) and have many nice properties, some of which we discuss in §9.8.

9.7 Skewness and Kurtosis

Though not as used as the mean and variance, the third and fourth moments have names and uses too. **Skewness** and **kurtosis** are the third and fourth *centered* moments, respectively. These are the first moments at which you can see the shape of the distribution, and they play a key role in how sums of random variables converge to the normal distribution in the Central Limit Theorem.

TABLE 9.4.

The joint density of two random variables X and Y; note X and Y have means of zero, their covariance is 0 but the random variables are not independent.

	$Pr(Y = -2)$	$Pr(Y = -1)$	$Pr(Y = 1)$	$Pr(Y = 2)$	
$Pr(X = -2)$	1/8	0	0	1/8	1/4
$Pr(X = -1)$	0	1/8	1/8	0	1/4
$Pr(X = 1)$	0	1/8	1/8	0	1/4
$Pr(X = 2)$	1/8	0	0	1/8	1/4
	1/4	1/4	1/4	1/4	

Skewness measures the asymmetry of a distribution. Thus the normal distribution has a skewness of zero, as the normal distribution is perfectly symmetric. If our distribution is unimodal, meaning that there is only one maximum point in our probability mass function, then the value of skewness can tell us which side of the distribution has a fatter or longer tail; so, if the skewness is negative, the tail on the left side is fatter or longer than the tail on the right side, and similarly, if the skewness is positive, the tail on the right side is fatter or longer than the tail on the left side. Understanding skewness, consequently, tells us where fluctuations in the mean will lie.

Kurtosis, on the other hand, measures how the data peaks or flattens out, with respect to the normal distribution. A data set with low kurtosis will have somewhat of a plateau at the mean; as an extreme example, a distribution with very low kurtosis would be the uniform distribution. However, a distribution with high kurtosis will have a very sharp point at the mean, have steep drop-offs on both sides, and relatively fat tails. As a source of comparison, the kurtosis of the standard normal distribution is three. See Exercises 9.10.10 and 9.10.11 for inequalities involving the skewness and kurtosis.

9.8 Covariances

As independent variables have a covariance of 0, the covariance is a measure of how a change to one variable effects changes in another. However, it is important to note that a covariance of zero does not always mean that two variables are independent (see Table 9.4 and Exercise 9.10.27). Covariance measures the level of linear association between two variables. A covariance above zero shows that two variables are positively related, while a covariance below zero means they are inversely related.

A term you will often hear alongside covariance is **correlation**, where the correlation ρ is defined by

$$\rho = \frac{\text{Cov}(X, Y)}{\sigma_X \sigma_Y}.$$

Correlation is a standardized version of the covariance, and we always have $\rho \in [-1, 1]$ (see Exercise 9.10.28). It represents the strength of the linear association between two variables. The closer the correlation is to -1 or 1, the stronger the linear association.

For any random variable X and Y (discrete or continuous) with means μ_X and μ_Y, the covariance of X and Y can be calculated by

$$\text{Cov}(X, Y) = \mathbb{E}[XY] - \mu_X \mu_Y;$$

notice how similar this is to our formula for computing variances. The proof follows from linearity of expectation:

$$\mathrm{Cov}(X, Y) = \mathbb{E}[(X - \mu_X)(Y - \mu_Y)]$$

$$= \mathbb{E}[XY - \mu_Y X - \mu_X Y + \mu_Y \mu_X]$$

$$= \mathbb{E}[XY] - \mu_X \mathbb{E}[X] - \mu_X \mathbb{E}[Y] + \mathbb{E}[\mu_X \mu_Y]$$

$$= \mathbb{E}[XY] - \mu_X \mu_Y - \mu_X \mu_Y + \mu_X \mu_Y$$

$$= \mathbb{E}[XY] - \mu_X \mu_Y.$$

9.9 Summary

We met one of the most important concepts in the course, that of the expectation of a random variable. We discussed several of its properties. The most important is linearity; we saw how that led to much simpler proofs (i.e., it removed the need of writing out long arguments with the density functions explicitly given).

The quote from *Ghostbusters* is particularly appropriate:

> *Winston Zeddemore*: We had the tools! We had
>   the talent!
> *Peter Venkman*: It's Miller Time!

All three parts matter. First, we need the right tools. Without the right tools, we can't make any progress; expectation is a powerful tool to add to our arsenal. Second, it's not enough to have the tools; we also need to know how to use them. For example, we regrouped terms in order to be able to recognize simpler expressions. A great example of the power in **rewriting algebra** is

$$\mathbb{E}\left[\left((X_1 + X_2) - (\mu_{X_1} + \mu_{X_2})\right)^2\right] = \mathbb{E}\left[\left((X_1 - \mu_{X_1}) + (X_2 - \mu_{X_2})\right)^2\right].$$

One of the hardest parts of mathematics is learning how to do algebra well; if you rewrite expressions in a good way, you can often see the connections and get a great hint at how to continue. The last line of the couplet refers to a (very old!) advertising campaign for Miller beer, and not to me. I don't recommend having a drink to celebrate a problem well done, but after mastering a long technical argument, some celebration is justified!

9.10 Exercises

Exercise 9.10.1 *Find if you can, or say why you cannot, the first five Taylor coefficients of (a)* $\log(1 - u)$ *at* $u = 0$; *(b)* $\log(1 - u^2)$ *at* $u = 0$; *(c)* $x \sin(1/x)$ *at* $x = 0$.

Exercise 9.10.2 *In §9.2 we found* $\sum_{k=0}^{n} k^2 \binom{n}{k} 2^{-n}$ *equaled a quadratic in n. What if we replace* k^2 *with* k^3? *Is this also a polynomial in n, and if so, what polynomial?*

Exercise 9.10.3 *Show that* $\int_0^1 dx/x$ *is infinite by evaluating* $\lim_{\epsilon \to 0^+} \int_\epsilon^1 dx/x$; *the notation* $\epsilon \to 0^+$ *means* ϵ *converges to zero through positive values.*

Exercise 9.10.4 *Prove the derivative of* $\arctan(x)$ *is* $1/(1+x^2)$. *(Hint: Apply the chain rule to* $\tan(\arctan(x)) = x$.)

Exercise 9.10.5 *Show that* $\lim_{A,B\to\infty} \int_{-A}^{B} \frac{\sin x}{x} dx$ *exists no matter how A and B tend to infinity. This is* not *the case if we replace the integrand with* $\left|\frac{\sin x}{x}\right|$.

Exercise 9.10.6 *Let X be a random variable and* $g_1, \ldots, g_n$ *continuous functions. Prove that* $\mathbb{E}[a_1 g_1(X) + \cdots + a_n g_n(X)] = \sum_{k=1}^{n} a_k \mathbb{E}[g_k(X)]$, *given that all expectations are finite. Why do we need to assume these quantities are finite?*

Exercise 9.10.7 *In §9.5 we commented about how often proofs start with definitions. This is the case with many calculus proofs. Using the definition of the derivative, prove the sum rule and the product rule: if f and g are differentiable functions, then* $(f(x) + g(x))' = f'(x) + g'(x)$ *and* $(f(x)g(x))' = f'(x)g(x) + f(x)g'(x)$.

Exercise 9.10.8 *Let X be a discrete random variable. Prove or disprove:* $\mathbb{E}[1/X] = 1/\mathbb{E}[X]$.

For the next four exercises, here are some good distributions to use.

- Uniform: $f_X(x) = 1$ for $0 \le x \le 1$ and 0 otherwise.
- Exponential: $f_X(x) = \exp(-x)$ for $x \ge 0$ and 0 otherwise.
- Laplace: $f_X(x) = \exp(-|x|)/2$.
- Chi-square: $f_X(x) = x^{\nu/2-1} \exp(-x/2)/2^{\nu/2}\Gamma(\nu/2)$ for $x \ge 0$ and 0 otherwise, where $\nu > 0$ and $\Gamma(s)$ is the Gamma function.
- Normal: $f_X(x) = (2\pi)^{-1/2} \exp(-x^2/2)$.

Exercise 9.10.9 *Calculate the first four centered moments of some common distributions (such as the ones above). Try to find a lower bound for the kurtosis in terms of the skewness and other centered moments. In other words, can you generalize* $\mathbb{E}[X^2] \ge \mathbb{E}[X]^2$?

Exercise 9.10.10 *Fix a random variable X whose first* $k \ge 4$ *moments are finite, and let* $\mu_k = \mathbb{E}[(X - \mu)^k]$. *Verify the inequality* $(\mu_4/\sigma^4) \ge (\mu_3/\sigma^3)^2 + 1$, *where* σ *is the standard deviation, for several common distributions. Notice that the* μ_k/σ *are unitless and are the natural quantities to study.*

Exercise 9.10.11 *Prove the inequality from the previous exercise.*

Exercise 9.10.12 *Do you think the last two exercises generalize and give an inequality for the sixth centered moment in terms of the lower centered moments? Investigate!*

Exercise 9.10.13 *Consider a standard deck of 52 cards. Assume the deck has been thoroughly shuffled, so all 52! possible orderings are equally likely. We pick cards one at a time (without replacement) until we get two cards of the same suit. What is the expected number of cards drawn before two of the same suit are drawn? What is the variance of this number?*

Exercise 9.10.14 *Consider the sum of two random variables* $Y = X_1 + X_2$. *Does the mean of Y depend on whether or not* X_1 *and* X_2 *are independent? Given that* $var(X_1) = var(X_2) = a$, *what are the maximum and minimum values for* $var(Y)$?

Exercise 9.10.15 *Given independent random variables X and Y, show that* $\mathbb{E}[XY] = \mathbb{E}[X]\mathbb{E}[Y]$.

Exercise 9.10.16 *Let $X_1, \ldots, X_n$ be independent, identically distributed random variables that have zero probability of taking on a non-positive value. Prove $\mathbb{E}[(X_1 + \cdots + X_m)/(X_1 + \cdots + X_n)] = m/n$ for $1 \le m \le n$. Does this result seem surprising?*

Exercise 9.10.17 *The standard normal variable has density $\frac{1}{\sqrt{2\pi}}\exp(-x^2/2)$. Find the first four moments.*

Exercise 9.10.18 *Show $\Pr((X, Y) = (m, n)) = \frac{1}{2^n}\frac{1}{2^m}$ for $m, n \in \{1, 2, 3, \ldots\}$ is a joint probability density function.*

Exercise 9.10.19 *Find the marginal distributions for $\text{Prob}((X, Y) = (m, n)) = \frac{1}{2^n}\frac{1}{2^m}$ $m, n \in \{1, 2, 3, \ldots\}$. Are X and Y independent?*

Exercise 9.10.20 *A variable is uniformly chosen on the square $[0, 1]^2$. Given that this variable satisfies $x^2 + y^2 \le 1$, find the marginal distribution on X. Find the marginal distribution on Y. Are X and Y independent with the additional information that $x^2 + y^2 \le 1$? What about without it?*

Exercise 9.10.21 *Calculate the second and third moments of X when $X \sim \text{Bin}(n, p)$ (a binomial random variable with parameters n and p).*

Exercise 9.10.22 *Imagine you are in the library. You have always believed that you spend an average amount of time in the library for a college student. You decide to verify this belief by walking through the library and asking each person how much time they spend there. You discover that the average person you surveyed spends significantly more time in the library than you. Does this necessarily contradict your belief?*

Exercise 9.10.23 *If instead you found that the average person you surveyed spends significantly less time than you in the library, would that contradict your belief that you spend an average amount of time in the library?*

Exercise 9.10.24 *Assume the population of the school is uniformly distributed, spending between 1 and 5 hours a day studying in the library. Further, when a person is in the library is independent of when all other people are in the library. At a given time in the library, what do you expect to be the distribution for the number of hours the people currently in the library spend in the library?*

Exercise 9.10.25 *What is the mean number of hours spent working in the library by the whole student body? What is the mean number of hours spent working in the library by the people in the library, assuming independence?*

Exercise 9.10.26 *Explain how diversification in investments helps to maximize expected return for a given expected level of variance in return, assuming the random variables associated with returns on investments are independent. (In practice, this turns out to not be a great assumption.)*

Exercise 9.10.27 *Prove the random variables X and Y in Table 9.4 are not independent. You can either do the direct calculation, or look at the magnitudes of X and Y where the joint probability is non-zero.*

Exercise 9.10.28 *Prove the correlation coefficient is at most 1 in absolute value. (Hint: The Cauchy-Schwarz inequality from §B.6 might be useful.)*

Exercise 9.10.29 *Prove no matter what units we assign to X and Y, the correlation coefficient is unitless.*

Exercise 9.10.30 *Let $Y = 100X + X^2$, where X is uniform on $[-1, 1]$. Calculate the covariance of X and Y and their correlation coefficient.*

In World War II, the Allies wanted a good method for estimating the number of German tanks in production. They knew the Germans numbered their tanks sequentially from 1 to some unknown N. They captured k tanks, and the maximum observed serial number was m. Unfortunately, the number of tanks captured, k, was not large enough that the allies could reasonably expect that $m \approx N$ and it was necessary to inflate m to obtain a more accurate estimate. Consider M to be a random variable, corresponding to the largest tank observed. One of the methods explored to estimate the number of German tanks is discussed below. We give a proof of this in §12.7.

Exercise 9.10.31 *Calculate the probability that $m = n$ for a given k and N.*

Exercise 9.10.32 *Find the mean value of m for a given N and k (this takes a lot of careful algebra). Solve the equation for N to get a formula for N in terms of m and k.*

Exercise 9.10.33 *Write code that simulates observing k tanks from a population of N tanks. Test the effectiveness of the formula you derived at estimating N given the k observed tanks.*

Exercise 9.10.34 *Generalize the German tank problem and assume the tanks are sequentially numbered from N_1 to N_2. We do not know those values, but we are able to observe k tank serial numbers. If the smallest observed is m_1 and the largest is m_2, estimate the number of tanks produced.*

Exercise 9.10.35 *Assume we choose a point uniformly in the unit disk by choosing r uniformly in $[0, 1]$ and θ uniformly in $[0, 2\pi)$; is this the same as choosing x uniformly in $[-1, 1]$ and then y uniformly in $[-\sqrt{1 - x^2}, \sqrt{1 - x^2}]$? If not, what is the joint density of the pairs (x, y) arising from how we chose our (r, θ)?*

CHAPTER 10 _____

Tools: Convolutions and Changing Variables

Time is a storm in which we are all lost. Only inside the convolutions of the storm itself shall we find our directions.
— WILLIAM CARLOS WILLIAMS, *Selected Essays of William Carlos Williams* (1954)

We've had great success in studying random variables in earlier chapters. The most natural question to ask about a random variable is what is its density. Unfortunately, it's not easy to pass from the densities of X and Y to the density of their sum $X + Y$, and things only get worse as we add more and more random variables. As we haven't talked too much about *why* we want to add many random variables, let's say a few words about that now, which will highlight the need for a way to find these new probability density functions. Discovering how to do this is one of the two major themes of this chapter.

As we can easily fill several chapters with applications of adding random variables (and we will!), I'll stick to one general example here. One of the great lessons of science is to reduce one complicated problem into many simpler problems. In chemistry classes you learn to break compounds into the constituent atoms. In number theory you learn to break integers into products of primes. There are many other examples we could give, all illustrating the same principle: break a complicated object into simpler constituents. We can do this in probability. For example, we can understand the result of rolling n fair die by understanding the roll of one die and then combining. Similarly we can understand the behavior of tossing n fair coins by understanding just one toss.

Of course, this principle is valuable for more than just tossing coins and die. Imagine we're trying to understand the behavior of consumers. Perhaps we want to know the demands for movies, or perhaps we're designing schedules for airlines, or sending products to markets. We can try to understand the likelihood of different behavior for individuals, and then aggregate them together.

The point is that there's a real need to understand sums of random variables. We made some progress in §9.5, where Theorem 9.5.1 gave us formulas for expected values of combinations of random variables in terms of the expected values of the random variables. We saw, and will see in greater detail in Chapter 19, that knowing the moments of a distribution provides clues as to its behavior. Of course, it would be better to know the distribution! This is where convolutions enter the picture—they provide a great way to get the probability density function of a sum of random variables, precisely the information we want to know!

We remarked earlier that finding the density of sums of independent random variables is one of the two major themes of the chapter. The other is consequences of the Change of Variables Formula. While these two topics may seem worlds apart, there is a natural connection which argues for their sharing a chapter. The Change of Variables Formula allows us to pass from the density of one random variable to that of another, related random variable. Specifically, if the random variable X has density f_X, and g is a nice function, we can find the density of $Y = g(X)$ in terms of f_X and g. Note how similar this is to our convolutions. In each case, we're trying to find the density of a random variable from the densities of its constituent pieces.

10.1 Convolutions: Definitions and Properties

It's possible to study sums of random variables without knowing convolutions, but it won't be pleasant! Convolutions are designed to facilitate finding densities of sums. We'll spend a good part of Chapter 19 on generating functions, seeing how to use convolutions. In fact, convolutions play a key role in our proof of the Central Limit Theorem, one of the gems of not only probability but also all of mathematics!

As they're so important, it's not surprising that there's a deep theory about convolutions. For now, we'll just look at the definition and some of the more basic, but very useful, properties, and save the remaining material for applications of convolutions to specific distributions and the proof of the Central Limit Theorem.

Here's the basic framework. We have a random variable X with density f_X and another random variable Y with density f_Y, and we want to know what the density of $Z = X + Y$ is. If this is all the information we know, sadly we're out of luck! The problem, as always, arises when X and Y are not independent.

Let's assume

$$f_X(x) = \begin{cases} 1 & \text{if } -1/2 \le x \le 1/2 \\ 0 & \text{otherwise} \end{cases}$$

and

$$f_Y(y) = \begin{cases} 1 & \text{if } -1/2 \le y \le 1/2 \\ 0 & \text{otherwise.} \end{cases}$$

Note that X and Y have the same density function, as do X and $-X$! If we let $Y = -X$, then $Z = X + Y$ is always zero, while if $Y = X$ then $Z = X + Y$ is just $2X$, and

$$f_{2X}(z) = \begin{cases} 1/2 & \text{if } -1 \le z \le 1 \\ 0 & \text{otherwise.} \end{cases}$$

There's a big lesson here: knowing f_X and f_Y isn't enough information to determine f_{X+Y}. Fortunately the situation is completely different if additionally we know X and Y are independent. In this case, the world is nice once again, and we have a very nice, explicit formula for f_Z.

Before stating the main result, we first need some notation and a few preliminary items.

Definition 10.1.1: *The **convolution** of independent continuous random variables X and Y on $\mathbb{R}$ with densities f_X and f_Y is denoted $f_X * f_Y$, and is given by*

$$(f_X * f_Y)(z) = \int_{-\infty}^{\infty} f_X(t) f_Y(z - t) dt.$$

If X and Y are discrete, we have

$$(f_X * f_Y)(z) = \sum_{n} f_X(x_n) f_Y(z - x_n);$$

note of course that $f_Y(z - x_n)$ is zero unless $z - x_n$ is one of the values where Y has positive probability (i.e., one of the special points y_m).

The convolution of two random variables has many wonderful properties, including the following theorem.

Theorem 10.1.2: *Let X and Y be continuous or discrete independent random variables on $\mathbb{R}$ with densities f_X and f_Y. If $Z = X + Y$, then*

$$f_Z(z) = (f_X * f_Y)(z).$$

Further, convolution is commutative: $f_X * f_Y = f_Y * f_X$.

Proof: We'll give the proof when everything is continuous; the discrete case is similar.

We first show the claim about the density of Z. Let f_Z be the probability density function for Z, and let F_Z be its cumulative distribution function. We can always find the density by differentiating the cumulative distribution function. We have

$$F_Z(z) = \text{Prob}(Z \le z).$$

How can we compute this probability? Well, let's say X takes on the value t. Since we want $Z = X + Y$ to be at most z, we have Y is at most $z - t$. From the definition, the probability of this is just $F_Y(z - t) = \text{Prob}(Y \le z - t)$. We let t range over all possible values of X, and find

$$F_Z(z) = \int_{t=-\infty}^{\infty} f_X(t) F_Y(z - t) dt.$$

We now differentiate under the integral sign. In a math class, this must be justified, though frequently instructors either forget to do so, or choose to hide the fact that this must be justified. The additional exercises at the end of this chapter give some cases where interchanges cannot be done. Fortunately in our case it's legal to interchange the integral and the derivative; for a statement of when you can do this, see §B.2.1. We have

$$f_Z(z) = \frac{d}{dz} \int_{t=-\infty}^{\infty} f_X(t) F_Y(z-t) dt$$

$$= \int_{t=-\infty}^{\infty} \frac{d}{dz} \left[f_X(t) F_Y(z-t) \right] dt$$

$$= \int_{t=-\infty}^{\infty} f_X(t) \frac{d}{dz} F_Y(z-t) dt$$

$$= \int_{t=-\infty}^{\infty} f_X(t) f_Y(z-t) dt$$

$$= (f_X * f_Y)(z).$$

For the second claim, the proof is trivial if f_X and f_Y are probability densities. We've just shown that the density of Z is $f_X * f_Y$ when Z is the sum of the independent random variables X and Y. As addition is commutative, $Y + X$ also equals Z, and thus $f_{X+Y} = f_{Y+X}$. Though we don't need it, convolution is commutative for any two functions, not just non-negative functions integrating to 1. □

I *really* like the proof of the second claim above. It's a beautiful example of the power of looking at things the right way. We *could* prove $f_X * f_Y = f_Y * f_X$ by writing out what each equals, and then changing variables to show they are equivalent; however, it's much better to note that addition is commutative. This is typical of a variety of problems—you can solve them with lots of algebra, but often there's a better way.

After stating such a technical result and wading through the proof, let's see how helpful it is in practice. Imagine X and Y are independent random variables with common density

$$f(t) = \begin{cases} 1 & \text{if } -1/2 \le t \le 1/2 \\ 0 & \text{otherwise.} \end{cases}$$

We can find the density of $Z = X + Y$ by evaluating

$$\int_{-\infty}^{\infty} f(t) f(z-t) dt.$$

Unfortunately, we need to be *very* careful. **One of the most common mistakes made by students is to replace $f(t)$ with 1 and $f(z-t)$ with 1 and have t range from $-1/2$ to $1/2$; this is not valid!** What's wrong with this substitution? The problem is that the function $f(u)$ equals 1 only when $-1/2 \le u \le 1/2$; if $f(t) = 1$ this means $-1/2 \le t \le 1/2$, while if $f(z-t) = 1$ this means $-1/2 \le z-t \le 1/2$ or equivalently $z - 1/2 \le t \le z + 1/2$. We thus have to break the integral into cases, which we'll do when we study sums of uniform random variables in §13.1.2. Specifically, we'll have the case when $|z| > 1$ and $|z| \le 1$.

Okay, the above is a bit disheartening. It would be nice to have just one single integral to evaluate, and not a piecewise definition. Perhaps the problem is that our original function was defined piecewise. What if we take a function with the same definition for all inputs? While this does help a bit, there's a fundamental difficulty: integration is hard! While we can always differentiate combinations of elementary functions, it's rare for a general integral involving these functions to have a simple, closed form expression. Thus, as much as we would like to have great expressions for convolutions, we're often out of luck.

Taking a step back, we'll see all is not lost. Convolutions *are* still useful, even if we don't always get a nice closed form answer for the density. What else can we do with them? It turns out we can deduce a lot of properties about the sums of the random variables from the convolution, so it's still a good tool. We start these studies in Chapter 19, leading up to a proof of the Central Limit Theorem. In fact, in Chapter 21 we'll see that there is a beautiful operation called the Fourier transform, and the Fourier transform of a convolution is the product of the Fourier transforms. We will exploit the consequences of this to convert difficult convolution integrations into simple multiplications (though at the end of the day we must invert back!), and thus some of the disadvantages of convolutions disappear.

We'll look at convolutions in much greater detail when we study special random variables in the next few chapters, but at this point they look like a mixed blessing. Yes, they give us an explicit formula for the density, but the resulting integral isn't so easy to evaluate. What's the point of having an expression for the answer if you can't evaluate it!

One point is that necessity is the mother of invention; it's a terrible taunt to have an answer staring us in the face that is unusable. Such situations are what have inspired many researchers to develop techniques for evaluating such integrals. The other answer, as we've mentioned above, is that for many problems in mathematics it's enough to just know something exists or that there is a formula. Amazingly, in Chapter 20 we'll see that the existence of this expansion is basically all we need to prove the Central Limit Theorem!

10.2 Convolutions: Die Example

10.2.1 Theoretical Calculation

Let's do a few examples of convolutions. For our first, consider rolling two fair die. We assume the outcomes of the two rolls are independent of each other. We take X to be the number rolled on the first die, and Y the number on the second. We thus have

$$f_X(k) = f_Y(k) = \begin{cases} 1/6 & \text{if } k \in \{1, 2, 3, 4, 5, 6\} \\ 0 & \text{otherwise.} \end{cases}$$

We found the density for $X + Y$ in §7.2 by enumerating the 36 pairs and seeing what $X + Y$ was for each. While we can do this when we roll two die, this becomes impractical as the number of die increase. Fortunately, the convolutions can help us navigate the algebra.

In[11]:= `Convolve[PDF[DiscreteUniformDistribution[{1, 6}], x],`
`PDF[DiscreteUniformDistribution[{1, 6}], x], x, y]`

Out[11]= $\begin{cases} \frac{1}{3} - \frac{y}{36} & 7 < y < 12 \\ \frac{1}{36}(-2+y) & 2 < y \le 7 \\ 0 & \text{True} \end{cases}$

Figure 10.1. Code for convolving two discrete uniform random variables, on $\{1, \ldots, 6\}$.

From Theorem 10.1.2 and the definition of convolution, if $Z = X + Y$ then

$$f_Z(z) = (f_X * f_Y)(z) = \sum_{k=-\infty}^{\infty} f_X(k) f_Y(z - k).$$

As f_X and f_Y are zero unless their argument is in $\{1, \ldots, 6\}$, we must simultaneously have

$$k \in \{1, \ldots, 6\} \quad \text{and} \quad z - k \in \{1, \ldots, 6\}.$$

Alright, what does this imply? First, $1 \le k \le 6$ and k is an integer, so we can restrict the sum. Second, z has to be an integer. Third, no matter what k we choose we can always find some valid z; of course, this isn't how we should view it. It's better to look at z as given, and then figure out what possible k we may take. The two conditions imply

$$\{z - 6, z - 5, z - 4, z - 3, z - 2, z - 1\} \cap \{1, 2, 3, 4, 5, 6\}.$$

For example, if $z = 2$ then only $k = 1$ works, while if $z = 8$ then $k = 2, 3, 4, 5,$ and 6 work.

We're thus left with eleven sums to evaluate, one for each integer z from 2 to 12. For example,

$$f_Z(8) = \sum_{k=2}^{6} f_X(k) f_Y(8 - k) = \sum_{k=2}^{6} \frac{1}{6} \cdot \frac{1}{6} = \frac{5}{36}.$$

Continuing in this way, we see

$$f_Z(k) = \begin{cases} \sum_{k=1}^{z-1} \frac{1}{36} = \frac{z-1}{36} & \text{if } z \in \{2, \ldots, 7\} \\ \sum_{k=z-6}^{6} \frac{1}{36} = \frac{13-z}{36} & \text{if } z \in \{7, \ldots, 12\} \\ 0 & \text{otherwise.} \end{cases}$$

10.2.2 Convolution Code

It's been awhile since we've included code to investigate. Mathematica has pre-defined functions to do convolutions (though you might prefer to write your own function so you can have more control over the output). Here it is, with the resulting output.

```
In[12]:= Convolve[ Convolve[PDF[DiscreteUniformDistribution[{1, 6}], x],
         PDF[DiscreteUniformDistribution[{1, 6}], x], x, y],
      Convolve[PDF[DiscreteUniformDistribution[{1, 6}], t],
         PDF[DiscreteUniformDistribution[{1, 6}],t], t, y], y, z]
```

$$
\text{Out[12]=}\begin{cases}
-\dfrac{(-24+z)\,(-9+z)^2}{3888} & z == 14 \\[2mm]
\dfrac{2852-924\,z+96\,z^2-3\,z^3}{7776} & 9 < z < 14 \\[2mm]
\dfrac{13\,824-1728\,z+72\,z^2-z^3}{7776} & 19 < z < 24 \\[2mm]
\dfrac{-10\,724+1578\,z-72\,z^2+z^3}{7776} & z == 19 \\[2mm]
\dfrac{-64+48z-12\,z^2+z^3}{7776} & 4 < z \le 9 \\[2mm]
\dfrac{-13\,612+2604\,z-156\,z^2+3\,z^3}{7776} & 14 < z \le 19 \\[2mm]
0 & \text{True}
\end{cases}
$$

Figure 10.2. Code for convolving four discrete uniform random variables on $\{1, \dots, 6\}$.

To do the sum of four die is possible; we still get a closed form solution, but it takes longer to run and, not surprisingly, the output is significantly more involved!

The complexity of the above strongly suggests we need a better approach. Unfortunately, there are many times in life when the answer is a mess of algebra, and this is one of those times! I leave it to you to try and find what the sum of eight die would be (I need to save pages for other material and cannot afford the space to print it out). The complexity we've just seen, however, suggests that perhaps we are looking at the wrong problem. What we are doing is finding *exactly* the probability of sums of rolls. In many applications it suffices to have a good approximation; we don't need the actual value, just a close approximation. This leads to the Central Limit Theorem, one of the highlights of any course. We'll see that for nice random variables, if we sum more and more independent, identically distributed random variables, then the answer converges to being normally distributed. You can see the normal shape emerge by plotting the above (or read ahead to Figure 10.3); we'll spend a lot of time on the Central Limit Theorem later in the book.

10.3 Convolutions of Several Variables

Building on our success with the sum of two *independent* random variables, let's explore what happens when we sum three or more *independent* random variables. Let X_i be a random variable with density f_{X_i}. We know from Theorem 10.1.2 that if U and V are independent random variables, then $f_{U+V}(z) = (f_U * f_V)(z)$. Can we use this to figure out $f_{X_1+X_2+X_3}$, or more generally, $f_{X_1+\dots+X_n}$? Yes!

> **Theorem 10.3.1 (Sums of independent random variables)**: *Let $X_1, X_2, \dots, X_n$ be independent random variables with densities $f_{X_1}, f_{X_2}, \dots, f_{X_n}$. Then*
>
> $$f_{X_1+\dots+X_n}(z) = (f_{X_1} * f_{X_2} * \dots * f_{X_n})(z),$$
>
> *where*
>
> $$(f_1 * f_2 * \dots * f_n)(z) = (f_1 * (f_2 * \dots * (f_{n-2} * (f_{n-1} * f_n)) \dots))(z).$$

We've already shown that convolution is commutative, which means $f * g = g * f$. It's also associative: $(f * g) * h = f * (g * h)$. Remember that convolution takes two functions as inputs and returns one function as the output. We can't immediately take the convolution of three functions. So, if we write $f * g * h$, we need to carefully explain what this means. Since convolutions always take two inputs, there are two ways to interpret this: it's $(f * g) * h$, or it's $f * (g * h)$. Fortunately, since convolutions are associative these two expressions are equal, and it doesn't matter what we write. While it's possible to prove associativity directly, for us there's no need. The reason is that we only need theorems about convolutions of probability density functions, and there's an elegant trick that gives us associativity almost for free.

Proof of Theorem 10.3.1: We'll do the case when $n = 3$; the general case follows similarly.

So, let's consider $Z = X_1 + X_2 + X_3$. We write this as $Z = (X_1 + X_2) + X_3$. The advantage of doing this is that we know the density of the sum of two independent random variables is the convolution of the densities (note that since X_3 is independent of X_1 and X_2, it's also independent of their sum). We thus get

$$f_Z(z) = (f_{X_1+X_2} * f_{X_3})(z).$$

Now we use Theorem 10.1.2, which tells us that $f_{X_1+X_2} = f_{X_1} * f_{X_2}$. Substituting this in, we get

$$f_Z(z) = \left(\left(f_{X_1} * f_{X_2} \right) * f_{X_3} \right)(z).$$

Of course, instead of writing $Z = (X_1 + X_2) + X_3$ we could have written $Z = X_1 + (X_2 + X_3)$. If we do this, we first get

$$f_Z(z) = (f_{X_1} * f_{X_2+X_3})(z).$$

Now we use $f_{X_2+X_3} = f_{X_2} * f_{X_3}$ to find

$$f_Z(z) = \left(f_{X_1} * \left(f_{X_2} * f_{X_3} \right) \right)(z).$$

If we continue to argue along these lines, we see it doesn't matter how we group the convolutions. We could even mix up the order of the functions, as $X_1 + X_2 + X_3 = X_2 + X_3 + X_1$ (and so on).

For four independent random variables, we group as $X_1 + (X_2 + (X_3 + X_4))$, which leads to a density of $f_{X_1} * (f_{X_2} * (f_{X_3} * f_{X_4}))$. □

The argument above is one of my favorites. Notice the complete lack of integrals or sums. We're getting the answer by grouping terms, and exploring the consequence of what we're doing. I call this method **proof by grouping**; see Appendix A.3 for additional examples.

 Let's return to the dice from §10.2. The next question illustrates a truly wonderful perspective. We can figure out the pdf of the number rolled from three independent die. We know the density for one roll of a die, and for two, so all we need to do is convolve. Unfortunately, this is easier said then done, as the algebra gets a bit messy.

We'll leave the headache of three die to you, and continue to four. What's nice about four is there's *another* way to group, which we haven't mentioned yet. Instead of grouping four as $X_1 + (X_2 + (X_3 + X_4))$, a *much* better way to do the problem is to group it as $(X_1 + X_2) + (X_3 + X_4)$. Why is this so much better? We still have three additions, right? Yes, *and no!* Look at the second approach again: $(X_1 + X_2) + (X_3 + X_4)$. Notice that the first addition and the last addition are the same, and the densities $f_{X_1+X_2} = f_{X_1} * f_{X_2}$ equal $f_{X_3+X_4} = f_{X_3} * f_{X_4}$. So, even though it's technically three additions, two of them are the same. This is far better than $X_1 + (X_2 + (X_3 + X_4))$, where all the additions are distinct.

If we move up to the sum of eight independent random variables with the same distribution, it's even more pronounced. Writing the sum as

$$((X_1 + X_2) + (X_3 + X_4)) + ((X_5 + X_6) + (X_7 + X_8)),$$

the savings is even larger. We have the sum of two copies four times, then the sum of these sums twice, and then the sum of those once. In other words, we have three different kinds of convolutions to find, which is much better than the 7 distinct convolutions we get if we use the naive grouping

$$X_1 + (X_2 + (X_3 + (X_4 + (X_5 + (X_6 + (X_7 + X_8)))))).$$

While in some sense the observations above don't matter, that's the wrong lesson to learn. Yes, we could compute the pdfs by chugging away and not using intelligent grouping, but we're only hurting ourselves. One of my goals in life is to minimize the amount of tedious algebra I need to do; grouping is a great way to keep the algebra to a manageable lesson.

To see this, let's now look at the sum of the numbers rolled on four independent die. Let X_1, X_2, X_3, and X_4 denote the values rolled on the four independent die. We know their distributions, and we know the densities of $X_1 + X_2$ and $X_3 + X_4$:

$$f_{X_1+X_2}(u) = f_{X_3+X_4}(u) = \begin{cases} \dfrac{u-1}{36} & \text{if } u \in \{2, \dots, 7\} \\ \dfrac{13-u}{36} & \text{if } u \in \{7, \dots, 12\} \\ 0 & \text{otherwise.} \end{cases}$$

We can now find the probabilities of rolling a 4, 5, 6, ..., 24.

Let $Z = X_1 + X_2 + X_3 + X_4$. We have $f_Z(z) = \left(f_{X_1+X_2} * f_{X_3+X_4}\right)(z)$, with the two densities on the right non-zero only for inputs from 2 to 12. For example,

$$f_Z(6) = \sum_{k=2}^{12} f_{X_1+X_2}(k) f_{X_3+X_4}(6-k)$$

$$= \sum_{k=2}^{4} f_{X_1+X_2}(k) f_{X_3+X_4}(6-k)$$

$$= \frac{2-1}{36}\frac{4-1}{36} + \frac{3-1}{36}\frac{3-1}{36} + \frac{4-1}{36}\frac{2-1}{36} = \frac{10}{1296} = \frac{5}{648}.$$

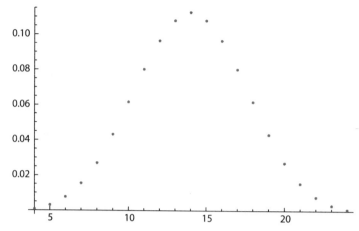

Figure 10.3. The probability density function for the sum of four rolls of independent, fair die.

We can similarly find the other 20 values; this is a lot better than writing down all $6^4 = 1296$ possible outcomes from rolling four fair die. We display the results in Figure 10.3.

If you've heard of the Central Limit Theorem, look at Figure 10.3 through that prism—you should begin to see the bell curve or the Gaussian trying to come out.

10.4 Change of Variable Formula: Statement

We now shift gears and turn to our second theme of the chapter. In the first part, we found the density for the sum of independent random variables in terms of the densities of the summands. Now, we look at the densities of random variables dependent on each other in a very nice way. Our major tool for this study is the Change of Variable Formula. This is one of the most important theorems in calculus. In one variable it's essentially just the chain rule; it becomes more complicated when we have several variables. For now we're looking just at densities that are functions of one variable, so we'll hold off on the more general case. If you're interested, see http://web.williams.edu/Mathematics/sjmiller/public_html/probabilitylifesaver/ for a proof and examples.

Here's a situation that arises all the time. Let's say we have a continuous random variable X with probability density function f_X. If g is a "nice" function, then surely we should be able to figure out the probability density function of $Y = g(X)$, and the answer should involve some combination of f_X and g. The Change of Variables Formula tells us what the relation is, and specifies which g are nice.

> **Theorem 10.4.1 (Change of Variables Formula):** *Let X be a continuous random variable with density f_X, and assume that there is an interval $I \subset \mathbb{R}$ such that $f_X(x) = 0$ whenever $x \notin I$ (in other words, the only non-zero values of X occur in I, which might be the entire real line). Let $g : I \to \mathbb{R}$ be a differentiable function with inverse h, and assume the derivative of g is either always positive or always negative in I, except at finitely many points where it may be zero. If we let $Y = g(X)$, then*
>
> $$f_Y(y) = f_X(h(y)) \cdot |h'(y)|.$$

Before proving the theorem, we'll discuss what it says and do an example, deferring to the end of the section to provide a proof. Some of the conditions (especially the differentiability ones) will become clearer when we see the proof.

There's a lot going on in the statement above, so let's parse it slowly.

- The first is that we have some nice interval I where X is defined. While I might be all of $\mathbb{R}$, frequently it's a subset. The reason it's worth introducing this smaller interval is that we're going to need our function g to have several nice properties if we want to compute f_Y; by restricting our study to I we often can get away with just requiring g to be nice on I and *not* on all of $\mathbb{R}$.

- The next issue is that we want g to be differentiable. That isn't too horrible of a restriction; most of the common transformations are differentiable, but *not* all. For example, if we let X be a random variable with $I = [-1, 1]$ then $g(X) = |X|$ would not be an acceptable choice, as the absolute value function isn't differentiable at the origin. If, however, $I = [2, 3]$ then we could take $g(X) = |X|$.

- The last condition is that we want the derivative of g to be either always positive or always negative. This implies that g is either strictly increasing (if the derivative is positive) or strictly decreasing (if the derivative is negative). As a consequence, we find that each value of X is associated with a unique value of Y, and vice versa. What if the derivative vanishes at a few points? It doesn't really matter; so long as the derivative is of the same sign immediately before and after, the function will still be either strictly increasing or decreasing. For example, consider $g(x) = x^3$ on $[-1, 1]$. Note $g'(x) = 3x^2$, which is positive everywhere except at $x = 0$, where it vanishes. The function g is still strictly increasing on $[-1, 1]$.

- Remember that if h is the inverse function to g, then $h(g(x)) = x$ and $g(h(y)) = y$. Using the chain rule, we find a nice relation between the derivatives of g and h. Differentiating

$$g(h(y)) = y$$

with respect to y yields

$$g'(h(y)) \cdot h'(y) = 1,$$

or

$$h'(y) = \frac{1}{g'(h(y))}.$$

Thus, if we know the derivative of g then we know the derivative of h. In practice, it's sometimes easier to explicitly compute h and differentiate that, but if it's a pain to differentiate h then we can determine h' by knowing g'. Note that, no matter what, we do need to find the function h as the formula requires us to evaluate f_X at $h(y)$.

- Finally, let's check and see if the formula is reasonable. The simplest thing to do is see if it's non-negative and integrates to 1. We now see that the absolute value sign is essential; if g' is negative then so too is h', and if we forget the absolute values then the density of Y would be negative at some points! Let's also make sure that it integrates to 1. For convenience, let's just look at the case when h' is positive, so we may drop the absolute values. In this case the interval $I = [a, b]$ is

mapped to $[g(a), g(b)]$, and thus

$$
\int_{g(a)}^{g(b)} f_Y(y)dy = \int_{g(a)}^{g(b)} f_X(h(y))h'(y)dy
$$

$$
= \int_{g(h(a))}^{g(h(b))} f_X(u)du \quad \text{(by } u - \text{substitution)}
$$

$$
= \int_a^b f_X(u)du
$$

$$
= F_X(b) - F_X(a) = 1
$$

(the last equality follows from the fact that F_X is the cdf of X, and since $a \leq X \leq B$ we have $F_X(b) = 1$ and $F_X(a) = 0$). Thus the proposed density does integrate to 1. If we had g' were negative, the calculation would be similar, except now $I = [a, b]$ would be mapped to $[g(b), g(a)]$, since g is decreasing and therefore flips the order.

Let's do an example. Imagine X has the density function

$$
f_X(x) = \begin{cases} 1/2 & \text{if } 0 \leq x \leq 2 \\ 0 & \text{otherwise,} \end{cases}
$$

and

$$
g(X) = X^2.
$$

Let's go through the list.

1. The interval I is just $[0, 2]$.
2. The derivative of $g(x) = x^2$ is just $g'(x) = 2x$, which is positive everywhere except at $x = 0$.
3. The inverse function is $h(y) = \sqrt{y}$, as $h(g(x)) = \sqrt{x^2} = x$ and $g(h(y)) = (\sqrt{y})^2 = y$. Remember that our interval is $[0, 2]$, so we're only taking positive square-roots.
4. As $h(y) = \sqrt{y}$, $h'(y) = \frac{1}{2}y^{-1/2}$. Alternatively, since $h'(y) = 1/g'(h(y))$ and $g'(x) = 2x$, we find $h'(y) = \frac{1}{2}y^{-1/2}$.

We now use the Change of Variables Formula, and find

$$
f_Y(y) = f_X(h(y)) \cdot |h'(y)|.
$$

We have $f_X(u) = 0$ unless $0 \leq u \leq 2$, and $h(y) = \sqrt{y}$. Thus $f_Y(y) = 0$ unless $0 \leq \sqrt{y} \leq 2$, or $0 \leq y \leq 4$. For such y, $f_X(h(y)) = 1/2$, and $h'(y) = 1/2\sqrt{y}$. Combining

everything gives

$$f_Y(y) = \begin{cases} \dfrac{1}{4\sqrt{y}} & \text{if } 0 \le y \le 4 \\ 0 & \text{otherwise.} \end{cases}$$

As a check, let's make sure the proposed density in the problem above is reasonable. It's clearly non-negative; does it integrate to 1? We only need to integrate from 0 to 4, as that's where the proposed density is non-zero. As

$$\int_0^4 f_Y(y)dy = \int_0^4 \frac{dy}{4\sqrt{y}} = \left. \frac{\sqrt{y}}{2} \right|_0^4 = 1,$$

we do observe that it integrates to 1. Of course this isn't a proof that we've done the algebra correctly, but it's reassuring.

Our answer above is quite interesting. Our original density was well-behaved everywhere, and our mapping function $g(x) = x^2$ is very nice; however, our new random variable $Y = g(X)$ has a density that is infinite at $y = 0$! As strange as this may seem, it's not inconsistent with our theory. The important point to note is that the density f_Y is only "weakly" infinite at the origin; namely, even though it blows up, it blows up at a slow enough rate as $y \to 0$ that the resulting integrals are still finite.

10.5 Change of Variables Formula: Proof

We now turn to the proof of the Change of Variables Formula. The main idea is to use the cumulative distribution function. This is such a good way of proving results that we'll give the method a (fairly obvious) name: the **Method of the Cumulative Distribution Function**.

Proof of the Change of Variables Formula: We have X with density f_X defined on an interval I, and $Y = g(X)$ for some nice function g whose derivative is always positive (except for finitely many points) with inverse function h (so $g(h(y)) = y$ and $h(g(x)) = x$). The cumulative distribution function F_Y of Y is simply the probability that Y takes on a value of y or less, and the density of Y is just the derivative of the cumulative distribution function. Thus, if we can find F_Y and if we can differentiate it, then we'll know the density f_Y.

It's very easy to make a mistake in the proof, so we're going to go very slowly and put in every detail. Imagine our interval $I = [a, b]$. Then statements like $X \le x$ translate to $a \le X \le x$. Consider how the interval I is mapped by g. We have $a \mapsto g(a)$ and $b \mapsto g(b)$. You should be tempted to say that the interval I thus maps to $[g(a), g(b)]$, but this may be wrong! If g' is positive this is true, as g is then an increasing function. If, however, g' is negative then g is decreasing, and $g(b) < g(a)$; in this case, I maps to $[g(b), g(a)]$. This is the cause for the absolute value in the statement of the Change of Variables Formula.

Case 1: Let's assume first that g' is positive, so I maps to $[g(a), g(b)]$. Then

$$F_Y(y) = \text{Prob}(Y \leq y)$$

$$= \text{Prob}(g(a) \leq Y \leq y)$$

$$= \text{Prob}(g(a) \leq g(X) \leq y)$$

$$= \text{Prob}(a \leq X \leq g^{-1}(y)),$$

because if $g(a) \leq g(X) \leq y$ then this is the same as $a \leq g^{-1}(g(X)) \leq g^{-1}(y)$. But we're just denoting the function g^{-1} by h, so the condition becomes $a \leq X \leq h(y)$. We couldn't make this argument if g didn't have a nice inverse; we'll comment more on this after the proof. Continuing, we have

$$F_Y(y) = \text{Prob}(a \leq X \leq h(y))$$

$$= F_X(h(y)) \quad \text{(by definition of } F_X).$$

We now differentiate to get f_Y, using the chain rule and recalling that $F_X' = f_X$. We get

$$f_Y(y) = F_X'(h(y)) \cdot h'(y) = f_X(h(y)) \cdot h'(y).$$

As h' and g' are of the same sign, $h'(y)$ is positive and we may write

$$f_Y(y) = f_X(h(y)) \cdot |h'(y)|.$$

Case 2: Assume now that g' is negative, so I maps to $[g(b), g(a)]$. Now $Y \leq y$ translates to $g(b) \leq Y \leq y$, and we find

$$F_Y(y) = \text{Prob}(Y \leq y)$$

$$= \text{Prob}(g(b) \leq Y \leq y)$$

$$= \text{Prob}(g(b) \leq g(X) \leq y)$$

$$= \text{Prob}(g^{-1}(y) \leq X \leq b),$$

because if $g(b) \leq g(X) \leq y$ then this is the same as $g^{-1}(y) \leq g^{-1}(g(X)) \leq b$. Why did we flip the order? The reason is that g is a decreasing function, as is g^{-1}; applying g or g^{-1} flips the relations. As before, we're just denoting the function g^{-1} by h, so the condition becomes $h(y) \leq X \leq b$. Continuing, we have

$$F_Y(y) = \text{Prob}(h(y) \leq X \leq b)$$

$$= \text{Prob}(a \leq X \leq b) - \text{Prob}(a \leq X \leq h(y))$$

$$= 1 - F_X(h(y)) \quad \text{(by definition of } F_X).$$

We now differentiate to get f_Y, using the chain rule and recalling that $F_X' = f_X$. We obtain

$$f_Y(y) = -F_X'(h(y)) \cdot h'(y) = -f_X(h(y)) \cdot h'(y);$$

however, as h' and g' are of the same sign, $h'(y)$ is negative. Thus $-h'(y) = |h'(y)|$, and we obtain

$$f_Y(y) = f_X(h(y)) \cdot |h'(y)|.$$

Note this is exactly the same answer as the previous case. $\qquad\square$

As the proof was long, it's worth it to step back and soak it in. Here are a few points worth highlighting:

- It was annoying having to deal with the case of g' being negative, but the argument was essentially the same. The main idea is to convert a statement about the cumulative distribution function of Y being less than a certain number to a related statement about the cumulative distribution function of X being less than another related number.
- We comment briefly on why it's so important for g to be one-to-one and onto. Imagine we have $g(X) = X^2$ with $I = [-1, 1]$. In this case the endpoints of the interval $[-1, 1]$ are both mapped to the same point, 1! This would result in us integrating over a point, and the integral over a point is zero. Here's another way to think about it. In order to have an inverse function, given any input there should be a unique inverse. In this situation, if we tried to go backwards from $g(X) = 1/4$ we would get either $X = 1/2$ or $X = -1/2$. By only looking at g that are one-to-one and onto we ensure that the inverse exists.
- If you look carefully at the proof, you might notice an interesting absence. The cumulative distribution function of X, F_X, plays a big role in the arguments, but we never have to write it down. Why? Immediately after it appears, we differentiate it and get f_X. It's extremely fortunate that this happens. Remember, integration is *hard*! It's rare to have a function which has a nice, closed form expression for its integral. Thus, we're very lucky that we don't need a formula for F_X.

We end by summarizing the procedure for using the Change of Variable Formula.

Method of the Cumulative Distribution Function: Let X be a random variable with density f_X whose density is non-zero on some interval I, and let $Y = g(X)$ where $g : I \to \mathbb{R}$ is a differentiable function with inverse h. Assume the derivative of g is either always positive or always negative in I, except at finitely many points where it may vanish. To find the density f_Y:

1. Identify the interval I where the random variable X is defined.
2. Prove the function g has a derivative that is always positive or always negative (except, of course, at potentially finitely many points).
3. Determine the inverse function $h(y)$, where $g(h(y)) = y$ and $h(g(x)) = x$.
4. Determine $h'(y)$, either by directly differentiating h or using the relation $h'(y) = 1/g'(h(y))$.
5. The density of Y is $f_Y(y) = f_X(h(y))|h'(y)|$.

For applications, sometimes it's better *not* to memorize a formula (such as the Change of Variables Formula), but to remember the idea and essentially re-derive it on the spot.

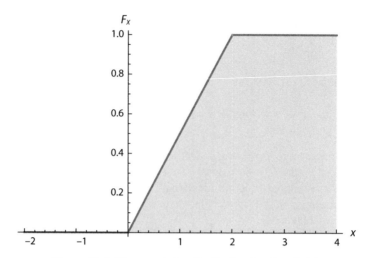

Figure 10.4. The cumulative distribution function $F_X(x)$.

For example, let's consider our old friend

$$f_X(x) = \begin{cases} 1/2 & \text{if } 0 \le x \le 2 \\ 0 & \text{otherwise.} \end{cases}$$

Note the cumulative distribution function is

$$F_X(x) = \begin{cases} 0 & \text{if } x \le 0 \\ x/2 & \text{if } 0 \le x \le 2 \\ 1 & \text{if } x \ge 2, \end{cases}$$

which is pictured in Figure 10.4; we get this by integrating f_X. We again want to find $Y = g(X)$ with $g(X) = X^2$. Here's another way to write the calculation.

We find the cdf of Y and then differentiate that to get the pdf of Y. The cdf is easily found if $y \le 0$ (it's zero) or $y \ge 4$ (it's 1). Thus, we concentrate on $0 \le y \le 4$, where

$$\begin{aligned} F_Y(y) &= \text{Prob}(Y \le y) \\ &= \text{Prob}(X^2 \le y) \\ &= \text{Prob}(-\sqrt{y} \le X \le \sqrt{y}) \\ &= \text{Prob}(0 \le X \le \sqrt{y}) \quad \text{as } X \text{ is non-negative} \\ &= F_X(\sqrt{y}). \end{aligned}$$

There are two ways to go from here. As we know the cdf of X, we can replace $F_X(\sqrt{y})$ with $\sqrt{y}/2$. We thus get, for $0 \le y \le 4$, that

$$F_Y(y) = \frac{\sqrt{y}}{2} \quad \text{which implies} \quad f_Y(y) = \frac{1}{4\sqrt{y}}.$$

Note this agrees with our earlier result from §10.4.

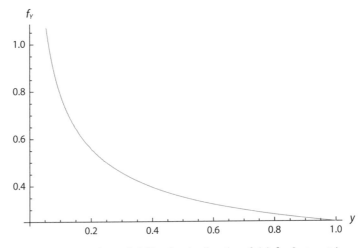

Figure 10.5. The probability density function $f_Y(y)$ for $0 \le y \le 1$.

Alternatively, if we don't have the cdf of X written down, we could differentiate with the chain rule, again finding for $0 \le y \le 1$ that

$$f_Y(y) \;=\; \frac{d}{dy} F_X(\sqrt{y}) \;=\; F'_X(\sqrt{y}) \frac{d}{dy} \sqrt{y} \;=\; f_X(\sqrt{y}) \frac{1}{2} y^{-1/2} \;=\; \frac{1}{2} \cdot \frac{1}{2\sqrt{y}} \;=\; \frac{1}{4\sqrt{y}},$$

shown in Figure 10.5.

We get exactly the same answer as before. If you can remember the formula, you can immediately jump to the answer. That's great and it's a bit faster, but there's always the danger of a loss of understanding. Formulas shouldn't just be memorized; you should remember why they're true. That's why it might be easier to remember that you can get pdfs from cdfs, and essentially re-derive the Change of Variables Formula every time you need it.

Method of the Cumulative Distribution Function (alternative formulation):
Let X be a random variable with density f_X and $Y = g(X)$ for a differentiable function g. For convenience assume $g'(x) \ge 0$. To find f_Y:

1. Express the cdf of Y in terms of X and g: $F_Y(y) = \mathrm{Prob}(Y \le y) = \mathrm{Prob}(g(X) \le y)$.
2. Replace the inequalities involving $g(X)$ with inequalities involving X by inverting. For example, we might have $g(X) \ge 0$, which is implicit in the lines above, and we would get $F_Y(y) = \mathrm{Prob}(g^{-1}(0) \le X \le g^{-1}(y)) = F_X(g^{-1}(y)) - F_X(g^{-1}(0))$.
3. Take the derivative, using the chain rule: $f_Y(y) = f_X\left(g^{-1}(y)\right) \frac{d}{dy} g^{-1}(y)$.

10.6 Appendix: Products and Quotients of Random Variables

So far we have studied the theory of sums and differences of random variables, but of course there are many other ways we can combine quantities, such as multiplication and division. Are there nice formulas for the density of XY and X/Y in terms of the densities of X and Y when the two random variables are independent?

There are, and the reason is our Pavlovian response: whenever we see a product (or a quotient) we should take a logarithm. Thus instead of studying XY we could instead study $U + V$ where $U = \log X$ and $V = \log Y$ (and similarly for the quotient). In other words, in some sense there is no need to develop a theory for products as we can derive everything we need from the theory of sums and change of variables; however, it's often easier to isolate a formula for a special case than to do the conversions back and forth. Thus, in the following subsections we'll analyze the general theory and look at an example or two.

10.6.1 Density of a Product

Let X and Y be independent, non-negative random variables with densities f_X and f_Y, cumulative distribution functions F_X and F_Y, and let $Z = XY$. Then

$$f_Z(z) = \int_{t=0}^{\infty} f_X(t) f_Y(z/t) \frac{dt}{t}.$$

The formula above is very close to the formula for the density of a sum of independent random variables. There we evaluated the two densities at t and $z - t$, as that sums to z, and then integrated with respect to the measure dt. The difference here is that we have a measure dt/t and not dt. There is a lot of deep theory behind this. Briefly, dt is invariant under additive transformations (if $w = t + \alpha$ then $dw = dt$), while dt/t is invariant under multiplicative transformations (if $w = \alpha t$ then $dw/w = dt/t$).

We show why this is true by similar arguments as before. We again start with the definition. Note that we assumed X, Y is non-negative to simplify the bounds of integration (we don't have to worry about the product of two negative numbers changing sign).

$$F_Z(z) := \text{Prob}(Z \le z)$$

$$= \int_{x=0}^{\infty} \int_{y=0}^{z/x} f_X(x) f_Y(y) dy dx$$

$$= \int_{x=0}^{\infty} f_X(x) [F_Y(z/x) - F_Y(0)] \, dx$$

$$f_Z(z) = \frac{d}{dz} \int_{x=0}^{\infty} f_X(x) [F_Y(z/x) - F_Y(0)] \, dx$$

$$= \int_{x=0}^{\infty} f_X(x) \frac{d}{dz} F_Y(z/x) dx$$

$$= \int_{x=0}^{\infty} f_X(x) f_Y(z/x) \frac{d}{dz} \left(\frac{z}{x} \right) dx$$

$$= \int_{x=0}^{\infty} f_X(x) f_Y(z/x) \frac{dx}{x};$$

the claim now follows by replacing x with t as the dummy variable of integration.
Warning: one needs to justify interchanging the integration and the differentiation.

10.6.2 Density of a Quotient

Let X and Y be independent, non-negative random variables with densities f_X and f_Y, cumulative distribution functions F_X and F_Y, and let $Z = X/Y$. Then

$$f_Z(z) = z^{-2} \int_{t=0}^{\infty} f_X(x) f_Y(x/z) x \, dx.$$

The proof follows very similarly as before; we include it as it provides another opportunity to discuss this method. The big difference at first is the bounds of integration.

$$F_Z(z) := \text{Prob}(Z \le z)$$

$$= \int_{x=0}^{\infty} \int_{y=x/z}^{\infty} f_X(x) f_Y(y) dy dx$$

$$= \int_{x=0}^{\infty} f_X(x) [F_Y(\infty) - F_Y(x/z)] \, dx$$

$$f_Z(z) = \frac{d}{dz} \int_{x=0}^{\infty} f_X(x) [1 - F_Y(x/z)] \, dx$$

$$= - \int_{x=0}^{\infty} f_X(x) \frac{d}{dz} F_Y(x/z) dx$$

$$= - \int_{x=0}^{\infty} f_X(x) f_Y(x/z) \frac{d}{dz} \left(\frac{x}{z} \right) dx$$

$$= z^{-2} \int_{x=0}^{\infty} f_X(x) f_Y(x/z) x \, dx;$$

the claim now follows by replacing x with t as the dummy variable of integration. Not surprisingly, the density of a product and a quotient are closely related. . . .

10.6.3 Example: Quotient of Exponentials

The following example leads to a very surprising result. Fix a $\lambda > 0$ and let $X, Y \sim$ Exp(λ) be two independent exponentially distributed random variables (so their densities are $\lambda^{-1} \exp(-t/\lambda)$ for non-negative t and zero otherwise; see Chapter 13 for more on exponential distributions), and set $Z = X/Y$. We use the results from §10.6.2 to find the density of Z:

$$f_Z(z) = z^{-2} \int_{t=0}^{\infty} f_X(t) f_Y(t/z) t \, dt$$

$$= \lambda^{-2} z^{-2} \int_{t=0}^{\infty} \exp(-t/\lambda) \exp(-(t/z)/\lambda) t \, dt$$

$$= \lambda^{-2} z^{-2} \int_{t=0}^{\infty} \exp(-t/(\lambda(1+1/z)^{-1})) t \, dt,$$

where the last simplification of the exponential arguments comes from

$$\frac{t}{\lambda} + \frac{t/z}{\lambda} = \frac{t(1+1/z)}{\lambda} = \frac{t}{\lambda(1+1/z)^{-1}}.$$

The reason we want to write it this way is that we can recognize the integrand as almost an exponential random variable, where now the parameter is

$$\omega = \lambda(1+1/z)^{-1}.$$

Thus, multiplying by 1 in the form of ω/ω (so the integrand is exactly the mean of an exponential random variable with parameter ω), we find

$$f_Z(z) = \lambda^{-2} z^{-2} \omega \int_{t=0}^{\infty} t \frac{1}{\omega} \exp(-t/\omega) dt.$$

The last integral can be done by parts; we'll see later in Chapter 13 that it is the mean of an exponential random variable with parameter ω, which is ω. Substituting this yields

$$f_Z(z) = \lambda^{-2} z^{-2} \omega^2 = z^{-2}(1+1/z)^{-2} = \frac{1}{(1+z)^2}. \tag{10.1}$$

The answer above has a lot of fascinating features. The most important is that it doesn't depend on the exponential parameter λ! This seems shocking—could we have made a mistake? Let's do some quick checks. Our proposed density is clearly nonnegative: does it integrate to 1? Yes, as

$$\int_0^{\infty} \frac{dz}{(1+z)^2} = \int_1^{\infty} \frac{du}{u^2} = \left. \frac{1}{u} \right|_{\infty}^{1} = 1. \tag{10.2}$$

Thus our answer is at least a probability density function.

We can try to do some simulations to see if the answer depends on λ. Here is some simple code to simulate a large number of ratios and calculate the sample mean.

```
ratioexp[lambda_, num_] := Module[{},
  sum = 0;
  For[n  = 1, n  <= num, n++,
    {
      x = Random[ExponentialDistribution[lambda]];
      y = Random[ExponentialDistribution[lambda]];
      sum = sum + (x/y);
    }];
  Print["Average is ", sum/num];
  ];
```

Running this for large numbers (on the order of hundreds of thousands to millions) of ratios for various λ's gave fairly consistent answers most of the time, but not always. Should you be surprised that sometimes some simulations are very different than others?

10.7 Summary

This chapter is concerned with a fundamental problem in probability: how can we understand a complicated random variable in terms of simpler ones? We saw two ways such a problem could appear. The second was through a change of variable, and we were able to dispatch that problem quickly and somewhat easily. In fact, in finding the density of $Y = g(X)$ in terms of f_X and g, we finally saw a real application of the cumulative distribution function. My favorite part of this method is that we don't need to actually *compute* the cdf F_X. All we need is that it exists, and we then differentiate it and return to f_X. This is pure math at its best—we only need to know it exists, not it's exact form!

The other situation was when we had a random variable as a sum of independent random variables (though we also handled products and quotients in the appendix). Many probability classes are designed to reach the Central Limit Theorem, which describes the limiting behavior of more and more sums of independent, identically distributed random variables. It's amazing how frequently such a situation arises. We saw that writing down exact expressions for these sums quickly becomes difficult. Fortunately, there is much in common with the other theme of this chapter. We're again in a situation where it's enough to know something exists and have a general formula, even if it doesn't appear particularly useful. The following quote from William Carlos Williams (*Selected Essays*, 1954) is very appropriate, at least if we replace "time" with "sums of independent random variables": "Time is a storm in which we are all lost. Only inside the convolutions of the storm itself shall we find our directions."

Convolutions provide a wonderful handle on sums of random variables. While a true appreciation needs to wait till our proof of the Central Limit Theorem (Chapter 20), already we can see hope. Convolutions provide a starting point for understanding these sums.

10.8 Exercises

Exercise 10.8.1   *Give an example of random variables X and Y with density functions f and g, respectively, so that the density of $X + Y$ is not $(f * g)(z)$.*

Exercise 10.8.2   *Consider a fair die with n faces. Assuming all rolls are independent, find the density for the sum of two rolls, for the sum of three rolls, and the sum of four rolls.*

Exercise 10.8.3   *A probability distribution is infinitely divisible if it can be written as the sum of arbitrarily many independent, identically distributed random variables. Find an example of an infinitely divisible random variable.*

Exercise 10.8.4   *The Poisson distribution is a common distribution that will be discussed in greater detail in subsequent chapters. It has density $P(X = n) = \lambda^n e^{-\lambda}/n!$ for n in the non-negative integers and a given $\lambda > 0$. If X is a Poisson variable with parameter λ_X and Y is a Poisson variable with parameter Y show that $X + Y$ is a Poisson variable with parameter $\lambda_X + \lambda_Y$.*

Exercise 10.8.5   *If X and Y are independent, identical exponential variables, that is, they have density*

$$f_X(x) = \begin{cases} \dfrac{1}{\lambda} e^{-x/\lambda} & \text{if } x \geq 0 \\ \\ 0 & \text{otherwise} \end{cases}$$

with parameter λ, find the density function for $\log(XY)$.

Exercise 10.8.6   *The product of two continuous independent random variables, X and Y with probability densities f_X and f_Y, is given by $\int_{-\infty}^{\infty} f_x(x) f_y(z/x) \frac{1}{|x|} dx$. Find the density for the product of two independent uniform random variables with positive probability on the interval $(0, 1)$.*

Exercise 10.8.7   *The normal distribution with parameters μ and σ has density $f(x) = \frac{1}{\sqrt{2\pi\sigma^2}} e^{-(x-\mu)^2/2\sigma^2}$. Show that the sum of two normal variables is also a normal variable.*

Exercise 10.8.8   *Another well-known distribution is the log normal distribution. As the name suggests, if X is a normal random variable, $\log X$ follows a log normal distribution. Use the result from the previous exercise to show that the product of two log normal variables is also log normal.*

Exercise 10.8.9   *We have to be very careful about interchanging a limit and an integral. Let*

$$f_n(x) = \begin{cases} n - |x - n| & \text{if } \dfrac{1}{n} \leq x \leq \dfrac{3}{n} \\ \\ 0 & \text{otherwise.} \end{cases}$$

Prove $\lim_{n\to\infty} \int_0^\infty f_n(x)dx \neq \int_0^\infty \lim_{n\to\infty} f_n(x)dx$.

Exercise 10.8.10   *The previous example illustrated the dangers of interchanging a limit and an integral. This one shows that we cannot always interchange orders of integration. For simplicity, we give a sequence a_{mn} such that $\sum_m(\sum_n a_{m,n}) \neq \sum_n(\sum_m a_{m,n})$; as a nice exercise find an analogue with integrals. For $m, n \geq 0$ let*

$$a_{m,n} = \begin{cases} 1 & \text{if } n = m \\ -1 & \text{if } n = m + 1 \\ 0 & \text{otherwise.} \end{cases}$$

Show that the two different orders of summation yield different answers (the reason for this is that the sum of the absolute value of the terms diverges).

Exercise 10.8.11 *We proved that convolutions are associative by grouping. We've seen this technique many times. Show that if we know the derivative of a sum of two functions is the sum of the two derivatives then the result immediately extends to derivatives of sums of three functions. More generally, it holds for any* finite *sum, though the derivative of an infinite sum need not be the sum of the derivatives.*

Exercise 10.8.12 *We saw that grouping sums of random variables together can simplify the algebra. For example, with 8 independent rolls of a fair die we grouped as follows:*

$$((X_1 + X_2) + (X_3 + X_4)) + ((X_5 + X_6) + (X_7 + X_8)).$$

This is very similar to the ***method of repeated squaring****, which plays a key role in making many cryptographic systems (such as RSA) practical (see for example Chapters 7 and 8 of [CM]). Naively we would expect it would require 99 multiplications to compute x^{100}; show by intelligently grouping that it can be done in under ten!*

Exercise 10.8.13 *We used the relation $g(h(y)) = y$ to express the derivative of g in terms of the derivative of h. This is a powerful approach, and is used to pass from knowledge of one function's derivative to another one. For example, assume you know that the derivative of e^x is e^x; use that to prove the derivative of $\ln x$ is $1/x$. For a more exotic example (which is related to the Cauchy distribution), find the derivative of $\arctan(x)$ given that the derivative of $\tan(x)$ is $1/\cos^2(x)$.*

Exercise 10.8.14 *We discussed at great length that we need our maps g in the Change of Variables Formula to be one-to-one and onto. Prove that such functions have a unique inverse. Specifically, let $g : I \to J$ be a one-to-one and onto map. Prove that there is a unique inverse $h : J \to I$. Show that this is not the case if either g is not one-to-one or g is not onto.*

Exercise 10.8.15 *Let X be the random variables associated to the number of 1's in six rolls of a fair die. Let $Y = \sqrt{x}$. Find the probability density function for Y.*

Exercise 10.8.16 *Let X be a random variable with pdf $f_X = 1/2$ on the interval $[-1, 1]$. Let $Y = X^2$. Find the pdf for Y.*

Exercise 10.8.17 *Imagine an insurance company is trying to figure out how many of the people carrying its plan will make claims. Each person makes a claim with probability p. Use convolutions to find the probability distribution if the insurance company has 100 policy holders.*

Exercise 10.8.18 *Prove that $(f * g)' = f' * g = g' * f$, where the prime denotes differentiation; assume all integrals converge absolutely.*

Exercise 10.8.19 *Prove that the integral of the convolution of two functions is the product of the integrals of both of the individual functions (in the case when both functions are pdf's all the integrals are just 1). (Advanced: Does the claim hold for all pairs of functions?)*

Exercise 10.8.20 *Write code that will take many samples from a random variable with probability density function $f_X(x) = 1$ on $[0, 1]$. Plot a histogram showing the approximate pdf for Y, where $Y = e^X$ and X is a random variable with pdf f_X.*

Exercise 10.8.21   *Find* Y *where* $Y = e^X$ *and* X *is a random variable with pdf* f_X. *Plot the pdf for* Y. *Compare this plot to the histogram produced in the previous exercise.*

Exercise 10.8.22   *In studying the ratio of two independent exponential random variables with parameter* λ, *we found the density was* $f_Z(z) = 1/(1 + z)^2$. *If you were to do lots of simulations for several different choices of* λ, *would you expect the means of the simulations to be roughly the same all the time? Why or why not?*

Exercise 10.8.23   *Let* Z *be a standard normal random variable,* V *be a chi-square random with* v *degrees of freedom, and assume* Z *and* V *are independent. Let* $T = Z/\sqrt{V/v}$. *Calculate the density of* T. *See Chapter 14 for properties of normal random variables, and Chapter 16 for properties of chi-square random variables. We say* T *has the (Student's)* t-**distribution** *with* v *degrees of freedom. This distribution surfaces in many statistics problems, especially in comparing differences in sample means.*

Exercise 10.8.24   *Let* X_i *be a chi-square distribution with* d_i *degrees of freedom (see Chapter 16). Compute the density of* $(X_1/d_1)/(X_2/d_2)$. *This is called the* F-**distribution**, *and plays an important role in many statistics problems such as hypothesis testing and analysis of variance, and leads to the* F-**test**.

Assume we can simulate a uniform random variable U *and independently a random variable* X *with a nice density* f_X. *If there is a positive* $M \geq 1$ *such that* $0 \leq f_Y(x) \leq M f_X(x)$ *for all* x *and* $\int_{-\infty}^{\infty} f_Y(x)dx = 1$, *then we can simulate a random variable* Y *with density* f_Y. *While the inverse cdf method allows us to easily generate from distributions such as the Cauchy or an exponential, it does not allow us to generate normal random variables and thus the importance of this method should be clear. The exercises below sketch the proof.*

Exercise 10.8.25   *Let* $h(x) = f_Y(x)/M f_X(x)$. *Prove that* $0 \leq h(x) \leq 1$ *and*

$$\mathrm{Prob}(U \leq X) = \int_{x=-\infty}^{\infty} \int_{u=0}^{1} h(x)du \cdot f_X(x)dx = \frac{1}{M} \int_{-\infty}^{\infty} f_Y(x)dx.$$

Exercise 10.8.26   *If a draw has* $u \leq x$ *then set* $Y = X$, *otherwise make no assignment to* Y *and continue drawing until this condition is met. Prove that the cdf of* Y *is*

$$\mathrm{Prob}(Y \leq y) = \mathrm{Prob}(X \leq y \text{ and } U \leq h(X))/\mathrm{Prob}(U \leq h(x)).$$

Show this implies the pdf of Y *is* f_Y, *as desired.*

Exercise 10.8.27   *Apply the above method to simulate a standard normal random variable from two uniform random variables and an exponential random variable. We use one uniform to determine the sign, and the other and the exponential to simulate half a normal. Note an exponential is easy to simulate as it has a nice inverse cdf. In particular, find the smallest* M *such that* $2e^{-x^2}/\sqrt{2\pi} \leq e^{-x}$ *(we double the standard normal's density as we are looking at the absolute value of a normal random variable).*

Exercise 10.8.28   *Why do we want* M *to be as small as possible in the above method? Must such an* M *always exist for a pair* (f_X, f_Y)? *If not give a counterexample, if yes give a proof.*

Exercise 10.8.29   *Find the expected value of* M, *and give an interpretation of what it represents.*

CHAPTER 11 _____

Tools: Differentiating Identities

Big fleas have little fleas,
Upon their backs to bite 'em,
And little fleas have lesser fleas,
and so, ad infinitum.
—"THE SIPHONAPTERA", *a nursery rhyme based on Jonathan Swift's*
  "On Poetry: A Rhapsody," (1733)

"Give us the tools, and we will finish the job."
— WINSTON CHURCHILL, *broadcast speech, February* (1941)

We've come to the last of our tool chapters. After this, we'll have everything we need to analyze the standard distributions, which we then do. There is no single "right" way to order the material in a book. As the main goal of this book is to supplement any standard probability book, I've made the decision to collect the various techniques together. All the theory is developed at once, and only after do we apply it (again and again and again) to different special distributions.

The drawback of this method, of course, is the paucity of examples at our disposal when developing the theory. We did a lot with coin flips and rolling die as these can be stated fairly quickly. To provide a balance, we'll take a different approach in this chapter. After we describe the general theory, we'll do several examples from some of the most important standard distributions. If you're using this book as a supplement, there's a very good chance you've already seen these distributions. If you haven't seen them, you can quickly glance ahead to those later chapters and read a little bit about them if you wish. If you prefer not to, don't worry. You don't need to know anything about those probability distributions to read the parts in this chapter. All that matters is that these are not busywork exercises—the last few examples are deliberately chosen from important random variables, but you can forget that and just view those sections as interesting in their own right.

The purpose of this chapter is to explain the Method of Differentiating Identities. Identities are the bread and butter of mathematics. They are the building blocks for developing sweeping theories about all sorts of things. Not surprisingly, it's usually a lot of work to prove an identity. Often we have to do something clever. Students typically can follow the proofs line by line, but learning how to make that creative leap and initiate the line of argument takes a lot longer to learn; one of the goals of this book in general and this chapter in particular is to help you with this leap.

Since identities are important and it's often hard to come up with a new one, surely any method to generate new identities from old identities should be a welcome addition to any course! Let's consider the following sum:

$$\frac{1}{2} + \frac{2}{4} + \frac{3}{8} + \frac{4}{16} + \frac{5}{32} + \frac{6}{64} + \cdots + = \sum_{n=0}^{\infty} \frac{n}{2^n}.$$

Is there a simple expression for this sum? It's almost a geometric series; what prevents it from being a geometric series is that we have n in the numerator of our fraction instead of a 1. If the n weren't present, it would be a geometric series with ratio $1/2$.

In the next section we'll show there's actually a simple way to get the sum of $n/2^n$ from the sum of $1/2^n$. It involves differentiation, and we'll be able to get a formula for our sum from differentiating a certain geometric series. This is just one example of a very general method for tackling a problem, the **Method of Differentiating Identities**.

We'll describe the method in detail in a moment, but a few words about why this is such a good idea are in order. As we've said above, it typically is hard to generate an identity. Thus, any method that creates new identities from current identities is sure to be useful; one that creates *infinitely many* new identities is particularly welcome! Differentiation isn't that hard, and if we can prove an identity with a parameter on both sides, differentiating with respect to that parameter creates a new identity! As with much of mathematics, often the hardest part in the method is the algebra, specifically finding a nice way of simplifying the resulting expressions. We'll talk a lot about some general tips to help clean up formulas.

11.1 Geometric Series Example

Rather than formalize the Method of Differentiating Identities, let's jump right in and do an example. In the course of studying this we'll see what the general method should be. We'll tackle two variants of our example; by doing this we'll see some of the issues that arise when we try to clean up the algebra in our formulas.

Assume, for some reason (perhaps because of the tantalizing simplicity of the expression), that we want to evaluate

$$\frac{1}{1} + \frac{2}{2} + \frac{3}{4} + \frac{4}{8} + \frac{5}{16} + \frac{6}{32} + \frac{7}{64} + \cdots = \sum_{n=0}^{\infty} \frac{n}{2^{n-1}}. \qquad (11.1)$$

The first thing we should do is make sure the series converges. It does by the comparison test from calculus. For n large, compare $\frac{n}{2^n}$ to $\frac{1}{(3/2)^n}$. We plot the first few partial sums in Figure 11.1, which suggests that the limit is 4. Let's try to prove that.

We haven't said much about the Method of Differentiating Identities. All that we've mentioned so far is that we need to have a free parameter on both sides of an identity and we then differentiate. Note that our series seems to be related to a geometric series. The first thing that comes to mind is

$$1 + \frac{1}{2} + \frac{1}{4} + \frac{1}{8} + \frac{1}{16} + \cdots + = \sum_{n=0}^{\infty} \frac{1}{2^n} = \frac{1}{1 - 1/2} = 2.$$

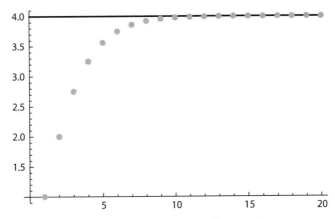

Figure 11.1. The first 20 partial sums of the series $\sum_{n=1}^{\infty} n/2^{n-1}$. Note that the limit appears to be 4.

While this is correct, differentiating can't give us anything useful as there are no variables anywhere to be seen! We just have two expressions that are equal.

Instead, we try **Abstraction**. Rather than consider our problem in isolation, we **generalize** and consider it as one of a multitude of problems, all of which have a similar form. Frequently in math it's easier to tackle the general case than a specific instance. As this probably seems a little strange, let's give a few reasons why this might be the case.

- First, if we're looking at just one specific instance, we might be misled by the actual numbers that appear. In other words, for this problem we might think the number 2 is somehow important in the solution, as 2 appears everywhere.

- The second reason is that we often have access to more powerful techniques when we generalize. We'll elaborate on this in a moment, but the essential idea is the following. Rather than looking at one special sum, we'll study a family of sums. We'll have a parameter at our disposal, and now we can use results from continuity and differentiation (in other words, calculus) to try and extract information.

So, instead of studying the sum in (11.1), let's consider the more general case

$$\sum_{n=0}^{\infty} n \cdot x^{n-1}. \tag{11.2}$$

Again, by the comparison test, one can show this series converges, this time for $|x| < 1$. If we didn't have the n in the front of the product in (11.2), the series would be easily summable: the geometric series formula (see §1.2) gives

$$\sum_{n=0}^{\infty} x^n = \frac{1}{1-x}. \tag{11.3}$$

Now we can begin to see the power of abstraction. Before we had an identity involving sums of $1/2^n$; now we have an identity involving sums of x^n. The point is this identity holds for all x with $|x| < 1$. We have a continuum of choices of x. We can vary x and the identity is still true. This opens up the possibility of differentiating both sides with respect to x. Why would we want to do that? Well, the derivative of x^n is just nx^{n-1}; if we then take $x = 1/2$ we recover our sum from (11.1).

Actually, we need to be able to do more than just differentiate both sides of the above equation; we also need to interchange the order of summation and differentiation. There's no problem taking the derivative of $1/(1-x)$, but we do need to appeal to some results from analysis to support switching the derivative and the summation on the left-hand side. The reason is that when we prove the derivative of a sum is the sum of the derivatives in a calculus class, it's only for a *finite* sum; we need to be careful with infinite sums. This point is often glossed over or worse, completely ignored, in a first course. It's fine not to do the technical details, but you should be aware that some rigor is needed. We provide some of this in §B.2.2.

So, accepting that we can interchange the derivative and the sum, we find

$$\frac{d}{dx} \sum_{n=0}^{\infty} x^n = \frac{d}{dx} \frac{1}{1-x}$$

$$\sum_{n=0}^{\infty} \frac{d}{dx} x^n = \frac{1}{(1-x)^2}$$

$$\sum_{n=0}^{\infty} n x^{n-1} = \frac{1}{(1-x)^2}.$$

Now all we have to do is take $x = 1/2$ above to solve the original problem. For this problem, as long as $|x| < 1$ we can justify interchanging the order of summation and differentiation. Taking $x = 1/2$, we see our sum equals 4 as claimed.

Let's look at a slight variant of this problem. Suppose now we want to find

$$\frac{1}{2} + \frac{2}{4} + \frac{3}{8} + \frac{4}{16} + \frac{5}{32} + \frac{6}{64} + \cdots + = \sum_{n=0}^{\infty} \frac{n}{2^n}.$$

There are two ways to attack this. The first is to note that it's just 1/2 of the sum in the previous problem, and so all we need to do is multiply our answer by 1/2 (which tells us the answer is just 2). Another way is to apply the operator $x \frac{d}{dx}$ to both sides of the geometric series formula instead of $\frac{d}{dx}$. The advantage of this is that it keeps the power of x unchanged when we differentiate. When we're dealing with just one derivative it doesn't really matter, but in many applications in probability we need to differentiate multiple times, and this will make a difference.

For our problem, applying $x \frac{d}{dx}$ to (11.3) gives

$$\sum_{n=0}^{\infty} n x^n = \frac{x}{(1-x)^2};$$

if we take $x = 1/2$ we find the sum is 2.

Let's end with a more interesting variant. Consider

$$\sum_{n=0}^{\infty} \frac{n^2}{2^n}.$$

Is there a nice formula for this sum? We start with the geometric series formula:

$$\sum_{n=0}^{\infty} x^n = \frac{1}{1-x}.$$

We apply $x\frac{d}{dx}$ to both sides; the reason we have a factor of x is to prevent the power of x from being lowered. We have

$$\sum_{n=0}^{\infty} nx^n = \frac{x}{(1-x)^2}.$$

We now apply $x\frac{d}{dx}$ again, obtaining (after applying the quotient rule and doing some algebra)

$$\sum_{n=0}^{\infty} n^2 x^n = \frac{x(1+x)}{(1-x)^3}.$$

If we take $x = 1/2$ we regain our desired sum, and see that it equals 6.

This example is a wonderful advocate for the method. With just a little more work, we were able to replace the n with the n^2, deriving another new identity. The motivation for this chapter's quote is now hopefully clear.

11.2 Method of Differentiating Identities

Inspired by the geometric series example, we state a useful version of the Method of Differentiating Identities. It's not the most general version, but it's sufficiently general to apply to many cases.

Method of Differentiating Identities: Let $\alpha, \beta, \gamma, \ldots, \omega$ be some parameters. Assume

$$\sum_{n=n_{\min}}^{n_{\max}} f(n; \alpha, \beta, \ldots, \omega) = g(\alpha, \beta, \ldots, \omega),$$

where f and g are differentiable functions with respect to α. Then

$$\sum_{n=n_{\min}}^{n_{\max}} \frac{\partial f(n; \alpha, \beta, \ldots, \omega)}{\partial \alpha} = \frac{\partial g(\alpha, \beta, \ldots, \omega)}{\partial \alpha},$$

provided that f has sufficient decay to justify the interchange of summation and differentiation.

The easiest case is when $n_{\min}$ and $n_{\max}$ above are finite, as then it's trivial to justify switching orders. Note that there are two powerful results from analysis that show up in this method and its application. The first, as we've just mentioned, is interchanging a sum (possibly infinite) and a derivative. Interchanging sums or a sum and a derivative is a common technique in math and physics. The second item to note, as discussed earlier, is that we're not considering just one identity, but rather a family of identities for the parameter of interest. It's essential that we have a continuum of values, as otherwise we can't use calculus.

In the next few sections we'll apply this method to calculate moments for some of the standard random variables. If you haven't seen these random variables before, you can go to the corresponding chapter and skim to learn more, but that isn't necessary to appreciate what happens. Our goal below is to show how differentiating identities gives us a better path through the algebra than brute force attacks to find the moments.

11.3 Applications to Binomial Random Variables

Consider a binomial distribution with n trials, where each trial has probability p of being a success (coded as 1) and probability $1 - p$ of being a failure (coded as 0). This is the same as n tosses of a coin with probability p of heads and $1 - p$ of tails. If X is the number of heads, then

$$\text{Prob}(X = k) = \begin{cases} \binom{n}{k} p^k (1 - p)^{n-k} & \text{if } k \in \{0, 1, \ldots, n\} \\ \\ 0 & \text{otherwise.} \end{cases}$$

This is a Binomial Random Variable; we discuss these in §12.2. To prove this is a random variable, we must show the probabilities are non-negative and sum to 1. The non-negativity is easy, while the sum follows from the Binomial Theorem (see §A.2.3), which says

$$(x + y)^n = \sum_{k=1}^{n} \binom{n}{k} x^k y^{n-k}.$$

Taking $x = p$ and $y = 1 - p$ proves the probabilities sum to 1.

Now that we've shown that this is a probability distribution, the first question to ask is what's its mean (the next question is what's the variance). We've seen ways of answering these questions by straightforward calculation; we now show how to get the same answer using differentiating identities.

Similar to the geometric series formulas above, we need a free parameter to differentiate with respect to. In this problem, that parameter will be p, the probability of getting a head on each flip. Even if a problem gives a particular value for p, it's easier to derive formulas for arbitrary p and then set p equal to the given value at the end. This allows us to use the tools of calculus.

Thus to study binomial random variables we should consider

$$(p + q)^n = \sum_{k=0}^{n} \binom{n}{k} p^k q^{n-k}. \tag{11.4}$$

If we take $p \in [0, 1]$ and $q = 1 - p$, then we have a binomial distribution and $(p + q)^n = 1$. We now differentiate the above with respect to p. While we eventually set $q = 1 - p$, for now we consider p and q as independent variables. The reason is that if we took $q = 1 - p$, the derivative with respect to p would no longer vanish when applied to q. In fact, if we took $q = 1 - p$ then our sum is just 1, and hence its derivative would be 0. We thus see the importance of having p and q initially unrelated.

Calculating the mean: We now calculate the mean, which is

$$\mathbb{E}[X] = \sum_{k=0}^{n} k \cdot \binom{n}{k} p^k (1-p)^{n-k}.$$

The Method of Differentiating Identities will not only give us this expression from the binomial formula, but will also give us what it equals!

It turns out that, while we could determine the mean by differentiating with respect to p (in other words, acting on both sides of (11.4) with $\partial/\partial p$), the resulting algebra is a little easier if we instead apply the operator $p\frac{\partial}{\partial p}$. The advantage of this is that we don't change the powers of p and q in our expressions. We find

$$p\frac{\partial}{\partial p} \left(\sum_{k=0}^{n} \binom{n}{k} p^k q^{n-k} \right) = p\frac{\partial}{\partial p} (p+q)^n$$

$$p \sum_{k=0}^{n} \binom{n}{k} k p^{k-1} q^{n-k} = p \cdot n(p+q)^{n-1}$$

$$\sum_{k=0}^{n} k \binom{n}{k} p^k q^{n-k} = np(p+q)^{n-1};$$

interchanging the differentiation and summation is trivial to justify because we have a finite sum. The expected number of successes (when each trial has probability p of success) is obtained by now setting $q = 1 - p$, which yields

$$\sum_{k=0}^{n} k \binom{n}{k} p^k (1-p)^n = np. \tag{11.5}$$

Note that we waited till the very end to set $q = 1 - p$; if we had made this substitution earlier we would end up with $0 = 0$, which is not that useful! As the left-hand side of the above equation is the definition of the mean, we see the mean is just np. It's always nice when our answers match our intuition: if we have n independent flips of a coin and on each flip the probability of getting a head is p, then we expect np heads.

Calculating the variance: To determine the variance, we differentiate again. There are several choices we could make. We could apply the operator $p\frac{\partial}{\partial p}$ again, or we could apply the operator $p^2 \frac{\partial^2}{\partial p^2}$ to the Binomial Theorem, (11.4). While both lead to the right answer for the variance, the algebra is a little easier if we apply $p\frac{\partial}{\partial p}$ twice.

Remember $X \sim \text{Bin}(n, p)$ (this notation means that X is a binomial random variable with parameters n and p). There are two formulas for its variance:

$$\text{Var}(X) = \mathbb{E}[(X - \mu_X)^2] = \sum_{k=0}^{n} (k - np)^2 \cdot \binom{n}{k} p^k (1-p)^{n-k}$$

$$\text{Var}(X) = \mathbb{E}[X^2] - \mathbb{E}[X]^2 = \sum_{k=0}^{n} k^2 \cdot \binom{n}{k} p^k (1-p)^{n-k} - (np)^2.$$

We'll see in a moment it's easier to use the second formula than the first.
We start with

$$\sum_{k=0}^{n} \binom{n}{k} p^k q^{n-k} = (p+q)^n.$$

From our work on the mean, we know that applying $p \frac{\partial}{\partial p}$ once yields

$$\sum_{k=0}^{n} k \binom{n}{k} p^k q^{n-k} = p \cdot n(p+q)^{n-1}.$$

Applying $p \frac{\partial}{\partial p}$ again gives

$$\sum_{k=0}^{n} k^2 \binom{n}{k} p^k q^{n-k} = p \left[1 \cdot n(p+q)^{n-1} + p \cdot n(n-1)(p+q)^{n-2} \right].$$

If we take $q = 1 - p$ the above becomes

$$\sum_{k=0}^{n} k^2 \binom{n}{k} p^k (1-p)^{n-k} = np + n(n-1)p^2;$$

further, note the left-hand side is just $\mathbb{E}[X^2]$. Thus

$$\text{Var}(X) = \mathbb{E}[X^2] - \mathbb{E}[X]^2$$

$$= \sum_{k=0}^{n} k^2 \binom{n}{k} p^k q^{n-k} - (np)^2$$

$$= np + n^2 p^2 - np^2 - n^2 p^2$$

$$= np - np^2 = np(1-p).$$

As our purpose is to learn how to use the Method of Differentiating Identities well, we cannot understate how useful it is to find ways to make the algebra nice. Instead of applying $p \frac{\partial}{\partial p}$ twice let's see what happens if we apply the operator $p^2 \frac{\partial^2}{\partial p^2}$ to (11.4). This operator is nicer for differentiation, but as we'll see leads to messier algebra. We get

$$p^2 \frac{\partial^2}{\partial p^2} \left(\sum_{k=0}^{n} \binom{n}{k} p^k q^{n-k} \right) = p^2 \frac{\partial^2}{\partial p^2} (p+q)^n.$$

After some simple algebra we find

$$\sum_{k=0}^{n} k(k-1) \binom{n}{k} p^k q^{n-k} = p^2 \cdot n(n-1)(p+q)^{n-2}. \tag{11.6}$$

Unfortunately, to find the variance we need to study

$$\sum_{k=0}^{n} (k-\mu)^2 \binom{n}{k} p^k q^{n-k},$$

where $\mu = np$ is the mean of the binomial random variable X, or

$$\sum_{k=0}^{n} k^2 \binom{n}{k} p^k q^{n-k} - (np)^2.$$

It's not a huge deal to go from what we have in (11.6) to the variance. We can determine the variance from $\mathbb{E}[X^2] - \mathbb{E}[X]^2$ and write $k(k-1)$ as $k^2 - k$; note the sum of k^2 will be $\mathbb{E}[X^2]$ while the sum of k will be the mean. Thus

$$n(n-1)p^2(p+q)^{n-2} = \sum_{k=0}^{n} k^2 \binom{n}{k} p^k q^{n-k} - \sum_{k=0}^{n} k \binom{n}{k} p^k q^{n-k}.$$

But we've already determined the second sum—it's just np when $q = 1 - p$. Setting $q = 1 - p$ we thus find

$$\sum_{k=0}^{n} k^2 \binom{n}{k} p^k (1-p)^{n-k} = n(n-1)p^2 + np = n^2 p^2 + np(1-p). \qquad (11.7)$$

Therefore the variance is just

$$\begin{aligned}
\text{Var}(X) &= \sum_{k=0}^{n} k^2 \binom{n}{k} p^k (1-p)^{n-k} - \left(\sum_{k=0}^{n} k \binom{n}{k} p^k (1-p)^{n-k} \right)^2 \\
&= n^2 p^2 + np(1-p) - (np)^2 \\
&= np(1-p),
\end{aligned}$$

which agrees with what we found above.

As a final remark, consider again (11.5) and (11.7). If we set $p = q = 1/2$ and then move those factors to the right-hand side, we obtain

$$\sum_{k=0}^{n} k \binom{n}{k} = 2^n, \quad \sum_{k=0}^{n} k^2 \binom{n}{k} = n(n+1)2^{n-2}.$$

Thus, we can find nice expressions for sums of products of binomial coefficients and their indices. It's interesting to note that even if we only want to evaluate sums of integers or rationals, we need to have *continuous* variables so that we can use the tools of calculus.

11.4 Applications to Normal Random Variables

By $X \sim N(\mu, \sigma^2)$ we mean that X is normally distributed with mean μ and variance σ^2, so its density is

$$f_X(x) = \frac{1}{\sqrt{2\pi\sigma^2}} e^{-(x-\mu)^2/2\sigma^2}.$$

We discuss in great length in Chapter 14 how we calculate the integral to prove this is a probability distribution. After proving it's a probability distribution, the next questions what are the mean, the variance, and more generally all of the moments?

Let's concentrate on the standard normal, whose density function is

$$(2\pi)^{-1/2} \exp(-x^2/2).$$

We are thus reduced to calculating

$$M(k) = \int_{-\infty}^{\infty} x^k \cdot \frac{1}{\sqrt{2\pi}} e^{-x^2/2} dx.$$

The integral is clearly zero for k odd, as we're integrating an odd function over a symmetric region. (Note the normal decays so rapidly that all the integrals exist.) There are at least two natural ways to handle even k: brute force integration and using differentiating identities.

Standard Approach: Let's do the brute force first, which is through induction and integration by parts. Consider the variance; since the mean is zero, the variance is just

$$\int_{-\infty}^{\infty} x^2 \cdot \frac{1}{\sqrt{2\pi}} e^{-x^2/2} dx.$$

To integrate by parts, we need to choose values for u and dv. While at first we might think the natural choices are either $u = x^2$ or $dv = x^2 dx$, if we try either we run into problems. The reason is that there is no nice anti-derivative for $e^{-x^2/2}$. Fortunately, all is not lost. The function $e^{-x^2/2}$ is *screaming* to us that it wants to be considered with a factor of x, as then it *will* have a nice anti-derivative. Thus we try

$$u = x, \quad dv = \frac{1}{\sqrt{2\pi}} e^{-x^2/2} x dx.$$

This leads to $du = dx$ and $v = -(2\pi)^{-1/2} e^{-x^2/2}$. Thus we find

$$M(2) = uv \Big|_{-\infty}^{\infty} + \int_{-\infty}^{\infty} \frac{1}{\sqrt{2\pi}} e^{-x^2/2} dx = I(0) = 1.$$

We have thus shown that the second moment is 1!

More generally, assume we know $M(2k) = (2k - 1)!!$, where the double factorial means we only consider odd terms of the factorial. Then we proceed as above, and to compute $M(2k + 2)$ when we integrate by parts we set $u = x^{2k+1}$, so $du = (2k + 1)x^{2k} dx$. The boundary term vanishes when evaluated at $\pm\infty$, and we find

$$M(2k + 2) = (2k + 1) \int_{-\infty}^{\infty} x^{2k} \frac{1}{\sqrt{2\pi}} e^{-x^2/2} dx$$

$$= (2k + 1)M(2k) = (2k + 1)(2k - 1)!! = (2k + 1)!!.$$

Differentiating Identities Approach: We now show how to calculate the moments through differentiating identities. It seems strange to talk about differentiating identities here, as

$$M(2k) = \int_{-\infty}^{\infty} x^{2k} \frac{1}{\sqrt{2\pi}} e^{-x^2/2} dx$$

has no free parameter! Remember **Abstraction**; rather than looking at this specific density let's look at a family of densities with a free parameter.

We begin with the fact that

$$1 = \int_{-\infty}^{\infty} \frac{1}{\sqrt{2\pi\sigma^2}} e^{-x^2/2\sigma^2} dx;$$

this is just the statement that the above is the probability density for a normal distribution with mean 0 and variance σ^2. Moving σ to the other side gives

$$\sigma = \int_{-\infty}^{\infty} \frac{1}{\sqrt{2\pi}} e^{-x^2/2\sigma^2} dx.$$

We keep applying $\sigma^3 \frac{d}{d\sigma}$ to both sides. Why do we multiply by σ^3? The reason is that the differentiation hits $-x^2\sigma^{-2}/2$, and thus brings down a factor of $x^2\sigma^{-3}$. Hence if we multiply by σ^3, we keep everything nice (i.e., we don't change the power of σ on the right-hand side). Differentiating once gives

$$\sigma^3 \frac{d}{d\sigma} \sigma = \sigma^3 \frac{d}{d\sigma} \int_{-\infty}^{\infty} \frac{1}{\sqrt{2\pi}} e^{-x^2/2\sigma^2} dx.$$

We interchange the derivative and the integration, and note that $\frac{d}{d\sigma} e^{-x^2/2\sigma^2}$ is $e^{-x^2/2\sigma^2}$. x^2, and thus we get

$$\sigma^3 \cdot 1 = \int_{-\infty}^{\infty} x^2 \cdot \frac{1}{\sqrt{2\pi}} e^{-x^2/2\sigma^2} dx.$$

Letting

$$I(k;\sigma) = \int_{-\infty}^{\infty} x^k \cdot \frac{1}{\sqrt{2\pi}} e^{-x^2/2\sigma^2} dx,$$

we've just shown that

$$\sigma^3 = I(2;\sigma);$$

further, there is a simple relation between the integrals $I(k;\sigma)$ and the moments of the standard normal, $M(k)$:

$$I(k;1) = M(k).$$

More generally, $I(k;\sigma)/\sigma$ is the kth moment of a normal random variable with mean 0 and variance σ^2. The reason for dividing by $\sigma = \sqrt{\sigma^2}$ is that our integral only has the factor $1/\sqrt{2\pi}$ and not $1/\sqrt{2\pi\sigma^2}$.

We showed

$$1 \cdot \sigma^3 = I(2;\sigma) \quad \text{and} \quad I(k;1) = M(k);$$

we'll see in a moment that it's convenient to write $1 \cdot \sigma^3$ and not just σ^3. Applying $\sigma^3 \frac{d}{d\sigma}$ to both sides yields

$$\sigma^3 \cdot 3 \cdot 1\sigma^2 = \int_{-\infty}^{\infty} x^2 \cdot x^2 \frac{1}{\sqrt{2\pi}} e^{-x^2/2\sigma^2} dx = I(4;\sigma),$$

or equivalently

$$3 \cdot 1 \cdot \sigma^5 = I(4; \sigma).$$

We again apply $\sigma^3 \frac{d}{d\sigma}$ to both sides, and find

$$\sigma^3 \cdot 5 \cdot 3 \cdot 1\sigma^4 = 5 \cdot 3 \cdot 1 \cdot \sigma^7 = I(6; \sigma);$$

note the left-hand side is $5!! \cdot \sigma^7$. We're now just a short induction away from proving the formula for the even moments. If we have

$$(2k - 1)!! \cdot \sigma^{2k+1} = I(2k; \sigma),$$

applying $\sigma^3 \frac{d}{d\sigma}$ one more time gives

$$\sigma^3 \cdot (2k + 1) \cdot (2k - 1)!! \cdot \sigma^{2k} = I(2k + 2; \sigma).$$

Taking $\sigma = 1$ yields

$$(2k + 1)!! = I(2k + 2, 1) = M(2k + 2),$$

which proves the formula for the moments.

Note for this problem that while differentiating identities is quite useful, it was not immediately apparent what identity we needed to use! One way to narrow down the options is to realize we cannot work with just the probability density we care about; if we were to do this, we wouldn't be able to take derivatives! If we were to use $\frac{d}{d\sigma}$ instead of $\sigma^3 \frac{d}{d\sigma}$ then there would be a lot more algebra to do, as we would now need the product rule to compute the derivatives. This is yet another example of how difficulties in math problems boil down to problems in algebra, and we want to reduce the amount of algebra we do by cleverly choosing which operator to use in the differentiation.

11.5 Applications to Exponential Random Variables

We can apply our Method of Differentiating Identities to almost any distribution to compute its mean, variance, or more generally any moment. What are the restrictions? We need the distribution to sit inside a family of densities which depend on several parameters, at least one of which is amenable to differentiation. Let's unwind this phrase. We'll assume our density has finite mean and variance. Then we can write our density as $f(x; \theta_1, \ldots, \theta_\ell)$, where $\theta_1, \ldots, \theta_\ell$ are the parameters. We can write the kth moment as

$$\int_{-\infty}^{\infty} x^k f(x; \theta_1, \ldots, \theta_\ell) dx.$$

Starting from

$$1 = \int_{-\infty}^{\infty} f(x; \theta_1, \ldots, \theta_\ell) dx,$$

the goal is to apply differentiable operators and end up with the formula for the kth moment. We'll illustrate this with another example.

Consider an exponential random variable. If $X \sim \text{Exp}(\lambda)$ then its density is

$$f_X(x) = \begin{cases} \lambda^{-1} e^{-x/\lambda} & \text{if } x \geq 0 \\ 0 & \text{otherwise;} \end{cases}$$

note that some books define the exponential differently, with $1/\lambda$ playing the role of λ. (*Aside: I prefer this way, as it leads to an exponential random variable with parameter λ having a mean of λ. Alternatively, another way to look at it is that we're measuring x and λ in the same units. We cannot exponentiate a quantity that has units, and the ratio x/λ is unitless. The alternative notation has λx, and now x and λ are in different "units."*)

We start off with the fact that the density integrates to 1:

$$1 = \int_0^\infty e^{-x/\lambda} \frac{dx}{\lambda};$$

even if we only care about the standard exponential (when $\lambda = 1$), we still need to have a free parameter and an entire family, since if we didn't then we wouldn't be able to differentiate! To keep things nice, this suggests multiplying both sides by λ, yielding

$$\lambda = \int_0^\infty e^{-x/\lambda} dx. \tag{11.8}$$

If we apply $d/d\lambda$ to both sides, when it hits the right hand side we see exponent $-x/\lambda$ contributes a x/λ^2. This suggests that the nicer operator to study is $\lambda^2 \frac{d}{d\lambda}$. If we use this, then applying it to both sides of (11.8) gives

$$\lambda^2 \cdot 1 = \lambda^2 \cdot \int_0^\infty \frac{x}{\lambda^2} e^{-x/\lambda} dx = \int_0^\infty x e^{-x/\lambda} dx.$$

If we want the mean of the standard exponential, we just take $\lambda = 1$ and we're done. What if we want the mean of the exponential with parameter λ? In this case, we see that the integral on the right-hand side isn't quite what we want; the mean is

$$\int_0^\infty x \cdot \frac{1}{\lambda} e^{-x/\lambda} dx,$$

while we have

$$\int_0^\infty x \cdot e^{-x/\lambda} dx.$$

The solution, of course, is to just divide both sides by λ, giving

$$\lambda = \int_0^\infty x \cdot e^{-x/\lambda} \frac{dx}{\lambda}.$$

What about the variance? As the mean isn't zero (it's λ), this is a little harder than the calculation for the standard normal. We use $\text{Var}(X) = \mathbb{E}[X^2] - \mathbb{E}[X]^2$. Applying $\lambda^2 \frac{d}{d\lambda}$ once to (11.8) yields

$$\lambda^2 = \int_0^\infty x e^{-x/\lambda} dx.$$

Applying it again gives

$$\lambda^2 \cdot 2\lambda = \int_0^\infty x^2 e^{-x/\lambda} dx.$$

Dividing both sides by λ yields

$$2\lambda^2 = \int_0^\infty x^2 e^{-x/\lambda} \frac{dx}{d\lambda} = \mathbb{E}[X^2];$$

thus $\mathbb{E}[X^2] = 2\lambda^2$. As the mean is $\mathbb{E}[X] = \lambda$, we find

$$\text{Var}(X) = \mathbb{E}[X^2] - \mathbb{E}[X]^2 = 2\lambda^2 - \lambda^2 = \lambda^2,$$

the variance of the exponential. We could of course have found the variance directly by integrating by parts; it's up to you which method you prefer.

As a nice exercise, continue this method to find the third and fourth moments. The easiest way to find the third moment is probably to use

$$\begin{aligned}
\mathbb{E}[(X - \mu)^3] &= \mathbb{E}[X^3 - 3X^2\mu + 3X\mu^2 - \mu^3] \\
&= \mathbb{E}[X^3] - 3\mathbb{E}[X^2]\mu + 3\mathbb{E}[X]\mu^2 - \mu^3,
\end{aligned}$$

where for an exponential random variable with parameter λ the mean $\mu = \mathbb{E}[X]$ is just λ. We already know $\mathbb{E}[X^2]$, so to find the third moment we just need to compute $\mathbb{E}[X^3]$.

11.6 Summary

The amusing quote at the beginning of the chapter, inspired by a poem of Jonathan Swift, beautifully describes the scope of differentiating identities:

> Big fleas have little fleas,
> Upon their backs to bite 'em,
> And little fleas have lesser fleas,
> and so, ad infinitum.

It's a process that never ends. Like the little fleas that have lesser fleas, and so on ad infinitum, once we have an identity we can keep generating more and more and more, indefinitely. It's extremely important that we can do so. We talked in Chapter 9 about how the moments of a probability distribution are like the Taylor coefficients of a function, and knowledge of these translates into knowledge of our density. As the process never ends, we get formulas for *all* the moments. It's spectacular seeing how much more old friends such as the geometric series formula or the Binomial Theorem can give.

11.7 Exercises

Exercise 11.7.1   *The double angle formula for sine is* $\sin(2x) = 2\sin(x)\cos(x)$. *Use this to find the double angle formula for cosine.*

Exercise 11.7.2   *Explain the error in the following statement: Given the equation* $ax^2 + bx + c = 0$, *differentiating both sides twice with respect to* x *gives* $2a = 0$, *so* $a = 0$. *Differentiating the remaining terms once with respect to* x *implies* $b = 0$. *Therefore,* c *also equals 0 for all* x.

Exercise 11.7.3   *Prove that for any polynomial* $a_0 + a_1 x + a_2 x^2 + \cdots + a_n x^n$ *to be identically 0 that* $a_0 = a_1 = a_2 = \cdots = a_n = 0$.

Exercise 11.7.4   *Earlier in the chapter, we began with the Binomial Theorem as the identity which we differentiated. Prove the Binomial Theorem.*

Exercise 11.7.5   *Derive a formula for* $(x_1 + x_2 + \cdots + x_m)^n$. *Include a proof of your formula.*

Exercise 11.7.6   *The Taylor series for* $\sin x$ *near 0 is*

$$\sin x = \sum_{n=1}^{\infty} (-1)^{n-1} \cdot \frac{x^{2n-1}}{(2n-1)!}.$$

Use differentiating identities to find the Taylor expansion for $\cos x$ *near 0.*

Exercise 11.7.7   *Without explicitly finding the Taylor series, use the properties of* e^x *to show that the Taylor series for* e^x *is* $\sum_{n=0}^{\infty} x^n / n!$.

Exercise 11.7.8   *Show using the Taylor series that* $\sum_{n=1}^{\infty} (-1)^{n-1}/n = \log 2$.

Exercise 11.7.9   *The Taylor series for* $\frac{1}{1-x}$ *is* $\sum_{n=0}^{\infty} x^n$ *on the interval* $(-1, 1)$. *The series converges uniformly on this interval (so the derivative of a sum is the sum of the derivatives). What is the Taylor series for* $\frac{x}{(1-x)^2}$ *on the same interval?*

Exercise 11.7.10   *Returning to* $\sum_{n=0}^{\infty} x^n$, *notice that if you split the sum into* $n < N$ *and* $n \geq N$ *that the sum over* $n \geq N$ *is another geometric series, and thus we can convert the infinite sum to a finite sum! Do this and justify taking the derivative term by term of the infinite sum.*

Exercise 11.7.11   *On the interval* $[0, 2\pi]$, $\sum_{n=1}^{\infty} \frac{\cos(nx)}{n^2} = \frac{3x^2 - 6\pi x + 2\pi^2}{12}$. *Use this identity to show that* $\sum_{n=1}^{\infty} \frac{1}{n^2} = \frac{\pi^2}{6}$.

Exercise 11.7.12   *Use the Taylor series to justify that* $e^{ix} = \cos(x) + i\sin(x)$.

Exercise 11.7.13   *Show that the identity mentioned in the previous exercise implies the Pythagorean identity:* $(\sin^2 x + \cos^2 x = 1)$.

Exercise 11.7.14   *The Poisson distribution is a probability function given by*

$$\text{Prob}(X = n) = \begin{cases} \lambda^n e^{-\lambda}/n! & \text{if } n \in \{0, 1, 2, \dots\} \\ 0 & \text{otherwise.} \end{cases}$$

Find the mean of the Poisson distribution.

Exercise 11.7.15   *Find the variance of the Poisson distribution.*

Exercise 11.7.16   *Find the variance of the negative binomial distribution.*

Exercise 11.7.17   *Find a closed form solution for $\sum_{n=0}^{\infty} \frac{n!}{(n-k)!} x^{n-k}$.*

Exercise 11.7.18   *Does the sum in the previous exercise converge for all k? Justify your answer.*

Exercise 11.7.19   *Imagine that inflation is fixed at 5% a year, that is, a dollar now is worth \$(1/1.05) a year from now. Consider a perpetuity that pays out one dollar this year and 50 cents more than the previous year each year forever. What is the value of this perpetuity in today's dollars?*

Exercise 11.7.20   *Imagine a pendulum with radius 1 dropped from a 30 degree angle. Due to friction, each swing does not go up as far as the previous. We will say that each time the pendulum moves $\frac{n^2}{2^n}$, where n is the number of swings as far horizontally as it did on its original swing. Write down an expression for the total distance the pendulum swings. Use the Taylor series to approximate this sum.*

Exercise 11.7.21   *Graph $\sum_{i=0}^{n} 2^{-i}$ for $n \in \{1, 2, \ldots, 50\}$ and the line $y = 2$ on the same plot. Comment on the convergence of the series.*

PART III
SPECIAL DISTRIBUTIONS

CHAPTER 12 _____

Discrete Distributions

Life defies our phrases, it is infinitely continuous and subtle and shaded, whilst our verbal terms are discrete, rude and few.
— WILLIAM JAMES, *The Varieties of Religious Experience* (1902)

The purpose of this chapter is to introduce you to some of the most common and important discrete distributions (in later chapters we'll consider continuous ones). These distributions are popular for two very good reasons. First, they happen to describe many natural and mathematical phenomena very well. Second, they are mathematically tractable, and we can do many calculations with them, from computing means and variances to sums of independent random variables. As you continue in your courses and career, you'll experience again and again just how important tractability is. Sadly, most situations are *not* nice. It's rare to get a nice, closed form expression for the answer in terms of the parameters of the problem. Why is that such an important goal? If you can get a solution in terms of the parameters, then you can quickly see how changes to these values affect the answer.

For example, economics models are frequently very complicated, as numerous complex effects must be incorporated. Sadly, it's often not possible to get an answer in terms of the input parameters, and we're forced to resort to millions of simulations. What this means is that if a parameter changes, we need to run new simulations. This is computationally expensive. If instead we could get a closed form answer, all we would have to do is substitute in a new value.

We begin with the Bernoulli distribution. We then explore several of its generalizations, and then end with a few other important choices. Some of these distributions should be a little familiar, as we saw special cases of them in Chapter 7. The reason we briefly met them is that it's painful to describe probability without any meaningful examples; in the sections below, we'll delve more deeply into the theory and applications.

12.1 The Bernoulli Distribution

The simplest discrete random variable always takes on the same value. Of course, this means there's no mystery. Thus, our story begins with the *second* simplest choice: it takes on two values (which we can scale to make 0 and 1).

> **Bernoulli Distribution**: X has a **Bernoulli distribution** with parameter $p \in [0, 1]$ if $\text{Prob}(X = 1) = p$ and $\text{Prob}(X = 0) = 1 - p$. We view the outcome 1 as a **success**, and 0 as a **failure**. We write $X \sim \text{Bern}(p)$. We also call X a **binary indicator random variable**.

Remember, random variables are always real-valued. If we're tossing a coin with probability p of heads, our random variable can't take on the values "Head" and "Tail"; these are elements of the outcome space, not numbers. We must associate a number to each possibility. The advantage, of course, is that we can add numbers together (what is the sum of two heads and a tail?). Instead of viewing the outcome as heads or tails, we could imagine people vote for a candidate or against them, it rains or it doesn't, our team wins or it loses, and so on. Whenever you meet a random variable, very quickly you should compute its mean and variance. These give a lot of information.

> **Theorem**: *If $X \sim \text{Bern}(p)$ then its mean μ_X equals p and its variance σ_X^2 is $p(1 - p)$.*

This follows from the definitions of the mean and variance:

$$\mu_X = 1 \cdot p + 0 \cdot (1 - p) = p$$
$$\sigma_X^2 = (1 - \mu_X)^2 \cdot p + (0 - \mu_X)^2 \cdot (1 - p)$$
$$= (1 - p)^2 \cdot p + (-p)^2 (1 - p)$$
$$= p(1 - p) \cdot (1 - p + p) = p(1 - p).$$

There's not much more to do with *one* Bernoulli random variable; however, the situation is drastically changed if we either consider multiple coins or if we keep flipping the same coin repeatedly. We explore the associated random variables in the next section.

12.2 The Binomial Distribution

There are two ways to view the binomial distribution. One is that we have n independent coins, each with probability p of success. We flip them all simultaneously, and record the number of heads. Alternatively, we could flip one coin n times, and record the number of heads. Both perspectives are useful. They're equivalent as we're assuming the tosses of the coins are independent. Since the tosses are independent, we can't tell the difference from n different coins being tossed, or one coin being tossed n times.

Binomial Distribution: Let n be a positive integer and let $p \in [0, 1]$. Then X has the **binomial distribution** with parameters n and p if

$$\text{Prob}(X = k) = \begin{cases} \binom{n}{k} p^k (1-p)^{n-k} & \text{if } k \in \{0, 1, \ldots, n\} \\ 0 & \text{otherwise.} \end{cases}$$

We write $X \sim \text{Bin}(n, p)$. The mean of X is np and the variance is $np(1-p)$.

We need to make sure that this is in fact a probability distribution. There are two items to check: (1) all the probabilities must be non-negative, and (2) the sum of the probabilities must equal 1. The first condition is easily verified, and the name of the distribution actually suggests the way to prove the second. We're calling this the **binomial** distribution. Where else have we heard the word binomial? Well, there's binomial coefficients, which are part of the definition of our probabilities. There's also the **Binomial Theorem**, which we review in §A.2.3. The Binomial Theorem says

$$(x + y)^n = \sum_{k=0}^{n} \binom{n}{k} x^k y^{n-k}.$$

For us, we have

$$\sum_{k=0}^{n} \text{Prob}(X = k) = \sum_{k=0}^{n} \binom{n}{k} p^k (1-p)^{n-k} = (1 - (1-p))^n = 1.$$

Thus the second condition holds, and we do have a probability distribution.

By now it should be automatic to compute the mean and the variance of any random variable you meet, which is why I included their values in the statement above. There's many ways to find these. If you go directly from the definition, you need to find

$$\mu_X = \sum_{k=0}^{n} k \binom{n}{k} p^k (1-p)^{n-k}, \quad \sigma_X^2 = \sum_{k=0}^{n} (k - \mu_X)^2 \binom{n}{k} p^k (1-p)^{n-k}.$$

These might appear to be unpleasant sums, as they involve binomial coefficients; however, it turns out there are powerful techniques to evaluate them. One of my favorites is **differentiating identities**, which we used to find the mean and variance in §11.3.

What I want to do now is showcase another approach, **linearity of expectation**. The idea is the following: let's break up our complicated random variable into a sum of simpler, independent random variables. If $X \sim \text{Bin}(n, p)$ and $X_1, \ldots, X_n$ are independent $\text{Bern}(p)$ random variables, then

$$X = X_1 + \cdots + X_n.$$

Why is a $\text{Bin}(n, p)$ random variable the same as a sum of n independent $\text{Bern}(p)$ random variables? It's essentially a **proof by grouping** again; we can either view all n coins as being tossed together and the number of heads counted all at once, or we can view the results of the n tosses one at a time and then add.

As each X_i is Bern(p), the means are $\mu_{X_i} = p$ and the variances are $\sigma^2_{X_i} = p(1-p)$. We now use **linearity of expectation** (see §9.5 as well as Theorem 9.6.2): the expected value of a sum of independent random variables is the sum of the expected values, and the variance of a sum of independent random variables is the sum of the variances. Thus

$$\begin{aligned}
\mu_X &= \mathbb{E}[X] \\
&= \mathbb{E}[X_1 + \cdots + X_n] \\
&= \mathbb{E}[X_1] + \cdots + \mathbb{E}[X_n] \\
&= p + \cdots + p = np,
\end{aligned}$$

and

$$\begin{aligned}
\sigma^2_X = \mathrm{Var}(X) &= \mathrm{Var}(X_1 + \cdots + X_n) \\
&= \mathrm{Var}(X_1) + \cdots + \mathrm{Var}(X_n) \\
&= p(1-p) + \cdots + p(1-p) = np(1-p).
\end{aligned}$$

It's worth taking a minute to celebrate what we've just accomplished—we've found a way to avoid evaluating sums of binomial coefficients (multiplied by a variety of factors!). We reduced a hard sum to n simpler sums. This is yet another example of one of our guiding principles: *It is often better to do many simpler problems than one hard problem.*

Actually, though, there's another way to view what we've done. One of the most powerful techniques in probability and combinatorics is to **compute something two different ways**. Typically one of the calculations is easier than the other, and this allows us to find the other. There's an added benefit to using linearity of expectation: we have just proved some theorems about binomial sums. The mean calculation gives

$$\sum_{k=0}^{n} k \binom{n}{k} p^k (1-p)^{n-k} = np,$$

while the variance calculation gives

$$\mathrm{Var}(X) = \sum_{k=0}^{n} (k - np)^2 \binom{n}{k} p^k (1-p)^{n-k}.$$

We've used this idea many times—if we can calculate something two different ways, all we have to do is evaluate one of them to determine the other.

We can solve all sorts of problems now involving binomial random variables. Let's return to gambling. Assume four independent dice are rolled, and a player bets on any number from 1 to 6. If the number that the player bet on appears k times, where k ranges from 1 to 4, then the player wins k dollars. If the number does not appear, the player loses 1 dollar. Should the player play this game?

To figure out if we should play, we want to know the *expected outcome*. That phrase is a clue on what we need to find. If we play this game many times over, do we expect to have won or lost money? We can answer this by calculating the expected value. If we let X be the amount we win or lose when we roll 4 independent, fair die, then the possible values for X are -1 (if we roll our number zero times), 1 (if we roll our number once), 2 (if we roll our number twice), and so on up to 4. We now compute the probability of each of these happening:

$$\text{Prob}(X = -1) = \binom{4}{0}\left(\frac{1}{6}\right)^0\left(\frac{5}{6}\right)^4 = \frac{625}{1296}$$

$$\text{Prob}(X = 1) = \binom{4}{1}\left(\frac{1}{6}\right)^1\left(\frac{5}{6}\right)^3 = \frac{125}{324}$$

$$\text{Prob}(X = 2) = \binom{4}{2}\left(\frac{1}{6}\right)^2\left(\frac{5}{6}\right)^2 = \frac{25}{216}$$

$$\text{Prob}(X = 3) = \binom{4}{3}\left(\frac{1}{6}\right)^3\left(\frac{5}{6}\right)^1 = \frac{5}{324}$$

$$\text{Prob}(X = 4) = \binom{4}{4}\left(\frac{1}{6}\right)^4\left(\frac{5}{6}\right)^0 = \frac{1}{1296}.$$

We can now find the expected value of X:

$$E[X] = (-1)\cdot\frac{625}{1296} + 1\cdot\frac{125}{324} + 2\cdot\frac{25}{216} + 3\cdot\frac{5}{324} + 4\cdot\frac{1}{1296} = \frac{239}{1296}.$$

As the expected value is positive, it's to our advantage to play. One way to interpret the above is that, *on average*, after every 1296 games played we expect to be 239 dollars richer.

The following code allows us to play this game and check our calculation.

```
diegame[numdo_] := Module[{},
  winnings = 0; (* keep track of how much have won *)
  For[n  = 1, n <= numdo, n++, (* start of main loop *)
   {
     (* RandomInteger[{1,6}] uniformly chooses a number from 1 to 6 *)
     (* Without loss of generality can assume our chosen number is 1 *)
     (* We do four times and save the number of 1s to numroll *)
     numroll = Sum[If[RandomInteger[{1, 6}] == 1, 1, 0], {i, 1, 4}];
     If[numroll == 0, winnings = winnings - 1,
      winnings = winnings + numroll]; (* adjust winnings accordingly *)
    }]; (* end of n loop *)
  Print["Expected value is 239/1296 or about ", 239/1296.0, "."];
  Print["Average winnings is ", 1.0 winnings / numdo, "."];
  ]
```

Doing 100,000,000 trials (it's a very simple program so we can do a large number of runs) yielded:

```
Expected value is 239/1296 or about 0.184414.
Average winnings is 0.184213.
```

12.3 The Multinomial Distribution

There are several ways to generalize the Bernoulli distribution. Our first choice led to the binomial distribution. We use a binomial distribution to study a situation where we have multiple trials with two possible outcomes, with the probabilities of each respective outcome the same for each trial and all of the trials independent. If you want, you can imagine tossing coins, or perhaps people voting in an election with only two alternatives. It's obviously quite limiting to only have two options, which suggests a natural further generalization to the **multinomial distribution**. Like the binomial distribution, the multinomial distribution considers multiple independent trials with the probabilities of respective outcomes the same for each trial. However, the multinomial distribution gives the probability of different outcomes when we have more than two possible outcomes for each trial. This is useful, as sometimes in life there actually are more than two possibilities!

Suppose that we have n trials and k mutually exclusive outcomes with probabilities $p_1, p_2, \ldots, p_k$. We let $f(x_1, x_2, \ldots, x_k)$ be the probability of having x_i outcomes of each corresponding type, for $1 \le i \le k$. Obviously, we must have $x_1 + x_2 + \cdots + x_k = n$. To compute $f(x_1, x_2, \ldots, x_k)$, we first note that the probability of getting these numbers of outcomes in a specific, particular order is $p_1^{x_1} p_2^{x_2} \cdots p_k^{x_k}$. We now compute the number of orders in which our combination of numbers of outcomes is attainable. The x_1 outcomes of the first type can be chosen in $\binom{n}{x_1}$ ways, the x_2 outcomes of the second type can be chosen in $\binom{n-x_1}{x_2}$ ways, and so on up to the x_k outcomes of type k which can be chosen in $\binom{n-x_1-x_2-\cdots-x_{k-1}}{x_k}$ ways. The total number of orderings is therefore

$$\binom{n}{x_1}\binom{n-x_1}{x_2}\cdots\binom{n-x_1-\cdots-x_{k-1}}{x_k}$$

$$= \frac{n!}{(n-x_1)!x_1!} \cdot \frac{(n-x_1)!}{(n-x_1-x_2)!x_2!} \cdots \frac{(n-x_1-\cdots-x_{k-1})!}{(n-x_1-\cdots-x_k)!x_k!}.$$

The product telescopes and we're left with

$$\frac{n!}{x_1!x_2!\cdots x_k!}. \tag{12.1}$$

The expression (12.1) is called a **multinomial coefficient** and is often denoted

$$\binom{n}{x_1, x_2, \ldots, x_k}.$$

We met these coefficients in §6.2.2 when we studied rearranging letters of a word to make new words; see that section for an alternative derivation of the expansion above. Using the multinomial coefficient, we can see that

$$f(x_1, x_2, ..., x_n) = \frac{n!}{x_1! x_2! \cdots x_k!} p_1^{x_1} p_2^{x_2} \cdots p_k^{x_k}.$$

This is the multinomial distribution. We often write $f(x_1, x_2, ..., x_n; p_1, p_2, ..., p_k)$ to emphasize the dependence on the parameters (the convention is to put parameters after the semicolon).

Of course, we really should make sure it's a probability distribution by showing the sum of the probabilities is 1. I encourage you to roll up your sleeves and do so. Alternatively, you can argue it *must* sum to one on combinatorial grounds. We have a set of exclusive and exhaustive options. Finally, in the next paragraph we derive the distribution by grouping a bunch of binomial distributions. As each of those is a distribution, so too is the multinomial.

One can also derive the multinomial distribution by repeated uses of the Binomial Theorem and the **method of grouping**. For example, if $k = 3$ there are three outcomes, say A, B, and C. We may amalgamate B and C and consider the case of two outcomes: A and not A. If we let p_1 equal the probability of A and $1 - p_1$ the probability of not A, we find the probability of x_1 outcomes being A and $n - x_1$ outcomes being not A is just

$$\binom{n}{x_1} p_1^{x_1} (1 - p_1)^{n - x_1}.$$

Let p_2 be the probability of outcome B, and p_3 the probability of outcome C. *Given A does not occur*, the probability that B occurs is $\frac{p_2}{p_2 + p_3}$; the probability that C occurs is $\frac{p_3}{p_2 + p_3}$. Note these are *conditional probabilities*, and they sum to 1 since $\frac{p_2}{p_2 + p_3} + \frac{p_3}{p_2 + p_3} = 1$.

Thus the probability that x_1 outcomes are A, x_2 are B, and $x_3 = n - x_1 - x_2$ are C is

$$\binom{n}{x_1} p_1^{x_1} \left[\binom{n - x_1}{x_2} \left(\frac{p_2}{p_2 + p_3} \right)^{x_2} \left(\frac{p_3}{p_2 + p_3} \right)^{n_1 - x_1 - x_2} \right] (1 - p_1)^{n - x_1};$$

note that if we sum the bracketed quantity over x_2 from 0 to $n - x_1$ that we get 1 by the Binomial Theorem.; this is no accident and is related to our expansion (we started with A and not A).

We can simplify this a lot further by noting $1 - p_1 = p_2 + p_3$ and $\binom{n}{x_1}\binom{n - x_1}{x_2} = \frac{n!}{x_1! x_2! x_3!}$, which gives

$$\frac{n!}{x_1! x_2! x_3!} p_1^{x_1} p_2^{x_2} p_3^{n_1 - x_1 - x_2},$$

which agrees with what we found above. Isolating what we've discovered gives the following.

Multinomial distribution and coefficients: Let n, k be positive integers, and let $p_1, p_2, \ldots, p_n \in [0, 1]$ be such that $p_1 + \cdots + p_n = 1$. Let $x_1, \ldots, x_n \in \{0, 1, \ldots, n\}$ be such that $x_1 + \cdots + x_n = n$. The corresponding **multinomial coefficient** is

$$\binom{n}{x_1, x_2, \ldots, x_k} = \frac{n!}{x_1! x_2! \cdots x_k!},$$

and all other choices of the x_i's evaluate to zero. The **multinomial distribution** with parameters n, k and $p_1, \ldots, p_k$ is non-zero only for such $(x_1, \ldots, x_k)$, where the density is

$$\binom{n}{x_1, x_2, \ldots, x_k} p_1^{x_1} p_2 x_2 \cdots p_k^{x_k}.$$

We write $X \sim$ Multinomial$(n, k, p_1, \ldots, p_k)$.

The multinomial distribution is different than the other distributions we've seen. Instead of getting *one* value, we get an n-tuple. We can understand its behavior by viewing it as binomial distributions. For example, if

$$X \sim \text{Multinomial}(n, k, p_1, \ldots, p_k)$$

we can form the random variable X_i, which is the number of occurrences of outcome i. We have $X_i \sim \text{Bin}(n, p_i)$; now all we care about is whether we get outcome i (which we do with probability p_i) or not (which happens with probability $1 - p_i$). Thus $\mu_{X_i} = n p_i$ and $\sigma_{X_i}^2 = n_i p_i (1 - p_i)$.

The following problem is one of the standard ones in the subject. We consider a cookie jar filled with three types of delicious, delectable cookies: chocolate chip, snickerdoodle, and sugar. We assume there are so many cookies that we can consider removing a few as insignificant, and will not change the probability of getting another cookie. The probability of drawing a chocolate chip is 45%, the probability of choosing a snickerdoodle is 30%, and the chance of drawing out a sugar cookie is 25%. What is the chance of getting a mix of 3 chocolate chip cookies, 2 snickerdoodles, and 1 sugar cookie in a random drawing of 6 cookies? If you are uncomfortable with this approximation (as drawing a chocolate chip cookie must affect the probability that the next one is chocolate chip), we can instead consider this a problem of **sampling with replacement**; that means we pull out six cookies one at a time, recording their type and then immediately returning them to the jar.

Solution: Not surprisingly for a problem in the multinomial section, this is an example of a multinomial distribution. We have $p_1 = .45$, $p_2 = .30$, and $p_3 = .25$. Our values are $n = 6$, $x_1 = 3$, $x_2 = 2$, and $x_3 = 1$. We have

$$\Pr(X_1 = 3, X_2 = 2, X_3 = 1) = \frac{6!}{3! 2! 1!}(0.45)^3 (0.30)^2 (0.25)^1 \approx 0.123.$$

We give some additional problems in the exercises at the end of this chapter (for example, Exercise 12.8.29).

12.4 The Geometric Distribution

Our next distribution is another generalization of a Bernoulli random variable. Recall a Bernoulli random variable assigns a probability of p to a success and $1 - p$ to a failure. What we do now is we keep playing the game until we get our first success, and we let the random variable X denote the time we wait until we reach our first success. We call this a **geometric random variable**.

> **Geometric distribution**: Let $p \in [0, 1]$. A random variable X has the geometric distribution with parameter p if
>
> $$\text{Prob}(X = n) = \begin{cases} p(1 - p)^{n-1} & \text{if } n \in \{1, 2, 3, \dots\} \\ 0 & \text{otherwise.} \end{cases}$$
>
> We write $X \sim \text{Geom}(p)$. The mean is $1/p$ and the variance is $\frac{1-p}{p^2}$.

Our first task is to prove that this is indeed a distribution. The probabilities are non-negative, so we're left with showing they sum to 1. We use the geometric series formula (which we discussed and proved in §1.2): if $|r| < 1$ then $\sum_{n=0}^{\infty} r^n = \frac{1}{1-r}$. We have

$$\sum_{n=1}^{\infty} \text{Prob}(X = n) = \sum_{n=1}^{\infty} p(1 - p)^{n-1}$$
$$= p \sum_{m=0}^{\infty} (1 - p)^m$$
$$= p \cdot \frac{1}{1 - (1 - p)} = 1.$$

We're done—it's a probability distribution!

We now turn to the mean, which is more involved. A great way to find the mean is to use the **Method of Differentiating Identities**. We did this in §11.1, calculating the mean relatively painlessly. We give another proof now. This proof is very similar in spirit to our analysis of the basketball game in §1.2. The key idea is exploiting a **memoryless process** (or, if you prefer, taking advantage of some symmetries of the problem).

Let's denote the mean by μ_X. I claim that

$$\mu_X = 1 \cdot p + (\mu_X + 1) \cdot (1 - p).$$

Where did this formula come from? We can derive it by using a very similar argument as the one we used for the Basketball Problem in §1.2. The probability we stop after one toss is p. If we stop here, the value of our random variable is 1. Thus, the $1 \cdot p$ is due to the fact that with probability p our random variable takes on the value 1. What about the other factor? This is the more interesting one. Imagine the first toss was a failure. This happens with probability $1 - p$; however, once the failure happens it's like we've started the game all over. We're choosing to denote the mean by μ_X, so from this point onward we expect to wait μ_X tosses until we stop. Thus the contribution to the mean is $(\mu_X + 1)(1 - p)$; the $+1$ is because the first toss was a failure, and the $1 - p$ is the probability we end up in this case. We now solve for μ_X, and find

$$(1 - (1 - p))\,\mu_X \;=\; p + (1 - p),$$

which implies $p\mu_X = 1$ or $\mu_X = 1/p$, exactly as claimed!

It's always nice to have different proofs of the same result. Sometimes one perspective is more useful than another for a problem, or just clicks better with us. As a nice exercise, use this approach to find the variance. I think it's easier to find the variance through differentiating identities, and the formula $\sigma_X^2 = \mathbb{E}[X^2] - \mathbb{E}[X]^2$.

Assume we have a jar with p purple balls and y yellow balls (while we can use any letter to represent the variables, it's nice to choose letters that have a clear association with the words, as it makes it easier to remember what is what). We randomly draw balls, and after each ball is drawn we put it back in the jar before the next one is taken. What is the probability that at least m draws are needed before the first purple ball is drawn?

Let D denote the number of draws needed to get a purple ball. Since we're sampling with replacement, the probability of getting a purple ball is the same for each draw: it's always $p/(p + y)$. We find

$$
\begin{aligned}
\text{Prob}(D \geq m) \;&=\; \frac{p}{p+y} \cdot \left(\frac{y}{p+y}\right)^{m-1} + \frac{p}{p+y} \cdot \left(\frac{y}{p+y}\right)^{m} + \cdots \\
&=\; \frac{p}{p+y} \sum_{i=m}^{\infty} \left(\frac{y}{p+y}\right)^{i-1} \\
&=\; \frac{p}{p+y} \cdot \left(\frac{y}{p+y}\right)^{m-1} \Big/ \left(1 - \frac{y}{p+y}\right) \\
&=\; \left(\frac{y}{p+y}\right)^{m-1}.
\end{aligned}
$$

12.5 The Negative Binomial Distribution

Warning: if you thought the definition of an exponential random variable was bad, this is worse. The problem is different books have different conventions for the definition of the negative binomial; some have the support start at 0 and others at 1, some use p for the probability of success and some for failure. So, be warned when you compare our formulas to those in other books (or in computer packages!). You need to carefully check what their normalization is.

By highlighting one word, we suggest the next generalization. We started with a Bernoulli random variable, say tossing a coin that lands on heads (a success) with probability p, and on tails (a failure) with probability $1 - p$. We got a geometric random variable by repeatedly tossing the coin until we got our *first* success, and then we stopped. Hopefully it seems natural to consider tossing until we get our *second* success, or *third*, or more generally our r^{th} success and then stopping. We can thus create a random variable X which counts how long we need to wait until we have r successes.

Let's think a bit about its density. We must have $\text{Prob}(X = n) = 0$ if n is not an integer or if $n \leq r - 1$. This is because X can only take on integer values, and we need at least r tosses if we're to have r successes. We're left with finding the probabilities for $n \geq r$. Let's consider such n. We have exactly r successes among the n tosses. Further, the last toss *must* be a success, as otherwise we would have had r successes among $n - 1$ (or fewer) tosses, and we're determining the probability that it took exactly n tosses to get exactly r successes. To recap: $n \geq r$, we have exactly r successes among the first n tosses and the last toss is a success. This means there are exactly $r - 1$ successes among the first $n - 1$. There are $\binom{n-1}{r-1}$ ways to choose $r - 1$ of the first $n - 1$ tosses to be a success. Each of these configurations has r successes and $n - r$ failures, for a probability of $p^r (1 - p)^{n-r}$ of occurring. Putting everything together, we get

$$\text{Prob}(X = n) \;=\; \begin{cases} \binom{n-1}{r-1} p^r (1 - p)^{n-r} & \text{if } n \in \{r, r + 1, r + 2, \dots\} \\ 0 & \text{otherwise.} \end{cases}$$

Unfortunately, this is *not* the definition of the negative binomial distribution, though it is very close to it. Frequently in mathematics there are several equally valid choices one can make for a notation or a definition. Instead of counting successes, (some) people instead count failures as for many applications this perspective is more natural. We want to know how long a piece of equipment will last; in that case, keeping track of how many failures is more natural than counting successes. Also, instead of counting how many tosses we have before getting r failures, we count how many successes we have before getting r failures. We denote the number of successes by k, which has to be a non-negative integer. As we remarked at the start of the section, it is annoying that there are many slightly different formulations, but that's how it is and you have to be careful when looking at other sources.

We argue in the same manner as before. If we are to have exactly k successes and r failures, then there were $k + r$ tosses. Further, the last toss *must* have been a failure, so we have exactly k successes among the first $k + r - 1$ tosses. The probability of having any string with k successes and r failures is $p^k (1 - p)^r$, giving a probability of exactly k successes before getting the rth failure is $\binom{k+r-1}{k} p^k (1 - p)^r$. This is the **negative binomial**.

Negative binomial distribution: Let r be a positive integer and $p \in [0, 1]$. Let X be the random variable counting the number of successes obtained from tossing a coin with probability p of a success (a head) until getting exactly r failures (tails). X has the **negative binomial distribution** with parameters r and p. Its density is

$$\text{Prob}(X = k) \;=\; \begin{cases} \binom{k+r-1}{k} p^k (1 - p)^r & \text{if } k \in \{0, 1, 2, \dots\} \\ 0 & \text{otherwise.} \end{cases}$$

We write $X \sim \text{NegBin}(r, p)$. The mean is $\frac{pr}{1-p}$ and the variance is $\frac{pr}{(1-p)^2}$.

It's possible to prove that these probabilities sum to 1 by actually summing them. A great way to do that is to use the general Binomial Theorem, where the exponent need not be a positive integer. We elected *not* to pursue that approach here and instead gave a **Proof by Story** (see §A.6 for more on proofs by stories).

 Whenever you see a complicated formula, if you can find a special case to check, *do so!* There is one case that we can check. Imagine $r = 1$ and let the probability of success p equal $1 - q$ (so $1 - p$, the probability of a failure, equals q). In this case we're counting the number of successes before a coin with probability q of a failure (a tail) gives us a failure (i.e., lands on a tail). For integer $k \geq 0$, the probability is just

$$\binom{k+1-1}{k}(1-q)^k (1-(1-q))^1 = q(1-q)^k.$$

This should look familiar. It's *almost* the density of a geometric random variable with probability q of success. The difference is now k represents the number of successes, *not* the total number of tosses. Let $\ell = k + 1$ denote the total number of tosses, with L the corresponding random variable. Then for ℓ a non-negative integer we have

$$\text{Prob}(L = \ell) = q(1-q)^{\ell-1},$$

which *is* the same as a geometric random variable.

The purpose of this calculation was to show you how to test the plausibility of a calculation. It's encouraging that everything worked out. It's nice to see that the negative binomial is a generalization of a geometric random variable. Of course, it's important to keep things in perspective. We shouldn't get *too* excited about our success here. After all, this was a very special case. A lot of terms canceled because $r = 1$. So, while this definitely supports the claim that we have a probability distribution, it's not a proof.

Okay, it's time to find the mean! We have a daunting task before us, as we need to find

$$\mu_X = \sum_{k=0}^{\infty} k \binom{k+r-1}{k} p^k (1-p)^r.$$

If we stare at this awhile, eventually we might combine k and p^k to get kp^k. This suggests taking $\frac{d}{dp}$ (or perhaps $p\frac{d}{dp}$), as this will bring down a k. In other words, perhaps **differentiating identities** is a good way to go. As we've remarked numerous times before, it's wonderful to have a closed form expression, but they're often hard to find. Frequently we need to be very clever in how we **rewrite the algebra** to reach an expression we can fruitfully manipulate.

We start with the only relation we know, namely that the sum is 1, and go from there.

$$1 = \sum_{k=0}^{\infty} \binom{k+r-1}{k} p^k (1-p)^r$$

$$p\frac{d}{dp} 1 = p\frac{d}{dp} \sum_{k=0}^{\infty} \binom{k+r-1}{k} p^k (1-p)^r$$

$$0 = \sum_{k=0}^{\infty} \binom{k+r-1}{k} p \frac{d}{dp} \left(p^k (1-p)^r \right)$$

$$= \sum_{k=0}^{\infty} \binom{k+r-1}{k} p \left(kp^{k-1}(1-p)^r - rp^k(1-p)^{r-1} \right)$$

$$= \sum_{k=0}^{\infty} k \binom{k+r-1}{k} p^k (1-p)^r - rp \sum_{k=0}^{\infty} \binom{k+r-1}{k} p^k (1-p)^{r-1}.$$

We now bring the negative term over to the other side, obtaining

$$rp \sum_{k=0}^{\infty} \binom{k+r-1}{k} p^k (1-p)^{r-1} = \sum_{k=0}^{\infty} k \binom{k+r-1}{k} p^k (1-p)^r$$

$$\frac{rp}{1-p} \sum_{k=0}^{\infty} \binom{k+r-1}{k} p^k (1-p)^r = \sum_{k=0}^{\infty} k \binom{k+r-1}{k} p^k (1-p)^r.$$

Note, however, that the sum on the left is just 1 (this is because the negative binomial is a probability distribution), and the sum on the right is just the mean. Thus

$$\mu_X = \frac{rp}{1-p}$$

as claimed.

A similar calculation, which I encourage you to do, gives the variance.

I like the above proof through differentiating identities. If you think about it, the only identity we know is that the sum is 1, so it makes sense to start differentiating that. Is there another way to get the mean? Fortunately, yes! It's a bit more involved, but it's similar in spirit to the memoryless proof we gave for the geometric distribution's mean, so let's give it a try.

Our goal is to show that $\mu_X = \frac{rp}{1-p}$ if $X \sim \text{NegBin}(r, p)$. We proceed by induction on r. We first do the base case. If $r = 1$, we showed earlier that $\text{NegBin}(1, p)$ has almost the same density as $\text{Geom}(1 - p)$; they're shifted by 1. Explicitly, if X is the negative binomial and G is the geometric, then $X + 1 = G$. As the mean of a geometric with parameter $1 - p$ is $\frac{1}{1-p}$, this means the mean of $\text{NegBin}(1, p)$ is $\frac{1}{1-p} - 1 = \frac{p}{1-p}$. This proves the base case.

We now assume the result is true for negative binomials with parameters r and p, and show it holds for parameters $r + 1$ and p. Let $\mu_{r,p}$ denote the mean of $\text{NegBin}(r, p)$, and $\mu_{r+1,p}$ the mean of $\text{NegBin}(r + 1, p)$. Arguing as we did when we calculated the mean of a geometric random variable, we get

$$\mu_{r+1,p} = (\mu_{r+1,p} + 1)p + \mu_{r,p}(1 - p).$$

Why? Remember we want exactly $r + 1$ failures, and we are looking for the mean number *of successes!* If the first item is a success (which happens with probability p), we still need $r + 1$ failures. It's as if we started the process now, except we have to remember to add 1 for the turn we just took which gave us a success. If the first toss is a failure, we now only need r more. The expected number here is $\mu_{r,p}$. Note we *do not* add 1 like we did before, because we're counting the number of successes, and here the first toss was a failure. By the inductive hypothesis, $\mu_{r,p} = \frac{rp}{1-p}$, and substituting gives

$$\mu_{r+1,p} = (\mu_{r+1,p} + 1)p + \mu_{r,p}(1-p)$$
$$(1-p)\mu_{r+1,p} = p + rp$$
$$\mu_{r+1,p} = \frac{(r+1)p}{1-p}$$

as claimed, completing the proof.

Remark: *When I was writing this second proof, I made a mistake. I didn't have* $\mu_{r,p}(1-p)$ *but* $(\mu_{r,p}+1)(1-p)$. *This led to* $(1-p)\mu_{r+1,p} = 1+rp$ *instead of* $p+rp$. *What's nice is that I could see that I would've been fine if I had a* $p+rp$ *instead of* $1+rp$. *This allowed me to work backwards and figure out where the error was. Everyone makes mistakes. It's easy to forget something, or to mess up some algebra. If, however, you have a sense of what the answer is, you can often use that to keep yourself on track.*

The negative binomial distribution can be used for determining the probability of more than one loss in a series. For example, if we roll a die and consider a "1" as a "loss," we can calculate the number of rolls it takes before we encounter a certain number of failures. So if we want to see seven 1's (for some reason), we can determine, on average, how many rolls it will take before we see seven 1's in our outcomes. This is helpful when, say, we are calculating the number of days before a machine breaks down completely, or how many games it takes to lose the series championships.

 A great example is the effectiveness of warehouse equipment. An expensive machine is bought for a paper factory, and the manager wants to know how long the machine will work until it breaks down. If the machine has a 98% rate of working each day, independent of another day, and it takes 5 breakdowns (these are days when the machine is *not* working) before the machine cannot be used again, how many days can we expect to use the machine before it breaks down completely?

Solution: This is an example of a random variable with a negative binomial distribution; we say *a* and not *the* as there are many negative binomial distributions; we need to find the values of the parameters to determine which one it is. From the problem, we see that $r = 5$ and $p = 0.98$. Thus $\Pr(X = k) = \binom{k+r-1}{k}(p^k)(1-p)^r$, and from our formulas we have the expected value of the number of successes is

$$\mu = \frac{rp}{1-p} = \frac{5(0.98)}{0.02} = 245.$$

12.6 The Poisson Distribution

So far all the discrete distributions we've studied are related somehow to the Bernoulli distribution. Our next example is related as well, but the connection is a bit more tenuous. While we can just state the definition of a Poisson random variable, it turns out it can be defined as a limit of binomial random variables with parameters n and p such that as $n \to \infty$, $np \to \lambda$. (I dislike this notation in introductory courses, as p looks like a fixed number. I prefer to write p_n in such circumstances, and then $np_n \to \lambda$. I encourage the reader to try this in Exercise 12.8.20.)

> **Poisson distribution**: Let $\lambda > 0$. Then X is a **Poisson random variable** with parameter λ if
> $$\text{Prob}(X = n) = \begin{cases} \lambda^n e^{-\lambda}/n! & \text{if } n \in \{0, 1, 2, \dots\} \\ 0 & \text{otherwise.} \end{cases}$$
> We write $X \sim \text{Pois}(\lambda)$. The mean and the variance are both λ.

As always, we start by showing it's a probability distribution. The probabilities are clearly non-negative. To see that they sum to 1, we use the Taylor series of the exponential function:

$$e^x = \sum_{n=0}^{\infty} \frac{x^n}{n!}.$$

We have

$$\sum_{n=0}^{\infty} \text{Prob}(X = n) = \sum_{n=0}^{\infty} \frac{\lambda^n e^{-\lambda}}{n!}$$

$$= e^{-\lambda} \sum_{n=0}^{\infty} \frac{\lambda^n}{n!}$$

$$= e^{-\lambda} e^{\lambda} = 1.$$

Now, onward to the mean! We again use the **Method of Differentiating Identities**. The only identity we have is that the sum of the probabilities is 1. We have

$$1 = \sum_{n=0}^{\infty} \frac{\lambda^n e^{-\lambda}}{n!}$$

$$\lambda \frac{d}{d\lambda} 1 = \lambda \frac{d}{d\lambda} \sum_{n=0}^{\infty} \frac{\lambda^n e^{-\lambda}}{n!}$$

$$0 = \sum_{n=0}^{\infty} \frac{\lambda}{n!} \frac{d}{d\lambda} \left(\lambda^n e^{-\lambda} \right)$$

$$= \sum_{n=0}^{\infty} \frac{\lambda}{n!} \left(n\lambda^{n-1} e^{-\lambda} - \lambda^n e^{-\lambda} \right)$$

$$= \sum_{n=0}^{\infty} n \frac{\lambda^n e^{-\lambda}}{n!} - \lambda \sum_{n=0}^{\infty} \frac{\lambda^n e^{-\lambda}}{n!}.$$

The last sum on the right is 1, as this is just the sum of the probabilities of a Poisson random variable. Further, the first sum in the last line is just μ_X. Therefore

$$\mu_X = \lambda,$$

completing the proof.

A similar calculation gives the variance; I urge you to do that to make sure you have the method down cold.

 Assume that the number of customers who enter the post office today is given by a Poisson distribution with $\lambda = 1/3$. What is the probability that at least one customer enters the post office today?

Solution: If we let X denote the number of customers who enter the post office today, we are looking for $\text{Prob}(X \geq 1)$. To compute this directly we'd have to find the probability that X equals 1, or 2, or 3, and so on. This requires us to analyze a lot of cases, and then execute an infinite sum. In situations like this it's often *much* easier to calculate the probability that something *doesn't* happen, and then the probability it *does* happen is just one minus the probability it didn't occur. The probability no one enters is just the probability that X equals zero, which is $(1/3)^0 e^{-1/3}/0! = e^{-1/3}$. Thus

$$\text{Prob}(X \geq 1) = 1 - \text{Prob}(X = 0) = 1 - e^{-1/3} \approx 0.283.$$

As always, if you can write some simple code to check your answer it's worth doing. Many systems have predefined functions to allow you to sample from standard random variables; you want to be able to take advantage of this.

```
poissonpostoffice[numdo_] := Module[{},
  count = 0; (* counts number of successes *)
  For[n = 1, n <= numdo, n++,
  {
    (* randomly generate from Poisson(1/3) *)
    x = Random[PoissonDistribution[1/3]];
    If[x > 0, count = count + 1];
    }]; (* end of n loop *)
  Print["Theory predicts probability customer is ",
   100. (1 - Exp[-1/3]), "."];
  Print["Observed probability is ", 100. count/numdo, "."];
  ]
```

Running 10,000,000 trials (I'm stuck in an airport and have plenty of time) gives:

```
Theory predicts probability customer is 28.3469.
Observed probability is 28.3239.
```

 We now turn to the other type of problem. Imagine the company IKEA knows from past experience that the number of beds sold at a particular store each day has a Poisson distribution with $\lambda = 1/2$. What is a good approximation for the probability that this store sells no more than 3 beds today?

For the previous problem we found the probability that our event didn't happen, and subtracted that from 1. We did this to avoid the infinite sum. For this problem, as there are only four cases to consider (selling 0, 1, 2, or 3 beds) we evaluate them directly. Letting B be the number of beds that the store sells today, we find

$$\text{Prob}(B \leq 3) = \left(\frac{1}{2}\right)^0 \frac{e^{-1/2}}{0!} + \left(\frac{1}{2}\right)^1 \frac{e^{-1/2}}{1!} + \left(\frac{1}{2}\right)^2 \frac{e^{-1/2}}{2!} + \left(\frac{1}{2}\right)^3 \frac{e^{-1/2}}{3!}$$
$$\approx .998.$$

We end with an advanced remark about Poisson random variables. At first it might seem strange that the mean and the variance are both λ, as that might suggest to you that the mean and the variance have the same units (they do not; the mean and the standard deviation have the same units in general). What is going on? Remember the density is $\Pr(X = n) = e^{-\lambda}\lambda^n/n!$. If λ had any units, then $e^{-\lambda}$ would make no sense, as that is just $1 + \lambda + \lambda^2/2! + \lambda^3/3! + \cdots$ and we would need all the different expressions to have the same units. This is absurd, and leads to λ having the same units as λ^2.

12.7 The Discrete Uniform Distribution

It's time for one last generalization of a Bernoulli random variable. Remember there are two outcomes for a Bernoulli distribution. Let's consider the special case when $p = 1/2$, so the two outcomes are equally probable. A natural generalization is to n outcomes, each happening with probability $1/n$. (Note this is the same as a multinomial random variable where there is just one result, and the probabilities are all equal.)

Discrete uniform distribution: Let $\{a_1, a_2, \ldots, a_n\}$ be a finite set. X is a **discrete random variable** if

$$\text{Prob}(X = a) = \begin{cases} 1/n & \text{if } a \in \{a_1, a_2, \ldots, a_n\} \\ 0 & \text{otherwise.} \end{cases}$$

The most important case is when the set is $\{a, a+1, a+2, \ldots, a+n-1\}$. In that case, the mean is $a + \frac{n-1}{2}$ and the variance is $\frac{n^2-1}{12}$.

The mean is easy to find: we just add everything up and divide by n. We can, however, make our lives even easier if we stop and think about the algebra for a moment. If we take $a = 0$ our numbers are much nicer; we can do this so long as we add a back at the end to the mean. Thus the mean is

$$a + \frac{1}{n}\sum_{k=0}^{n-1} k = a + \frac{1}{n}\frac{(n-1)n}{2} = a + \frac{n-1}{2},$$

where we used the formula for sums of consecutive integers (see §A.2.1). The variance is handled similarly, with the key ingredient being the formula for the sums of squares of consecutive integers (see Exercise A.2.2).

While the two most common examples of a discrete uniform distribution are a fair coin and a fair die, there are others. My favorite is its occurrence in the German Tank Problem. The Germans had the serial numbers for their items consecutive. That ordering was useful to the Germans. For example, one could have any part with middle digits 01 be made in January, 02 in February, and so on, and thus by looking at the serial number quickly know how old a piece is and when it might be due to be replaced. Unfortunately for them but fortunately for the Allies, by looking at the serial numbers of destroyed tanks the Allies were able to accurately estimate how many tanks they could expect to face in the field!

The German Tank Problem: In World War II, the Western Allies attempted to statistically estimate the number of Panzer V tanks produced by the Germans. To do this, they collected serial numbers of destroyed tanks, analyzed the tank wheels, and estimated how many wheel molds were in use at the time. Using this information, the Allies predicted that about 270 Panzers were produced just before D-Day; in reality, 276 tanks were produced. We describe below the mathematics behind this incredibly accurate and valuable estimate.

This method of analysis predicts the maximum of a discrete uniform distribution from sampling without replacement. From its application in World War II, it is now known colloquially as the "German Tank Problem." We discuss a simpler problem where the serial numbers range from 1 to N and we are trying to find N; in the real problem it's slightly harder as we don't know the lowest number, and I leave that to you as Exercise 12.8.31.

We first set some notation. Let N be the number of tanks produced, with N as the largest serial number and 1 the smallest. We assume we record k serial numbers, with m the highest serial number observed. Our goal is to try and estimate the unknown N given our value of m and our number of observations k. Clearly our estimate must return a value at least as large as m, and almost surely larger. The difficulty is how much larger should our optimal estimate be? There are many approaches to this problem. In the one below, we calculate the expected value for M (the random variable representing the largest of k observations of a sample without replacement from $\{1, 2, \ldots, N\}$) in terms of N, and then reverse it to take our observed value of m to estimate N.

Thus we want $\Pr(M = m)$; note this is really a conditional probability as it depends on k and N, but to simplify notation we shall suppress these quantities as they are fixed in the discussion. We start by noting there are $\binom{N}{k}$ ways to choose k distinct numbers from $\{1, \ldots, N\}$. Next, we claim that the number of k-tuples where m is the largest number is $\binom{m-1}{k}$. Why? We are choosing k numbers from N where we *must* choose m and we *cannot* choose anything larger. This means that we have to choose $k - 1$ distinct numbers from $\{1, 2, \ldots, m - 1\}$, and the claim follows. We have therefore shown, assuming we have N tanks, that

$$\Pr(M = m) \;=\; \frac{\binom{m-1}{k-1}}{\binom{N}{k}} \quad \text{and} \quad \sum_{m=k}^{N} \frac{\binom{m-1}{k-1}}{\binom{N}{k}} \;=\; 1. \tag{12.7}$$

This last sum will be very important in a moment; it follows from the fact that we have a probability distribution and that m ranges from k to N.

We now calculate the expected value of M, the random variable denoting the largest observed tank value (note that M must be at least as large as k). We find

$$\mathbb{E}[M] \;=\; \sum_{m=k}^{N} m \Pr(M = m) \;=\; \sum_{m=k}^{N} m \frac{\binom{m-1}{k-1}}{\binom{N}{k}}$$

$$= \sum_{m=k}^{N} m \frac{\frac{(m-1)!}{(k-1)!(m-k)!}}{\frac{N!}{k!(N-k)!}}$$

$$= \sum_{m=k}^{N} \frac{\frac{m!}{(k-1)!(m-k)!}}{\frac{N!}{k!(N-k)!}}.$$

The challenge now is to simplify the sum above. When we wrote the formula for the probability M equals m in (12.7), remember we emphasized that the sum of these probabilities equals 1. This is always true for a probability distribution, and noting that allows us to evaluate our final sum for $\mathbb{E}[M]$ above! The reason is that what we have looks a lot like the sum of the probabilities where the maximum observed serial number is $m + 1$ and we have $k + 1$ observations. We now multiply by 1 (one of my favorite techniques) in a clever way, and then invoke the Theory of Normalization Constants to claim a subset of our sum is 1. If you haven't seen calculations like this before it can be very confusing; often people follow line by line but miss the big picture. To help you see what's going on, we include far more steps than is normally done in an attempt to highlight the ideas and techniques so that you can do similar arguments in the future. Our goal will be to rewrite the sum as a sum of the probability of observing a largest number of u from a sample of size $k + 1$ from $N + 1$. Why these choices? It's because in the formula for the expected value of M we have m times $\binom{m-1}{k-1}$; this suggests combining these two factors and multiplying in a clever way to get something nice times $\binom{m}{k} = \binom{m+1-1}{k+1-1}$, which is related to the density we mentioned.

It's time to do the algebra. We find

$$\mathbb{E}[M] = \sum_{m=k}^{N} \frac{\frac{m!}{(k-1)!(m-k)!}\frac{k}{k}}{\frac{N!}{k!(N-k)!}\frac{N+1}{N+1}\frac{k+1}{k+1}}$$

$$= \sum_{m=k}^{N} \frac{\binom{m}{k}k}{\binom{N+1}{k+1}\frac{k+1}{N+1}}$$

$$= \frac{k}{k+1}(N+1)\sum_{m=k}^{N} \frac{\binom{m}{k}}{\binom{N+1}{k+1}}$$

$$= \frac{k}{k+1}(N+1)\sum_{m=k}^{N} \frac{\binom{m+1-1}{k+1-1}}{\binom{N+1}{k+1}}$$

$$= \frac{k}{k+1}(N+1)\sum_{u=k+1}^{N+1} \frac{\binom{u-1}{k+1-1}}{\binom{N+1}{k+1}} = \frac{k}{k+1}(N+1)\cdot 1,$$

as our last sum is just summing over the probabilities that the largest tank has value u when we have a sample of $k + 1$ tanks from $N + 1$! It is worth thinking about this last part very carefully—this observation allows us to bypass some difficult combinatorial sums, as the sum of a density must be 1!

We thus have

$$\mathbb{E}[M] = \frac{k}{k+1}(N+1),$$

which yields the estimate

$$N = \frac{k+1}{k}m - 1;$$

as $m \geq k$ our estimate is at least as large as m. For m large, we essentially estimate N by inflating m by a factor of $\frac{k+1}{k}$. Not surprisingly, the larger k the less inflation we use.

Table 12.1.

Comparison of our statistical estimation to conventional estimates. From http://en.wikipedia.org/wiki/German_tank_problem.

Month	Statistical estimate	Intelligence estimate	German records
June 1940	169	1,000	122
June 1941	244	1,550	271
August 1942	327	1,550	342

How well did this work? Table 12.1 (from Wikipedia's page) shows that it was significantly better than conventional estimates.

Hopefully it's clear why this is one of my favorite problems. It has incredible real-world significance, and it involves a lot of great math and techniques. Thus, even if you don't care about this problem, it's a great review of many concepts.

There is a lot more that one can do for this problem. For example, we could find confidence intervals. In other words, given m and k, give a range of N where you're 95% sure the true value lives. Thus this is only the beginning of a fascinating story, and I encourage you to read on.

12.8 Exercises

Exercise 12.8.1   *Which of the following distributions can be symmetric: Poisson, geometric, discrete uniform, binomial, negative binomial, Bernoulli? What parameters make them symmetric?*

Exercise 12.8.2   *Find* $\mathrm{Prob}(X + Y = n)$ *for X and Y independent negative binomial random variables with parameters (r_X, p) and (r_Y, p).*

Exercise 12.8.3   *Find the CDF for a uniform discrete variable with positive probabilities assigned to the set $\{1, 2, 3, \ldots, n\}$.*

Exercise 12.8.4   *Find the CDF for a geometric random variable with parameter p.*

Exercise 12.8.5   *Find $\mathbb{E}[1/X]$ for a geometric distribution with parameter p.*

Exercise 12.8.6   *Imagine we flip a fair coin 100 times. What do we expect the sum of the position of the first head and the number of heads to be approximately? Is the true expected value higher, equal, or lower to that?*

Exercise 12.8.7   *Consider the game "Rock, Paper, Scissors" in which each player picks rock, paper, or scissors independently, each with probability 1/3. All you need to know about the game for this exercise is that if both players pick the same option the game continues, or else it ends and there is a winner. How many turns do we expect the game to take?*

Exercise 12.8.8   *Show without using convolutions that the sum of n geometric distributions with parameter p is a negative binomial distribution with parameters n, p. You may assume that the sum of two negative binomials is a negative binomial with parameters $n_1 + n_2, p$.*

Exercise 12.8.9   *Imagine we are taking a random walk starting at 0 and moving right 1 with probability p. How many turns do we expect for it to take us to reach 10? What about 100?*

Exercise 12.8.10 *Repeat the previous exercise, but this time on the ith possible move we move up i (passing 10 counts as reaching 10).*

Exercise 12.8.11 *If instead we move up with probability p and down with probability $1 - p$ what is the probability we return to point 0 after n steps? (Hint: Consider odd and even n separately.)*

Exercise 12.8.12 *Show that in the random walk described in the last exercise we will eventually cross the origin with probability 1 as we let the number of steps increase to infinity as long as p does not equal 0 or 1 (in fact, we will cross the origin arbitrarily many times).*

Exercise 12.8.13 *Why can we not use a geometric distribution to model our wait time for picking an ace in a deck of cards?*

Exercise 12.8.14 *Imagine we have a deck with 1000 cards numbered 1 through 250 in each suit. Would a geometric model be a good approximation for how long it takes us to get a 2? What about how long it takes us to get a red card?*

Exercise 12.8.15 *Suppose we have a random variable, X, which is distributed according to a geometric distribution with parameter p. Consider another random variable Y which is also distributed according to a geometric distribution with parameter p'. What value of p' maximizes $P(X = Y)$ for a fixed p?*

Exercise 12.8.16 *Find the probability distribution for the sum of 5 independent random variables that take on the value 3 with probability p and 5 with probability $1 - p$.*

Exercise 12.8.17 *Let X be a geometric random variable. Show that*

$$\text{Prob}(X > x + a | X > a) \; = \; \text{Prob}(X > x).$$

Exercise 12.8.18 *Find the probability distribution for the sum of 5 independent uniform random variables that are supported on $\{1, 2, 3\}$.*

Exercise 12.8.19 *Imagine you are the commissioner of a new sports league and are trying to decide how many games should be in the final series. Team A wins each game with probability p and Team B wins with probability q. Assume $p > q$, $p + q = 1$, and the outcome of each game is independent of the previous games. The series must have an odd number of games. In a series with n games, the first team to win $\frac{n+1}{2}$ games wins the series. Write down an inequality to solve for how many games must be in the series for the team with higher probability of winning to win at least 90% of the time.*

Exercise 12.8.20 *Consider a sequence of binomial random variables X_n, where X_n has probability p_n of success, $1 - p_n$ of failure, n outcomes, and as $n \to \infty$, $np_n \to \lambda$. Prove that as $n \to \infty$ the probability $X_n = k$ for any fixed k approaches that of a Poisson random variable with parameter λ.*

Exercise 12.8.21 *Imagine we continually flip a fair coin until we get a tail. We assign a value of i to the ith head for each head. What is the expected total sum?*

Exercise 12.8.22 *Find the third moment and the third centered moment for a geometric distribution with parameter p.*

Exercise 12.8.23 *Find the third centered moment (skewness) for a Poisson distribution with parameter λ.*

Exercise 12.8.24 *Find the median of a geometric distribution with parameter p.*

Exercise 12.8.25 *Sam is working on a politician's campaign. It is known that the true percentage of people supporting the politician is 52%, and the true percentage of people supporting her opponent is 48%. However, voter turnout is only 50% regardless of which candidate the person plans to vote for. There are 3000 people in the district. Estimate the probability Sam's candidate wins.*

Exercise 12.8.26 *Plot the cumulative density functions for binomial random variables with parameters $n = 10$ and $p = .2, .4, .6, .8$. (It might take a couple times to get the shading right so that it is easily readable.)*

Exercise 12.8.27 *Consider a grocery store with 5 horizontal aisles and 2 vertical aisles.*

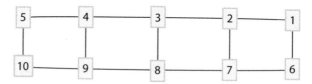

Imagine that you and your friend become separated and neither of you knows where the other is. You are at location 1. Each move consists of you randomly selecting an adjacent vertex to walk to, with equal probability assigned to each immediately accessible vertex. For example, if you are at vertex 9, you would move to vertex 10 with probability 1/3, to vertex 8 with probability 1/3 and to vertex 4 with probability 1/3. How many moves does it take on average for you to reach vertex 10, where your friend is standing? For example, if you are at vertex 9, you would move to vertex 10 with probability 1/3, to vertex 8 with probability 1/3 and to vertex 4 with probability 1/2. Use a simulation to model this situation.

Exercise 12.8.28 *Repeat the previous exercise, but assume this time you and your friend take turns moving, and each move is randomly selected from all of your possible move options.*

Exercise 12.8.29 *Candidate A is expected to receive 65% of the votes, Candidate B is expected to receive 20% of the votes, Candidate C is expected to receive 10% of the votes, and Candidate D expects, then, 5% of the population votes. If a sample size of 100 is taken, what is the probability that the group is comprised of 50 votes for A, 30 votes for B, 15 votes for C, and 5 votes for D?*

Exercise 12.8.30 *Write a program to randomly choose an integer N and then sample without replacement k elements from $\{1, 2, \ldots, N\}$. Compare the estimate for N from the German Tank Problem, $\frac{k+1}{k}m - 1$.*

Exercise 12.8.31 *Consider the more general case of the German Tank Problem, where now the serial numbers are chosen from $\{N_{\min}, N_{\min} + 1, \ldots, N_{\max}\}$. If the largest observed tank has serial number $m_{\max}$ and the smallest is $m_{\min}$, what is your estimate for $N_{\max} - N_{\min}$? (Note: Do you want to try to estimate $N_{\max}$ and $N_{\min}$ directly, or their difference? Does it matter? At the end of the day all we care about is how separated they are, but perhaps, similar to Lagrange multipliers and the parameter λ, sometimes it is easier to calculate more than is needed.)*

CHAPTER 13

Continuous Random Variables: Uniform and Exponential

We now use the tools from Chapters 9 to 11 to study different continuous random variables. We'll see that many random variables that at first look quite different can be shown to live in a "family." This means that there are some parameters whose values we get to choose, and different choices lead to the different random variables, often with similar properties. We've already seen how useful it can be to have a free parameter (see Chapter 11 and differentiating identities).

In this chapter we look at two of the most important distributions, the uniform and the exponential. We show how to compute the moments (especially the mean and variance), sums of such random variables, and finally discuss how to generate random numbers following these distributions.

13.1 The Uniform Distribution

It's time to look at some specific distributions! While different textbooks often list the "common" continuous random variables in a different order (and, in fact, might even have slight disagreements as to which densities should and should not be on the list), most begin with a study of the uniform distribution. In some sense it's the simplest and most natural. It's the continuous analogue of flipping a fair coin or rolling a fair die.

Uniform Distribution: We say a random variable X has the uniform distribution on $[a, b]$ (with $-\infty < a < b < \infty$) if

$$f_X(x) = \begin{cases} \frac{1}{b-a} & \text{if } a \leq x \leq b \\ 0 & \text{otherwise.} \end{cases}$$

We denote this by writing $X \sim \text{Unif}(a, b)$.

We'll prove a lot of properties of the uniform distribution. It's a bit misleading to write "the" uniform distribution, of course, as there are infinitely many. Fortunately, it's easy to write down the a and b dependence for many of the properties, and thus we can study "all" the uniform distributions simultaneously. Or, alternatively, we can **standardize** the random variable: if $X \sim \text{Unif}(a, b)$ and $U \sim \text{Unif}(0, 1)$ then $X = (b - a)U + a$, which means if we understand a uniform random variable on $[0, 1]$ we can easily transfer that knowledge to any other.

13.1.1 Mean and Variance

Whenever you meet a distribution, the first thing you should determine is its mean and variance, so let's get to it!

Let $X \sim \text{Unif}(a, b)$. Then the mean μ_X and variance σ_X^2 are

$$\mu_X = \frac{b + a}{2} \quad \text{and} \quad \sigma_X^2 = \frac{(b - a)^2}{12}.$$

In the special case that $[a, b] = [0, 1]$ the mean is $1/2$ and the variance is $1/12$, while if $[a, b] = [-1/2, 1/2]$ the mean is 0 and the variance is $1/12$. Another important case is when $[a, b] = [-\sqrt{3}, \sqrt{3}]$, as in this case the mean is 0 and the variance is 1.

The proof follows from straightforward integration. We'll do this for arbitrary a and b so that we can handle all the uniform distributions at once; if you're more comfortable with specific numbers, just mimic these arguments with your favorite values of a and b; we've listed three of the most important.

The mean is

$$\mu_X = \int_{-\infty}^{\infty} x f_X(x) dx$$

$$= \int_{a}^{b} x \cdot \frac{1}{b - a} dx$$

$$= \frac{1}{b - a} \frac{x^2}{2} \Big|_{a}^{b}$$

$$= \frac{1}{b - a} \frac{b^2 - a^2}{2}$$

$$= \frac{1}{b - a} \frac{(b - a)(b + a)}{2}$$

$$= \frac{b + a}{2}.$$

We now turn to computing the variance. There are two approaches commonly taken: compute $\mathbb{E}[(X - \mu_X)^2]$ or compute $\mathbb{E}[X^2] - \mathbb{E}[X]^2$. The calculations are fairly similar, both leading to standard Calculus I integrals. Let's do the first approach. We won't substitute the value of μ_X until the end, when we need to do some algebra to isolate a nice answer from our computation. We have

$$\sigma_X^2 = \int_{-\infty}^{\infty} (x - \mu_X)^2 f_X(x) dx$$

$$= \int_a^b (x - \mu_X)^2 \cdot \frac{1}{b-a} dx$$

$$= \frac{1}{b-a} \left. \frac{(x - \mu_X)^3}{3} \right|_a^b$$

$$= \frac{1}{b-a} \frac{(b - \mu_X)^3 - (a - \mu_X)^3}{3}.$$

At this point, it's impossible to go further without knowing that $\mu_X = (b+a)/2$, which means $b - \mu_X = (b-a)/2$ and $a - \mu_X = -(b-a)/2$. We now have to roll up our sleeves and do some algebra (or, what is more likely in the twenty-first century, have some computer program do the algebra for us!). We find

$$\sigma_X^2 = \frac{1}{3(b-a)} \left[\left(\frac{b-a}{2} \right)^3 - \left(-\frac{b-a}{2} \right)^3 \right]$$

$$= \frac{1}{24(b-a)} \left[(b-a)^3 + (b-a)^3 \right]$$

$$= \frac{(b-a)^2}{12}.$$

The rest of the claims follow from substituting these special values into the formulas. The final choice, namely $[a, b] = [-\sqrt{3}, \sqrt{3}]$, was chosen as this gives a mean 0, variance 1 uniform random variable. As there are a lot of applications where we want to standardize our variables to have mean 0 and variance 1, we record this for the future.

Our Pavlovian response now kicks in—we have an answer, is it reasonable? The mean is readily checked. It's halfway between a and b, which makes sense. Further, if we double a and b then the mean doubles, exactly as we would predict.

What about the variance? There are a lot of nice things about our final formula. The first is that it only depends on the difference of b and a. A moment's reflection shows that we could have predicted this. If we add 100 to both a and b, we change the mean but we don't change the magnitude of the fluctuations about the mean. The next item to observe is that it depends on the square of $b - a$. This has two nice features. First, it ensures that the variance is non-negative. Second, the variance has the correct units (if a and b are in meters, the variance is in meters-squared, and hence the standard deviation is in meters). Finally, the standard deviation, which is the square-root of the variance, is $(b-a)/\sqrt{12}$. Notice that this is less than $b - a$. As the furthest apart two values can be is $b - a$, the standard deviation *should* be smaller than that. Again, if you have the opportunity to do a simple test to make sure your answer is reasonable, it's worth doing this. In addition to checking for errors, it often helps build intuition.

13.1.2 Sums of Uniform Random Variables

There's one more question we'd like to answer about uniform random variables: What is the distribution of sums of uniform random variables? For simplicity, let's assume X and Y are uniform on $[0, 1]$; if we can handle this case then we should be able to generalize to any uniform random variable.

Let X and Y be independent, Unif(0, 1) random variables. Then the density of $Z = X + Y$ is

$$f_Z(z) = \begin{cases} z & \text{if } 0 \le z \le 1 \\ 2 - z & \text{if } 1 \le z \le 2 \\ 0 & \text{otherwise.} \end{cases}$$

The Theory of Convolutions (Chapter 10) gives us the answer: it's just

$$f_Z(z) = \int_{-\infty}^{\infty} f(t)f(z - t)dt, \quad f(u) = \begin{cases} 1 & \text{if } 0 \le u \le 1 \\ 0 & \text{otherwise.} \end{cases} \tag{13.1}$$

We just have to roll up our sleeves and evaluate this integral. A little thought shows that there are three ranges of z to consider: $z \le 0$, $0 \le z \le 2$, and $z \ge 2$. Why these three? Well, our uniform random variables live in $[0, 1]$. Thus the density can only be non-zero for $z \in [0, 2]$, and it becomes natural to break up the calculation to these three regions. We'll show that the probability is zero for $z \le 0$; a similar argument shows the probability is zero when $z \ge 2$. After showing this, we'll be left with the hard part, namely $0 \le z \le 2$.

Let's assume $z \le 0$. In (13.1) the integrand is $f(t)f(z - t)$. In order for the integrand to be non-zero, we need $0 \le t \le 1$ and $0 \le z - t \le 1$, and the latter is the same as $z - 1 \le t \le z$. If $z \le 0$ then the second condition forces $t \le 0$, and thus the first condition cannot be satisfied. Thus, when $z \le 0$, the integrand is always zero and hence the probability is zero.

We are now left with what happens for $0 \le z \le 2$. As before, the integrand is zero unless $0 \le t \le 1$ and $z - 1 \le t \le z$. As $0 \le z \le 2$, for each z there are always some t such that both conditions are met. Don't be upset, but it's easiest if we break this case into subcases! Arg—subcases of cases! Is this really necessary? Sadly, yes. While the first condition is nice, the second condition has z-dependence. We know the first condition is met, so $0 \le t \le 1$. If $0 \le z \le 1$ then $z - 1 \le t$ is trivially satisfied, as $z - 1 \le 0$; in this case, it's the upper bound that is non-trivial, as that forces $t \le z$. If instead $z \ge 1$ then the upper bound $t \le z$ is trivially satisfied and it's the lower bound of $z - 1 \le t$ that is non-trivial. Thus, it's easier to break things into cases.

We'll just do the case $0 \le z \le 1$. The analysis of the other case follows similarly, and we'll leave that to you (though we'll give a few words after our analysis of how to immediately deduce that subcase from this subcase).

To recap, we're reduced to considering the subcase where $0 \le z \le 1$. We need to integrate over all t such that $0 \le t \le 1$ and $z - 1 \le t \le z$. The first condition

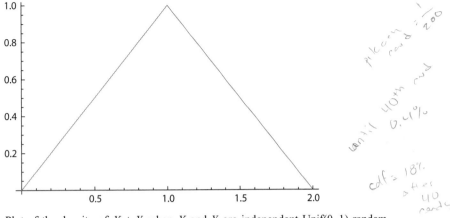

Figure 13.1. Plot of the density of $X + Y$ when X and Y are independent Unif(0, 1) random variables.

restricts $t \in [0, 1]$, while the second condition further restricts us to have $t \in [0, z]$. Both conditions must hold, so the only t where the integrand $f(t)f(z - t)$ is non-zero are $t \in [0, z]$. We thus find

$$f_Z(z) = \int_0^z 1 \cdot 1 dt = z.$$

As promised, we add a few words on how we could quickly get that the answer is $2 - z$ for the subcase $z \geq 1$ without resorting to the integration. The trick is to note that $\widetilde{X} = 1 - X$ and $\widetilde{Y} = 1 - Y$ have the *same* densities as X and Y! This is because our densities are symmetric about 1/2. Why is this a good observation? Well, $\widetilde{X}$ and $\widetilde{Y}$ are also both Unif(0, 1), and when $\widetilde{X} + \widetilde{Y} \in [0, 1]$ then $X + Y = 2 - (\widetilde{X} + \widetilde{Y}) \in [1, 2]$. Our arguments above show that the probability that $\widetilde{X} + \widetilde{Y} = u \in [0, 1]$ is just u. If we write $u = 2 - z$ for $z \in [1, 2]$, we then see that the probability that $X + Y = z$ is just $2 - z$ as claimed! These **symmetric arguments** are very powerful, and frequently can save you lots of tedious, duplicate calculations.

We plot the density of $X + Y$ (when both are independent Unif(0, 1) random variables) in Figure 13.1.

Notice the result is a nice triangular function! If we think back to sums of two fair die, this result shouldn't be surprising. We had a triangular function there, with the middle event of a 7 happening with probability 6/36, linearly decreasing to the extremes of 2 and 12 happening with probabilities 1/36 (the probability of getting r is $\frac{6 - |7 - r|}{12}$, a triangle centered at 7).

What if we convolve again? We plot the answer in Figure 13.2, as well as the result of summing eight independent Unif(0, 1) random variables, and compare that to the corresponding normal distribution.

One of the biggest results in probability is that sums of "nice" independent random variables converge to being normally distributed. We can begin to see that with just eight uniform random variables; the corresponding normal (which has the same mean and variance) is an excellent fit!

While it's possible to explicitly write down the density for the sum of an arbitrary number of independent uniform random variables, the resulting density is only piecewise continuous, and as the number of summands increases so too does the number of different regions.

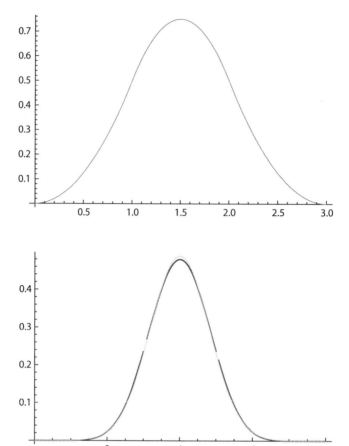

Figure 13.2. Plot of the density of $X_1 + \cdots + X_n$ when the X_i's are independent Unif(0, 1) random variables: (top) $n = 3$; (bottom) $n = 8$, and also a normal distribution with the same mean and variance.

13.1.3 Examples

 We now reap the benefits of all our work. Let's calculate the probability that if X and Y are independent Unif(0, 1) random variables then $Z = X + Y$ is between 1/2 and 3/2. We know the density from above; it's $f_Z(z) = z$ for $0 \le z \le 1$ and $2 - z$ for $1 \le z \le 2$; as the form of the density changes at 1, it suggests we break the integral into two parts. Thus

$$\text{Prob}(1/2 \le Z \le 3/2) = \int_{1/2}^{3/2} f_Z(z)dz$$

$$= \int_{1/2}^{1} f_Z(z)dz + \int_{1}^{3/2} f_Z(z)dz$$

$$= \int_{1/2}^{1} z\,dz + \int_{1}^{3/2} (2 - z)dz$$

$$= \left.\frac{z^2}{2}\right|_{1/2}^{1} + \left.\frac{-(2-z)^2}{2}\right|_{1}^{3/2}$$

$$= \left(\frac{1}{2} - \frac{1}{8}\right) + \left(-\frac{1}{8} + \frac{1}{2}\right) = \frac{3}{4}.$$

Let's do one more example. Assume now $\widetilde{X}$ and $\widetilde{Y}$ are independent Unif(1, 3), and we want to find the probability that $\widetilde{Z} = \widetilde{X} + \widetilde{Y}$ is in [3, 5]. We could of course just mimic the above calculations, but there is a faster way! We can reduce this to the problem we've just solved by a change of variables (see §10.4), justifying all the time we spent in that section developing the Change of Variables Formula.

We have $Z = X + Y$ with X, Y independent Unif(0, 1), and $\widetilde{Z} = \widetilde{X} + \widetilde{Y}$ with $\widetilde{X}$, $\widetilde{Y}$ independent Unif(1, 3). We have

$$\widetilde{X} = 2X + 1$$

$$\widetilde{Y} = 2Y + 1$$

$$\widetilde{Z} = 2(X + Y) + 2 = 2Z + 2.$$

We thus have $\widetilde{Z} = g(Z)$ with $g(z) = 2z + 2$ on the interval $I = [0, 2]$. Note $g'(z) = 2$ so g is strictly increasing, and the inverse function is $h(\widetilde{z}) = (\widetilde{z} - 2)/2$. How did we find this? We have $\widetilde{z} = g(z) = 2z + 2$; we now solve for z in terms of $\widetilde{z}$. The claim now follows from algebra. We can easily find $h'(\widetilde{z})$; it's just $1/2$.

According to the Change of Variables Formula,

$$f_{\widetilde{Z}}(\widetilde{z}) = f_Z(h(\widetilde{z})) \cdot h'(\widetilde{z}).$$

It's very important to remember the h' factor above; we're not just substituting this into the density (we'll remark on this more in a moment). From above, $f_Z(z) = 0$ if $z \le 0$ or $z \ge 2$, and is z for $0 \le z \le 1$ and $2 - z$ for $1 \le z \le 2$. If $0 \le h(\widetilde{z}) \le 1$ then $g(0) \le \widetilde{z} \le g(1)$ or $2 \le \widetilde{z} \le 4$; similarly if $1 \le h(\widetilde{z}) \le 2$ then $4 \le \widetilde{z} \le 6$. We thus find

$$f_{\widetilde{Z}}(\widetilde{z}) = \begin{cases} \dfrac{\widetilde{z} - 2}{4} & \text{if } 2 \le \widetilde{z} \le 4 \\[2mm] 1 - \dfrac{\widetilde{z} - 2}{4} & \text{if } 4 \le \widetilde{z} \le 6 \\[2mm] 0 & \text{otherwise.} \end{cases}$$

Now that we have the density, we can find the desired probability by direct integration.

Actually, there is no need to write down the density $f_{\widetilde{Z}}$ explicitly; we can change variables and work directly with f_Z. We have

$$\text{Prob}(3 \le \widetilde{Z} \le 5) = \text{Prob}(3 \le 2Z + 2 \le 5)$$

$$= \text{Prob}(1/2 \le Z \le 3/2)$$

$$= \int_{1/2}^{3/2} f_Z(z)\,dz;$$

this, however, is exactly the integral we did in the previous example, and thus the answer is again just 3/4.

It's worth dwelling a bit on the h' factor in the chain rule; it's omission is one of the most common mistakes. Imagine we have a function $A(x) = f(h(x))$. The chain rule says $A'(x) = f'(h(x)) \cdot h'(x)$. Most people remember to evaluate f' at $h(x)$ (as f is evaluated at $h(x)$, it makes sense its derivative is evaluated there as well). The difficulty is remembering the $h'(x)$. If $h(x) = 2x$, you can think of this as we're traveling at twice the speed, and thus our derivative should have this factor of 2 as well. If that argument doesn't help, here's another one. Imagine $f(x) = x^n$ and $h(x) = x$. Then $A(x) = x^n$ and $A'(x) = nx^{n-1}\frac{d}{dx}x$; the last factor is just 1, so we don't see it if $h(x) = x$, but it's there.

13.1.4 Generating Random Numbers Uniformly

There's a pattern to the various sections of this chapter, though it shouldn't yet be apparent as we've only looked at one distribution. What we'll do is for each distribution we'll first calculate its mean and variance, then examine sums of independent identically distributed random variables, move on to some examples and applications, and finally conclude with a brief discussion as to how one would generate random variables from such a distribution in practice.

This subsection will be shorter than the others, as we assume the existence of a random number generator which can give us a Unif(0, 1) random variable. In a math book, it's so easy to get caught up with conditions such as "let X be uniformly distributed on $[0, 1]$" that we forget about what it would take to actually *do* this in practice. It turns out that if we can generate a random variable that is truly Unif(0, 1), then we can generate many other continuous random variables with ease by the Change of Variables Formula. We'll discuss this in greater detail for the other sections; for now, we'll content ourselves with a brief discussion of generating from the uniform distribution on $[0, 1]$. For more on generating random numbers from various distributions, see http://www.random.org/, especially the background at http://www.random.org/randomness/.

Here's a nice link to a webpage by Chris Wetzel, http://faculty.rhodes.edu/wetzel/random/intro.html, which in addition to providing background reading also gives you the option to generate what you believe are random sequences, and have the computer run some of the standard tests.

We end by briefly thinking about how a computer might generate a uniform random variable on $[0, 1]$. If we had a perfectly fair coin, we could flip it n times and let $a_i = 1$ if the ith toss is a head and a 0 if it's a tail. We then form the number

$$\frac{a_1}{2} + \frac{a_2}{2^2} + \frac{a_3}{2^3} + \cdots + \frac{a_n}{2^n} \in [0, 1].$$

If our coin were fair, this process would generate a random number uniformly from the set

$$\left\{0, \frac{1}{2^n}, \frac{2}{2^n}, \frac{3}{2^n}, \ldots, \frac{2^n - 1}{2^n}\right\}.$$

Sadly, we cannot just send $n \to \infty$, as we're working on a computer. We may only do finitely many steps in a finite amount of time. Now, if n is quite large, this discrete set is for all practical purposes indistinguishable from a uniform distribution on $[0, 1]$.

Imagine, for example, that $n = 100,000$. Then $2^{100,000} > 10^{30,000}$, and it's hard to imagine us being able to do any physical measurement accurate to 30,000 orders of magnitude! That said, just because we can't measure anything this accurately does not mean that there is *no* difference between this approximation and a truly uniform random variable. There are instances in mathematics and physics where, *in the limit*, there are profound differences between rational and irrational numbers, even if the irrational numbers are extremely close (in some sense) to a rational number. For an interesting example, look up **Lissajous figures** from physics. For another, look up **Kronecker's theorem** in number theory/ergodic theory (see for example [MT-B]). It says that if α is irrational then $n\alpha$ mod 1 is equidistributed; the **Liouville number** $\sum_{n=1}^{\infty} 10^{-n!}$ is irrational (actually it's transcendental), but it is *very* hard to distinguish it from its rational approximations.

13.2 The Exponential Distribution

In the previous section we looked at uniform random variables. Now we turn our sights to exponential random variables. There are some new features that emerge here that we didn't have to worry about with uniform random variables. The most important difference is that these random variables will not have compact support. What this means is that there is no finite interval such that all of the probability lies in that interval. We'll have to deal with infinities, and a little more work is needed to justify some of the (extremely reasonable appearing) steps.

Exponential Distribution: We say a random variable X has the exponential distribution with parameter $\lambda > 0$ if

$$f_X(x) = \begin{cases} \frac{1}{\lambda} e^{-x/\lambda} & \text{if } x \geq 0 \\ 0 & \text{otherwise.} \end{cases}$$

We denote this by $X \sim \text{Exp}(\lambda)$.
Note: Unfortunately, this notation isn't standard; some books use $e^{-\lambda x}\lambda$ for the exponential random variable with parameter λ; these two definitions are of course related, but one should choose a definition and stick with it.

We now go through the same litany of steps and results as in the random variables of the previous section.

13.2.1 Mean and Variance

Let $X \sim \text{Exp}(\lambda)$. Then the mean μ_X and variance σ_X^2 are

$$\mu_X = \lambda \quad \text{and} \quad \sigma_X^2 = \lambda^2.$$

In the special case $\lambda = 1$, the mean and the variance are both 1.

Similar to the uniform distribution, the proof follows from lots of integrating. We find

$$\mu_X = \int_{-\infty}^{\infty} x f_X(x) dx = \int_0^{\infty} x \frac{1}{\lambda} e^{-x/\lambda} dx.$$

We're now left with finding the best way to handle this integral. The first thing we should note is that the λ dependence is nice. If we let $t = x/\lambda$ so $dt = dx/\lambda$, we can pull out the λ dependence, finding

$$\mu_X = \lambda \int_0^{\infty} t e^{-t} dt.$$

What's really nice about this is that we can see how all the different exponential distributions are related. The λ dependence is really quite minor; the difficulty comes down to evaluating one integral. If we know the Gamma function (see Chapter 15), we would notice that this integral is just $\Gamma(2) = 1$; we'll discuss this approach later.

The more common approach, of course, is to integrate by parts, which says

$$\int_0^{\infty} u\, dv = uv \Big|_0^{\infty} - \int_0^{\infty} v\, du.$$

We take $u = t, du = dt, dv = e^{-t} dt$, and $v = -e^{-t}$, which gives

$$\begin{aligned}
\mu_X &= \lambda \left[uv \Big|_0^{\infty} - \int_0^{\infty} v\, du \right] \\
&= \lambda \left[-t e^{-t} \Big|_0^{\infty} + \int_0^{\infty} e^{-t} dt \right] \\
&= \lambda \left[(-0 + 0) - e^{-t} \Big|_0^{\infty} \right] \\
&= \lambda [0 + 1] = \lambda.
\end{aligned}$$

Certain parts of this calculation often cause confusion, especially if it's been a while since you've done calculus. The first, of course, is deciding to use integration by parts, and then after making this decision determining how to choose u and v. It's very natural to integrate by parts. The reason is that we know how to integrate each factor separately; the hope is that integration by parts will help us deal with their product. There are two choices: we could take $u = t$ and $dv = e^{-t} dt$, or $u = e^{-t}$ and $dv = t dt$. Which choice is better? The first. The reason is that if we take the second choice, then after integration by parts we have a t^2. This is the wrong direction to go; we started with an exponential times a polynomial of degree 1 and ended with an exponential times a polynomial of degree 2! We want the power of the polynomial to be going down and not up, and thus we take $u = t$.

There's another part of our work that is screaming for justification, namely the claim

$$-t e^{-t} \Big|_0^{\infty} = 0.$$

Half of this is no problem; taking $t = 0$ clearly gives $-0 e^{-0} = 0$; however, how do we make sense of $-\infty e^{-\infty}$? The trick is to remember **L'Hôpital's Rule**, which says that if $\lim_{x \to x_0} f(x)/g(x)$ is $0/0$ or ∞/∞ and f and g are differentiable then $\lim_{x \to x_0} f(x)/g(x) = \lim_{x \to x_0} f'(x)/g'(x)$ (and if *that* is still $0/0$ or ∞/∞, we L'Hôpital again and again and again until we no longer have one of these two fractions); note that x_0 may be a finite number or infinity. Using that, we have

$$\lim_{t \to \infty} te^{-t} = \lim_{t \to \infty} \frac{t}{e^t} = \lim_{t \to \infty} \frac{1}{e^t} = 0.$$

We have proved that the mean is λ; also, we have shown that

$$\int_0^\infty te^{-t}dt = 1. \tag{13.2}$$

This fact will be very important in determining the variance and is worthy of being isolated. Note there is a nice probabilistic interpretation, namely the mean of an exponential random variable with parameter 1 is just 1.

The variance is computed similarly. We have

$$\sigma_X^2 = \int_{-\infty}^\infty (x - \mu_X)^2 f_X(x)dx$$

$$= \int_0^\infty (x - \lambda)^2 \frac{1}{\lambda} e^{-x/\lambda}dx$$

$$= \int_0^\infty \lambda^2 \left(\frac{x}{\lambda} - 1\right)^2 e^{-x/\lambda} \frac{dx}{\lambda}$$

$$= \lambda^2 \int_0^\infty (t - 1)^2 e^{-u}dt,$$

where we changed variables by setting $t = x/\lambda$.

There are several ways to proceed. The most direct is to just bite the bullet, roll up our sleeves, and prepare ourselves for *two* integrations by parts. If we take $u = (t-1)^2$ and $dv = e^{-t}dt$, then if we integrate by parts once we'll be left with something nice times a te^{-t}, which we know how to evaluate from our mean calculation. Now, to the details. We have

$$u = (t - 1)^2 \quad \text{and} \quad dv = e^{-t}dt,$$

so

$$du = 2(t - 1)dt \quad \text{and} \quad v = -e^{-t}.$$

As

$$\int_0^\infty u\,dv = uv\Big|_0^\infty - \int_0^\infty v\,du,$$

we find

$$\sigma_X^2 = \lambda^2 \left[(t-1)^2(-e^{-t})\Big|_0^\infty + \int_0^\infty 2(t-1)e^{-t}dt\right]$$

$$= \lambda^2 \left[1 + 2\int_0^\infty (t-1)e^{-t}dt\right],$$

where we used L'Hôpital's rule to see that

$$\lim_{t\to\infty}(t-1)e^{-t} \;=\; \lim_{t\to\infty}\frac{t-1}{e^t} \;=\; \lim_{t\to\infty}\frac{1}{e^t} \;=\; 0.$$

Sadly, we're still not done; we have another integral to do! One option, of course, is to integrate by parts again, setting $u = t - 1$ and $dv = e^{-t}dt$. There is another way; we can expand the remaining integral and notice that it's just the difference of the mean (of an exponential random variable with parameter 1) and the integral of the probability density of an exponential random variable with parameter 1. The first integral was done in (13.2) and was shown to equal 1, while the second is just 1 (as all probability densities integrate to 1). Thus, the remaining integral vanishes.

If you're not comfortable with this approach, you should of course just do the integration by parts again. Whenever possible, it's nice to be able to **reduce a calculation to something you've already done**, as this cuts down on how much of your life you have to spend doing algebra or integration!

As promised, we now discuss another approach to this problem. It involves the **Gamma function** $\Gamma(s)$. We discuss this function in great detail in Chapter 15; for now, what matters most is that

$$\Gamma(s) \;=\; \int_0^{\infty} e^t t^{s-1}dt \quad \text{for } \mathrm{Re}(s) > 0$$

and

$$\Gamma(n+1) \;=\; n! \quad \text{for } n \text{ a positive integer.}$$

Using these two facts, we can easily find the variance. It's just

$$\sigma_X^2 \;=\; \lambda^2 \int_0^{\infty} (t-1)^2 e^{-t}dt$$

$$= \;\lambda^2 \int_0^{\infty} (t^2 - 2t + 1)e^{-t}dt$$

$$= \;\lambda^2 \left[\int_0^{\infty} t^2 e^{-t}dt - 2\int_0^{\infty} te^{-t}dt + \int_0^{\infty} e^{-t}dt \right]$$

$$= \;\lambda^2 \left[\Gamma(3) - 2\Gamma(2) + \Gamma(1) \right].$$

We'll show why this is true for one of the three terms, and leave the other two to you:

$$\int_0^{\infty} t^2 e^{-t}dt \;=\; \int_0^{\infty} e^{-t} t^{3-1}dt \;=\; \Gamma(3).$$

Using the second property of the Gamma function, we find

$$\sigma_X^2 \;=\; \lambda^2 \left[2! - 2\cdot 1! + 0! \right] \;=\; \lambda^2.$$

Of course, we ended up with the same answer, but now the heavy lifting is being done for us in the background; specifically, we're reaping the dividends of learning about the Gamma function. I wanted to show you this approach as it's a great example of a general principle: *do a calculation in general, and then pluck off the cases you need by specializing the parameter.* This is similar to differentiating polynomials. No one memorizes or looks at the derivative of say $f(x) = 1701x^{1017} - 24601x^{123} + 314x^{15} - 2718$. What we do is we learn what the derivative of x^n is, we learn the derivative of the sum is the sum of the derivatives, and then use these facts to immediately get the derivative of $f(x)$. It's a similar story here—by spending time doing computations for the Gamma function once, we can reap the benefits later.

13.2.2 Sums of Exponential Random Variables

Amazingly, there is a nice, clean formula for the sum of n independent exponential random variables. Not surprisingly, it's also one of the "named" distributions, though it isn't as well-known as many of the other densities you'll meet in a first course. It's called the Erlang distribution. It originally surfaced in Erlang's work analyzing the distribution of telephone calls and is now used extensively in queuing theory.

Erlang Distribution: Let $X_1, \ldots, X_n$ be n independent identically distributed exponential random variables with parameter $\lambda > 0$. Then the density of $X = X_1 + \cdots + X_n$ is

$$f_X(x) = \begin{cases} \dfrac{x^{n-1}e^{-x/\lambda}}{\lambda^n(n-1)!} & \text{if } x \geq 0 \\ \\ 0 & \text{otherwise.} \end{cases}$$

We say any random variable with this density f_X is an Erlang distribution with parameters λ and n; it has a mean of $n\lambda$ and a variance of $n\lambda^2$.

Warning! Similar to the exponential, some references use $1/\lambda$ instead of λ, so you have to be careful and check before using any reference

---

We'll prove the formula when $n = 2$. If you want some practice to test how well this material clicks, you should do Exercise 13.3.22 and generalize these arguments for any n. There is a huge advantage in knowing what the answer is—this can frequently be used to help guide you in terms of how to manage the algebra. After reading the computation below, you should attempt the general case by induction.

Now to the details for the $n = 2$ case. We again use the fact that the density function of the sum of two independent random variables is given by the convolution integral. Let

$$f(x) = \begin{cases} e^{-x} & \text{if } x \geq 0 \\ 0 & \text{otherwise} \end{cases}$$

denote the density of the standard exponential random variable. Then $X = X_1 + X_2$ has density

$$f_X(x) = \int_{-\infty}^{\infty} f(t)f(x-t)dt.$$

At this point, remembering the nightmarish answers we found in the case of sums of uniform random variables, you should be quite nervous and worried about the algebra zoo that seems ready to unfold; fortunately, this case is actually *much* easier! The reason is that the uniform had *two* points where the functional form of the definition changed, while the exponential has only one. This means we have fewer cases to analyze, and the algebra won't be too involved.

For the integrand to be non-zero, we need $t \geq 0$ and $x - t \geq 0$. If $x < 0$ this is impossible. Great—we're already halfway done! If now $x \geq 0$, then our two conditions combine to give $0 \leq t \leq x$, which is a nice interval. We find, for $x \geq 0$, that

$$
\begin{aligned}
f_X(x) &= \int_0^x f(t) f(x - t) dt \\
&= \int_0^x e^{-t} e^{-(x-t)} dt \\
&= \int_0^x e^{-t} e^{-x} e^t dt \\
&= e^{-x} \int_0^x dt = x e^{-x}.
\end{aligned}
$$

Unlike the uniform case, we have a very nice explicit formula for the convolution of n independent, identically distributed random variables. It's thus natural to ask what the answer looks like as $n \to \infty$. Later in the book (Chapter 20) we'll prove the Central Limit Theorem, which implies that as $n \to \infty$ these sums become normally distributed (again, see Chapter 14 for more on the normal distribution). The normal distribution with mean μ_X and variance σ_X^2 has density

$$
\frac{1}{\sqrt{2\pi\sigma_X^2}} e^{-(x-\mu_X)^2/2\sigma_X^2}.
$$

It therefore isn't enough to say that our distribution approaches a normal distribution; we must also specify the mean and the variance. Not surprisingly, the correct choices are the mean and the variance of the Erlang distribution. In other words, if n is large then the density of the Erlang distribution with parameters λ and n should be approximately that of a normal distribution with mean $n\lambda$ and variance $n\lambda^2$.

We plot the density of $X + Y$ (when both are independent Exp(1) random variables) in Figure 13.3.

We can continue, and in Figure 13.4 we plot the density of the sum of eight and thirty independent standard exponential random variables, and compare this to the corresponding normal distribution. Note that the fit isn't good at all for eight but is okay (though not great) for thirty.

Why is the convergence here so much slower than for sums of uniform random variables? We'll see an explanation when we prove the Central Limit Theorem. The quick and simple reason is the following: the uniform distribution is symmetric about its mean, so its centered third moment vanishes; this isn't true for the exponential distribution. We chose $n = 30$ as often books say that once n reaches 30, the normal is a good approximation, and leave it to you to decide if the fit looks good.

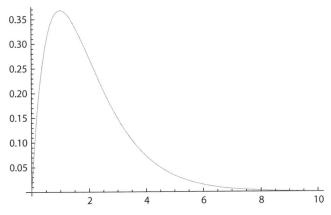

Figure 13.3. Plot of the density of $X + Y$ when X and Y are independent Exp(1) random variables.

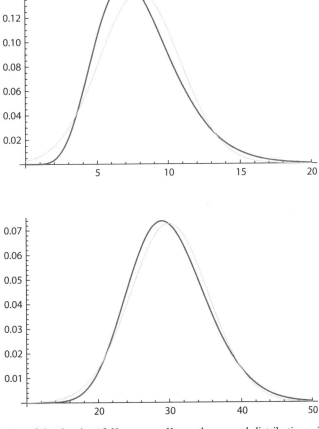

Figure 13.4. Plot of the density of $X_1 + \cdots + X_8$ vs. the normal distribution with mean 8 and variance 8 (top) and $X_1 + \cdots + X_{30}$ vs. the normal distribution with mean 30 and variance 30 (bottom) when the X_i's are independent Exp(1) random variables.

13.2.3 Examples and Applications of Exponential Random Variables

The exponential distribution occurs naturally when describing the lengths of the inter-arrival times in a homogeneous Poisson process, which are described in Chapter 12.6. The exponential distribution may be viewed as the continuous counterpart to the geometric distribution, which describes the number of Bernoulli trials necessary for a discrete process to change state. In contrast, the exponential distribution describes the time for a continuous process to change state.

In the real world, the assumption of a constant rate (or probability per unit time) is rarely satisfied. For example, scoring goals in soccer has been shown to approximately be a Poisson process. The rate of goal scoring differs, however, according to factors such as the teams playing, the weather conditions, etc. If we focus on a certain game, between teams of constant skill and in constant weather conditions, the exponential distribution can be used as a good approximate model for the time until the next goal is scored.

Example: Let's assume the time until the next goal is scored can be modeled by an exponential distribution with a parameter of 30 minutes. What is the chance that there are no goals scored in the first half (a half in soccer is 45 minutes long)?

Solution: The probability density function is

$$f_X(x) = \begin{cases} \dfrac{1}{30}e^{-x/30} & \text{if } x \geq 0 \\ 0 & \text{otherwise.} \end{cases}$$

Thus, the answer to our question is

$$\int_{45}^{\infty} \frac{1}{30}e^{-x/30}dx = -e^{-x/30}\Big|_{45}^{\infty} = e^{-3/2} \approx 22.3\%.$$

Example: Another example of a process well modeled by a Poisson distribution is the action potentials emitted by a neuron. As a result, we can model the time between action potentials with an exponential distribution. Suppose you have a group of ten independent neurons which each activate their action potentials on average every 12 milliseconds. What is the probability that the sum of the time it takes each of the ten neurons to fire is less than .1 seconds? (One second is equal to 1000 milliseconds, so .1 seconds is 100 milliseconds.)

Solution: To solve this, we will use the Erlang distribution with $\lambda = 12$ and $n = 10$ integrating from $x = 0$ to $x = 100$:

$$\int_0^{100} \frac{x^{10-1}e^{-x/12}}{12^{10}(10-1)!}dx.$$

By integrating by parts nine times, we come to a final answer of

$$1 - \frac{1002136941077}{357128352e^{25/3}} \approx 32.55\%;$$

however, you should have noticed that we did not include the calculations for this solution. There is a reason for this: integrating by parts nine times is painful! While we could use a computer to calculate this integral—and indeed, this is what we did—what we would really like to do in general, as stated in the previous section, is to approximate the Erlang distribution as a normal distribution with a mean of $n\lambda$ and a variance of $n\lambda^2$.

13.2.4 Generating Random Numbers from Exponential Distributions

It's one thing to develop a mathematical theory; it's another to implement it. For example, we can of course say, "Choose a real number from the standard exponential distribution"; however, how would we do this in practice?

We assume we have a way of choosing numbers randomly from the uniform distribution on $[0, 1]$. As discussed in §13.1.4, this isn't a horrible assumption. Amazingly, if we can generate random numbers uniformly on $[0, 1]$ then we can generate a multitude of random variables easily, specifically, any random variable where we have an explicit formula for the cumulative distribution function. We describe the method, and then discuss why it works.

Cumulative Distribution Method to Generate Random Numbers: Let X be a random variable with density f_X and cumulative distribution function F_X. If Y is a uniform random variable on $[0, 1]$, then

$$X = F_X^{-1}(Y).$$

This is also known as the **inverse transform sampling** or the **inverse transformation method**.

The above is absolutely miraculous! All we need to do is compute the inverse cumulative distribution function and, assuming we can generate random numbers uniformly in $[0, 1]$, we're done! This is truly amazing! Of course, we're left with two questions.

1. Why does this method work?
2. How often does a random variable have a nice, closed form expression for its inverse cumulative distribution function?

We'll tackle the first question in detail, and for now content ourselves with showing that the exponential has a nice inverse cumulative distribution function.

So, why does this method work? Let's set $Z = F_X^{-1}(Y)$. Right now we have no idea why this should work; we're merely guessing it and exploring the consequences. We now do the same arguments we always do to find probabilities. In particular, if we can find the cumulative distribution function of Z then we can find its density by taking derivatives. *Very important* in the proof is the fact that F_X is a non-decreasing function, as is F_X^{-1}. This allows us to pass from $-\infty \leq F_X^{-1}(Y) \leq z$ to $F_X(-\infty) \leq Y \leq F_X(z)$ (and since $F_X(-\infty) = 0$, it becomes $0 \leq Y \leq F_X(z)$). We have

$$\text{Prob}(Z \leq z) = \text{Prob}(-\infty \leq F_X^{-1}(Y) \leq z)$$

$$= \text{Prob}(0 \leq Y \leq F_X(z))$$

$$= \int_0^{F_X(z)} 1 dy = F_X(z);$$

thus Z has the same cumulative distribution function as X, and taking derivatives we see its density also equals the density of X. This forces $Z = X$ (it's actually enough that they have the same cumulative distribution functions; however, it's often more natural to think about densities, and so we showed the densities were equal as well).

Okay, the method works, but how did anyone ever think of this? One of the goals of this book is to help you think probabilistically (or mathematically). We don't want to present a chain of algebra and leave it at that; we want to *motivate* that chain of algebra so you can think of similar things to try for future problems. So, let's go back to what it means to generate numbers from a given distribution. If we're to generate random numbers with X's distribution, then the probability we generate a number in $[x, x + \Delta x]$ must be $F_X(x + \Delta x) - F_X(x)$. If Y is a uniform random variable on $[0, 1]$, then the probability Y takes on a value in an interval $[a, b] \subset [0, 1]$ is just $b - a$. This suggests the following decision rule: generate Y, and if the value is in the interval $[F_X(x), F_X(x + \Delta)]$ then take the value x for X. As $\Delta x \to 0$, the interval shrinks to a point, and we're basically saying that the associated value of X to Y satisfies $y = F_X(x)$, or $x = F_X^{-1}(y)$.

Let's look at an exponential random variable with parameter λ. The density is

$$f_X(x) = \begin{cases} \dfrac{1}{\lambda} e^{-x/\lambda} & \text{if } x \geq 0 \\ 0 & \text{otherwise.} \end{cases}$$

We just need to integrate to get the cumulative distribution function. Clearly this is zero for $x \leq 0$, while for $x \geq 0$ we have

$$F_X(x) = \int_0^x f_X(t)dt$$

$$= \int_0^x \frac{1}{\lambda} e^{-t/\lambda} dt$$

$$= \int_0^{x/\lambda} e^{-u} du$$

$$= -e^{-t} \Big|_0^{x/\lambda} = 1 - e^{-x/\lambda}.$$

Thus for $x \geq 0$ we have

$$F_X(x) = 1 - e^{-x/\lambda}.$$

We now invert: if $F_X(x) = y$ then

$$y = 1 - e^{-x/\lambda}$$

$$e^{-x/\lambda} = 1 - y$$

$$-\frac{x}{\lambda} = \log(1 - y)$$

$$x = -\lambda \log(1 - y) = F_X^{-1}(y)$$

(to check that this is the inverse of F_X at y, direct calculation shows $F_X(F_X^{-1}(y)) = y$).

Is our answer reasonable? Again, asking this question should be second nature for you by now. It should be automatic. We *always* test to see if a formula is reasonable. Fortunately, this formula has many things we can check. First, note it does depend on λ. If there were no λ dependence then we would understandably be suspicious, because this is supposed to equal the density of an exponential random variable with parameter λ! What else can we test? The next thing to ask is whether or not the logarithm is well-defined. The answer is yes; as Y is uniformly distributed on $[0, 1]$, $0 \leq y \leq 1$ and hence the logarithm is well-defined. (Okay, there's a problem if $y = 1$, but that happens with zero probability.) Finally, our initial reaction should be one of horror at seeing a minus sign, as probability densities are non-negative! Everything is fine. Remember that $0 \leq 1 - y \leq 1$, thus we're taking the logarithm of a number less than 1, and this is *negative*. We *need* that minus sign to make the final expression positive.

In addition to checking the reasonableness of the answer by looking at the formula, we can of course also resort to some numerical simulations. In Figure 13.5 we compare the results of simulating 10,000 values from a uniform distribution on $[0, 1]$ and converting with the standard exponential. Not surprisingly, if we increase the number of points the fit improves. We're using bin sizes of .01, and with 100,000 values the fit with the exponential is quite good.

13.3 Exercises

Exercise 13.3.1 *Why can't we have a uniform random variable on $(0, \infty)$?*

Exercise 13.3.2 *Find the variance of $X \sim \text{Unif}(a, b)$ by computing $\mathbb{E}[X^2] - \mathbb{E}[X]^2$. The hardest part of the variance calculation is doing the algebra simplification—is this approach easier than the other one?*

Exercise 13.3.3 *Give a dimensional analysis argument that $X \sim \text{Unif}(a, b)$. There must exist a constant C independent of a and b such that $\text{Var}(X) = C(b - a)^2$. What bounds can you deduce for C?*

Exercise 13.3.4 *Instead of using the symmetry argument to get the density of $\Pr(Z = z)$ ($Z = X + Y$ with X, Y independent uniform random variables on $[0, 1]$), do the integration and confirm our earlier answer.*

Exercise 13.3.5 *Let $X_1, X_2, \ldots$ be independent uniform random variables on $[0, 1]$. Building on our knowledge of the densities of X_1 and $X_1 + X_2$, is it easier to find the density of $X_1 + X_2 + X_3$ or $X_1 + X_2 + X_3 + X_4$ (or are the two equally difficult)? Find these two densities.*

Exercise 13.3.6 *We calculated the mean of $X \sim \text{Unif}(a, b)$ by direct integration. We can avoid finding anti-derivatives by noting that the density is symmetry about the midpoint, $\frac{b+a}{2}$. Use one of the most important techniques in mathematics, **adding zero**, to show*

$$
\mathbb{E}[X] = \int_a^b \left(x - \frac{b+a}{2} + \frac{b+a}{2} \right) \frac{dx}{b-a}
$$

$$
= \int_a^b \left(x - \frac{b+a}{2} \right) \frac{dx}{b-a} + \frac{b+a}{2} \int_a^b \int_a^b \frac{dx}{b-a}.
$$

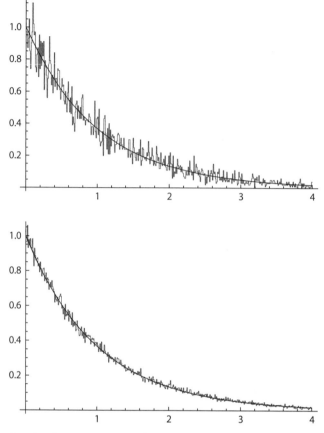

Figure 13.5. Result of using the cumulative distribution method to generate 10,000 (top) and 100,000 (bottom) randomly chosen numbers from the standard exponential distribution, compared with the standard exponential distribution.

Complete the analysis to determine the mean. Can you use a similar argument to quickly find the variance?

Exercise 13.3.7 *We calculated the variance of $X \sim \text{Exp}(\lambda)$ by finding $\mathbb{E}[(X - \mu_X)^2]$; compute it using $\mathbb{E}[X]^2 - \mathbb{E}[X]^2$. Which integration do you think is easier (or are they roughly the same)?*

Exercise 13.3.8 *Find $\mathbb{E}[\sqrt{X}]$ if X is an exponential variable with parameter λ.*

Exercise 13.3.9 *Find $\mathbb{E}[1/X]$ if X is a uniform variable on the interval $[0, n]$.*

Exercise 13.3.10 *Find all the moments for a uniform variable on the interval (a, b).*

Exercise 13.3.11 *Find all the moments for an exponential variable with parameter λ.*

Exercise 13.3.12 *Find $\text{Prob}(X > 3)$ where X is an exponential random variable with parameter 1.*

Exercise 13.3.13 *Find $\text{Prob}(X > 2)$ or $(X > 4)$ where X is an exponential with parameter 1.*

Exercise 13.3.14   *Between what values of x does the middle 50% of an exponential random variable X with parameter λ lie? In other words, find a_λ and b_λ such that* $\Pr(X \le a_\lambda) = \Pr(X \ge b_\lambda) = .25$.

Exercise 13.3.15   *If X is a uniform random variable with positive probability on the interval [0, n], find the probability density function for e^X.*

Exercise 13.3.16   *Find the **median** of an exponential random variable X with parameter λ; remember the median $\widetilde{\mu}$ for a random variable with a continuous density is the point such that $\Pr(X \le \widetilde{\mu}) = \Pr(X \ge \widetilde{\mu}) = 1/2$. Is the median greater than or less than the mean? By how much?*

Exercise 13.3.17   *Building on the previous exercise, find a piecewise continuous probability density such that the set of numbers satisfying the definition of the median above is an interval.*

Exercise 13.3.18   *If X and Y are independent uniform random variables on the interval [0, n], find the pdf for $\min(X, Y)$ and $\max(X, Y)$.*

Exercise 13.3.19   *If X is a uniform random variable on the interval [0, n] and Y is an exponential random variable with parameter λ, find the pdf for $\min(X, Y)$ and $\max(X, Y)$.*

Exercise 13.3.20   *If X is an exponential random variable with parameter λ_X and Y is an exponential random variable with parameter λ_Y find the pdf for $\min(X, Y)$ and $\max(X, Y)$.*

Exercise 13.3.21   *Let X and Y be independent uniform random variables on [0, 1]. Let Z be the uniform random variable from $\min(X, Y)$ to $\max(X, Y)$. What can you say about Z? What is its expected value? Its variance?*

Exercise 13.3.22   *We derived the density of the Erlang distribution when n = 2; compute it for as many general n and values of the parameter you can.*

Exercise 13.3.23   *Find $\mathbb{E}[e^{tX}]$, where X is an exponential random variable with parameter λ and $t < 1/\lambda$.*

Exercise 13.3.24   *Calculate the mean and the variance of an Erlang distribution with parameters λ and n.*

Exercise 13.3.25   *Are you surprised that the prediction in Figure 13.4 seems to be undershooting the probability? Why or why not?*

Exercise 13.3.26   *Let X be an exponential random variable. Show that $\text{Prob}(X > x + a | X > a) = \text{Prob}(X > x)$.*

Exercise 13.3.27   *Why is the exponential distribution useful for estimating how long it takes certain types of objects to break down or fail? When does it not work for estimating when an object will break down?*

Exercise 13.3.28   *The probability that a particle has a given lifetime is an exponential with parameter $\frac{t}{\sqrt{1-v^2/c^2}}$, where t is the mean lifetime of the particle at rest, v is the speed at which the particle is moving, and c is the speed of light, about 3 hundred million meters/sec. Given that a neutron has a mean lifetime at rest of around 880sec and is moving quickly, at around 14 million meters/sec, what is the probability that it lives more than ten minutes?*

Exercise 13.3.29   *Plot the sum of an exponential random variable with $\lambda = 1$ and a uniform variable on the interval $[0, n]$ for $n \in \{1, 2, 3, 4, 5\}$.*

Exercise 13.3.30   *Evaluate the following limits if they exist, and if they don't say why they don't: (a) $\lim_{x \to 0}(e^x - 1)/x$, (b) $\lim_{x \to 0}(\cos^2 x - 1)/x^3$, (c) $\lim_{x \to 1}(x^2 - 1)/(x^3 - 1)$, (d) $\lim_{x \to \infty} x^{2004}/e^{x/2001}$, (e) $\lim_{x \to \infty} x^{1/2007}/e^{\log(\log(x))}$, and (f) $\lim_{x \to 0} x \log(x)$.*

Exercise 13.3.31   *A calculus student is asked to evaluate $\lim_{x \to 0} \frac{\sin(x)}{x}$. They claim it is easy; as the limit is $0/0$ we can use L'Hôpital's rule, and thus the limit equals $\lim_{x \to 0} \frac{\cos(x)}{1} = 1$. Is there anything wrong or dangerous with this argument?*

Exercise 13.3.32   *Tim and Lisa live in Albany, NY, and are taking their kids to the Boston Red Sox game at Fenway Park. It takes a random time uniformly distributed between 150 and 240 minutes to drive from their home to Fenway. If the game is at 7:00 PM and they leave at 3:50 PM, find the probability that they arrive at the game on time.*

Exercise 13.3.33   *A slot machine has two independent random number generators that generate a uniformly distributed number between 0 and 80. You win the jackpot if the numbers sum to less than 20 or more than 140. What is the probability that you win the jackpot?*

Exercise 13.3.34   *Jimmy works at a local ice cream shop, where customers arriving and ordering ice cream is a Poisson process, averaging 15 customers per hour. Realizing he ran out of Rocky Road ice cream, he decides to travel to the store to buy more, a 5 minute round trip. What is the probability that Jimmy is able to successfully return before a customer arrives?*

Exercise 13.3.35   *The time between car crashes in Odwalla, MD, can be accurately modeled by an exponential distribution averaging 2 days between each crash. What is the probability that there are at least 4 crashes in the next week?*

CHAPTER 14

Continuous Random Variables: The Normal Distribution

Unquestionably one of the most important distributions, not just in probability but in all of mathematics and science, is the normal distribution. To a large extent, this is due to the Central Limit Theorem, which says that in numerous cases sums of independent random variables converge to being normally distributed. These conditions are frequently weak and are met in many theoretical and practical problems. We'll cover the Central Limit Theorem in detail in Chapter 20; our purpose here is to introduce the normal distribution and many of its properties. We first record the definition.

> **Normal Distribution**: A random variable X is normally distributed with mean μ and variance σ^2 if its density is
>
> $$ f_X(x) = \frac{1}{\sqrt{2\pi\sigma^2}} e^{-(x-\mu)^2/2\sigma^2}; $$
>
> we write $X \sim N(\mu, \sigma^2)$. This density is so important we also say X is a **Gaussian random variable** (with mean μ and variance σ^2). If X has the **standard normal distribution** then $X \sim N(0, 1)$. If X is normally distributed we sometimes say it follows the **bell curve**.

Note that in $N(\mu, \sigma^2)$ the second argument refers to the variance; thus if $X \sim N(0, 4)$ then X is normally distributed with mean 0 and variance 4, and thus the standard deviation is 2. We plot three normal distributions in Figure 14.1.

Unlike the other random variables we've looked at, it isn't at all clear that the normal distribution is a distribution! It's clearly non-negative, but does it integrate to 1? Sadly, the normal distribution is a lot harder to work with than the other densities we've studied to date. The reason is that there's no elementary function which is the anti-derivative of its density. This means we can't write down the cumulative distribution function, and must resort to series expansions and numerical approximations.

As bad as this is, things could be a lot worse. Amazingly, even though there is no nice, closed form expression for its cumulative distribution function, there is a very

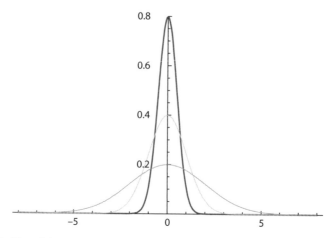

Figure 14.1. Plot of three normal distributions, all with mean zero and variances 1/2, 1, and 2.

elementary way to show the area under the curve is 1. We'll first show two different ways to see this, then follow the same path as the other chapters (calculating the mean and variance, looking at sums of normal random variables, and so on).

14.1 Determining the Normalization Constant

The **Theory of Normalization Constants** is one of the most beautiful parts of the subject. It tells us that we can take any non-negative function that has a finite integral and convert it to a probability density by simple multiplication. At this stage in the game, it's not immediately clear why this is so important and useful, and not just a simple obvious fact. It becomes very helpful when we have functions of several variables and we integrate out some variables but keep others. We'll discuss this in greater detail in §16.3.2; for now, we'll state the method and then apply it to the normal density.

> **Theory of Normalization Constants**: Let g be a non-negative, real-valued function with
>
> $$c = \int_{-\infty}^{\infty} g(x)dx > 0.$$
>
> If $c < \infty$ then $f(x) = g(x)/c$ is a probability density.

The proof is a follow-your-nose argument; namely, we just have to check the two conditions of a probability function. The non-negativity is easy; since $g(x) \geq 0$ and $c > 0$, $g(x)/c \geq 0$. We are left with showing that f integrates to 1. We have

$$\int_{-\infty}^{\infty} f(x)dx = \int_{-\infty}^{\infty} \frac{g(x)}{c}dx = \frac{1}{c}\int_{-\infty}^{\infty} g(x)dx = \frac{c}{c} = 1.$$

Let's return to the normal distribution, with density

$$f(x) = \frac{1}{\sqrt{2\pi\sigma^2}} e^{-(x-\mu)^2/2\sigma^2}.$$

Clearly this density is non-negative. If we can just show it has a finite, non-zero integral then we know there is some normalization constant that will make this a probability distribution. We won't know exactly what that constant is, just that it exists.

Its integral is also non-negative. Let's find an upper bound. We'll make several changes of variables to simplify the integration. Also, we'll use that the integral of an even function about a symmetric region is equal to twice the integral over half the region (see §A.4). Towards the end we split the integral in two, as we use different functions to bound e^{-v^2} depending on v's size (see Exercise 14.6.1). We find

$$\int_{-\infty}^{\infty} \frac{1}{\sqrt{2\pi\sigma^2}} e^{-(x-\mu)^2/2\sigma^2} dx = \frac{1}{\sqrt{2\pi\sigma^2}} \cdot 2 \int_{\mu}^{\infty} e^{-(x-\mu)^2/2\sigma^2} dx$$

$$= \frac{2}{\sqrt{2\pi\sigma^2}} \int_{0}^{\infty} e^{-u^2/2\sigma^2} du$$

$$= \frac{2}{\sqrt{2\pi\sigma^2}} \int_{0}^{\infty} e^{-u^2/2\sigma^2} \sigma\sqrt{2} \frac{du}{\sigma\sqrt{2}}$$

$$= \frac{2}{\sqrt{\pi}} \int_{0}^{\infty} e^{-v^2} dv$$

$$= \frac{2}{\sqrt{\pi}} \left[\int_{0}^{1} e^{-v^2} dv + \int_{1}^{\infty} e^{-v^2} dv \right]$$

$$< \frac{2}{\sqrt{\pi}} \left[1 + \int_{1}^{\infty} e^{-v} dv \right]$$

$$= \frac{2}{\sqrt{\pi}} \left[1 + (-e^{-v}) \Big|_{1}^{\infty} \right]$$

$$= \frac{2}{\sqrt{\pi}} \left[1 + e^{-1} \right] \approx 1.54349 < \infty.$$

Thus the integral is finite for all μ and σ, and therefore can be rescaled to be a probability density.

 As all we need here is that the integral is finite, though very shortly we'll need the actual value, we can be crude in our approximation. This is a valuable skill to learn, namely trying to find the **order of magnitude of a quantity**. This is a *very* hard integral to evaluate, as there is no closed form anti-derivative. We'll see in a moment that it does equal 1, and thus our upper bound of about 1.54349 is pretty good.

Now that we have showed that it *can* be rescaled, we need to prove that the scale factor is 1. In other words, we need to show that including the factor of $1/\sqrt{2\pi\sigma^2}$ is *precisely* what we need to make this a probability density. There are two ways to do this. The first is to use a wonderful trick from multivariable calculus. You should be a bit confused after reading this, as we only have one variable of integration—where does multivariable calculus come in? The idea is that x^2 looks like half of $x^2 + y^2$, which in polar coordinates is just r^2. The trick is to square the integral so we can use circles, which is probably one of the strangest sentences you'll see.

Things will look clearer after some details. Let's set

$$I(\mu, \sigma) = \int_{-\infty}^{\infty} \frac{1}{\sqrt{2\pi\sigma^2}} e^{-(x-\mu)^2/2\sigma^2} dx.$$

The first thing to note is that

$$I(\mu, \sigma) = I(0, 1)$$

for all μ, σ. This is just changing variables:

$$I(\mu, \sigma) = \int_{-\infty}^{\infty} \frac{1}{\sqrt{2\pi\sigma^2}} e^{-(x-\mu)^2/2\sigma^2} dx$$

$$= \frac{1}{\sqrt{2\pi\sigma^2}} \int_{-\infty}^{\infty} e^{-(u/\sigma)^2/2} \sigma \frac{dx}{\sigma}$$

$$= \frac{1}{\sqrt{2\pi}} \int_{-\infty}^{\infty} e^{-x^2/2} dx = I(0, 1).$$

Unlike the last argument (where we only wanted to show the integral is finite), here it's more convenient to leave the factor of 2 inside the exponential.

Okay, so we've reduced the problem to showing $I(0, 1) = \sqrt{2\pi}$. Why should this be true? The answer actually gives us a clue as to how to proceed. This is an extremely clever idea; do not be disheartened and worry that you won't be able to think of something like this, as seeing such connections becomes easier with experience. Well, whenever we see a π we should think circles, and since we have variables we should think polar coordinates:

$$x = r\cos\theta, \quad y = r\sin\theta, \quad r \in [0, \infty), \quad \theta \in [0, 2\pi), \quad dxdy = rdrd\theta.$$

Now it's time for the trick:

$$I(0, 1) = \int_{-\infty}^{\infty} \frac{1}{\sqrt{2\pi}} e^{-x^2/2} dx = \int_{-\infty}^{\infty} \frac{1}{\sqrt{2\pi}} e^{-y^2/2} dy.$$

This works because x and y are just dummy variables. We now consider $I(0, 1)^2$, writing one of the I's as an integral over x and another as an integral over y. This is the truly remarkable, clever idea (called by some the **polar coordinate trick**). We can't use the same dummy variable for the two integrals, as that could cause confusion; however, using two different letters should make us think of certain tricks from multivariable calculus. Basically, whenever you can't solve an integral with two or more variables, try one of two things: (1) Fubini's theorem (to change the order of integration), or (2) change variables.

For us, Fubini's theorem can't help, but changing variables from the Cartesian coordinates x, y to the polar coordinates r, θ does. (Another way to view what we're doing is to note we have $e^{-x^2}dx$; while this is hard to integrate, $e^{-x^2}xdx$ is easy. Thus we need to somehow insert an x. Obviously we can't just insert variables, but changing to polar will have $dxdy$ go to $rdrd\theta$ and lead to tractable integration.) We find

$$
\begin{aligned}
I(0, 1)^2 &= \int_{-\infty}^{\infty} \frac{1}{\sqrt{2\pi}} e^{-x^2/2} dx \int_{-\infty}^{\infty} \frac{1}{\sqrt{2\pi}} e^{-y^2/2} dy \\
&= \frac{1}{2\pi} \int_{x=-\infty}^{\infty} \int_{y=-\infty}^{\infty} e^{-(x^2+y^2)/2} dxdy \\
&= \frac{1}{2\pi} \int_{r=0}^{\infty} \int_{\theta=0}^{2\pi} e^{-r^2/2} rdrd\theta \\
&= \frac{1}{2\pi} \int_{r=0}^{\infty} e^{-r^2/2} rdr \int_{\theta=0}^{2\pi} d\theta \\
&= \frac{1}{2\pi} \left[-e^{-r^2/2} \Big|_0^{\infty} \right] \cdot \left[\theta \Big|_0^{2\pi} \right] = \frac{1}{2\pi} \cdot 1 \cdot 2\pi = 1,
\end{aligned}
$$

which proves that it *is* a probability density (as $I(0, 1)$ is clearly positive, it must equal 1).

 The above calculation is an example of how frustrating math can be. Line by line, the argument isn't hard to follow. The difficulty is reproducing the insight of this proof in a similar problem, i.e., learning how to think outside the box and square the integral. Whomever first thought of the polar coordinate trick deserves our praise for a truly original approach. The best advice I can give is to do *lots and lots* of math problems. The more problems you do, the more ways you see how to solve something, growing the techniques and tricks of your arsenal. It takes awhile to gain fluency, but until you hear the language you don't have a chance. A lot of mathematics is pattern recognition, seeing how to look at something the right way. The more you do, the easier it becomes.

14.2 Mean and Variance

Let $X \sim N(\mu, \sigma^2)$. This means X is a normal random variable with density:

$$
f(x) = \frac{1}{\sqrt{2\pi\sigma^2}} e^{-(x-\mu)^2/2\sigma^2}.
$$

In the previous section, we showed that this is a density because it integrates to 1. It's time to determine its mean, which our notation suggests is μ (and similarly we should be thinking the variance is σ^2). Let's rewrite it in a very suggestive manner:

$$f(x) = \frac{1}{\sqrt{2\pi\sigma^2}} e^{-(x-\mu)^2/2\sigma^2}$$

$$= \frac{1}{\sqrt{2\pi}} \exp\left(-\frac{1}{2}\left(\frac{x-\mu}{\sigma}\right)^2\right) \frac{1}{\sigma}.$$

We see the density depends on $(x-\mu)/\sigma$, suggesting that the mean is μ and it has been adjusted so that the standard deviation is σ. Why? To see the mean is μ, notice the probability of being $\mu + a$ is the same as being $\mu - a$, so our density is symmetric about $x = \mu$, and thus μ is the mean. We'll now prove these facts, justifying our notation of writing a random variable with this density as $N(\mu, \sigma^2)$.

 First, the mean. We must be a little careful here; we'll write μ_X for the mean. The reason we do this is that we already have the symbol μ appearing in our density, and if we wrote μ for the mean we could potentially have the same symbol taking on two different values. We'll show in a little bit that $\mu_X = \mu$, but right now *we do not know that these are the same!*

As our distribution is symmetric about μ, it's natural to guess that the mean is μ. Let's prove this; from the definition of the mean we have

$$\mu_X = \int_{-\infty}^{\infty} x \cdot \frac{1}{\sqrt{2\pi\sigma^2}} e^{-(x-\mu)^2/2\sigma^2} dx.$$

There are a lot of ways to proceed; the following is one of my favorite tricks. See how the density only depends on $x - \mu$? This suggests that we should change variables by shifting and set $u = x - \mu$, so $x = u + \mu$ and $dx = du$. This gives

$$\mu_X = \int_{-\infty}^{\infty} (u + \mu) \cdot \frac{1}{\sqrt{2\pi\sigma^2}} e^{-u^2/2\sigma^2} du$$

$$= \int_{-\infty}^{\infty} u \cdot \frac{1}{\sqrt{2\pi\sigma^2}} e^{-(x-\mu)^2/2\sigma^2} du + \mu \int_{-\infty}^{\infty} \frac{1}{\sqrt{2\pi\sigma^2}} e^{-(x-\mu)^2/2\sigma^2} du.$$

Note the second integral is very friendly; it's just 1 as it's the integral of a probability density! Thus,

$$\mu_X = \int_{-\infty}^{\infty} u \cdot \frac{1}{\sqrt{2\pi\sigma^2}} e^{-(x-\mu)^2/2\sigma^2} du + \mu.$$

We just need to show that the first integral vanishes. The easiest way to do this is to note that we're integrating an odd function about a symmetric region.

 If we don't see this, progress can still be made. We can try to integrate by parts; however, if we aren't careful we'll make one of the most common mistakes in probability! **We will deliberately be a little careless and make a mistake—see if you**

can catch it! We have

$$\int_{-\infty}^{\infty} u \cdot \frac{1}{\sqrt{2\pi\sigma^2}} e^{-u^2/2\sigma^2} du = \int_{-\infty}^{\infty} \frac{1}{\sqrt{2\pi}} e^{-u^2/2\sigma^2} \frac{u\,du}{\sigma}.$$

We now change variables. If we let $t = u^2/2\sigma^2$, then $dt = u\,du/\sigma$ and we have

$$\int_{-\infty}^{\infty} u \cdot \frac{1}{\sqrt{2\pi\sigma^2}} e^{-u^2/2\sigma^2} du = \int_{0}^{\infty} \frac{1}{\sqrt{2\pi}} e^{-t} dt$$

$$= \frac{1}{\sqrt{2\pi}} \left[-e^{-t}\right]_{t=0}^{\infty}$$

$$= \frac{1}{\sqrt{2\pi}}.$$

Ack! What gives? We're integrating an odd function over a symmetric region—how could this be positive? The reason is that we made one of the most common mistakes in doing probability integrals—**we didn't change the bounds of integration properly!** Remember u ran from $-\infty$ to ∞. If we let $t = u^2/2\sigma^2$, then t runs from ∞ to ∞. This is absurd! The problem is our change of variables isn't invertible on $(-\infty, \infty)$, as it maps both $(-\infty, 0]$ and $[0, \infty)$ onto $[0, \infty)$. Thus, what we need to do is split the integral into two regions, such that our change of variables is invertible in each region.

Let's try this again, but this time we'll be a bit more careful. We have

$$\int_{-\infty}^{\infty} u \cdot \frac{1}{\sqrt{2\pi\sigma^2}} e^{-u^2/2\sigma^2} du$$

$$= \int_{-\infty}^{0} u \cdot \frac{1}{\sqrt{2\pi\sigma^2}} e^{-u^2/2\sigma^2} du + \int_{0}^{\infty} u \cdot \frac{1}{\sqrt{2\pi\sigma^2}} e^{-u^2/2\sigma^2} du$$

$$= \int_{-\infty}^{0} \frac{1}{\sqrt{2\pi}} e^{-u^2/2\sigma^2} \frac{u\,du}{\sigma} + \int_{0}^{\infty} \frac{1}{\sqrt{2\pi}} e^{-u^2/2\sigma^2} \frac{u\,du}{\sigma}$$

$$= \int_{\infty}^{0} \frac{1}{\sqrt{2\pi}} e^{-t} dt + \int_{0}^{\infty} \frac{1}{\sqrt{2\pi}} e^{-t} dt.$$

There are two ways to proceed. One is of course to just evaluate the two integrals directly. If we do this, we'll see that everything cancels and we get zero. The other way is to note that $\int_{b}^{a} g(t)dt = -\int_{a}^{b} g(t)dt$, as when we go through the integral backwards we count the area oppositely. Using this, we see

$$\int_{-\infty}^{\infty} u \cdot \frac{1}{\sqrt{2\pi\sigma^2}} e^{-u^2/2\sigma^2} du$$

$$= -\int_{0}^{\infty} \frac{1}{\sqrt{2\pi\sigma^2}} e^{-t} dt + \int_{0}^{\infty} \frac{1}{\sqrt{2\pi\sigma^2}} e^{-t} dt = 0.$$

This now completes the proof that the mean *is* μ; in other words, $\mu_X = \mu$.

We now turn to the variance. Again, we expect it to be σ^2, but as we don't know that yet, we must be careful not to use the same symbol for two different purposes in the same equation. Thus, we'll write σ_X^2 for the variance now, and not be too surprised when we find $\sigma_X^2 = \sigma^2$.

We find the variance by integrating $(x - \mu_X)^2$ times the density. Fortunately we know $\mu_X = \mu$, so the variance is just

$$\sigma_X^2 = \int_{-\infty}^{\infty} (x - \mu)^2 \frac{1}{\sqrt{2\pi\sigma^2}} e^{-(x-\mu)^2/2\sigma^2} dx.$$

Let's change variables to simplify this mess. We take $w = (x - \mu)/\sigma$, so $dw = dx/\sigma$ and $(x - \mu)^2 = \sigma^2 w^2$. As we're doing a *linear* change of variables, the bounds of integration will change nicely; we had x run from $-\infty$ to ∞, and w will also run from $-\infty$ to ∞, giving us

$$\sigma_X^2 = \int_{-\infty}^{\infty} \sigma^2 w^2 \frac{1}{\sqrt{2\pi\sigma^2}} e^{-w^2/2} \sigma \, dw$$

$$= \sigma^2 \int_{-\infty}^{\infty} w^2 e^{-w^2/2} dw.$$

We now integrate by parts. We have to split $w^2 \exp(-w^2/2)dx$ into $u \, dv$. It's natural to try $u = w^2$ and $dv = \exp(-w^2/2)dw$, but this won't work. It's natural because we replace the $u = w^2$ term with $du = 2w \, dw$ when we integrate by parts, which lowers the power of w; the difficulty is that we need to find v, and there is no closed form for the anti-derivative of $\exp(-w^2/2)$. We thus need to be a bit more clever here than for a typical integration by parts problem. Let's look at the terms a bit more. The u term should involve the w factor, as we want to lower the polynomial (which is causing trouble). The $\exp(-w^2/2)$ factor should be part of the dv term; however, as we need to integrate it, we should have one factor of w with it. The reason is this gives us a function where there *is* a nice anti-derivative.

So, we take

$$u = w, \quad dv = \exp(-w^2/2)w \, dw,$$

which implies

$$du = dw, \quad v = -\exp(-w^2/2).$$

We obtain

$$\sigma_X^2 = \sigma^2 \left[uv \Big|_{-\infty}^{\infty} - \int_{-\infty}^{\infty} v \, du \right]$$

$$= \sigma^2 \left[w \exp(-w^2/2) \Big|_{-\infty}^{\infty} + \int_{-\infty}^{\infty} \exp(-w^2/2)dw \right] = \sigma^2,$$

where a lot happened in the last line. First, the integral is just 1; this is the same integral we saw when proving our proposed density was valid (i.e., that it integrated to 1). Second, the exponential decay is so great that $w \exp(-w^2/2) \to 0$ as $w \to \pm\infty$.

There are many ways to see this. We can write it as $w/\exp(w^2/2)$ and then recall that for t positive we have $\exp(t) > 1 + t$. This is true because we're just truncating the series expansion of the exponential function. This gives us $w/\exp(w^2/2) < w/(1 + w^2/2)$, which clearly goes to zero. Alternatively, we can use **L'Hôpital's rule**: if f and g are continuously differentiable functions with $\lim_{x\to x_0} f(x) = \lim_{x\to x_0} g(x)$ both 0 or both ∞, then $\lim_{x\to x_0} f(x)/g(x) = \lim_{x\to x_0} f'(x)/g'(x)$. In our case, this means

$$\lim_{w\to\infty} \frac{w}{\exp(w^2/2)} = \lim_{w\to\infty} \frac{1}{w\exp(w^2/2)} = 0.$$

The above calculations highlight many of the pitfalls in the subject. It's easy to get the bounds wrong, or to sometimes miss a good change of variables. The more calculus you remember, the easier these problems will be.

14.3 Sums of Normal Random Variables

Unfortunately, most of the time the theorems we wish were true turn out to be false. For example, imagine how easy it would be to find derivatives if the derivative of a product were the product of the derivatives, or even better, if the derivative of a quotient were the quotient of the derivatives. Sadly, of course, neither is true.

What is the best we could wish for in probability? Say X and Y are independent random variables with the same shape, but possibly different means and variances. Let's say they both come from the *Dream Distribution*, with means μ_X and μ_Y and variances σ_X^2 and σ_Y^2. If we look at $X + Y$, what would be a reasonable hope? We know the mean of a sum is the sum of the means, and the variance of the sum of independent random variables is the sum of the variances. Thus, no matter what, $X + Y$ must have mean $\mu_{X+Y} = \mu_X + \mu_Y$ and variance $\sigma_{X+Y}^2 = \sigma_X^2 + \sigma_Y^2$. Wouldn't it be wonderful if the *shape* of $X + Y$ were the same as the shape of X and Y? Such distributions are called **stable distributions**.

We know that sums of uniform random variables are not uniformly distributed; nor are sums of exponential random variables exponentially distributed. Our luck, however, changes if we look at sums of normal random variables, as these *are* normally distributed.

Sums of Normal Random Variables: If $X \sim N(\mu_X, \sigma_X^2)$ and $Y \sim N(\mu_Y, \sigma_Y^2)$ are two independent normal random variables (with means μ_X and μ_Y and variances σ_X^2 and σ_Y^2), then $X + Y \sim N(\mu_X + \mu_Y, \sigma_X^2 + \sigma_Y^2)$; i.e., the sum is normally distributed with mean $\mu_X + \mu_Y$ and variance $\sigma_X^2 + \sigma_Y^2$.
More generally, if we have independent normally distributed random variables $X_i \sim N(\mu_i, \sigma_i^2)$, then

$$X_1 + \cdots + X_n \sim N(\mu_1 + \cdots + \mu_n, \sigma_1^2 + \cdots + \sigma_n^2).$$

Just in case we didn't have enough reasons to love the normal distribution, here's one more! Think back to the nightmare of finding the probability density for sums of uniform random variables. If only all distributions were this easy!

There are many ways to prove our claim. We'll attack it in the most straightforward manner possible, by directly computing the probability density for $X + Y$. It turns out that if we can do it for the sum of two normals, through mathematical induction we can do it for an arbitrary sum. After we do the computation we'll step back and see that a lot of it was unnecessary!

Derivation of the formula for the sum of two independent normal random variables: As always, we use the Theory of Convolutions to find the density for $X + Y$. Letting f_X be the density for X, f_Y the density for Y and f_{X+Y} the density of $X + Y$, we have

$$f_X(x) = \frac{1}{\sqrt{2\pi\sigma_X^2}} \exp(-(x - \mu_X)^2/2\sigma_X^2)$$

$$f_Y(y) = \frac{1}{\sqrt{2\pi\sigma_Y^2}} \exp(-(y - \mu_Y)^2/2\sigma_Y^2)$$

$$f_{X+Y}(z) = \int_{-\infty}^{\infty} f_X(t) f_Y(z - t) dt.$$

We now "follow our nose"—we substitute for the two densities and do some algebra to make the resulting integral nice. We find

$$f_{X+Y}(z) = \int_{-\infty}^{\infty} \frac{1}{\sqrt{2\pi\sigma_X^2}} \exp(-(t - \mu_X)^2/2\sigma_X^2)$$

$$\cdot \frac{1}{\sqrt{2\pi\sigma_Y^2}} \exp(-(z - t - \mu_Y)^2/2\sigma_Y^2) dt$$

$$= \frac{1}{2\pi\sigma_X\sigma_Y} \int_{-\infty}^{\infty} \exp\left(-\frac{(t - \mu_X)^2\sigma_Y^2 + (z - t - \mu_Y)^2\sigma_X^2}{2\sigma_X^2\sigma_Y^2}\right) dt.$$

There are two natural approaches at this point. The first is to just push onward and evaluate the integral. The problem with integrals like this is that there are so many different parameters floating around the problem that it becomes very intimidating. I want you to feel comfortable attacking calculations like this, so let's spend some time really thinking about how to continue.

One method that often shines light on a problem is to consider a special case first. The easiest case, of course, is when $\mu_X = \mu_Y = 0$ and $\sigma_X^2 = \sigma_Y^2 = 1$. In this case a lot of the alphabet soup above vanishes; the hope is that if we can understand this case, we can then see how to do the general case.

14.3.1 Case 1: $\mu_X = \mu_Y = 0$ and $\sigma_X^2 = \sigma_Y^2 = 1$

In this case, we have

$$f_{X+Y}(z) = \frac{1}{2\pi} \int_{-\infty}^{\infty} \exp\left(-\frac{t^2 + (z - t)^2}{2}\right) dt.$$

The most natural thing to do is expand everything out and get a polynomial in t. As

$$t^2 + (z - t)^2 = t^2 + t^2 - 2zt + z^2$$
$$= 2t^2 - 2zt + z^2,$$

we find that

$$f_{X+Y}(z) = \frac{1}{2\pi} \int_{-\infty}^{\infty} \exp\left(-\frac{2t^2 - 2zt + z^2}{2}\right) dt.$$

It's time for a change of variable; this is the hardest step in the algebra, as what we need to do is **complete the square** by adding zero. **Adding zero** is one of the most important methods in math, but it takes a long time to become truly proficient at doing nothing! The idea is that if we add zero we haven't changed the value of the expression, but we have possibly simplified the algebra (and **rewriting the algebra** can lead to expressions easier to analyze). We have

$$2t^2 - 2zt + z^2 = 2(t^2 - zt) + z^2$$
$$= 2\left(t^2 - zt + \frac{z^2}{4} - \frac{z^2}{4}\right) + z^2$$
$$= 2\left(\left(t - \frac{z}{2}\right)^2 - \frac{z^2}{4}\right) + z^2$$
$$= 2\left(t - \frac{z}{2}\right)^2 - \frac{z^2}{2} + z^2$$
$$= 2\left(t - \frac{z}{2}\right)^2 + \frac{z^2}{2}.$$

We substitute back, and we find

$$f_{X+Y}(z) = \frac{1}{2\pi} \int_{-\infty}^{\infty} \exp\left(-\frac{2\left(t - \frac{z}{2}\right)^2 + \frac{z^2}{2}}{2}\right) dt$$
$$= \frac{1}{2\pi} \exp\left(-\frac{z^2}{4}\right) \int_{-\infty}^{\infty} \exp\left(-\left(t - \frac{z}{2}\right)^2\right) dt.$$

This integral is screaming for us to do a change of variables. If we let $u = t - \frac{z}{2}$, it simplifies to

$$f_{X+Y}(z) = \frac{1}{2\pi} \exp\left(-\frac{z^2}{4}\right) \int_{-\infty}^{\infty} \exp(-u^2) du.$$

Fortunately, we've already computed this u-integral earlier in the chapter! When we proved the normal distribution integrates to 1, we showed

$$\frac{1}{\sqrt{2\pi\sigma^2}} \int_{-\infty}^{\infty} \exp\left(-(x-\mu)^2/2\sigma^2\right) dx = 1,$$

as this is the density of a normal with mean μ and variance σ^2. Rearranging gives

$$\int_{-\infty}^{\infty} \exp\left(-\frac{(x-\mu)^2}{2\sigma^2}\right) dx = \sqrt{2\pi\sigma^2}, \qquad (14.1)$$

and thus the u-integral above is just $\sqrt{\pi}$. Using this value for the integral, we see

$$f_{X+Y}(z) = \frac{1}{2\pi} \exp\left(-\frac{z^2}{4}\right) \cdot \sqrt{\pi}$$

$$= \frac{1}{2\sqrt{\pi}} \exp\left(-\frac{z^2}{4}\right).$$

Note this is looking an awful lot like a normal. To detect the mean and the variance of a normal, recall the exponential part should look like $-(z-\mu)^2/2\sigma^2$. Thus this suggests that we have a normal with mean 0 and variance 2. If this is the case, the normalization constant outside should be $1/\sqrt{2\pi \cdot 2} = 1/2\sqrt{\pi}$, which is exactly what we have! Thus, writing our answer in a far more suggestive manner, we see that we've shown

$$f_{X+Y}(z) = \frac{1}{\sqrt{2\pi \cdot \sqrt{2}^2}} \exp\left(-\frac{(z-0)^2}{2 \cdot \sqrt{2}^2}\right);$$

in other words, in this case the sum of two independent normals with mean 0 and variance 1 is just a normal with mean 0 and variance $1+1=2$.

 Any time we have algebra taking more than a page it's worth stopping and thinking about what we've done. The key idea is that we've already shown the function

$$\frac{1}{\sqrt{2\pi\sigma^2}} \exp\left(-\frac{(x-\mu)^2}{2\sigma^2}\right)$$

is a density, and thus integrates to 1. We have

$$f_{X+Y}(z) = \int_{-\infty}^{\infty} f_X(t) f_Y(z-t).$$

This gives us an integral of a quadratic in t whose coefficients are at most quadratics in z. By completing the square we got a nice quadratic in t, of the form $2(t-z/2)^2 + z^2/2$. By changing variables with $u = t - z/2$, the new expression in the exponential is $u^2 + z^2/4$. The result of all of this is that we have *separated* the u and the z variables; we can pull the z variables outside the integral, and the u-integral is *just some constant independent of z*. As we've pulled out an exponential of a quadratic in z, we can *see* the Gaussian starting to emerge! All we need to do is find the z-dependence; the rest becomes a constant after integration.

 Here's another interpretation. The **Theory of Normalization Constants** uses the fact that the density integrates to 1 to get the normalization constant. This means that *if* we know we have a probability density *and* the dependence on the variable is the same as the dependence on the variable for some other known density, *then* we can just read off the normalization constant without doing the integration! In particular, let's revisit (14.1). The u-integration gives us some constant, say C, and thus we have

$$f_{X+Y}(z) = \frac{C}{2\pi} e^{-z^2/4}.$$

As f_{X+Y} is a density, it must integrate to 1. Looking at the exponential factor, we see we can write it as $\exp(-(z-0)^2/2 \cdot 2)$. Thus we must have a normal distribution with mean 0 and variance 2, and $C = \sqrt{\pi}$ (as the normalization constant for this Gaussian is $1/\sqrt{2\pi \cdot 2}$).

14.3.2 Case 2: General μ_X, μ_Y and σ_X^2, σ_Y^2

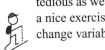 The general case follows similarly; the only difference is the algebra is a bit more tedious as we have more symbols present. The general ideas, however, are the same. As a nice exercise, modify the algebra to get this case too. The easiest way to proceed is to change variables in the integration to take into account the means and variances.

Derivation of the formula for the sum of n independent normal random variables: The general case follows from the first by induction, or if you wish, from massively **grouping parentheses (proof by grouping)**. This is another common technique in math, where if you know something holds for two items, you can extend it to any finite number. See §A.3 for a discussion of this powerful method.

How does this help us for sums of independent normal random variables? We essentially just mirror the above. Now the key input is that if $X \sim N(\mu_X, \sigma_X^2)$ and $Y \sim N(\mu_Y, \sigma_Y^2)$ then $X + Y \sim N(\mu_X + \mu_Y, \sigma_X^2 + \sigma_Y^2)$. In a slight abuse of notation, we'll write this as

$$N(\mu_X, \sigma_X^2) + N(\mu_Y, \sigma_Y^2) = N(\mu_X + \mu_Y, \sigma_X^2 + \sigma_Y^2).$$

As

$$X_1 + X_2 + X_3 = (X_1 + X_2) + X_3,$$

we see

$$N(\mu_1, \sigma_1^2) + N(\mu_2, \sigma_2^2) + N(\mu_3, \sigma_3^2) = \left[N(\mu_1, \sigma_1^2) + N(\mu_2, \sigma_2^2)\right] + N(\mu_3, \sigma_3^2)$$
$$= N(\mu_1 + \mu_2, \sigma_1^2 + \sigma_2^2) + N(\mu_3, \sigma_3^2)$$
$$= N(\mu_1 + \mu_2 + \mu_3, \sigma_1^2 + \sigma_2^2 + \sigma_3^2).$$

If we have more than three independent normals, it's just lather, rinse, repeat. For example, if $n = 4$ we could use the grouping

$$(((X_1 + X_2) + X_3) + X_4).$$

As the sum of two independent normal random variables is so important, we give one more argument to determine the density of the sum. Again, for simplicity, let's assume our two independent normal random variables X and Y have mean 0 and variance 1. The idea is to use a bit of **divine inspiration**. This is a great way to prove results in mathematics, but it has the drawback of requiring you to know or intuit the answer beforehand. This isn't so unreasonable. We know that $X + Y$ has mean 0 and variance 2. Perhaps we've run some numerical experiments, and the plot we see looks normal. The *only* possible normal distribution would be $N(0, 2)$, so let's try that. In other words, if we divide $f_{X+Y}(z)$ by the density of a normal with mean 0 and variance 2, we should get 1. We'll denote this normal's density (i.e., a normal with mean 0 and variance 2) by

$$f_{0,2}(z) = \frac{1}{\sqrt{2\pi}} \exp(-z^2/2 \cdot 2).$$

Now, on to the algebra! We have

$$\frac{f_{X+Y}(z)}{f_{0,2}(z)} = \frac{1}{f_{0,2}(z)} \cdot \frac{1}{2\pi} \int_{-\infty}^{\infty} \exp\left(-\frac{t^2 + (z-t)^2}{2}\right) dt$$

$$= \sqrt{2\pi \cdot 2} \exp\left(\frac{z^2}{2 \cdot 2}\right) \cdot \frac{1}{2\pi} \int_{-\infty}^{\infty} \exp\left(-\frac{t^2 + (z-t)^2}{2}\right) dt$$

$$= \frac{1}{\sqrt{\pi}} \int_{-\infty}^{\infty} \exp\left(\frac{z^2}{4} - \frac{t^2 + (z-t)^2}{2}\right) dt$$

$$= \frac{1}{\sqrt{\pi}} \int_{-\infty}^{\infty} \exp\left(-\frac{2(t^2 + t^2 - 2zt + z^2) - z^2}{4}\right) dt$$

$$= \frac{1}{\sqrt{\pi}} \int_{-\infty}^{\infty} \exp\left(-\frac{4t^2 - 4zt + z^2}{4}\right) dt$$

$$= \frac{1}{\sqrt{\pi}} \int_{-\infty}^{\infty} \exp\left(-\frac{4(t^2 - zt) + z^2}{4}\right) dt.$$

We want an integral we can evaluate. We thus complete the square (for the t quadratic), which will convert the integral to one we know. We find

$$\frac{f_{X+Y}(z)}{f_{0,2}(z)} = \frac{1}{\sqrt{\pi}} \int_{-\infty}^{\infty} \exp\left(-\frac{4(t^2 - zt + z^2/4 - z^2/4) + z^2}{4}\right) dt$$

$$= \frac{1}{\sqrt{\pi}} \int_{-\infty}^{\infty} \exp\left(-\frac{4(t - z/2)^2 - z^2 + z^2}{4}\right) dt$$

$$= \frac{1}{\sqrt{\pi}} \int_{-\infty}^{\infty} \exp\left(-\left(t - \frac{z}{2}\right)^2\right) dt.$$

We can evaluate the integral using (14.1), which gives us the answer for integrals of exponentials of quadratics. For us it looks like the integrand of an exponential with mean $\mu = z/2$ and variance $\sigma^2 = 1/2$, and thus the answer is $\sqrt{2\pi\sigma^2} = \sqrt{2\pi/2} = \sqrt{\pi}$. Substituting this in for the integral, we obtain

$$\frac{f_{X+Y}(z)}{f_{0,2}(z)} = \frac{1}{\sqrt{\pi}} \cdot \sqrt{\pi} = 1,$$

or equivalently

$$f_{X+Y}(z) = f_{0,2}(z) = \frac{1}{\sqrt{2\pi \cdot \sqrt{2}^2}} \exp\left(-\frac{z^2}{2 \cdot \sqrt{2}^2}\right),$$

exactly what we wished to prove!

Obviously, there's a lot of similarity with the algebra here and our first approach. The difference is we tried to use our intuition as to what the answer should be to guide us in attacking the algebra. The *hardest* part of much of mathematics is learning the right way to look at algebra. If you can find the right perspective, you can often see interesting relationships.

14.3.3 Sums of Two Normals: Faster Algebra

Let's look at sums of two normal random variables one more time. This is the problem that keeps on giving, and we can find yet another way to do integration without integrating. There is a terrific technique to be mastered here: often if we can get close to the correct answer we can correct our work and arrive at it exactly, with *much* less work expended than if we attacked it directly. We'll call this the **correcting the guess method**.

Remember if $X \sim N(0, \sigma_X^2)$ and $Y \sim N(0, \sigma_Y^2)$ are two independent Gaussians and $Z = X + Y$ then the pdf of Z, $f_Z(z)$, satisfies

$$f_Z(z) = \int_{t=-\infty}^{\infty} \frac{1}{\sqrt{2\pi\sigma_X^2}} e^{-t^2/2\sigma_X^2} \frac{1}{\sqrt{2\pi\sigma_X^2}} e^{-(z-t)^2/2\sigma_Y^2} dy.$$

Further, as X and Y are independent the mean of Z must be zero and its variance must be $\sigma_X^2 + \sigma_Y^2$. Thus *if* we knew that Z were a normal, *then* the only normal it could be would be $N(0, \sigma_X^2 + \sigma_Y^2)$. We thus concentrate our efforts on showing that Z is a normal, as once we show that, all the constants are immediately determined for us for free!

Notice the argument of the exponential in $f_Z(z)$ is

$$-\left[\frac{t^2}{2\sigma_X^2} + \frac{(z-t)^2}{2\sigma_Y^2}\right].$$

We can expand and simplify, and there must be constants, depending on σ_X and σ_Y, such that the argument of the exponential is

$$-az^2 - b(t - cz)^2;$$

we get this by adding zero to complete the square with respect to t, and then collecting the z^2 piece. Explicitly, if we have $-\alpha t^2 - \beta(z-t)^2$ we note this equals

$$-\beta z^2 - (\alpha + \beta)\left[t^2 - 2\frac{\beta z}{\alpha + \beta} + \left(\frac{\beta z}{\alpha + \beta}\right)^2 - \left(\frac{\beta z}{\alpha + \beta}\right)^2\right] = -az^2 - b(t - cz)^2.$$

How does this help us? We see that

$$f_Z(z) = \int_{t=-\infty}^{\infty} \frac{1}{2\pi\sigma_X\sigma_Y} e^{-az^2} e^{-b(t-cz)^2} dt.$$

We change variables and let $u = t - cz$. Pulling $\exp(-az^2)$ outside the integration yields

$$f_Z(z) = e^{-az^2} \left[\frac{1}{2\pi\sigma_X\sigma_Y} \int_{u=-\infty}^{\infty} e^{-bu^2} du \right] = Ce^{-az^2}$$

for some C depending on σ_X and σ_Y.

Now the miraculous simplifications happen. Note the density is that of a Gaussian with mean 0 and variance $\frac{1}{2a}$, as

$$C \exp(-az^2) = C \exp\left(-\frac{z^2}{2\frac{1}{2a}} \right).$$

As we know the variance of Z is $\sigma_Z^2 = \sigma_X^2 + \sigma_Y^2$, we must have $a = \frac{1}{2(\sigma_X^2+\sigma_Y^2)}$. Further, as we have a Gaussian of mean 0 and variance 1, the normalization constant must be $1/\sqrt{2\pi\sigma_Z^2}$, and thus we find that

$$f_Z(z) = \frac{1}{\sqrt{2\pi(\sigma_X^2+\sigma_Y^2)}} \exp\left(-x^2/2(\sigma_X^2+\sigma_Y^2) \right).$$

It's worth pausing and reflecting on this approach. When we first attacked this problem we did all the algebra fully. It turns out *we don't need to know what the coefficients are when we complete the square*; all that matters is that there are *some* coefficients. When the dust settles, since we know the mean and variance of Z, we can immediately correct our approximation!

14.4 Generating Random Numbers from Normal Distributions

While the title of this section is generating random numbers, we're going to do a lot more. Below we'll learn how to do series expansions to approximate probabilities. These series expansions can be used for many things, from calculating likelihood of events to generating random numbers from a given distribution.

That said, our object is to generate some numbers from the standard normal distribution, given a random number generator that chooses numbers uniformly in [0, 1]. We'd like to use the cumulative distribution method. It worked so well for other cases like the uniform and the exponential distributions, so why not here as well? Unfortunately, the cumulative distribution method is much harder to use for normal distributions. This hardly seems fair, as it's such a nice method and this is one of the most important cases. What goes wrong? The problem goes back to the battle between Calculus I (differentiation) and Calculus II (integration). No matter how evil your professor is, they cannot give you a differentiation problem involving the standard functions that is beyond your ability. Imagine, for example, you are asked to find the derivative of

$$g(x) = \left(\log \left(\cos \left(x^2 + 1 \right) \exp \left(x + \sqrt{\sin x} \right) \right) \right)^2.$$

Even though this is quite intimidating, we can find the derivative by repeated applications of the Rules of Differentiation, and the fact that all the elementary functions have known derivatives. In this case, we see we can write $g(x)$ as

$$g(x) = (A(x))^2, \quad A(x) = \log \left(\cos \left(x^2 + 1 \right) \exp \left(x + \sqrt{\sin x} \right) \right).$$

This isn't bad at all; it's just the power rule, and we have

$$g'(x) = 2A(x)A'(x).$$

Now we turn to finding $A'(x)$. We can write *that* as

$$A(x) = \log B(x), \quad B(x) = \cos \left(x^2 + 1 \right) \exp \left(x + \sqrt{\sin x} \right),$$

and we just use the chain rule to see

$$A'(x) = \frac{1}{B(x)} \cdot B'(x).$$

The point is that if we're patient (as well as *very* careful), we'll always end up with the answer. Each step decreases the complexity of the mess, and eventually we'll have no more functions to differentiate. Patience pays off! It may not be pleasurable, but there's no mystery in how to proceed.

Integration is much harder. Our integration rules can help us, but it's a completely different game. Think of differentiation as a very mechanical procedure. Check and see if it's a sum of two functions. If it is, use the sum rule. If not, see if it's a difference. A product. A composition. You get the point. We can just methodically go through our list, and eventually we'll get it. For integration we have no procedure to fall back on that always works. Most functions, in fact, don't have a nice, closed form integral. Sometimes a function does have a nice integral, but it requires a trick or insight to see it. For example, if $f(x) = \log x$ then $F(x) = x \log x - x$ is its anti-derivative. While it's hard to find this, it's easy to verify. (As a fun exercise, find the "nice" anti-derivative for $f(x) = x^5 \log^2 x$.) That said, there is *no* nice expression for the integral of $f(x) = \sin x \cdot \log x$.

You can probably guess where all this is going: there is no nice expression for the integral of the normal's density. This means that we don't have a nice formula for its cumulative distribution function, which is a huge blow to using the cumulative distribution method to generate random numbers from a normal distribution.

What can we do? We have to resort to approximations. One way is to numerically integrate the normal's density, and use that to approximate the cumulative distribution function. While this won't be exact, if we choose a sufficiently small step size, the error can be made arbitrarily small. If you remember methods such as Simpson's rule from Calculus II, this is a great time to use them!

The other possibility is to try and numerically approximate the cumulative distribution function. If we let $\phi(x)$ denote the density of the standard normal, its cumulative distribution function is

$$\Phi(x) = \int_{-\infty}^{x} \phi(t) dy = \int_{-\infty}^{x} \frac{1}{\sqrt{2\pi}} e^{-t^2/2} dt.$$

Ack—how can we ever hope to integrate this! Well, let's think. What's the most natural thing to do? We do know the series expansion for the exponential. So, let's put that in and see what we get. We won't worry about convergence or justifying any of the interchanges. Let's just play fast and loose with the math and see what happens in an attempt to shed some light on what this function $\Phi(x)$ looks like. We find

$$\Phi(x) = \frac{1}{\sqrt{2\pi}} \int_{-\infty}^{x} \sum_{n=0}^{\infty} \frac{1}{n!} \left(-\frac{t^2}{2}\right)^n dt$$

$$= \frac{1}{\sqrt{2\pi}} \sum_{n=0}^{\infty} \frac{(-1)^n}{2^n n!} \int_{-\infty}^{x} t^{2n} dt.$$

We are irretrievably up the creek now—there is no way to make this work. Why? Even if we can justify switching the orders of the integral and the sum (which in general is a good thing to try), we end up with a horrible integral. The problem is that the integral of t^{2n} from $-\infty$ to x is *always* infinite, and thus our expression is going to be an alternating sum of positive and negative infinities. You should have been taught that there are several expressions that you must avoid and never, ever do; do not try to make sense of $\infty - \infty$ or $\infty \cdot 0$ or ∞/∞.

This seems to doom our approach to finding a nice series expansion for $\Phi(t)$. All is not lost, however. As you read on, remember that the issue was that we were integrating from $-\infty$ to x; there was no problem in evaluating the integrals at x, only at $-\infty$. There *is* a way to get a very nice, tractable expression. It involves exploiting some symmetry in the problem. **Exploiting symmetries** is a powerful way to assault a variety of intractable problems; see §A.4 for a longer discussion. Let's record everything we know about $\Phi(x)$. Well, $\lim_{x \to -\infty} \Phi(x) = 0$ and $\lim_{x \to \infty} \Phi(x) = 1$. This isn't that much; this is true for all cumulative distribution functions. Fortunately, there are two more very useful facts. The first is

$$\Phi(0) = 1/2.$$

This just means that half the probability comes before 0, and half after. The reason this is true is that our probability density $\phi(t)$ is an even function which is symmetric about zero. Thus, half of the probability falls before 0, and half after.

We're in luck, as even more is true. As our density $\phi(t)$ is symmetric about 0, it suffices to know how to compute $\int_0^x \phi(t) dt$ for any fixed, positive x; if we can do that, we can find $\Phi(x)$ easily for any input!

How does this work? There are two cases. First assume our input is non-negative, so we write $x \geq 0$. Then

$$\Phi(x) = \int_{-\infty}^{x} \phi(t)dt$$

$$= \int_{-\infty}^{0} \phi(t)dt + \int_{0}^{x} \phi(t)dt$$

$$= \frac{1}{2} + \int_{0}^{x} \phi(t)dt;$$

thus in this case, our claim holds.

What if we had negative input? Let's write it as $-x$, with $x \geq 0$. Then

$$\Phi(-x) = \int_{-\infty}^{-x} \phi(t)dt$$

$$= \int_{-\infty}^{0} \phi(t)dt - \int_{-x}^{0} \phi(t)dt$$

$$= \frac{1}{2} - \int_{-x}^{0} \phi(t)dt$$

$$= \frac{1}{2} - \int_{0}^{x} \phi(t)dt.$$

This means we can compute Φ at any input so long as we can compute $\int_{0}^{x} \phi(t)dt$ for x positive. What's the advantage of all this? This observation allows us to use our series expansion *without* the danger of integrating over an infinite interval. One must be extremely careful when dealing with infinities—if you can avoid them, so much the better! We'll now find a nice series expansion. Again, the reason we can do this is that there is an enormous amount of symmetry of the density function to exploit. Now, to the details. Let's take $x \geq 0$. Then

$$\Phi(x) = \int_{-\infty}^{x} \phi(t)dt$$

$$= \int_{-\infty}^{0} \phi(t)dt + \int_{0}^{x} \phi(t)dt$$

$$= \frac{1}{2} + \int_{0}^{x} \frac{1}{\sqrt{2\pi}} e^{-t^2/2} dt$$

$$= \frac{1}{2} + \frac{1}{\sqrt{2\pi}} \int_{0}^{x} \sum_{n=0}^{\infty} \frac{1}{n!} \left(-\frac{t^2}{2}\right)^n$$

$$= \frac{1}{2} + \frac{1}{\sqrt{2\pi}} \sum_{n=0}^{\infty} \frac{(-1)^n}{2^n n!} \int_{0}^{x} t^{2n} dt.$$

Note we're now integrating a well-behaved function, t^{2n}, over a finite interval. Though the values of these integrals grow, they're hit by the factor $1/2^n n!$, which decays so rapidly that the resulting sum converges. Explicitly, we see that

$$\Phi(x) = \frac{1}{2} + \frac{1}{\sqrt{2\pi}} \sum_{n=0}^{\infty} \frac{(-1)^n}{2^n n!} \frac{x^{2n+1}}{2n+1} \quad \text{if } x \ge 0.$$

Similarly, we find

$$\Phi(-x) = \frac{1}{2} - \int_0^x \phi(t)dt = \frac{1}{2} - \frac{1}{\sqrt{2\pi}} \sum_{n=0}^{\infty} \frac{(-1)^n}{2^n n!} \frac{x^{2n+1}}{2n+1} \quad \text{if } x \ge 0.$$

Let's test and see how good of an approximation this gives by computing the probability a standard normal random variable takes on a value between -2 and 2. This is

$$\text{Prob}(-2 \le X \le 2) = \int_{-2}^{2} \phi(t)dt$$

$$= \int_{-\infty}^{2} \phi(t)dt - \int_{-\infty}^{-2} \phi(t)dt$$

$$= \Phi(2) - \Phi(-2)$$

$$= \frac{2}{\sqrt{2\pi}} \sum_{n=0}^{\infty} \frac{(-1)^n}{2^n n!} \frac{2^{2n+1}}{2n+1}$$

$$= \frac{2}{\sqrt{2\pi}} \sum_{n=0}^{\infty} \frac{(-1)^n 2^{n+1}}{n!(2n+1)},$$

where this comes from subtracting (14.4) from (14.2). Taking the first five terms of the expansion gives us

$$\text{Prob}(-2 \le X \le 2) \approx \frac{2}{\sqrt{2\pi}} \left[2 - \frac{4}{3} + \frac{4}{5} - \frac{8}{21} + \frac{4}{27} \right] \approx 0.98448;$$

the correct answer being about 0.954499736. We see with just five terms we get a pretty good approximation, and with very little work. The reason the approximation isn't better is we got a factor of $2^{2n+1}/(2n+1)$ when we integrated; this is so large that when we divide by $2^n n!$ we have $2^{n+1}/n!(2n+1)$. While eventually $n!$ grows far faster than 2^n, for small n the two are of comparable size, and thus the first several terms have significant contributions to the probability. If we went up to 10 terms, our approximation is 0.954481375. With just 10 terms, our error is about 0.000018361, which is quite small!

In Figure 14.2 we plot some series approximations to $\Phi(2) - \Phi(-2)$, the probability our standard normal random variable takes on a value between -2 and 2. Note that very quickly we have a good fit. With a little bit of work, we can study the series expansion

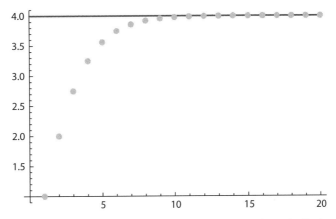

Figure 14.2. Plot of convergence of series approximation to $\Phi(2) - \Phi(-2)$, the probability a standard normal takes on a value in $[-2, 2]$. We see that 6 terms suffice to give a great approximation.

and see how many terms we need for a good approximation. This is possible as our series expansion alternates between being positive and negative, and eventually each term is less in absolute value than the previous. This means truncating at a positive term eventually gives an upper bound, while having the last kept term be a negative one yields a lower bound.

Though we couldn't solve the original problem, namely come up with a simple, exact formula for the integral of $\phi(t)$, with a little bit of work we can approximate the integrals to whatever level of accuracy we need, and often without having to take too many terms. Try to do as well as we've done using Simpson's rule or any other numerical approximation to the integral—the series expansion is far more pleasant! The reason we can get such a nice result is that we can expand the integrand as an infinite sum, and integrate term by term. Up to here, our answer is exact; the only error comes when we truncate, but our series is so rapidly converging that this isn't too bad.

No introduction to Φ would be complete without discussing the error function. Frequently when mathematicians encounter an integral they can't evaluate they give it a name. This doesn't mean that we can now evaluate the integral; rather, it means that this integral occurs so frequently it's worthy of a name, and almost surely this means someone has tabulated many of its values or computed its series expansion. As the normal distribution is so important to probability, it's not surprising that its integral has been named. Unfortunately, the way it was named isn't what you would expect. It's called the **error function**, denoted Erf(x), and it is

$$\text{Erf}(x) := \frac{2}{\sqrt{\pi}} \int_0^x e^{-t^2} dt.$$

Sadly, we're stuck with this notation. This is the integral of a normal with variance $1/2$ from 0 to x. While it would be nice to have variance 1, this choice was made so that the argument of the exponential is nice ($-t^2$ instead of $-t^2/2$). With a little bit of algebra and some change of variables, we can relate our series expansions to this. Let's do the case we considered above, finding the probability that our standard normal takes on a

value between -2 and 2. We have

$$\text{Prob}(-2 \leq X \leq 2) = \Phi(2) - \Phi(-2)$$

$$= \int_{-2}^{2} \phi(t)dt$$

$$= 2\int_{0}^{2} \phi(t)dt$$

$$= 2\int_{0}^{2} \frac{1}{\sqrt{2\pi}} e^{-t^2/2}dt$$

$$= 2\int_{0}^{2/\sqrt{2}} \frac{1}{\sqrt{2\pi}} e^{-u^2}\sqrt{2}du$$

$$= \frac{2}{\sqrt{\pi}} \int_{0}^{\sqrt{2}} e^{-u^2}du$$

$$= \text{Erf}\left(\sqrt{2}\right).$$

If instead of the interval -2 to 2 we had $-x$ to x, all that would change is that we'd evaluate at $x/\sqrt{2}$ and not $2/\sqrt{2} = \sqrt{2}$. We've thus found the series expansion for the error function!

If Φ is the cumulative distribution function of the standard normal, and $\text{Erf}(x)$ is the error function, then

$$\Phi(x) - \Phi(-x) = \text{Erf}\left(\frac{x}{\sqrt{2}}\right) = \frac{2}{\sqrt{2\pi}} \sum_{n=0}^{\infty} \frac{(-1)^n x^{n+1}}{n!(2n+1)}.$$

14.5 Examples and the Central Limit Theorem

In this section we'll see how to use our results on the normal distribution to compute probabilities. Such calculations are very important in statistics, and you'll see them again if you do Chapter 22.

Let X be a normal random variable with $\mu = 5$ and $\sigma^2 = 16$. Find (a) $\text{Prob}(2 < X < 7)$ and (b) $\text{Prob}(X > 0)$.

We proceed by **standardizing** our random variable; this means that we change variables and convert the calculations to ones concerning a standard normal, which has mean 0 and variance 1. This allows us to use tables of values of the standard normal distribution; in other words, it suffices to tabulate the behavior of *one* normal distribution to be able to handle them all (for more on this see the index entries on the change of base formula, or Exercise 14.6.4).

We adjust by sending $X \to \frac{X-\mu}{\sigma}$. We denote the standard normal by Z.
For (a), we have

$$\text{Prob}(2 < X < 7) = \text{Prob}\left(\frac{2-5}{4} < \frac{X-5}{4} < \frac{7-5}{4}\right)$$

$$= \text{Prob}\left(\frac{-3}{4} < Z < \frac{1}{2}\right)$$

$$= \Phi(1/2) - \Phi(-3/4)$$

$$= \Phi(1/2) - \left(1 - \frac{3}{4}\right) \approx 0.4649,$$

while for (b) we find

$$\text{Prob}(X > 0) = \text{Prob}\left(\frac{X-5}{4} > \frac{0-5}{4}\right)$$

$$= \text{Prob}\left(Z > -\frac{5}{4}\right)$$

$$= 1 - \Phi\left(-\frac{5}{4}\right) = \Phi\left(\frac{5}{4}\right).$$

$$\approx 0.8944.$$

You might have noticed we only have a table of z-values, and that it is only for the standard normal distribution. As in our previous example, any normal distribution can be standardized. If we have a random variable with mean μ and variance δ^2, it can be standardized by sending $X \to \frac{X-\mu}{\sigma}$, where we can then use the table. For this reason we only need one table for all distributions.

14.6 Exercises

Exercise 14.6.1   *We split the integral involving e^{-v^2} (in bounding the integral of a normal) into two regions, $[0, 1]$ and $[1, \infty)$, and used the function 1 as a bound on the first interval and e^{-v} on the second. Could we have bounded it by e^{-v} everywhere? Why or why not?*

Exercise 14.6.2   *Which of the following distributions can be well approximated by a normal distribution for certain values of their parameters: continuous uniform, exponential, Erlang, Poisson, geometric, binomial, negative binomial, chi-square?*

Exercise 14.6.3   *(Product of Perpendicular Lines) Here is a fun example where $\infty \cdot 0 = -1$. Consider two perpendicular lines going through the origin. If the two lines are not in the direction of the coordinate axes, show that if one has slope m then the other has slope $-1/m$; thus the product of the two slopes is -1. In the limit as the two lines align with the axis, while at each moment the product of the slopes is -1 the x-axis has slope 0 and the y-axis has slope ∞. Thus, in this case, it is natural to say $\infty \cdot 0 = -1$.*

Exercise 14.6.4   *The process of **standardizing** a random variable is similar to something you might have seen with tables of logarithms (or probably not, as very few* *people use tables these days, as even phones can take logarithms). In the "old" days there were look-up tables to compute logarithms (and in the "older" days some people's jobs were to compute these values). It takes a lot of time to create these tables, and you don't want to have to carry them all with you. Fortunately, if you can find logarithms in* one *base you can find them in* any *base! This is because of the **change of base formula**, which says* $\log_c x = \log_b x / \log_b c$ *(thus, if we know logarithms base b, we can easily find them in any other base). Prove this formula.*

Exercise 14.6.5   *Prove that the normal distribution is symmetric about the line* $x = \mu$.

Exercise 14.6.6   *Consider the density of the standard normal distribution,* $f(x) = \frac{1}{\sqrt{2\pi}} e^{\frac{-x^2}{2}}$. *Find* $f'(x)$ *in terms of* $f(x)$.

Exercise 14.6.7   *Find the second derivative of the pdf of the standard normal distribution,* $f''(x)$ *in terms of* $f(x)$.

Exercise 14.6.8   *Find all the odd centered moments for a normal distribution with parameters* (μ, σ^2).

Exercise 14.6.9   *Find all the even centered moments for a normal distribution with parameters* (μ, σ^2).

Exercise 14.6.10   *Discuss a method for generating random numbers from a standard normal distribution using that the sum of independent identically distributed random variables tends to be normal.*

Exercise 14.6.11   *Use the Taylor approximation to estimate the probability that a random variable selected from a standard normal distribution falls within a standard deviation of the mean to 4 decimal places. Justify the precision of your approximation.*

Exercise 14.6.12   *Must variables be independent for the Central Limit Theorem to apply? Explain.*

Exercise 14.6.13   *Let X be a normal variable with* $\mu = 3$ *and* $\sigma^2 = 25$. *Find* $Prob(5 < X < 7)$ *using standardization.*

Exercise 14.6.14   *If X is normally distributed and* $Prob(2 < X < 3) = \frac{1}{5}$ *and* $Prob(X < 0) = \frac{1}{4}$, *what are* μ *and* σ^2?

Exercise 14.6.15   *If X is normally distributed and* $Prob(X < 1) = \frac{1}{4}$, *what is the smallest possible value of* μ?

Exercise 14.6.16   *For what* λ *is the area within two standard deviations of the mean of a Poisson within a factor of 1.3 of the area within two standard deviations of the mean in a normal distribution?*

Exercise 14.6.17   *What constraints must be placed on a binomial distribution so that the area inside one standard deviation of the mean is within a factor of 1.2 of the area within one standard deviation of the normal model?*

Exercise 14.6.18   *Why are errors in measurement and production often believed to be normally distributed?*

Exercise 14.6.19   *Let X and Y be independent normally distributed random variables with mean 0 and variance 1. Find the pdf for* $\max(X - Y, Y - X)$.

Exercise 14.6.20   *Let X be a standard normal random variable. Find the pdf for X^2.*

Exercise 14.6.21   *Show that the product of two independent normal distribution random variable is also a normal distribution random variable.*

Exercise 14.6.22   *What is the joint probability density function for two independent normal distributions $X_1 \sim N(\mu_1, \sigma_1^2)$ and $X_2 \sim N(\mu_2, \sigma_2^2)$?*

Exercise 14.6.23   *In physical chemistry it is often desirable to know the location of a particle. Consider the one-dimensional case. We will assume, as is often done, that this can be modeled by a random walk, with probability of moving in either direction a distance of 1 unit equal to 1/2 after each unit of time. Fortunately, each unit of time is very small since particles move quickly. Explain why a normal model is useful for finding the probability of the particle being in a certain location after a sufficiently long amount of time.*

Exercise 14.6.24   *What is your best guess for the location of a particle that moves as described in the previous exercise? What is your best guess for how far the particle is from its initial location?*

Exercise 14.6.25   *One method for numerically approximating the area of a region is to place the region in a larger rectangle and "throw darts" or more precisely, pick points from the larger rectangle with uniform probability. If you keep track of how many darts land in the region of interest as a percentage of the darts thrown, this should converge to the area of the region divided by the area of the rectangle as the number of darts increase. This technique is known as **Monte Carlo integration**. Use Monte Carlo integration to numerically approximate the area within one standard deviation of a normal curve.*

Exercise 14.6.26   *Use Monte Carlo integration to approximate the probability a random variable sampled from a standard normal distribution is less than 2 or greater than 3.*

Exercise 14.6.27   *If the sum of two linearly independent copies of a random variable, after possibly scaling the resulting mean and variance, has the same distribution as the summands we say the initial distribution is **stable**. Our results in this chapter show that the normal distribution is stable. What about other distributions such as uniform, exponential, or binomial? What about the Cauchy distribution, which has density $f_X(x) = \frac{1}{\pi}(1 + x^2)^{-1}$?*

Exercise 14.6.28   *Suppose X is a normal distribution with mean μ and variance σ. Prove that $X = \alpha X + \beta$ has mean $(\alpha\sigma + \beta)$ and variance $(\alpha\sigma)$.*

Exercise 14.6.29   *At Infinity Co., the temperatures of its rooms follow a normal distribution with a mean of 55 degrees Fahrenheit and a standard deviation of 2 degrees. Given a random room, what is the probability of its temperature being between 55 and 65 degrees Fahrenheit?*

Exercise 14.6.30   *You are hiring for your company, Infinity Co., and your job as the only employee on floor B1 is to conduct phone interviews with potential employees. Each interviewee was told to call at 6:00 PM sharp. Your company values early birds, and your instructions are to only pass callers who call between 5:45 PM and 5:55 PM. Suppose the time the candidates call approximate a continuous normal distribution with* *a mean of 6:00 PM and a standard deviation of 5 minutes. If you must hire 15 applicants, approximately how many total callers were there?*

Exercise 14.6.31   *You are an employee at Infinity Co., and you work in the basement on floor B1 (or floor −1). The building has an infinite number of floors with range $(-\infty, \infty)$. You're waiting for the elevator so you can return home, but on this day the elevator makes random stops along the elevator shaft. Unfortunately, your floor is one of the only two floors without an emergency stairwell or exit. In accordance with safety regulations, the elevator door will only open if the elevator overshot or undershot a floor by a margin of one-fifth of a floor. For example, the elevator door will only open on floor 44 if the elevator stops in [43.8, 44.2]. If the elevator stops at 44.5, no door on any floor will open. Suppose the stops of the elevator form a normal distribution with a mean of 45 and standard deviation of 10. If each stop takes one minute, how long would you expect to wait before the elevator reaches floor B1?*

Exercise 14.6.32   *Your friend works at Infinity Co. on floor B24, and the building has an infinite number of floors with range $(-\infty, \infty)$. Unfortunately, his floor is one of the only two floors in the building without an emergency stairwell or exit. He has an important meeting in an hour on floor 291, and if he is not there on time his relations with the company will be immediately terminated. The conditions of the elevator are the same as in the previous exercise. Your friend complains to Human Resources, and requests elevator service. After consulting with the company manager, they agree to help your friend, but there are conditions. They will directly teleport the friend to the meeting, but only if either one of two conditions are met: if the elevator arrives on your friend's floor, or the elevator stops two times in a row on the same floor, and this floor is lower than the 30th floor. A stop is only counted if the elevator door on that floor opens (see the previous exercise). For example, the elevator is only counted as stopping on floor 25 if the number selected is in the range [24.8, 25.2]. Suppose the stops of the elevator form a normal distribution with a mean of 45 and standard deviation of 10, and each elevator stop takes one minute. What is the probability that your friend makes it before he is late and consequently terminated?*

Exercise 14.6.33   *Following the previous question: Suppose the company manager at Infinity Co. was in a generous mood, and programmed the elevator so it would reroll if the random number generator returned any integer within [30, 100]. The building has an infinite number of floors going upward and downward.*

1. *How much would your friend's odds improve? Why? (Hint: Use the Method of **Divine Inspiration**.)*

2. *The generous company manager at Infinity Co. wishes to extend his generosity infinitely. He will now have the elevator reroll if the random number generator returned any integer within [30, ∞]. How would your friend's odds change?*

3. *It was suggested to the generous company manager that he would have the elevator reroll instead on the interval [45.9, 46], but the manager refused on the grounds that it would decrease the probability of success. Assuming the company manager was well versed in probability, what is his definition of success?*

Exercise 14.6.34   *The Infinity Co. building where you work on floor B1 has an infinite number of floors going upward and downward. You seem to be stuck indefinitely on your floor. Seeking escape, you find a teleporter in the closet and instantly make it to floor B24. At this point, the Human Resources is alerted, and, concerned, they teleport two* *agents: to floors B1 and to B29. They also close all emergency exits and stairwells. You and the agent on B29 have teleporters that will teleport you according to a standard normal distribution with mean +2 floors and a standard deviation of 2. The agent on B1 will teleport with a mean of −2 floors and a standard deviation of 2. Note that although logistically, if you are teleported to floor 2.5, you really are on floor 2, the teleporter will calculate the next leap as if you were on floor 2.5. That is, the probability function for the next floor would have a distribution with mean (2.5 + 2 = 4.5) and a standard deviation of 2 floors. From this point on, you and the agents teleport simultaneously once every thirty seconds. If at any point you are on the same floor as an agent, your teleporter will be disabled, and you will be brought to the office of Human Resources for paperwork. If you make it to the first floor before then, you are free to go home.*

1. *How likely are you to go home?*
2. *How likely are you to be caught and be brought to the office of Human Resources for paperwork? Which agent is more likely to catch you?*
3. *How long do you expect to last before you're either caught or free?*

CHAPTER 15 _____

The Gamma Function and Related Distributions

In this chapter we'll explore some of the strange and wonderful properties of the Gamma function $\Gamma(s)$.

For $s > 0$ (or actually $\Re(s) > 0$), the **Gamma function** $\Gamma(s)$ is

$$\Gamma(s) := \int_0^\infty e^{-x} x^{s-1} dx = \int_0^\infty e^{-x} x^s \frac{dx}{x}.$$

There are countless integrals or functions we can define. Just looking at it, there's nothing that suggests it's one of the most important functions in all of mathematics, appearing throughout probability and statistics (and many other fields), but it is. We'll see where it occurs and why, and discuss many of its most important properties. If you can't wait, before reading on evaluate the integral for $s = 1, 2, 3$, and 4, and try and figure out the pattern. Also try to figure out why we rewrote $x^{s-1} dx$ as $x^s dx / x$.

15.1 Existence of $\Gamma(s)$

Looking at the definition of $\Gamma(s)$, it's natural to ask: *Why do we have restrictions on s?* Whenever you're given an integrand, you must make sure it's well-behaved before you can conclude the integral exists. The purpose of this section is to highlight some useful techniques to investigate integrals. Frequently there are two trouble points to check, near $x = 0$ and near $x = \pm\infty$ (okay, three points).

For example, consider the function $f(x) = x^{-1/2}$ on the interval $[0, \infty)$. This function blows up at the origin, but only mildly. Its integral is $2x^{1/2}$, and this is integrable near the origin. This just means that

$$\lim_{\epsilon \to 0} \int_\epsilon^1 x^{-1/2} dx$$

exists and is finite. Unfortunately, even though this function is tending to zero, it approaches zero so slowly for large x that it's not integrable on $[0, \infty)$. The problem

is that integrals such as

$$\lim_{B \to \infty} \int_1^B x^{-1/2} dx$$

are infinite. Can the reverse problem happen, namely our function decays fast enough for large x but blows up too rapidly for small x? Sure—consider $g(x) = 1/x^2$. Note g has a nice integral:

$$G(x) = \int g(x) dx = \int \frac{dx}{x^2} = -\frac{1}{x}.$$

Now the integral over large x is fine and finite, being just

$$\lim_{B \to \infty} \int_1^B g(x) dx = \lim_{B \to \infty} -\frac{1}{x}\Big|_1^B = \lim_{B \to \infty} \left[1 - \frac{1}{B}\right] < \infty;$$

however, the integral over small x blows up:

$$\lim_{\epsilon \to \infty} \int_\epsilon^1 g(x) dx = \lim_{\epsilon \to 0} -\frac{1}{x}\Big|_\epsilon^1 = \lim_{\epsilon \to 0} \left[\frac{1}{\epsilon} - 1\right] = \infty.$$

So it's possible for a positive function to fail to be integrable because it decays too slowly for large x, or it blows up too rapidly for small x. As a rule of thumb, if as $x \to \infty$ a function is decaying faster than $1/x^{1+\epsilon}$ for any epsilon, then the integral at infinity will be finite. For small x, if as $x \to 0$ the function is blowing up slower than $x^{-1+\epsilon}$ then the integral at 0 will be okay near zero. You should always do tests like this, and get a sense for when things will exist and be well-defined.

Returning to the Gamma function, let's make sure it's well-defined for any $s > 0$. The integrand is $e^{-x} x^{s-1}$. As $x \to \infty$, the factor x^{s-1} is growing polynomially but the term e^{-x} is decaying exponentially, and thus their product decays rapidly. If we want to be a bit more careful and rigorous, we can argue as follows: choose some integer $M > s + 1701$ (we put in a large number to alert you to the fact that the actual value of our number does not matter). We clearly have $e^x > x^M/M!$, as this is just one term in the Taylor series expansion of e^x (all terms have a positive contribution as $x > 0$). Thus $e^{-x} < M!/x^M$, and the integral for large x is finite and well-behaved, as it's bounded by

$$\int_1^B e^{-x} x^{s-1} dx \leq \int_1^B M! x^{-M} x^{s-1} dx$$

$$= \int M! \int_1^B x^{s-M-1}$$

$$= M! \frac{x^{s-M}}{s-M}\Big|_1^B$$

$$= \frac{M!}{s-M} \left[\frac{1}{B^{M-s}} - 1\right].$$

Remember, our goal is not just to understand the Gamma function, but to understand functions in general. Thus, it's important to get a sense of what techniques are available, and when a method has a chance of succeeding. Our approach above was a very good choice. We know e^x grows very rapidly, so e^{-x} decays quickly. We're **borrowing some of the decay** from e^{-x} to handle the x^{s-1} piece; **borrowing decay** is a great technique to bound the behavior of integrals.

What about the other issue, near $x = 0$? Well, near $x = 0$ the function e^{-x} is bounded; its largest value is when $x = 0$ so it's at most 1. Thus

$$\int_0^1 e^{-x} x^{s-1} dx \ \leq \ \int_0^1 1 \cdot x^{s-1} dx \ = \ \left. \frac{x^s}{s} \right|_0^1 \ = \ \frac{1}{s}.$$

We've shown everything is fine for $s > 0$; what if $s \leq 0$? Could these values be permissible as well? The same type of argument as above shows that there are no problems when x is large. Unfortunately, it's a different story for small x. For $x \leq 1$ we clearly have $e^{-x} \geq 1/e$; before we had an upper bound to show the integral was okay, now we need a lower bound to show it blows up. Thus our integrand is at least as large as x^{s-1}/e. If $s \leq 0$, this is no longer integrable on $[0, 1]$. For definiteness, let's do $s = -2$. Then we have

$$\int_0^\infty e^{-x} x^{-3} dx \ \geq \ \int_0^\infty \frac{1}{e} x^{-3} dx \ = \ \left. -\frac{1}{e} x^{-2} \right|_0^1 \ = \ \infty,$$

and this blows up.

The arguments above can (and should!) be used every time you meet an integral. Even though our analysis hasn't suggested a reason why anyone would *care* about the Gamma function, we at least know that it's well-defined and exists for all $s > 0$. In the next section we'll show how to make sense of Gamma for all values of s. This should be a bit alarming—we've just spent this section talking about being careful and making sure we only use integrals where they are well-defined, and now we want to talk about putting in values such as $s = 1/2$? Obviously, whatever we do, it won't be anything as simple as just plugging $s = 1/2$ into the formula.

If you're interested, $\Gamma(1/2) = \sqrt{\pi}$—we'll prove this soon!

If you're looking for a fun integral, explore whether or not $\int_0^\infty f(x) dx$ exists, where

$$f(x) \ = \ \begin{cases} \dfrac{1}{(x + 1) \log^2(x + 1)} & \text{if } x > 0 \\ 0 & \text{otherwise.} \end{cases}$$

Is the integral fine at infinity? At zero?

15.2 The Functional Equation of $\Gamma(s)$

We turn to *the* most important property of $\Gamma(s)$. This property allows us to make sense of *any* value of s as input, such as the $s = 1/2$ of the last section. Obviously this can't mean just naively throwing in any s in the definition, though many good mathematicians have accidentally done so. What we're going to see is the **Analytic (or Meromorphic) Continuation**. The gist of this is that we can take a function f that makes sense in one region and extend its definition to a function g defined on a larger region in such a way that our new function g agrees with f where they are both defined, but g is defined for more points.

The following absurdity is a great example. What is

$$1 + 2 + 4 + 8 + 16 + 32 + 64 + \cdots?$$

Well, we're adding all the powers of 2, thus it's clearly infinity, right? Wrong—the "natural" meaning for this sum is -1! A sum of infinitely many positive terms is negative? What's going on here?

This example comes from something you've probably seen many times, a geometric series. If we take the sum

$$1 + r + r^2 + r^3 + r^4 + r^5 + r^6 + \cdots$$

then, *so long as* $|r| < 1$, the sum is just $\frac{1}{1-r}$. There are many ways to see this. The most common, as well as one of the most boring, is to let

$$S_n = 1 + r + \cdots + r^n.$$

If we look at $S_n - r S_n$, almost all the terms cancel; we're left with

$$S_n - r S_n = 1 - r^{n+1}.$$

We factor the left-hand side as $(1 - r)S_n$, and then dividing both sides by $1 - r$ gives

$$S_n = \frac{1 - r^{n+1}}{1 - r}.$$

If $|r| < 1$ then $\lim_{n \to \infty} r^n = 0$, and thus taking limits gives

$$\sum_{m=0}^{\infty} r^m = \lim_{n \to \infty} S_n = \lim_{n \to \infty} \frac{1 - r^{n+1}}{1 - r} = \frac{1}{1 - r}.$$

This is known as the **geometric series formula**, and is used in a variety of problems. See §1.2 for a more entertaining derivation.

Let's rewrite the above. The summation notation is nice and compact, but that's not what we want right now—we want to really see what's going on. We have

$$1 + r + r^2 + r^3 + r^4 + r^5 + r^6 + \cdots = \frac{1}{1 - r}, \quad |r| < 1.$$

Note the left-hand side makes sense only for $|r| < 1$, but the right-hand side makes sense for *all* values of r other than 1! We say the right-hand side is an **analytic continuation** of the left, with a **pole** at $s = 1$ (poles are where our functions blow-up).

Let's define the function

$$f(x) = 1 + x + x^2 + x^3 + x^4 + x^5 + x^6 + \cdots .$$

For $|x| < 1$ we also have

$$f(x) = \frac{1}{1-x}.$$

We're now ready for the big question: what's $f(2)$? If we use the second definition, it's just $\frac{1}{1-2} = -1$, while if we use the first definition it's that strange sum of all the powers of 2. *THIS* is the sense in which we mean the sum of all the powers of 2 is -1. We don't mean plugging in 2 for the series expansion; instead, we evaluate the extended function at 2.

It's now time to apply these techniques to the Gamma function. We'll show, using integration by parts, that $\Gamma(s)$ can be extended for all s (or at least for all s except the negative integers and zero). Before doing the general case, let's do a few representative examples to see why integration by parts is such a good thing to do, and to get a feeling for the Gamma function's behavior. Recall

$$\Gamma(s) = \int_0^\infty e^{-x} x^{s-1} dx, \quad s > 0.$$

The easiest value of s to take is $s = 1$, as then the x^{s-1} term becomes the harmless $x^0 = 1$. In this case, we have

$$\Gamma(1) = \int_0^\infty e^{-x} dx = -e^{-x} \Big|_0^\infty = -0 + 1 = 1.$$

Building on our success, what's the next easiest value of s to take? A little experimentation suggests we try $s = 2$. This makes x^{s-1} equal x, a nice integer power. We find

$$\Gamma(2) = \int_0^\infty e^{-x} x \, dx.$$

Now we can begin to see why integration by parts will play such an important role. If we let $u = x$ and $dv = e^{-x} dx$, then $du = dx$ and $v = -e^{-x}$, then we'll see great progress—we start with needing to integrate xe^{-x} and after integration by parts we're left with having to do e^{-x}, a wonderful savings. Putting in the details, we find

$$\Gamma(2) = uv \Big|_0^\infty - \int_0^\infty v \, du = -xe^{-x} \Big|_0^\infty + \int_0^\infty e^{-x} dx.$$

The boundary term vanishes (it's clearly zero at zero; use L'Hôpital's rule to evaluate it at ∞, giving $\lim_{x \to \infty} \frac{x}{e^x} = \lim_{x \to \infty} \frac{1}{e^x} = 0$), while the other integral is just $\Gamma(1)$. We've thus shown that

$$\Gamma(2) = \Gamma(1);$$

however, it's more enlightening to write this in a slightly different way. We took $u = x$ and then said $du = dx$; let's write it as $u = x^1$ and $du = 1dx$. This leads us to

$$\Gamma(2) = 1 \cdot \Gamma(1).$$

At this point you should be skeptical—does it really matter? Anything times 1 is just itself! It does matter, and should remind you of our work with binomial coefficients and combinatorics. If we were to calculate $\Gamma(3)$, we would find it equals $2 \cdot \Gamma(2)$, and if we then progressed to $\Gamma(4)$ we would see it's just $3 \cdot \Gamma(3)$. This pattern suggests $\Gamma(s + 1) = s\Gamma(s)$, which we now prove.

Proof that $\Gamma(s + 1) = s\Gamma(s)$ *for* $\Re(s) > 0$. We have

$$\Gamma(s + 1) = \int_0^\infty e^{-x} x^{s+1-1} dx = \int_0^\infty e^{-x} x^s dx.$$

We now integrate by parts. Let $u = x^s$ and $dv = e^{-x} dx$; we're basically forced to do it this way as e^{-x} has a nice integral, and by setting $u = x^s$ when we differentiate the power of our polynomial goes down, leading to a simpler integral. We thus have

$$u = x^s, \quad du = sx^{s-1} dx, \quad dv = e^{-x} dx, \quad v = -e^{-x},$$

which gives

$$\Gamma(s + 1) = -x^s e^{-x} \Big|_0^\infty + \int_0^\infty e^{-x} sx^{s-1} dx$$

$$= 0 + s \int_0^\infty e^{-x} x^{s-1} dx = s\Gamma(s),$$

completing the proof. □

This relation is so important it's worth isolating it, and giving it a name.

Functional equation of $\Gamma(s)$: The Gamma function satisfies

$$\Gamma(s + 1) = s\Gamma(s).$$

This allows us to extend the Gamma function to all s. We call the extension the Gamma function as well, and it's well-defined and finite for all s save the negative integers and zero.

Let's return to the example from the previous section. Later we'll prove that $\Gamma(1/2) = \sqrt{\pi}$. For now we assume we know this, and show how we can figure out what $\Gamma(-3/2)$ should be. From the functional equation, $\Gamma(s + 1) = s\Gamma(s)$. We can rewrite this as $\Gamma(s) = s^{-1}\Gamma(s + 1)$, and we can now use this to "walk up" from $s = -3/2$, where we don't know the value, to $s = 1/2$, where we assume we do. We have

$$\Gamma\left(-\frac{3}{2}\right) = -\frac{2}{3}\Gamma\left(-\frac{1}{2}\right) = -\frac{2}{3} \cdot (-2)\Gamma\left(\frac{1}{2}\right) = \frac{4\sqrt{\pi}}{3}.$$

This is the power of the functional equation—it allows us to define the Gamma function essentially everywhere, so long as we know its values for $s > 0$ (or more generally for $\Re(s) > 0$). Why are zero and the negative integers special? Well, let's look at $\Gamma(0)$:

$$\Gamma(0) = \int_0^\infty e^{-x} x^{0-1} dx = \int_0^\infty e^{-x} x^{-1} dx.$$

The problem is that this isn't integrable. While it decays very rapidly for large x, for small x it looks like $1/x$. The details are:

$$\lim_{\epsilon \to 0} \int_\epsilon^1 e^{-x} x^{-1} dx \geq \frac{1}{e} \lim_{\epsilon \to 0} \int_\epsilon^1 \frac{dx}{x} = \frac{1}{e} \lim_{\epsilon \to 0} \log x \Big|_\epsilon^1 = \lim_{\epsilon \to 0} - \log \epsilon = \infty.$$

Thus $\Gamma(0)$ is undefined, and hence by the functional equation it's also undefined for all the negative integers.

15.3 The Factorial Function and $\Gamma(s)$

In the last section we showed that $\Gamma(s)$ satisfies the functional equation $\Gamma(s + 1) = s\Gamma(s)$. This is reminiscent of a relation obeyed by a better known function, the **factorial function**. Remember

$$n! = n \cdot (n - 1) \cdot (n - 2) \cdots 3 \cdot 2 \cdot 1;$$

we write this in a more suggestive way as

$$n! = n \cdot (n - 1)!.$$

Note how similar this looks to the relationship satisfied by $\Gamma(s)$. It's not a coincidence—the Gamma function is a generalization of the factorial function!

> $\Gamma(s)$ **and the Factorial Function**: If n is a non-negative integer, then $\Gamma(n + 1) = n!$. Thus the Gamma function is an extension of the factorial function.

We've shown that $\Gamma(1) = 1$, $\Gamma(2) = 1$, $\Gamma(3) = 2$, and so on. We can interpret this as $\Gamma(n) = (n - 1)!$ for $n \in \{1, 2, 3\}$; however, applying the functional equation allows us to extend this equality to all n. We proceed by induction. Proofs by induction have two steps, the base case (where you show it holds in some special instance) and the inductive step (where you assume it holds for n and then show that it holds for $n + 1$). See §A.2 for a review and additional examples of this technique.

We've already done the base case, as we've checked $\Gamma(1) = 0!$. (This is probably one of the few times in your life when you are grammatically correct to end a sentence with an exclamation point *and* a period. It's a good idea not to use another exclamation point for excitement, as the !!, called the double factorial, has a meaning in probability too!) We checked a few more cases than we needed. Typically that's a good strategy when doing inductive proofs. By getting your hands dirty and working out a few cases in detail, you often get a better sense of what's going on, and you can see the pattern. Remember, we initially wrote $\Gamma(2) = \Gamma(1)$, but after some thought (as well as years of experience) we rewrote it as $\Gamma(2) = \mathbf{1} \cdot \Gamma(1)$.

We now turn to the inductive step. We assume $\Gamma(n) = (n - 1)!$, and we must show $\Gamma(n + 1) = n!$. From the functional equation, $\Gamma(n + 1) = n\Gamma(n)$; but by the inductive step $\Gamma(n) = (n - 1)!$. Combining gives $\Gamma(n + 1) = n(n - 1)!$, which is just $n!$, or what we needed to show. This completes the proof. □

We now have two different ways to calculate say 1020!. The first is to do the multiplications out: $1020 \cdot 1019 \cdot 1018 \cdots$. The second is to look at the corresponding integral:

$$1020! = \Gamma(1021) = \int_0^\infty e^{-x} x^{1020} dx.$$

There are advantages to both methods; I want to discuss some of the benefits of the integral approach, as this is definitely not what most people have seen. Integration is hard; most students don't see it until late in high school or college. We all know how to multiply numbers—we'be been doing this since grade school. Thus, why make our lives difficult by converting a simple multiplication problem to an integral?

The reason is a general principle of mathematics—often by looking at things in a different way, from a higher level, new features emerge that you can exploit. Also, once we write it as an integral we have a lot more tools in our arsenal; we can use results from integration theory and from analysis to study this. We do this in Chapter 18, and see just how much we can learn about the factorial function by recasting it as an integral.

 Remark: The relation in this section is so important it's worth one last look before moving on. In the early chapters of the book we did a lot with combinatorics and probability. The factorial function was almost always lurking in the background, either directly through multiplicative trees of probabilities, or indirectly through binomial coefficients (recall $\binom{n}{k}$, the number of ways of choosing k objects from n when order doesn't matter, is $n!/k!(n-k)!$). This section connects the factorial function to the Gamma function, and suggests the possibility of a greater understanding through calculus and real analysis.

15.4 Special Values of $\Gamma(s)$

We know that $\Gamma(s+1) = s!$ whenever s is a non-negative integer. Are there other choices of s that are important, and if so, what are they? In other words, we've just generalized the factorial function. What was the point? It may be that the non-integral values are just curiosities that don't really matter, and the entire point might be to have the tools of calculus and analysis available to study $n!$. This, however, is most emphatically *not* the case. Some of these other values are very important in probability; in a bit of foreshadowing, we'll say they play a *central* role in the subject.

So, what are the important values for s? Because of the functional equation, once we know $\Gamma(1)$ we know the Gamma function at all non-negative integers, which gives us all the factorials. So 1 is an important choice of s. What should we look at next? The simplest number after the integers are the half-integers, those of the form $n/2$ where n is an integer. The simplest one which isn't an integer is 1/2. We'll now see that $s = 1/2$ is also very important.

One of the most important, if not the most important, distribution is the normal distribution (see Chapter 14 for a detailed tour). We say X is normally distributed with mean μ and variance σ^2, written $X \sim N(\mu, \sigma^2)$, if the density function is

$$f_{\mu,\sigma}(x) = \frac{1}{\sqrt{2\pi\sigma^2}} e^{-(x-\mu)^2/2\sigma^2}.$$

Looking at this density, we see there are two parts. There's the exponential part, and the constant factor of $1/\sqrt{2\pi\sigma^2}$. Because the exponential function decays so rapidly, the integral is finite and thus, if appropriately normalized, we will have a probability density. The hard part is determining just what this integral is. Let $g(x) = e^{-(x-\mu)^2/2\sigma^2}$. As g decays rapidly and is never negative, it can be rescaled to integrate to one and hence become a probability density. That scale factor is just $1/c$, where

$$c = \int_{-\infty}^{\infty} e^{-(x-\mu)^2/2\sigma^2}.$$

In Chapter 14 we saw numerous applications and uses of the normal distribution. It's not hard to make an argument that it's important, and thus we *need* to know the value of this integral. That said, why is this in the Gamma function chapter?

The reason is that, with a little bit of algebra and some change of variables, we'll see that this integral is just $\sqrt{2}\Gamma(1/2)\sigma^2$. We might as well assume $\mu = 0$ and $\sigma = 1$ (if not, then step 1 is just to change variables and let $t = \frac{x-\mu}{\sigma}$). So let's look at

$$I := \int_{-\infty}^{\infty} e^{-x^2/2}dx = 2\int_0^{\infty} e^{-x^2/2}dx,$$

where we **exploited the symmetry** to reduce the integration to be from 0 to infinity (see §A.4). This only vaguely looks related to the Gamma function. The Gamma function is the integral of e^{-x} times a polynomial in x, while here we have the exponential of $-x^2/2$. Looking at this, we see that there's a natural change of variable to try to make our integral look like the Gamma function at some special point. We try $u = x^2/2$, as this is the only way we'll end up with the exponential of the negative of our variable. We want to find dx in terms of u and du for the change of variables, thus we rewrite $u = x^2/2$ as $x = (2u)^{1/2}$, which gives $dx = (2u)^{-1/2}du$. Plugging all of these in, we see

$$I = 2\int_0^{\infty} e^{-u}(2u)^{-1/2}du = \sqrt{2}\int_0^{\infty} e^{-u}u^{-1/2}du.$$

We're almost done—this does look very close to the Gamma function. There are just two issues: one trivial and one minor. The first is that we're using the letter u instead of x, but that's fine as we can use whatever letter we want for our variable. The second is that $\Gamma(s)$ involves a factor of u^{s-1} and we have $u^{-1/2}$. This is easily fixed; we just write

$$u^{-\frac{1}{2}} = u^{\frac{1}{2}-\frac{1}{2}-\frac{1}{2}} = u^{\frac{1}{2}-1};$$

we just **added zero**, one of the most useful things to do in mathematics. (It takes awhile to learn how to "do nothing" well, which is why we keep pointing this out.) Thus

$$I = \sqrt{2}\int_0^{\infty} e^{-u}u^{\frac{1}{2}-1}du = \sqrt{2}\Gamma(1/2).$$

We did it—we've found another value of s that's important. Now we just need a way to find out what $\Gamma(1/2)$ equals! We could of course just go back to the standard normal's density and do the polar coordinate trick (see §14.1); however, it's possible to evaluate this directly by using the cosecant identity.

Cosecant identity: If s is not an integer, then

$$\Gamma(s)\Gamma(1-s) = \pi\csc(\pi s) = \frac{\pi}{\sin(\pi s)}.$$

We'll give a few different proofs in §15.8; note it implies that $\Gamma(1/2) = \sqrt{\pi}$.

Remark: It's worth remarking again why we chose to study $s = 1/2$. We had already mastered the Gamma function at the positive integers, and we needed to figure out what to study next. It's best to walk before running. Before running to $s = \sqrt{2}$ or $s = \pi$, it's a good idea to try the simplest numbers remaining. So, what's the simplest numbers that aren't positive integers? Those that are almost *positive integers. This thought process led us to the half-integers, and I hope you see that these are natural items to investigate.*

15.5 The Beta Function and the Gamma Function

The **Beta function** is defined by

$$B(a,b) = \int_0^1 t^{a-1}(1-t)^{b-1}dt, \quad a, b > 0.$$

Note the similarities with the Gamma function; both involve the integration variable raised to a parameter minus 1. It turns out this isn't just a coincidence or a stretch of the imagination, but rather these two functions are intimately connected.

Fundamental Relation of the Beta Function: For $a, b > 0$ we have

$$B(a,b) := \int_0^1 t^{a-1}(1-t)^{b-1}dt = \frac{\Gamma(a)\Gamma(b)}{\Gamma(a+b)}.$$

With a little bit of algebra, we can rearrange the above and find

$$\frac{\Gamma(a+b)}{\Gamma(a)\Gamma(b)} \int_0^1 t^{a-1}(1-t)^{b-1}dt = 1;$$

this means that we've discovered a new density, the density of the **Beta distribution**.

Beta distribution: Let $a, b > 0$. If X is a random variable with the **Beta distribution** with parameters a and b, then its density is

$$f_{a,b} = \begin{cases} \dfrac{\Gamma(a+b)}{\Gamma(a)\Gamma(b)}t^{a-1}(1-t)^{b-1}dt & \text{if } 0 \le t \le 1 \\ 0 & \text{otherwise.} \end{cases}$$

We write $X \sim B(a,b)$.

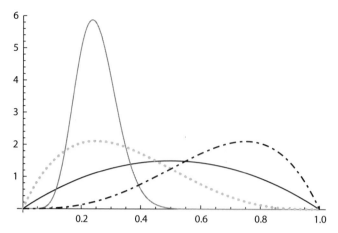

Figure 15.1. Plots of Beta densities for (a, b) equal to $(2, 2)$, $(2, 4)$, $(4, 2)$, $(3, 10)$, and $(10, 30)$.

We'll discuss this distribution in a bit more detail in §15.7. For now we'll just say briefly that it's an important family of densities as often our input is between 0 and 1, and the two parameters a and b give us a lot of freedom in creating "one-hump" distributions (namely densities that go up and then go down). We plot several of these densities in Figure 15.1.

15.5.1 Proof of the Fundamental Relation

We prove the fundamental relation of the Beta function. While this is an important result, remember that our purpose in doing so is to help you see how to attack problems like this. Multiplying both sides by $\Gamma(a + b)$, we see that we must prove

$$\Gamma(a)\Gamma(b) \quad \text{and} \quad \Gamma(a + b) \int_0^1 t^{a-1}(1 - t)^{b-1} dt$$

are equal. There are two ways to do this; we can either work with the product of the Gamma functions, or expand the $\Gamma(a + b)$ term and combine it with the other integral.

Let's try working with the product of the Gamma functions. Note that we can use the integral representation freely, as we've assumed $a, b > 0$. We'll argue along the lines of our first proof of the cosecant identity (see §15.8.1), and we find

$$\Gamma(a)\Gamma(b) = \int_0^\infty e^{-x} x^{a-1} dx \int_0^\infty e^{-y} y^{b-1} dy$$

$$= \int_{y=0}^\infty \int_{x=0}^\infty e^{-(x+y)} x^{a-1} y^{b-1} dx dy.$$

Remember, we can't change the order of integration, as that won't gain us anything as the two variables are not mixed. Our only remaining option is to change variables. We've fixed y and are integrating with respect to x. Let's try $x = yu$ so $dx = y du$; this at least mixes things up, and turns out to be a good choice for many problems. We find

$$\Gamma(a)\Gamma(b) = \int_{y=0}^{\infty} \left[\int_{u=0}^{\infty} e^{-(1+y)u}(yu)^{a-1} y^{b-1} y\, du \right] dy$$

$$= \int_{y=0}^{\infty} \int_{u=0}^{\infty} y^{a+b-1} u^{a-1} e^{-(1+u)y}\, du\, dy$$

$$= \int_{u=0}^{\infty} \int_{y=0}^{\infty} y^{a+b-1} u^{a-1} e^{-(1+u)y}\, dy\, du.$$

We've changed variables and then switched the order of integration. So right now we're fixing u and then integrating with respect to y. For u fixed, consider the change of variables $t = (1 + u)y$. This is a good choice, and a somewhat reasonable one to try. We need to get a $\Gamma(a + b)$ somehow. For that, we want something like the exponential of the negative of one of our variables. Right now we have $e^{-(1+u)y}$, which isn't of the desired form. By letting $t = (1 + u)y$, however, it now becomes e^{-t}. Again, what drives this change of variables is trying to get something looking like $\Gamma(a + b)$; *note how useful it is to have a sense of what the answer is!*

Anyway, if $t = (1 + u)y$ then $dy = dt/(1 + u)$ and our integral becomes

$$\Gamma(a)\Gamma(b) = \int_{u=0}^{\infty} \int_{t=0}^{\infty} \left(\frac{t}{1+u} \right)^{a+b-1} u^{a-1} e^{-t} \frac{1}{1+u}\, dt\, du$$

$$= \int_{u=0}^{\infty} \left(\frac{u}{1+u} \right)^{a-1} \left(\frac{1}{1+u} \right)^{b+1} \left[\int_{t=0}^{\infty} e^{-t} t^{a+b-1}\, dt \right] du$$

$$= \Gamma(a + b) \int_{u=0}^{\infty} \left(\frac{u}{1+u} \right)^{a-1} \left(\frac{1}{1+u} \right)^{b+1} du,$$

where we used the definition of the Gamma function to replace the t-integral with $\Gamma(a + b)$. We're definitely making progress—we've found the $\Gamma(a + b)$ factor.

We should also comment on how we wrote the algebra above. We combined everything that was to the $a - 1$ power together, and what was left was to the $b + 1$ power. Again, this is a promising sign; we're trying to show that this equals $\Gamma(a + b)$ times an integral involving x^{a-1} and $(1 - x)^{b-1}$; it's not exactly this, but it's close. (You might be a bit worried that we have a $b + 1$ and not a $b - 1$—it'll work out after yet another change of variables.) So, looking at what we have and again comparing it with where we want to go, what's the next change of variables? Let's try $\tau = \frac{u}{1+u}$, so $1 - \tau = \frac{1}{1+u}$ and $d\tau = \frac{du}{(1+u)^2}$ (by the quotient rule), or $du = (1 + u)^2 d\tau = \frac{d\tau}{(1-\tau)^2}$. Since $u : 0 \to \infty$, we have $\tau : 0 \to 1$,

$$\Gamma(a)\Gamma(b) = \Gamma(a + b) \int_0^1 \tau^{a-1}(1 - \tau)^{b+1} \frac{d\tau}{(1 - \tau)^2}$$

$$= \Gamma(a + b) \int_0^1 \tau^{a-1}(1 - \tau)^{b-1}\, d\tau,$$

which is what we needed to show! Why did we set τ equal to $\frac{u}{1+u}$? Remember we're trying to get the Beta integral, which involves integrating the product of τ (which is less than one) to a power times one minus τ to another power. As u ranges from 0 to ∞, $\frac{u}{1+u}$ runs from 0 to 1. This suggests that $\tau = \frac{u}{1+u}$ could be a useful change of variables.

Remark: As always, after going through a long proof we should stop, pause, and think about what we did and why. There were several change of variables and an interchange of orders of integration. As we've already discussed why these changes of variables are reasonable, we won't rehash that here. Instead, we'll talk one more time about how useful it is to know the answer. If you can guess the answer somehow, that can provide great insight as to what to do. For this problem, knowing we wanted to find a factor of $\Gamma(a + b)$ helped us make the change of variables to fix the exponential. And knowing we wanted factors like a variable to the $a - 1$ power suggested the change of variables $\tau = \frac{u}{1+u}$.

15.5.2 The Fundamental Relation and $\Gamma(1/2)$

We give yet another derivation of $\Gamma(1/2)$, this time using properties of the Beta function. Taking $a = b = 1/2$ gives

$$\Gamma\left(\frac{1}{2}\right)\Gamma\left(\frac{1}{2}\right) = \Gamma\left(\frac{1}{2} + \frac{1}{2}\right)\int_0^1 t^{1/2-1}(1 - t)^{1/2-1}dt$$

$$= \Gamma(1)\int_0^1 t^{-1/2}(1 - t)^{-1/2}dt.$$

As always, the question becomes: what's the right change of variables? If we think back to our studies of the Gamma function and the cosecant identity, we have $\Gamma(1/2)^2$ was supposed to be $\pi/\sin(\pi/2)$. This is telling us that trig functions should play a big role, so perhaps we want to do something to facilitate using trig functions or trig substitution. If so, one possibility is to take $t = u^2$. This makes the factor $(1 - t)^{-1/2}$ equal to $(1 - u^2)^{-1/2}$, which is ideally suited for a trig substitution.

Now for the details. We set $t = u^2$ or $u = t^{1/2}$, so $du = dt/2t^{1/2}$ or $t^{-1/2}dt = 2du$; we write it like this as we have a $t^{-1/2}dt$ already! The bounds of integration are still 0 to 1, and we find

$$\Gamma\left(\frac{1}{2}\right)^2 = \int_0^1 (1 - u^2)^{-1/2}2du.$$

We now use trig substitution. Take $u = \sin\theta$, $du = \cos\theta d\theta$, so $u : 0 \to 1$ becomes $\theta : 0 \to \pi/2$ (we chose $u = \sin\theta$ over $u = \cos\theta$ as this way the bounds of integration become 0 to $\pi/2$ and not $\pi/2$ to 0, though of course either approach is fine). We now have

$$\Gamma\left(\frac{1}{2}\right)^2 = 2\int_0^{\pi/2}(1 - \sin^2\theta)^{-1/2}\cos\theta d\theta$$

$$= 2\int_0^{\pi/2}\frac{\cos\theta d\theta}{(\cos^2\theta)^{1/2}}$$

$$= 2\int_0^{\pi/2}d\theta = 2\cdot\frac{\pi}{2} = \pi,$$

which gives us yet another way to see $\Gamma(1/2) = \sqrt{\pi}$.

15.6 The Normal Distribution and the Gamma Function

It would be irresponsible to cover the Gamma function without mentioning some of the other connections with the normal distribution. The three most important integrals related to the standard normal are

$$1 = \int_{-\infty}^{\infty} \frac{1}{\sqrt{2\pi}} e^{-x^2/2} dx$$

$$0 = \int_{-\infty}^{\infty} x \cdot \frac{1}{\sqrt{2\pi}} e^{-x^2/2} dx$$

$$1 = \int_{-\infty}^{\infty} (x - 0)^2 \cdot \frac{1}{\sqrt{2\pi}} e^{-x^2/2} dx;$$

a fourth useful one is the $2m^{\text{th}}$ moment,

$$\mu_{2m} = \int_{-\infty}^{\infty} x^{2m} \cdot \frac{1}{\sqrt{2\pi}} e^{-x^2/2} dx = (2m - 1)!!,$$

where the **double factorial** means we take every other term until we reach 2 or 1 (so $5!! = 5 \cdot 3 \cdot 1$ while $6!! = 6 \cdot 4 \cdot 2$); we don't bother recording the odd moments as these are trivially zero.

The mean is easily understood—we're integrating an odd function over a symmetric region, and as our integrand decays very fast the integral converges and is zero. The other ones are a bit harder, and we had to do a lot of work to show the Gaussian's density did in fact integrate to 1, and that the variance was 1.

If, and this is a big if, we know the Gamma function very well, then we can immediately get any even moment. All we have to do is a little change of variables. We have

$$\mu_{2m} = \int_{-\infty}^{\infty} x^{2m} \cdot \frac{1}{\sqrt{2\pi}} e^{-x^2/2} dx = 2 \int_0^{\infty} x^{2m} \cdot \frac{1}{\sqrt{2\pi}} e^{-x^2/2} dx.$$

How should we change variables? Looking at the definition of the Gamma function, we see that it has a term e^{-u}; our exponential term is $e^{-x^2/2}$. This suggests that we set $u = x^2/2$, which implies $x = (2u)^{1/2}$, so

$$x^{2m} = 2^m u^m, \quad dx = \frac{du}{\sqrt{2u}}.$$

Doing this gives

$$\mu_{2m} = \frac{2}{\sqrt{2\pi}} \int_0^{\infty} 2^m u^m e^{-u} \frac{du}{\sqrt{2u}} = \frac{2^m}{\sqrt{\pi}} \int_0^{\infty} u^{m-\frac{1}{2}} e^{-u} du.$$

We now do a nice trick: we add zero. Remember **adding zero** is one of the most powerful tools in our arsenal. We *almost* have the definition of the Gamma function, but we need to have u^{s-1} and we have $u^{m-\frac{1}{2}}$. Thus we'll write

$$u^{m-\frac{1}{2}} = u^{m+\frac{1}{2}-1},$$

implying that $s = m + \frac{1}{2}$. The integral above is now just $\Gamma\left(m + \frac{1}{2}\right)$, and we end up with

$$\mu_{2m} = \frac{2^m}{\sqrt{\pi}}\,\Gamma\left(m + \frac{1}{2}\right).$$

We see now why there was a big *if* before; we have answers for the moments, but unless you really know a lot about the Gamma function, the answers don't look that useful. For example, if we take $m = 0$ we get the area under the curve. This is supposed to be 1; our formula tells us it's $2^0\Gamma(1/2)/\sqrt{\pi}$. All works as $\Gamma(1/2) = \sqrt{\pi}$; in fact, if we didn't know $\Gamma(1/2) = \sqrt{\pi}$ we could use the polar trick evaluation to provide yet another proof of this important fact! What about the variance? That requires us to take $m = 1$ (remember we're looking at the $2m^{\text{th}}$ moment). In this case, we find $2^1\Gamma(3/2)/\sqrt{\pi}$, and (as you surely have guessed) we do have $\Gamma(3/2) = \sqrt{\pi}/2$.

The Gamma function satisfies a lot of beautiful properties. We showed in §15.2 from integrating by parts that $\Gamma(s + 1) = s\Gamma(s)$, at least if $s > 0$. We gave several proofs that $\Gamma(1/2) = \sqrt{\pi}$. Using these two facts, it's a nice exercise to show that $\Gamma(m + 1/2) = \frac{(2m-1)!!}{2^m}\Gamma(1/2)$, and thus $\mu_{2m} = (2m - 1)!!$.

15.7 Families of Random Variables

We could easily fill up many more chapters by going through all the different, important distributions in general. Even if we restricted ourselves to distributions related to the Gamma function we would still have many more chapters to write. Instead of doing that, what we'll do instead is discuss one such distribution in greater detail (the chi-square distribution, the subject of Chapter 16), and briefly comment on a few here.

We've already talked about the Beta distribution in §15.5. We give two other **families** of densities (we'll explain the terminology in a bit).

Gamma and Weibull Distributions: A random variable X has the **Gamma distribution** with (positive) parameters k and σ if its density is

$$f_{k,\sigma}(x) = \begin{cases} \dfrac{1}{\Gamma(k)\sigma^k}x^{k-1}e^{-x/\sigma} & \text{if } x \geq 0 \\ 0 & \text{otherwise.} \end{cases}$$

We call k the **shape** parameter and σ the **scale** parameter, and write $X \sim \Gamma(k, \sigma)$ or $X \sim \text{Gamma}(k, \sigma)$.

> A random variable X has the **Weibull** distribution with (positive) parameters k and σ if its density is
>
> $$f_{k,\sigma}(x) = \begin{cases} (k/\sigma)(x/\sigma)^{k-1}e^{-(x/\sigma)^k} & \text{if } x \geq 0 \\ 0 & \text{otherwise.} \end{cases}$$
>
> We call k the **shape** parameter and σ the **scale** parameter, and write $X \sim W(k, \sigma)$.

Note these two distributions are fundamentally different from each other and from the Beta distribution. All three densities have a polynomial factor, but the Gamma and Weibull have (different) exponential factors if $k \neq 1$ and are non-zero outside $[0, 1]$. What's particularly nice about these distributions is that we can vary the parameters and get different, but related, distributions. This leads us to the notion of a **family of distributions**. These are densities that are different specializations of parameters. In practice, we frequently have reason to believe a natural or mathematical phenomenon is modeled by some distribution, but with unknown values of the parameters. We then try to figure out the value of these parameters, either through mathematical analysis or through statistical inference.

One of my favorite examples of this is some work I did with the Weibull distribution to provide a theoretical justification for a formula used to predict a baseball team's winning percentage knowing just its average runs scored and allowed per game (see [Mil], or http://www.youtube.com/watch?v=gFDly_6qOn4 for a lecture on the subject). It turns out that, for appropriate choices of parameters, a Weibull distribution does an excellent job fitting the runs scored and allowed data.

The more distributions you know, the more likely you are to make such a connection. I strongly urge you to read the paper; it's a nice application of basic probability and mathematical modeling (and some elementary statistics). I started by exploring what would happen if the runs scored or allowed were drawn from an exponential distribution (the density is proportional to $e^{-x/\sigma}$) and a Rayleigh distribution (the density is proportional to $xe^{-x^2/2\sigma^2}$). I knew about these distributions from physics, and saw I could get a nice answer, but not a perfect one. Then, inspiration hit: I noticed these two densities were of the form $x^{k-1}e^{-x^k/\lambda}$. They sat inside a family, and by choosing "good" values for k and λ I could get both good fits with the real-world data, as well as have mathematically tractable integrals. This is how I learned about the Weibull distribution.

The Weibull distribution is used in many problems involving survival analysis; there are similarly applications of the Beta and Gamma distributions (Wikipedia and a Google search will quickly yield many examples). Again, the point is to have families of distributions on your radar. The bigger your tool chest, the better job you can do modeling.

15.8 Appendix: Cosecant Identity Proofs

Books have entire chapters on the various identities satisfied by the Gamma function. In this section we'll concentrate on one that's particularly well-suited to our investigation of $\Gamma(1/2)$, namely, the cosecant identity.

Cosecant identity: If s is not an integer, then

$$\Gamma(s)\Gamma(1-s) = \pi\csc(\pi s) = \frac{\pi}{\sin(\pi s)}.$$

Before proving this, let's take a moment to use this to finish our study. For almost all s the cosecant identity relates two values, Gamma at s and Gamma at $1-s$; if you know one of these values, you know the other. Unfortunately, this means that in order for this identity to be useful, we have to know at least one of the two values. Unless, of course, we make the *very* special choice of taking $s = 1/2$. As $1/2 = 1 - 1/2$, the two values are the same, and we find

$$\Gamma(1/2)^2 = \Gamma(1/2)\Gamma(1/2) = \frac{\pi}{\sin(\pi/2)} = \pi;$$

taking square-roots gives $\Gamma(1/2) = \sqrt{\pi}$. We're quite fortunate that the very special value happens to be the value we wanted earlier!

In the following subsections I'll give various proofs of the cosecant identity. If all you care about is using it, you can of course skip this; however, if you read on you'll get some insight as to how people come up with formulas like this, and how they prove them. The arguments will become involved in places, but I'll try to point out why we're doing what we're doing, so that if you come across a situation like this in the future, a new situation where you are the first one looking at a problem and there's no handy guidebook available, you'll have some tools for your studies.

15.8.1 The Cosecant Identity: First Proof

Proof of the cosecant identity: We've seen the cosecant identity is useful; now let's see a proof. How should we try to prove this? Well, one side is $\Gamma(s)\Gamma(1-s)$. Both of these numbers can be represented as integrals. So this quantity is really a double integral. Whenever you have a double integral, you should start thinking about changing variables or changing the order of integration, or maybe even both! The point is using the integral formulations gives us a starting point. This argument might not work, but it's something to *try* (and, for many math problems, one of the hardest things is just figuring out where to begin).

What we are about to write looks like it does what we have decided to do, but there's *two* subtle mistakes:

$$\Gamma(s)\Gamma(1-s) = \int_0^\infty e^{-x}x^{s-1}dx \cdot \int_0^\infty e^{-x}x^{1-s-1}dx$$

$$= \int_0^\infty e^{-x}x^{s-1} \cdot e^{-x}x^{1-s-1}dx. \tag{15.1}$$

Why is this wrong? The first expression is the integral representation of $\Gamma(s)$; the second expression is the integral representation of $\Gamma(1-s)$, so their product is $\Gamma(s)\Gamma(1-s)$ and then just collect terms.... Unfortunately, **NO!** The problem is that we used the same dummy variable for both integrations. We can't write it as one integral—we had two integrations, each with a dx, and then ended up with just one dx. This is one of the most common mistakes students make. By not using a different letter for the variables in each integration, we accidentally combined them and went from a double integral to a single integral.

We should use two different letters, which in a fit of creativity we'll take to be x and y. Then

$$\Gamma(s)\Gamma(1-s) = \int_0^\infty e^{-x}x^{s-1}dx \cdot \int_0^\infty e^{-y}y^{1-s-1}dy$$

$$= \int_{y=0}^\infty \int_{x=0}^\infty e^{-x}x^{s-1}e^{-y}y^{-s}dxdy.$$

While the result we're gunning for, the cosecant formula, is beautiful and important, even more important (and far more useful!) is to learn how to attack problems like this. There aren't that many options for dealing with a double integral. You can integrate as given, but in this case that would be a bad idea as we would just get back the product of the Gamma functions. What else can we do? We can switch the orders of integration. Unfortunately, that too isn't any help; switching orders can only help us if the two variables are mingled in the integral, and that isn't the case now. Here, the two variables aren't seeing each other; if we switch the order of integration, we haven't really changed anything. Only one option remains: we need to change variables.

This is the hardest part of the proof. We have to figure out a good change of variables. Let's look at the first possible choice. We have $x^{s-1}y^{-s} = (x/y)^{s-1}y^{-1}$ (we could have written it as $(x/y)^s x^{-1}$, but since the definition of the Gamma function involves a variable to the $s-1$ power, we try this first). Perhaps a good change of variables would be to let $u = x/y$? If we do this, we fix y, and then for fixed y we set $u = x/y$, giving $du = dx/y$. The $1/y$ is encouraging, as we had an extra y earlier. This leads to

$$\Gamma(s)\Gamma(1-s) = \int_{y=0}^\infty e^{-y}\left[\int_{u=0}^\infty e^{-uy}u^{s-1}du\right]dy.$$

We now switch orders of integration. That gives

$$\Gamma(s)\Gamma(1-s) = \int_{u=0}^\infty u^{s-1}\left[\int_{y=0}^\infty e^{-(u+1)y}dy\right]du$$

$$= \int_{u=0}^\infty u^{s-1}\left[-\frac{e^{-(u+1)y}}{u+1}\Big|_0^\infty\right]du$$

$$= \int_{u=0}^\infty u^{s-1}\frac{1}{u+1}du = \int_{u=0}^\infty \frac{u^{s-1}}{u+1}du.$$

Warning: We have to be very careful above, and make sure the interchange is justified. Remember earlier in the chapter when we had a long discussion about the importance of making sure an integral makes sense? The integrand above is $\frac{u^{s-1}}{u+1}$. It has to decay sufficiently rapidly as $u \to \infty$ and it cannot blow up too quickly as $u \to 0$ if the integral is to be finite. If you work out what this entails, it forces $s \in (0, 1)$; if $s \le 0$ then it blows up too rapidly near 0, while if $s \ge 1$ it doesn't decay fast enough at infinity.

In hindsight, this restriction isn't surprising, and in fact we should have expected it. Why? Remember earlier in the proof we remarked that there were **two** mistakes in (15.1); if you were really alert, you would have noticed we only mentioned **one** mistake! What is the missing mistake? We used the integral representation of the Gamma function. That is only valid when the argument is positive. Thus we need $s > 0$ and $1 - s > 0$; these two inequalities force $s \in (0, 1)$. If you didn't catch this mistake this time, don't worry about it; just be aware of the danger in the future. This is one of the most common errors made (by both students and researchers.) It's so easy to take a formula that works in some cases and accidentally use it in a place where it's not valid.

Alright. For now, let's restrict ourselves to taking $s \in (0, 1)$. We leave it as an exercise to show that if the relationship holds for $s \in (0, 1)$ then it holds for all s. (Hint: Keep using the functional equation of the Gamma function.) It's easy to see how $\csc(\pi s)$ or $\sin(\pi s)$ changes if we increase s by 1; the Gamma pieces follow with a bit more work.

Now we really can say

$$\Gamma(s)\Gamma(1 - s) = \int_0^\infty \frac{u^{s-1}}{u + 1} \, du. \tag{15.2}$$

What next? Well, we have two factors, u^{s-1} and $\frac{1}{u+1}$. Note the second looks like the sum of a geometric series with ratio $-u$. To see that, we can write $\frac{1}{u+1}$ as $\frac{1}{1-(-u)}$, which is the sum of a geometric series with ratio 1 (so long as $|u| < 1$). Admittedly, this isn't going to be an obvious identification at first, but the more math you do, the more experience you gain and the easier it's to recognize patterns. We know $\sum_{n=0}^\infty r^n = \frac{1}{1-r}$, so all we have to do is take $r = -u$.

We must be careful—we're about to make the same mistake again, namely using a formula where it isn't applicable. It's very easy to fall into this trap. Fortunately, there's a way around it. We split the integral into two parts, the first part is when $u \in [0, 1]$ and the second when $u \in [1, \infty]$. In the second part we'll then change variables by setting $v = 1/u$ and do a geometric series expansion there. **Splitting an integral** is another useful technique to master. It allows us to break a complicated problem up into simpler ones, ones where we have more results at our disposal to attack it. We need to do something like this as we're searching for a Taylor series expansion. We want to get rid of an infinity and replace it with something we know.

For the second integral, we'll make the change $v = 1/u$. This gives $dv = -du/u^2$ or $du = -v^2 dv$ (since $1/u^2 = v^2$), and the bounds of integration go from being $u : 1 \to \infty$ to $v : 1 \to 0$ (we'll then use the negative sign to switch the order of integration to the more common $v : 0 \to 1$). Continuing onward, we have

$$\Gamma(s)\Gamma(1 - s) = \int_0^1 \frac{u^{s-1}}{u + 1} \, du + \int_1^\infty \frac{u^{s-1}}{u + 1} \, du$$

$$= \int_0^1 \frac{u^{s-1}}{u + 1} \, du - \int_1^0 \frac{(1/v)^{s-1}}{(1/v) + 1} \, v^2 dv$$

$$= \int_0^1 \frac{u^{s-1}}{u + 1} \, du + \int_0^1 \frac{v^{-s}}{v + 1} \, dv.$$

Note how similar the two expressions are (and are the same at the very special value of $s = 1/2$). We now use the geometric series formula, and then we'll interchange the integral and the sum. Everything can be justified (see §B.2) because $s \in (0, 1)$, so all the integrals exist and are well-behaved, giving

$$\Gamma(s)\Gamma(1 - s) = \int_0^1 u^{s-1} \sum_{n=0}^{\infty} (-1)^n u^n \, du + \int_0^1 v^{-s} \sum_{m=0}^{\infty} (-1)^m v^m \, dv$$

$$= \sum_{n=0}^{\infty} (-1)^n \int_0^1 u^{s-1+n} \, du + \sum_{m=0}^{\infty} (-1)^m \int_0^1 v^{m-s} \, dv$$

$$= \sum_{n=0}^{\infty} (-1)^n \frac{u^{s+n}}{n+s} \Big|_0^1 + \sum_{m=0}^{\infty} (-1)^m \frac{v^{m+1-s}}{m+1-s} \Big|_0^1$$

$$= \sum_{n=0}^{\infty} (-1)^n \frac{1}{n+s} + \sum_{m=0}^{\infty} (-1)^m \frac{1}{m+1-s}.$$

Note we used two different letters for the different sums. While we could have used the letter n twice, it's a good habit to use different letters. What happens now is that we'll adjust the counting a bit to easily combine them.

The two sums look very similar. They both look like a power of negative one divided by either $k + s$ or $k - s$. Let's rewrite both sums in terms of k. The first sum has one extra term, which we'll pull out. In the first sum we'll set $k = n$, while in the second we'll set $k = m + 1$ (so $(-1)^m$ becomes $(-1)^{k-1} = (-1)^{k+1}$). We get

$$\Gamma(s)\Gamma(1 - s) = \frac{1}{s} + \sum_{k=1}^{\infty} (-1)^k \frac{1}{k+s} + \sum_{k=1}^{\infty} (-1)^{k+1} \frac{1}{k-s}$$

$$= \frac{1}{s} + \sum_{k=1}^{\infty} (-1)^k \left[\frac{1}{k+s} - \frac{1}{k-s} \right]$$

$$= \frac{1}{s} + \sum_{k=1}^{\infty} (-1)^k \frac{2s}{k^2 - s^2}$$

$$= \frac{1}{s} - \sum_{k=1}^{\infty} (-1)^k \frac{-2s}{k^2 - s^2}.$$

It may not look like it, but we've just finished the proof. The problem is recognizing the above is $\pi \csc(\pi s) = \pi / \sin(\pi s)$. This is typically proved in a complex analysis course; see for instance [SS2].

We can at least see it's reasonable. We're claiming

$$\frac{\pi}{\sin(\pi s)} = \frac{1}{s} - \sum_{k=1}^{\infty} (-1)^k \frac{2s}{k^2 - s^2}.$$

If s is an integer then $\sin(\pi s) = 0$ and thus the left-hand side is infinite, while exactly one of the terms on the right-hand side blows up. This at least shows our answer is reasonable. Or mostly reasonable. It seems likely that our sum is $c/\sin(\pi s)$ for some c, but it isn't clear that c equals π. Fortunately, there's even a way to get that, but it involves knowing a bit more about certain special sums. If we take $s = 1/2$ then the sum becomes

$$\frac{1}{1/2} - \sum_{k=1}^{\infty} (-1)^k \frac{1}{k^2 - (1/2)^2} = 2 - \sum_{k=1}^{\infty} \frac{(-1)^k}{k^2 - 1/4}$$

$$= 2 - \sum_{k=1}^{\infty} \frac{(-1)^k 4}{4k^2 - 1}$$

$$= 2 - 4 \sum_{k=1}^{\infty} \frac{(-1)^k}{(2k-1)(2k+1)}$$

$$= 2 - 4 \sum_{k=1}^{\infty} \frac{(-1)^k}{2} \left(\frac{1}{2k-1} - \frac{1}{2k+1} \right)$$

$$= 2 + 2 \left(\frac{1}{1} - \frac{1}{3} \right) - 2 \left(\frac{1}{3} - \frac{1}{5} \right) + \cdots$$

$$= 4 \left(1 - \frac{1}{3} + \frac{1}{5} - \cdots \right).$$

As the alternating sum of the reciprocals of the odd numbers is $\pi/4$, this proves our constant c is π as claimed. This is the **Gregory-Leibniz formula** for π, which completes our analysis (see Exercise 15.10.21 for a sketch of how to prove this).

 Remark: This was a long proof, but there were a lot of good ideas in it. At the end, we tried to check the reasonableness of our formula by looking at special values. This is a great idea, but it's only as useful as our ability to find special values. Knowing the Gregory-Leibniz formula allowed us to verify the claim at $s = 1/2$, which fortunately is the value we care about most!

15.8.2 The Cosecant Identity: Second Proof

We already have a proof of the cosecant identity for the Gamma function—why do we need another? For us, the main reason is educational. The goal of this book is not to teach you how to answer one specific problem at one moment in your life, but rather to give you the tools to solve a variety of new problems whenever you encounter them. Because of that, it's worth seeing multiple proofs as different approaches emphasize different aspects of the problem, or generalize better for other questions.

Let's go back to the setup. We had $s \in (0, 1)$ and

$$\Gamma(s)\Gamma(1-s) = \int_0^\infty e^{-x}x^{s-1}dx \cdot \int_0^\infty e^{-y}y^{1-s-1}dy$$

$$= \int_{y=0}^\infty \int_{x=0}^\infty e^{-x}x^{s-1}e^{-y}y^{-s}dxdy.$$

We've already talked about what our options are. We can't integrate it as is, or we'll just get back the two Gamma functions. We can't change the order of integration, as the x and y variables are not mingled and thus changing the order of integration won't really change the problem. The only thing left to do is change variables.

Before we set $u = x/y$. We were led to this because we saw $x^{s-1}y^{-s} = (x/y)^{s-1}y^{-1}$, and thus it's not unreasonable to set $u = x/y$. Are there any other "good" choices for a change of variable? There is, but it's not surprising if you don't see it. It's our old friend, polar coordinates.

It should seem a little strange to use polar coordinates here. After all, we use those for problems with radial and angular symmetry. We use them for integrating over circular regions. **NONE** of this is happening here! That said, I think a good case can be made for trying the **polar coordinate trick**.

- First, we don't know that many change of variables; we do know polar coordinates, so we might as well try it.
- Second, we're trying to show the answer is $\pi \csc(\pi s) = \pi/\sin(\pi s)$. The answer involves the sine function, so perhaps this suggests we should try polar coordinates.

At the end of the day, a method either works or it doesn't. We hope the above at least motivates why we're trying this here, and can provide guidance for you in the future.

Recall for polar coordinates we have the following relations:

$$x = r\cos\theta, \quad y = r\sin\theta, \quad dxdy = rdrd\theta.$$

What are the bounds of integration? We're integrating over the upper right quadrant, $x, y : 0 \to \infty$. In polar coordinates it becomes $r : 0 \to \infty$ and $\theta : 0 \to \pi/2$. Our integral now becomes

$$\Gamma(s)\Gamma(1-s) = \int_{\theta=0}^{\pi/2} \int_{r=0}^\infty e^{-r\cos\theta}(r\cos\theta)^{s-1}e^{-r\sin\theta}(r\sin\theta)^{-s}rdrd\theta$$

$$= \int_{\theta=0}^{\pi/2} \int_{r=0}^\infty e^{-r(\cos\theta+\sin\theta)}\left(\frac{\cos\theta}{\sin\theta}\right)^{s-1}\frac{1}{\sin\theta}drd\theta$$

$$= \int_{\theta=0}^{\pi/2} \left(\frac{\cos\theta}{\sin\theta}\right)^{s-1}\frac{1}{\sin\theta}\left[\int_{r=0}^\infty e^{-r(\cos\theta+\sin\theta)}dr\right]d\theta$$

$$= \int_{\theta=0}^{\pi/2} \left(\frac{\cos\theta}{\sin\theta} \right)^{s-1} \frac{1}{\sin\theta} \left[-\frac{e^{-r(\cos\theta+\sin\theta)}}{\cos\theta + \sin\theta} \right]_0^\infty d\theta$$

$$= \int_{\theta=0}^{\pi/2} \left(\frac{\cos\theta}{\sin\theta} \right)^{s-1} \frac{1}{\sin\theta} \frac{1}{\cos\theta + \sin\theta} \, d\theta.$$

It doesn't look like we've made much progress, but we're just one little change of variables away from a great simplification. Note that a lot of the integrand only depends on $\cos\theta/\sin\theta = \operatorname{ctan}\theta$ (the cotangent of θ). If we do make the change of variables $u = \operatorname{ctan}\theta$ then $du = -\csc^2\theta = -1/\sin^2\theta$; if you don't remember this formula, you can get it by the quotient rule:

$$\operatorname{ctan}'(\theta) = \left(\frac{\cos\theta}{\sin\theta} \right)' = \frac{\cos'\theta \sin\theta - \sin'\theta \cos\theta}{\sin^2\theta}$$

$$= \frac{-\sin^2\theta - \cos^2\theta}{\sin^2\theta} = -\frac{1}{\sin^2\theta}.$$

Now things are looking really promising; our proposed change of variables needs a $1/\sin^2\theta$, and we already have a $1/\sin\theta$ in the integrand. We get the other by writing

$$\frac{1}{\cos\theta + \sin\theta} = \frac{1}{\sin\theta} \frac{1}{(\cos\theta/\sin\theta) + 1} = \frac{1}{\sin\theta} \frac{1}{\operatorname{ctan}\theta + 1}.$$

All that remains is to find the bounds of integration. If $u = \operatorname{ctan}\theta = \cos\theta/\sin\theta$, then $\theta : 0 \to \pi/2$ corresponds to $u : \infty \to 0$ (don't worry that we're integrating from infinity to zero—we have a minus sign floating around, and that will flip the order of integration).

Putting all the pieces together, we find

$$\Gamma(s)\Gamma(1-s) = \int_{\theta=0}^{\pi/2} \frac{\operatorname{ctan}^{s-1}\theta}{\operatorname{ctan}\theta + 1} \frac{d\theta}{\sin^2\theta}$$

$$= \int_{u=\infty}^0 \frac{u^{s-1}}{u+1}(-du) = \int_0^\infty \frac{u^{s-1}}{u+1} du.$$

This integral should look familiar—it's exactly the integral we saw in the previous section, in Equation (15.2). Thus from here onward we can just follow the steps in that section.

 Remark: A lot of students freeze when they first see a difficult math problem. Why varies from student to student, but a common refrain is: "I didn't know where to start." For those who feel that way, this should be comforting. There are (at least!) two different change of variables we can do, both leading to a solution for the problem. As you continue in math you'll see again and again that there are many different approaches you can take. Don't be afraid to try something. Work with it for awhile and see how it goes. If it isn't promising you can always backtrack and try something else.

15.8.3 The Cosecant Identity: Special Case $s = 1/2$

While obviously we want to prove the cosecant formula for arbitrary s, *the* most important choice of s is clearly $s = 1/2$. We need $\Gamma(1/2)$ in order to write down the density functions for normal distributions, and to compute its moments. Thus, while it would be nice to have a formula for any s, it's still cause for celebration if we can handle just $s = 1/2$.

Remember in (15.2) that we showed

$$\Gamma(s)\Gamma(1-s) = \int_0^\infty \frac{u^{s-1}}{u+1}\, du.$$

Taking $s = 1/2$ gives

$$\Gamma(1/2)^2 = \int_0^\infty \frac{u^{-1/2}}{1+u}\, du.$$

We're going to solve this with a highly non-obvious change of variable. Let's state it first, see how it works, and then discuss why this is a reasonable thing to try. Here it is: take $u = z^2$, so $z = u^{1/2}$ and $dz = du/2\sqrt{u}$. Note how beautifully this fits with our integral. We have a $u^{-1/2}du$ term already, which becomes $2dz$. Substituting gives

$$\Gamma(1/2)^2 = \int_0^\infty \frac{2dz}{1+z^2} = 2\int_0^\infty \frac{dz}{1+z^2}.$$

Looking at this integral, you should think of the trigonometric substitutions from calculus. Whenever you see $1 - z^2$ you should try $z = \sin\theta$ or $z = \cos\theta$; when you see $1 + z^2$ you should try $z = \tan\theta$. Let's make this change of variables. The reason it's so useful is the Pythagorean formula

$$\sin^2\theta + \cos^2\theta = 1$$

becomes, on dividing both sides by $\cos^2\theta$,

$$\tan^2\theta + 1 = \frac{1}{\cos^2\theta} = \sec^2\theta.$$

Letting $z = \tan\theta$ means we replace $1 + z^2$ with $\sec^2\theta$. Further, $dz = \sec^2\theta\, d\theta$ (if you don't remember this, just use the quotient rule applied to $\tan\theta = \sin\theta/\cos\theta$). As $z : 0 \to \infty$, we have $\theta : 0 \to \pi/2$. Collecting everything gives

$$\Gamma(1/2)^2 = 2\int_0^{\pi/2} \frac{1}{\sec^2\theta}\sec^2\theta\, d\theta$$

$$= 2\int_0^{\pi/2} d\theta = 2\frac{\pi}{2} = \pi,$$

and if $\Gamma(1/2)^2 = \pi$ then $\Gamma(1/2) = \sqrt{\pi}$ as claimed!

And there we have it: a correct, elementary proof that $\Gamma(1/2) = \sqrt{\pi}$. You should be able to follow the proof line by line, but that's not the point of mathematics. The point is to *see* why the author is choosing to do these steps so that you too could create a proof like this.

There were two changes of variables. The first was replacing u with z^2, and the second was replacing z with $\tan\theta$. The two changes are related. How can anyone be expected to think of these? To be honest, when writing this chapter I had to consult my notes from teaching a similar course several years ago. I remembered that somehow tangents came into the problem, but couldn't remember the exact trick I used so long ago. It's not easy. It takes time, but the more you do, the more patterns you can detect. We have a $1 + u$ in the denominator; we know how to handle terms such as $1 + z^2$ through trig substitution. As the cosecant identity involves trig functions, that suggests this could be a fruitful avenue to explore. It's not a guarantee, but we might as well try it and see where it leads.

Flush with our success, the most natural thing to try next are these substitutions for general s. If we do this, we would find

$$\Gamma(s)\Gamma(1-s) = \int_0^\infty \frac{z^{2s-2}}{1+z^2} \, 2z \, dz$$

$$= 2 \int_0^\infty \frac{z^{2s-1}}{1+z^2} \, dz$$

$$= 2 \int_0^{\pi/2} \frac{\tan^{2s-1}\theta}{\sec^2\theta} \, \sec^2\theta \, d\theta$$

$$= 2 \int_0^{\pi/2} \tan^{2s-1}\theta \, d\theta.$$

We now see how special $s = 1/2$ is. For this, and only for this, value does the integrand collapse to just being the constant function 1, which is easily integrated. Any other choice of s forces us to have to find integrals of powers of the tangent function, which is no easy task! Formulas do exist; for example,

$$\int \tan^{1/2}\theta \, d\theta$$

$$= \frac{1}{2\sqrt{2}} \Big[-2\arctan\big(1 - \sqrt{2}\sqrt{\tan\theta}\big) + 2\arctan\big(1 + \sqrt{2}\sqrt{\tan\theta}\big)$$

$$+ \log\big(1 - \sqrt{2}\sqrt{\tan\theta} + \tan\theta\big) - \log\big(1 + \sqrt{2}\sqrt{\tan\theta} + \tan\theta\big) \Big].$$

Remark: If we remember that the derivative of $\arctan(z)$ *is* $\frac{1}{1+z^2}$, *we can avoid the* $z = \tan\theta$ *substitution and directly evaluate* $\int_0^\infty \frac{1}{1+z^2} dz$ *as* $\arctan(\infty) - \arctan(0) = \pi/2$. *One of the best ways to see this is to note that if* $f(g(x)) = x$, *then by the chain rule* $f'(g(x))g'(x) = 1$, *or* $g'(x) = 1/f'(g(x))$. *Use this relation with* $g(x) = \arctan(x)$ *and* $f(x) = \tan(x)$ *to find the derivative of* $\arctan(x)$. *The difficult part is drawing the correct right triangle to get the nice expression for* $f'(g(x))$.

15.9 Cauchy Distribution

This chapter is on the Gamma function and related distributions. Our final entry at first glance appears to have no relation, but its normalization constant is immediately computable by using the cosecant identity at $s = 1/2$; in particular, we looked at exactly this integral in §15.8.3.

Cauchy Distribution: A random variable X has the standard Cauchy distribution if its density is

$$f_X(x) = \frac{1}{\pi} \frac{1}{1 + x^2}.$$

This random variable has no mean and infinite variance.

As remarked, we computed the normalization constant in §15.8.3; another way to interpret that calculation is that the anti-derivative of $1/(1 + x^2)$ is $\arctan(x)$, and thus

$$\int_{-\infty}^{\infty} \frac{dx}{1 + x^2} = 2 \int_0^{\infty} \frac{dx}{1 + x^2} = 2 \left[\arctan(\infty) - \arctan(0) \right] = 2 \left[\frac{\pi}{2} - 0 \right] = 0;$$

thus we must multiply by $1/\pi$ to have the integral equal 1.

Why does the Cauchy distribution have no mean? Aren't we investigating

$$\int_{-\infty}^{\infty} x \frac{dx}{\pi(1 + x^2)},$$

and shouldn't the integral of an odd function over a symmetric region vanish? The problem, as always, is that we're dealing with an infinity—you must *always* be careful when infinities enter the scene. The correct definition is that an improper integral like this exists if and only if

$$\lim_{A,B \to \infty} \int_{-A}^{B} x \frac{dx}{\pi(1 + x^2)}$$

exists and takes on the same value no matter how A and B tend to infinity. Unfortunately in our case the value depends on how they tend to infinity. For example, if $B = A$ then we get zero, while if $B = 2A$ the contribution from $-A$ to A is zero and we get

$$\lim_{A \to \infty} \int_{-A}^{2A} x \frac{dx}{\pi(1 + x^2)} = \lim_{A \to \infty} \int_A^{2A} \frac{x \, dx}{\pi(1 + x^2)}.$$

For A very large, $x/(1 + x^2)$ is approximately $1/x$ (it is at least $1/2x$ and at most $1/x$ if you wish to make the argument more precise), and thus our integral is approximately

$$\lim_{A \to \infty} \int_A^{2A} \frac{dx}{\pi x} = \frac{1}{\pi} \left[\log(2A) - \log(A) \right] = \frac{\log(2)}{\pi} \neq 0.$$

Thus the value of the integral for the mean depends on the path we take to infinity, and the mean does not exist (though see Exercise 15.10.23).

The situation is even worse for the variance, which is clearly infinite. Note for $x \geq 2016$ we have $x^2/(1+x^2) \geq 1/2$, and thus

$$\int_{-\infty}^{\infty} x^2 \frac{dx}{\pi(1+x^2)} \geq \int_{2016}^{\infty} \frac{dx}{2\pi} = \infty.$$

The Cauchy distribution is one of the most important distributions, and you must master and remember it. Because it has no mean and infinite variance, it has very different behavior than other random variables which we have studied. Thus, if you're trying to see if a result holds for all densities, it's great to test your conjecture on the Cauchy distribution as well as the "nicer" distributions (such as uniform, exponential, Gaussian, ...).

It also plays a major role in some economics theories. A simple variant of the **random walk hypothesis** asserts stock motions can be well modeled by tossing independent coins, and thus the Central Limit Theorem and Gaussian behavior emerge. Data, however, suggests that there are more days with large swings than would be predicted by such a theory, and one needs to work with a distribution with larger variance than the Gaussian. It turns out that the Cauchy and Gaussian densities can be placed in a common family with a varying parameter; interestingly both are also stable distributions (see Exercise 15.10.22). For more on such applications see the work of Fama and Mandelbrot [Fa1, Fa2, Man, ManHu].

15.10 Exercises

Exercise 15.10.1 *Find $\Gamma(3/2)$.*

Exercise 15.10.2 *Find $\Gamma(-1/2)$.*

Exercise 15.10.3 *Prove $\Gamma(m+1/2) = \frac{(2m-1)!!}{2^m}\Gamma(1/2)$, and thus $\mu_{2m} = (2m-1)!!$.*

Exercise 15.10.4 *Find $\Gamma(1/2 - m)$ for positive integers m.*

Exercise 15.10.5 *Prove that if the relationship $\Gamma(s)\Gamma(1-s) = \pi \csc(\pi s)$ holds for $s \in (0, 1)$ then it holds for all s (or at least all s that are not integers, as if s is an integer then we have to interpret the equality among two infinities).*

Exercise 15.10.6 *For what values of a and b is the Beta distribution symmetric about its mean?*

Exercise 15.10.7 *Find the mean of the Beta distribution.*

Exercise 15.10.8 *Find the mean of the Weibull distribution.*

Exercise 15.10.9 *Find the mean of the Gamma distribution with parameters k and σ.*

Exercise 15.10.10 *Find the variance of the Gamma distribution with parameters k and σ.*

Exercise 15.10.11 *Find $\mathbb{E}[X^n]$ for X a Gamma variable with parameters k and σ.*

Exercise 15.10.12 *Comment on the relationship between the Gamma distribution and the Erlang distribution.*

Exercise 15.10.13 *For what x does the Gamma distribution take on its maximum value given parameters k and θ? Use that to try and estimate n!.*

Exercise 15.10.14 *Justify that the kth smallest variable in a list of n identical, independent random variables $X_1, X_2, \ldots, X_k$ each with density f and cdf F is*

$$nf(x)\binom{n-1}{k-1}(F(x))^{k-1}(1-F(x))^{n-k}.$$

Exercise 15.10.15 *Use the formula from the previous exercise to show that the kth smallest of a set of n uniform variables on (0,1) can be modeled with a Beta distribution. Find the appropriate parameters for the Beta distribution.*

Exercise 15.10.16 *The incomplete lower Gamma function is defined by*

$$\gamma(s,x) = \int_0^x e^{-x}x^{s-1}dx.$$

Find a recurrence relation relating $\gamma(s,x)$ to $\gamma(s-1,x)$.

Exercise 15.10.17 *Prove the **Gregory-Leibniz formula** for π by evaluating $\int_0^1 \frac{dx}{1+x^2}$ two ways: (1) use the geometric series formula to expand, interchange the summation and the integral, and integrate term by term (all this must be justified); (2) use the derivative of $\arctan x$ is $1/(1+x^2)$.*

Exercise 15.10.18 *Due to the Gamma distribution's relationship to the exponential distribution and the exponential distribution's "memorylessness," Gamma distributions are useful in measuring Web server traffic. Particularly, the time at which the kth person connects can be modeled by a Gamma distribution with shape parameter k and some scale parameter σ. If $\sigma = 1/10$, find the probability that the 100th person connects within an hour of the start time.*

Exercise 15.10.19 *Wind speed is well approximated by a Weibull distribution. Find the probability of the wind speed being over 20 if the wind speed in a given area is a Weibull variable with shape parameter 2 and scale parameter 10.*

Exercise 15.10.20 *Plot the Beta distribution with parameters $a = 2, b = 3$. Shade in the area under the curve corresponding to $P(.2 < X < .6)$. (There are several ways to do this, some look better than others. Play around with it some.)*

Exercise 15.10.21 *Prove the **Gregory-Leibniz formula**. (Hint: Use the derivative of $\arctan(x)$ is $\frac{1}{1+x^2}$ to get $\int_0^1 \frac{1}{1+x^2}dx = \frac{\pi}{4}$. Write $1+x^2$ as $1-(-x^2)$, and expand $\frac{1}{1+x^2}$ using the geometric series formula with $r = -x^2$. Justify interchanging the sum and the integral, integrate term by term, and smile!)*

Exercise 15.10.22 *We say X is a **generalized Cauchy distribution** with parameters a and b if its density is*

$$f_X(x) = \frac{1}{b\pi}\frac{1}{1+(x-a)^2/b^2},$$

and we write $X \sim \text{Cauchy}(a,b)$ (note that other books use slightly different notation, so be careful). Prove that if $X \sim \text{Cauchy}(a_1, b_1)$ and $Y \sim \text{Cauchy}(a_2, b_2)$ then there *are constants a_3, b_3 such that $X + Y \sim \mathrm{Cauchy}(a_3, b_3)$, and determine these constants in terms of a_1, a_2, b_1, b_2; thus the Cauchy distribution is* **stable**. *(Warning: this is a challenging exercise if you don't know complex analysis; if you know the residue theorem you can give a straightforward argument.)*

Exercise 15.10.23 *Let $X \sim \mathrm{Cauchy}(a, b)$ (see the previous exercise). Though the mean does not exist, the median does, where the median $\widetilde{\mu}$ is the value such that*

$$\int_{-\infty}^{\widetilde{\mu}} f_X(x)dx \;=\; \int_{\widetilde{\mu}}^{\infty} f_X(x)dx.$$

Find the median in terms of a and b.

Exercise 15.10.24 *Generalize Exercise 15.10.22 to densities of the form $C(a, b, k)/(1 + |x - a|^k/b^k)$, where $k, b > 0$ and $x \in \mathbb{R}$. (Warning: this exercise is very hard without complex analysis.)*

CHAPTER 16 _____

The Chi-square Distribution

The chi-square distribution is one of the most important in statistics. It arises all the time in hypothesis testing. It's defined as follows.

Chi-square distribution: If X is a chi-square distribution with $v \geq 0$ degrees of freedom, then X has density

$$f(x) = \begin{cases} \dfrac{1}{2^{v/2}\Gamma(v/2)} x^{(v/2-1)} e^{-x/2} & \text{if } x \geq 0 \\ 0 & \text{otherwise.} \end{cases}$$

We write $X \sim \chi^2(v)$ to denote this.

The density is clearly non-negative, but does it integrate to 1? Before integrating, we first recall the definition of the Gamma function (see Chapter 15 for a detailed tour of this function):

$$\Gamma(s) = \int_0^\infty e^{-t} t^{s-1} dt.$$

As there is a $\Gamma(v/2)$ in the density it should come as no surprise that the Gamma function plays a big role in understanding the chi-square distribution.

Let's calculate the integral:

$$\int_0^\infty \frac{1}{2^{v/2}\Gamma(v/2)} x^{v/2-1} e^{-x/2} dx = \frac{1}{2^{v/2}\Gamma(v/2)} \int_0^\infty e^{-x/2} x^{v/2-1} dx.$$

Notice how close the integral is to the definition of the Gamma function; the only difference is that we have the exponential of $-x/2$, and in the definition of the Gamma function we just have $-x$. This suggests we change variables to $t = x/2$, so $x = 2t$ and

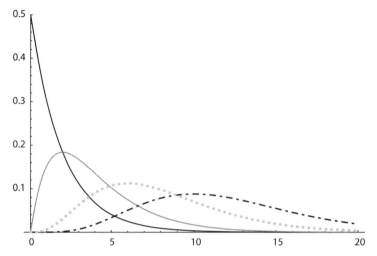

Figure 16.1. Plot of chi-square distributions with $\nu \in \{1, 2, 3, 5, 10, 20\}$; as the degree of freedom increases, the location of the bump moves rightward.

$dx = 2dt$, and we get

$$\int_0^\infty \frac{1}{2^{\nu/2}\Gamma(\nu/2)} x^{\nu/2-1} e^{-x/2} dx = \frac{1}{2^{\nu/2}\Gamma(\nu/2)} \int_0^\infty e^{-t} (2t)^{\nu/2-1} 2dt$$

$$= \frac{2^{\nu/2}}{2^{\nu/2}\Gamma(\nu/2)} \int_0^\infty e^{-t} t^{\nu/2-1} dt$$

$$= \frac{2^{\nu/2}\Gamma(\nu/2)}{2^{\nu/2}\Gamma(\nu/2)} = 1,$$

proving it's a probability distribution.

In the next section we'll discuss where the chi-square distribution comes from, then continue with a summary of some of its properties and applications. Before doing this, though, it's worth commenting on the choice of nomenclature. For each choice of $\nu > 0$ we get a different distribution. The most important choices, as we'll see below, are when ν is a positive integer. We need some way to refer to the parameter; the convention is to call ν the **degrees of freedom** (of the chi-square distribution). The number of degrees of freedom of a chi-square distribution is very important because it dramatically changes the shape of the distribution, as you can see in Figure 16.1

The term degrees of freedom is formally defined as the number of values in the statistic that are free to vary. If there are N observations, the number of degrees of freedom is normally $N - 1$ or N. See Exercise 16.5.21 for more on this concept.

There are many hypothesis tests in statistics whose test statistics have a chi-square distribution under the null hypothesis. Such hypothesis tests that reference test statistics to a chi-square distribution are applicable to categorical data. The advantage of a chi-square test is that it is a non-parametric test. Specifically, this means that it makes no assumptions about the underlying population distribution from which the data was drawn. The trade-off then, is that a chi-square test is less statistically powerful than its parametric counterpart.

16.1 Origin of the Chi-square Distribution

While we can of course take any non-negative function that integrates to 1 as a probability density, most of these aren't worth studying. Why is the chi-square distribution so important? The reason is that it arises from sums of independent, normally distributed random variables.

Let $X \sim N(0, 1)$; this means X is a random variable with the standard normal distribution (see Chapter 14). Thus X is a Gaussian with mean zero and variance 1. The big result follows.

Relation between Chi-square and Normal Random Variables: If $X \sim N(0, 1)$ then $X^2 \sim \chi^2(1)$.

This is just the tip of the iceberg; a massive generalization exists, which we'll see in a few moments. First, though, let's prove the square of the standard normal has a chi-square distribution with one degree of freedom. The easiest way to do this is through the **cumulative distribution function method** (described in detail in §10.5). We let $Y = X^2$; X has density

$$f_X(x) = \frac{1}{\sqrt{2\pi}} e^{-x^2/2}$$

and cumulative distribution function

$$F_X(x) = \text{Prob}(X \le x).$$

If we know the cumulative distribution function of Y, then we can find the density of Y by taking its derivative. For $y \ge 0$ we have

$$F_Y(y) = \text{Prob}(Y \le y)$$

$$= \text{Prob}(X^2 \le y)$$

$$= \text{Prob}(-\sqrt{y} \le X \le \sqrt{y}).$$

We had to be careful here; the condition $X^2 \le y$ isn't $X \le \sqrt{y}$ but rather $-\sqrt{y} \le X \le \sqrt{y}$, as the square of a negative number is a positive number. Continuing, we see

$$F_Y(y) = \int_{-\sqrt{y}}^{\sqrt{y}} \frac{1}{\sqrt{2\pi}} e^{-t^2/2} dt = F_X(\sqrt{y}) - F_X(-\sqrt{y}).$$

The calculation is finished by using the chain rule: if $A(y) = B(C(y))$ then $A'(y) = B'(C(y))C'(y)$. Thus

$$f_Y(y) = F_Y'(y)$$

$$= F_X'(\sqrt{y}) \cdot (\sqrt{y})' - F_X'(-\sqrt{y}) \cdot (-\sqrt{y})'$$

$$= f_X(\sqrt{y}) \cdot \frac{1}{2\sqrt{y}} - f_X(-\sqrt{y}) \cdot \frac{-1}{2\sqrt{y}}$$

$$= \frac{f_X(\sqrt{y})}{\sqrt{y}}$$

$$= \frac{1}{\sqrt{2\pi}} e^{-y/2} y^{-1/2}$$

$$= \frac{1}{2^{1/2}\Gamma(1/2)} y^{(1/2-1)} e^{-y/2},$$

where we used $\Gamma(1/2) = \sqrt{\pi}$ (which we proved many times in Chapter 15); this completes the proof. □

We note that we could have done the calculation in a slightly different manner, but of course we'd end up at the same place. We could have simplified the integral by using the fact that it's symmetric, and thus write

$$F_Y(y) = 2 \int_0^{\sqrt{y}} \frac{1}{\sqrt{2\pi}} e^{-t^2/2} dt = 2 \left(F_X(\sqrt{y}) - F_X(0) \right).$$

There's one other item worth mentioning. When we use the **cumulative distribution function approach** to find the density, note that we *never* need to know a nice formula for F_X. All we need to know is f_X. This is very important, as many distributions (such as the standard normal) have a nice expression for their densities, but as integration is hard we have no nice formulas for their cumulative distribution function. It's quite amazing, really. We integrate and get the cumulative distribution function, but then we immediately differentiate and return to the densities.

We now have an interpretation for a chi-square distribution with one degree of freedom—it's the square of a standard normal random variable. Is there a similar interpretation for other chi-square distributions? Yes! To find it, just read on.

16.2 Mean and Variance of $X \sim \chi^2(1)$

Now let's calculate the expected value (or mean) of a chi-square random variable. We'll do the simplest case first, and assume we have $k = 1$ degrees of freedom. We must evaluate

$$\mathbb{E}[X] = \int_0^\infty x f(x) dx = \int_0^\infty \frac{x}{2^{(1/2)}\Gamma(1/2)} x^{\frac{1}{2}-1} e^{-x/2} dx.$$

Bringing the constants outside of the integral sign and combining the powers of x yield

$$\mathbb{E}[X] = \frac{1}{2^{(1/2)}\Gamma(1/2)} \int_0^\infty x^{1/2} e^{-x/2} dx.$$

This form very closely resembles that of the Gamma function. However, we have e raised to the power of $-x/2$, instead of just to the power of $-x$. This problem is easily remedied with a change of variables. Consider the same change of variables used earlier:

$t = x/2$. In this case, $x = 2t$ and $dx = 2dt$. Our integral becomes

$$\mathbb{E}[X] = \frac{1}{2^{(1/2)}\Gamma(1/2)} \int_0^\infty 2t^{1/2}e^{-t}2dt.$$

Pulling our constants outside of our integral sign and allowing for cancelation, we get

$$\mathbb{E}[X] = \frac{2}{\Gamma(1/2)} \int_0^\infty t^{1/2}e^{-t}2dt.$$

This should look *very* familiar. If $t^{1/2}$ is rewritten as $t^{\frac{3}{2}-1}$, then the form above is exactly the definition of the Gamma function evaluated at 3/2 (as we have reason to believe there is a chi-square distribution lurking, it is natural to write the exponent of t as something minus one, which suggested this algebra). Thus, our integral simplifies to

$$\frac{2\Gamma(3/2)}{\Gamma(1/2)}.$$

From the functional equation of the Gamma function, we know $\Gamma(s+1) = s\Gamma(s)$. Taking $s = 1/2$ and substituting yields

$$\mathbb{E}[X] = \frac{2\Gamma(3/2)}{\Gamma(1/2)} = \frac{2(\frac{1}{2})\Gamma(1/2)}{\Gamma(1/2)} = 1,$$

and thus we've shown that the expected value of a chi-square random variable with $k = 1$ degrees of freedom is 1.

Now, let's calculate the variance of a chi-square random variable with $k = 1$ degrees of freedom. As

$$\text{Var}(X) = \mathbb{E}[X^2] - \mathbb{E}[X]^2,$$

we just need to find $\mathbb{E}[X^2]$:

$$\mathbb{E}[X^2] = \int_0^\infty x^2 f(x)\,dx = \int_0^\infty \frac{x^2}{2^{(1/2)}\Gamma(1/2)} x^{\frac{1}{2}-1}e^{-x/2}\,dx.$$

Simplifying the integrand and moving the constants outside of the integral signs gives

$$\mathbb{E}[X^2] = \frac{1}{2^{(1/2)}\Gamma(1/2)} \int_0^\infty x^{3/2}e^{-x/2}dx.$$

Let's make the same change of variables as we did in the expectation calculation: $t = x/2$. In this case, $x = 2t$ and $dx = 2dt$. With this substitution, our integral becomes

$$\mathbb{E}[X^2] = \frac{1}{2^{(1/2)}\Gamma(1/2)} \int_0^\infty 2t^{3/2}e^{-t}2dt.$$

Now, if we write $t^{3/2}$ as $t^{\frac{5}{2}-1}$ and move the constants outside of the integral sign, then our integral is identical to $\Gamma(5/2)$. And so, further simplification yields

$$\mathbb{E}[X^2] = \frac{2^2\Gamma(5/2)}{\Gamma(1/2)}.$$

We again appeal to the functional equation for the Gamma function, this time taking $s = 3/2$, and obtain

$$\mathbb{E}[X^2] = \frac{2^2\Gamma(5/2)}{\Gamma(1/2)} = \frac{2^2(\frac{3}{2})\Gamma(3/2)}{\Gamma(1/2)} = \frac{2^2(\frac{3}{2})(\frac{1}{2})\Gamma(1/2)}{\Gamma(1/2)} = 3.$$

All that's left to obtain $\text{Var}(X)$ is to subtract off the square of the mean, which we showed was 1. Thus the variance equals $3 - 1^2$, or 2.

We only did the mean and the variance for a chi-square random variable with one degree of freedom, but the calculation is similar for higher moments or other ν, and I encourage you to try doing some of these before reading on.

16.3 Chi-square Distributions and Sums of Normal Random Variables

In this section we'll give an interpretation for chi-square distributions with k degrees of freedom (for k a positive integer). The answer turns out to be related to sums of independent normal random variables. This shouldn't be too surprising, as we saw in §16.1 a connection between normal and chi-square random variables. This relation is why the chi-square distribution is so important. We'll elaborate on *why* this makes the chi-square distribution important in the next section; for now, we concentrate on showing the relation.

Chi-square distributions and sums of normal random variables: Let k be a positive integer, and $X_1, \ldots, X_k$ independent standard normal random variables; this means each $X_i \sim N(0, 1)$. Then if $Y_k = X_1^2 + \cdots + X_k^2$, $Y_k \sim \chi^2(k)$. More generally, let $Y_{\nu_1}, \ldots, Y_{\nu_m}$ be m independent chi-square random variables, where $Y_{\nu_i} \sim \chi^2(\nu_i)$. Then $Y = Y_{\nu_1} + \cdots + Y_{\nu_m}$ is a chi-square random variable with $\nu_1 + \cdots + \nu_m$ degrees of freedom.

As always, there are several different proofs we can take. We'll do a few of them, as each of these illustrate different important concepts in probability. In particular, we could skip many of the subsections below if we just prove the following special case.

If $Y_{\nu_1} \sim \chi^2(\nu_1)$ and $Y_{\nu_2} \sim \chi^2(\nu_2)$ are two independent, chi-square random variables, then $Y_{\nu_1} + Y_{\nu_2} \sim \chi^2(\nu_1 + \nu_2)$.

Why does this one special case prove *all* of our claims? Again, it's due to appropriately **grouping parentheses**. We've seen this method many, many times (see

§A.3 for more examples). We can save ourselves a lot of work by becoming proficient in this; you definitely want to add this to your bag of tricks. The reason is that we break our complicated case into lots of simpler cases, and frequently it is easier to do lots of easy calculations and combine than one hard one. What's really going on is a proof by induction is lurking in the background.

Let's illustrate with two examples. First, let's assume X_1, X_2, X_3 are independent $N(0, 1)$ random variables. We are assuming that the sum of two independent chi-square random variables is a chi-square random variable whose degree of freedom equals the sum of the two degrees of freedom. We have X_1^2, X_2^2, and X_3^2 are all independent $\chi^2(1)$ random variables. We have

$$X_1^2 + X_2^2 + X_3^2 = (X_1^2 + X_2^2) + X_3^2.$$

By our claim, the sum of the first two is just a chi-square random variable with two degrees of freedom; let's denote it by Y_2. We then have

$$X_1^2 + X_2^2 + X_3^2 = Y_2 + X_3^2;$$

however, we again have just a sum of two independent chi-square random variables. We can thus use our claim again, and we see that $Y_2 + X_3^2$ is $\chi^2(3)$. We've shown that the sum of the squares of three standard normals is a chi-square with 3 degrees of freedom. The general case for sums of squares of standard normals follows similarly.

What about sums of general chi-square random variables? The argument is quite similar, and we give just a sketch. Using what is hopefully obvious notation (all random variables are independent, and Y_ν is always a chi-square distribution with ν degrees of freedom):

$$
\begin{aligned}
Y_{\nu_1} + Y_{\nu_2} + Y_{\nu_3} + \cdots + Y_{\nu_m} &= \left(Y_{\nu_1} + Y_{\nu_2}\right) + Y_{\nu_3} + \cdots + Y_{\nu_m} \\
&= Y_{\nu_1 + \nu_2} + Y_{\nu_3} + \cdots + Y_{\nu_m} \\
&= \left(Y_{\nu_1 + \nu_2} + Y_{\nu_3}\right) + Y_{\nu_4} + \cdots + Y_{\nu_m} \\
&= Y_{\nu_1 + \nu_2 + \nu_3} + Y_{\nu_4} + \cdots + Y_{\nu_m}.
\end{aligned}
$$

If we continue arguing like this, eventually we get $Y_{\nu_1 + \cdots + \nu_m}$.

So, everything comes down to showing $\chi^2(\nu_1) + \chi^2(\nu_2) = \chi^2(\nu_1 + \nu_2)$. We won't jump straight to the proof of this, but rather we'll do a few special cases first. Remember, the goal is not to finish this book prepared to only do a few fixed problems, but rather have the skills to attack a variety of questions you've never seen. Thus, we frequently include some additional proofs that could be skipped if we wanted the most concise introduction to probability possible; however, it's worth the time reading these to build your toolbag.

In the next subsections we give proofs by direct integration in §16.3.1, by the Change of Variables Theorem in §16.3.2, and finally by convolutions. If all you care about is the proof of the claim, you can skip to the two convolution sections (§16.3.3 and §16.3.4); however, there are a lot of great items and perspectives in the other approaches. It's worth the time it takes to digest the material. In particular, in §16.3.2 we develop the Theory of Normalization Constants and see how to integrate without integrating!

16.3.1 Sums of Squares by Direct Integration

We showed that if $X \sim N(0, 1)$ then $X^2 \sim \chi^2(1)$. What about the sum of the squares of two independent standard normal distributions? We again calculate the cumulative distribution function and then differentiate. Let $Y = X_1^2 + X_2^2$ with each $X_i \sim N(0, 1)$ and X_1, X_2 independent. We find

$$
\begin{aligned}
F_Y(y) &= \text{Prob}(Y \leq y) \\
&= \text{Prob}(X_1^2 + X_2^2 \leq y) \\
&= \iint_{x_1^2 + x_2^2 \leq y} \frac{1}{\sqrt{2\pi}} e^{-x_1^2/2} \frac{1}{\sqrt{2\pi}} e^{-x_2^2/2} dx_1 dx_2 \\
&= \iint_{x_1^2 + x_2^2 \leq y} \frac{1}{2\pi} e^{-(x_1^2 + x_2^2)/2} dx_1 dx_2.
\end{aligned}
$$

By now you should see the next step coming. We're integrating over the disk of radius y; this problem is *screaming* at us to change variables to polar coordinates. Further, just look at the integrand. There is no dependence on the angle θ; there is only radial dependence.

We now switch to polar coordinates, setting $x_1 = r \cos \theta$ and $x_2 = r \sin \theta$. The Change of Variables Formula (see the review online for the special case of polar coordinates) gives $dx_1 dx_2 = r dr d\theta$. What are the bounds of integration? We have $\theta : 0 \to 2\pi$, while $r : 0 \to \sqrt{y}$ as the circle has radius $\sqrt{y}$ and not y. We get

$$
\begin{aligned}
F_Y(y) &= \int_{\theta=0}^{2\pi} \int_{r=0}^{\sqrt{y}} \frac{1}{2\pi} e^{-r^2/2} r \, dr \, d\theta \\
&= \int_{\theta=0}^{2\pi} \frac{d\theta}{2\pi} \int_{r=0}^{\sqrt{y}} e^{-r^2/2} r \, dr \\
&= \left(2\pi \cdot \frac{1}{2\pi} \right) \cdot \left[-e^{-r^2/2} \right]_0^{\sqrt{y}} \\
&= 1 \cdot \left(1 - e^{-y/2} \right).
\end{aligned}
$$

Now that we know the cumulative distribution function $F_Y(y)$, the density is simply the derivative. Thus we finally obtain

$$
f_Y(y) = \frac{1}{2} e^{-y/2},
$$

which by inspection is the density of the chi-square distribution with two degrees of freedom.

16.3.2 Sums of Squares by the Change of Variables Theorem

Given the amount of work it took to evaluate the sum of the squares of two standard normal distributions, we're justified in being a little afraid of the calculation for the sum of n squares. It seems like we will need to know the change of variable formula for n-dimensional Cartesian coordinates to n-dimensional spherical coordinates! This section is a bit more advanced, and we assume the reader is familiar with the multivariable Change of Variables Theorem (which we review online). *You can safely skip this section and move on to the convolution approach, though I do encourage you to at least skim these arguments.* The reason is that the Method of Normalization Constants, which we use in the convolution approach, also surfaces here; the more times you see an advanced concept, the easier it is. Another reason is that the change of variable approach leads to a lot of great results; for example, in the course of our analysis we get for free the area of the n-dimensional sphere! Finally, this section is the natural generalization of the previous section on direct integration, and thus logically it makes sense to see how far we can push those calculations. That said, feel free to jump to the convolution sections and return to this section later. As one of my mentors Serge Lang once remarked in one of his books, it's a shame that material must be totally ordered along the page axis, as different people would prefer different orders here!

Amazingly, though there are nice formulas for generalized spherical coordinates, we do not need to know them because we can exploit a method known as the **Theory of Normalization Constants**. We know we can represent $x_1, \ldots, x_k$ through the radius r and $k-1$ angles $\theta_1, \ldots, \theta_{k-1}$. We have relations of the form

$$x_1 = r h_1(\theta_1, \ldots, \theta_{k-1})$$

$$\vdots$$

$$x_k = r h_k(\theta_1, \ldots, \theta_{k-1}).$$

Just to make sure we're all on the same page: there *are* formulas for the different h_i's. We could go this route and use those explicit formulas for our calculations; however, the reason for this section is to show you how to bypass those horrendous calculations by looking at the problem the right way. We're still going to use the Change of Variables Theorem. It may seem a bit surprising that we're going to use it when we don't even know the functions involved in the change of variables, but that's the power and beauty of this method!

Let's recall the Change of Variables Theorem.

Change of Variables Theorem: Let V and W be bounded open sets in $\mathbb{R}^k$. Let $h : V \to W$ be a 1–1 and onto map, given by

$$h(u_1, \ldots, u_k) = (h_1(u_1, \ldots, u_k), \ldots, h_k(u_1, \ldots, u_k)).$$

Let $f : W \to \mathbb{R}$ be a continuous, bounded function. Then

$$\int \cdots \int_W f(x_1, \ldots, x_k) dx_1 \cdots dx_k$$

$$= \int \cdots \int_V f(h(u_1, \ldots, u_k)) J(u_1, \ldots, u_v) du_1 \cdots du_k,$$

where J is the Jacobian

$$J = \begin{vmatrix} \frac{\partial h_1}{\partial u_1} & \cdots & \frac{\partial h_1}{\partial u_k} \\ \vdots & \ddots & \vdots \\ \frac{\partial h_k}{\partial u_1} & \cdots & \frac{\partial h_k}{\partial u_k} \end{vmatrix}.$$

If we're to use this theorem, it does seem like we need to know the change of variable functions h_i. Here's the slick trick that gets us out of that painful computation. We need to figure out how the volume element $dx_1 \cdots dx_k$ changes; we clearly have

$$dx_1 \cdots dx_k = \mathcal{G}(r, \theta_1, \ldots, \theta_{k-1}) dr d\theta_1 \cdots d\theta_{k-1};$$

there's no math involved here, we're simply using the letter $\mathcal{G}$ to denote the answer. Now we input some math. We must have

$$\mathcal{G}(r, \theta_1, \ldots, \theta_{k-1}) = r^{k-1} \mathcal{C}(\theta_1, \ldots, \theta_{k-1}).$$

Why? This follows from unit analysis. In two dimensions we have $dx_1 dx_2 \mapsto r dr d\theta$ and in three dimensions it's $dx_1 dx_2 dx_3 \mapsto r^2 \sin\theta_1 dr d\theta_1 d\theta_2$. Note that we have the radius to a power one less than the number of variables. This is because the angular variables are unitless, and thus the units of $dr d\theta_1 \cdots d\theta_{k-1}$ are meters (say), while $dx_1 \cdots dx_k$ has units of meters^k. Thus we need the factor r^{k-1}. We have therefore shown that there is some complicated function $\mathcal{C}$ such that

$$dx_1 \cdots dx_k = r^{k-1} \mathcal{C}(\theta_1, \ldots, \theta_{k-1}) dr d\theta_1 \cdots d\theta_{k-1}.$$

We now return to our problem. Let $Y = X_1^2 + \cdots + X_k^2$. We again use the cumulative distribution function technique and find

$$F_Y(y) = \text{Prob}(X_1^2 + \cdots + X_k^2 \leq y)$$

$$= \int \cdots \int_{x_1^2 + \cdots + x_k^2 \leq y} \frac{1}{\sqrt{2\pi}} e^{-x_1^2/2} \cdots \frac{1}{\sqrt{2\pi}} e^{-x_k^2/2} dx_1 \cdots dx_k$$

$$= \int \cdots \int_{x_1^2 + \cdots + x_k^2 \leq y} \frac{1}{(2\pi)^{k/2}} e^{-(x_1^2 + \cdots + x_k^2)/2} dx_1 \cdots dx_k.$$

We now change variables. We don't care what the angular integrations are over, so we just denote those by ℓ_i to u_i (for lower and upper bound):

$$F_Y(y) = \int_{r=0}^{\sqrt{y}} \int_{\theta_1=\ell_1}^{u_1} \cdots \int_{\theta_{k-1}=\ell_{k-1}}^{u_{k-1}} \frac{e^{-r^2/2} r^{k-1} \mathcal{C}(\theta_1, \ldots, \theta_{k-1})}{(2\pi)^{k/2}} dr d\theta_1 \cdots d\theta_{k-1}.$$

Actually, while what we wrote is correct it could have been wrong. We assumed that the integration is over some $(k-1)$-dimensional box of angles. We should really have ℓ_{k-1} and u_{k-1} functions of the angles θ_1 through θ_{k-2}, and then ℓ_{k-2} and u_{k-2} functions of the angles θ_1 through θ_{k-3}, and so on. All that changes, fortunately, is the bounds of integration above, and the next line is the same.

We integrate over the $k - 1$ angles; the answer is independent of r and y, and we denote it by C_k (it does depend on the number of angular variables). Hence

$$F_Y(y) = C_k \int_{r=0}^{\sqrt{y}} e^{-r^2/2} r^{k-1} dr.$$

Let $h(r) = C_k e^{-r^2/2} r^{k-1}$ and $H(r)$ be its anti-derivative. Then

$$F_Y(y) = H(\sqrt{y}) - H(0).$$

We take the derivative and finally (almost) obtain the density:

$$f_Y(y) = H'(\sqrt{y}) \frac{1}{2\sqrt{y}} = \frac{C_k}{2} e^{-y/2} y^{\frac{k}{2}-1}.$$

Why do we say "almost" above? The problem is we still have the constant C_k, which we *should* have determined by doing the angular integrations but did not. Thus we do not have the final answer; fortunately, it's trivial to compute C_k now. This seems absurd—how can we compute C_k now? Shouldn't we have computed it earlier? And, if we're going to compute it, shouldn't we figure out what the change of variable formulas are for going from Cartesian to spherical?

The reason we can evaluate it so easily is that $Y = X_1^2 + \cdots + X_k^2$ is a random variable; **therefore its density must integrate to 1!** We know from above the formula for the density of a chi-square random variable with k degrees of freedom; using y for the dummy variable it's just (for $y \geq 0$)

$$\frac{1}{2^{k/2} \Gamma(k/2)} y^{\frac{k}{2}-1} e^{-y/2}.$$

Note this has exactly the same y-dependence as our part, and thus the normalization constants must match up!

This is a very important problem, without a doubt one of the most important in the book. While there are other ways to compute this answer by doing more direct computations, I prefer this approach as it illustrates the power of the **Theory of Normalization Constants***. It's incredible how it allows us to bypass certain painful computations. This arises all the time in higher mathematics, especially in Random Matrix Theory.*

Remark 16.3.1: *If we hadn't been given the probability density function for a chi-square with k degrees of freedom, we could still have found the value of C_k by noting that $f_Y(y)$ integrates to 1. We need to use the Gamma function, which is defined by*

$$\Gamma(s) = \int_0^{\infty} e^{-x} x^{s-1} dx.$$

Returning to our problem, we have

$$1 = \int_0^\infty f_Y(y)dy = \frac{C_k}{2} \int_0^\infty e^{-y/2} y^{\frac{k}{2}-1} dy.$$

We change variables, letting $x = y/2$ so $dy = 2dx$ and find

$$1 = \frac{C_k}{2} \int_0^\infty e^{-x} 2^{\frac{k}{2}-1} x^{\frac{k}{2}-1} 2dx = \frac{C_k}{2} \cdot 2^{\frac{k}{2}} \Gamma\left(\frac{k}{2}\right),$$

which implies

$$\frac{C_k}{2} = \frac{1}{2^{k/2}\Gamma(k/2)}.$$

Remark 16.3.2: *Whenever we see a new method, it's worth exploring how far we can push it. What else can we glean from the above analysis? Implicit in our computation is the "surface area" of the n-**dimensional sphere**! Remember our volume element became*

$$r^{k-1} \mathcal{C}(\theta_1, \ldots, \theta_{k-1}) dr d\theta_1 \cdots d\theta_{k-1},$$

and we showed

$$\int_{\theta_1=\ell_1}^{u_1} \cdots \int_{\theta_{k-1}=\ell_{k-1}}^{u_{k-1}} \frac{1}{(2\pi)^{k/2}} \mathcal{C}(\theta_1, \ldots, \theta_{k-1}) d\theta_1 \cdots d\theta_{k-1} = C_k.$$

Using our value for C_k above, we find

$$\int_{\theta_1=\ell_1}^{u_1} \cdots \int_{\theta_{k-1}=\ell_{k-1}}^{u_{k-1}} \mathcal{C}(\theta_1, \ldots, \theta_{k-1}) d\theta_1 \cdots d\theta_{k-1} = \frac{2(2\pi)^{k/2}}{2^{k/2}\Gamma(k/2)} = \frac{2 \cdot \pi^{k/2}}{\Gamma(k/2)}.$$

We claim that this is the surface area of the n-dimensional sphere. Why? We were integrating a function that depended only on the radius; thus we may consider our change of variables as partitioning the n-dimensional sphere of radius $\sqrt{y}$ into a collection of shells of radii ranging from 0 to $\sqrt{y}$. What does this formula give for specific n? We find

$$n = 2 : 2\pi$$

$$n = 3 : 4\pi$$

$$n = 4 : 2\pi^2;$$

except for the last, the previous two are well-known as the perimeter of the unit circle and the surface area of the unit sphere.

As a nice exercise, push the method further and compute the volume of the *n*-dimensional sphere. It's interesting to look at the ratio of the volume of a unit *n*-sphere divided by the volume of an *n*-dimensional cube with sides of length 2, as this tells us what percentage of the *n*-cube is filled by the *n*-sphere.

16.3.3 Sums of Squares by Convolution

In this section we give yet another proof that sums of squares of standard normal random variables have a chi-square distribution. The spirit of this section is very much in line with the Change of Variables approach. Specifically, our purpose is to show you how you can frequently bypass a lot of tedious calculations by appealing to the **Theory of Normalization Constants**.

In this section, we use the **Theory of Convolutions** to analyze the sum of independent random variables. Recall that if X_1 has density f_1, X_2 has density f_2, and if X_1 and X_2 are independent then the density of their sum is

$$f_{X_1+X_2}(x) = \int_{-\infty}^{\infty} f_1(t)f_2(x-t)dt.$$

We discuss convolutions in great detail in Chapters 10 and 19; here we content ourselves on a few quick words as to why this is reasonable. Imagine we want $X_1 + X_2$ to equal x. How can this happen? Well, if $X_1 = t$ then clearly X_2 must be $x - t$. The probability that $X_1 = t$ is just $f_1(t)$, while the probability that $X_2 = x - t$ is $f_2(x-t)$. Multiplying these two together gives the probability that both occur. We then integrate over all values of t, and this accounts for all possibilities. We denote the convolution of f_1 and f_2 by $f_1 * f_2$; thus

$$(f_1 * f_2)(x) = \int_{-\infty}^{\infty} f_1(t)f_2(x-t)dt.$$

Recall a random variable has a chi-square distribution with d degrees of freedom if it has density

$$f_d(x) = \begin{cases} \dfrac{1}{2^{d/2}\Gamma(d/2)}x^{\frac{d}{2}-1}e^{-x/2} & \text{if } x \geq 0 \\ 0 & \text{otherwise,} \end{cases}$$

where Γ is the Gamma function (the generalization of the factorial function), which is given by

$$\Gamma(s) = \int_0^{\infty} x^{s-1}e^{-x}dx.$$

We know that if X_i has the standard normal distribution, then X_i^2 has the chi-square distribution with 1 degree of freedom. We write c_d for the normalization constant of the chi-square distribution with d degrees of freedom.

Base Case: We first consider the case of the sum of two chi-square distributions, each with 1 degree of freedom. The density is

$$(f_1 * f_1)(x) = \int_{-\infty}^{\infty} f_1(t)f_1(x-t)dt$$

$$= \int_0^x c_1 t^{-1/2}e^{-t/2} \cdot c_1(x-t)^{-1/2}e^{-(x-t)/2}dt.$$

The range of integration stops at x as $f_1(x - t)$ is zero if the argument is negative. Simplifying yields

$$(f_1 * f_1)(x) = c_1^2 e^{-x/2} \int_0^x t^{-1/2} (x - t)^{-1/2} dt.$$

There are two ways to proceed. The first is to try and evaluate this integral directly. It may be possible to do this through brute force, but it won't be pleasant. Note that the final answer *must* be a probability distribution. Thus, **we don't need to figure out the integral exactly; it's enough to determine the x dependence!** The reason is that if we know the x-dependence, then we get the normalization constant by integrating $(f_1 * f_1)(x)$ with respect to x and setting the result equal to 1. This is an extremely clever and powerful idea; if you can learn to do this correctly, you can save yourself from many painful integrals.

Let's look at the t-integral. The only place x occurs is in $(x - t)^{-1/2}$. If we let $t = ux$ then we get $(x - ux)^{-1/2} = x^{-1/2}(1 - u)^{-1/2}$, which allows us to pull out the x dependence completely! Thus, lets make the following clever change of variables: set $t = ux$ and $dt = xdu$. As t runs from 0 to x we have u runs from 0 to 1. This yields

$$(f_1 * f_1)(x) = c_1^2 e^{-x/2} \int_0^1 (xu)^{-1/2} (x - xu)^{-1/2} x \, du$$

$$= c_1^2 e^{-x/2} \frac{x}{x^{1/2} x^{1/2}} \int_0^1 u^{-1/2} (1 - u)^{-1/2} du.$$

The u-integral can be done in closed form, as it's proportional to integrating the Beta density with parameters $\alpha = \beta = 1/2$ (see §15.5). We can also do some clever change of variables; however, there's no need! Letting $\mathcal{C}_1$ denote the value of the u-integral, we see

$$(f_1 * f_1)(x) = \begin{cases} \mathcal{C}_1 c_1^2 e^{-x/2} & \text{if } x \geq 0 \\ 0 & \text{otherwise.} \end{cases}$$

For this to be a probability distribution, the integral must be 1, which implies $\mathcal{C}_1 c_1^2 = 1/2$ (because $\int_0^\infty e^{-x/2} dx = 2$). Again, we emphasize that while we could have computed $\mathcal{C}_1$ by brute force, *there was no need*. To show that we have a chi-square distribution with 2 degrees of freedom, it suffices to show that we have the correct x-dependence, as then the normalization constants *must* match.

Induction Step: We now turn to the general case. We proceed by induction. We've already handled the base case; now we must show $X_1^2 + \cdots + X_{n+1}^2$ is a chi-square distribution with $n + 1$ degrees of freedom. By induction $X_1^2 + \cdots + X_n^2$ is a chi-square distribution with n degrees of freedom. Calling the normalization constants c_n and c_1 again, we see that

$$(f_1 * \cdots * f_1)(x) = \int_{-\infty}^\infty f_n(t) f_1(x - t) dt$$

$$= \int_0^x c_n t^{\frac{n}{2}-1} e^{-t/2} \cdot c_1 (x - t)^{-\frac{1}{2}} e^{-(x-t)/2} dt.$$

The exponential factors combine to give $e^{-x/2}$, and we again set $t = ux$ and $dt = xdu$, and find

$$(f_1 * \cdots * f_1)(x) = c_n c_1 e^{-x/2} \int_0^1 (xu)^{\frac{n}{2}-1}(x - xu)^{-\frac{1}{2}} x\,du$$

$$= c_n c_1 e^{-x/2} x^{\frac{n}{2}-1} x^{-\frac{1}{2}} x \int_0^1 u^{\frac{n}{2}-1}(1-u)^{-\frac{1}{2}}\,du$$

$$= c_n c_1 x^{\frac{n+1}{2}-1} e^{-x/2} \int_0^1 u^{\frac{n}{2}-1}(1-u)^{-\frac{1}{2}}\,du.$$

Again, it's possible to evaluate the u-integral in closed form (it's essentially the integral of a Beta density with parameters $n/2$ and $1/2$); however, all that matters is that it has no x-dependence. Calling this integral C_n, we find

$$(f_1 * \cdots * f_1)(x) = \begin{cases} C_n c_n c_1 x^{\frac{n+1}{2}-1} e^{-x/2} & \text{if } x \geq 0 \\ 0 & \text{otherwise.} \end{cases}$$

Note the x-dependence is exactly that of the chi-square distribution with $n + 1$ degrees of freedom, and thus the normalization constant $C_n c_n c_1$ *must* equal the normalization constant of the chi-square with $n + 1$ degrees of freedom. We emphasize again that we could have computed this constant by brute force, but that there was again no need!

16.3.4 Sums of Chi-square Random Variables

We've concentrated on giving the proofs that sums of squares of standard normal random variables have a chi-square distribution. These proofs are nice and illustrate a lot of great techniques; however, from a minimalist point of view they are not needed. All we need to do is show that if $Y_1 \sim \chi^2(\nu_1)$ and $Y_2 \sim \chi^2(\nu_2)$ are two independent chi-square random variables then $Y_1 + Y_2 \sim \chi^2(\nu_1 + \nu_2)$. If ν_1 and ν_2 are integers, we can reduce this to sums of squares of standard normal random variables with parentheses,

$$\left(X_1^2 + \cdots + X_{\nu_1}^2\right) + \left(X_{\nu_1+1}^2 + \cdots + X_{\nu_1+\nu_2}^2\right),$$

and then use generalized spherical coordinates.

What if ν_1 and ν_2 are not integers? It turns out we can just follow the convolution approach in §16.3.3. The reason is that our proof *never* used the assumption that they were integers, and thus the argument holds in greater generality. As this is such an important problem we'll give the general proof.

Letting the densities be f_{ν_1} and f_{ν_2}, the density of $Y = Y_{\nu_1} + Y_{\nu_2}$ is

$$f_Y(y) = (f_{\nu_1} * \cdots * f_{\nu_2})(y) = \int_{-\infty}^{\infty} f_{\nu_1}(t) f_{\nu_2}(y - t)\,dt$$

$$= \int_0^y c_{\nu_1} t^{\frac{\nu_1}{2}-1} e^{-t/2} \cdot c_{\nu_2}(y - t)^{\frac{\nu_2}{2}-1} e^{-(y-t)/2}\,dt.$$

The exponential factors combine to give $e^{-y/2}$, and we again set $t = uy$ and $dt = ydu$, and find

$$f_Y(y) = c_{v_1}c_{v_2}e^{-y/2}\int_0^1 (yu)^{\frac{v_1}{2}-1}(y-yu)^{\frac{v_2}{2}-1}ydu$$

$$= c_{v_1}c_{v_2}e^{-y/2}y^{\frac{v_1}{2}-1}y^{\frac{v_2}{2}-1}y\int_0^1 u^{\frac{v_1}{2}-1}(1-u)^{\frac{v_2}{2}-1}du$$

$$= c_nc_1y^{\frac{v_1+v_2}{2}-1}e^{-y/2}\int_0^1 u^{\frac{v_1}{2}-1}(1-u)^{\frac{v_2}{2}-1}du.$$

Again, it's possible to evaluate the u-integral in closed form as it's essentially the integral of a Beta density with parameters $n/2$ and $1/2$ (see §15.5); however, all that matters is that it has no y-dependence. Calling this integral C_{v_1,v_2}, we find

$$f_Y(y) = \begin{cases} C_{v_1,v_2}c_{v_1}c_{v_2}y^{\frac{v_1+v_2}{2}-1}e^{-y/2} & \text{if } y \geq 0 \\ 0 & \text{otherwise.} \end{cases}$$

Note the y-dependence is exactly that of the chi-square distribution with $v_1 + v_2$ degrees of freedom, and thus the normalization constant $C_{v_1,v_2}c_{v_1}c_{v_2}$ must equal the normalization constant of the chi-square with $v_1 + v_2$ degrees of freedom.

It's worth emphasizing again (and again and again!) that the **Theory of Normalization Constants** allows us to avoid some very difficult integrals. The normalization constant for a chi-square random variable with v degrees of freedom was easily found; it's just a simple expression involving the Gamma function. While it might not be easy for us to evaluate the Gamma function, from a theoretical point of view that doesn't matter—we have named the normalization constant. We don't have to evaluate the u-integral!

 It's worth noting that we can turn these arguments around and use them to derive the fundamental relation for the Beta function! We leave this as a nice exercise for the reader (see Exercise 16.5.22).

16.4 Summary

We introduced another important family of probability distributions, the chi-square distribution. It has one free variable, and calculating quantities associated to it are related to the Gamma function. We saw that the sum of independent chi-square random variables is also a chi-square random variable. Thus the chi-square distribution is a **stable distribution**, which means that the shape of the sum of independent copies is the same shape as the constituents. This is a very unusual property, and frequently does not hold. For example, sums of uniform variables are not uniform, nor are sums of exponential random variables exponential. There are other stable distributions (continuous ones include the normal and the Cauchy, while a discrete one is the Poisson); these are very useful for analysis as the functional form does not change.

We discussed several approaches to handling the integration, ranging from brute force to change of variables to sniffing out the functional dependence and using the

Theory of Normalization Constants. My personal favorite is the last, where we are essentially integrating without integrating!

We end this chapter with a brief introduction to applications of the chi-square distribution in statistics. The reason they play such an important role is that the square of a standard normal random variable is a chi-square random variable with 1 degree of freedom, and often one assumes errors are distributed normally. The sum of the error terms, however, are not incredibly useful as positive errors can cancel with negative errors. We could sum the absolute values of errors, but the absolute value function is not differentiable, and calculus is no longer available. The next simplest combination is to sum the squares of errors, and now we can see why the chi-square emerges. For more on this, see Chapter 24 on the Method of Least Squares.

There are three different types of chi-square tests that are very important in application. See §22.5 for an explanation of a **chi-square goodness of fit test**. This measures whether a set of categorical data came from a certain discrete distribution. Another is the **chi-square test of homogeneity**: it is applied to a single categorical variable from more than one distinct population, and tests to see if frequency counts of that variable are statistically different among the populations. Or, worded differently, the chi-square test of homogeneity tests if the frequency counts are distributed identically in the different populations. The other test is the **chi-square test of independence**: it is applied to multiple categorical variables from the same population, and determines if the association between the variables within the same population is statistically significant. Worded differently, the chi-square test of independence tests to see if independence exists between our variables.

The test of goodness of fit, of homogeneity, and of independence all calculate their test statistic in the same way. Usually, the data is organized into a contingency table, which is divided by the categorical variables of interest. The chi-square test statistic is calculated by summing together the squared difference between the observed and expected frequency counts in each cell of the contingency table, and then dividing by the expected frequency count::

$$\sum_{i=1}^{n} \frac{(\text{Observed}_i - \text{Expected}_i)^2}{\text{Expected}_i}.$$

The chi-square test of homogeneity and the chi-square test of independence are both incredibly useful in the field of genetics. The latter test is particularly in determining the degree of genetic linkage between genes on the same chromosome. See Exercise 16.5.16 for an application of a chi-square test of goodness of fit, and Exercise 16.5.17 for an application of the chi-square test of independence.

16.5 Exercises

Exercise 16.5.1   *Find the mean and variance of a chi-square distribution with k degrees of freedom.*

Exercise 16.5.2   *Find the skewness of a chi-square distribution with k degrees of freedom.*

Exercise 16.5.3   *What is the pdf for a chi-square variable with parameter $k = 2$? How does this relate to other distributions we have seen?*

Exercise 16.5.4   *Find the cdf for a chi-square variable with parameter $k = 2$.*

Exercise 16.5.5   *Is the chi-square distribution symmetric for any k? If so, about what point and for what k?*

Exercise 16.5.6   *Show that in the limit as k goes to infinity, the chi-square approaches being symmetric about k.*

Exercise 16.5.7   *If X is a chi-square variable with parameter k and $Y = \sqrt{X}$, find the probability density function for Y.*

Exercise 16.5.8   *You are measuring the distance between two points. If errors in both directions are independent, identical, and normally distributed random variables with parameters μ and σ, find the probability distribution for the total error.*

Exercise 16.5.9   *The most common method for fitting a line to data involves trying to minimize the square of the distance between each point and the fitted line. Explain why squaring the difference makes sense instead of just using the raw errors or taking the absolute values. What is a potential drawback? For more on this see Chapter 24.*

Exercise 16.5.10   *Explain how the chi-square and Gamma distributions are related.*

Exercise 16.5.11   *For those knowing some statistics, explain how a chi-square distribution could be used to check for the normality of errors in a linear regression.*

Exercise 16.5.12   *For those knowing some statistics, write a paragraph justifying the methodology of a chi-square test of homogeneity from a probabilistic standpoint (i.e., why is the test statistic a chi-square random variable, and why are the degrees of freedom the number of levels of the categorical variable minus 1 times the number of populations).*

Exercise 16.5.13   *Imagine we are throwing a dart at a dartboard and the amount we miss the center in each direction is distributed according to the standard normal distribution. Find the probability density function for the distance a dart hits from the origin. What if instead of having the errors arise from the standard normal distribution they come from a normal distribution with mean 0 and variance 2?*

Exercise 16.5.14   *Imagine we have two dartboards, one of which is the two-dimensional dartboard from the previous exercise and the other is a straight line. Again our errors in each direction are distributed according to the standard normal distribution. What is the probability that we are closer to the center of the two-dimensional dartboard than the one-dimensional dartboard?*

Exercise 16.5.15   *More generally, find the probability density function for the distance from the center of an n-dimensional dartboard in which the errors in each direction are normally distributed. How does the mean of the total distance change as n increases?*

Exercise 16.5.16   *For those knowing some statistics, chi-square distributions are commonly used in genetics. Assume that we perform a test cross in which we cross a purely bred red flower and a purely bred white flower. If 25 flowers are white, 35 are pink (a cross between red and white), and 20 are red, do you reject the hypothesis that the flowers exhibit monogenetic incomplete dominance resulting in a 1:2:1 ratio of phenotypes?*

Grade	Row 1	Row 2	Row 3	Row 4
A	8	6	7	6
B	13	13	13	14
C	6	9	5	8
D	8	7	10	7

Exercise 16.5.17   *Given the data in the table, is there sufficient evidence to conclude that in which row a person sits in a probability class is related to her overall performance in the class? Assume the probability class is drawn out of the larger student body population and is a representative sample.*

Exercise 16.5.18   *If you purchase a stock that returns a profit that can be modeled by a random variable $X \sim N(.1, .1)$ in the first year and will return at the same rate as it did in the first year in the second year, what is the probability that the stock has increased by at least 30% after the first two years?*

Exercise 16.5.19   *Find the median of chi-square distributions with 2, 4, 6, and 8 degrees of freedom. Plot the points and draw a regression line through them to get an approximate formula for the median of the chi-square distribution as a function of the degrees of freedom.*

Exercise 16.5.20   *Randomly select 1000 numbers from a chi-square distribution with 1 degree of freedom. Plot the frequency of each first digit. Repeat this with a chi-square with 10 degrees of freedom. Compare the histograms.*

Exercise 16.5.21   *Say we wanted to determine if a die was fair or not using a chi-square test of homogeneity and a significance level of $\alpha = .1$. We roll it 60 times, and get the following outcomes:*

Die Value	1	2	3	4	5	6
Observed	7	11	9	10	11	10
Expected	10	10	10	10	10	10

Let's calculate the test statistic: $\sum_{i=1}^{n} \frac{(\text{Observed}_i - \text{Expected}_i)^2}{\text{Expected}_i} = \frac{(7-10^2)}{10} + \frac{(11-10)^2}{10} + \frac{(9-10)^2}{10}$
$+ \frac{(10-10)^2}{10} + \frac{(11-10)^2}{10} + \frac{(10-10)^2}{10} = 1.2$. *Is this test statistic unreasonable given our assumption that the die is fair? Let's compare it to the following p-table for a chi-square random variable with 5 degrees of freedom; the probabilities are the probability of getting a value as large or larger than the test statistic in the next line.*

Probability	.90	.50	.10	.05	.01
Test Statistic	1.61	4.35	9.24	11.07	15.09

Our p-table tells us that our test statistic of 1.2 falls below the critical value of 1.61 at an $\alpha = .1$ significance level. Thus, we do not have enough statistical evidence to suggest that our die is not fair. Or, in other words, we fail to reject the null hypothesis that the die is a fair one.
What if instead we have the following results?

Die Value	1	2	3	4	5	6
Observed	6	13	9	13	5	12
Expected	10	10	10	10	10	10

Exercise 16.5.22   *Use the arguments of §16.3.4 to derive the fundamental relation for the Beta function.*

PART IV
LIMIT THEOREMS

CHAPTER 17 _____

Inequalities and Laws of Large Numbers

In a perfect world, we could easily and precisely answer any question. In practice, however, it's often too difficult to exactly calculate the probabilities that arise. This is due to a variety of reasons. Sometimes the initial distribution is hard to work with, while other times it's because we want to know what happens in the limit. Fortunately, for many problems we don't need to know the exact answer, and a good range for the answer suffices. For example, think of weight limits in an elevator. If it says the maximum capacity is 1200 pounds, while the true value is probably close to 1200 it's unlikely that 1199 pounds can safely be done but 1201 pounds will cause it to plummet. The point of this number is to give us a decent ballpark for our quantity of interest.

The purpose of this chapter is to introduce you to several different inequalities for probabilities. Not surprisingly, the more we know about the underlying distribution, the more we can say. We'll prove results both for individual random variables as well as for limiting values of standardized sums. Tackling the sums is a bit tricky, as there are many different ways we can view the limiting process. This requires us to introduce various notions of convergence; these parts of the chapter use more analysis than most other places in the book.

17.1 Inequalities

Markov's inequality requires very little input. All we assume about our random variable X is that it has finite mean and is non-negative. If the average value isn't even defined, we shouldn't expect to say anything meaningful about probabilities, so that assumption below is very weak. The advantage of a weak assumption is that it applies to many different situations; the disadvantage is that because it's applicable in so many cases, its consequences aren't strong. The second assumption, namely that our random variable is non-negative, is initially not as clear and thus we spend some time explaining why this is reasonable to include.

There are infinitely many inequalities we could prove; most, however, will be useless. Our goal is to find an inequality that uses readily available information and leads to non-trivial inferences. To this end, let's explore bounding the probability of large values given only information about the mean.

For example, what can we say about $\mathrm{Prob}(X \geq a)$, given knowledge about $\mu = \mathbb{E}[X]$? As we're trying to get a feel for what's possible, let's assume $\mu = 0$ to simplify the algebra, and for definiteness we'll take $a = 2$. Thus we'll explore bounds on $\mathrm{Prob}(X \geq 2)$. Of course, we know this must be at least 0 and at most 1 as it is a probability; the question is whether or not we can say anything more.

What we're going to do is choose different distributions for X and see how these choices impact $\mathrm{Prob}(X \geq 2)$. Our first choice is to take $X \sim \mathrm{Unif}(-1, 1)$; this has mean 0 (as needed), and $\mathrm{Prob}(X \geq 2) = 0$. In fact, $X \sim \mathrm{Unif}(-c, c)$ will give $\mathrm{Prob}(X \geq 2) = 0$ for any $c \leq 2$.

What happens if we increase c beyond 2? Then $\mathrm{Prob}(X \geq 2)$ becomes positive, and rises to just less than 1/2 as $c \to \infty$; the actual probability is $\mathrm{Prob}(X \geq 2) = \frac{c-2}{2c} = \frac{1}{2} - \frac{1}{c}$ for $X \sim \mathrm{Unif}(-c, c)$ and $c \geq 2$. While this might suggest $\mathrm{Prob}(X \geq 2)$ is bounded by 1/2 and strictly less, we can make this probability larger. For example, if

$$f_X(x) = \begin{cases} \dfrac{1}{2} & \text{if } |x| \in [2, 3] \\[2mm] 0 & \text{otherwise,} \end{cases}$$

then $\mathrm{Prob}(X \geq 2) = 1/2$.

Can we choose a random variable where the probability is greater than 1/2 and the mean is still zero? Interestingly, we can and we can drive that probability as close to 1 as we desire. Consider a discrete random variable (though as a nice exercise you should tweak and find a continuous one) where

$$\mathrm{Prob}(X = 2) = p, \quad \mathrm{Prob}(X = -b) = 1 - p$$

for some positive b. *I chose to write $-b$ and not b as I like my unknowns to be positive if possible, so I can just look at the sign and get some information.* As we want the mean to be zero, we need

$$2p - b(1 - p) = 0, \quad \text{or} \quad b = \frac{2p}{1 - p}.$$

Thus as long as $p < 1$, we can find a random variable X where $\mathrm{Prob}(X \geq 2) = p$.

What does all this mean? It means that as we range over all possible random variables X which have mean 0, $\mathrm{Prob}(X \geq 2)$ ranges from 0 to as close to 1 as we wish. Thus it is impossible to get a non-trivial bound for $\mathrm{Prob}(X \geq 2)$, given *only* knowledge about its mean. If we are going to make progress in estimating this probability, we have to make additional assumptions on X.

Our examples above suggest a natural restriction. In order to get $\mathrm{Prob}(X \geq 2)$ close to 1, we needed it to take a large negative value with a small probability. If we somehow knew that X couldn't be too negative, then these examples would not be available and *perhaps* we would have a non-trivial bound on $\mathrm{Prob}(X \geq 2)$. This turns out to be the case, but of course until we do the analysis it's possible that there were other random variables out there that would have satisfied this additional constraint and also had $\mathrm{Prob}(X \geq 2)$ tending to 1.

In the next section we'll state Markov's inequality, and hopefully our discussion here adequately motivates why it has the assumptions it does. In addition to assuming we know the mean of X, we also assume $X \geq 0$; in other words, X has zero probability of taking on negative values. This eliminates our family of densities which allowed us to drive $\text{Prob}(X \geq 2)$ as close to 1 as we wish.

If all you care about is knowing the statement of Markov's inequality, this section was useless and you could have skipped it and jumped straight to the result; however, the thought process here is useful for a variety of problems. The idea is to find what are reasonable constraints, to get a sense of what you should try to prove, to learn what has a chance of being true. We've seen that it is unreasonable to expect a good bound knowing just the mean—we need more information. There are thus several ways to go. Markov's inequality takes the route of restricting to non-negative random variables. While this avoids the examples where $\text{Prob}(X \geq 2)$ tended to 1, there are other alternatives. We could also have assumed a bound on the variance of X, which would then restrict how spread out our distribution could be and would also lead to non-trivial bounds on probabilities; we take this approach in Chebyshev's inequality.

17.2 Markov's Inequality

From our discussion above, we saw that we cannot get a non-trivial bound for $\text{Prob}(X \geq a)$ given only knowledge of the mean of X. Thus we need to assume more; our first result comes from restricting to non-negative random variables.

Markov's inequality: Let X be a non-negative random variable with finite mean $\mathbb{E}[X]$ (this means $\text{Prob}(X < 0) = 0$). Then for any positive a we have

$$\text{Prob}(X \geq a) \leq \frac{\mathbb{E}[X]}{a}.$$

Some authors write μ_X for $\mathbb{E}[X]$. An alternative formulation is

$$\text{Prob}(X < a) \geq 1 - \frac{\mathbb{E}[X]}{a}.$$

Before discussing the proof, it's worth looking at the claim and seeing if it's reasonable. Let's give X some units; say it's measuring heights. Whatever units X has, a has too, and thus $\mathbb{E}[X]/a$ is a unitless quantity. This is good, as it's supposed to be a probability.

What other checks can we do? There are several particularly good choices for a. If $a < \mathbb{E}[X]$, Markov's inequality tells us the probability that X is at most a is at most some number exceeding 1. As probabilities are at most 1, this provides absolutely no information. If we think about it, though, we *shouldn't* expect to get anything useful in this case. Perhaps X equals a with probability 1, in which case this probability is zero. Or, going for the opposite extreme, let p be any number less than 1. We could have X equals 0 with probability p and X equals $\mathbb{E}[X]/(1 - p)$ with probability $1 - p$. This gives us a non-negative random variable with mean $\mathbb{E}[X]$, but for any a less than the mean the probability can be as close to 1 as we desire!

The inequality is only useful once a exceeds $\mathbb{E}[X]$. This makes sense, and is similar to the **Intermediate Value Theorem** (see §B.1). If the only non-zero probabilities are for values larger than $\mathbb{E}[X]$, then X's mean would have to be larger than $\mathbb{E}[X]$, which can't happen. Thus, Prob$(X > a) < 1$ for all $a > \mathbb{E}[X]$; the meat of Markov's inequality is quantifying just how much less than 1 this is (as a function of a).

As you read the proof, try to see where we use non-negativity. We should need to use it somewhere, as it's a condition in the theorem! Typically either these conditions are essential, or their inclusion simplifies the proof. It's always a great idea to remove a condition and see if you can find a counterexample.

Proof of Markov's inequality: The proof uses two nice tricks. The first is that if $x \geq a$ then $x/a \geq 1$. We start by unwinding the definition of Prob$(X \geq a)$, and then use this observation to rewrite 1. For convenience we assume our random variable is continuous with probability density function f_X, and thus use integrals for the probabilities; a similar proof holds if X is discrete. We have

$$\text{Prob}(X \geq a) = \int_{x=a}^{\infty} f_X(x)dx$$

$$= \int_{x=a}^{\infty} 1 \cdot f_X(x)dx$$

$$\leq \int_{x=a}^{\infty} \frac{x}{a} f_X(x)dx$$

$$= \frac{1}{a} \int_{x=a}^{\infty} x f_X(x)dx.$$

Our last integral is *almost* the definition of the mean of X. The only reason it isn't is that $\mathbb{E}[X]$ is the integral of $x f_X(x)$ from $-\infty$ to ∞, and we're only integrating from a to ∞. We can easily fix this. As our integrand is positive, we can extend the integration down to 0; this either increases the value of the integral, or leaves it unchanged. We don't have to extend the integration to $-\infty$, as X is a non-negative random variable and is thus less than 0 with zero probability. We get

$$\text{Prob}(X \geq a) \leq \frac{1}{a} \int_{x=0}^{\infty} x f_X(x)dx = \frac{\mathbb{E}[X]}{a},$$

which proves the claim. $\qquad \square$

So, where did we use non-negativity? We used it to say

$$\mathbb{E}[X] = \int_0^{\infty} x f_X(x)dx;$$

in other words, we don't have to do any integration below 0. Why is this important? We need to make sure the inequalities go the right way. We used

$$\int_{x=a}^{\infty} x f_X(x)dx \leq \int_{x=0}^{\infty} x f_X(x)dx;$$

this is true as the integrand $x f_X(x)$ is non-negative in this range. If, however, we were to go below zero then the integrand can be negative. In that case, the integral could become smaller, and we would have the inequality go the wrong way.

For example, imagine X is 1 with probability 1/2 and -1 with probability 1/2. Then $\mathbb{E}[X] = 0$. If the conclusion of Markov's inequality held, we'd find $\text{Prob}(X \geq a) \leq 0$ for any positive a. This is clearly absurd (just take $a = 1/2$ or 1).

Now that we've seen a proof, let's do an example. *Imagine the mean US income is $60,000. What's the probability a household chosen at random has an income of at least $120,000? Of at least $1,000,000?*

As stated, we don't have enough information to solve this problem. Maybe there's a few very rich people and everyone else earns essentially nothing. Or, the opposite extreme, maybe everyone makes close to the average. Without knowing more about how incomes are distributed, we can't get an exact answer. We can, however, get some bounds on the answer by using Markov's inequality. To use this, we need a non-negative random variable with finite mean. If we assume that no household has a negative income then we're fine, as the other condition is met (the mean is $60,000, which is finite).

Thus the probability of an income of at least $120,000 is at most $60000/120000 = 1/2$; or, at most half the population makes twice the mean. What about the millionaire's club? The probability of being a millionaire is at most $60000/1000000 = .06$, or at most 6% of the households.

We can generalize our example and extract the following bound.

Let X be a non-negative random variable with finite mean $\mathbb{E}[X]$. Then the probability of being at least ℓ times the mean is at most $1/\ell$:

$$\text{Prob}(X \geq \ell\mathbb{E}[X]) \leq \frac{1}{\ell}.$$

Unfortunately this is the best we can do with our limited information. So long as our random variable has finite mean and is non-negative, the probability of being 100 or more times the mean is at most 1/100 or 1%. Of course, in many problems the true probability is *magnitudes* less than this. This is an excessively high overestimate at times. This suggests, of course, the next step: incorporate more information and get a better bound! We do this in the next section.

Problem: *Play with some non-negative random variables with finite means. Which ones are the "closest" to having the inequality in Markov's inequality as an equality for as many a's as possible? This is a very open question. You need to figure out how you want to measure closeness, as well as which a's you care about. It's good to explore a few somewhat vague problems; try a few different distributions and see what you need to do to have high probabilities of being far above the mean.*

17.3 Chebyshev's Inequality

17.3.1 Statement

Markov's inequality bounded probabilities by using just one input—the mean of the random variable. We saw this was not enough, and in order to get a non-trivial result we needed to assume more. One approach, which is what we took for Markov's inequality, was to assume our random variable was non-negative. In this section we take another tack, which not only leads to another inequality but to a better inequality. The reason we get a better result is that we are going to incorporate a lot more information on our random variable than simply it is non-negative.

The next logical choice is to assume a bound on the standard deviation (or the variance, as knowing one allows us to easily find the other). The standard deviation tells us the scale of fluctuations about the mean; there's good reason to hope that using this will lead to better, tighter bounds. This leads to Chebyshev's inequality. It's one of my favorite results for a variety of reasons. The first is that the conditions are very weak, which means you can use it for almost any distribution that you would encounter. Of course, there's a price for that—because the conditions are so weak, its consequences aren't as strong as other results. The second is more personal: I'm a horrible speller, and Chebyshev holds the record for most acceptable transliterations into English. Wolfram's *MathWorld* lists 40 possibilities; popular ones include Chebyshov, Chebishev, Chebysheff, Tschebischeff, Tschebyshev, Tschebyscheff, and Tschebyschef.

Theorem 17.3.1 (Chebyshev's inequality): *Let X be a random variable with finite mean μ_X and finite variance σ_X^2. Then for any $k > 0$ we have*

$$\text{Prob}(|X - \mu_X| \geq k\sigma_X) \leq \frac{1}{k^2}.$$

Some authors write $\mathbb{E}[X]$ for μ_X. This means that the probability of obtaining a value at least k standard deviations from the mean is at most $1/k^2$. A useful, alternative formulation is

$$\text{Prob}(|X - \mu_X| < k\sigma_X) > 1 - \frac{1}{k^2}.$$

Before proving Chebyshev's inequality, let's discuss its statement and examine some implications. The first, as always, is to assign and check units. If X is measuring a height (say in meters), then $|X - \mu_X| \geq k\sigma_X$ involves three height measurements. We're seeing how far we are from the mean. We're also using the most natural scale for such a comparison. As the standard deviation measures the size of the fluctuations about the mean, it's reasonable to count how many standard deviations a value is from the mean.

Another good test is to look at different choices of k. Note that if $k \leq 1$ then the probability is at least 1 (i.e., at least 100%). In other words, we get *no* useful information as every event happens with probability at most 1. Let's consider two different random variables. For the first we have *all* the mass at 0, while for the second we have half at 10^{100} and half at -10^{100}. Both have a mean of zero. In the first case there is a 100% probability of being within k standard deviations for any $k < 1$, while for the second the probability is zero (the variance is

$$(10^{100} - 0)^2 \cdot \frac{1}{2} + (-10^{100} - 0)^2 \cdot \frac{1}{2} = 10^{200},$$

making the standard deviation 10^{100}). This example shows we can't expect to get any information for $k < 1$, as essentially anything can happen.

The story is very different once $k > 1$, as now we *do* get some information. The larger k, the smaller the probability of being k or more standard deviations from the mean.

In Markov's inequality we saw the probability of being ℓ times the mean was at most $1/\ell$. What does Chebyshev tell us? Unfortunately, Chebyshev can't directly answer that. The reason is in Chebyshev's inequality we measure our distance from the mean by counting how many standard deviations we must travel, *not* by the magnitude. Thus, if the standard deviation is σ_X, the mean is μ_X, and we are at least ℓ times the mean, then we need $|X - \mu_X| \geq \ell\mu_X - \mu_X$. Thus we're $(\ell - 1)\mu_X$ away. We write

$$(\ell - 1)\mu_X = k\sigma_X,$$

and find

$$k = \frac{(\ell - 1)\mu_X}{\sigma_X}.$$

Note the units of μ_X and σ_X cancel, and as needed k is a unitless quantity. We now substitute this into Chebyshev, and finally obtain that the probability of being at least ℓ times the mean is at most $\sigma_X^2/(\mu_X^2(\ell - 1)^2)$.

How does this compare to Markov? Again, we can't directly compare these two results as Chebyshev requires us to input the standard deviation; however, we can investigate what happens as ℓ gets large. Markov's inequality tells us the probability of being at least ℓ times the mean falls off at least as fast as $1/\ell$, while Chebyshev tells us that there's some constant C such that the probability is decaying like $C/(\ell - 1)^2$ (and if ℓ is really large this is essentially just C/ℓ^2). This gives us a sense of the strength of Chebyshev's inequality relative to Markov's; for large ℓ Chebyshev will *always* do better.

17.3.2 Proof

It's now time to prove it. Any result this important deserves at least two different proofs. Our first shows how Chebyshev follows immediately from Markov's inequality, while the second proof is a more direct one. It's similar in spirit to our proof of Markov's inequality (which shouldn't be too surprising, as we'll first show how Markov implies Chebyshev). The proofs below are deliberately long-winded; the goal as always is not to come up with the shortest proof possible, but rather to highlight how one should approach problems like this as training for the future.

Proof of Chebyshev's inequality from Markov's inequality: Let X be a random variable with mean μ_X and standard deviation σ_X. We have a hint in terms of how to prove this: we're supposed to use Markov's inequality. We therefore need to find a random variable that's non-negative and has finite mean, and it should be related to the quantity we're investigating for Chebyshev.

A good first attempt is $Y = |X - \mu_X|$. This is non-negative, but it's not immediately clear what the mean is. We'd have to evaluate

$$\int_{y=0}^{\infty} y f_Y(y)dy,$$

and then figure out what $f_Y(y)$ is in terms of $f_X(x)$. This involves integrating an absolute value, which is hard. Also, it's not at all clear how the standard deviation enters. This suggests our next attempt, $W = (X - \mu_X)^2$. The advantage of this is that squaring is much nicer than absolute values, and when we try to find the mean of W we'll see the variance of X emerge (see Chapter 24 on the Method of Least Squares for more on the advantages of squaring over absolute values). We have

$$\mathbb{E}[W] = \int_{w=0}^{\infty} w f_W(w) dw = \int_{x=-\infty}^{\infty} (x - \mu_X)^2 f_X(x) dx = \sigma_X^2,$$

where the last equality follows from the definition of the variance.

We've shown $W = (X - \mu_X)^2$ is non-negative, and its mean is σ_X^2 which, by our assumptions on X, is finite. We now apply Markov's inequality to W. We have

$$\text{Prob}(W \geq a) \leq \frac{\mathbb{E}[W]}{a}.$$

All we need to do is choose values for a. Substituting gives

$$\text{Prob}((X - \mu_X)^2 \geq a) \leq \frac{\sigma_X^2}{a}, \quad \text{or} \quad \text{Prob}(|X - \mu_X| \geq \sqrt{a}) \leq \frac{\sigma_X^2}{a}.$$

At this point, the statement of Chebyshev's theorem rescues us and tells us what a should be: we just take $\sqrt{a} = k\sigma_X$ (so $a = k^2 \sigma_X^2$). This gives

$$\text{Prob}(|X - \mu_X| \geq k\sigma_X) \leq \frac{\sigma_X^2}{k^2 \sigma_X^2} = \frac{1}{k^2},$$

completing the proof. □

My goal in the first proof is to highlight the thought process. If we have the bright idea to use Markov's inequality, how can we rearrange what we have so that we can use it? The goal is to see our goal and use that to influence how we attack the algebra.

Our second proof shares many features with the proof of Markov's inequality. This shouldn't be too surprising, as in some sense we're basically reproving some of Markov. Instead of using $x/a \geq 1$ for $x \geq a$, now we use $(\frac{x - \mu_X}{k\sigma_X})^2 \geq 1$ when $|x - \mu_X| \geq k\sigma_X$.

Direct proof of Chebyshev's inequality: Let f_X be the probability density function of X. We assume X is a continuous random variable, though a similar proof holds in the discrete case. We have

$$\text{Prob}(|X - \mu_X| \geq k\sigma_X) = \int_{x:|x-\mu_X| \geq k\sigma_X} 1 \cdot f_X(x) dx$$

$$\leq \int_{x:|x-\mu_X| \geq k\sigma_X} \left(\frac{x - \mu_X}{k\sigma_X}\right)^2 \cdot f_X(x) dx$$

$$= \frac{1}{k^2 \sigma_X^2} \int_{x:|x-\mu_X| \geq k\sigma_X} (x - \mu_X)^2 f_X(x) dx$$

$$\leq \frac{1}{k^2 \sigma_X^2} \int_{x=-\infty}^{\infty} (x - \mu_X)^2 f_X(x) dx$$

$$= \frac{1}{k^2 \sigma_X^2} \cdot \sigma_X^2 = \frac{1}{k^2},$$

completing the proof. □

Note above that we could extend the integration to be over all of x as the integrand $(x - \mu_X)^2 f_X(x)$ is always non-negative. This is different than the integrand of $x f_X(x)$ that we had in the proof of Markov's inequality. We had to be very careful there, and couldn't extend the integral below $x = 0$.

17.3.3 Normal and Uniform Examples

Let's do a problem for comparison purposes. Let X be a standard normal random variable, which means it has mean 0 and standard deviation 1. According to Chebyshev's inequality, the probability of being k standard deviations from the mean is at most $1/k^2$. Thus, there is at most a 1/4 or 25% chance of X being 2 standard deviations or more from the mean (i.e., $|X| \geq 2$), at most 1/25 or 4% chance of X being at least 5 standard deviations from the mean, and at most a 1/100 or 1% chance of X being 10 or more standard deviations from the mean. These are just *upper bounds* for the probabilities. How close are they? The actual values are approximately 4.55%, .0000573%, and $1.52 \cdot 10^{-21}\%$.

While these values are all below the bounds we get from Chebyshev's inequality, we see that they are *far* below. What happened? The problem is Chebyshev's theorem is being asked to do too much. Remember we have very weak conditions. This is great, as it means we can apply Chebyshev in a large number of situations. The problem, however, is that the result can't be that strong as it has to apply to so many cases. We need to use finer properties of our distribution to get better bounds.

In particular, this means our quest for good bounds must continue. This will naturally lead to our studies about sums and limiting behavior of random variables, which in many situations are governed by Central Limit Theorem laws. Before we turn to these, it's worth revisiting the bounds from Markov and Chebyshev for probability densities that decay significantly slower than the standard normal.

Let $X \sim \text{Unif}(0, 1)$ be the uniform distribution on $[0, 1]$, and let $Y \sim \text{Exp}(1)$ be an exponential random variable with density $f_Y(y) = e^{-y}$ for $y \geq 0$ (and 0 otherwise). Both of these random variables are non-negative. The means are

$$\mathbb{E}[X] = \int_0^1 x \cdot 1 dx = \frac{x^2}{2}\Big|_0^1 = \frac{1}{2} - 0 = \frac{1}{2}$$

$$\mathbb{E}[Y] = \int_0^\infty y e^{-y} dy$$

$$= -y e^{-y}\Big|_0^\infty + \int_0^\infty e^{-y} dy$$

$$= (0 - 0) - e^{-y}\Big|_0^\infty = 1,$$

while a similar calculation (which we don't do) yields the variances are

$$\sigma_X^2 = \int_0^1 \left(x - \frac{1}{2}\right)^2 dx = \frac{1}{12}$$

$$\sigma_Y^2 = \int_0^\infty (y - 1)^2 e^{-y} dy = 1.$$

We now compute the bounds Markov and Chebyshev give and compare to the actual value. Let's examine $X \geq .95$ and $Y \geq 4$. *Warning: I'm deliberately making one of the most common mistakes below—see if you can find it as you read on!*

- Markov: We use $\text{Prob}(X \geq a) \leq \mathbb{E}[X]/a$. So the probability $X \geq .95$ is at most $\mathbb{E}[X]/.95$, or $.5/.9 \approx .55$; similarly the probability $Y \geq 4$ is at most $1/4 = .25$. It's very easy to use Markov's inequality!

- Chebyshev: Here we need to write our desired probability in the form $\text{Prob}(|X - \mathbb{E}[X]| \geq k\sigma_X)$, and this is at most $1/k^2$. Therefore the probability $X \geq .95$ is at most $1/k^2$, where $.95 - \mathbb{E}[X] = k\sigma_X$. Using our values from above, we see $k = (.95 - .5)/(1/12) = 27/5$. Thus the probability is at most $1/(27/5)^2$, which is $25/729 \approx .034$. We really found the probability that $|X - .5| \geq .45$; unfortunately Chebyshev includes both $X \geq .95$ as well as $X \leq .05$. For Y, we note that if Y is at least 4 then Y is at least 3 standard deviations from the mean (so $k = 3$), and thus the probability that Y is at least 4 is at most $1/3^2 \approx .11$.

- The actual probabilities are $.05$ for $X \geq .95$ and $1/e^4 \approx .018$ for $Y \geq 4$.

So, where's the mistake? The upper bounds we get from Markov's inequality are all greater than the actual answer, so these numbers pass the first test. This is *not* the case with our Chebyshev bounds; the upper bound for the uniform is smaller than the actual probability! Oops! What did we do wrong? The problem was in determining k. It's correct to say $.95 - .5 = k\sigma_X$. The problem is that it's the variance, σ_X^2, that's $1/12$. We can't use $1/12$ for σ_X; we need to use $\sigma_X = \sqrt{1/12}$. Using that, we find $k = (.95 - .5)/\sqrt{1/12} \approx 1.56$. Thus the upper bound is $1/k^2$, and that's $100/243 \approx .41$. This is still smaller than the bound we get from Markov, but it's significantly above the truth.

Not surprisingly, Markov is easier to use but in these examples Chebyshev is closer to the truth.

As a nice challenge, try to find a random variable X with a probability density f_X such that Chebyshev's inequality is as close to an equality as possible. For example, can you find a random variable such that $\text{Prob}(|X - \mu_X| \geq k\sigma_X) = 1/k^2$ for all integer $k \geq 1$?

17.3.4 Exponential Example

Let's do two more examples to compare the Markov and Chebyshev inequalities with each other and the truth. For our first test, let $X \sim \text{Exp}(1)$, so $f_X(x) = e^{-x}$ for $x \geq 0$, and 0 otherwise. The mean $\mu = 1$ and the variance $\sigma^2 = 1$; these are easily seen as the moments of the standard exponential random variable are given by values of the Gamma function:

$$\mathbb{E}[X^k] = \int_0^\infty x^k e^{-x} dx = \int_0^\infty e^{-x} x^{(k+1)-1} dx = \Gamma(k+1) = k!$$

(we use $\sigma^2 = \mathbb{E}[X^2] - \mathbb{E}[X]^2$).

- Markov: $\text{Prob}(X \geq 4) \leq \mathbb{E}[X]/4 = 1/4 = .25$.
- Chebyshev: $\text{Prob}(X \geq 4) = \text{Prob}(|X - 1| \geq 3 \cdot 1) \leq 1/3^3 \approx .111111$.
- Answer: $\text{Prob}(X \geq 4) = \int_4^\infty e^{-x} dx = e^{-4} \approx 0.0183156$.

Not surprisingly Chebyshev does better than Markov, but both are far from the truth (Chebyshev is off by about a factor of 6).

17.4 The Boole and Bonferroni Inequalities

The purpose of this section is to isolate some useful inequalities from the **Method of Inclusion-Exclusion** (see Chapter 5, especially §5.2). The Method (also known as the Principle) of Inclusion-Exclusion is exact; however, by "truncating" the sums we can get upper and lower bounds. Before giving the truncations, let's recall the statement.

Method of Inclusion-Exclusion: Let $A_1, \ldots, A_n$ be a finite collection of sets. Then

$$\text{Prob}\left(\bigcup_{i=1}^{n} A_i\right) = \sum_{1 \leq i_1 \leq n} \text{Prob}(A_{i_1}) - \sum_{1 \leq i_1 < i_2 \leq n} \text{Prob}(A_{i_1} \cap A_{i_2})$$

$$+ \sum_{1 \leq i_1 < i_2 < i_3 \leq n} \text{Prob}(A_{i_1} \cap A_{i_2} \cap A_{i_3})$$

$$- \cdots + (-1)^n \text{Prob}(A_1 \cap \cdots \cap A_n).$$

The proof was by careful bookkeeping. We kept track of how often an element x was counted, and saw that it occurs the same number of times on both sides. If we go back through the proof, we can discover what happens if we truncate the sum on the right. If we stop after a positive sign, we've possibly counted some elements multiple times and *not* adjusted the overcounting by removing. Thus, if we stop after a positive sign, the resulting answer is an upper bound on the probability of the union of the A_i's. It's reasonable that this is an upper bound, as the next term is negative, which suggests we might be too large.

Similarly, we could stop after a negative sign. In this case, we've just avoided overcounting. Now, however, the danger is that we may have removed some elements too many times. In this case, we're now potentially undercounting, and we get a lower bound for the true probability of the union. Again, this is reasonable as the next term is positive, suggesting we removed too much.

We now extract our new inequalities.

Boole's inequality: We have

$$\text{Prob}\left(\bigcup_{i=1}^{n} A_i\right) \leq \sum_{i=1}^{n} \text{Prob}(A_i);$$

this result is still true if we have countably many A_i's.

It's useful to introduce some notation before stating the Bonferroni inequalities. Let

$$S_k = \sum_{1 \leq i_1 < i_2 < \cdots < i_k \leq n} \text{Prob}(A_{i_1} \cap A_{i_2} \cap \cdots \cap A_{i_k}).$$

In words, S_k is the sum of the probabilities of all intersections of k of the A_i's; note there are $\binom{n}{k}$ summands.

> **Bonferroni's inequality**: With S_k as above and positive integers ℓ and m we have
>
> $$\sum_{k=1}^{2\ell}(-1)^{k-1}S_k \;\le\; \mathrm{Prob}\left(\bigcup_{i=1}^{n}A_i\right) \;\le\; \sum_{k=1}^{2m-1}(-1)^{k-1}S_k.$$

For example, let's compute bounds on the probability that at least one person is dealt a hand of 13 cards (from a standard deck) that is all one suit. If we let A_i be the event that person i has a one-suited hand ($1 \le i \le 4$), then

$$\mathrm{Prob}(A_i) \;=\; \frac{\binom{4}{1}\binom{13}{13}}{\binom{52}{13}}$$

(remember, there are four suits in a deck, and thus four different ways to be one-suited). The event $A_i \cap A_j$ means i and j have one-suited hands; for $i \ne j$ we have

$$\mathrm{Prob}(A_i \cap A_j) \;=\; \frac{4 \cdot 3 \cdot \binom{13}{13}\binom{13}{13}}{\binom{52}{26}};$$

this is because there are $4 \cdot 3 = 12$ ways to assign a suit to i and a different suit to j (and it matters who gets which suit). We find

$$S_1 \;=\; \sum_{i=1}^{4}\mathrm{Prob}(A_i) \;=\; 4 \cdot \frac{\binom{4}{1}\binom{13}{13}}{\binom{52}{13}} \;=\; \frac{1}{39688347475}$$

$$S_2 \;=\; \sum_{1\le i<j\le 4}\mathrm{Prob}(A_i \cap A_j) \;=\; \binom{4}{2}\frac{4 \cdot 3 \cdot \binom{13}{13}\binom{13}{13}}{\binom{52}{26}} \;=\; \frac{1}{82653088824684}.$$

Using the simplest version of Bonferroni's inequality, we get

$$S_1 - S_2 \;\le\; \mathrm{Prob}\left(\bigcup_{i=1}^{4}A_i\right) \;\le\; S_1,$$

with S_1 and S_2 as above. If you want to see the actual numbers, it's

$$.0000000000251842 \;\le\; \mathrm{Prob}\left(\bigcup_{i=1}^{4}A_i\right) \;\le\; .0000000000251964.$$

What's particularly nice about our argument above is that we get an answer quickly, and not only are our bounds close to the true answer, but we can quantify just how close they are (the upper and lower bounds differ by S_2, which is much smaller than the main term S_1)!

17.5 Types of Convergence

Before we can state the Weak and Strong Laws of Large Numbers in §17.6, we need to describe the different types of convergence. The Weak and the Strong Laws deal with the behavior of sums of independent random variables (appropriately normalized). It turns out that, in many cases, there is a limiting behavior. We want to discuss the various ways our sums can approach a result. The discussion below is meant to just introduce the terminology and help make it understandable; to do the subject justice requires some familiarity with analysis, which we aren't assuming.

17.5.1 Convergence in Distribution

> **Convergence in distribution (or weak convergence)**: Let $X, X_1, X_2, \ldots$ be random variables with cumulative distribution functions $F, F_1, F_2, \ldots$. Let $\mathcal{C}$ be the set of all real numbers where F is continuous. The sequence of random variables $X_1, X_2, \ldots$ **converges in distribution** (or **converges weakly**) to the random variable X if $\lim_{n \to \infty} F_n(x) = F(x)$ for all $x \in \mathcal{C}$. In other words, if F is continuous at x then the limit of the cumulative distribution functions at x equals $F(x)$. Popular ways of notating this are $X_n \xrightarrow{d} X$ or $X_n \xrightarrow{D} X$. If we know the type of the random variable X, we sometimes write that instead. Thus we might have $X_n \xrightarrow{d} N(0, 1)$ to indicate convergence to the standard normal, or $X_n \xrightarrow{d} \text{Exp}(2)$ for convergence to an exponential random variable with parameter 2.

It turns out this is the weakest type of convergence; it's implied by all others.

Let's look at an example. Let X_n be a uniform random variable, with density

$$f_n(x) = \begin{cases} \dfrac{1}{n} & \text{if } x \in \left\{0, \frac{1}{n}, \frac{2}{n}, \ldots, \frac{n-2}{n}, \frac{n-1}{n}\right\} \\ 0 & \text{otherwise.} \end{cases}$$

The subscript n is to highlight the fact that we have n non-zero values, which are uniformly spaced; see Figure 17.1 for plots of the pdf and the cdf.

- If we look at the densities, then for every $x \in [0, 1]$ we have $\lim_{n \to \infty} f_n(x) = 0$. The easiest way to see this is that $f_n(x) = 0$ except for n special choices of x, where it equals $1/n$. As n grows, this shrinks to zero.
- The situation is drastically different with the CDFs. For $x \in [0, 1]$, $F_n(x)$ is $\frac{1}{n}$ times the number of points of $\{0, 1/n, 2/n, \ldots, (n-1)/n\}$ that are at most x. This is because the CDF gives us the probability of getting a value of at most x. For $x \in [0, 1]$,

$$F_n(x) = \sum_{\substack{0 \le k \le n-1 \\ k/n \le x}} \frac{1}{n} = \frac{\lfloor nx \rfloor}{n},$$

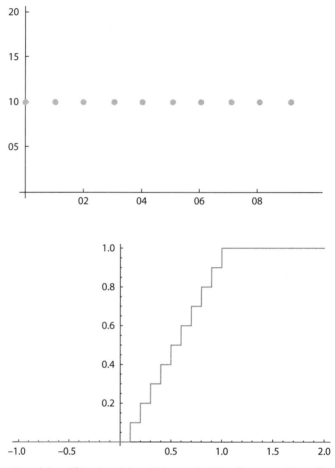

Figure 17.1. Plot of the pdf (top) and the cdf (bottom) of the discrete uniform distribution when $n = 10$.

where the **floor function** $\lfloor y \rfloor$ gives the greatest integer at most y. As the floor of y differs from y by at most 1, we have $nx - 1 \leq \lfloor nx \rfloor \leq nx$, and thus

$$x - \frac{1}{n} \leq F_n(x) \leq x.$$

Taking the limit as $n \to \infty$ we see $\lim_{n \to \infty} F_n(x) = x$, which is the cumulative distribution function of a uniform random variable on $[0, 1]$.

We've shown that $X_n \overset{d}{\to} \text{Unif}(0, 1)$. We also see why this is sometimes called weak convergence, as the density functions do not converge to the density of the uniform. At first this should be disappointing; however, for many purposes it's not too bad. Remember, we can calculate probabilities of events from the cumulative distribution function. For example, the probability we're in an interval $[a, b]$ is $F(b) - F(a)$. Thus, we can still do a lot if we know the cdfs. That said, our example clearly indicates the need and desire for a stronger convergence.

17.5.2 Convergence in Probability

Convergence in probability: Let $X, X_1, X_2, \ldots$ be random variables. We say the sequence $\{X_n\}_{n=1}^{\infty}$ **converges in probability** to X if for every $\epsilon > 0$ we have

$$\lim_{n \to \infty} \text{Prob}(|X_n - X| \geq \epsilon) = 0.$$

We denote this by $X_n \overset{p}{\to} X$ or $X_n \overset{P}{\to} X$.

In this book, we're only going to look at convergence in probability when X is a constant random variable. This means that there's some c such that $\text{Prob}(X = c) = 1$. We often denote this distribution by δ_c.

Let $X_n \sim N(1701, 1/n^2)$ (so X_n is normally distributed with mean 1701 and variance $1/n^2$). We claim that $X_n \overset{p}{\to} \delta_{1701}$. To prove this, we must show that, given any $\epsilon > 0$,

$$\lim_{n \to \infty} \text{Prob}(|X_n - 1701| \geq \epsilon) = 0$$

(we replaced X with 1701, as X equals 1701 with probability 1). Notice that this looks a lot like Chebyshev's inequality, which states

$$\text{Prob}(|X_n - 1701| \geq k\sigma_n) \leq \frac{1}{k^2}.$$

To use this, we just have to find k so that $k\sigma_n = \epsilon$. As $\sigma_n = 1/n$, we have $k = n\epsilon$. Thus

$$\text{Prob}(|X_n - 1701| \geq \epsilon) \leq \frac{1}{n^2 \epsilon^2}.$$

As ϵ is fixed and as $n \to \infty$, we see the above tends to zero, proving convergence in probability.

17.5.3 Almost Sure and Sure Convergence

The final two convergence notions require a greater analysis background. We'll just state them for completeness.

Almost sure convergence: Let $(\Omega, \mathcal{F}, \text{Prob})$ be a probability space and let $X, X_1, X_2, \ldots$ be random variables. We say $\{X_n\}_{n=1}^{\infty}$ **converges almost surely** (or **almost everywhere** or **with probability 1**) to X if

$$\text{Prob}(\omega \in \Omega : \lim_{n \to \infty} X_n(\omega) = X(\omega)) = 1.$$

We denote this by $X_n \overset{a.s.}{\to} X$.

> **Sure convergence**: Let $(\Omega, \mathcal{F}, \text{Prob})$ be a probability space and let $X, X_1, X_2, \ldots$ be random variables. We say $\{X_n\}_{n=1}^{\infty}$ **converges surely** to X if for all $\omega \in \Omega$ we have
>
> $$\lim_{n \to \infty} X_n(\omega) = X(\omega).$$

17.6 Weak and Strong Laws of Large Numbers

The two results here describe convergence of a very special sequence of random variables. If $X_1, X_2, \ldots$ are random variables, let

$$\bar{X}_n = \frac{X_1 + \cdots + X_n}{n}$$

be the random variable that's the average of the first n random variables. The Central Limit Theorem (see Theorem 20.2.2) states that, under appropriate assumptions about the X_i's, the standardized random variables $Z_n = (\bar{X}_n - \mathbb{E}[\bar{X}_n])/\text{StDev}(\bar{X}_n)$ converge to the standard normal. We'll always assume our random variables X_i are identically distributed with finite mean μ. If we don't standardize, we find $\bar{X}_n$ converges to being a Gaussian with mean μ and variance tending to zero. This means that $\bar{X}_n$ is very close to being the constant function μ; the added bonus of the Central Limit Theorem is that it discusses *how* the convergence happens (in particular, it describes the fluctuations).

In this section we discuss two weaker results which are, nevertheless, sufficient for many applications. These are the Weak and Strong Laws of Large Numbers. These results merely state that, in some sense, $\bar{X}_n$ converges to μ. We give up a detailed description of the fluctuations. The advantage is the proof is much shorter than the Central Limit Theorem, and is doable in a first course on probability.

> **Weak Law of Large Numbers**: Let $X_1, X_2, \ldots$ be independent identically distributed random variables with mean μ, and let $\bar{X}_n = \frac{X_1 + \cdots + X_n}{n}$. Then $\bar{X}_n \overset{p}{\to} \mu$ (i.e., $\bar{X}_n$ converges in probability to μ).

Proof: To simplify the proof, we make an additional assumption. We assume the random variables all have a finite variance σ^2. Doing this allows us to give a very simple proof that highlights what's going on. We have

$$\mathbb{E}[\bar{X}_n] = \mathbb{E}\left[\frac{X_1 + \cdots + X_n}{n}\right] = \frac{1}{n} \sum_{k=1}^{n} \mathbb{E}[X_k] = \frac{n\mu}{n} = \mu$$

$$\text{Var}(\bar{X}_n) = \text{Var}\left(\frac{X_1 + \cdots + X_n}{n}\right) = \frac{1}{n^2} \sum_{k=1}^{n} \text{Var}(X_k) = \frac{n\sigma^2}{n^2} = \frac{\sigma^2}{n}.$$

We thus see that the expected value of $\bar{X}_n$ is μ and the variance is σ^2/n, which tends to zero as $n \to \infty$. We want to show convergence to the constant function μ in probability. This means given any $\epsilon > 0$ we have

$$\lim_{n \to \infty} \text{Prob}(|\bar{X}_n - \mu| \geq \epsilon) = 0.$$

This looks a lot like Chebyshev's inequality, which states

$$\lim_{n \to \infty} \text{Prob}(|\bar{X}_n - \mu_{\bar{X}_n}| \geq k\sigma_{\bar{X}_n}) \leq \frac{1}{k^2}.$$

All we have to do is substitute. We have $\mu_{\bar{X}_n} = \mu$, $\sigma_{\bar{X}_n} = \sigma/\sqrt{n}$ and thus if $k\sigma_{\bar{X}_n} = \epsilon$ then $k = \epsilon\sqrt{n}/\sigma$. This gives

$$\lim_{n \to \infty} \text{Prob}(|\bar{X}_n - \mu| \geq \epsilon) \leq \frac{\sigma^2}{n\epsilon^2}.$$

As $n \to \infty$ this probability tends to zero, completing the proof. $\square$

It's worth pausing and looking at the chain of consequences. Markov's inequality gave us Chebyshev's inequality, which just gave us the Weak Law of Large Numbers. A great way to view this result is that if we keep sampling independently from a fixed distribution, then with probability tending to 1 the sample average is arbitrarily close to the population mean μ.

One of the great uses of this result is to estimate the mean of a distribution through sampling.

Not surprisingly, if there's a weak law there's also a strong law. The proof is more involved, and we'll content ourselves with just stating it.

Strong Law of Large Numbers: Let $X_1, X_2, \ldots$ be independent identically distributed random variables with mean μ, and set $\bar{X}_n = \frac{X_1 + \cdots + X_n}{n}$. Then $\bar{X}_n \overset{a.s.}{\to} \mu$.

17.7 Exercises

Exercise 17.7.1   *Prove the Markov inequality for the discrete case.*

Exercise 17.7.2   *Prove Chebyshev's inequality for the discrete case.*

Exercise 17.7.3   *Find a distribution such that Markov's inequality holds as an equality. Is there one distribution where it holds for all a, or do you need a different distribution for each a?*

Exercise 17.7.4   *Find a distribution such that Chebyshev's inequality holds as an equality. Is there one distribution where it holds for all k, or do you need a different distribution for each k?*

Exercise 17.7.5   *Results from the IQ test are fixed to have a mean of 100 and standard deviation of 15. Find a bound for the percentage of people with IQs above 140. Since IQ scores are also designed to be normally distributed, use this to find an estimate of the true percentage of people with IQ scores above 140. Compare the results.*

Exercise 17.7.6   *If the class average on an exam is 80% in a class of 30 students, what is the maximum number of students who scored above a 95%? If the standard deviation on the exam was 10%, what is the maximum number of students who scored above 95%?*

Exercise 17.7.7   *If a coin is flipped 100 times, find an upper bound on the probability it lands heads every time using Chebyshev's inequality. Compare this to the true probability.*

Exercise 17.7.8   *The random variable X has a **Laplace distribution** with parameters μ and b, written $X \sim$ Laplace(μ, b), if*

$$f_X(x) = \frac{1}{2b} \exp\left(-|x - \mu|/b\right).$$

Assume $Y \sim$ Laplace$(0, 1)$. What is the probability X is at least seven times its mean? Find the exact answer, and bound this probability using Chebyshev's inequality. Can you obtain a bound for this answer by using Markov's inequality?

Exercise 17.7.9   *Assume a random variable X has its first 2m moments finite. Use these to bound* Prob$(|X - \mu| \geq k\sigma)$. *What is the best bound you can obtain?*

Exercise 17.7.10   *Test how well your bounds from the previous exercise do for various distributions (say uniform, exponential, Laplace); compare your bounds to what you get from Chebyshev's inequality, and what the truth is, for various k.*

Exercise 17.7.11   *We can generalize a Cauchy distribution to obtain random variables that have a fixed number of moments existing. We say X has a **generalized Cauchy distribution** with parameters μ and $m > 1/2$ if the density is*

$$f_X(x) = \frac{m}{2\pi \, \csc(\pi/m)} \frac{1}{1 + |x - \mu|^m};$$

you can derive the normalization constant (at least for many m) through complex analysis and contour integration, though Mathematica or Wolfram/Alpha can compute it as well:

```
Integrate[2/(1 + x^(m)), {x, 0, Infinity},
    Assumptions -> Element[m, Integers]]
```

(if you've never used the Assumption feature in Mathematica, it's a great one to know, and cuts down enormously on the clutter of the response).

   Consider your bounds from Exercise 17.7.9, and apply those to $X \sim$ Cauchy$(0, m)$ for $m \in \{4, 5, 6\}$; how do your bounds compare to the actual probabilities?

Exercise 17.7.12   *Give upper and lower bounds for the probability that at least one person is dealt a straight in a hand of 5 cards.*

Exercise 17.7.13   *Let X_n be a random variable with CDF defined as $F_n(x) = 1 - (1 - \frac{x}{n})^n$ if $x \geq 0$, and 0 otherwise. Prove that $X_n \xrightarrow{d} Exp(1)$.*

Exercise 17.7.14   *Let $X_1, X_2, \ldots$ be independent, identically distributed exponential random variables with parameter λ, prove that $\bar{X}_n \xrightarrow{a.s.} \lambda$.*

Exercise 17.7.15   *Show that convergence in probability implies convergence in distribution.*

Exercise 17.7.16   *Show that almost sure convergence implies convergence in probability.*

Exercise 17.7.17   *If the league average of a baseball league with 20 teams is 50% of the games played are won with a standard deviation of 10%, what is the maximum number of teams that can have a perfect season?*

Exercise 17.7.18   *If a library has a budget of $25,000 to buy books, and a book costs $15 on average, with a standard deviation of $5, can we find an upper bound for the number of books the library can purchase? Explain why or why not.*

Exercise 17.7.19   *One of the conditions of the Weak Law of Large Numbers is that the variables must be independent. Explain where this is used in the proof. Find a counterexample if the variables are not independent.*

Exercise 17.7.20   *Show that the Strong Law of Large Numbers implies the Weak Law of Large Numbers.*

Exercise 17.7.21   *Find bounds on the probability that the sum of 100 six-sided dice rolls is at least 400.*

Exercise 17.7.22   *If $X_1, X_2, \ldots, X_n$ are n identically distributed independent random variables with mean μ, find $\mathbb{E}[\frac{X_1+X_2+\cdots+X_n}{n}]$.*

Exercise 17.7.23   *The Cantor set is constructed by starting with the interval [0,1] and removing the middle third, then recursively removing the middle third from each remaining segment. Let C_n be the probability that a number uniformly chosen from [0,1] is in the remaining interval. Show that C_n converges in probability to 0.*

Exercise 17.7.24   *What makes the Cantor set particularly interesting is that it contains uncountably infinite points despite its other properties (see the previous exercise). Find a surjective mapping from the Cantor set to the interval [0,1]. That is, find an expression that relates each real number in [0,1] to at least one number in the Cantor set. (Hint: Look at the numbers in the Cantor set in base 3, and try to map them to the real numbers in base 2.)*

Exercise 17.7.25   *Is it possible to assign a uniform probability to selecting an element from the Cantor set? If so, describe a method for doing this.*

Exercise 17.7.26   *Plot C_n for $n \in \{1, 2, \ldots, 20\}$ against 0. Comment on its rate of convergence to 0.*

Exercise 17.7.27   *If the fourth moment is finite must the first, second, and third moments be finite?*

Exercise 17.7.28   *Assume X has a finite fourth moment; find a generalization of Chebyshev's theorem that uses that moment.*

Exercise 17.7.29   *The Yankees are having a down year. In this down year, they only have a 20% chance of beating the Red Sox in any given game. They play 18 games against the Red Sox this year. Using Markov's and Chebyshev's inequalities, find upper bounds on the chances that the Red Sox win every game against the Yankees this year. Is there enough information to find the true probability? Why or why not? If there is, find it.*

Exercise 17.7.30   *A couple has a $100,000 budget for a new condo. The average condo costs $40,000 and has a standard deviation of $5,000. Using Chebyshev's inequality, find an upper bound on the probability that the couple will find a condo that is out of their budget. Is there enough information to find the true probability? Why or why not? If there is, find it.*

Exercise 17.7.31   *You are dealt 5 cards from a fair standard shuffled deck of 52 cards. Using Markov's inequality, what is an upper bound on the probability of getting 4 or more hearts in a hand of 5 cards? Is there enough information to find the true probability? Why or why not? If there is, find it.*

Exercise 17.7.32   *We have 50 dice. Using Markov's inequality and Chebyshev's inequality, find an upper bound on the probability of rolling at least 25 sixes when rolling these 50 fair die. Is there enough information to find the true probability? Why or why not? If there is, find it.*

Exercise 17.7.33   *Let X_n be a geometric random variable with parameter $\lambda_n = \lambda/n$, and set $Y_n = \frac{1}{n} X_n$. Prove that Y_n converges to being $\mathrm{Exp}(\lambda)$. Why must we study Y_n and not X_n?*

Exercise 17.7.34   *Let $\mathcal{R}(\mu, \sigma^2)$ be a random variable with mean μ and finite variance σ^2, and let $X_n \sim \mathcal{R}(\mu, \sigma^2/n^2)$. Show $X_n \overset{p}{\to} \delta_\mu$.*

CHAPTER 18 —————————

Stirling's Formula

It's possible to study probabilities without seeing **factorials**, but you have to do a lot of work to avoid them. Remember that $n!$ is the product of the first n positive integers, with the convention that $0! = 1$. Thus $3! = 3 \cdot 2 \cdot 1 = 6$ and $4! = 4 \cdot 3! = 24$. There's a nice combinatorial interpretation: $n!$ is the number of ways to arrange n people in a line when order matters (there are n choices for the first person, then $n - 1$ for the second, and so on). With this point of view, we interpret $0! = 1$ as meaning there is just one way to arrange nothing!

Factorials arise everywhere, sometimes in an obvious manner and sometimes in a hidden one. The first instance of factorials in probability is arranging objects when order matters, followed quickly by arranging when order does not matter (the binomial coefficients, $\binom{n}{k} = \frac{n!}{k!(n-k)!}$, are the number of ways to choose k objects from n when order does not matter).

There are, however, less obvious occurrences of the factorial function. Perhaps the best hidden one occurs in the density function of the standard normal, $\frac{1}{\sqrt{2\pi}} \exp(-x^2/2)$. It turns out that $\sqrt{\pi} = (-1/2)!$. You should be hearing alarm bells upon reading this; after all, we've defined the factorial function for integer input, and now we're taking the factorial, not just of a negative number, but of a negative rational! What does this mean? How do we interpret the factorial of $-1/2$? What does it mean to ask about the number of ways to order -1/2 people?

The answer to this is through the Gamma function, $\Gamma(s)$.

Gamma function: The Gamma function $\Gamma(s)$ is

$$\Gamma(s) = \int_0^\infty e^{-x} x^{s-1} dx, \quad \Re(s) > 0.$$

Some authors write this as

$$\Gamma(s) = \int_0^\infty e^{-x} x^s \frac{dx}{x}.$$

There's no difference in the two expressions, though they look different. Writing dx/x emphasizes how nicely the measure transforms under rescaling: if we send x to $u = ax$ for any fixed a, then $dx/x = du/u$.

It turns out the Gamma function generalizes the factorial function: if n is a non-negative integer then $\Gamma(n + 1) = n!$. We described this and other properties of the Gamma function in Chapter 15, including several proofs that $\Gamma(1/2) = \sqrt{\pi}$.

The purpose of this chapter is to describe Stirling's formula for approximating $n!$ for large n, and discuss some applications.

Stirling's formula: As $n \to \infty$, we have

$$n! \approx n^n e^{-n} \sqrt{2\pi n};$$

by this we mean

$$\lim_{n \to \infty} \frac{n!}{n^n e^{-n} \sqrt{2\pi n}} = 1.$$

More precisely, we have the following series expansion:

$$n! = n^n e^{-n} \sqrt{2\pi n} \left(1 + \frac{1}{12n} + \frac{1}{288n^2} - \frac{139}{51840n^3} - \cdots \right).$$

Whenever you see a formula, you should always try some simple tests to see if it's reasonable. What kind of tests can we apply? Well, Stirling's formula claims that $n! \approx n^n e^{-n} \sqrt{2\pi n}$. As $n! = n(n - 1) \cdots 1$, clearly $n! \le n^n$. This is consistent with Stirling, though if Stirling is correct it's too crude as it's too large by approximately e^n. What about a lower bound? Well, clearly $n! \ge n(n - 1) \cdots \frac{n}{2}$ (we'll assume $n/2$ is even for convenience), so $n! \ge (n/2)^{n/2}$. While this is a lower bound, it's a poor one, as it looks like $n^{n/2}2^{-n/2}$; the power of n should be n and not $n/2$ according to Stirling. There's a bit of an art to finding good, elementary upper and lower bounds. This is a more advanced topic and requires a "feel" for how to proceed; nevertheless, it's a great skill to develop. So as not to interrupt the flow, we'll hold off on these more elementary bounds for now, and wait till §18.5 to show you how close one can get to Stirling just by knowing how to count!

What other checks can we do? Well, $(n + 1)!/n! = n + 1$; let's see what Stirling gives:

$$\frac{(n + 1)!}{n!} \approx \frac{(n + 1)^{n+1}e^{-(n+1)}\sqrt{2\pi(n + 1)}}{n^n e^{-n}\sqrt{2\pi n}}$$

$$= (n + 1) \cdot \left(\frac{n + 1}{n} \right)^n \cdot \frac{1}{e} \cdot \sqrt{\frac{n + 1}{n}}$$

$$= (n + 1)\left(1 + \frac{1}{n} \right)^n \frac{1}{e}\sqrt{1 + \frac{1}{n}}.$$

As $n \to \infty$, $(1 + 1/n)^n \to e$ (this is the definition of e; see the end of §B.3) and $\sqrt{1 + 1/n} \to 1$; thus our approximation above is basically $(n + 1) \cdot e \cdot \frac{1}{e} \cdot 1$, which is $n + 1$ as needed for consistency!

While the arguments above are not proofs, they're almost as valuable. It's essential to be able to look at a formula and get a feel for what it's saying, and for whether or not it's true. By some simple inspection, we can get upper and lower bounds that sandwich Stirling's formula. Moreover, Stirling's formula is consistent with $(n + 1)! = (n + 1)n!$. This gives us reason to believe we're on the right track, especially when we see how close our ratio was to $n + 1$.

Remark: While we give a full proof of Stirling's formula in §18.6, in an introductory course you shouldn't be fixated on the rigorous proof. What really matters is (1) being able to use the result, and (2) having a sense of why it's true. I'm a huge fan of ballpark estimates. Numbers and functions shouldn't be mysterious; we should have a feeling for their values. That's why this chapter has so many arguments giving good approximations to Stirling's formula. The time you spend reading these is time well spent; you'll get a sense of what's going on without being overwhelmed with all the technical details. Of course, it's great to know a proof, and proving results distinguishes us from other fields. That's why we do give a full proof as well, and remark on what else is needed to flesh out some of our sketches.

18.1 Stirling's Formula and Probabilities

Before getting bogged down in the technical details of the proofs of Stirling's formula, let's first spend some time seeing how it can be used. Our first problem is inspired by a phenomenon that surprises many students. Imagine we have a fair coin, so it lands on heads half the time and tails half the time. If we flip it $2n$ times, we expect to get n heads; thus if we flip it two million times we expect one million heads. It turns out that as $n \to \infty$, the probability of getting *exactly* n heads in $2n$ tosses of the fair coin tends to zero!

This result can seem quite startling. The expected value from $2n$ tosses is n heads, and the probability of getting n heads tends to zero? What's going on? If this is the expected value, shouldn't it be very likely? The key to understanding this problem is to note that while the expected value is n heads in $2n$ tosses, the standard deviation is $\sqrt{n/2}$. Returning to the case of two million tosses, we expect one million heads with fluctuations of the size 700. If now we tossed the coin two trillion times, we would expect one trillion heads and fluctuations on the order of 700,000. As n increases, the "window" about the mean where outcomes are likely is growing like the standard deviation, i.e., it's growing like $\sqrt{n}$ (up to some constants). Thus the probability is being shared among more and more values, and thus it makes sense that the probabilities of individual, specific outcomes (like exactly n tosses) is going down. In summary: the probability of exactly n heads in $2n$ tosses is going down as the probability is being spread over a wider and wider set. We now use Stirling's formula to quantify just how rapidly this probability decays.

For large n, we can painlessly approximate $n!$ with Stirling's formula. Remember we're trying to answer: *What is the probability of getting exactly n heads in 2n tosses of a fair coin?*

It's very easy to write down the answer: it's simply

$$\text{Prob(exactly } n \text{ heads in } 2n \text{ tosses)} = \binom{2n}{n}\left(\frac{1}{2}\right)^n\left(\frac{1}{2}\right)^n = \binom{2n}{n}\frac{1}{2^{2n}}.$$

The reason is that each of the 2^{2n} strings of heads and tails are equally likely, and there are $\binom{2n}{n}$ strings with exactly n heads. If it helps, think of the second $(1/2)^n$ as $(1-1/2)^{2n-n}$. How big is $\binom{2n}{n}$?

It's now Stirling's formula to the rescue. We have

$$\binom{2n}{n} = \frac{(2n)!}{n!n!}$$

$$\approx \frac{(2n)^{2n}e^{-2n}\sqrt{2\pi \cdot 2n}}{n^n e^{-n}\sqrt{2\pi n} \cdot n^n e^{-n}\sqrt{2\pi n}}$$

$$= \frac{2^{2n}}{\sqrt{\pi n}};$$

thus the probability of exactly n heads is

$$\binom{2n}{n}\frac{1}{2^{2n}} \approx \frac{1}{\sqrt{\pi n}}.$$

This means that if $n = 100$ there is a little less than a 6% chance of getting half heads, while if $n = 1,000,000$ the probability falls to less than .06%.

The above exercise was actually discussed in the 2008 presidential primary season. Clinton and Obama each received 6,001 votes in the democratic primary in Syracuse, NY. While there were 12,346 votes cast in the primary, for simplicity let's assume that there were just 12,002, and ask what is the probability of a tie. If we assume each candidate is equally likely to get any vote, the answer is just $\binom{12,002}{6,001}/2^{12,002}$. The exact answer is approximately 0.0072829; using our approximation from above we would estimate it as

$$\frac{1}{\sqrt{\pi \cdot 6001}} \approx 0.00728305,$$

which is fairly close to the true answer.

While we found the probability is a little less than 1%, some news outlets reported that the probability was about one in a million, with some going so far as to say it was "almost impossible." Why are there such different answers? It all depends on how you model the problem. If we assume the two candidates are equally likely to get each vote, then we get an answer of a little less than 1%. If, however, we take into account other bits of information, then the story changes. For example, Clinton was then a senator from NY, and it should be expected that she'd do better in her current home state. In fact, she won the state overall with 57.37% of the vote to Obama's 40.32%. Again for ease of analysis, let's say Clinton had 57% and Obama 43%. If we use these numbers, we no longer assume each voter is equally likely to vote for Clinton or Obama, and we find the probability of a tie is just $\binom{12002}{6001}.57^{6001}.43^{6001}$. Using Stirling's formula, we found $\binom{2n}{n} \approx 2^{2n}/\sqrt{\pi n}$. Thus, under these assumptions, the probability of a tie is approximately

$$\frac{2^{12002}}{\sqrt{\pi \cdot 6001}} \cdot .57^{6001} .43^{6001} \approx 1.877 \cdot 10^{-54}.$$

Note how widely different the probabilities are depending on our assumption of what is to be expected!

18.2 Stirling's Formula and Convergence of Series

Another great application of Stirling's formula is to help us determine for what x certain series converge. We have lots of powerful tests from calculus for this purpose, such as the ratio, root, and integral test; we review these in §B.3; however, we can frequently avoid having to use these tests and instead use the simpler comparison test by applying Stirling's formula.

For our first example, let's take

$$e^x = 1 + x + \frac{x^2}{2!} + \frac{x^3}{3!} + \cdots = \sum_{n=0}^{\infty} \frac{x^n}{n!}.$$

We want to find out for which x it converges. Using the ratio test we would see that it converges for *any* choice of x. We could use the root test, but that requires us to know a little bit about the growth of $n!$; fortunately Stirling's formula gives us this information. We have

$$n!^{1/n} \approx \left(n^n e^{-n} \sqrt{2\pi n}\right)^{1/n} = \frac{n}{e};$$

as this tends to infinity we find $(1/n!)^{1/n}$ tends to zero, and thus by the root test the radius of convergence is infinity (i.e., the series converges for all x).

While the above is a method to determine that the series for e^x converges for all x, it's quite unsatisfying. In addition to using Stirling's formula we also had to use the powerful root test. Is it possible to avoid using the root test, and determine that the series always converges *just* by using Stirling's formula? The answer is yes, as we now show.

We wish to show the series for e^x converges for all x. The idea is that $n!$ grows so rapidly that, no matter what x we take, $x^n/n!$ rapidly tends to zero. If we can show that, for all n sufficiently large, $|x^n/n!| < r(x)^n$ for some $r(x)$ less than 1, then the series for e^x converges by the comparison test; we wrote $r(x)$ to emphasize that the bound may depend on x. Explicitly, let's say our inequality holds for all $n \geq N(x)$. To determine if a series converges, it's enough to study the tail, as the finite number of summands in the beginning don't affect convergence. We have

$$\left| \sum_{n=N(x)}^{\infty} \frac{x^n}{n!} \right| \leq \sum_{n=N(x)}^{\infty} r(x)^n = \frac{r(x)^{N(x)}}{1 - r(x)}.$$

We are thus reduced to proving that $|x^n/n!| \le r(x)^n < 1$ for all n large for some $r(x) < 1$. Plugging in Stirling's formula, we see that $x^n/n!$ looks like

$$\frac{x^n}{n^n e^{-n}\sqrt{2\pi n}} = \frac{1}{\sqrt{2\pi n}}\left(\frac{ex}{n}\right)^n \le \left(\frac{ex}{n}\right)^n.$$

For any fixed x, once $n \ge 2ex + 1$ then $|ex/n| \le 1/2$, which completes the proof.

We can use this type of argument to prove many important series converge. Let's consider the moment generating function of the standard normal (for the definition and properties of the moment generating function, see Chapter 19). The moments of the standard normal are readily determined; the nth moment is

$$\mu_n = \int_{-\infty}^{\infty} x^n \frac{1}{\sqrt{2\pi}} e^{-x^2/2}dx = \begin{cases} (2m-1)!! & \text{if } n = 2m \text{ is even} \\ 0 & \text{if } n = 2m+1 \text{ is odd,} \end{cases}$$

where the **double factorial** means take every other term until you reach 2 or 1. Thus $4!! = 4 \cdot 2$, $5!! = 5 \cdot 3 \cdot 1$, $6!! = 6 \cdot 4 \cdot 2$, and so on. The moment generating function, $M_X(t)$, is

$$M_X(t) = \sum_{n=0}^{\infty} \frac{\mu_n}{n!}t^n = \sum_{m=0}^{\infty} \frac{(2m-1)!!}{(2m)!}t^{2m}.$$

We need to understand how rapidly $(2m-1)!!t^{2m}/(2m)!$ is decaying. We have

$$\frac{(2m-1)!!}{(2m)!} = \frac{(2m-1)!!}{(2m)!! \cdot (2m-1)!!} = \frac{1}{2m \cdot (2m-2)\cdots 2} = \frac{1}{2^m m!}.$$

Thus, arguing as before, we see the series converges for all t. In fact, we can even identify what it converges to:

$$M_X(t) = \sum_{m=0}^{\infty} \frac{t^{2m}}{2^m m!} = \sum_{m=0}^{\infty} \frac{(t^2/2)^m}{m!} = e^{t^2/2}.$$

18.3 From Stirling to the Central Limit Theorem

This section becomes a bit technical, and can safely be skipped; however, if you spend the time mastering it you'll learn some very useful techniques for attacking and estimating probabilities, and see a very common pitfall and learn how to avoid it.

The Central Limit Theorem is one of the gems of probability, saying the sum of nice independent random variables converges to being normally distributed as the number of summands grows. As a powerful application of Stirling's formula, we'll show it implies the Central Limit Theorem for the special case when the random variables $X_1, \dots, X_{2N}$ are all binomial random variables with parameter $p = 1/2$. It's technically easiest if we normalize these by

$$
\text{Prob}(X_i = n) =
\begin{cases}
1/2 & \text{if } n = 1 \\
1/2 & \text{if } n = -1 \\
0 & \text{otherwise.}
\end{cases}
\tag{18.1}
$$

Think of this as a 1 for a head and a -1 for a tail; this normalization allows us to say that the expected value of the sum $X_1 + \cdots + X_{2N}$ is 0. As we may interpret X_i as whether or not the ith toss is a head or a tail, this sum is the expected number of excess heads to tails (which is zero).

Let $X_1, \ldots, X_{2N}$ be independent binomial random variables with probability density given by (18.1). Then the mean is zero as $1 \cdot (1/2) + (-1) \cdot (1/2) = 0$, and the variance of each is

$$
\sigma^2 = (1 - 0)^2 \cdot \frac{1}{2} + (-1 - 0)^2 \cdot \frac{1}{2} = 1.
$$

Finally, we set

$$
S_{2N} = X_1 + \cdots + X_{2N}.
$$

Its mean is zero. This follows from

$$
\mathbb{E}[S_{2N}] = \mathbb{E}[X_1] + \cdots + \mathbb{E}[X_{2N}] = 0 + \cdots + 0 = 0.
$$

Similarly, we see the variance of S_{2N} is $2N$. We therefore expect S_{2N} to be on the order of 0, with fluctuations on the order of $\sqrt{2N}$.

Let's consider the distribution of S_{2N}. We first note that the probability that $S_{2N} = 2k + 1$ is zero. This is because S_{2N} equals the number of heads minus the number of tails, which is always even: if we have k heads and $2N - k$ tails then S_{2N} equals $2N - 2k$.

The probability that S_{2N} equals $2k$ is just $\binom{2N}{N+k}(\frac{1}{2})^{N+k}(\frac{1}{2})^{N-k}$. This is because for S_{2N} to equal $2k$, we need $2k$ more 1's (heads) than -1's (tails), and the number of 1's and -1's add to $2N$. Thus we have $N + k$ heads (1's) and $N - k$ tails (-1's). There are 2^{2N} strings of 1's and -1's, $\binom{2N}{N+k}$ have exactly $N + k$ heads and $N - k$ tails, and the probability of each string is $(\frac{1}{2})^{2N}$. We wrote $(\frac{1}{2})^{N+k}(\frac{1}{2})^{N-k}$ to show how to handle the more general case when there is a probability p of heads and $1 - p$ of tails.

We now use Stirling's formula to approximate $\binom{2N}{N+k}$. We find

$$
\binom{2N}{N+k} \approx \frac{(2N)^{2N} e^{-2N} \sqrt{2\pi \cdot 2N}}{(N+k)^{N+k} e^{-(N+k)} \sqrt{2\pi(N+k)} (N-k)^{N-k} e^{-(N-k)} \sqrt{2\pi(N-k)}}
$$

$$
= \frac{(2N)^{2N}}{(N+k)^{N+k}(N-k)^{N-k}} \sqrt{\frac{N}{\pi(N+k)(N-k)}}
$$

$$
= \frac{2^{2N}}{\sqrt{\pi N}} \frac{1}{(1 + \frac{k}{N})^{N+\frac{1}{2}+k}(1 - \frac{k}{N})^{N+\frac{1}{2}-k}}.
$$

The rest of the argument is just doing some algebra to show that this converges to a normal distribution. There is, unfortunately, a very common trap people frequently fall into when dealing with factors such as these. To help you avoid these in the future, we'll describe this common error first and then finish the proof.

We would like to use the definition of e^x (see the end of §B.3) to deduce that as $N \to \infty$, $\left(1 + \frac{w}{N}\right)^N \approx e^w$; unfortunately, we must be a little more careful as the values of k we consider grow with N. For example, we might believe that $(1 + \frac{k}{N})^N \to e^k$ and $(1 - \frac{k}{N})^N \to e^{-k}$, so these factors cancel. As k is small relative to N we may ignore the factors of $1/2$, and then say

$$\left(1 + \frac{k}{N}\right)^k = \left(1 + \frac{k}{N}\right)^{N \cdot \frac{k}{N}} \to e^{k^2/N};$$

similarly, $(1 - \frac{k}{N})^{-k} \to e^{k^2/N}$. Thus we would claim (*and we shall see later in Lemma 18.3.1 that this claim is in error!*) that

$$\left(1 + \frac{k}{N}\right)^{N + \frac{1}{2} + k} \left(1 - \frac{k}{N}\right)^{N + \frac{1}{2} - k} \to e^{2k^2/N}.$$

We show that $\left(1 + \frac{k}{N}\right)^{N + \frac{1}{2} + k} \left(1 - \frac{k}{N}\right)^{N + \frac{1}{2} - k} \to e^{k^2/N}$. The importance of this calculation is that it highlights how crucial *rates* of convergence are. While it's true that the main terms of $(1 \pm \frac{k}{N})^N$ are $e^{\pm k}$, the error terms (in the convergence) are quite important, and yield large secondary terms when k is a power of N. What happens here is that the secondary terms from these two factors reinforce each other. Another way of putting it is that one factor tends to infinity while the other tends to zero. Remember that $\infty \cdot 0$ is one of our undefined expressions; it can be anything depending on how rapidly the terms grow and decay; we'll say more about this at the end of the section.

The short of it is that we cannot, sadly, just use $\left(1 + \frac{w}{N}\right)^N \approx e^w$. We need to be more careful. The correct approach is to take the logarithms of the two factors, Taylor expand the logarithms, and then exponentiate. This allows us to better keep track of the error terms.

Before doing all of this, we need to know roughly what range of k will be important. As the standard deviation is $\sqrt{2N}$, we expect that the only k's that really matter are those within a few standard deviations from 0; equivalently, k's up to a bit more than $\sqrt{2N}$. We can carefully quantify exactly how large we need to study k by using Chebyshev's inequality (Theorem 17.3.1). From this we learn that we need only study k where $|k|$ is at most $N^{\frac{1}{2}+\epsilon}$. This is because the standard deviation of S_{2N} is $\sqrt{2N}$. We then have

$$\text{Prob}(|S_{2N} - 0| \geq (2N)^{1/2+\epsilon}) \leq \frac{1}{(2N)^{2\epsilon}},$$

because $(2N)^{1/2+\epsilon} = (2N)^{\epsilon} \text{StDev}(S_{2N})$. Thus it suffices to analyze the probability that $S_{2N} = 2k$ for $|k| \leq N^{1/2+1/9}$.

We now come to the promised lemma which tells us what the right value is for the product; the proof will show us how we should attack problems like this in general.

Lemma 18.3.1: : *For any $\epsilon \leq 1/9$, for $N \to \infty$ with $|k| \leq (2N)^{1/2+\epsilon}$, we have*

$$\left(1 + \frac{k}{N}\right)^{N+\frac{1}{2}+k} \left(1 - \frac{k}{N}\right)^{N+\frac{1}{2}-k} \longrightarrow e^{k^2/N} e^{O(N^{-1/6})}.$$

Proof: Recall that for $|x| < 1$,

$$\log(1+x) = \sum_{n=1}^{\infty} \frac{(-1)^{n+1} x^n}{n}.$$

As we're assuming $k \leq (2N)^{1/2+\epsilon}$, note that any term below of size k^2/N^2, k^3/N^2, or k^4/N^3 will be negligible. Thus if we define

$$P_{k,N} := \left(1 + \frac{k}{N}\right)^{N+\frac{1}{2}+k} \left(1 - \frac{k}{N}\right)^{N+\frac{1}{2}-k}$$

then using the big-Oh notation from §B.4 we find

$$
\begin{aligned}
\log P_{k,N} &= \left(N + \frac{1}{2} + k\right) \log \left(1 + \frac{k}{N}\right)^{N+\frac{1}{2}+k} \\
&\quad + \left(N + \frac{1}{2} - k\right) \log \left(1 - \frac{k}{N}\right)^{N+\frac{1}{2}-k} \\
&= \left(N + \frac{1}{2} + k\right) \left(\frac{k}{N} - \frac{k^2}{2N^2} + O\left(\frac{k^3}{N^3}\right)\right) \\
&\quad + \left(N + \frac{1}{2} - k\right) \left(-\frac{k}{N} - \frac{k^2}{2N^2} + O\left(\frac{k^3}{N^3}\right)\right) \\
&= \frac{2k^2}{N} - 2\left(N + \frac{1}{2}\right) \frac{k^2}{2N^2} + O\left(\frac{k^3}{N^2} + \frac{k^4}{N^3}\right) \\
&= \frac{k^2}{N} + O\left(\frac{k^2}{N^2} + \frac{k^3}{N^2} + \frac{k^4}{N^3}\right).
\end{aligned}
$$

As $k \leq (2N)^{1/2+\epsilon}$, for $\epsilon < 1/9$ the big-Oh term is dominated by $N^{-1/6}$, and we finally obtain that

$$P_{k,N} = e^{k^2/N} e^{O(N^{-1/6})},$$

which completes the proof. $\qquad\square$

We now finish the proof of S_{2N} converging to a Gaussian. Combining Lemma 18.3.1 with (18.2) yields

$$\binom{2N}{N+k}\frac{1}{2^{2N}} \approx \frac{1}{\sqrt{\pi N}} e^{-k^2/N}$$

(the careful analysis in the lemma alerted us to the existence of the factor $e^{-k^2/N}$, which our fast and loose calculations missed). The proof of the Central Limit Theorem in this case is completed by some simple algebra. We're studying $S_{2N} = 2k$, so we should replace k^2 with $(2k)^2/4$. Similarly, since the variance of S_{2N} is $2N$, we should replace N with $(2N)/2$. While these may seem like unimportant algebra tricks, it's very useful to become comfortable at doing this. By doing such small adjustments we make it easier to compare our expression with its conjectured value. We find

$$\text{Prob}(S_{2N} = 2k) = \binom{2N}{N+k}\frac{1}{2^{2N}} \approx \frac{2}{\sqrt{2\pi \cdot (2N)}} e^{-(2k)^2/2(2N)}.$$

Remember S_{2N} is never odd. The factor of 2 in the numerator of the normalization constant above reflects this fact, namely the contribution from the probability that S_{2N} is even is twice as large as we would expect, because it has to account for the fact that the probability that S_{2N} is odd is zero. Thus it looks like a Gaussian with mean 0 and variance $2N$. For N large such a Gaussian is slowly varying, and integrating from $2k$ to $2k + 2$ is basically $2/\sqrt{2\pi(2N)} \cdot \exp -(2k)^2/2(2N)$. □

As our proof was long, let's spend some time going over the key points. We were fortunate in that we had an explicit formula for the probability, and that formula involved binomial coefficients. We used Chebyshev's inequality to limit which probabilities we had to investigate. We then expanded using Stirling's formula, and did some algebra to make our expression look like a Gaussian.

For a nice challenge: Can you generalize the above arguments to handle the case when $p \neq 1/2$.

18.4 Integral Test and the Poor Man's Stirling

Using the integral test from calculus, we'll show the poor man's Stirling.

Poor man's Stirling: Let $n \geq 3$ be a positive integer. Then

$$n^n e^{-n} \cdot e \leq n! \leq n^n e^{-n} \cdot en.$$

As we've remarked throughout the book, whenever we see a formula our first response should be to test its reasonableness. Before going through the proof, let's compare this to Stirling's formula, which says $n! \approx n^n e^{-n}\sqrt{2\pi n}$. We correctly identify the main factor of $n^n e^{-n}$, but we miss the factor of $\sqrt{2\pi n}$. It is interesting to note how much we miss by. Our lower bound is just e while our upper bound is en. If we take the

geometric mean of these two (recall the geometric mean of x and y is $\sqrt{xy}$), $\sqrt{e \cdot en}$, we get $e\sqrt{n}$. This is approximately $2.718\sqrt{n}$, which is remarkably close to the true answer, which is $\sqrt{2\pi n} \approx 2.5063\sqrt{n}$. With just a bit more work, we could actually get Stirling's formula by arguing along the lines below. We're going to use the integral test from calculus; what we would need for a full proof is a slightly better version with more control on the error terms. As the goal is to highlight the method and ideas, I prefer not adding these extra details. They clutter the proof just a little bit; after you read the rest of this section you can look up the **Euler–Maclaurin formula** and extract a complete proof without too much more work.

The gist of the above discussion is that our answer is quite close. The arguments below are fairly elementary. They involve the integral test from calculus, and the Taylor series expansion of $\log(1 + x)$, which is just

$$\log(1 + x) = x - \frac{x^2}{2} + \frac{x^3}{3} - \cdots \quad (\text{if } |x| < 1).$$

While the algebra grows a bit at the end, it's important not to let that distract from the main idea, which is that we can approximate a sum very well and fairly easily through an integral. The difficulty is when we want to quantify how close the integral is to our sum; this requires us to do some slightly tedious bookkeeping. We go through the details of this proof to highlight how you can use these techniques to attack problems. We'll recap the arguments afterwards, emphasizing what you should take away from all of this.

Now, to the proof! Let $P = n!$ (we use the letter P because *product* starts with P). It's very hard to look at a product and have a sense of what it's saying. This is in part because our experiences in previous classes are with sums, not products. For example, in calculus we encounter Riemann sums all the time; I've never seen a Riemann product!

We thus want to convert our problem on $n!$ to a related one where we have more experience. The natural thing to do is to take the logarithm of both sides. The reason is that the logarithm of a product is the sum of the logarithms. This is excellent general advice: you should have a Pavlovian response to **take a logarithm** whenever you run into a product.

Returning to our problem, we have

$$\log P = \log n! = \log 1 + \log 2 + \cdots + \log n = \sum_{k=1}^{n} \log k.$$

We want to approximate this sum with an integral. Note that if $f(x) = \log x$ then this function is increasing for $x \geq 1$. We claim this means

$$\int_1^n \log t\, dt \leq \sum_{k=1}^{n} \log k \leq \int_2^{n+1} \log t\, dt.$$

This follows by looking at the upper and lower sums; note this is the same type of argument as you've seen in proving the Fundamental Theorem of Calculus (the old

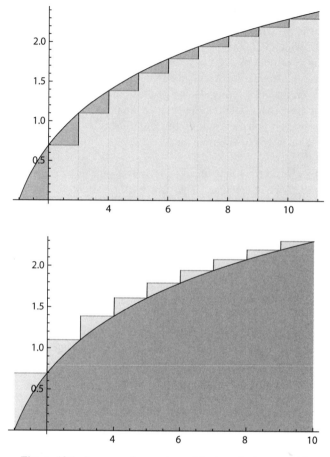

Figure 18.1. Lower and upper bound for $\log n!$ when $n = 10$.

upper and lower sum approach); see Figure 18.1. This is probably the most annoying part of the argument, getting the bounds for the integrals correct.

We now come to the hardest part of the argument. We need to know what is the integral of $\log t$. This isn't one of the standard functions, but it turns out to have a relatively simple anti-derivative, namely $t \log t - t$. While it's very hard to find a typical anti-derivative, it's very straightforward to check and make sure it works—all we have to do is take the derivative. We can now use this to approximate $\log n!$. We find

$$(t \log t - t)\Big|_{t=1}^{n} \leq \log n! \leq (t \log t - t)\Big|_{t=2}^{n+1}$$
$$n \log n - n + 1 \leq \log n! \leq (n+1)\log(n+1) - (n+1) - (2 \log 2 - 2).$$

We'll study the lower bound first. From

$$n \log n - n + 1 \leq \log n!,$$

we find after exponentiating that

$$e^{n \log n - n + 1} = n^n e^{-n} \cdot e \leq n!.$$

What about the upper bound? In the argument below we'll have the logarithm of quantities involving n. We have a Taylor expansion for the logarithm near 0, *not* near infinity. Thus we want to rewrite expressions to allow us to use what we have. In particular, we'll rewrite $\log(n + 1)$ as $\log(n(1 + 1/n))$, which by the log laws is $\log n + \log(1 + 1/n)$. This last expression is now ideally suited for a Taylor expansion, using $\log(1 + u) = u - u^2/2 + u^3/3 - \cdots$.

We thus find

$$(n + 1)\log(n + 1) - n + 1 - 2\log 2$$
$$= (n + 1)\log\left(n\left(1 + \frac{1}{n}\right)\right) - n + 1 - 2\log 2$$
$$= (n + 1)\log n + (n + 1)\log\left(1 + \frac{1}{n}\right) - n + 1 - 2\log 2$$
$$= n\log n + \log n - n + (n + 1)\left(\frac{1}{n} - \frac{1}{2n^2} + \frac{1}{3n^2} - \cdots\right) + 1 - 2\log 2$$
$$\leq n\log n + \log n - n + \frac{n + 1}{n} + 1 - 2\log 2,$$

where the last follows from two facts: (1) we have an alternating series, and so truncating after a positive term overestimates the sum; (2) if $n \geq 3$ then $\frac{n+1}{n} + 1 - 2\log 2 < 1$. Exponentiating gives

$$n! \leq e^{n\log n - n + \log n + 1} = n^n e^{-n} \cdot en.$$

Putting the two bounds together, we've shown that

$$n^n e^{-n} \cdot e \leq n! \leq n^n e^{-n} \cdot en,$$

which is what we wanted to show.

Remark: Note our arguments were entirely elementary, and introduce many nice and powerful techniques that can be used for a variety of problems. We replace a sum with an integral. We replace a complicated function, $\log(n + 1)$, with its Taylor expansion. We note that for an alternating series, if we truncate after a positive term we overestimate. Finally, and most importantly, we've seen how to get a handle on products—we should take logarithms and convert the product to a sum!

We end with a few words about how we found the anti-derivative of $\log t$ is $t \log t - t$. We want a function whose derivative is $\log t$; unfortunately, we don't know any nice functions with that as a derivative. All is not lost, however. We know the product rule, so if we try $g(t) = t \log t$ we know that when the differentiation hits the t factor, then we get $\log t$. Unfortunately, $g(t) = \log t + t\frac{1}{t} = \log t + 1$. We thus see the derivative is off by a little bit. We need to decrease the value of our derivative by 1. This means we need to subtract from $g(t)$ a function whose derivative is 1. That's easy—we just subtract t, and this is how we found $t \log t - t$.

Now that you've seen the integration by parts attack, read about the Euler-Maclaurin formula. This creates tighter upper and lower bounds, and you can use it to prove Stirling.

18.5 Elementary Approaches towards Stirling's Formula

In the last section we used the integral test to get very good upper and lower bounds for $n!$. While the bounds are great, we did have to use calculus twice. Once was in applying the integral test, and the other was being inspired to see that the anti-derivative of $\log t$ is $t \log t - t$; while it's easy to check this by differentiation, if you haven't seen such relations before it looks like it's a real challenge to find.

In this section we'll present an entirely elementary approach to estimates of Stirling's formula. We'll mostly avoid using calculus, instead just counting in a clever manner. *You may safely skip this section; however, the following arguments highlight a great way to look at estimation, and these ideas will almost surely be useful for a variety of other problems that you'll encounter over the years.* As you go further in mathematics, you start to really appreciate techniques and methods over facts; that's why I've spent a few pages expanding on these counting arguments in great length. Enjoy!

18.5.1 Dyadic Decompositions

One of the most useful skills you can develop is the knack for how to approximate well a very complicated quantity. While we can often resort to a computer for brute force calculations to get a feel for the answer, there are times when the parameter dependence is so wild that this isn't realistic. Thus, it's very useful to learn how to look at a problem and glean something about the behavior as parameters change.

Stirling's formula provides a wonderful testing ground for some of these methods. Remember it says that $n! \sim n^n e^{-n} \sqrt{2\pi n}$ as n tends to infinity. We've already seen how to get a reasonable upper bound without too much work; what about a good lower bound? Our first attempt yielded 1^n, which was quite poor; now we show a truly remarkable approach that leads to a very good lower bound with very little work. We'll recover our lower bound of $n^{n/2} 2^{-n/2}$, but we'll explain more how to view the calculation leading to it.

To simplify the presentation, we assume that $n = 2^N$ for some N; if we don't make this assumption, we need to use floor functions throughout, or do a bit more arguing (which we'll do later). Using the floor function makes the formulas and the analysis look more complicated, and the key insight, the main idea, is now buried in the unenlightening algebra which is required to make the statements true. I prefer to concentrate on this special case as I can then highlight the method without being bogged down in details.

The idea is to use **dyadic decompositions**. This is a powerful idea, and occurs throughout mathematics. It works as follows: we break the initial problem into two smaller problems, work on those, and then combine those results. This is very similar to recursion: we're building better bounds for a big problem by doing work on two smaller problems. Of course, nothing prevents us from stopping at one division: we can take the two smaller intervals, subdivide each of those, and apply the idea again.

As this is such an important concept, let's work slowly and carefully through its application here. Our goal is to bound $n! = n(n-1) \cdots 2 \cdot 1$. As each factor is at least

1 and at most n, we start with the trivial bound

$$1^n \le n! \le n^n.$$

Notice the *enormous* spread between our upper and lower bounds. The problem is our set $I_0 := \{1, 2, \ldots, n\}$ is very large as $n \to \infty$, and thus it is horrible trying to find *one* upper bound for each factor, and *one* lower bound for each. The idea behind dyadic decompositions is to break this large interval into smaller ones, where the bounds are better, then put them together.

Explicitly, let's split our set in half:

$$\mathcal{S}_0 = \{1, 2, \ldots, n\} = \{1, 2, \ldots, n/2\} \cup \{n/2 + 1, n/2 + 2, \ldots, n\} := \mathcal{S}_1 \cup \mathcal{S}_2.$$

In the first interval, each term is at least 1 and at most $n/2$, and thus we obtain

$$1^{n/2} \le 1 \cdot 2 \cdots (n/2 - 1)(n/2) \le (n/2)^{n/2}.$$

Similarly in the second interval each term is at least $n/2 + 1$, though we'll use $n/2$ as a lower bound as that makes the algebra cleaner, and at most n. Thus we find

$$(n/2)^{n/2} \le (n/2 + 1)(n/2 + 2) \cdots (n - 1)n \le n^{n/2}.$$

Notice that we're still just using the trivial idea of bounding each term by the smallest or largest; the gain comes from the fact that the sets $\mathcal{S}_1$ and $\mathcal{S}_2$ are each half the size of the original set $\mathcal{S}_0$. Thus the upper and lower bounds are much better, as these sets have less variation. Multiplying the two lower (respectively, upper) bounds together gives a lower (respectively, upper) bound for $n!$:

$$1^{n/2}(n/2)^{n/2} \le [1 \cdot 2 \cdots (n/2)] [(n/2 + 1)(n/2 + 2) \cdots n] \le (n/2)^{n/2} n^{n/2},$$

which simplifies to

$$n^{n/2}\sqrt{2}^{-n} \le n! \le n^n \sqrt{2}^{-n}.$$

Notice how much better this is than our original trivial bound of $1 \le n! \le n^n$; the upper bound is very close (we have a $\sqrt{2}^{-n}$ instead of an $e^{-n}\sqrt{2\pi n}$), while the lower bound is significantly closer.

We now use the advice from shampoo: **lather, rinse, repeat**. We can break $\mathcal{S}_1$ and $\mathcal{S}_2$ into two smaller intervals, argue as above, and then break those new intervals further (though in practice we'll do something slightly different). We do all this in the next subsection; our purpose here was to introduce the method slowly and describe why it works so well. Briefly, the success is from a delicate balancing act. If we make things too small, there is no variation and no approximation—the numbers are what they are; if we have things too large, there is too much variation and the bounds are trivial. We need to find a happy medium between the two.

18.5.2 Lower Bounds towards Stirling: I

We continue our elementary attack on $n!$, and build on the dyadic decomposition idea from the previous subsection. Instead of breaking each smaller set in half, what we will do is just break the earlier set (the one with smaller numbers). We thus end up with sets of different size, getting a chain of sets where each is half the size of the previous.

Explicitly, we study the factors of $n!$ in the intervals $I_1 = (n/2, n]$, $I_2 = (n/4, n/2]$, $I_3 = (n/8, n/4]$, ..., $I_N = (1, 2)$. Note on I_k that each of the $n/2^k$ factors is at least $n/2^k$. Thus

$$n! = \prod_{k=1}^{N} \prod_{m \in I_k} m$$

$$\geq \prod_{k=1}^{N} \left(\frac{n}{2^k}\right)^{n/2^k}$$

$$= n^{n/2+n/4+n/8+\cdots+n/2^N} 2^{-n/2} 4^{-n/4} 8^{-n/8} \cdots (2^N)^{-n/2^N}.$$

Let's look at each factor above slowly and carefully. Note the powers of n *almost* sum to n; they would if we just add $n/2^N = 1$ (since we're assuming $n = 2^N$). Remember, though, that $n = 2^N$; there is thus no harm in multiplying by $(n/2^N)^{n/2^N}$ as this is just 1[1] (**multiplying by one** is a powerful technique; see §A.12 for more applications of this method). We now have $n!$ is greater than

$$n^{n/2+n/4+n/8+\cdots+n/2^N+n/2^N} 2^{-n/2} 4^{-n/4} 8^{-n/8} \cdots (2^N)^{-n/2^N} (2)^{-n/2^N}.$$

Thus the n-terms give n^n. What of the sum of the powers of 2? That's just

$$2^{-n/2} 4^{-n/4} 8^{-n/8} \cdots (2^N)^{-n/2^N} \cdot 2^{-n/2^N} = 2^{-n(1/2+2/4+3/8+\cdots N/2^N)} 2^{-2^N/2^N}$$

$$> 2^{-n\left(\sum_{k=0}^{N} k/2^k\right)} 2^{-2^N/2^N}$$

$$\geq 2^{-n\left(\sum_{k=0}^{\infty} k/2^k\right)} 2^{-1}$$

$$= 2^{-2n-1} = \frac{1}{2} 4^{-n}.$$

To see this, we use the following wonderful identity:

$$\sum_{k=0}^{\infty} k x^k = \frac{x}{(1-x)^2};$$

for a proof, see §11.1 (on differentiating identities involving the geometric series formula).

Putting everything together, we find

$$n! \geq \frac{1}{2} n^n 4^{-n},$$

which compares favorably to the truth, which is $n^n e^{-n}$. It's definitely much better than our first lower bound of $n^{n/2} 2^{-n/2}$.

As with many things in life, we can get a better result if we're willing to do a little more work. For example, consider the interval $I_1 = (n/2, n]$. We can pair elements at the beginning and the end: n and $n/2 + 1$, $n - 1$ and $n/2 + 2$, $n - 2$ and $n/2 + 3$, and so on until $3n/4$ and $3n/4 + 1$; for example, if we have the interval $(8, 16]$ then the pairs are: $(16,9)$, $(15,10)$, $(14,11)$, and $(13,12)$. We now use one of the gold standard problems from calculus: if we want to maximize xy given that $x + y = L$ then the maximum occurs when $x = y = L/2$. This is frequently referred to as the Farmer Bob (or Brown) problem, and is given the riveting interpretation that if we're trying to find the rectangular pen that encloses the maximum area for his cows to graze given that the perimeter is L, then the answer is a square pen. Thus of all our pairs, the one that has the largest product is $3n/4$ with $3n/4 + 1$, and the smallest is n and $n/2 + 1$, which has a product exceeding $n^2/2$. We therefore decrease the product of all elements in I_1 by replacing each product with $\sqrt{n^2/2} = n/\sqrt{2}$. Thus a little thought gives us that

$$n \cdot (n - 1) \cdots \frac{3n}{4} \cdots \left(\frac{n}{2} + 1\right) \cdot \frac{n}{2} \geq \left(\frac{n}{\sqrt{2}}\right)^{n/2} = \left(\frac{n\sqrt{2}}{2}\right)^{n/2},$$

a nice improvement over $(n/2)^{n/2}$, and this didn't require too much additional work!

We now do a similar analysis on I_2; again the worst case is from the pair $n/2$ and $n/4 + 1$ which has a product exceeding $n^2/8$. Arguing as before, we find

$$\prod_{m \in I_2} m \geq \left(\frac{n}{\sqrt{8}}\right)^{n/4} = \left(\frac{n}{2\sqrt{2}}\right)^{n/4} = \left(\frac{n\sqrt{2}}{4}\right)^{n/4}.$$

At this point hopefully the pattern is becoming clear. We have almost exactly what we had before; the only difference is that we have a $n\sqrt{2}$ in the numerator each time instead of just an n. This leads to very minor changes in the algebra, and we find

$$n! \geq \frac{1}{2}(n\sqrt{2})^n 4^{-n} = \frac{1}{2}n^n(2\sqrt{2})^{-n}.$$

Notice how close we are to $n^n e^{-n}$, as $2\sqrt{2} \approx 2.82843$, which is just a shade larger than $e \approx 2.71828$. It's amazing how close our analysis has brought us to Stirling; we're within striking distance of it!

We end this section on elementary questions with a few things for you to try.

- Can you modify the above argument to get a reasonably good upper bound for $n!$?

- After reading the above argument, you should be wondering exactly how far can we push things. What if we didn't do a dyadic decomposition; what if instead we did say a triadic: $(2n/3, n]$, $(4n/9, 2n/3]$, ... ? Maybe powers of 2 are nice, so perhaps instead of thirds we should do fourths? Or perhaps fix an r and look at $(rn, n]$, $(r^2n, rn]$, ... for some universal constant r. Using this and the pairing method described above, what is the largest lower bound attainable? In other words, what value of r maximizes the lower bound for the product?

Our proof in this section was *almost* entirely elementary. We used calculus in one step: we needed to know that $\sum_{k=0}^{\infty} kx^k$ equals $x/(1-x)^2$. Fortunately it's possible to prove this result *without* resorting to calculus. All we need is our work on memoryless processes from the basketball game of §1.2. I'll outline the argument in Exercise 18.8.19.

18.5.3 Lower Bounds toward Stirling II

We continue seeing just how far we can push elementary arguments. Of course, in some sense there is no need to do this; there are more powerful approaches that yield better results with less work. As this is true, we're left with the natural, nagging question: *why spend time reading this?*

There are several reasons for giving these arguments. Even though they're weaker than what we can prove, they need less machinery. To prove Stirling's formula, or good bounds towards it, requires results from calculus, real and complex analysis; it's nice to see what we can do just from basic properties of the integers. Second, there are numerous problems where we just need some simple bound. By carefully going through these pages, you'll get a sense of how to generate such elementary bounds, which we hope will help you in something later in life.

Again, the rest of the material in this subsection is advanced and not needed in the rest of the book. You may safely skip it, but I urge you to at least skim these arguments.

We now generalize our argument showing that $n! > (n/4)^n$ for $n = 2^N$ to any integer n; in other words, it was harmless assuming n had the special form $n = 2^N$. Suppose $2^k < n < 2^{k+1}$. Then we can write $n = 2^k + m$ for some positive $m < 2^k$, and use our previous result to conclude

$$n! = n \cdot (n-1) \cdots (2^k + 1) \cdot (2^k)! > (2^k)^m \cdot (2^k)! > (2^k)^m \cdot (2^k/4)^{2^k}.$$

Our goal, then, is to prove that this quantity is greater than $(n/4)^n$. Here's one possible method: write

$$2^{km} \cdot (2^k/4)^{2^k} = (n/4)^{\alpha}.$$

If $\alpha > n$, then we're done. Taking logarithms, we find

$$k \cdot m \cdot \log 2 + 2^k \cdot \log(2)(k-2) = \alpha(\log(n) - 2\log 2).$$

Solving for α gives

$$\alpha = \frac{k \cdot m \cdot \log 2 + 2^k \cdot \log(2)(k-2)}{\log(n) - 2\log 2}.$$

Remember, we want to show that $\alpha > n$. Substituting in our prior expression $n = 2^k + m$, this is equivalent to showing

$$\frac{k \cdot m \cdot \log 2 + 2^k \cdot \log(2)(k-2)}{\log(2^k + m) - 2\log 2} > 2^k + m.$$

So long as $2^k + m > 4$, the denominator is positive, so we may multiply through without altering the inequality:

$$\log(2)(k(2^k + m) - 2^{k+1}) > (2^k + m)\log(2^k + m) - \log(2)2^{k+1} - 2m\log 2.$$

With a bit of algebra, we can turn this into a nicer expression:

$$\log(2^k)(2^k + m) > (2^k + m)(\log(2^k + m) - 2m\log 2)$$
$$2m\log 2 > (2^k + m)\log(1 + m/2^k)$$
$$2\log 2 > (1 + 2^k/m)\log(1 + m/2^k).$$

Let's write $t = m/2^k$. Then showing that $\alpha > n$ is equivalent to showing

$$2\log 2 > (1 + 1/t)\log(1 + t)$$

for $t \in (0, 1)$. Why $(0, 1)$? Since we know $0 < m < 2^k$, then $0 < m/2^k < 1$, so t is always between 0 and 1. While we're only really interested in whether this equation holds when t is of the form $m/2^k$, if we can prove it for all t in $(0,1)$, then it automatically holds for the special values we care about. Letting $f(t) = (1 + 1/t)\log(1 + t)$, we see $f'(t) = (t - \log(1 + t))/t^2$, which is positive for all $t > 0$ (fun exercise: show that the limit as t approaches 0 of $f'(t)$ is $1/2$). Since $f(1) = 2\log 2$, we see that $f(t) < 2\log 2$ for all $t \in (0, 1)$. Therefore $\alpha > n$, so $n! > (n/4)^n$ for all integer n.

18.5.4 Lower Bounds towards Stirling: III

Again, this subsection may safely be skipped; it's the last in our chain of seeing just how far elementary arguments can be pushed. Reading this is a great way to see how to do such arguments, and if you continue in probability and mathematics there is a good chance you'll have to argue along these lines someday.

We've given a few proofs now showing that $n! > (n/4)^n$ for any integer n. However, we know that Stirling's formula tells us that $n! > (n/e)^n$. Why have we been messing around with 4, then, and where does e come into play? The following sketch doesn't *prove* that $n! > (n/e)^n$, but hints suggestively that e might enter into our equations.

In our previous arguments we've taken n and then broken the number line up into the following intervals: $\{[n, n/2), [n/2, n/4), \dots\}$. The issue with this approach is that $[n, n/2)$ is a pretty big interval, so we lose a fair amount of information by approximating $n \cdot (n - 1) \cdots \frac{n}{2}$ by $(n/2)^{n/2}$. It would be better if we could use a smaller interval. Therefore, let's think about using some ratio $r < 1$, and suppose $n = (1/r)^k$. We would like to divide the number line into $\{[n, rn), [rn, r^2n), \dots\}$, although the problem we run into is that $r^\ell n$ isn't always going to be an integer for every integer $\ell < k$. Putting that issue aside for now (*this is why this isn't a proof!*), let's proceed as we typically do: having broken up the number line, we want to say that $n!$ is greater than the product of the smallest numbers in each interval raised to the number of integers in that interval:

$$n! > (rn)^{(1-r)n}(r^2n)^{r\cdot(1-r)n} \cdot (r^3n)^{r^2\cdot(1-r)n} \cdots (r^k \cdot n)^{r^{k-1}\cdot(1-r)n}.$$

Since $r^{k+m}n < 1$ for all $m > 1$, we can extend this product to infinity:

$$n! > (rn)^{(1-r)n}(r^2n)^{r\cdot(1-r)n} \cdot (r^3n)^{r^2\cdot(1-r)n} \cdots (r^k \cdot n)^{r^{k-1}\cdot(1-r)n} \cdots .$$

While this lowers our value, it shouldn't change it too much. The reason is that $\lim_{x\to 0} x^x = 1$. Let's simplify this a bit. Looking at the n terms, we have

$$n^{(1-r+r-r^2+r^2-\cdots)n} = n^n$$

because the sum **telescopes**. Looking at the r terms we see

$$
\begin{aligned}
r^{n(1-r)(1+2r+3r^2+\cdots)} &= r^{n(1-r)/r(r+2r^2+3r^3+\cdots)} \\
&= r^{n(1-r)/r\cdot r/(1-r)^2} \\
&= r^{n/(1-r)},
\end{aligned}
$$

where in the third step we use the identity

$$\sum_{k=1}^{\infty} kr^k = \frac{r}{(1-r)^2};$$

remember we used this identity earlier as well! Combining the two terms, we have

$$n! > (r^{1/(1-r)}n)^n.$$

To make this inequality as strong as possible, we want to find the largest possible value of $r^{1/(1-r)}$ for $r \in (0, 1)$. Substituting $x = 1/(1-r)$, this becomes: what is the limit as $x \to \infty$ of $(1 - 1/x)^x$? Hopefully you've encountered this limit before; the first exposure to it is often from continuously compounded interest. It's just e^{-1} (see §B.3). There are two definitions of e^x, one as a series and one as this limit. Thus we see that this argument gives a heuristic proof (remember we only looked at special n that were a power of r) that $n! > (n/e)^n$.

18.6 Stationary Phase and Stirling

Any result as important as Stirling's formula deserves multiple proofs. The proof below is a modification from Eric W. Weisstein's post "Stirling's Approximation" (See [We]).

To prove the theorem, we'll use the identity

$$n! = \Gamma(n + 1) = \int_0^\infty e^{-x}x^n\, dx. \tag{18.2}$$

A review of the Gamma function, including a proof that it does generalize the factorial function, can be found in Chapter 15. The reason we're doing this is we're replacing the discrete sequence of integer factorials with an integral of a continuous function. This expands our tool set, and we can throw some powerful results at it.

 In order to get an approximation, we want to find where the integrand is largest. Because of the exponential factor in the integrand, we take a logarithm before differentiating. We can do this as maximizing a positive function $f(x)$ is equivalent to maximizing $\log f(x)$. There are many useful versions of this principle: in calculus it's often easier to minimize the square of the distance rather than the distance (as this avoids the square-root function).

We find

$$\frac{d}{dx}\log(e^{-x}x^n) = \frac{d}{dx}(-x + n\log x) = \frac{n}{x} - 1.$$

The maximum value of the integrand is therefore seen to occur only for $x = n$. The exponential factor shrinks much more quickly than the growth of x^n, so we *assume* that the only significant contribution to the integral comes from $x = n + \alpha$ with $|\alpha|$ much smaller then n. We are not going to say anything further about this assumption; the purpose here is to provide another argument in support of Stirling's formula, not to provide the proof in all its glory and gory details. You are, of course, encouraged to explore the contributions from the other x and make this analysis rigorous. We have

$$\log x = \log(n + \alpha) = \log n + \log\left(1 + \frac{\alpha}{n}\right).$$

We now expand the second term using the Taylor series for $\log(1 + u)$ to find

$$\log(n + \alpha) = \log n + \frac{\alpha}{n} - \frac{1}{2}\frac{\alpha^2}{n^2} + \cdots.$$

Therefore

$$\log(x^n e^{-x}) = n\log x - x \approx n\left(\log n + \frac{\alpha}{n} - \frac{1}{2}\frac{\alpha^2}{n^2}\right) - (n + \alpha)$$

$$= n\log n - n - \frac{\alpha^2}{2n^2}.$$

It follows that

$$x^n e^{-x} \approx \exp\left(n\log n - n - \frac{\alpha^2}{2n^2}\right) = n^n e^{-n} \cdot \exp\left(-\frac{\alpha^2}{2n^2}\right),$$

with α smaller than n. Returning to the integral expression for $n!$ of (18.2), we have

$$n! = \int_0^\infty e^{-x}x^n\, dx$$

$$\approx \int_{-n}^\infty n^n e^{-n} \cdot \exp\left(-\frac{\alpha^2}{2n^2}\right)\, d\alpha$$

$$\approx n^n e^{-n} \cdot \int_{-\infty}^\infty \exp\left(-\frac{\alpha^2}{2n^2}\right)\, d\alpha.$$

In the last step, we rely on the fact that the integrand is very small for $\alpha < -n$. The integral is the same as the one we would obtain in integrating a normal density with mean 0 and variance $\sqrt{n}$. Its value is $\sqrt{2\pi n}$. We thus have

$$n! \approx n^n e^{-n} \sqrt{2\pi n},$$

which is the statement of the theorem. $\qquad\qquad\square$

This proof is an example of a very powerful method, the **method of stationary phase** (see for example [SS2]). This is a wonderful tool to approximate highly oscillatory integrals by looking for the source of most of the contribution. Note that most of the contribution occurs from x near the maximum. I particularly like how the factor $\sqrt{2\pi}$ emerges in the proof. It's the normalization constant of the standard normal. Completing the circle, remember we say this is related to $\Gamma(1/2) = \sqrt{\pi}$.

18.7 The Central Limit Theorem and Stirling

We end this chapter with one more proof of Stirling's formula. We continue our trend of needing more and more input. Our first proof was very elementary, essentially just the integral test. The second was more complicated, needing the Gamma function. Our final approach involves an application of the Central Limit Theorem, one of the gems of probability. We prove the Central Limit Theorem in Chapter 20; if you haven't seen the proof you can go there, or take the result on faith for now and read up on it later.

The idea of the proof is to apply the Central Limit Theorem to a sum of independent, identically distributed Poisson random variables with parameter 1. We have an explicit formula for this probability, as a sum of appropriately normalized Poissonian random variables is itself Poissonian (see §19.1 for a proof). We also know what this probability is (or at least approximately is) by the Central Limit Theorem. Equating the two gives Stirling's formula.

Remember:

1. X has a Poisson distribution with parameter λ means

$$\text{Prob}(X = n) = \begin{cases} \frac{\lambda^n e^{-\lambda}}{n!} & \text{if } n \geq 0 \text{ is an integer} \\ 0 & \text{otherwise.} \end{cases}$$

2. If $X_1, \ldots, X_N$ are independent, identically distributed random variables with mean μ, variance σ^2 and a little more (such as the third moment is finite, or the moment generating function exists), then $X_1 + \cdots + X_N$ converges to being normally distributed with mean $n\mu$ and variance $n\sigma^2$.

Let's flesh out the details. An important and useful fact about Poisson random variables is that the sum of n independent identically distributed Poisson random variables with parameter λ is a Poisson random variable with parameter $n\lambda$. As the mass function for a Poisson random variable Y with parameter λ is $\text{Prob}(Y = m) = \lambda^m e^{-\lambda}/m!$ for m a non-negative integer and 0 otherwise. Thus the probability density of $X_1 + \cdots + X_n$ is

$$f(m) = \begin{cases} n^m e^{-n}/m! & \text{if } m \text{ is a non-negative integer} \\ 0 & \text{otherwise.} \end{cases}$$

For n large, $X_1 + \cdots + X_n$ (in addition to being a Poisson random variable with parameter n) is by the Central Limit Theorem approximately normal with mean $n \cdot 1$ and variance n (as the mean and variance of a Poisson random variable with parameter λ is λ for each). We must be a bit careful due to the discreteness of the values taken on by $X_1 + \cdots + X_n$; however, a little inspection shows that the Central Limit Theorem allows us to approximate the probability $n - \frac{1}{2} \le X_1 + \cdots + X_n \le n + \frac{1}{2}$ with

$$\int_{n-\frac{1}{2}}^{n+\frac{1}{2}} \frac{1}{\sqrt{2\pi n}} \exp\left(-(x-n)^2/2n\right) dx = \frac{1}{\sqrt{2\pi n}} \int_{-1/2}^{1/2} e^{-t^2/2n} dt.$$

As n is large, we may approximate $e^{-t^2/2n}$ with the zeroth term of its Taylor series expansion about $t = 0$, which is 1. Thus

$$\text{Prob}\left(n - \frac{1}{2} \le X_1 + \cdots + X_n \le n + \frac{1}{2}\right) \approx \frac{1}{\sqrt{2\pi n}} \cdot 1 \cdot 1,$$

where the second 1 is from the length of the interval; however, we can easily calculate the left-hand side, as this is just the probability our Poisson random variable $X_1 + \cdots + X_n$ with parameter n takes on the value n; this is $n^n e^{-n}/n!$. We thus find

$$\frac{n^n e^{-n}}{n!} \approx \frac{1}{\sqrt{2\pi n}} \implies n! \approx n^n e^{-n} \sqrt{2\pi n}.$$

One of the most common mistakes in this approach is forgetting that the Poisson is discrete and the standard normal is continuous. Thus, to approximate the Poisson's mass at n, we should integrate the continuous density of the standard normal from $n - 1/2$ to $n + 1/2$. It's just fortuitous that, for this problem, we get the same answer if we forget about the integral.

We should note that the above argument is not quite an independent proof of Stirling's formula. The problem is that when we use the Central Limit Theorem, we need to have some control over the error term. That said, the above should be a very convincing argument. It's also interesting as it's the second proof we've given where the factor $\sqrt{2\pi}$ is coming from an integral involving a normal distribution.

18.8 Exercises

Exercise 18.8.1   *For large n, does Stirling's formula over or under approximate n!? Check your answer numerically.*

Exercise 18.8.2   *For large n, find a formula for* $\log(n!)$.

Exercise 18.8.3   *Approximate 200!. What is the percent error on the approximate?*

Exercise 18.8.4   *How many error terms must you keep track of to get your error on the approximation of 200! under one millionth of a percent?*

Exercise 18.8.5   *Approximate* $\binom{500}{499}$ *using Stirling's formula to approximate **all** the factorial. What went wrong in this estimate? What would be a better way to approximate* $\binom{n}{k}$ *where n is large, but k or $n - k$ is not?*

Exercise 18.8.6   *Imagine there are 200 people in a school cafeteria. The cafeteria has 20 circular tables, each of which can seat 10 kids. How many distinct seating arrangements are there given that each table is distinct and who each person is sitting next to at the table matters?*

Exercise 18.8.7   *Repeat the previous exercise, but now assume the tables are indistinguishable.*

Exercise 18.8.8   *Now assume the tables are arranged in a giant circle. The tables are not distinct, except it matters which tables are next to each other table, since the kids don't mind shouting across tables to talk to nearby friends. Find approximately how many distinct arrangements there are.*

Exercise 18.8.9   *Use dyadic decomposition to get an upper bound on n!.*

Exercise 18.8.10   *Find the least upper bound that can be achieved using decompositions.*

Exercise 18.8.11   *Show that $\lim_{n\to\infty} n^{1/n} = 1$.*

Exercise 18.8.12   *Show that $\lim_{n\to\infty} \frac{n!}{x^n}$ diverges for all x.*

Exercise 18.8.13   *Find $\lim_{n\to\infty} \sum_{k=1}^{n} \frac{\log(n)-\log(k)}{n}$.*

Exercise 18.8.14   *Find an approximation for $\lim_{n\to\infty} \sum_{k=1}^{n} \log(k)^{1/k}$.*

Exercise 18.8.15   *The Taylor series for arcsine is $\sin^{-1}(x) = \sum_{n=0}^{\infty} \frac{(2n)!}{4^n(n!)^2(2n+1)} x^{2n+1}$. Use Stirling's formula to find the radius of convergence for this series.*

Exercise 18.8.16   *Find good upper and lower bounds on $n!!$ for large n. Does it matter if n is even or odd?*

Exercise 18.8.17   *One of the most useful inequalities in mathematics is the **arithmetic mean–geometric mean (AMGM)**. Prove that if $x_1, x_2 \geq 0$ then $\frac{x_1+x_2}{2} \geq \sqrt{x_1 x_2}$. More generally, prove $\frac{x_1+\cdots+x_n}{n} \geq \sqrt{x_1 \cdots x_n}$.*

Exercise 18.8.18   *Find, if you can, the anti-derivatives of $t^2 \log t$ and $t \log^2 t$.*

Exercise 18.8.19   *Fill in the details to elementarily prove $\sum_{k=0}^{\infty} kx^k = x/(1-x)^2$.*

1. *Let $S(x) = \sum_{k=0}^{\infty} kx^k$. Clearly $S(x) = \frac{1}{1-x} \sum_{k=0}^{\infty} kx^k(1-x)$. If we have a coin with probability x of tails and $1-x$ of heads, then $x^k(1-x)$ is the probability that the first head is on the $(k+1)^{st}$ toss. This is almost the expected value of how long we must wait for the first head. The division by the extra factor of $1-x$ is easily fixed, but we have a factor of k and not $k+1$ in the summand. Fix this by writing k as $(k+1) - 1$, and then showing $S(x) = \frac{1}{1-r} \sum_{k=0}^{\infty}(k+1)x^k(1-x) - \frac{1}{1-x}$.*

2. *Let W be the random variable of how many tosses of a coin with probability $1-x$ of heads are needed before the first head. If μ_W is the expected value of W, then $S(x) = \frac{\mu_W}{1-x} - \frac{1}{1-x}$.*

3. Prove that $\mu_W = (1 - x) + x(\mu_W + 1)$. (Hint: If we get a head on the first toss we are done; if we get a tail it is as if we've started the tossing over but with one extra tail). Simple algebra yields $\mu_W = \frac{1}{1-x}$.

4. Substitute μ_W into our expression for $S(x)$ and deduce $S(x) = \frac{x}{(1-x)^2}$.

Exercise 18.8.20   Show that $\lim_{x \to 0^+} x^x = 1$.

Exercise 18.8.21   Plot the difference between the factorial function and Stirling's approximation.

Exercise 18.8.22   Plot the difference between the the Gamma function (shifted by 1) and Stirling's approximation.

CHAPTER 19 _____

Generating Functions and Convolutions

A common complaint in mathematics is the ubiquitous: *I can follow the proof when you do it line by line, but how could anyone ever think of doing this!* Of all the areas in probability, one of the most appropriate for sounding this complaint is in generating functions. At first glance, it seems like it's making our lives needlessly complex; however, at the end of the chapter you'll have learned how many different problems generating functions solve. Further, the time you spend learning these techniques will continue to pay dividends as you continue your studies, as these are used not just in probability, but throughout mathematical physics.

The reason for their helpfulness is that they allow us to package neatly a lot of information about a problem. You should be skeptical as to whether or not this is worthwhile; however, we'll see time and time again that this new viewpoint simplifies the algebra we need to do. We'll give several motivating examples from previous courses of how a change in viewpoint can save you hours of labor, and then describe many of the properties and applications of generating functions. While there are many problems where it's quite difficult to find and use the correct generating function, a lot of useful problems can be handled with a small bag of tricks. Thus, be patient as you read on—the time you spend mastering this material will help you for years to come.

In probability, the most important use of generating functions is to understand moments of random variables. As we know, the moments tell us about the shape of the distribution. A very powerful application of this is in proving the Central Limit Theorem, which tells us that, in many cases, the sum of independent random variables tends towards a Gaussian as the number of summands grows. We'll devote Chapter 20 to this theorem (which shouldn't be surprising—anything given the name "Central" should be expected to play a prominent role in a course).

19.1 Motivation

Frequently in mathematics we encounter complex data sets, and then do operations on it to make it even more complex! For example, imagine the first data set is the probabilities that the random variable X_1 takes on given values, and the second set is the probabilities of another random variable X_2 taking on given values. From these we can, painfully through brute force, determine the probabilities of $X_1 + X_2$ equaling anything; however, if at all possible we would like to avoid these tedious computations. Below we'll study this problem in great detail in the special case that our two random variables have Poisson distributions (see §12.6 for properties of Poisson random variables). We'll solve the problem completely in this case, but the solution will be unsatisfying. The problem is we need to have some moments of divine inspiration in how to handle the algebra. The purpose of this example is to set the stage: we will introduce generating functions to automate the algebra.

Let's consider the case when X_1 has the Poisson distribution with parameter 5 and X_2 is a Poisson with parameter 7. This means

$$\text{Prob}(X_1 = m) = 5^m e^{-5}/m!$$

$$\text{Prob}(X_2 = n) = 7^n e^{-7}/n!,$$

where m and n range over the non-negative integers. If k is a non-negative integer, then the probability that $X_1 + X_2 = k$ can be found by looking at all the different ways two non-negative integers can add to k. Clearly X_1 must take on a value between 0 and k; if it's ℓ then we must have X_2 equaling $k - \ell$. As our random variables are independent, the probability this happens is just the product of the probability that X_1 is ℓ and the probability that X_2 is $k - \ell$. If we now sum over ℓ we get the probability that $X_1 + X_2$ is k:

$$\text{Prob}(X_1 + X_2 = k) = \sum_{\ell=0}^{k} \text{Prob}(X_1 = \ell)\text{Prob}(X_2 = k - \ell)$$

$$= \sum_{\ell=0}^{k} \frac{5^\ell e^{-5}}{\ell!} \cdot \frac{7^{k-\ell} e^{-7}}{(k - \ell)!}.$$

For general sums of random variables, it would be hard to write this in a more illuminating manner; however, we're lucky for sums of Poisson random variables *if we happen to think of the following sequence of simplifications!*

1. First, note that we have a factor of $1/\ell!(k - \ell)!$. This is almost $\binom{k}{\ell}$, which is $k!/\ell!(k - \ell)!$. We do one of the most useful tricks in mathematics, we **multiply cleverly by 1** (see §A.12 for more examples), where we write 1 as $k!/k!$. Thus this factor becomes $\binom{k}{\ell}/k!$. As our sum is over ℓ, we may pull the $1/k!$ outside the ℓ-sum.

2. The e^{-5} and e^{-7} inside the sum don't depend on ℓ, so we may pull them out, giving us an e^{-12}.

3. We now have $\frac{e^{-12}}{k!} \sum_{\ell=0}^{k} \binom{k}{\ell} 5^\ell 7^{k-\ell}$. Recalling the Binomial Theorem (Theorem A.2.7), we see the ℓ-sum is just $(5 + 7)^k$, or just 12^k.

Putting all the pieces together, we find

$$\text{Prob}(X_1 + X_2 = k) = \frac{12^k e^{-12}}{k!};$$

note this is the probability density for a Poisson random variable with parameter 12 (and $12 = 5 + 7$). There's nothing special about 5 and 7 in the argument above. Working more generally, we see the sum of two Poisson random variables with parameters λ_1 and λ_2 is a Poisson random variable with parameter $\lambda_1 + \lambda_2$.

This argument can be generalized. Using induction (or cleverly grouped parentheses), we find

---

Sums of Poisson random variables: The sum of n independent Poisson random variables with parameters $\lambda_1, \ldots, \lambda_n$ is a Poisson random variable with parameter $\lambda_1 + \cdots + \lambda_n$.

We were fortunate in this case in that we found a "natural" way to manipulate the algebra so that we could recognize the answer. What would happen if we considered other sums of random variables? We want a procedure that will work in general, which will *not* require us to see these clever algebra tricks.

Fortunately, there is such an approach. It's the theory of generating functions. We'll first describe what generating functions are (there are several variants; depending on what you are studying, some versions are more useful than others), and then show some applications.

19.2 Definition

We now define the generating function of a sequence. Though the most common applications are when the terms in the sequence are probabilities of different events or moments of distributions, a generating function can be defined for any sequence. In this section we'll define generating functions and give an example of their utility. Later on we'll apply what we learn to probability by either (1) taking the a_n's below to be the probability that a discrete random variable taking only non-negative integer values is n, or (2) taking the a_n's to be the moments of a random variable.

Definition 19.2.1 (Generating Function): *Given a sequence* $\{a_n\}_{n=0}^{\infty}$, *we define its generating function by*

$$G_a(s) = \sum_{n=0}^{\infty} a_n s^n$$

for all s where the sum converges.

The standard convention is to use the letter s for the variable; however, it's just a dummy variable and we could use any letter: s, x, or even a ☺. Just looking at this definition, there's no reason to believe that we've made any progress in studying anything. We want to understand a sequence $\{a_n\}_{n=0}^{\infty}$—how can it possibly help to make an infinite series out of these! The reason is that frequently there's a simple, closed form expression for $G_a(s)$, and from this simple expression we can derive many properties of the a_n's with ease!

 Let's do an example. This example is long, but it's worth the time as it highlights many of the points of generating functions, and why they're so useful. Almost everyone has seen the **Fibonacci numbers**, defined by $F_0 = 0$, $F_1 = 1$, and in general $F_n = F_{n-1} + F_{n-2}$. The first few terms are $0, 1, 1, 2, 3, 5, 8, 13, \ldots$. These numbers have many wonderful properties. They occur throughout nature, from pine cones to branchings in trees (and of course to counting rabbits). They have applications in computer science, and generalizations arise in gambling theory (we'll discuss that application in Chapter 23). In principle, there are no mysteries about the Fibonacci numbers, as we have an explicit formula that allows us to compute any term in the sequence; in practice, this formula is clearly not useful for large n. While we can compute $F_{10} = 55$, it would be tedious to find $F_{100} = 354{,}224{,}848{,}179{,}261{,}915{,}075$, while computing F_{2011} with pen and paper is cause for alarm, as there are over 400 digits!

We now show how generating functions allow us to determine *any* Fibonacci number without having to compute *any* of the previous terms! The generating function is

$$G_F(s) = \sum_{n=0}^{\infty} F_n s^n.$$

We isolate the $n = 0$ and $n = 1$ terms, and for $n \geq 2$ we use the defining recurrence $F_n = F_{n-1} + F_{n-2}$ and find

$$G_F(s) = F_0 + F_1 s + \sum_{n=2}^{\infty} (F_{n-1} + F_{n-2}) s^n$$

$$= 0 + s + \sum_{n=2}^{\infty} F_{n-1} s^n + \sum_{n=2}^{\infty} F_{n-2} s^n.$$

Notice the last two sums are almost our original generating function—they differ in having the wrong power of s, and the sums don't start at $n = 0$. We can fix this by pulling out some powers of s and then relabeling the summation; this is the hardest part of the argument, but after many examples it does eventually start to appear as a natural thing to do:

$$G_F(s) = s + s \sum_{n=2}^{\infty} F_{n-1} s^{n-1} + s^2 \sum_{n=2}^{\infty} F_{n-2} s^{n-2}$$

$$= s + s \sum_{m=1}^{\infty} F_m s^m + s^2 \sum_{m=0}^{\infty} F_m s^m.$$

As $F_0 = 0$, we may extend the first sum to also be from $m = 0$. The two sums above are just $G_F(s)$, and thus we find

$$G_F(s) = s + s G_F(s) + s^2 G_F(s).$$

We now use the quadratic formula, and find

$$G_F(s) = \frac{s}{1 - s - s^2}. \tag{19.1}$$

Great—we've determined the generating function for the Fibonacci numbers: *How does this help us?* The reason we've made so much progress, though it doesn't appear as if we have, is that the left-hand side and right-hand side of (19.1) are both functions of s. On the left-hand side, the coefficient of s^n is just F_n; thus the coefficient of s^n on the right-hand side must also be F_n. That said, it's not at all clear what the coefficient of s^n is on the right hand side. One natural idea is to try and expand using the geometric series:

$$\frac{1}{1-(s+s^2)} = \sum_{k=0}^{\infty}(s+s^2)^k = \sum_{k=0}^{\infty}\sum_{\ell=0}^{k}\binom{k}{\ell}s^\ell(s^2)^{k-\ell},$$

which gives

$$\frac{s}{1-s-s^2} = \sum_{k=0}^{\infty}\sum_{\ell=0}^{\infty}\binom{k}{\ell}s^{2k-\ell+1};$$

it's not easy to look at this and collect powers of s (but it's a nice exercise and leads to an interesting formula for the Fibonacci numbers!)

Fortunately there's a better way of looking at the right-hand side. It goes back to one of the most disliked integration methods from calculus: **partial fractions**. Not surprisingly, there are good reasons your calculus professors taught this; in addition to being useful here, partial fractions also arise in solving certain differential equations. We factor $1-s-s^2$ as $(1-As)(1-Bs) = 1-(A+B)s+ABs^2$, and then write

$$\frac{s}{1-s-s^2} = \frac{a}{1-As} + \frac{b}{1-Bs},$$

and then use the geometric series to expand each fraction. It's because we want to use the geometric series formula that we write it as $(1-As)(1-Bs)$ and not $-(s-C)(s-D)$; for the geometric series formula we want the denominator to look like 1 minus something small. Note if $|s| < \min(1/|A|, 1/|B|)$ then $|As|$ and $|Bs|$ are less than 1, and we can use the formula.

A little algebra (or the quadratic formula) gives the values for A and B. We have $A+B=1$ and $AB=-1$. Thus $B=-1/A$ and $A-1/A=1$, or $A^2-A-1=0$. Therefore $A = \frac{1\pm\sqrt{5}}{2}$. We take the positive sign, and simple algebra then gives $B = \frac{1-\sqrt{5}}{2}$ (if we had taken the minus sign, the roles of A and B would just be reversed).

We now find a and b:

$$\frac{s}{1-s-s^2} = \frac{a}{1-As} + \frac{b}{1-Bs} = \frac{a+b-(aB+bA)s}{(1-As)(1-Bs)}.$$

Note the above is an equality, and it must hold for all values of s. As the denominators are the same, the only way this can happen is if the two numerators are equal. Each numerator is a polynomial in s; there's only one way these two polynomials can be equal for every choice of s—they must be the same polynomial, which means they must have the same coefficients.

Looking at the constant term, we find $a+b=0$, so $b=-a$. We now consider the coefficients of the s term. We now need $-(aB+bA)$ to equal 1. Using our values for A and B and the fact that $b=-a$ gives

$$-a\frac{1-\sqrt{5}}{2} + a\frac{1+\sqrt{5}}{2} = 1,$$

or $a = 1/\sqrt{5}$ and thus $b = -1/\sqrt{5}$. We've proved

$$G_F(s) = \frac{s}{1 - s - s^2} = \frac{1}{\sqrt{5}} \frac{1}{1 - As} - \frac{1}{\sqrt{5}} \frac{1}{1 - Bs}.$$

We now expand with the geometric series, and see

$$G_F(s) = \frac{1}{\sqrt{5}} \sum_{n=0}^{\infty} A^n s^n - \frac{1}{\sqrt{5}} \sum_{n=0}^{\infty} B^n s^n$$

$$= \sum_{n=0}^{\infty} \left[\frac{1}{\sqrt{5}} \left(\frac{1 + \sqrt{5}}{2} \right)^n - \frac{1}{\sqrt{5}} \left(\frac{1 - \sqrt{5}}{2} \right)^n \right] s^n.$$

We've found and proved the desired formula for the nth Fibonacci number.

Binet's formula: Let $\{F_n\}_{n=0}^{\infty}$ denote the Fibonacci series, with $F_0 = 0$, $F_1 = 1$, and $F_{n+2} = F_{n+1} + F_n$. Then

$$F_n = \frac{1}{\sqrt{5}} \left(\frac{1 + \sqrt{5}}{2} \right)^n - \frac{1}{\sqrt{5}} \left(\frac{1 - \sqrt{5}}{2} \right)^n.$$

Binet's formula is spectacular. We can now jump to *any* term in the sequence without calculating all the previous terms! I've always been amazed by it. The Fibonacci numbers are integers, and this expression involves division and square-roots, yet somehow it all works out to be an integer.

After such a long argument, it's a good idea to go back and see what we've done. We started with a relation for the Fibonacci numbers. While we could use it to find any term, it would be time-consuming. We bundled the Fibonacci numbers into a generating function $G_F(s)$. The miracle is that there's a nice closed form expression for $G_F(s)$, and from that we can deduce a nice formula for the Fibonacci numbers.

It's worth emphasizing the miracle that occurred, namely that $G_F(s)$ is nice. If we were to take a random sequence of numbers for the a_n's, this would not happen. Fortunately in many problems of interest, when the a_n's are related to probabilistic items we care about, there will be a nice form for the generating function.

The rest of this section may be safely skipped; however, as miracles are rare, it's worth trying to understand why one just happened. We're trying to answer why it's worth constructing a generating function. After all, if it's just equivalent to our original sequence of data, what have we gained? Were we just really lucky with the Fibonacci numbers, or do we expect this to happen again? Their most important advantage is that generating functions help simplify the algebra we'll encounter in probability calculations. We can't stress too strongly how useful it is in life to minimize the algebra you need to do. In addition to being a frequent source for errors, the more elaborate an expression is, the harder it is to see patterns and connections. Simplifying algebra is a great aid in illuminating connections, and often leads to enormous computational savings.

We give two examples to remind you how useful it can be to simplify algebra. The first is from calculus and involves telescoping series.

Consider the following addition problem: evaluate

$$12 - 7$$
$$+ \quad 45 - 12$$
$$+ \quad 231 - 45$$
$$+ \, 7981 - 231$$
$$+ \, 9812 - 7981.$$

The "natural" way to do this is to do evaluate each line and then add; if we do this we get

$$5 + 33 + 186 + 7750 + 1831 = 9805$$

(or at least that's what I got on my calculator). A much faster way to do this is to regroup (see §A.3 for additional instances of **proofs by grouping**); we have a $+12$ and a -12, and so these terms cancel. Similarly we have a $+45$ and a -45, so these terms cancel. In the end we're left with

$$9812 - 7 \; = \; 9805,$$

a much simpler problem! One of the most important applications of telescoping series is in the proof of the Fundamental Theorem of Calculus, where they're used to show the area under the curve $y = f(x)$ from $x = a$ to b is given by $F(b) - F(a)$, where F is any anti-derivative of f.

We turn to linear algebra for our second example; if you haven't seen eigenvalues and eigenvectors don't worry, as we won't use this later in the book but merely provide it as another illustration of the utility of simplifying algebra. Consider the matrix

$$A \; = \; \begin{pmatrix} 1 & 0 \\ 1 & 1 \end{pmatrix};$$

what is A^{100}? If your probability (or linear algebra) grade depended on you getting this right, you would be in good shape. So long as you don't make any algebra errors, after a lot of brute force computations (namely 99 matrix multiplications!) you'll find

$$A^{100} \; = \; \begin{pmatrix} 218922995834555169026 & 354224848179261915075 \\ 354224848179261915075 & 573147844013817084101 \end{pmatrix}.$$

We can find this answer much faster if we diagonalize A. The eigenvalues of A are $\varphi = \frac{1+\sqrt{5}}{2}$ and $-1/\varphi$, with corresponding eigenvectors

$$\vec{v}_1 = \begin{pmatrix} -1 + \varphi \\ 1 \end{pmatrix} \quad \text{and} \quad \vec{v}_2 = \begin{pmatrix} -1 - 1/\varphi \\ 1 \end{pmatrix}$$

(remember $\vec{v}$ is an **eigenvector** of the matrix A with **eigenvalue** λ if $A\vec{v} = \lambda\vec{v}$; in other words, applying A to $\vec{v}$ doesn't change the direction—it just rescales its length). Letting $S = (\vec{v}_1\ \vec{v}_2)$ and $\Lambda = \begin{pmatrix} \varphi & 0 \\ 0 & -1/\varphi \end{pmatrix}$, we see $A = S\Lambda S^{-1}$. The key observation is that $S^{-1}S = I$, the 2×2 identity matrix. Thus

$$A^2 = (S\Lambda S^{-1})(S\Lambda S^{-1}) = S\Lambda(S^{-1}S)\Lambda S^{-1} = S\Lambda^2 S^{-1};$$

more generally,

$$A^n = S\Lambda^n S^{-1}.$$

If we only care about finding A^2, this is significantly more work; however, there's a lot of savings if n is large. Note how similar this is to the telescoping example, with all the $S^{-1}S$ terms canceling.

As you might have guessed, this is not a randomly chosen matrix! This matrix arises in another approach to solving the Fibonacci relation $F_{n+1} = F_n + F_{n-1}$ (with $F_0 = 0$, $F_1 = 1$). If we let

$$\vec{v}_0 = \begin{pmatrix} 0 \\ 1 \end{pmatrix} \quad \text{and} \quad \vec{v}_n = \begin{pmatrix} F_n \\ F_{n+1} \end{pmatrix},$$

then $\vec{v}_n = A^n \vec{v}_0$. Thus, if we know A^n, we can quickly compute any Fibonacci number without having to determine its predecessors. This gives an alternative derivation of **Binet's formula**.

19.3 Uniqueness and Convergence of Generating Functions

Depending on the sequence $\{a_n\}_{n=0}^{\infty}$, it's possible for the generating function $G_a(s)$ to exist for all s, for only some s, or sadly only $s = 0$ (as $G_s(0) = a_0$, this isn't really saying much!).

 Consider the following examples.

1. The simplest case is when $a_0 = 1$ and all other $a_n = 0$, which leads to $G_a(s) = 1$. More generally, if a_n is zero except for finitely many n then $G_a(s)$ is a polynomial.

2. If $a_n = 1$ for all n then $G_a(s) = \sum_{n=0}^{\infty} s^n = \frac{1}{1-s}$ by the geometric series formula. Of course, we need $|s| < 1$ in order to use the geometric series formula; for larger s, the series doesn't converge.

3. If $a_n = 1/n!$, then $G_a(s) = \sum_{n=0}^{\infty} s^n/n!$. This is the definition of e^s, and hence $G_a(s)$ exists for all s.

4. If $a_n = 2^n$, then $G_a(s) = \sum_{n=0}^{\infty} 2^n s^n = \sum_{n=0}^{\infty} (2s)^n$. This is a geometric series with ratio $2s$; the series converges for $|2s| < 1$ and diverges if $|2s| > 1$. Thus $G_a(s) = (1 - 2s)^{-1}$ if $|s| < 1/2$.

5. If $a_n = n!$, a little inspection shows $G_a(s)$ diverges for any $|s| > 0$. Probably the easiest way to see that this series diverges is to note that for any fixed $s \neq 0$, for all n sufficiently large we have $n!|s| > 1$; as the terms in the series don't tend to zero, the series can't converge. Using Stirling's formula (see Chapter 18) we can get a good estimate on how large n must be for $n!|s| > 1$. Stirling's formula states that $n! \sim (n/e)^n \sqrt{2\pi n}$, so $n!|s|^n > (n|s|/e)^n$, which doesn't go to zero as whenever $n > e/|s|$ we have $n!|s^n| > 1$.

If we're given a sequence $\{a_n\}_{n=0}^{\infty}$, then clearly we know its generating function (it may not be *easy* to write down a closed form expression for $G_a(s)$, but we do have a formula for it). The converse is also true: if we know a generating function $G_a(s)$ (which converges for $|s| < \delta$ for some r), then we can recover the original sequence. This is easy if we can differentiate $G_a(s)$ arbitrarily many times, as then $a_n = \frac{1}{n!} \frac{d^n G_a(s)}{ds^n}$. This result is extremely important; as we'll use it frequently later, it's worth isolating as a theorem.

Theorem 19.3.1 (Uniqueness of generating functions of sequences): *Let $\{a_n\}_{n=0}^{\infty}$ and $\{b_n\}_{n=0}^{\infty}$ be two sequences of numbers with generating functions $G_a(s)$ and $G_b(s)$ which converge for $|s| < \delta$. Then the two sequences are equal (i.e., $a_i = b_i$ for all i) if and only if $G_a(s) = G_b(s)$ for all $|s| < \delta$. We may recover the sequence from the generating function by differentiating: $a_n = \frac{1}{n!} \frac{d^n G_a(s)}{ds^n}$.*

Proof: Clearly if $a_i = b_i$ then $G_a(s) = G_b(s)$. For the other direction, if we can differentiate arbitrarily many times, we find $a_i = \frac{1}{i!} \frac{d^i G_a(s)}{ds^i}$ and $b_i = \frac{1}{i!} \frac{d^i G_b(s)}{ds^i}$; as $G_a(s) = G_b(s)$, their derivatives are equal and thus $a_i = b_i$. □

 Remark 19.3.2: *The division by $n!$ is a little annoying; later we'll see a related generating function that doesn't have this factor. If we don't want to differentiate, we can still determine the coefficients from the generating function. Clearly we can get a_0 by setting $s = 0$. We can then find a_1 by looking at $(G_a(s) - a_0)/s$ and sending s to zero in this expression; continuing in this manner we can find any a_m. Note, of course, how similar this is to differentiating!*

We end with a quick caveat to the reader: just because we've written down the generating function, it doesn't mean that it makes sense! Unfortunately it's possible that the resulting sum doesn't converge for any value of s (other than $s = 0$, of course, which trivially converges). Fortunately the generating functions that arise in probability frequently (but not always) converge, at least for some s; we'll discuss this in much greater detail later. There are many tests to determine whether or not a series converges or diverges, and we summarize four of the more popular and powerful (ratio, root, comparison, and integral) in Appendix B.3.

In the next section we show how generating functions behave nicely with convolution, and from this we'll finally get some examples of why generating functions are so useful in probability.

19.4 Convolutions I: Discrete Random Variables

Above we introduced generating functions. We gave a few examples, we talked about how to see where it converges and diverges; however, we haven't seen why they're such a powerful tool in probability. We correct that now. After defining some notation, we'll return to the problem from the motivation section, namely determining the density of the sum of two random variables. The main result is that generating functions allow us to readily determine probability densities.

First, however, we need some notation.

Definition 19.4.1 (Convolution of sequences): *If we have two sequences $\{a_m\}_{m=0}^{\infty}$ and $\{b_n\}_{n=0}^{\infty}$, we define their convolution to be the new sequence $\{c_k\}_{k=0}^{\infty}$ given by*

$$c_k = a_0 b_k + a_1 b_{k-1} + \cdots + a_{k-1} b_1 + a_k b_0 = \sum_{\ell=0}^{k} a_\ell b_{k-\ell}.$$

We frequently write this as $c = a * b$.

This definition arises from multiplying polynomials; if $f(x) = \sum_{m=0}^{\infty} a_m x^m$ and $g(x) = \sum_{n=0}^{\infty} b_n x^n$, then assuming everything converges we have

$$h(x) = f(x)g(x) = \sum_{k=0}^{\infty} c_k x^k,$$

with $c = a * b$. For example, if $f(x) = 2 + 3x - 4x^2$ and $g(x) = 5 - x + x^3$, then $f(x)g(x) = 10 + 13x - 23x^2 + 6x^3 + 3x^4 - 4x^5$. According to our definition, c_2 should equal:

$$a_0 b_2 + a_1 b_1 + a_2 b_0 = 2 \cdot 0 + 3 \cdot (-1) + (-4) \cdot 5 = -23,$$

which is exactly what we get from multiplying $f(x)$ and $g(x)$.

Lemma 19.4.2: *Let $G_a(s)$ be the generating function for $\{a_m\}_{m=0}^{\infty}$ and $G_b(s)$ the generating function for $\{b_n\}_{n=0}^{\infty}$. Then the generating function of $c = a * b$ is $G_c(s) = G_a(s)G_b(s)$.*

We can now give a nice application of how generating functions can simplify algebra: *What is $\sum_{m=0}^{n} \binom{n}{m}^2$?* If we evaluate this sum for small values of n we find that when $n = 1$ the sum is 1, when $n = 2$ it's 6, when $n = 3$ it is 20, then 70, and then 252. We might realize that the answer seems to be $\binom{2n}{n}$, but even if we notice this, how would we prove it? A natural idea is to try induction. We could write $\binom{n}{m}^2$ as $\left(\binom{n-1}{m-1} + \binom{n-1}{m} \right)^2$ (noting that we have to be careful when $m = 0$). If we expand the square we get two sums similar to the initial sum but with an $n-1$ instead of an n, which we would know by induction; the difficulty is that we have the cross term $\binom{n-1}{m-1}\binom{n-1}{m}$ to evaluate, which requires some effort to get this to look like something nice times something like $\binom{n-1}{\ell}^2$.

Using generating functions, the answer just pops out. Let $a = \{a_m\}_{m=0}^n$, where $a_m = \binom{n}{m}$. Thus

$$G_a(s) = \sum_{m=0}^{n} \binom{n}{m} s^m = \sum_{m=0}^{n} \binom{n}{m} s^m 1^{n-m} = (1+s)^n$$

(when we have binomial sums such as this, it's *very* useful to introduce factors such as 1^{n-m}, which facilitates using the Binomial Theorem, Theorem A.2.7).

Let $c = a * a$, so by Lemma 19.4.2 we have $G_c(s) = G_a(s)G_a(s) = G_a(s)^2$. At first this doesn't seem too useful, until we note that

$$c_n = \sum_{\ell=0}^{n} a_\ell a_{n-\ell} = \sum_{\ell=0}^{n} \binom{n}{\ell}\binom{n}{n-\ell} = \sum_{\ell=0}^{n} \binom{n}{\ell}^2$$

as $\binom{n}{n-\ell} = \binom{n}{\ell}$. Thus the answer to our problem is c_n. We don't know c_n, but we *do* know its generating function, *and the entire point of this exercise is to show that sometimes it's more useful to know one and deduce the other.* We have

$$\sum_{k=0}^{2n} c_k s^k = G_c(s) = G_a(s)^2 = (1+s)^n \cdot (1+s)^n = (1+s)^{2n} = \sum_{k=0}^{2n} \binom{2n}{k} s^k,$$

where the last equality is just the Binomial Theorem. Thus $c_n = \binom{2n}{n}$ as claimed.

While we've found an example where it's easier to study the problem through generating functions, some things are unsatisfying about this example. The first is we still needed to have some combinatorial expertise, noting $\binom{n}{\ell} = \binom{n}{n-\ell}$. This is minor for two reasons. First, this is one of the most important properties of binomial coefficients (the number of ways of choosing ℓ people from n people when order doesn't matter is the same as the number of ways of excluding $n-\ell$). The second is more severe: *why would one ever consider convolving our sequence a with itself to solve this problem!*

The answer to the second objection is that convolutions arise all the time in probability, and thus it's natural to study any process which is nice with respect to convolution. To see this, we define the probability generating function.

Definition 19.4.3 (Probability generating function): *Let X be a discrete random variable taking on values in the integers. Let $G_X(s)$ be the generating function to $\{a_m\}_{m=-\infty}^{\infty}$ with $a_m = \mathrm{Prob}(X = m)$. Then $G_X(s)$ is called the probability generating function. If X is only non-zero at the integers, a very useful way of computing $G_X(s)$ is to note that*

$$G_X(s) = \mathbb{E}[s^X] = \sum_{m=-\infty}^{\infty} s^m \mathrm{Prob}(X = m).$$

More generally, if the probabilities are non-zero on an at most countable set $\{x_m\}$, then

$$G_X(s) = E[s^X] = \sum_m s^{x_m} \text{Prob}(X = x_m).$$

The function $G_X(s)$ can be a bit more complicated than the other generating functions we've seen if X takes on negative values; if this is the case, we're no longer guaranteed that $G_X(0)$ makes sense! One way we can get around this problem is by restricting to s with $0 < \alpha < |s| < \beta$ for some α, β; another is to restrict ourselves to random variables that are never negative, and thus this issue can't arise! We concentrate on the latter. While this does restrict the distributions we may study a bit, so many of the common, important probability distributions (Bernoulli, geometric, Poisson, negative binomial, ...) of Chapter 12 take on non-negative integer values that we have a wealth of examples and applications.

We can now state one of the most important results for probability generating functions.

Theorem 19.4.4: Let $X_1, \ldots, X_n$ be independent *discrete random variables taking on non-negative integer values*, with corresponding probability generating functions $G_{X_1}(s), \ldots, G_{X_n}(s)$. Then

$$G_{X_1 + \cdots + X_n}(s) = G_{X_1}(s) \cdots G_{X_n}(s).$$

Proof: This is one of the cornerstone results in the subject; you should keep reading the proof until it completely sinks in. We'll do the case when $n = 2$ in full detail, and leave arbitrary n for you.

Basically, all we need to do is unwind the definitions. We have

$$\text{Prob}(X_1 + X_2 = k) = \sum_{\ell=0}^{\infty} \text{Prob}(X_1 = \ell)\text{Prob}(X_2 = k - \ell).$$

If we let $a_m = \text{Prob}(X_1 = m)$, $b_n = \text{Prob}(X_2 = n)$, and $c_k = \text{Prob}(X_1 + X_2 = k)$, we see that $c = a * b$. Thus $G_c(s) = G_a(s)G_b(s)$, or equivalently, $G_{X_1+X_2}(s) = G_{X_1}(s)G_{X_2}(s)$.

What if now $n = 3$? It's another **proof by grouping** (see §A.3): write $X_1 + X_2 + X_3$ as $(X_1 + X_2) + X_3$. Using the $n = 2$ result *twice* we get

$$G_{X_1+X_2+X_3}(s) = G_{(X_1+X_2)+X_3}(s)$$

$$= G_{X_1+X_2}(s)G_{X_3}(s) = G_{X_1}(s)G_{X_2}(s)G_{X_3}(s).$$

A similar idea works for all n. $\qquad\square$

 Whenever you see a theorem, you should remove a hypothesis and ask if it's still true. Usually the answer is a resounding *NO!* (or, if true, the proof is usually significantly harder). In the theorem above, how important is it for the random variables to be independent? As an extreme example consider what would happen if $X_2 = -X_1$. Then $X_1 + X_2$ is identically zero, but $G_{X_1+X_2}(s) \neq G_{X_1}(s)G_{-X_1}(s)$.

The above shows why generating functions play such a central role in probability.

The density of the sum of independent discrete random variables is the convolution of their probabilities!

We can begin to see why generating functions are so useful. From Theorem 19.3.1 we know the generating function is unique, and from Theorem 19.4.4 we know that the generating function of the sum of random variables is the product of the generating functions. If we happen to recognize the resulting product, we can immediately glean the density function of the sum!

 Let's return to the problem from the motivation section, §19.1. We have two independent Poisson random variables, X_1 with parameter 5 and X_2 with parameter 7, and we want to understand $X_1 + X_2$. From Definition 19.5.1, the generating function of a Poisson random variable X with parameter λ is just

$$G_X(s) = \sum_{n=0}^{\infty} \text{Prob}(X = n)s^n$$

$$= \sum_{n=0}^{\infty} \frac{\lambda^n e^{-\lambda}}{n!} s^n$$

$$= e^{-\lambda} \sum_{n=0}^{\infty} \frac{(\lambda s)^n}{n!}$$

$$= e^{-\lambda} e^{\lambda s} = e^{\lambda(s-1)},$$

where we used the exponential function's series expansion: $e^u = \sum_{n=0}^{\infty} u^n/n!$. Thus

$$G_{X_1} = e^{5(s-1)}, \quad G_{X_2} = e^{7(s-1)}.$$

From Theorem 19.4.4 we have

$$G_{X_1+X_2}(s) = G_{X_1}(s)G_{X_2}(s)$$

$$= e^{5(s-1)} \cdot e^{7(s-1)}$$

$$= e^{12(s-1)};$$

however, note that $e^{12(s-1)}$ is just the generating function of a Poisson random variable with parameter 12. As Theorem 19.3.1 tells us generating functions are unique, we can now deduce that $X_1 + X_2$ is a Poisson random variable with parameter 12.

In the above example, note how much easier it was to understand $X_1 + X_2$ by using properties of generating functions than from doing the algebra directly. We tackled the algebra in §19.1; while we solved the problem, we had to make several clever choices in the analysis. The arguments are far more straightforward when we use generating functions. We'll do more examples of this later, and even study cousins of generating functions that makes the algebra even easier, namely the moment generating functions and the characteristic functions.

19.5 Convolutions II: Continuous Random Variables

Fortunately the same arguments that analyzed the discrete case can be easily adapted to handle continuous random variables. Essentially the only difference is writing integrals rather than sums. (There's a few subtle, technical difficulties with integration, which we'll briefly mention.) While a general random variable need not be purely discrete or continuous, for most problems our random variables are one or the other. Frequently books adopt the convention that a sum could also mean an integral, or an integral could mean a sum. This allows them greater flexibility in writing as one notation can refer to either case.

Let's now adjust our notation and study the case of generating functions for continuous random variables.

Definition 19.5.1 (Probability generating function): *Let X be a continuous random variable with density f. Then*

$$G_X(s) = \int_{-\infty}^{\infty} s^x f(x) dx$$

is the probability generating function of X.

Let's compute some generating functions of continuous random variables. If we let X be an exponential with parameter λ, we have its density is

$$f(x) = \begin{cases} \frac{1}{\lambda} \exp(-x/\lambda) & \text{if } x \geq 0 \\ 0 & \text{otherwise.} \end{cases}$$

(Note that there's unfortunately a difference in opinion among authors as to what the exponential density should be; some books use this notation while others would use $\lambda \exp(-\lambda x)$; I prefer the first choice as this way an exponential random variable with parameter λ has mean λ and not mean $1/\lambda$.) The generating function is thus

$$G_X(s) = \int_0^{\infty} s^x \frac{1}{\lambda} \exp(-x/\lambda) dx = \frac{1}{\lambda} \int_0^{\infty} \exp(x \log s) \exp(-x/\lambda) dx.$$

Notice we rewrote s^x as $\exp(x \log s)$. While we can see these two expressions are the same by taking logarithms, why did we do this? Remember s is fixed and x is the integration variable. If instead of s^x we had e^x then we could just combine the two exponential factors into one factor. (Later we'll look at a close cousin of these probability generating functions, which yields easier algebra to solve.) The point of this substitution is to make the integral easier to evaluate. Since $s = e^{\log s}$,

$$s^x = \left(e^{\log s}\right)^x = e^{x \log s}$$

by the laws of exponents.

We now continue with the integration, and find

$$G_X(s) = \frac{1}{\lambda} \int_0^\infty \exp\left(-x\left(\frac{1}{\lambda} - \log s\right)\right) dx$$

$$= \frac{1}{\lambda} \frac{1}{\frac{1}{\lambda} - \log s} \int_0^\infty \exp\left(-x\left(\frac{1}{\lambda} - \log s\right)\right) \left(\frac{1}{\lambda} - \log s\right) dx.$$

We now assume $\frac{1}{\lambda} - \log s > 0$. If this holds, the integrand exponentially decays and we can evaluate the integral; if it were less than zero then the integrand would exponentially grow and the integral wouldn't exist. We get

$$G_X(s) = \frac{1}{1 - \lambda \log s} \int_0^\infty \exp(-u) du = \frac{1}{1 - \lambda \log s}.$$

It's worth discussing these algebra tricks again, as the goal is to get you to the point where you can do these on related problems. The first was rewriting s^x. As this is so important, let's discuss it once more to really drive home the point. Remember s is fixed, and we're integrating with respect to x. We have an exponential in x coming from the density of X; it's thus natural to rewrite s^x as $e^{x \log s} = \exp(x \log s)$ and combine exponentials. We then change variables and perform the integration, which brings up the second algebra item worth highlighting. It's essential to look carefully at the integration. We wrote the argument of the exponential as $-x\left(\frac{1}{\lambda} - \log s\right)$; if $\frac{1}{\lambda} > \log s$ then the integration makes sense, as we're integrating an exponential with negative argument from 0 to infinity, and that will converge. If, however, $\frac{1}{\lambda} \le \log s$, then the integral would diverge. Thus the generating function of the exponential with parameter λ is

$$G_X(s) = \begin{cases} (1 - \lambda \log s)^{-1} & \text{if } \log s < \frac{1}{\lambda} \\ \text{undefined} & \text{otherwise.} \end{cases}$$

Looking at our answer for the generating function, we see it has a marked change in behavior when $\log s = \frac{1}{\lambda}$, which supports our observation that our arguments only work for small s.

Definition 19.5.2 (Convolution of functions): *The convolution of two functions* f_1 *and* f_2, *denoted* $f_1 * f_2$, *is*

$$(f_1 * f_2)(x) = \int_{-\infty}^\infty f_1(t) f_2(x - t) dt.$$

If the f_i's *are densities then the integral converges.*

Note that this is a natural generalization of the convolution of two sequences, where $c_k = \sum a_\ell b_{k-\ell}$ becomes $(f_1 * f_2)(x) = \int f_1(t)f_2(x - t)$. In the discrete case, the two indices summed to the new index $(k = \ell + (k - \ell))$, while in the continuous case the sum of the two arguments is the new argument $(x = t + (x - t))$. Not surprisingly, we have similar results for continuous random variables as we had for discrete ones. The most important result follows.

Theorem 19.5.3 (Sums of continuous random variables): *The probability density function of the sum of independent continuous random variables is the convolution of their probability density functions. In particular, if $X_1, \ldots, X_n$ have densities $f_1, \ldots, f_n$, then the density of $X_1 + \cdots + X_n$ is $f_1 * f_2 * \cdots * f_n$.*

Proof: While the argument is almost identical, it's worth giving it again as this is such an important point. We'll just do the case of two random variables, as the same grouping argument as before allows us to go from two random variables to n.

Let X_1 and X_2 be continuous random variables with densities f_1 and f_2, and set $X = X_1 + X_2$. Consider the convolution of their densities:

$$(f_1 * f_2)(x) = \int_{-\infty}^{\infty} f_1(t)f_2(x - t)dt.$$

Note that if we want $X_1 + X_2 = x$, then $X_1 = t$ for some t and X_2 is then forced to be $x - t$. Thus this integral gives the probability density for $X_1 + X_2$, which we denote by f. In other words,

$$f(x) = (f_1 * f_2)(x) = \int_{-\infty}^{\infty} f_1(t)f_2(x - t)dt.$$

We check that f is a density. As f_1 and f_2 are densities, they're non-negative and thus the integral defining $f(x)$ is clearly non-negative. We must show that if we integrate over all x that we get 1. We have

$$\int_{x=-\infty}^{\infty} f(x)dx = \int_{x=-\infty}^{\infty} \int_{t=-\infty}^{\infty} f_1(t)f_2(x - t)dt\,dx$$

$$= \int_{t=-\infty}^{\infty} f_1(t)\left[\int_{x=-\infty}^{\infty} f_2(x - t)dx\right] dt$$

$$= \int_{t=-\infty}^{\infty} f_1(t)\left[\int_{u=-\infty}^{\infty} f_2(u)du\right] dt$$

$$= \int_{t=-\infty}^{\infty} f_1(t) \cdot 1\,dt = 1.$$

In analysis classes we're constantly told to be careful about interchanging orders of integration. This is always permissible in probability theory as our densities take on non-negative values, and thus Fubini's theorem holds (see Theorem B.2.1 of Appendix B.2 for a statement and discussion on interchanging integrations).   □

We won't prove the result for n random variables, but we do need to say a few words about the meaning of $f_1 * f_2 * f_3$. There are two ways to interpret this, but fortunately we'll see below that they're the same. We have $(f_1 * f_2 * f_3)(x)$ is one of the following:

$$(f_1 * (f_2 * f_3))(x) = \int_{t=-\infty}^{\infty} f_1(t)(f_2 * f_3)(x - t)dt$$

$$= \int_{t=-\infty}^{\infty} \int_{u=-\infty}^{\infty} f_1(t)f_2(u)f_3(x - t - u)du\,dt$$

$$((f_1 * f_2) * f_3)(x) = \int_{w=-\infty}^{\infty} (f_1 * f_2)(w)f_3(x - w)dw$$

$$= \int_{w=-\infty}^{\infty} \int_{t=-\infty}^{\infty} f_1(t)f_2(w - t)f_3(x - w)dt\,dw$$

$$= \int_{t=-\infty}^{\infty} \int_{u=-\infty}^{\infty} f_1(t)f_2(u)f_3(x - t - u)du\,dt$$

(the last equality follows from switching the order of integration and making the substitution $w = u + t$); note our two interpretations are the same. The reason there are just two possibilities is that convolution is a binary operation: it takes two functions as input and gives one function as output. We have to group in pairs, so we can either first convolve the first two functions and then convolve with the third, or first convolve the last two functions and then convolve with the first. A similar proof shows that it doesn't matter how we place the parentheses to have a sequence of binary convolutions with n functions.

What this means is that **convolution is associative**. You may have heard this word in a few math classes. Maybe a professor said multiplication is associative, or addition (either of real number, complex numbers, or matrices). This might be the first time you see *why* people care about associativity. It tells us that convolution is well-defined; we can afford to be a bit careless or lazy and not write down the parentheses as it doesn't matter!

It's worth isolating the following result, as convolution has another nice property.

Theorem 19.5.4 (Commutativity of convolution): *The convolution of two sequences or functions is* **commutative***; in other words, $a * b = b * a$ or $f_1 * f_2 = f_2 * f_1$.*

Any result as important as this is worthy of several proofs. We first give a standard algebra proof, and then discuss some other ways to prove it.

Proof: The proof follows immediately from simple algebra. We'll do the case of continuous functions and leave the case of sequences as an exercise. We have

$$(f * g)(x) = \int_{t=-\infty}^{\infty} f(t)g(x - t)dt.$$

We now change variables and let $t = x - u$, so $dt = -du$. As t goes from $-\infty$ to ∞, u goes from ∞ to $-\infty$. Thus

$$(f * g)(x) = \int_{u=\infty}^{-\infty} f(x-u)g(x-(x-u))(-du) = \int_{u=-\infty}^{\infty} g(u)f(x-u)du$$
$$= (g * f)(u),$$

completing the proof. $\square$

Our second proof only works in the special case when the two functions or sequences are probability densities. In this case, there's a probabilistic interpretation for $f * g$: it's the density of the random variable $X_1 + X_2$. Similarly, $g * f$ is the density of $X_2 + X_1$. However, addition is **commutative**, and thus $X_1 + X_2 = X_2 + X_1$, which implies the corresponding densities $f * g$ and $g * f$ are the same. I really like this proof. The first proof is the result of some algebra. We change variables and after a little work we get the result. This proof feels nicer. We can "see" the commutativity of convolution coming from the commutativity of addition (i.e., it inherits the property). See §4.4 for another example of a result that can be proved through observing commutativity; this turns out to be a very powerful technique.

Let's give one last proof! We have

$$G_{f*g}(s) = G_f(s)G_g(s) = G_g(s)G_f(s) = G_{g*f}(s);$$

it's easy to justify the algebra above, as $G_f(s)$ and $G_g(s)$ are real numbers, and multiplication is commutative. We'd *like* to say that since the generating functions are the same, the densities of the associated random variables must be the same. Unfortunately, we haven't proved the generating function is unique in the continuous case. This requires a bit more analysis, so we'll stop this attempted proof here and leave this as a nice observation, simmering for the future.

Earlier we showed the sum of two independent Poisson random variables was Poisson—does a similar statement hold for exponential random variables? Let X_1 and X_2 be independent exponential random variables with parameters 5 and 7. From our analysis above, we know their respective generating functions are

$$G_{X_1}(s) = (1 - 5 \log s)^{-1}, \quad G_{X_2}(s) = (1 - 7 \log s)^{-1},$$

at least so long as $\log s \le \min\left(\frac{1}{5}, \frac{1}{7}\right)$. Letting $X = X_1 + X_2$, we have

$$G_X(s) = G_{X_1}(s)G_{X_2}(s)$$
$$= (1 - 5 \log s)^{-1}(1 - 7 \log s)^{-1}.$$

As there's *no* choice of λ such that this is the generating function of an exponential with parameter λ, we sadly conclude that the sum of two exponentials need not be an exponential.

It's not hard to write down the actual density. From the definition of the convolution, it's just

$$f_{X_1+X_2}(x) = \left(f_{X_1} * f_{X_2}\right)(x)$$

$$= \int_{-\infty}^{\infty} f_{X_1}(t) f_{X_2}(x-t) dt.$$

One of the most common mistakes students make in doing such convolutions is being careless when substituting for the two densities f_{X_1} and f_{X_2}. Remember these are the densities of exponential random variables and are zero if the argument is negative. Thus, we don't just replace $f_{X_2}(x-t)$ with $\frac{1}{7}\exp(-(x-t)/7)$; we can only do this replacement if the argument $x-t$ is non-negative.

Clearly the integral is zero unless $x \geq 0$: we have the sum of two non-negative random variables, and thus the sum must be non-negative too. As f_{X_1} is the density of an exponential with parameter 5, if $t < 0$ then the integrand vanishes. Similarly, if $t > x$ the second density vanishes, and thus we need only study $0 \leq t \leq x$:

$$f_{X_1+X_2}(x) = \int_0^x \frac{1}{5}\exp(-t/5)\frac{1}{7}\exp(-(x-t)/7)dt$$

$$= \frac{1}{35}\int_0^x \exp\left(-\frac{t}{5} - \frac{x}{7} + \frac{t}{7}\right)dt$$

$$= \frac{\exp(-x/7)}{35}\int_0^x \exp\left(-2t/35\right)dt$$

$$= \frac{\exp(-x/7)}{2}[1 - \exp(-2x/35)]$$

$$= \frac{\exp(-x/7) - \exp(-x/5)}{2}.$$

We have thus found the density for the sum, and this density is clearly not that of an exponential random variable.

19.6 Definition and Properties of Moment Generating Functions

In Remark 19.3.2 we commented that we can recover our sequence from the generating function through differentiation. In particular, if $a = \{a_m\}_{m=0}^{\infty}$ and $G_a(s) = a_m s^m$, then $a_m = \frac{1}{m!}\frac{d^m G_a(s)}{ds^m}$; however, the factor $1/m!$ is annoying. There's a related generating function that doesn't have this factor, the moment generating function. Its derivatives are not the probabilities of our random variable taking on given values, but actually it's not desirable to have that! The reason is that for continuous random variables, the probability of taking on any given value is just 0. Instead, as the name suggests, our new generating function will give us the moments of our density. Before defining it, we briefly recall the definition of moments.

Definition 19.6.1 (Moments): *Let X be a random variable with density f. Its kth **moment**, denoted μ'_k, is defined by*

$$\mu'_k := \sum_{m=0}^{\infty} x_m^k f(x_m)$$

if X is discrete, taking non-zero values only at the x_m's, and for continuous X by

$$\mu'_k := \int_{-\infty}^{\infty} x^k f(x)dx.$$

*In both cases we denote this as $\mu'_k = \mathbb{E}[X^k]$. We define the kth **centered moment**, μ_k, by $\mu_k := \mathbb{E}[(X - \mu'_1)^k]$. We frequently write μ for μ'_1 and σ^2 for μ_2.*

Whenever we deal with a discrete random variable, we let $\{x_m\}_{m=-\infty}^{\infty}$ or $\{x_m\}_{m=0}^{\infty}$ or $\{x_m\}_{m=1}^{\infty}$ denote the set of points where the probability density is non-zero. In most applications, we have $\{x_m\}_{m=0}^{\infty} = \{0, 1, 2, \dots\}$. We can now define the moment generating function.

Definition 19.6.2 (Moment generating function): *Let X be a random variable with density f. The moment generating function of X, denoted $M_X(t)$, is given by $M_X(t) = \mathbb{E}[e^{tX}]$. Explicitly, if X is discrete then*

$$M_X(t) \doteq \sum_{m=-\infty}^{\infty} e^{tx_m} f(x_m),$$

while if X is continuous then

$$M_X(t) = \int_{-\infty}^{\infty} e^{tx} f(x)dx.$$

Note $M_X(t) = G_X\left(e^t\right)$, or equivalently $G_X(s) = M_X(\log s)$.

Of course, it's not clear that $M_X(t)$ exists for any value of t. Frequently what happens is that it exists for some, but not all, t. Usually this is enough to allow us to deduce an amazing number of facts. We now collect many of the nice properties of the moment generating function, which show its usefulness in probability. We give a complete proof, as these properties are crucial in understanding the moment generating function proof of the Central Limit Theorem (which we give in Chapter 20). Many of these properties are similar to ones for generating functions, which shouldn't be too surprising as $M_X(t) = G_X(\log t)$. We'll see, however, that the moment generating function provides a better way of doing much of the algebra (i.e., this is a *really* good change of variables!).

Theorem 19.6.3: *Let X be a random variable with moments μ'_k.*

1. We have

$$M_X(t) = 1 + \mu'_1 t + \frac{\mu'_2 t^2}{2!} + \frac{\mu'_3 t^3}{3!} + \cdots \, ;$$

in particular, $\mu'_k = d^k M_X(t)/dt^k \big|_{t=0}$.

2. Let α and β be constants. Then

$$M_{\alpha X + \beta}(t) = e^{\beta t} M_X(\alpha t).$$

Useful special cases are $M_{X+\beta}(t) = e^{\beta t} M_X(t)$ and $M_{\alpha X}(t) = M_X(\alpha t)$; when proving the Central Limit Theorem, it's also useful to have $M_{(X+\beta)/\alpha}(t) = e^{\beta t/\alpha} M_X(t/\alpha)$.

3. Let X_1 and X_2 be independent random variables with moment generating functions $M_{X_1}(t)$ and $M_{X_2}(t)$ which converge for $|t| < \delta$. Then

$$M_{X_1+X_2}(t) = M_{X_1}(t) M_{X_2}(t).$$

More generally, if $X_1, \ldots, X_N$ are independent random variables with moment generating functions $M_{X_i}(t)$ which converge for $|t| < \delta$, then

$$M_{X_1+\cdots+X_N}(t) = M_{X_1}(t) M_{X_2}(t) \cdots M_{X_N}(t).$$

If the random variables all have the same moment generating function $M_X(t)$, then the right-hand side becomes $M_X(t)^N$.

Proof: For notational convenience, we only prove the claims when X is a continuous random variable with density f.

1. As the first claim is so important (this is the reason moment generating functions are studied, and the source of their name!) we provide two proofs. The two proofs are similar. Both require some results from analysis for general f; if we restrict to the common f seen in a probability class, everything is straightforward.

   For our first proof, we use the series expansion for the exponential function: $e^{tx} = \sum_{k=0}^\infty (tx)^k/k!$. We have

$$M_X(t) = \int_{-\infty}^\infty \sum_{k=0}^\infty \frac{x^k t^k}{k!} f(x) dx$$

$$= \sum_{k=0}^\infty \frac{t^k}{k!} \int_{-\infty}^\infty x^k f(x) dx;$$

the claim follows by noting the integral is just the definition of the kth moment μ'_k. Note this proof requires us to switch the order of an integral and a sum; while this can be justified if $M_X(t)$ converges for $|t| < \delta$ for some positive r, it's important to note that this *must* be justified. See Appendix B.2 for a discussion on when we can interchange orders.

For our second proof, differentiate $M_X(t)$ a total of k times. Arguing that the derivative of the integral is the integral of the derivative, and noting the only t-dependence in the integrand is the e^{tx} factor, we find

$$\frac{d^k M_X}{dt^k} = \int_{-\infty}^{\infty} \left[\frac{d^k e^{tx}}{dt^k}\right] f(x)dx = \int_{-\infty}^{\infty} x^k e^{tx} f(x)dx;$$

the claim now follows from taking $t = 0$ and recalling the definition of the moments. Note again that the proof reduces to a claim from analysis, this time on when we can interchange an integral and a derivative (see Theorem B.2.2 of Appendix B.2.1).

2. We now turn to the second claim. We have

$$M_{\alpha X + \beta}(t) = \int_{-\infty}^{\infty} e^{t(\alpha x + \beta)} f(x)dx$$

$$= e^{\beta t} \int_{-\infty}^{\infty} e^{t\alpha x} f(x)dx = e^{\beta t} M_X(\alpha t),$$

as the last integral is just the moment generating function evaluated at αt instead of t. The special cases now readily follow.

3. The third property follows from the fact that the expected value of independent random variables is the product of the expected values. If X_1 and X_2 are independent, so too is the pair e^{tX_1} and e^{tX_2} (remember t is fixed). Thus

$$M_{X_1 + X_2}(t) = \mathbb{E}[e^{t(X_1 + X_2)}]$$

$$= \mathbb{E}[e^{tX_1} e^{tX_2}]$$

$$= \mathbb{E}[e^{tX_1}]\mathbb{E}[e^{tX_2}] = M_{X_1}(t)M_{X_2}(t).$$

The case of n random variables follows similarly. $\qquad\square$

Let's do some examples where we compute moment generating functions and see how useful they can be.

Example 19.6.4: *Let X be a Poisson random variable with parameter λ and density f; this means that*

$$f(n) = \text{Prob}(X = n) = \frac{\lambda^n e^{-\lambda}}{n!}$$

for $n \geq 0$, *and* 0 *otherwise. The moment generating function is*

$$M_X(t) = \sum_{n=0}^{\infty} e^{tn} f(n)$$

$$= \sum_{n=0}^{\infty} e^{tn} \frac{\lambda^n e^{-\lambda}}{n!}$$

$$= e^{-\lambda} \sum_{n=0}^{\infty} \frac{\lambda^n e^{tn}}{n!}$$

$$= e^{-\lambda} \sum_{n=0}^{\infty} \frac{(\lambda e^t)^n}{n!}$$

$$= e^{-\lambda} e^{\lambda e^t} = e^{\lambda(e^t - 1)}.$$

From part (3) of Theorem 19.6.3, if X_1 *and* X_2 *are independent Poisson random variables with parameters* λ_1 *and* λ_2, *then*

$$M_{X_1+X_2}(t) = M_{X_1}(t) M_{X_2}(t)$$

$$= e^{\lambda_1(e^t-1)} e^{\lambda_2(e^t-1)}$$

$$= e^{(\lambda_1+\lambda_2)(e^t-1)}.$$

This is probably the first time you have ever seen an exponential of an exponential naturally arise in a problem! Note this is exactly the moment generating function of a Poisson random variable with parameter $\lambda_1 + \lambda_2$, obtained with significantly less work than the brute force approach! Does this imply that $X_1 + X_2$ is a Poisson random variable with parameter $\lambda_1 + \lambda_2$? Yes, because of the following theorem.

Theorem 19.6.5 (Uniqueness of moment generating functions for discrete random variables): *Let* X *and* Y *be discrete random variables taking on nonnegative integer values (i.e., they're non-zero only in* $\{0, 1, 2, \ldots\}$*) with moment generating functions* $M_X(t)$ *and* $M_Y(t)$, *each of which converges for* $|t| < \delta$. *Then* X *and* Y *have the same distribution if and only if there is an* $r > 0$ *such that* $M_X(t) = M_Y(t)$ *for* $|t| < r$.

In other words, discrete random variables are uniquely determined by their moment generating functions (if they converge).

Proof: One direction is trivial; namely, if X and Y have the same distribution then clearly $M_X(t) = M_Y(t)$. What about the other direction?

From Theorem 19.3.1, we know that two sequences $\{a_m\}_{m=0}^{\infty}$ and $\{b_n\}_{n=0}^{\infty}$ are equal if and only if their generating functions are equal. Let $a_m = \text{Prob}(X = m)$ and

$b_n = \text{Prob}(Y = n)$. The generating functions are (see Definition 19.2.1)

$$G_a(s) \;=\; \mathbb{E}[s^X] \;=\; \sum_{m=0}^{\infty} s^m \text{Prob}(X = m)$$

$$G_b(s) \;=\; \mathbb{E}[s^Y] \;=\; \sum_{n=0}^{\infty} s^n \text{Prob}(Y = n);$$

however, the generating functions are trivially related to the moment generating functions through

$$M_X(t) \;=\; \mathbb{E}[e^{tX}], \quad M_Y(t) \;=\; \mathbb{E}[e^{tY}].$$

If we let $s = e^t$, we find $G_a(e^t) = M_X(t)$ and $G_b(e^t) = M_Y(t)$; as $M_X(t) = M_Y(t)$, $G_a(e^t) = G_b(e^t)$. We now know that the generating functions are equal, and hence by Theorem 19.3.1 the corresponding sequences are equal. But this means $\text{Prob}(X = i) = \text{Prob}(Y = i)$ for all i, and so the two densities are the same. $\qquad\square$

There's a lot to remark about in the theorem above. It's *very* useful; it says that the moment generating function of a discrete random variable which is non-zero only at the non-negative integers *uniquely* determines the distribution! While there are *a lot* of hypotheses in this statement, these are fairly mild ones. Most of the discrete distributions we study and use are supported on the non-negative integers, so this isn't that restrictive an assumption. Arguing as in Remark 19.3.2, however, we can remove this hypothesis. Imagine first that the random variables only take on non-negative values, so we have

$$G_a(s) \;=\; \mathbb{E}[s^X] \;=\; \sum_{m=0}^{\infty} a_m s^{x_m}$$

$$G_b(s) \;=\; \mathbb{E}[s^Y] \;=\; \sum_{n=0}^{\infty} b_n s^{y_n}.$$

Without loss of generality, assume $x_0 \le y_0$. Let's explore the consequences of $G_a(s)$ equaling $G_b(s)$. As $G_a(s)/s^{x_0} = G_b(s)/s^{x_0}$ for all s, sending $s \to 0$ gives $a_0 = b_0 \lim_{s \to 0} s^{y_0 - x_0}$; as each $a_m \ne 0$, the only way this can hold is if $y_0 = x_0$ and $a_0 = b_0$. We continue in this manner (specifically, we play this game again, except now our two functions are $G_a(s) - a_0 s^{x_0}$ and $G_b(s) - a_0 s^{x_0}$).

We now return to Example 19.6.4. Using moment generating functions, we saw the sum of two Poisson random variables with parameters λ_1 and λ_2 had its moment generating function equal to $e^{(\lambda_1 + \lambda_2)(e^t - 1)}$. As the moment generating function of a Poisson random variable with parameter λ is just $e^{\lambda(e^t - 1)}$, by Theorem 19.6.5 we can now conclude that the sum of two Poisson random variables with parameters λ_1 and λ_2 is a Poisson random variable, with parameter equal to the $\lambda_1 + \lambda_2$.

We now consider a continuous example.

Let X be an exponentially distributed random variable with parameter λ, so its density function is $f(x) = \lambda^{-1} e^{-x/\lambda}$ for $x \ge 0$ and 0 otherwise. We can calculate its moment generating function:

$$M_X(t) = \int_0^\infty e^{tx} \cdot \frac{e^{-x/\lambda}}{\lambda} dx$$

$$= \frac{1}{\lambda} \int_0^\infty e^{-(\lambda^{-1}-t)x} dx.$$

We change variables by setting $u = (\lambda^{-1} - t)x$, so $dx = du/(\lambda^{-1} - t)$. As long as $\lambda^{-1} > t$ (in other words, so long as $t < 1/\lambda$) the exponential has a negative argument, and thus converges. We find

$$M_X(t) = \frac{1}{\lambda} \int_0^\infty e^{-u} \frac{du}{\lambda^{-1} - t} = \frac{1}{1 - \lambda t} \int_0^\infty e^{-u} du = (1 - \lambda t)^{-1}.$$

In our analysis we needed $t < 1/\lambda$; note that for such t, the resulting expression for $M_X(t)$ makes sense. (While $(1 - \lambda t)^{-1}$ makes sense for all $t \neq 1/\lambda$, clearly something is happening when t goes from below $1/\lambda$ to above.)

If X_i ($i \in \{1, 2\}$) are independent exponentially distributed random variables with parameters λ_i, from the example above and part (3) of Theorem 19.6.3 we find $M_{X_1+X_2}(t) = (1 - \lambda_1 t)^{-1}(1 - \lambda_2 t)^{-1}$. What does this imply about the distribution of $X_1 + X_2$? Is it anything nice? What if we restrict to the special case $\lambda_1 = \lambda_2 = \lambda$? Can we say anything here?

The following dream theorem would make life easy: *A probability distribution is uniquely determined by its moments.* This would be the natural analogue of Theorem 19.6.5 for continuous random variables. Is it true? **Sadly, this isn't always the case.**

> There exist distinct probability distributions which have the same moments. In other words, knowing all the moments doesn't always uniquely determine the probability distribution.

Example 19.6.6: *The standard examples given are the following two densities, defined for $x \geq 0$ by*

$$f_1(x) = \frac{1}{\sqrt{2\pi x^2}} e^{-(\log^2 x)/2}$$

$$f_2(x) = f_1(x)[1 + \sin(2\pi \log x)]. \tag{19.2}$$

It's a nice calculation to show that these two densities have the same moments; they're clearly different (see Figure 19.1).

What went wrong? It should seem absurd that two probability distributions could have the same moments without being the same. This example above isn't a mere annoying curiosity to be forgotten, but rather a warning as to how difficult and technical the subject really is. This example isn't an isolated problem, but rather indicative as to how strange and non-intuitive real valued functions can be. The king of all examples is

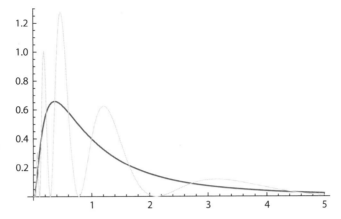

Figure 19.1. Plot of $f_1(x)$ and $f_2(x)$ from (19.2).

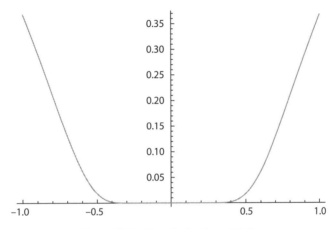

Figure 19.2. Plot of $g(x)$ from (19.3).

the function

$$g(x) = \begin{cases} \exp(-1/x^2) & \text{if } x \neq 0 \\ 0 & \text{otherwise.} \end{cases} \tag{19.3}$$

Looking at the plot (see Figure 19.2), we see the function is unbelievably flat near the origin. Using L'Hôpital's rule, one can show that *all* of the derivatives of g at the origin vanish. This has tremendous implications. Recall the Taylor series of a function h about the origin is just

$$h(x) = h(0) + h'(0) + \frac{h''(0)}{2!}x^2 + \frac{h'''(0)}{3!}x^3 + \cdots.$$

As all the derivatives of h vanish at the origin, the Taylor series is just the function which is identically zero; however, we know g is clearly not identically zero. We now come to the main point of this example. Even if you don't care about the function g, you surely must care about *some* other function. Perhaps it's the cosine function. While cosine has a nice Taylor series that converges for all x, unfortunately so too does $\cos x + g(x)$; in fact, as the derivatives of these two functions take on the same values at the origin, these two functions have the same Taylor series expansion about the origin.

Okay—the preceding argument tells us why we should care, but it's still a bit unsatisfying. Why are generating functions unique for discrete random variables but not continuous random variables. Haven't we been told, time and time again, that there's essentially no difference between these two cases but notation, where we write sums for one and integrals for another? While for many properties and problems this is true, there are some significant differences between the two types of random variables. We've seen one example already, in that discrete random variables can assign different probabilities to $[a, b]$ and $(a, b]$ while continuous random variables give the same probability for each. For discrete random variables living on the integers, the generating function is just weighted sums of all the probabilities. This is why it's uniquely determined. For continuous random variables the situation is quite different.

In the online supplemental chapter, "Complex Analysis and the Central Limit Theorem," we explore what goes wrong with the functions defined in Equation (19.2). After seeing what the problem is, we discuss what additional properties we need to assume to prevent such an occurrence. The solution involves results from complex analysis, which will tell us when a moment generating function (of a continuous random variable) uniquely determines a probability distribution.

19.7 Applications of Moment Generating Functions

We've only scratched the surface of the consequences of part (1) of Theorem 19.6.3. It's a truly amazing statement, justifying completely why we call $M_X(t) = \mathbb{E}[e^{tX}]$ the moment generating function. *If* (and admittedly this is a big if) we can compute the moment generating function of a random variable, then we can find any moment simply by taking derivatives. This is a really good trade, as it converts having to evaluate a sum or an integral to having to do a derivative. You should remember from calculus classes that differentiation is *much* easier than integration. It's impossible to give you a combination of the standard functions which you can't differentiate—if you keep applying the rules (constant, sum, difference, product, quotient, power, chain), you will eventually get the answer. On the other hand, it's trivial to give you functions which are very hard to integrate (there are elementary answers to $\int x \ln x \, dx$ or worse $\int x \ln^{1701} x \, dx$, though they're hard to find), as well as functions where there's *no* nice closed form expression for the integral. Sadly the most common example of this is $\int \exp(-x^2) \, dx$, which means we can't find simple formulas for the cumulative distribution function of normal random variables.

 Let's consider a Poisson random variable with parameter λ. To compute the mean μ we need to evaluate

$$\sum_{n=0}^{\infty} n \cdot \frac{\lambda^n e^{-\lambda}}{n!}, \tag{19.4}$$

while the variance is

$$\sum_{n=0}^{\infty} (n - \mu)^2 \cdot \frac{\lambda^n e^{-\lambda}}{n!}. \tag{19.5}$$

While it's possible to use the Method of Differentiating Identities (from Chapter 11) to find these, it's amazingly simple to compute these by differentiating the moment generating function. In Example 19.6.4 we showed the moment generating function is just

$$M_X(t) = e^{\lambda(e^t - 1)}.$$

Let's see how easily we can compute the mean and the variance. From part (1) of Theorem 19.6.3, we have the mean is

$$\begin{aligned} \mu &= \frac{d}{dt} M_X(t) \Big|_{t=0} \\ &= \frac{d}{dt} e^{\lambda(e^t - 1)} \Big|_{t=0} \\ &= \left[e^{\lambda(e^t - 1)} \cdot \lambda e^t \right]_{t=0} = \lambda. \end{aligned}$$

That was almost pleasant; it certainly was better than the sum in (19.4).

What about the variance? To find the variance, it suffices to find the second moment, which is just the second derivative of $M_X(t)$ at $t = 0$. We already know the first derivative, so we need only differentiate that again. We find

$$\begin{aligned} \mu_2' &= \frac{d^2}{dt^2} M_X(t) \Big|_{t=0} \\ &= \frac{d}{dt} \left[e^{\lambda(e^t - 1)} \lambda e^t \right]_{t=0} \\ &= \left[e^{\lambda(e^t - 1)} \lambda e^t \cdot \lambda e^t + e^{\lambda(e^t - 1)} \lambda e^t \cdot \lambda \right]_{t=0} = \lambda + \lambda^2. \end{aligned}$$

As the variance is the second moment minus the square of the mean, we find

$$\sigma^2 = \lambda + \lambda^2 - \lambda^2 = \lambda.$$

We had to use the product rule to find this; big deal—that's much better than the sum in (19.5).

As it is so important to be able to find moments, we give one final approach. We can directly expand the moment generating function into a power series. While the resulting expression often becomes complicated for even moderate powers of t, frequently it isn't too bad to isolate the coefficients of the t and t^2 terms, which are the most important ones. For example, let's consider again $M_X(t) = \exp(\lambda(e^t - 1))$. In all the equations below, the dots indicate higher powers of t. As $\exp(u) = 1 + u + u^2/2! + \cdots$, we find

$$M_X(t) = 1 + \left(\lambda(e^t - 1)\right) + \left(\lambda(e^t - 1)\right)^2/2! + \cdots.$$

To make further progress, we expand $e^t - 1$:

$$e^t - 1 = (1 + t + t^2/2! + \cdots) - 1 = t + t^2/2! + \cdots.$$

Substituting this into our expansion for $M_X(t)$ gives

$$\begin{aligned}
M_X(t) &= 1 + \lambda(t + t^2/2! + \cdots) + \lambda^2(t + t^2/2! + \cdots)^2/2! + \cdots \\
&= 1 + \lambda t + \lambda t^2/2! + \lambda^2 t^2/2 + \cdots \\
&= 1 + \lambda t + \frac{\lambda^2 + \lambda}{2}t^2 + \cdots = 1 + \mu_1' t + \frac{\mu_2'}{2!}t^2 + \cdots;
\end{aligned}$$

notice we've recovered our evaluations of μ_1' and μ_2', the first two moments. If we wanted the third moment, all we'd have to do is keep all terms up to t^3. Unfortunately, the more terms we need, the more involved the algebra; however, if all we care about are the first two moments then we can frequently isolate these without too much work.

 There are a million examples we can do, or at least a very large number. We'll conclude with another common distribution, the binomial distribution. Recall X has a binomial distribution with parameters n (a positive integer) and p (a real number in $[0, 1]$) if

$$\text{Prob}(X = k) = \begin{cases} \binom{n}{k}p^k(1 - p)^{n-k} & \text{if } k \in \{0, 1, \ldots, n\} \\ 0 & \text{otherwise.} \end{cases}$$

This means to find the mean we need to evaluate

$$\sum_{k=0}^{n} k\binom{n}{k}p^k(1 - p)^{n-k};$$

even in the special case $p = 1/2$ this is a formidable sum, requiring us to understand the sum of $k\binom{n}{k}$. While there are other ways to compute this and the variance, the generating function approach is extremely clean and easy.

We first find $M_X(t)$:

$$\begin{aligned}
M_X(t) &= \mathbb{E}[e^{tX}] \\
&= \sum_{k=0}^{n} e^{tk}\binom{n}{k}p^k(1 - p)^{n-k}
\end{aligned}$$

$$= \sum_{k=0}^{n} \binom{n}{k} e^{tk} p^k (1-p)^{n-k}$$

$$= \sum_{k=0}^{n} \binom{n}{k} (e^t p)^k (1-p)^{n-k}$$

$$= (e^t p + 1 - p)^n,$$

where the last line follows from the always useful Binomial Theorem (Theorem A.2.7). Note the key step in the analysis is doing algebra well, namely grouping e^{tk} and p^k together (this is yet another example of the powerful **grouping technique**). After enough examples, this becomes second nature. The reason is that both terms can be viewed as the kth power of something, and it therefore makes sense to move them next to each other. We then note that we have $(A+B)^n$ with the role of A played by $e^t p$ and the role of B by $1-p$.

To find the mean, we compute the first derivative at $t=0$, so

$$\mu = \frac{d}{dt} M_X(t) \Big|_{t=0}$$

$$= \frac{d}{dt} \left(e^t p + (1-p) \right)^n \Big|_{t=0}$$

$$= n \left(e^t p + (1-p) \right)^{n-1} \cdot e^t p \Big|_{t=0}$$

$$= n(p+1-p)^n p = np.$$

Similar to our previous example, the first step in finding the variance is to find the second moment, which is $M_X''(0)$. As we know $M_X'(t)$, we need only differentiate that one more time:

$$\mu_2' = \frac{d^2}{dt^2} M_X(t) \Big|_{t=0}$$

$$= \frac{d}{dt} \left[n \left(e^t p + (1-p) \right)^{n-1} e^t p \right]_{t=0}$$

$$= \left[n(n-1) \left(e^t p + (1-p) \right)^{n-2} e^t p \cdot e^t p + n \left(e^t p + (1-p) \right)^{n-1} e^t p \right]_{t=0}$$

$$= n(n-1)(p+1-p)^{n-2} p^2 + n(p+1-p)^{n-1} p = n^2 p^2 - np^2 + np.$$

We can now get the variance from simple algebra:

$$\sigma^2 = \mu_2' - \mu^2$$

$$= n^2 p^2 - np^2 + np - (np)^2$$

$$= -np^2 + np = np(1-p).$$

Rather than taking derivatives, we show how to get the first two moments by expanding the moment generating function as a power series in t. As we only need to keep terms up to t^2, we ignore t^3 and higher terms (indicating their presence by dots). We use $e^u = 1 + u + u^2/2! + \cdots$ and $(x + y)^n = x^n + \binom{n}{1}x^{n-1}y + \binom{n}{2}x^{n-2}y^2 + \cdots$ (note $\binom{n}{1} = n$ and $\binom{n}{2} = \frac{n(n-1)}{2}$). We find

$$
\begin{aligned}
M_X(t) &= \left(e^t p + 1 - p\right) \\
&= \left((1 + t + t^2/2 + \cdots)p + 1 - p\right)^n \\
&= \left(1 + (tp + t^2 p/2 + \cdots)\right)^n \\
&= 1 + n(tp + t^2 p/2 + \cdots) + \frac{n(n-1)}{2}(tp + t^2 p/2 + \cdots) + \cdots \\
&= 1 + npt + \left(\frac{np}{2} + \frac{n(n-1)p^2}{2}\right)t^2 + \cdots \\
&= 1 + np + \frac{n^2 p^2 - np^2 + np}{2!}t^2 + \cdots = 1 + \mu_1' t + \frac{\mu_2'}{2!}t^2 + \cdots,
\end{aligned}
$$

which agrees with the values we found for μ_1' and μ_2' above. Similar to our success with the moment generating function of a Poisson random variable, we're able to find the first two moments (though you should be a bit wary of finding the third).

These examples, and many more which you should try, hopefully explain why moment generating functions are worth the effort. As their name suggest, they provide a framework to find any moment. In general the higher moments become a little unwieldy as more and more terms arise from the differentiation; however, if we're patient enough we can always find the answer this way. We don't need to wait for flashes of insight or divine inspiration—it's very reassuring to know that patience and Calc I suffice.

19.8 Exercises

Exercise 19.8.1   *We showed*

$$
\frac{1}{1 - (s + s^2)} = \sum_{k=0}^{\infty}(s + s^2)^k = \sum_{k=0}^{\infty}\sum_{\ell=0}^{k}\binom{k}{\ell}s^\ell(s^2)^{k-\ell};
$$

and also

$$
\frac{s}{1 - s - s^2} = \sum_{n=0}^{\infty}\left[\frac{1}{\sqrt{5}}\left(\frac{1 + \sqrt{5}}{2}\right)^n - \frac{1}{\sqrt{5}}\left(\frac{1 - \sqrt{5}}{2}\right)^n\right]s^n.
$$

Use these two results to deduce a relation connecting sums of binomial coefficients and the Fibonacci numbers. Can you give a combinatorial interpretation for this?

Exercise 19.8.2   *Find the probability density function for $X - Y$, where X and Y are independent random variables following the standard normal distribution.*

Exercise 19.8.3 *Find the probability density function $X - Y$, where X and Y are independent, normal random variables with $X \sim N(\mu_X, \sigma_X)$ and $Y \sim N(\mu_Y, \sigma_Y)$.*

Exercise 19.8.4 *Find the probability density function $X - Y$, where X and Y are independent, with $X \sim \text{Exp}(\lambda_X)$ and $Y \sim \text{Exp}(\lambda_Y)$.*

Exercise 19.8.5 *Find the density of the sum of two exponential random variables with the same parameter λ.*

Exercise 19.8.6 *Find the moment generating function for a geometric random variable with parameter p.*

Exercise 19.8.7 *Find the moment generating function for a chi-square random variable with parameter k.*

Exercise 19.8.8 *Find a formula for the variance in terms of the moment generating function.*

Exercise 19.8.9 *The **Lucas numbers** $\{L_n\}$ are defined by $L_0 = 2, L_1 = 1, L_n = L_{n-1} + L_{n-2}$ (note they are closely related to the Fibonacci numbers). Use generating functions to find an explicit formula for the Lucas numbers.*

Exercise 19.8.10 *There are many definitions for the **Catalan numbers** $\{C_n\}$; one of my favorites is that C_n is the number of paths on an $n \times n$ grid from the lower left $(0, 0)$ to the upper right (n, n) such that (1) all paths are either one unit to the right or one unit up, and (2) at no time is the path above the main diagonal (thus if the path goes through (i, j) we have $i \geq j$). Find a recurrence relation for the Catalan numbers, and find an explicit formula for C_n.*

Exercise 19.8.11 *Find the generating function $G_C(s)$ for the Catalan numbers (defined in the previous exercise). For what values of s does it converge? What is $\sum_{n=0}^{\infty} C_0 / 4^n$?*

Exercise 19.8.12 *Explain the relationship between generating functions and Taylor series.*

Exercise 19.8.13 *Show that the variance being linear for independent random variables is consistent with the theory of moment generating functions and convolutions.*

Exercise 19.8.14 *Show that the following two densities, defined for $x \geq 0$ by $f_1(x) = \frac{1}{\sqrt{2\pi x^2}} e^{-(\log^2 x)/2}$ and $f_2(x) = f_1(x)[1 + \sin(2\pi \log x)]$, have the same moment generating functions, but are not equivalent. What are the consequences of this result?*

Exercise 19.8.15 *Find another example of two functions that have identical moment generating functions but are not identical.*

Exercise 19.8.16 *Prove algebraically that convolution is commutative in the discrete case.*

Exercise 19.8.17 *Prove that convolution is associative in general; specifically, if we have the convolution of n probability densities then it doesn't matter how we group them. For example, $(f_1 * f_2) * (f_3 * f_4)$ equals $(f_1 * (f_2 * f_3)) * f_4$.*

Exercise 19.8.18 *Find the density of the sum of two exponential random variables with the same parameter. Can you just modify the argument in the book from the case of distinct parameters? If not, consider the case where $\lambda_2 = \lambda$ and $\lambda_1 \to \lambda$. If you've* *studied recurrence relations, this approach of considering the limit of one parameter is similar to how recurrence relations with repeated roots are solved.*

Exercise 19.8.19 *Calculate the moments of the two densities from Example 19.6.6, and observe they are equal.*

Exercise 19.8.20 *Show all the derivatives of the function $g(x)$ from Equation (19.3) are zero at zero. (Hint: L'Hôpital's rule may be useful.)*

Exercise 19.8.21 *Let's revisit our method of finding moments by expanding the moment generating functions. (a) Find the third moment of a Poisson random variable with parameter λ. (b) Find the third moment of a binomial random variable with parameters n and p.*

Exercise 19.8.22 *Find the (unique) discrete random variable X whose moment generating function is*

$$M_X(x) = 1 + \frac{x^2}{2!} + \frac{x^4}{4!} + \frac{x^6}{6!} + \cdots = \cosh x.$$

CHAPTER 20 _____

Proof of the Central Limit Theorem

*Archimedes—Give me a long enough lever and a place to stand, and
I will move the earth.*

The Central Limit Theorem (frequently abbreviated CLT) is one of the true gems of probability. The hypotheses can be very weak and are frequently met in practice. What's so amazing is the universality of the result. Briefly, the sum of "nice" independent random variables converges to a Gaussian as the number of summands grows, with the mean and variance of the Gaussian the obvious candidates in terms of the means and variances of the independent random variables.

After quickly stating the Central Limit Theorem, we'll carefully and slowly work our way there by reviewing some concepts from earlier in the course. There are two parts to the Central Limit Theorem: the quantity studied, and what that quantity tends to. People frequently overlook the quantity studied and concentrate on the limiting behavior; we want you to see why this is such a natural and important object to study. We'll thus quickly review means and variances, and talk about the right way to standardize random variables.

There are numerous proofs of the Central Limit Theorem. Often the differences in the proofs are due to the amount assumed about the underlying random variables. In this chapter we'll concentrate on sums of independent identically distributed random variables whose moment generating functions converge in some neighborhood of the origin. While at first this might seem restrictive, most of the random variables we've encountered satisfy this restriction, and thus assume it will still leave us with a very useful theorem. In Chapter 21 and the online supplementary chapter "Complex Analysis and the Central Limit Theorem" we'll give proofs under weaker conditions, though doing so requires us to use some results from Fourier Analysis and complex analysis. If you continue in probability (or much of mathematics) you'll see a lot of these two fields, which is why we take the time to introduce these subjects.

20.1 Key Ideas of the Proof

Before spending *many* pages going through the technical details of the proof of the Central Limit Theorem, we're going to take a few moments and give a high level overview of what we're going to do, why this method of proof might work, why it might not, and what we need to assume so that it has a good chance of going through. Thus our goal in this section is not to dot all the i's, but rather to give a brief introduction to the method and motivate the algebra that follows.

We assume the reader has some familiarity with the statement of the Central Limit Theorem. For us, this means that if we have "nice" independent, identically distributed random variables X_i with mean μ and standard deviation σ, then the standardized random variable $Z_n = (X_1 + \cdots + X_n - n\mu)/\sigma\sqrt{n}$ (which has mean 0 and variance 1) converges in some sense to the standard normal random variable Z. Exactly how the sum converges depends on what we assume about the random variables, and in fact this is why we put *nice* in quotes earlier; to make the arguments rigorous, or to get a certain level of convergence, we need to assume certain properties on our random variables. It shouldn't be surprising that the more we assume, the stronger results we get; this is similar in spirit to the differences between Markov's inequality and Chebyshev's inequality in Chapter 17).

We use the moment generating functions of Chapter 19; thus given X we associate

$$M_X(t) = \mathbb{E}[e^{tX}] = \int_{-\infty}^{\infty} e^{tx} f_X(x)dx = \sum_{n=0}^{\infty} \frac{\mu_n' t^n}{n!}.$$

The easiest place to evaluate $M_X(t)$ is when $t = 0$; as f_X is a density we always have $M_X(0) = 0$. Our arguments will require $M_X(t)$ to exist in a small neighborhood centered at $t = 0$; unfortunately, sometimes there is no such neighborhood! For example, if we take a Cauchy random variable the integral diverges for any $t > 0$, and thus we cannot have an interval of convergence. Fortunately for most of the standard probability distributions this is not a strong condition, and $M_X(t)$ will converge for all $|t| < \delta$ for some $\delta > 0$.

The idea of the proof is the following: the way we show that Z_n converges to the standard normal random variable Z is to show that $M_{Z_n}(t) \to M_Z(t)$ for $|t| < \delta$. While it seems plausible that the convergence of the moment generating functions imply the convergence of the corresponding densities, this is the hardest step of the proof; to do it properly requires several major theorems in complex analysis. Instead of proving those theorems we'll content ourselves with stating them later in this chapter, and giving an argument below on *why* this is a natural theorem to expect.

Briefly, we want to show that

$$\lim_{n\to\infty} \int_{-\infty}^{\infty} e^{tx} f_{Z_n}(x)dx = \int_{-\infty}^{\infty} e^{tx} f_Z(x)dx \quad \text{implies} \quad \lim_{n\to\infty} f_{Z_n} = f_Z.$$

If we ignore the fact that we have a limit going on (which is a big *if*, as the limit of an integral is *not* always the integral of the limit—see Exercise 20.11.4), what we want is

$$\int_{-\infty}^{\infty} e^{tx} f(x)dx = \int_{-\infty}^{\infty} e^{tx} g(x)dx \quad \text{implies} \quad \lim_{n\to\infty} f = g.$$

Why should something like this be true? The idea is that the only way two functions f and g can have the same integral against a large class of test functions is if they are equal. Here we have infinitely many integrals that are equal, one for each test function e^{tx} for $-\delta < t < \delta$.

Instead of going through the complex analysis of why the equality of these integrals implies the equality of the functions, we study a simpler example and wave our hands, hoping that the analogue is compelling and convincing. Imagine we know

$$\int_{-\infty}^{\infty} h(x)f(x)dx = \int_{-\infty}^{\infty} h(x)g(x)dx \tag{20.1}$$

for all h; does this imply $f = g$? *Not necessarily*; if $f = g$ everywhere but at one point, where say $f(0) = g(0) + 1$, that will have no effect on the integrals. Thus let's assume f and g are continuous (you should view this assumption as an analogue to assuming the densities in the Central Limit Theorem are "nice"). Is that enough?

Amazingly, the answer is yes and all we need is the equality in (20.1) for h that are piecewise continuous. Let's assume the integrals are equal but there is some point where f and g differ. Without loss of generality we can assume the point is $x = 0$, and we might as well adjust our functions so $f(0) = 1$ and $g(0) = -1$. Since f and g are continuous there is a small interval about 0 where $f(x) > 1/2$ and $g(x) < -1/2$; let's call that interval $[-\eta, \eta]$. We now take

$$h_\eta(x) = \begin{cases} 1/\eta & \text{if } x \in [-\eta, \eta] \\ 0 & \text{otherwise.} \end{cases}$$

Note that

$$\int_{-\infty}^{\infty} h_\eta(x)f(x)dx \geq \int_{-\eta}^{\eta} \frac{1}{\eta} \cdot \frac{1}{2}dx = 1,$$

and a similar calculation shows the integral of h_η against g is less than or equal to -1. Thus (20.1) doesn't hold for all h, contradiction! Therefore we must have $f = g$!

We've ended our digression on motivation. It turns out this is a powerful technique in higher mathematics: one way to prove two functions are equal is to show they have the same integral against a large class of test functions. The difficulty, of course, is (1) proving that there is a large class of test functions giving rise to equal integrals, and (2) proving that these equalities force the functions to be equal. There must be some beef to these calculations. How do we prove integrals are the same when we don't know what function we're integrating! Remember one of the points of the CLT is that it's painful to use the exact formulas coming from convolutions for the densities. The answer is through limits; we don't show the integrals are equal, but rather that certain limits are equal. Of course, while this makes one problem easier it opens another can of worms, as now we have to deal with how limits and integrals interact (and as the exercises at the end of this chapter show, such as Exercise 20.11.4, the interaction can be bad!). 20.11.4

20.2 Statement of the Central Limit Theorem

Of the many distributions encountered in probability, perhaps the most important is the normal distribution. One way to measure how important a quantity or concept is to a subject is to count how many different names are used to refer to it; in this case, names include the normal distribution, the Gaussian distribution, and the bell curve.

> **Definition 20.2.1 (Normal distribution):** *A random variable X is normally distributed (or has the normal distribution, or is a Gaussian random variable) with mean μ and variance σ^2 if the density of X is*
>
> $$f(x) = \frac{1}{\sqrt{2\pi\sigma^2}} \exp\left(-\frac{(x-\mu)^2}{2\sigma^2}\right).$$
>
> *We often write $X \sim N(\mu, \sigma^2)$ to denote this. If $\mu = 0$ and $\sigma^2 = 1$, we say X has the standard normal distribution.*

There are many versions of the Central Limit Theorem. The differences range from the hypotheses assumed to the type of convergence obtained; not surprisingly, the more nice properties one assumes, the stronger the convergence or the simpler the proofs. We state a theorem which, while not the most general one possible, is easy to state and has hypotheses satisfied by most of the common distributions we encounter.

> **Theorem 20.2.2 (Central: Limit Theorem (CLT)):** *Let $X_1, \ldots, X_N$ be independent, identically distributed random variables whose moment generating functions converge for $|t| < \delta$ for some $\delta > 0$ (this implies all the moments exist and are finite). Denote the mean by μ and the variance by σ^2, let*
>
> $$\overline{X}_N = \frac{X_1 + \cdots + X_N}{N}$$
>
> *and set*
>
> $$Z_N = \frac{\overline{X}_N - \mu}{\sigma/\sqrt{N}}.$$
>
> *Then as $N \to \infty$, the distribution of Z_N converges to the standard normal (see Definition 20.2.1 for a statement).*

One way to interpret the above is as follows: imagine $X_1, \ldots, X_N$ are N independent measurements of some process or phenomenon. Then $\overline{X}_N$ is the average of the observed values. As the X_i's are drawn from a common distribution with mean μ, and as expectation is linear, we have

$$\mathbb{E}[\overline{X}_N] = \mathbb{E}\left[\frac{X_1 + \cdots + X_N}{N}\right] = \frac{1}{N}\sum_{n=1}^{N} \mathbb{E}[X_n] = \frac{1}{N} \cdot N\mu = \mu.$$

Since the X_n's are independent, the variance of $\overline{X}_N$ is

$$\text{Var}(\overline{X}_N) = \text{Var}\left(\frac{X_1 + \cdots + X_N}{N}\right) = \frac{1}{N^2}\sum_{n=1}^{N} \text{Var}(X_n) = \frac{1}{N^2} \cdot N\sigma^2 = \frac{\sigma^2}{N},$$

so the standard deviation of $\overline{X}_N$ is just $\sigma/\sqrt{N}$. Note as $N \to \infty$ the standard deviation of $\overline{X}_N$ tends to zero. This leads to the following interpretation: as we take more and more measurements, the distribution of the average value is living in a tighter and tighter band about the true mean. We chose to write $\overline{X}_N$ for the average value of the X_n's to emphasize that we have a sum of N random variables.

It's *very* important to be clear on what is becoming normally distributed. It is *not* the individual X_i's; these can have any "nice" distribution. What becomes normally distributed is the *average* of the X_i's. It is amazing that the distribution of the average seems to be independent of the shape of the distribution of the summands. This should feel wrong; how could it be that the average doesn't care what shape we draw our random variables from? In reality, there's no problem as the average *does* care about the shape of the initial distribution; however, it only weakly cares. By this we mean that the shape of the underlying distributions enter in determining the *rate* of convergence to a normal (the **Berry-Esseen theorem**).

20.3 Means, Variances, and Standard Deviations

Even though we've discussed means, variances, and standard deviations before, it's a good idea to spend a few minutes and look at them very carefully, as they'll play key roles in the rest of this chapter. As we've said before, it's very easy in a math course to fall into a state where you can follow proofs and arguments line by line, but have no idea why the author is arguing thusly. The more comfortable you are with the building block concepts, the more likely you are to understand the order of steps in a proof.

Recall that the **mean** μ and **variance** σ^2 of a random variable X with density f is given by

$$\mu = \mathbb{E}[X] = \int_{-\infty}^{\infty} x f(x) dx$$

$$\sigma^2 = \mathbb{E}[(X-\mu)^2] = \int_{-\infty}^{\infty} (x-\mu)^2 f(x) dx$$

if X is a continuous random variable, and

$$\mu = \mathbb{E}[X] = \sum_{n=1}^{\infty} x_n f(x_n)$$

$$\sigma^2 = \mathbb{E}[(X-\mu)^2] = \sum_{n=1}^{\infty} (x_i - \mu)^2 f(x_n)$$

if X is discrete. The **standard deviation** is the square-root of the variance.

We often write $\text{Var}(X)$ for the variance of X. The mean measures the average value of X, and the variance how spread out it is (the larger the variance, the more spread out the density).

Consider the following two data sets:

$$S_1 = \{0, 0, 0, 0, 0, 0, 0, 0, 0, 0, 100, 100, 100, 100, 100, 100, 100, 100, 100, 100\}$$
$$S_2 = \{50, 50, 50, 50, 50, 50, 50, 50, 50, 50, 50, 50, 50, 50, 50, 50, 50, 50, 50, 50\}.$$

Both data sets have a mean of 50, but the first is clearly more spread out than the second. If we try to compute the variances of these two sets, we run into a problem, namely: *what are the probabilities* $f(x_i)$? Unless there's information to the contrary, one typically assumes that all data points are equally likely. There are two ways now to determine the probabilities. Let's look at S_1. The first way is to treat each observation as a different measurement. In that case, we have $x_1 = x_2 = \cdots = x_{10} = 0$, all with probabilities $1/20$, and $x_{11} = x_{12} = \cdots x_{20} = 100$, all with probability $1/20$. Alternatively, we could consider $x_1 = 0$ with probability $1/2$ and $x_2 = 100$ with probability $1/2$. Note, however, that while the *number* of data points is different in the two interpretations, all computed quantities will be the same. For example, let's calculate the variance using both methods. Using the first, we find the variance is

$$\sum_{n=1}^{10}(0 - 50)^2 \cdot \frac{1}{20} + \sum_{n=11}^{20}(100 - 50)^2 \cdot \frac{1}{20} = 10 \cdot 50^2 \cdot \frac{1}{20} + 10 \cdot 50^2 \cdot \frac{1}{20} = 50^2,$$

while the second method gives

$$(0 - 50)^2 \cdot \frac{1}{2} + (100 - 50)^2 \cdot \frac{1}{2} = 50^2.$$

The second set, S_2, is significantly easier to compute. All values are the same, and thus the variance is clearly zero.

As a nice test, prove that the two methods *always* give the same answer for the mean and the variance, no matter what the x_i equal.

Not surprisingly, the second data set has significantly smaller variance than the first; however, there's something a bit unsettling about using the variance to quantify how spread out a data set is. The difficulty comes when our numbers have physical meaning. For example, imagine the two data sets are recording the wait time (in seconds) for a bank teller. Thus we either have a wait of 0 seconds, of 50 seconds, or of 100 seconds. In both banks the average waiting time of customers is the same; however, in the second bank all customers have the same experience, while in the first some are presumably very happy with no wait, while others are almost surely upset at a very long wait. This can be seen by noting the variance in the second set is zero while in the first it's $50^2 = 2500$; however, it's not quite right to say this. In this situation, there are *units* attached to the variance. As time is measured in seconds, the variance is measured in seconds-squared. To be honest, I have no idea what a seconds-squared is. I can imagine a meter-squared (area), but a seconds-squared? Yet this is *precisely* the unit that arises here. To see this, note that the x_i and μ are measured in seconds, the probabilities are unitless numbers, so the variance is a sum of expressions such as $(0\text{sec} - 50\text{sec})^2$, $(50\text{sec} - 50\text{sec})^2$, and $(100\text{sec} - 50\text{sec})^2$. Thus, the variance is measured in seconds-squared.

If I want to find out how long I need to wait, I'm expecting an answer such as "say 10 minutes, plus or minus a minute or two". I'm *not* expecting anyone to respond with "say 10 minutes, with a variance of one or four minutes-squared".

Fortunately, there's a simple solution to this problem; instead of reporting the variance, it's frequently more appropriate to report the standard deviation, which is the square-root of the variance.

Returning to our earlier example, we would say that for the first data set the mean wait time was 50 seconds, with a standard deviation of 50 seconds, while in the second it was also a mean wait time of 50 seconds, but with a standard deviation of 0 seconds.

The point of the above is that the standard deviation and the mean have the same units, while the variance and the mean do not; we always want to compare apples and apples (i.e., objects with the same dimensions). In fact, this is why the notation for the variance is σ^2, highlighting the fact that the quantity we will frequently care about is σ, its square-root. Similar to writing Var(X) for the variance of X, we occasionally write StDev(X) for its standard deviation.

20.4 Standardization

In the previous section we saw that the variance of a random variable isn't the right scale to look at fluctuations, as the units were wrong. In particular, if X is measured in seconds then the variance is in the physically mysterious unit of seconds-squared; it's the standard deviation that has the same units, and thus it's the standard deviation that we use to discuss how spread out a data set is.

Finding the correct scale or units to discuss a problem is very important. For example, imagine we have two sections of calculus with identical students in each but very different professors (admittedly this is not an entirely realistic situation as no two classes are identical; however, if the classes are large then this is approximately true). Let's assume one professor writes really easy exams, and another writes very challenging ones. If we're told that Hari from the first section has a 92 average and Daneel from the second section has an 84, who is the better student? Without more information, it's very hard to judge—how does a 92 in the "easier" section compare to an 84 in the "harder"?

Let's assume we know more about the two classes. Let's say that in section 1 (the one with the easier exams) the average grade is a 97 and the standard deviation is 1, while in section 2 the average grade is a 64 and the standard deviation is 10. Once we know this, it's clear that Daneel is the superior student (remember in this pretend example we're assuming the two classes are identical in terms of ability; the only difference is that one gives easier tests than the other). Hari is actually below average (by 5 standard deviations, a sizeable number), while Daneel is significantly above average (by 2 standard deviations).

We were warned about comparing apples and oranges, and that's what happened here—we have two different scales, and an 84 on one scale doesn't mean the same as an 84 on the other. To avoid problems like this (i.e., to compare apples and apples), we frequently standardize our data to have mean zero and variance 1. This puts different data sets on the same scale. This is done as follows.

Definition 20.4.1 (Standardization of a random variable): *Let X be a random variable with mean μ and standard deviation σ, both of which are finite.*

> *The standardization, Z, is defined by*
>
> $$Z := \frac{X - \mathbb{E}[X]}{\text{StDev}(X)} = \frac{X - \mu}{\sigma}.$$
>
> *Note that*
>
> $$\mathbb{E}[Z] = 0 \quad \text{and} \quad \text{StDev}(Z) = 1.$$

The standardization process we've discussed is quite natural; it rescales any "nice" random variable to a new one having mean 0 and variance 1. The only assumption we need is that it have finite mean and standard deviation. This a mild assumption, but not all distributions satisfy it. For example, consider the Cauchy distribution

$$f(x) = \frac{1}{\pi} \frac{1}{1 + x^2}.$$

It is debatable whether or not this distribution has a mean; it clearly doesn't have a finite variance. Why is the mean of this distribution problematic? It's because we have an improper integral where the integrand is sometimes positive and sometimes negative. This means *how* we go to infinity matters. For example,

$$\lim_{A \to \infty} \int_{-A}^{A} \frac{x\,dx}{\pi(1 + x^2)} = \lim_{A \to \infty} 0 = 0,$$

while

$$\lim_{A \to \infty} \int_{-A}^{2A} \frac{x\,dx}{\pi(1 + x^2)} = \lim_{A \to \infty} \frac{1}{\pi} \int_{A}^{2A} \frac{x\,dx}{1 + x^2},$$

and the last integral is, for A enormous, essentially $\int_{A}^{2A} dx/x = \log(2A) - \log(A) = \log(2)$. Thus, *how* we tend to infinity matters!

For completeness, let's go through how we would adjust a random variable to have mean zero and variance 1. We'll do the general case in all its gory details first, and then apply it to a friendly distribution. All we're really doing below is finding the density function attached to the process in Definition 20.4.1. Let's assume X is a continuous random variable with mean μ, variance σ^2, and density f_X. Thus

$$\int_{-\infty}^{\infty} f_X(x)dx = 1, \quad \int_{-\infty}^{\infty} x f_X(x) = \mu, \quad \int_{-\infty}^{\infty} (x - \mu)^2 f_X(x)dx = \sigma^2.$$

We first find a random variable Y that has mean zero and variance σ^2. Clearly $Y = X - \mu$, but what is its density? We can find the answer by using the **method of the cumulative distribution function** (see §10.5 for a review of the method), as the probability densities are just the derivative of the cumulative distribution function.

Writing F_X and F_Y for the cumulative distribution functions of X and Y, we find

$$
\begin{aligned}
F_Y(y) &= \text{Prob}(Y \le y) \\
&= \text{Prob}(X - \mu \le y) \\
&= \text{Prob}(X \le y + \mu) \\
&= \int_{-\infty}^{y+\mu} f_X(x)dx = F_X(y+\mu) - F_X(-\infty) = F_X(y+\mu).
\end{aligned}
$$

Differentiating gives

$$
f_Y(y) = F_X'(y+\mu)\frac{d}{dy}(y+\mu) = f_X(y+\mu) \cdot 1.
$$

Thus the density is just $f_Y(y) = f_X(y+\mu)$.

Let's check this. We should have $\int_{-\infty}^{\infty} yf_Y(y)dy = 0$. Well,

$$
\begin{aligned}
\int_{-\infty}^{\infty} yf_Y(y)dy &= \int_{-\infty}^{\infty} yf_X(y+\mu)dy \\
&= \int_{-\infty}^{\infty} (x - \mu)f_X(x)dx,
\end{aligned}
$$

where we changed variables by setting $x = y + \mu$ so $y = x - \mu$ and $dy = dx$ (note the bounds of integration are unchanged). Continuing and recalling that f_X is the density of X and X has mean μ, we find

$$
\begin{aligned}
\int_{-\infty}^{\infty} yf_Y(y)dy &= \int_{-\infty}^{\infty} xf_X(x)dx - \int_{-\infty}^{\infty} \mu f_X(x)dx \\
&= \mu - \mu \int_{-\infty}^{\infty} f_X(x)dx \\
&= \mu - \mu \cdot 1 = 0,
\end{aligned}
$$

where the last integral is 1 because f_X is a probability density.

How do we get the variance to equal 1? We don't want to change the mean, as that is now correct. Thus we simply want to rescale. As $\text{Var}(aU) = a^2\text{Var}(U)$ for any random variable, we just have to take $Z = Y/\sqrt{\text{Var}(Y)} = Y/\text{StDev}(Y)$ (note of course that $\text{StDev}(Y) = \text{StDev}(X)$). We then go through the cumulative distribution argument as before, skipping a few of the steps (which we recommend you do) We find

$$
\begin{aligned}
F_Z(z) &= \text{Prob}(Z \le z) \\
&= \text{Prob}(Y \le z\text{StDev}(Y)) \\
&= F_Y(z\text{StDev}(Y)) = F_X(z\text{StDev}(X) + \mu)
\end{aligned}
$$

(as $\text{StDev}(Y) = \text{StDev}(X)$). Differentiating and using the chain rule gives

$$
f_Z(z) = f_X(z\text{StDev}(X) + \mu) \cdot \text{StDev}(X).
$$

An explicit example should help. Let X be a uniform random variable on $[1, 3]$. Note the density is 1/2 on this interval and 0 otherwise, the mean is 2 and the variance is $\int_1^2 (x - 2)^2 \cdot \frac{1}{2} dx = 1/3$ (so the standard deviation is $1/\sqrt{3}$). Thus the density of the standardized random variable $Z = (X - \mu)/\text{StDev}(X)$ should be

$$f_Z(z) = f_X(z\,\text{StDev}(X) + \mu) \cdot \text{StDev}(X) = f_X\left(\frac{z}{\sqrt{3}} + 2\right) \cdot \frac{1}{\sqrt{3}},$$

or explicitly

$$f_Z(z) = \begin{cases} \frac{1}{2\sqrt{3}} & \text{if} -\sqrt{3} \le z \le \sqrt{3} \\ 0 & \text{otherwise.} \end{cases}$$

A quick calculation shows that this integrates to 1, has mean 0 and variance 1 too.

> We cannot stress enough how important and useful it is to standardize a random variable. We'll discuss this again below, but given any random variable X, sending X to $(X - \mathbb{E}[X])/\text{StDev}(X)$ is an extremely natural and often useful thing to do.

There is another benefit to all the effort we've put in to understanding standardization. Namely, this process suggests why something like the Central Limit Theorem is true! We'll discuss this in greater detail after stating the Central Limit Theorem, but we can begin to see why there might be universal behavior for sums of random variables under standardization. The reason is that, so long as the mean and variance are finite, *we can always change our units so that our random variable has mean zero and variance one.* As the moments of a distribution tell us about its shape, adopting this viewpoint shows that all "nice" random variables can be standardized so that, when so viewed, they have similar properties. It's not until we get to the third moment (or the fourth moment if the third moment is zero) that we start to see the true "shape" of the distribution. The Central Limit Theorem says that as we add more and more independent copies of the same random variable, these higher moments (the third and beyond) have less and less of an effect on the distribution of the sum; their main role becomes controlling the rate of convergence of the sum to the normal distribution.

20.5 Needed Moment Generating Function Results

Before proving the Central Limit Theorem, we'll analyze some special cases where the proof is simpler. As our hypotheses include statements about moment generating functions (discussed in §19.6), it should come as no surprise that we'll need to know the moment generating function of the standard normal.

> **Theorem 20.5.1 (Moment generating function of normal distributions)**: *Let X be a normal random variable with mean μ and variance σ^2. Its moment generating*

function is

$$M_X(t) = e^{\mu t + \frac{\sigma^2 t^2}{2}}.$$

In particular, if Z has the standard normal distribution, its moment generating function is

$$M_Z(t) = e^{t^2/2}.$$

Sketch of proof: While we could try to directly compute $M_X(t)$ through $M_X(t) = \mathbb{E}[e^{tX}]$, clearly we would much rather compute $M_Z(t) = \mathbb{E}[e^{tZ}]$. The reason is that Z has mean 0 and variance 1, and thus the numbers are a little cleaner. We could set up the equation for $M_X(t)$ and then do some change of variables, or we could note that we can deduce $M_X(t)$ from $M_Z(t)$ through part (2) of Theorem 19.6.3. Specifically, we have

$$Z = \frac{X - \mu}{\sigma},$$

or equivalently

$$X = \sigma Z + \mu.$$

We then use $M_{\alpha Z + \beta}(t) = e^{\beta t} M_Z(\alpha t)$.

Thus we are reduced to computing $M_Z(t)$, or

$$M_Z(t) = \mathbb{E}[e^{tZ}] = \int_{-\infty}^{\infty} e^{tz} \cdot \frac{e^{-z^2/2} dz}{\sqrt{2\pi}}.$$

We solve this by **completing the square**. The argument of the exponential is

$$tz - \frac{z^2}{2} = -\frac{z^2 - 2tz}{2} = -\frac{z^2 - 2tz + t^2 - t^2}{2} = -\frac{(z-t)^2}{2} + \frac{t^2}{2};$$

notice we **added zero** to simplify the algebra; this is a powerful technique (see §A.12 for more examples). Note the second term is independent of z, the variable of integration. We find

$$M_Z(t) = \int_{-\infty}^{\infty} \frac{1}{\sqrt{2\pi}} \exp\left(-\frac{(z-t)^2}{2} + \frac{t^2}{2}\right) dz$$

$$= e^{t^2/2} \int_{-\infty}^{\infty} \frac{e^{-(z-t)^2/2} dz}{\sqrt{2\pi}}$$

$$= e^{t^2/2} \int_{-\infty}^{\infty} \frac{e^{-u^2/2} du}{\sqrt{2\pi}} = e^{t^2/2},$$

as the last integral is 1 as it's the integral of the standard normal's density from $-\infty$ to ∞. $\qquad\square$

The main idea in our proof of the Central Limit Theorem is extremely easy to describe. We know the moment generating function of the standard normal. For our standardized sum of independent random variables, we can calculate its moment generating function. It's not too hard to show that, in many cases, as the number of summands tends to infinity the resulting moment generating function converges to the moment generating function of the standard normal. All that remains is to argue that if two densities have the same moment generating functions then they're equal, or more generally if a sequence of moment generating functions converges to the moment generating function of the standard normal, then the densities must converge to the density of the standard normal. Unfortunately such a result isn't always true—just see Example 19.6.6.

Fortunately all is not lost, and we can salvage this approach. What is needed are some technical conditions to assure us that this cannot happen. In other words, the natural state should be convergence of moment generating functions imply convergence of the corresponding densities, and if we assume just a little more about our distributions then dangers such as Example 19.6.6 are gone.

Not surprisingly, removing these technicalities is beyond the scope of a first course in probability. We really need results from complex analysis or Fourier Analysis. For those interested, we'll sketch the key ideas in Chapter 21; if you want even more see the online supplementary chapter "Complex Analysis and the Central Limit Theorem." For now, we'll simply state two black-box results to be freely used without any qualms below. These results involve the cumulative distribution function, so for completeness we review that definition first.

Definition 20.5.2: *Let F_X and G_Y be the cumulative distribution functions (cdf) of the random variables X and Y with densities f and g. This means*

$$F_X(x) = \int_{-\infty}^{x} f(t)dt$$

$$G_Y(y) = \int_{-\infty}^{y} g(v)dv.$$

Complex analysis (specifically the Laplace and Fourier inversion formulas) yields the following two *very* important and useful theorems for determining when we have enough information from the moments to uniquely determine a probability density.

Theorem 20.5.3: *Assume the moment generating functions $M_X(t)$ and $M_Y(t)$ exist in a neighborhood of zero (i.e., there's some δ such that both functions exist for $|t| < \delta$). If $M_X(t) = M_Y(t)$ in this neighborhood, then $F_X(u) = F_Y(u)$ for all u. As the densities are the derivatives of the cumulative distribution functions, we have $f = g$.*

Theorem 20.5.4: *Let $\{X_i\}_{i \in I}$ be a sequence of random variables with moment generating functions $M_{X_i}(t)$. Assume there's a $\delta > 0$ such that when $|t| < \delta$ we* have $\lim_{i \to \infty} M_{X_i}(t) = M_X(t)$ *for some moment generating function* $M_X(t)$, *and all moment generating functions converge for* $|t| < \delta$. *Then there exists a unique cumulative distribution function* F *whose moments are determined from* $M_X(t)$, *and for all* x *where* $F_X(x)$ *is continuous,* $\lim_{n \to \infty} F_{X_i}(x) = F_X(x)$.

20.6 Special Case: Sums of Poisson Random Variables

As a warm-up for the general proof of the Central Limit Theorem, we'll show that the normalized sum of Poisson random variables converges to the standard normal distribution. The proof includes the key ideas of the general case and involves moment generating functions. We know from Theorem 20.5.1 that the moment generating function of the standard normal is $e^{t^2/2}$. We computed the moment generating function of a Poisson random variable X with mean λ in Example 19.6.4. We showed that it's

$$M_X(t) = 1 + \mu t + \frac{\mu_2' t^2}{2!} + \cdots = e^{\lambda(e^t - 1)}.$$

Note the mean is λ and the variance is λ. To see this, we differentiate the moment generating function and then set $t = 0$. Remember that the expansion of the moment generating function involves the non-centered moments (the first moment, the second moment, and so on). While the mean is just the first moment, the variance isn't the second moment (unless the mean happens to be zero). In general the variance is the second moment minus the square of the mean. We find

$$\mu = \left. \frac{dM_X}{dt} \right|_{t=0} = \left. \left(\lambda e^t \cdot e^{\lambda(e^t - 1)} \right) \right|_{t=0} = \lambda$$

$$\mu_2' = \left. \frac{d^2 M_X}{dt^2} \right|_{t=0}$$

$$= \left. \left(\lambda e^t \cdot e^{\lambda(e^t - 1)} + \lambda^2 e^{2t} \cdot e^{\lambda(e^t - 1)} \right) \right|_{t=0} = \lambda + \lambda^2;$$

as

$$\sigma^2 = \mathbb{E}[(X - \mu)^2] = \mathbb{E}[X^2] - \mathbb{E}[X]^2,$$

we see that

$$\sigma^2 = (\lambda + \lambda^2) - \lambda^2 = \lambda.$$

Alternatively, we could have found the mean and variance by Taylor expanding the moment generating function. We would have

$$e^{\lambda(e^t - 1)} = 1 + \lambda(e^t - 1) + \frac{(\lambda(e^t - 1))^2}{2!} + \frac{(\lambda(e^t - 1))^3}{3!} + \cdots.$$

While at first this looks very complicated, we note that a Taylor expansion of $e^t - 1$ gives $t + t^2/2! + \cdots = t(1 + t/2 + \cdots)$; in other words, $(e^t - 1)^k$ is divisible by t^k. This means

$$e^{\lambda(e^t - 1)} = 1 + \lambda t \left(1 + \frac{t}{2} + \cdots\right) + \lambda^2 t^2 \frac{(1 + t/2 + \cdots)^2}{2!}$$

$$+ \lambda^3 t^3 \frac{(1 + t/2 + \cdots)^3}{3!} + \cdots$$

$$= 1 + \lambda t + \lambda \frac{t^2}{2} + \lambda^2 \frac{t^2}{2} + \text{terms in } t^3 \text{ or higher}$$

$$= 1 + \lambda t + \frac{(\lambda + \lambda^2)t^2}{2} + \cdots.$$

Thus, by knowing the Taylor series expansion of e^x, we can find the first two moments through algebra and avoid differentation; we leave it to the reader to determine which approach they like more (or hate less!).

Theorem 20.6.1: *Let $X_1, \ldots, X_N$ be independent Poisson random variables with parameter λ. Let*

$$\overline{X}_N = \frac{X_1 + \cdots + X_N}{N}.$$

As $N \to \infty$, $\overline{X}_N$ converges to a normal distribution with mean λ and variance λ.

Proof: We expect $\overline{X}_N$ to be approximately equal to the mean of the Poisson random variable, which in this case is λ. This follows from the linearity of expected value:

$$\mathbb{E}[\overline{X}_N] = \mathbb{E}\left[\frac{X_1 + \cdots + X_N}{N}\right] = \frac{1}{N} \sum_{n=1}^{N} \mathbb{E}[X_i] = \frac{1}{N} \cdot N\lambda = \lambda.$$

We write μ for the mean (and not λ); this keeps the argument a bit more general, and the resulting calculations will look like the general case for a bit longer if we do this.

Let σ^2 denote the variance of the X_n's (Poisson distributions with parameter λ). We know $\sigma = \sqrt{\lambda}$; however, we again choose to write σ below so that these calculations will look a lot like the general case. The variance of $\overline{X}_N$ is computed similarly; since the X_n are independent we have

$$\text{Var}(\overline{X}_N) = \text{Var}\left(\frac{X_1 + \cdots + X_N}{N}\right) = \frac{1}{N^2} \sum_{n=1}^{N} \text{Var}(X_n) = \frac{1}{N^2} \cdot N\sigma^2 = \frac{\sigma^2}{N}.$$

As always, the natural quantity to study is

$$Z_N = \frac{\overline{X}_N - \mathbb{E}[\overline{X}_N]}{\text{StDev}(\overline{X}_N)} = \frac{\frac{X_1 + \cdots + X_N}{N} - \mu}{\sigma/\sqrt{N}} = \frac{(X_1 + \cdots + X_N) - N\mu}{\sigma\sqrt{N}}.$$

We now use

$$M_{\frac{X+a}{b}}(t) = e^{at/b} M_X(t/b)$$

and the fact that the moment generating function of a sum of independent variables is the product of the moment generating functions (Theorem 19.6.3) to find the moment generating function of Z_N. We have

$$
\begin{aligned}
M_{Z_N}(t) &= M_{\frac{(X_1+\cdots+X_N)-N\mu}{\sigma\sqrt{N}}}(t) \\
&= M_{\sum_{n=1}^{N} \frac{X_n-\mu}{\sigma\sqrt{N}}}(t) \\
&= \prod_{n=1}^{N} M_{\frac{X_n-\mu}{\sigma\sqrt{N}}}(t) \\
&= \prod_{n=1}^{N} e^{\frac{-\mu t}{\sigma\sqrt{N}}} M_X\left(\frac{t}{\sigma\sqrt{N}}\right) \\
&= \prod_{n=1}^{N} e^{\frac{-\mu t}{\sigma\sqrt{N}}} e^{\mu\left(e^{\frac{t}{\sigma\sqrt{N}}}-1\right)},
\end{aligned}
\tag{20.2}
$$

where in the final step we take advantage of knowing the moment generating function of $M_X(t)$. We now Taylor expand the exponential, using

$$e^u = \sum_{k=0}^{\infty} \frac{u^k}{k!} = 1 + u + \frac{u^2}{2!} + \frac{u^3}{3!} + \cdots .$$

This is one of the most important Taylor expansions in the world; as the Central Limit Theorem involves a Gaussian and Gaussians involve the exponential function, it shouldn't be surprising that this Taylor series makes an appearance.

Thus the exponential in (20.2) is

$$e^{\frac{t}{\sigma\sqrt{N}}} = 1 + \frac{t}{\sigma\sqrt{N}} + \frac{t^2}{2\sigma^2 N} + \frac{t^3}{6\sigma^3 n\sqrt{N}} + \cdots .$$

The important thing to note is that after subtracting 1, the first piece is $\frac{t}{\sigma\sqrt{N}}$, the next piece is $\frac{t^2}{2\sigma^2 N}$, and the remaining pieces are dominated by a geometric series (starting with the cubed term) with $r = \frac{t}{\sigma\sqrt{N}}$. Thus, the contribution from all the other terms is of size at most some constant times $\frac{t^3}{n\sqrt{N}}$. For large n, this will be negligible, and we write errors like this as $O\left(\frac{t^3}{N\sqrt{N}}\right)$. This is called **big-Oh notation**. The technical definition is below, and see Exercise 20.11.36 for some examples.

Big-Oh Notation: $f(x) = O(g(x))$ (read: $f(x)$ is big-Oh of $g(x)$) means that there are constants C and x_0 such that whenever $x > x_0$, $|f(x)| \le Cg(x)$. In other words, from some point onward $|f(x)|$ is dominated by a constant times $g(x)$.

Thus, remembering that $\mu = \lambda$ and $\sigma^2 = \lambda$ for these Poisson random variables, (20.2) becomes

$$M_{Z_N}(t) = \prod_{n=1}^{N} e^{\frac{-\mu t}{\sigma\sqrt{N}}} \, e^{\lambda \cdot \left(\frac{t}{\sigma\sqrt{N}} + \frac{t^2}{2\sigma^2 N} + O\left(\frac{t^3}{N\sqrt{N}} \right) \right)}$$

$$= \prod_{n=1}^{N} e^{\frac{\mu t^2}{2\sigma^2 N} + O\left(\frac{t^3}{n\sqrt{N}} \right)}$$

$$= e^{\frac{t^2}{2} + O\left(\frac{t^3}{\sqrt{N}} \right)}$$

where the last line follows from the fact that we have a product over N identical terms. Thus, for all t, as $N \to \infty$ the moment generating function of Z_N tends to $e^{t^2/2}$, which is the moment generating function of the standard normal. The proof is completed by invoking Theorem 20.5.4, one of our black-box results from complex analysis, which states that if a sequence of moment generating functions which exist for $|t| < \delta$ converges to a moment generating function of a density, then the corresponding density converges to that density. In our case, this implies convergence to the standard normal. □

We only need to Taylor expand far enough to get the main term (which has a finite limit as $N \to \infty$) and then estimate the size of the error term (which tends to zero as $N \to \infty$). If we were to Taylor expand further and do a better job keeping track of the error terms, we would be able to prove far more. In addition to showing convergence to the standard normal, we would get results on how rapidly the convergence happens. Even though this was a special case, some features are visible here that reappear when we consider the general case. *Note that the higher moments of the distribution don't seem to matter; all we used was the first and second moments of X.* The higher moments *do* matter; their effect is to control the rate of convergence to the standard normal. They are felt in the $e^{O(t^3/\sqrt{N})}$ term.

20.7 Proof of the CLT for General Sums via MGF

We deliberately kept the proof of Theorem 20.6.1 (normalized sums of independent identically distributed Poisson random variables converge to the standard normal) as general as possible as long as possible for use in proving the full version of the Central Limit Theorem. Before jumping into a difficult proof, it's a great idea to try some special cases first. This allows you to build your intuition. It's often easier to do a specific case first, and get a sense of what might be important or helpful in the general case.

Proof of Theorem 20.2.2 (the Central Limit Theorem): Looking at the proof of Theorem 20.6.1, our arguments held for *any* distribution up until the last line of (20.2), where we finally used the fact that we had independent Poisson random variables by substituting for $M_X(t/\sigma\sqrt{N})$. This time, we can't substitute a specific expansion for $M_X(t/\sigma\sqrt{N})$ as we don't know M_X. We thus have

$$M_{Z_N}(t) = \prod_{n=1}^{N} e^{\frac{-\mu t}{\sigma\sqrt{N}}} M_X\left(\frac{t}{\sigma\sqrt{N}} \right) = e^{\frac{-\mu t\sqrt{N}}{\sigma}} M_X\left(\frac{t}{\sigma\sqrt{N}} \right)^N \qquad (20.3)$$

(as the random variables are identically distributed).

There are several ways to do the algebra to finish the proof; we chose the following approach as it emphasizes one of the most important tricks in mathematics. Namely, whenever you see a product you should *seriously* consider replacing it with a sum. The reason is we have lots of experience evaluating sums. We have formulas for special sums, and using Taylor series we can expand nice functions as sums. We don't really know that many products, or expansions of functions in terms of products.

How do we convert a product to a sum? We know the logarithm of a product is the sum of the logarithms. Thus, let's **take logarithms** of (20.3), and then when we're done analyzing it we just exponentiate. We find

$$\log M_{Z_N}(t) = -\frac{\mu t \sqrt{N}}{\sigma} + N \log M_X \left(\frac{t}{\sigma \sqrt{N}} \right). \tag{20.4}$$

Note the first term in the expansion above is of size $\sqrt{N}$ for fixed t; if it isn't canceled by something from the other term, the limit won't exist. Fortunately it is canceled, and all we will care about are terms up to size $1/N$. We need to be concerned with terms this small because we multiply by N; however, terms of size $1/N^{3/2}$ or smaller won't contribute in the limit as they're only multiplied by N, and thus are still small.

We know

$$M_X(t) = 1 + \mu t + \frac{\mu'_2 t^2}{2!} + \cdots = 1 + t \left(\mu + \frac{\mu'_2 t}{2} + \cdots \right).$$

We now use the Taylor series expansion for $\log(1 + u)$, which is

$$\log(1 + u) = u - \frac{u^2}{2} + \frac{u^3}{3!} - \cdots .$$

Combining the two gives

$$\log M_X(t) = t \left(\mu + \frac{\mu'_2 t}{2} + \cdots \right) - \frac{t^2 \left(\mu + \frac{\mu'_2 t}{2} + \cdots \right)^2}{2} + \cdots$$

$$= \mu t + \frac{\mu'_2 - \mu^2}{2} t^2 + \text{terms in } t^3 \text{ or higher.}$$

Thus

$$\log M_X(t) = \mu t + \frac{\mu'_2 - \mu^2}{2} t^2 + \text{terms in } t^3 \text{ or higher.}$$

But we do not want to evaluate M_X at t, but rather at $t/\sigma \sqrt{N}$. We find

$$\log M_X \left(\frac{t}{\sigma \sqrt{N}} \right) = \frac{\mu t}{\sigma \sqrt{N}} + \frac{\mu'_2 - \mu^2}{2} \frac{t^2}{\sigma^2 N} + \text{terms in } t^3/N^{3/2} \text{ or lower in } N.$$

Henceforth we'll denote these lower order terms by $O(N^{-3/2})$, and when we multiply these by N we'll denote the new error by $O(N^{-1/2})$.

The entire point of all of this is to simplify (20.4), the expansion for $\log M_{Z_N}(t)$. Collecting our pieces, we find

$$
\begin{aligned}
\log M_{Z_N}(t) &= -\frac{\mu t \sqrt{N}}{\sigma} + N\left(\frac{\mu t}{\sigma \sqrt{N}} + \frac{\mu_2' - \mu^2}{2}\frac{t^2}{\sigma^2 N} + O(N^{-3/2})\right) \\
&= -\frac{\mu t \sqrt{N}}{\sigma} + \frac{\mu t \sqrt{N}}{\sigma} + \frac{\mu_2' - \mu^2}{2}\frac{t^2}{\sigma^2} + O(N^{-1/2}) \\
&= \frac{t^2}{2} + O(N^{-1/2}).
\end{aligned}
$$

Why is the last step true? We have $\mu_2' - \mu^2$; this equals $\mathbb{E}[X^2] - \mathbb{E}[X]^2$, which is an alternate way of defining the variance. Thus $\mu_2' - \mu^2 = \sigma^2$, and the claim follows.

So, if $\log M_{Z_N}(t)$ is like $t^2/2 + O(N^{-1/2})$, then

$$
M_{Z_N}(t) = e^{\frac{t^2}{2} + O(N^{-1/2})}.
$$

Though we took a different route, we end in the same place as in the proof of Theorem 20.6.1. We again appeal to Theorem 20.5.4, one of our black-box results from complex analysis, which states that if a sequence of moment generating functions which exist for $|t| < \delta$ converges to a moment generating function of a density, then the corresponding density converges to that density. In our case, this implies convergence to the standard normal. $\qquad\square$

In the proof above, we have to be a bit careful in discarding the t^3 and higher terms, and this is where one uses the assumption that the moment generating function converges for $|t| < \delta$ to show the moments are not growing too rapidly, and thus these terms do not contribute in the limit. Make this argument rigorous; see Exercise 20.11.7.

20.8 Using the Central Limit Theorem

Most probability books (or at least the older ones) have tables with the values of the standard normal. For example, imagine we want to compute the probability that a random variable with the standard normal distribution is within one standard deviation of 0. We can just turn to the back of the book and grab this information; however, it's extremely unlikely you'll ever find a book with the tabulation of probabilities for a normally distributed random variable with mean $\sqrt{2}$ and variance π.

Why aren't there such tables? There are two reasons today. The first, of course, is that computers are very powerful and accessible, and thus the need for printed tables like the last one alluded to is greatly lessened, as a few lines of code will give us the answer. While this might be a satisfactory answer today, why were there no such tables before computers? Perhaps our example is a bit absurd, but what about a normally distributed random variable with mean 0 and variance 4; surely that must have occurred in someone's research?

The reason we don't need such tables is that if we know the probabilities for the standard normal, we can use those to compute the probabilities for *any* normally distributed random variable. For definiteness, imagine $W \sim N(3, 4)$, which means W has a mean of 3 and a variance of 4 (or a standard deviation of 2). Imagine we want to know the probability that $W \in [2, 10]$. We normalize W (see Definition 20.4.1) by setting

$$Z = \frac{W - \mathbb{E}[W]}{\mathrm{StDev}(W)} = \frac{W - 3}{2}. \tag{20.5}$$

Thus asking that $W \in [2, 10]$ is equivalent to asking Z to be in a certain interval. Which interval? Well, $W \in [2, 10]$ is the same as $Z \in [-1/2, 7/2]$. If we have a table of probabilities for the standard normal, we can now compute this probability, and hence find the probability that $W \in [2, 10]$.

This is wonderful! We only have *one* table of probabilities for *one* normally distributed random variable, as we can deduce the probabilities for any other with simple algebra. In the days before computers, this was a *very* important observation. It meant people needed only calculate *one* table of probabilities in order to study *any* normal distribution.

This is very similar to logarithm tables. Most books only had logarithms base e (sometimes base 10 was given, or perhaps base 2). Through a similar normalization process, if we have a table of logarithms in one base we can compute logarithms in any base. This is because of the following log law (commonly called the **change of base formula**): For any $b, c, x > 0$ we have

$$\log_c x = \frac{\log_b x}{\log_b c}.$$

While this is probably the most forgotten of the log laws, it's extremely useful and deserves to be remembered. Imagine we know logarithms base b. Then using the right-hand side of the above formula, we can compute the logarithm of any x base c. Thus it suffices to compile just one table of logarithms (as base e and base 10 are often both used, it might be a kindness to assemble both tables, but just one would suffice).

20.9 The Central Limit Theorem and Monte Carlo Integration

One of the biggest lies of integral calculus is that you can integrate. Specifically, that given a function you can find nice, closed form expression for its anti-derivative. Sadly, in general this is not possible; for example, even though e^{-x^2} is one of the most important functions in probability, we have to resort to an infinite series expansion (called the **error function**, described in §14.4) for its integral. As professors we have to work very hard to cook up functions whose integrals are nice.

This section is about one of the most important applications of the Central Limit Theorem: the **Monte Carlo Method**. There are many who say that this was the greatest achievement of mathematics in the 20 twentieth century; such a strong statement clearly requires some evidence in support. Why would people say this? It allows you to simulate very complex phenomena accurately and quickly, especially determining high dimensional integrals with remarkable precision. One of its earliest uses was in trying to understand how to build an atomic bomb and dealing with extremely complicated interactions. Such applications continue to this day; for example, most of the integrals in economics cannot be done in closed form, and we must resort to simulations.

We'll first describe the method and then see why it works. Imagine we have a unit square, and some region entirely contained in it; let's call that region A. Imagine we can easily determine if a point (x, y) is in A. For example, maybe there are some nice functions ff and g such that

$$A = \{(x, y) : f(x) \leq y \leq g(x), \ 0 \leq x \leq 1\}.$$

Then

$$\text{Area}(A) = \int_{x=0}^{1} [g(x) - f(x)] \, dx;$$

unfortunately evaluating this requires computing anti-derivatives!

The idea is incredibly simple, yet powerful. Assume we can choose points uniformly at random in the unit square. Then if we choose a large number of points, say N, a good approximation for the area of A should be the fraction of points that lie in A. This is often entertainingly described as throwing darts at a target, and seeing what fraction hit (we assume all darts hit the unit square; if a dart misses we just ignore it and throw again). Thus if we throw N dart,

$$\text{Area}(A) \approx \frac{\text{number of darts that hit } A}{N}.$$

Of course if A were contained in a 2×3 rectangle we would have to adjust for this by a simple rescaling, and we find for A contained in some nice region R that

$$\text{Area}(A) \approx \frac{\text{number of darts that hit } A}{N} \times \text{Area}(R).$$

We wrote $\approx$ for approximately equals above; how good is the approximation? This is where the Central Limit Theorem kicks in; for simplicity we assume R is the unit square. The probability a point chosen uniformly at random is inside A is $\text{Area}(A)$. As we choose our N points independently, if we let $X_i = 1$ if the ith point is in A and 0 otherwise, then we have N independent Bernoulli random variables with probability $p = \text{Area}(A)$ of success. Thus $(X_1 + \cdots + X_N)/N$ converges to being normally distributed as $N \to \infty$ by the Central Limit Theorem, with mean $\text{Area}(A)$ and standard deviation $\sqrt{\text{Area}(A)(1 - \text{Area}(A)}N^{-1/2}$. This is amazing: if N is on the order of a million, our error in estimating the area is on the order of one-thousandth. We give an example in Figure 20.1.

While we already know the formula for the area of the circle, this is meant to be a friendly example to illustrate the method. So long as we can easily generate points uniformly at random and determine if they're in the desired region, we can estimate the area well. Note that we are not limited to two dimensions; and thus 300+ dimensional regions (which arise in some financial models) are accessible.

20.10 Summary

In this chapter we reached the pinnacle of a good probability proof, a good, meaty discussion of why the Central Limit Theorem is true. Mathematicians don't throw

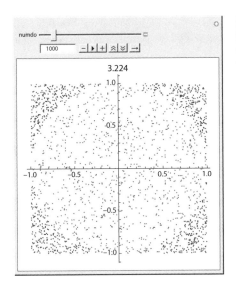
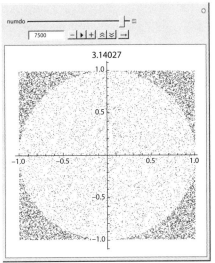

Figure 20.1. Monte Carlo simulations of the area of a unit circle: (left) 1000 points and an estimate of 3.224; (right) 7500 points and an estimate of 3.14027. The true value is $\pi \approx 3.14159$.

down adjectives willy-nilly. A theorem needs to be exceptionally important to merit the designation *fundamental*. By this point in your career, you've probably seen three fundamental theorems.

- **Fundamental Theorem of Arithmetic**. Any positive integer greater than 1 can be written uniquely as a product of primes (up to rearranging the order of the factors, of course). Why is this theorem so important? It means primes are the building blocks of the integers, just like the atoms are the building blocks of molecules in chemistry. It's also the basis for many cryptographic results, and is the starting point in building a very useful representation for the prime numbers in number theory (the Riemann zeta function).

- **Fundamental Theorem of Algebra**. If $f(x)$ is a polynomial with complex coefficients of degree n, then (counting multiplicities) f has exactly n complex roots. (Remember the **complex numbers** are all numbers of the form $z = x + iy$, with i equal to $\sqrt{-1}$ and x, y real.) This result is crucial in our understanding of functions and numbers. Note an integral polynomial need not have integral roots ($f(x) = 2x - 3$ has $3/2$ as a root) and a real polynomial need not have real roots ($f(x) = x^2 + 1$ has $i = \sqrt{-1}$ as a root), but, once we add $\sqrt{-1}$ (needed to solve $x^2 + 1 = 0$), we don't need to add anything else! We can solve any cubic, quartic, quintic, et cetera without having to add anything else! It turns out polynomial iteration (given some point x and a polynomial f, look at the sequence x, $f(x)$, $f(f(x))$, $f(f(f(x)))$, and so on) does a great job modeling many phenomena, and arises all the time in chaos theory and fractal geometry, which are used in everything from realistic special effects in the movies to powerful models for the stock market.

- **Fundamental Theorem of Calculus (FTC)**. We've already discussed this in §8.1: if F is the anti-derivative of f (so $F'(x) = f(x)$) then the area under the curve $y = f(x)$ from a to b is $F(b) - F(a)$. This is our starting point in understanding probabilities of continuous random variables. These probabilities can be interpreted as areas under a curve, and the FTC gives us a great way to attack them.

To this elite list we now add the Central Limit Theorem (CLT). Instead of calling it *fundamental* the convention is to say it's *central*, but the idea's the same. The needed conditions are fairly weak (and the ones we give can be weakened further with more work), which ensures that it's applicable in a variety of cases.

We based our discussion of the Central Limit Theorem's proof on moment generating functions. Archimedes remarked: "Give me a long enough lever and a place to stand, and I will move the earth." For us, generating functions are our lever, and they move the entire subject. They lead beautifully to the CLT. Moreover, even the method of proof is a joy to see. The generating function approach highlights why only the first two moments matter. As we've seen, any nice density can be standardized to have mean zero and variance 1. So, if only the first two moments matter, and the first two moments can always be taken to be 0 and 1, we *must* expect universal behavior!

We end with one last remark. Archimedes' quote has two suppositions. In addition to needing a lever, he asks for a place to stand. It's easy to forget about this, but it's a mistake to do so. We need to remember *why* the generating function approach works so well. To a large extent, it's because it's built on a beautiful, powerful theory, that of Fourier Analysis or complex analysis. We need some major results of these subjects to finish the proof. Without these results, we can't go from the limit of the moment generating function equaling the moment generating function of the standard normal to the density converges to that of the standard normal. As these foundations are so important (in probability and other fields), we'll explore them in some detail in Chapter 21 (if you want even more, see the online supplementary chapter "Complex Analysis and the Central Limit Theorem").

20.11 Exercises

Exercise 20.11.1   *Imagine f and g are two continuous functions and*

$$\int_{-\infty}^{\infty} h(x)f(x)dx \ = \ \int_{-\infty}^{\infty} h(x)g(x)dx$$

whenever $h(x)$ is constant on an interval of length 1. Must $f = g$? If not, can we at least deduce any common property between f and g?

Exercise 20.11.2   *In §20.1 we said that if $f(0) \neq g(0)$ we may assume without loss of generality that $f(0) = 1$ and $g(0) = -1$. Show that we may make such a transformation.*

Exercise 20.11.3   *In §20.1 we used a step function h_η which is not continuous. What if we only know the integrals in (20.1) are equal when h is continuous? Modify the construction of h_η to come up with continuous functions.*

Exercise 20.11.4   *Consider the triangle densities*

$$f_n(x) \ = \ \begin{cases} n - n|x - 2/n| & \text{for } 1/n \leq x \leq 3/n \\ 0 & \text{otherwise;} \end{cases}$$

these are densities as they are non-negative and integrate to 1 for each n. Prove

$$\lim_{n \to \infty} \int_{-\infty}^{\infty} f_n(x)dx \ \neq \ \int_{-\infty}^{\infty} \lim_{n \to \infty} f_n(x)dx,$$

even though all functions are continuous! The difficulty is similar to Fubini's theorem in multivariable calculus—what goes wrong here is that the functions f_n are not uniformly bounded with n.

Exercise 20.11.5   *Prove for any $r > 0$ that $(\log x)^r = O(x)$ and $x^r = O(e^x)$.*

Exercise 20.11.6   *Let $r > 0$. Is $x^r = O(\exp(\sqrt{\log x}))$? Prove your claim.*

Exercise 20.11.7   *In our proof of the Central Limit Theorem we look at the series expansion of $\log M_X(t/\sigma\sqrt{N})$ (for X whose moment generating function converges for $|t| < \delta$) and argue only the terms at most quadratic in t matter. Prove this rigorously. If you want, assume a slightly stronger condition on the moments, for example, $|\mu_n'| = O(n!^{1-\epsilon})$ for some $\epsilon > 0$.*

Exercise 20.11.8   *Find a formula for the probability density function of Z, where Z is the standardized form of an exponential random variable X with parameter λ.*

Exercise 20.11.9   *Find the formula for the probability density function of Z, where Z is the standardized form of a Poisson random variable X with parameter λ.*

Exercise 20.11.10   *Prove the CLT for the sum of exponential random variables without using the generalized argument.*

Exercise 20.11.11   *Imagine we have N identically distributed exponential and N identically distributed geometric random variables, all of which are independent. Show that $\frac{X_1+X_2+\cdots+X_N+Y_1+Y_2+\cdots+Y_N}{2N}$ converges to a normal distribution as N tends to infinity.*

Exercise 20.11.12   *What is the mean and variance of the normal distribution in the previous exercise. How do you standardize it?*

Exercise 20.11.13   *What if in Exercise 20.11.11 we have N exponential and M geometric random variables. How should we standardize the sum? Does the standardized sum converge to a normal as N and M tend to infinity? Does it matter how they tend to infinity?*

Exercise 20.11.14   *In Exercise 20.11.11 you were asked to prove the sum becomes normally distributed. One cannot directly appeal to the Central Limit Theorem as our random variables are not identically distributed; can you see an elegant way around that obstruction?*

Exercise 20.11.15   *In the previous exercises we saw that in some cases independent variables that are not identically distributed still approach normality when summed. Find a counterexample of independent random variables that shows this is not always the case (i.e., their sum does not approach a normal).*

Exercise 20.11.16   *Find an example to show that variables that are dependent do not necessarily approach a normal distribution when summed.*

Exercise 20.11.17   *Find an example of a discrete random variable with a probability distribution such that if independent variables identical to it are summed, it does not approach normality.*

Exercise 20.11.18 *Let $X \sim N(0, 1)$ and $Y \sim \text{Unif}(0, 1)$ be independent. Must $X + Y$ be normal?*

Exercise 20.11.19 *We define the* mod1 *function from real numbers to $[0, 1)$ by $x \bmod 1 = x - \lfloor x \rfloor$ (where $\lfloor x \rfloor$ is the* **floor function***, which gives the greatest integer at most x); thus $1701.24601 \bmod 1$ is $.24601$, while $-21.75 \bmod 1$ is $.25$. Given a random variable X define the new random variable X_{mod} to be $X \bmod 1$. If $X \sim \text{Unif}(0, 1)$ and Y is any continuous random variable, must $(X + Y) \bmod 1$ be uniform on $[0, 1)$?*

Exercise 20.11.20 *Find an example of a continuous random variable with a probability distribution such that if independent variables identical to it are summed, it does not approach normality.*

Exercise 20.11.21 *Consider a set $S = \{a_1, a_2, \ldots, a_n\}$ where the a_i's are not necessarily distinct. Let the distinct values be $B = \{b_1, b_2, \ldots, b_k\}$ with multiplicities $m_1, m_2, \ldots, m_k$ (so of the n terms in S, exactly m_1 equal b_1, exactly m_2 equal b_2, and so on). There are two ways to calculate the mean and the variance. First, we can study the set S directly, with each of the n elements having probability $1/n$; second, we can consider the set B where the probability of b_i is m_i/n. Prove these two approaches give the same mean and variance.*

Exercise 20.11.22 *We say a function $f(x)$ is* **odd** *if $f(-x) = -f(x)$. Prove that if a density f is odd then any odd moment, if it exists, is zero.*

Exercise 20.11.23 *Prove the* **change of base formula** *for logarithms: for any $b, c, x > 0$ we have $\log_c x = \log_b x / \log_b c$.*

Exercise 20.11.24 *Explain in detail why errors are often assumed to be normally distributed. Don't forget to address the criteria for the CLT to apply.*

Exercise 20.11.25 *You are conducting a survey of 1000 people in a small town. It asks them how they feel about tax rates on a scale of 1–10 (with 1 being they should be significantly lower, and 10 being they should be significantly higher). Do you expect your results to be normally distributed?*

Exercise 20.11.26 *You are conducting a survey of 1000 people in a small town. You ask each person to add up the digits in their address, their birthday, and their phone number and divide by the total number of digits. What do you expect the distribution of responses to look like?*

Exercise 20.11.27 *Particles in a fluid move according to a random walk, that is after a given period of time their movement in any direction is independent of their movement in the past. This is known as* **Brownian motion***. How can we model the one dimensional location of a particle after some time t relative to the initial location? How can we model the distance of a particle after some time t relative to the initial location in two dimensions?*

Exercise 20.11.28 *Use Mathematica's built-in animation functions to animate the repeated sum of uniform random variables as it begins to approach normality.*

Exercise 20.11.29 *Consider the generalized Cauchy random variables from Exercise 17.7.11 with $m = 4$, so the means and variances exist. Does the moment generating function exist? Compare sums of independent Cauchy random variables with* *sums of generalized Cauchy random variables with m = 4. Do either converge to being normally distributed? Why or why not? (Note: this might be a very hard question!)*

Exercise 20.11.30   *Let $Z \sim N(0, 1)$, and let I_p be an interval such that $\text{Prob}(Z \in I_\alpha) = p$. For a given $p < 1$, find the interval I_p of smallest length.*

Exercise 20.11.31   *Using notation as in the previous exercise, what range of p force the mean (which is 0) to be included in the interval I_p of smallest length?*

Exercise 20.11.32   *Let X be a random variable with $\mu = 10$ and $\sigma = 4$. A sample of size 100 is taken from this population. Find the probability that the sample mean of these 100 observations is less than 9.*

Exercise 20.11.33   *Let $A = \{0, 0, 1, 1, 1, 2, 2, 2, 2, 3, 3, 3, 3, 3\}$ with each element having probability $1/14$. Let $B = \{0, 1, 2, 3\}$ with $\text{Prob}(0) = 2/14$, $\text{Prob}(1) = 3/14$, $\text{Prob}(2) = 4/14$, and $\text{Prob}(3) = 5/14$. Show that A and B have the same mean and the same variance.*

Exercise 20.11.34   *At what moment (first moment, second moment, ...) do we start to see the shape of a distribution with a finite mean and a finite variance? Why not sooner?*

Exercise 20.11.35   *Let the number of car accidents in Albany each day follow a Poisson distribution with mean 2. Use the Central Limit Theorem to approximate the probability that there are at least 800 accidents in Albany in one year.*

Exercise 20.11.36   *Remember that $f(x) = O(g(x))$ if there is an x_0 and a $C > 0$ such that $|f(x)| \le Cg(x)$ for all $x \ge x_0$; if this holds we say f is **big-Oh** of g. Sometimes we might want to send x to zero, in which case it would be for all $|x| < x_0$. Prove the following:*

- *$x = O(x^2)$ as $x \to \infty$, and $x^2 = O(x)$ as $x \to 0$.*
- *$(\log x)^N = O(x^r)$ as $x \to \infty$ for any $N, r > 0$.*
- *$x^N = O(e^x)$ as $x \to \infty$ for any $N > 0$.*
- *$x = O(x/4)$ as $x \to \infty$ or as $x \to 0$.*
- *$\cosh(x) = O(\sinh(x))$ as $x \to \infty$, where $\cosh(x) = \frac{1}{2}(e^x + e^{-x})$ is the **hyperbolic cosine** and $\sinh(x) = \frac{1}{2}(e^x - e^{-x})$ is the **hyperbolic sine**.*

Exercise 20.11.37   *Using $i = \sqrt{-1}$ and*

$$e^x = \sum_{n=0}^{\infty} \frac{x^n}{n!} = 1 + x + \frac{x^2}{2!} + \frac{x^3}{3!} + \cdots,$$

prove $e^{ix} = \cos(x) + i \sin(x)$; note, the Taylor series for sine and cosine will be useful!

Exercise 20.11.38   **(Deriving Trig Identities)** *Building on the previous two exercises, use the relation $e^{ix} = \cos(x) + i\sin(x)$ (where $i = \sqrt{-1}$) to prove $\cos(x) = \frac{1}{2}(e^{ix} + e^{-ix})$ and $\sin(x) = \frac{1}{2i}(e^{ix} - e^{-ix})$, which allows us to interpret the hyperbolic trig functions as the normal trig functions at imaginary arguments! (So now you know what the cosine of i equals!) Using these relations and $e^{u+v} = e^u e^v$ (this must be proved; see §B.5), one can quickly prove all **trig identities**. For example*

$$\cos(x + y) + i\sin(x + y) = e^{i(x+y)}$$
$$= e^{ix}e^{iy} = (\cos(x) + i\sin(x))(\cos(y) + i\sin(y));$$

finish the proof by multiplying out the right-hand side and noting that two complex numbers are equal if and only if they have the same real and the same imaginary part.

Exercise 20.11.39   *Is $\cos(x) = O(\sin(x))$? Why or why not?*

Exercise 20.11.40   *Write a code to perform Monte Carlo simulations for regions inside the unit square bounded by two continuous functions satisfying $0 \leq f(x) \leq g(x) \leq 1$.*

Exercise 20.11.41   *Using your code from the previous exercise, estimate the area of the ellipse $(x/a)^2 + (y/b)^2 \leq 1$. Make a conjecture for the area as a function of a and b by varying these two parameters and simulating the area.*

Exercise 20.11.42   *Prove your conjecture from the previous exercise, and generalize to the ellipsoid $(x/a)^2 + (y/b)^2 + (z/c)^2 \leq 1$.*

Exercise 20.11.43   *The area of the unit circle can be written as $4 \int_0^1 \sqrt{1 - x^2} dx$. Is there a nice, closed form anti-derivative for the integral $\sqrt{1 - x^2}$? If yes, find it and compute the area of the unit circle.*

Exercise 20.11.44   *Let A be contained in the unit square. What bounds does Chebyshev's inequality give for our error estimation? What about the Central Limit Theorem?*

Exercise 20.11.45   *Let A be contained in the unit square. What is the value of its area that leads to the largest variance in our estimate of it? What leads to the smallest variance?*

CHAPTER 21

Fourier Analysis and the Central Limit Theorem

Any theorem as important as the Central Limit Theorem deserves more than one proof. Different proofs emphasize different aspects of the problem. Our first proof was based on properties of moment generating functions. It's a nice proof, and the main idea is easily explained (for "nice" distributions, if the moment generating function converges to the moment generating function of the standard normal, then the densities converge to the density of the standard normal). Unfortunately the proof uses some major results in complex analysis. We thus want to provide a proof that works under less restrictive conditions.

Sadly, the proof below won't be it. It too appeals to some black-box results from complex analysis. As this argument still requires us to assume results beyond the scope of the book, why do we bother giving this proof? There are many reasons. The first is that it introduces **integral transforms**, specifically the **Fourier transform**. Integral transforms in general and the Fourier transform in particular are ubiquitous in higher mathematics, and it never hurts to see them. The second reason is that our statement of the Central Limit Theorem is for functions whose moment generating functions exist in a neighborhood of the origin. There are plenty of densities that have finite first, second, and third moments but whose moment generating function doesn't exist. Consider for example a cousin of the Cauchy distribution,

$$f(x) = \frac{4\sin(\pi/8)}{\pi} \frac{1}{1+x^8}.$$

It shouldn't be apparent that this is a probability distribution. It is clearly non-negative and it decays rapidly enough as $|x| \to \infty$ so that the integral converges; however, it's not at all clear that it will integrate to 1, although the constant does have some nice features (it has an 8 in it, which could come from the power of x, and the normalization constant of the Cauchy distribution had a π in the denominator, which this does as well). For our purposes, *it doesn't matter!* Say we have the normalization constant wrong—who cares! There's some constant, let's call it C_8, such that $C_8/(1+x^8)$ is a probability density. While this will have finite mean, variance, and third moment, the eight moment is clearly infinite, as it's

$$\int_{-\infty}^{\infty} x^8 \frac{C_8}{1+x^8} dx;$$

the integrand is essentially C_8 for $|x|$ large, and thus the integral diverges. Similarly one can show all the larger even moments blow up, and hence the moment generating function can't converge in a neighborhood of the origin *as the moment generating function doesn't exist!*

This example shows us that our approach to the Central Limit Theorem is too restrictive, as it eliminates many nice distributions. (For example, the Cauchy distribution arises in Mandelbrot's work on fractal behavior of financial and commodities markets.) The moment generating function approach is fundamentally flawed; there's just no getting around the fact that some nice distributions don't have a moment generating function, and thus we can't do any argument that requires a moment generating function to exist! While the sums of independent Cauchy distributions do not approach normality, the cousin of it we mentioned above does. The key turns out to be having finite mean and variance.

One solution to this quandary is to study the Fourier transform of our density, which in probability is called the **characteristic function**. We'll see later that unlike the moment generating function, the characteristic function *always* exists, and is a very close analogue of the moment generating function. It has better properties (such as existence!) and is more amenable to analysis. This will allow us to adapt our previous proof. The ideas are similar, but the algebra is a little different.

The material in this chapter is thus a bit more advanced than many introductory probability courses. Most courses just don't have time to delve this deeply. While we won't prove everything we need, we'll provide enough details so that hopefully the big picture is clear, and give you a sense of some of what's waiting for you in future math classes.

21.1 Integral Transforms

Given a function $K(x, y)$ and an interval I (which is frequently $(-\infty, \infty)$ or $[0, \infty)$), we can construct a map from functions to functions as follows: send f to

$$(\mathcal{K}f)(y) := \int_I f(x)K(x, y)dx.$$

As the integrand depends on the two variables x and y and we only integrate out x, the result is a function of y. Obviously it doesn't matter what letters we use for the dummy variables; other common choices are $K(t, x)$ or $K(t, s)$ or $K(x, \xi)$. We call K the **kernel** and the new function the **integral transform** of f.

Integral transforms are useful for studying a variety of problems. Their utility stems from the fact that the related function leads to simpler algebra for the problem at hand. We define two of the most important integral transforms, the Laplace and the Fourier transforms.

Definition 21.1.1 (Laplace Transform): *Let $K(t, s) = e^{-ts}$. The Laplace transform of f, denoted $\mathcal{L}f$, is given by*

$$(\mathcal{L}f)(s) = \int_0^\infty f(t)e^{-st}dt.$$

Given a function g, its inverse Laplace transform, $\mathcal{L}^{-1}g$, is

$$(\mathcal{L}^{-1}g)(t) \;=\; \lim_{T\to\infty} \frac{1}{2\pi i} \int_{c-iT}^{c+iT} e^{st} g(s)\,ds \;=\; \lim_{T\to\infty} \frac{1}{2\pi i} \int_{-T}^{T} e^{(c+i\tau)t} g(c+i\tau)i\,d\tau.$$

Definition 21.1.2 (Fourier Transform (or Characteristic Function)): *Let* $K(x, y) = e^{-2\pi i x y}$. *The Fourier transform of* f, *denoted* $\mathcal{F}f$ *or* $\widehat{f}$, *is given by*

$$\widehat{f}(y) \;:=\; \int_{-\infty}^{\infty} f(x)e^{-2\pi i x y}\,dx,$$

where

$$e^{i\theta} \;:=\; \sum_{n=0}^{\infty} \frac{(i\theta)^n}{n!} \;=\; \cos\theta + i\sin\theta.$$

The inverse Fourier transform of g, *denoted* $\mathcal{F}^{-1}g$, *is*

$$(\mathcal{F}^{-1}g)(x) \;=\; \int_{-\infty}^{\infty} g(y)e^{2\pi i x y}\,dy.$$

Note other books define the Fourier transform differently, sometimes using $K(x, y) = e^{-ixy}$ *or* $K(x, y) = e^{-ixy}/\sqrt{2\pi}$.

The Laplace and Fourier transforms are related. If we let $s = 2\pi i y$ and consider functions $f(x)$ which vanish for $x \le 0$, we see the Laplace and Fourier transforms are equal.

While we have chosen to write the Fourier transform of f by

$$\widehat{f}(y) \;=\; \int_{-\infty}^{\infty} f(x)e^{-2\pi i x y}\,dx,$$

other books sadly might use a different notation. Some authors use e^{-ixy} or $e^{ixy}/\sqrt{2\pi}$ instead of $e^{-2\pi i x y}$, so always check the convention when you reference a book or use a program such as Mathematica. Why are there so many different notations? It turns out that different notations lead to cleaner algebra for different problems. For our purposes, this choice leads to the simplest algebra, which is why we use it.

Given a function f we can compute its transform. What about the other direction? If we are told g is the transform of some function f, can we recover f from knowing g? If yes, is the corresponding f unique? Notice how similar these questions are to the two black-box complex analysis theorems from Chapter 20. There we knew moment generating functions and wanted to recover densities. Fortunately, the answer to both questions turns out to be "yes", provided f and g satisfy certain nice conditions. A particularly nice set of functions to study is the Schwartz space.

Definition 21.1.3 (Schwartz space): *The Schwartz space, $\mathcal{S}(\mathbb{R})$, is the set of all infinitely differentiable functions f such that, for any non-negative integers m and n,*

$$\sup_{x \in \mathbb{R}} \left| (1+x^2)^m \frac{d^n f}{dx^n} \right| < \infty,$$

where $\sup_{x \in \mathbb{R}} |g(x)|$ is the smallest number B such that $|g(x)| \le B$ for all x (think "maximum value" whenever you see supremum).

Whenever we define a space or a set, it's worthwhile to show that it isn't empty! Let's show there are infinitely many Schwartz functions. We claim the Gaussians $f(x) = \frac{1}{\sqrt{2\pi\sigma^2}}\, e^{-(x-\mu)^2/2\sigma^2}$ are in $\mathcal{S}(\mathbb{R})$ for any $\mu, \sigma \in \mathbb{R}$. By a change of variables, it suffices to study the special case of $\mu = 0$ and $\sigma = 1$. Clearly the standard normal, $f(x) = \frac{1}{\sqrt{2\pi}}e^{-x^2/2}$, is infinitely differentiable. Its first few derivatives are

$$f'(x) = -x \cdot \frac{1}{\sqrt{2\pi}}e^{-x^2/2}$$

$$f''(x) = (x^2 - 1) \cdot \frac{1}{\sqrt{2\pi}}e^{-x^2/2}$$

$$f'''(x) = -(x^3 - 3x) \cdot \frac{1}{\sqrt{2\pi}}e^{-x^2/2}.$$

By induction, we can show that the nth derivative is a polynomial $p_n(x)$ of degree n times $\frac{1}{\sqrt{2\pi}}e^{-x^2/2}$. To show f is Schwartz, by Definition 21.1.3 we must show

$$\left| (1+x^2)^m \cdot p_n(x)\frac{1}{\sqrt{2\pi}}e^{-x^2/2} \right|$$

is bounded. This follows from the fact that the standard normal decays faster than any polynomial. Say we want to show $|x^m e^{-x^2/2}|$ is bounded. The claim is clear for $|x| \le 1$. What about larger $|x|$? By keeping just one term of the Taylor series expansion of the exponential function, we know $(x^2/2)^k / k! \le e^{x^2/2}$ for any k, so $e^{-x^2/2} \le k!2^k/x^{2k}$. Thus $|x^m e^{-x^2/2}| \le k!2^k/x^{2k-m}$, and if we choose $2k > m$ then this is bounded by $k!2^k$.

We now state the main result we need from complex analysis. It states precisely when the integral transform arises from a unique input. We only give the statement for the inverse Fourier transform—just stating the result for the Laplace transform requires a lot of new notation from complex analysis! A proof can be found in many books on complex analysis or Fourier Analysis (see for example [SS1, SS2]).

Theorem 21.1.4 (Inversion Theorem): *Let $f \in \mathcal{S}(\mathbb{R})$, the Schwartz space. Then*

$$f(x) = \int_{-\infty}^{\infty} \widehat{f}(y)e^{2\pi ixy}dy,$$

where $\widehat{f}$ is the Fourier transform of f. In particular, if f and g are Schwartz functions with the same Fourier transform, then $f(x) = g(x)$.

This interplay between a function and its transform will be very useful for us when we study probability distributions, as the moment generating function is an integral transform of the density! Recall the moment generating function is defined by $M_X(t) = \mathbb{E}[e^{tX}]$, which means

$$M_X(t) = \int_{-\infty}^{\infty} e^{tx} f(t)dt.$$

If $f(x) = 0$ for $x \leq 0$, *this is just the Laplace transform of f*. Alternatively, if we take $t = -2\pi i y$ then it's the Fourier transform of f. This is trivially related to (yet another!) generating function, the characteristic function of X.

Characteristic function: The characteristic function of a random variable X is

$$\phi(t) := \mathbb{E}[e^{itX}].$$

Unlike the moment generating function, if X has a continuous density then the characteristic function *always* exists for all t. Note the characteristic function is essentially the Fourier transform of the density: the Fourier transform is just $\phi(-2\pi t)$.

Why does the characteristic function always exist? Remember the density is a non-negative integrable function f. We have

$$|\phi(t)| = \left| \int_{x=-\infty}^{\infty} e^{itx} f(x)dx \right| \leq \int_{x=-\infty}^{\infty} |e^{itx}| f(x)dx = \int_{x=-\infty}^{\infty} f(x)dx = 1,$$

as $|e^{itx}| = 1$. The reason the absolute value of this exponential is 1 is the **Pythagorean theorem**. We have $e^{i\theta} = \cos\theta + i\sin\theta$ for any real θ (there are many ways to see this; one way is to compare the Taylor series expansions of e^{θ}, $\cos\theta$, and $\sin\theta$, noting $i = \sqrt{-1}$). If $z = a + ib$ is a complex number, then $|z|^2 = z\bar{z}$, where $\bar{z} = a - ib$ is the **complex conjugate**; we call $|z|$ the **length**, **absolute value**, or the **norm** of z. For us, we get

$$|e^{itx}|^2 = (\cos tx + i\sin tx)(\cos tx - i\sin tx) = \cos^2 tx + \sin^2 tx = 1.$$

This is an immense improvement over the moment generating function—the most important property an object can have is existence, so already we've made progress.

Furthermore, we see that the characteristic function is simply related to the moment generating function. What a difference, though, an i makes! The characteristic function and the Fourier transform are trivially related; they differ by rescaling the input by a factor of 2π. The rescaling from the characteristic function to the moment generating function is far more profound; the presence of the factor of i leads to *very* different behavior, and very different algebra.

We now see why these results from complex analysis will save the day. The inversion formulas above tell us that if our initial distribution is nice, then knowing the integral transform of the function is the same as knowing the function; in other words, knowing the integral transform uniquely determines the distribution.

 The following remark is a bit more advanced, and is meant to put these arguments into their proper context in the realm of analytic arguments. A function $f : \mathbb{R} \to \mathbb{C}$ has **compact support** if there's a finite closed interval $[a, b]$ such that for all $x \notin [a, b]$, $f(x) = 0$. Schwartz functions with compact support are extremely useful in many arguments. It can be shown that given any continuous function g on a finite closed interval $[a, b]$, there's a Schwartz function f with compact support arbitrarily close to g; i.e., for all $x \in [a, b]$, $|f(x) - g(x)| < \epsilon$. Similarly, given any such continuous function g, one can find a sum of **step functions** of intervals arbitrarily close to g in the same sense as above (a step function is a finite sum of characteristic functions of closed intervals). Often, to prove a result for step functions it suffices to prove the result for continuous functions, which is the same as proving the result for Schwartz functions. Schwartz functions are infinitely differentiable and as the Fourier inversion formula holds, we can pass to the Fourier transform space, which is sometimes easier to study.

21.2 Convolutions and Probability Theory

An important property of the Fourier transform is that it behaves nicely under **convolution**. Remember we denote the convolution of two functions f and g by $h = f * g$, where

$$h(x) = \int_{-\infty}^{\infty} f(t)g(x - t)dt = \int_I f(x - t)g(t)dt.$$

A natural question to ask is: what must we assume about f and g to ensure that the convolution exists? For our purposes, f and g will be probability densities. Thus they're non-negative and integrate to 1. While this is all we need to ensure that $h = f * g$ integrates to 1, it's not quite enough to guarantee that $f * g$ is finite. Let's first show it integrates to 1. Since our integrand is non-negative, we're allowed to switch the order of integration. Note for each x the integral is either non-negative or positive infinity. We have

$$\int_{x=-\infty}^{\infty} (f * g)(x)dx = \int_{x=-\infty}^{\infty} \int_{t=-\infty}^{\infty} f(t)g(x - t)dtdx$$

$$= \int_{t=-\infty}^{\infty} f(t) \left[\int_{x=-\infty}^{\infty} g(x - t)dx \right] dt.$$

The integral in brackets is 1. If you want, change variables and let $u = x - t, du = dx$. We're integrating a probability density from $-\infty$ to ∞; that's always 1. We're left with

$$\int_{x=-\infty}^{\infty} (f * g)(x)dx = \int_{t=-\infty}^{\infty} f(t)dt = 1,$$

again as the integral of a probability density from $-\infty$ to ∞ is always 1. This means our non-negative function $(f * g)(x)$ can only be zero on a set of measure (or length) 0. If you're not familiar with measure theory, no worries: here's another formulation. It means that for any M, the length of $\{x : (f * g)(x) > M\}$ is at most $1/M$, as otherwise the integral would exceed 1.

So, this proves that for almost all x we have $(f * g)(x)$ is finite. What must we assume about f and g so that the convolution is finite for *all* x? If we assume f and g are **square-integrable**, namely $\int_{-\infty}^{\infty} f(x)^2 dx$ and $\int_{-\infty}^{\infty} g(x)^2 dx$ are finite, then $f * g$ is well-behaved everywhere. We'll see shortly how this follows from the Cauchy-Schwarz inequality, which is proved in Appendix B.6.

Cauchy-Schwarz inequality: For complex-valued functions f and g,

$$\int_{-\infty}^{\infty} |f(x)g(x)|dx \leq \left(\int_{-\infty}^{\infty} |f(x)|^2 dx \right)^{1/2} \cdot \left(\int_{-\infty}^{\infty} |g(x)|^2 dx \right)^{1/2}.$$

Assuming f and g are square-integrable is very weak, and is met in all the standard densities we study. Even in situations where they're not square-integrable, often there are no problems. For example, if we take

$$f(x) = \begin{cases} \frac{1}{2\sqrt{x}} & \text{if } 0 < x \leq 1 \\ 0 & \text{otherwise,} \end{cases}$$

then f is integrable but not square-integrable, as $\int_0^1 dx/x$ blows up. That said, the convolution of f with itself is well-behaved. After "some" integration, you would find

$$(f * f)(y) = \begin{cases} \pi/4 & \text{if } 0 < y \leq 1 \\ (\text{arccsc}(\sqrt{y}) - \arctan(\sqrt{y-1}))/2 & \text{if } 1 < y < 2 \\ 0 & \text{otherwise.} \end{cases}$$

We now state a wonderful result. It is because of this that the Fourier transform is so prevalent in probability. It's such an important result that we provide a full proof.

Theorem 21.2.1 (Convolutions and the Fourier Transform): *Let f, g be continuous functions on $\mathbb{R}$. If $\int_{-\infty}^{\infty} |f(x)|^2 dx$ and $\int_{-\infty}^{\infty} |g(x)|^2 dx$ are finite then $h = f * g$ exists, and $\widehat{h}(y) = \widehat{f}(y)\widehat{g}(y)$. Thus the Fourier transform converts convolution to multiplication.*

Proof: We first show $h = f * g$ exists. We have

$$h(x) = (f * g)(x)$$
$$= \int_{-\infty}^{\infty} f(t)g(x - t)dt$$
$$|h(x)| \leq \int_{-\infty}^{\infty} |f(t)| \cdot |g(x - t)|dt$$
$$\leq \left(\int_{-\infty}^{\infty} |f(t)|^2 dt \right)^{1/2} \left(\int_{-\infty}^{\infty} |g(x - t)|^2 dt \right)^{1/2}$$

by the Cauchy-Schwarz inequality. As we're assuming f and g are square-integrable, both integrals are finite (for x fixed, as t runs from $-\infty$ to ∞ so too does $x - t$). We're not assuming f and g are densities; if we did, then the inequalities would be equalities as the densities are never negative.

Now that we know that the convolution h exists, we can explore its properties. Let's calculate its Fourier transform. This leads to a double integral (one integral from the definition of h, and then another from the definition of the Fourier transform). The fact that we'll have two integrals suggests how we'll handle it. Usually there are two things to try with a double integral: we can change variables, or we can interchange the order of integration. We'll interchange orders; this is justified as the integrals of the absolute value are finite, and we can appeal to the Fubini theorem (see Theorem B.2.1).

Before we change orders, however, we first cleverly **add zero** to facilitate the algebra (see §A.12 for more examples of this method). We'll see shortly an integral of $g(x - t)$ against the exponential $e^{-2\pi i x y}$. As we have g evaluated at $x - t$, we want $x - t$ in the exponential and not x. This suggests writing x as $x - t + t$. We find

$$\widehat{h}(y) = \int_{-\infty}^{\infty} h(x)e^{-2\pi i x y} dx$$

$$= \int_{-\infty}^{\infty}\int_{-\infty}^{\infty} f(t)g(x - t)e^{-2\pi i x y} dt dx$$

$$= \int_{-\infty}^{\infty}\int_{-\infty}^{\infty} f(t)g(x - t)e^{-2\pi i(x - t + t)y} dt dx$$

$$= \int_{t=-\infty}^{\infty} f(t)e^{-2\pi i t y}\left[\int_{x=-\infty}^{\infty} g(x - t)e^{-2\pi i(x-t)y} dx\right] dt$$

$$= \int_{t=-\infty}^{\infty} f(t)e^{-2\pi i t y}\left[\int_{u=-\infty}^{\infty} g(u)e^{-2\pi i u y} du\right] dt$$

$$= \int_{t=-\infty}^{\infty} f(t)e^{-2\pi i t y}\widehat{g}(y) dt = \widehat{f}(y)\widehat{g}(y),$$

where the last line is from the definition of the Fourier transform.   □

If for all $i = 1, 2, \ldots$ we have f_i is square-integrable, prove for all i and j that $\int_{-\infty}^{\infty} |f_i(x)f_j(x)| < \infty$. What about $f_1 * (f_2 * f_3)$ (and so on)? Prove $f_1 * (f_2 * f_3) = (f_1 * f_2) * f_3$. Therefore convolution is associative, and we may write $f_1 * \cdots * f_N$ for the convolution of N functions. If you're stuck, we discuss this in §19.5.

It's unusual to have two operations that essentially commute. We have the Fourier transform of a convolution is the product of the Fourier transforms; as convolution is like multiplication, this is saying that using this special type of multiplication, we can switch the orders of the operations. It is rare to have two operations satisfying such a rule. For example, $\sqrt{a + b}$ typically is *not* $\sqrt{a} + \sqrt{b}$.

The following lemma is the starting point to the Fourier analytic proof of the Central Limit Theorem.

Lemma 21.2.2: *Let X_1 and X_2 be two independent random variables with densities f and g. Assume f and g are square-integrable probability densities, so $\int_{-\infty}^{\infty} f(x)^2 dx$ and $\int_{-\infty}^{\infty} g(x)^2 dx$ are finite. Then $f * g$ is the probability density for $X_1 + X_2$. More generally, if $X_1, \ldots, X_N$ are independent random variables with square-integrable densities $p_1, \ldots, p_N$, then $p_1 * p_2 * \cdots * p_N$ is the density for $X_1 + \cdots + X_N$. (As convolution is commutative and associative, we don't have to be careful when writing $p_1 * p_2 * \cdots * p_N$.)*

Proof: The probability of $X_i \in [x, x + \Delta x]$ is $\int_x^{x+\Delta x} f(t)dt$, which is approximately $f(x)\Delta x$ when Δx is small (as the integrand is essentially constant). The probability that $X_1 + X_2 \in [x, x + \Delta x]$ is just

$$\int_{x_1=-\infty}^{\infty} \int_{x_2=x-x_1}^{x+\Delta x-x_1} f(x_1)g(x_2)dx_2 dx_1.$$

As $\Delta x \to 0$ we obtain the convolution $f * g$, and find

$$\text{Prob}(X_1 + X_2 \in [a, b]) = \int_a^b (f * g)(z)dz. \tag{21.1}$$

We must justify our use of the word "probability" in (21.1); namely, we must show $f * g$ is a probability density. Clearly $(f * g)(z) \geq 0$ as $f(x), g(x) \geq 0$. As we are assuming f and g are square-integrable,

$$\int_{-\infty}^{\infty} (f * g)(x)dx = \int_{-\infty}^{\infty} \int_{-\infty}^{\infty} f(x - y)g(y)dy\,dx$$

$$= \int_{-\infty}^{\infty} \int_{-\infty}^{\infty} f(x - y)g(y)dx\,dy$$

$$= \int_{-\infty}^{\infty} g(y) \left(\int_{-\infty}^{\infty} f(x - y)dx \right) dy$$

$$= \int_{-\infty}^{\infty} g(y) \left(\int_{-\infty}^{\infty} f(t)dt \right) dy.$$

As f and g are probability densities, these integrals are 1, completing the proof. $\quad\square$

Remark: We really don't need to assume the densities are square-integrable. The purpose of that assumption is to make sure the density of the sum of the random variables is finite everywhere. If we're willing to allow our density to be infinite at a few places, we can drop that assumption.

This section introduced a lot of material and results, but we can now begin to see the big picture. If we take N independent random variables with densities $p_1, \ldots, p_N$, then the sum has density $p = p_1 * \cdots * p_N$. While at first this equation looks frightening (what is the convolution of N exponential densities?), there's a remarkable simplification that happens. Using the Fourier transform of a convolution is the product of the Fourier transforms, we find $\widehat{p}(y) = \widehat{p}_1(y) \cdots \widehat{p}_N(y)$; in the special case when the random variables are identically distributed, this simplifies further to just $\widehat{p}_1(y)^N$. Now "all" (and, sadly, it's a big "all") we need to do to prove the Central Limit Theorem in the case when all the densities are equal is show that, as $N \to \infty$, $\widehat{p}_1(y)^N$ converges to the Fourier transform of something normally distributed (remember we haven't normalized our sum), and the inverse Fourier transform is uniquely determined and is normally distributed.

21.3 Proof of the Central Limit Theorem

We can now sketch the proof of the Central Limit Theorem. The version we prove is a bit more general than our earlier work. We no longer need to assume the moment generating function exists. To really grasp the nuts and bolts of this proof, we encourage you to provide the complete details to the series of problems below, each of which gives another needed input for the proof.

Theorem 21.3.1 (Central Limit Theorem): *Let $X_1, \ldots, X_N$ be independent, identically distributed random variables whose first three moments are finite and whose probability density decays sufficiently rapidly. Denote the mean by μ and the variance by σ^2, let*

$$\overline{X}_N = \frac{X_1 + \cdots + X_N}{N}$$

and set

$$Z_N = \frac{\overline{X}_N - \mu}{\sigma/\sqrt{N}}.$$

Then as $N \to \infty$, the distribution of Z_N converges to the standard normal.

We highlight the key steps, but we do not provide detailed justifications (which would require several standard lemmas about the Fourier transform; see for example [SS1]). Without loss of generality, we may consider the case where we have a probability density p on $\mathbb{R}$ that has mean zero and variance one (see §20.4). We assume the density decays sufficiently rapidly so that all convolution integrals that arise below converge.

Specifically, our density p satisfies

$$\int_{-\infty}^{\infty} x p(x)dx = 0, \quad \int_{-\infty}^{\infty} x^2 p(x)dx = 1, \quad \int_{-\infty}^{\infty} |x|^3 p(x)dx < \infty. \tag{21.2}$$

Assume $X_1, X_2, \ldots$ are independent identically distributed random variables drawn from p; thus, $\text{Prob}(X_i \in [a, b]) = \int_a^b p(x)dx$. Define $S_N = \sum_{i=1}^{N} X_i$. Recall the standard Gaussian (mean zero, variance one) has density $p(-x^2/2)/\sqrt{2\pi}$.

As we are assuming $\mu = 0$ and $\sigma = 1$, we have $Z_N = \frac{(X_1 + \cdots + X_N)/N}{1/\sqrt{N}} = \frac{X_1 + \cdots + X_N}{\sqrt{N}}$, so $Z_N = S_N/\sqrt{N}$. We must show $S_N/\sqrt{N}$ converges in probability to the standard

Gaussian:

$$\lim_{N \to \infty} \text{Prob}\left(\frac{S_N}{\sqrt{N}} \in [a, b]\right) = \frac{1}{\sqrt{2\pi}} \int_a^b e^{-\frac{x^2}{2}} dx.$$

We sketch the proof. The Fourier transform of p is

$$\widehat{p}(y) = \int_{-\infty}^{\infty} p(x)e^{-2\pi i xy} dx.$$

Clearly, $|\widehat{p}(y)| \le \int_{-\infty}^{\infty} p(x)dx = 1$, and $\widehat{p}(0) = \int_{-\infty}^{\infty} p(x)dx = 1$.

Claim 1: *One useful property of the Fourier transform is that the derivative of $\widehat{g}$ is the Fourier transform of $2\pi i x g(x)$; thus, differentiation (hard) is converted to multiplication (easy). Explicitly, show*

$$\widehat{g}'(y) = \int_{-\infty}^{\infty} 2\pi i x \cdot g(x)e^{-2\pi i xy} dx.$$

If g is a probability density, note $\widehat{g}'(0) = 2\pi i \mathbb{E}[x]$ and $\widehat{g}''(0) = -4\pi^2 \mathbb{E}[x^2]$.

The above claim shows why it's, at least potentially, natural to use the Fourier transform to analyze probability distributions. The mean and variance (and the higher moments) are simple multiples of the derivatives of $\widehat{p}$ at zero. By Claim 1, as p has mean zero and variance one, $\widehat{p}'(0) = 0$, $\widehat{p}''(0) = -4\pi^2$. We Taylor expand $\widehat{p}$ (we do not justify that such an expansion exists and converges; however, in most problems of interest this can be checked directly, and this is the reason we need technical conditions about the higher moments of p), and find near the origin that

$$\widehat{p}(y) = 1 + \frac{\widehat{p}''(0)}{2}y^2 + \cdots = 1 - 2\pi^2 y^2 + O(y^3). \tag{21.3}$$

Near the origin, the above shows $\widehat{p}$ looks like a concave down parabola. There is no y term as $\widehat{p}'(0) = 0$. Here $O(y^3)$ is **big-Oh notation** for an error at most on the order of y^3; see §20.6 for more on big-Oh notation.

From §21.2, we know:

- The probability that $X_1 + \cdots + X_N \in [a, b]$ is $\int_a^b (p * \cdots * p)(z)dz$.
- The Fourier transform converts convolution to multiplication. If $\text{FT}[f](y)$ denotes the Fourier transform of f evaluated at y, then we have

$$\text{FT}[p * \cdots * p](y) = \widehat{p}(y) \cdots \widehat{p}(y).$$

However, we do not want to study the distribution of $X_1 + \cdots + X_N = x$, but rather the distribution of $S_N = \frac{X_1 + \cdots + X_N}{\sqrt{N}} = x$.

Claim 2: *If $B(x) = A(cx)$ for some fixed $c \ne 0$, show $\widehat{B}(y) = \frac{1}{c}\widehat{A}\left(\frac{y}{c}\right)$.*

Claim 3: *Show that if the probability density of $X_1 + \cdots + X_N = x$ is $(p * \cdots * p)(x)$ (i.e., the distribution of the sum is given by $p * \cdots * p$), then the probability density of*

$\frac{X_1 + \cdots + X_N}{\sqrt{N}} = x$ is $(\sqrt{N}p * \cdots * \sqrt{N}p)(x\sqrt{N})$. *Using Claim 2, show*

$$FT\left[(\sqrt{N}p * \cdots * \sqrt{N}p)(x\sqrt{N})\right](y) = \left[\widehat{p}\left(\frac{y}{\sqrt{N}}\right)\right]^N.$$

The above claims allow us to determine the Fourier transform of the distribution of S_N. It's just $\left[\widehat{p}\left(\frac{y}{\sqrt{N}}\right)\right]^N$. We take the limit as $N \to \infty$ for **fixed** y. From (21.3), $\widehat{p}(y) = 1 - 2\pi^2 y^2 + O(y^3)$. Thus we have to study

$$\left[1 - \frac{2\pi^2 y^2}{N} + O\left(\frac{y^3}{N^{3/2}}\right)\right]^N.$$

For any fixed y, we have

$$\lim_{N \to \infty}\left[1 - \frac{2\pi^2 y^2}{N} + O\left(\frac{y^3}{N^{3/2}}\right)\right]^N = e^{-2\pi y^2}. \tag{21.4}$$

There are two definitions of e^x (see the end of §B.3); while we normally work with the infinite sum expansion, in this case the product formulation is far more useful:

$$e^x = \lim_{N \to \infty}\left(1 + \frac{x}{N}\right)^N$$

(you might recall this formula from compound interest). Of course, this isn't a fully rigorous proof. The problem is we don't have *exactly* the same setting as the definition, as we have the lower order error $O(y^3/N^{3/2})$. A great way to proceed rigorously is to take the logarithm of both sides of (21.4), and notice that as $N \to \infty$ the two sides are equal.

Claim 4: *Show that the Fourier transform of $e^{-2\pi y^2}$ at x is $\frac{1}{\sqrt{2\pi}} e^{-x^2/2}$. (Hint: This problem requires contour integration from complex analysis. If you haven't had a course in complex analysis, this is another black-box result to return to after taking more math).*

We would like to conclude that as the Fourier transform of the distribution of S_N converges to $e^{-2\pi y^2}$ and the Fourier transform of $e^{-2\pi y^2}$ is $\frac{1}{\sqrt{2\pi}} e^{-x^2/2}$, then the distribution of S_N equalling x converges to $\frac{1}{\sqrt{2\pi}} e^{-x^2/2}$. Justifying these statements requires some results from complex analysis. We refer the reader to [Fe] for the details, which completes the proof. □

The key point in the proof is that we used Fourier Analysis to study the sum of independent identically distributed random variables, as Fourier transforms convert convolution to multiplication. The universality is due to the fact that *only* terms up to the second order contribute in the Taylor expansions. Explicitly, for "nice" p the distribution of S_N converges to the standard Gaussian, independent of the fine structure of p. The fact that p has mean zero and variance one is really just a normalization to study all probability distributions on a similar scale; see Section 20.4.

The higher order terms are important in determining the *rate* of convergence in the Central Limit Theorem (see [Fe] for details and [KonMi] for an application to Benford's

Law).

Here are some good problems to think about.

- Modify the proof to deal with the case of p having mean μ and variance σ^2.
- For reasonable assumptions on p, estimate the rate of convergence to the Gaussian.
- Let p_1, p_2 be two probability densities satisfying (21.2). Consider $S_N = X_1 + \cdots + X_N$, where for each i, X_1 is equally likely to be drawn randomly from p_1 or p_2. Show the Central Limit Theorem is still true in this case. What if we instead had a fixed, finite number of such distributions $p_1, \ldots, p_k$, and for each i we draw X_i from p_j with probability q_j (of course, $q_1 + \cdots + q_k = 1$)?

21.4 Summary

There are, not surprisingly, many structural similarities with the Fourier Analysis proof and the moment generating function proof. If you forgive the pun, this is to be expected. Why? The moment generating function is $M_X(t) = \mathbb{E}[e^{tX}]$ while the Fourier transform (or the characteristic function) is $\mathbb{E}[e^{-2\pi iyX}]$. The two are thus related by $t \mapsto -2\pi iy$, but what a difference that i makes! The characteristic function exists for any density, which isn't the case for the moment generating function.

This relation sheds light on Claim 1. It should now be clear why derivatives of the Fourier transform are related to the moments of the density; it's because the Fourier transform is a very close cousin of the moment generating function, where the derivatives are the moments.

There's essentially an unlimited number of things one can do in math. We can define almost anything; the question is which definitions are useful, which definitions lead to good viewpoints. In Theorem 21.2.1 we saw the Fourier transform of a convolution is the product of the Fourier transforms. When studying sums of random variables, it's hard not to try to use convolutions, as that is the most natural way to find the density. As the Fourier transform interacts well with convolutions, it shouldn't be surprising that it enters the proof.

21.5 Exercises

Exercise 21.5.1   *Find sufficient conditions on f and g so that the Cauchy-Schwarz inequality holds as an equality. Try to find the weakest such conditions.*

Exercise 21.5.2   *Find the Fourier transform of $f(x) = e^{-|x|}$.*

Exercise 21.5.3   *Show that $\widehat{g}'(y) = \int_{-\infty}^{\infty} 2\pi ix \cdot g(x)e^{-2\pi ixy}dx$.*

Exercise 21.5.4   *Prove Claim 2, that if $B(x) = A(cx)$ for some fixed $c \neq 0$, then $\widehat{B}(y) = \frac{1}{c}\widehat{A}\left(\frac{y}{c}\right)$.*

Exercise 21.5.5   *Find constants C_{2m} such that $\int_{-\infty}^{\infty} \frac{C_{2m}}{1+x^{2m}}dx = 1$ for $m \in \{1, 2, 3, 4\}$. (Using techniques from complex analysis it's possible to find C_{2m} for all m.)*

Exercise 21.5.6   *Using the Taylor series of e^x, $\cos x$, and $\sin x$, "prove" that $e^{ix} = \cos x + i \sin x$.*

Exercise 21.5.7   *Show that $(\mathcal{L}^{-1}(\mathcal{L}f)(s))(x) = f(x)$.*

Exercise 21.5.8 *Show that $(\mathcal{F}^{-1}(\mathcal{F}f)(s))(x) = f(x)$.*

Exercise 21.5.9 *Show that any infinitely differentiable functions with compact support (that is, they are only non-zero on a finite interval) are in the Schwartz space.*

Exercise 21.5.10 *Explain how a function can be smooth and have compact support.*

Exercise 21.5.11 *Show that the Cauchy distribution is not in the Schwartz space.*

Exercise 21.5.12 *Use characteristic functions to show that the Cauchy distribution is strictly stable, that is, the sum of two identically distributed Cauchy distributions is a rescaled version of the original distribution.*

Exercise 21.5.13 *Let f be a non-negative, continuous function. Prove $\int_{-\infty}^{\infty} f_n(x)dx < \infty$ does not imply that $\lim_{x\to\infty} f(x) = 0$.*

Exercise 21.5.14 *Let $X_i \sim \text{Exp}(1)$, with associated density function $f_{X_i}(x) = e^{-x}$ for $x \geq 0$ and 0 otherwise. (a) Find $f_{X_1} * f_{X_2}$. (b) More generally, can you find a closed form expression for the convolution of 3, 4, or arbitrarily many such exponential functions? (c) What if the exponential random variables have different parameters, say λ_1 and λ_2?*

Exercise 21.5.15 *An inner product is a function that maps two "vectors" to a single value. We will denote the inner product of v_1 and v_2 as $\langle v_1, v_2 \rangle$. For such a function to be an inner product, it must have 3 properties: $\langle v_1, v_2 \rangle = \langle v_2, v_1 \rangle$ (we will restrict ourselves to real valued functions, the complex case is a bit different); linearity, that is $\langle (av_1 + bv_2), v_3 \rangle = a\langle v_1, v_3 \rangle + b\langle v_2, v_3 \rangle$; and finally, the inner product of a vector with itself is always non-negative and is 0 only if that vector is the 0 vector. Show that if we consider two random variable X and Y to be vectors, $\mathbb{E}[XY]$ is an inner product.*

Exercise 21.5.16 *The Cauchy-Schwarz inequality has a more general form: $|\langle x, y \rangle|^2 \leq \langle x, x \rangle \cdot \langle y, y \rangle$. Use this to show that the square of the covariance of X and Y is less than the product of their variances.*

Exercise 21.5.17 *Use characteristic functions to show that the sums of independent, identical exponential random variables are distributed according to the Erlang distribution.*

Exercise 21.5.18 *Find a function whose Fourier transform does not exist.*

Exercise 21.5.19 *Would you expect sums of exponential random variables with parameter 1 or uniformly distributed random variables on $[0, 1]$ to approach normality more rapidly?*

Exercise 21.5.20 *Plot sums of exponential random variables with parameter 1 and uniformly distributed random variables $[0, 1]$ and test your hypothesis from the previous exercise.*

PART V
ADDITIONAL TOPICS

CHAPTER 22 ————————————

Hypothesis Testing

If your experiment needs statistics, you ought to have done a better experiment.
— ATTRIBUTED TO ERNEST RUTHERFORD

In 2003–2004 I participated in a data analysis seminar at Ohio State. I remember one speaker mentioning that every day weather satellites beam down more information than is in the entire Library of Congress, and forecasters have only a few hours to analyze the data and make their predictions. The wealth of data available is one of the boons of the twenty-first century, as well as one of its greatest challenges. We ignore this data at our own peril. Frequently we have mathematical models for problems of interest, ranging from predicting the weather to choosing a professional sports team to judging the financial impact of regulations and laws to describing the fundamental particles and forces in physics. We have a hypothesis about some issue, gather data related to the issue, and see whether or not the data supports our belief, or contradicts it.

This leads us to the very important field of model testing, an important part of **statistics**. As this is a probability book and not a statistics one, our treatment must be brief; however, many classes are actually part probability and part statistics, and thus as a compromise I decided to have one long, semi-detailed statistics chapter. I *strongly* urge you to take a statistics course in the future. When you do, you'll encounter many different tests to determine whether or not the data supports your conjecture. Why can we trust these tests? Probability! The tests in this chapter are consequences of many of our probability results and theorems, and thus the material below is a great way to review what we have learned and see its utility.

Our point below is to introduce you to some of the major tests, and the reasons why they're true. This can't of course be a complete substitute for a full course devoted to statistics, but hopefully it will give you a good sense of what can be done with probability, and encourage you to continue your studies. The tests in the following section are beautiful and important applications of the earlier material in the book. In a first course, most of the examples are of the following form. We have some population where the quantity of interest is drawn from one of the standard distributions with some unknown parameter. We have an idea what the unknown parameter should equal. Our goal is to see if the data supports our claims for this parameter's value. For example, maybe we believe the wealth of people in America is exponentially distributed with $\lambda = \$60,000$. We then gather our data. It's unlikely that our data will be a perfect exponential with parameter \$60,000, and thus our goal is to quantify how close it is, and discuss the implications of what we observe. Or, more interestingly, maybe we believe wealth is exponentially distributed but we don't know what the parameter is; now we want to use our observations to find the best fit exponential (as well as see how good of a job it does!).

22.1 Z-tests

We first describe the null and alternative hypothesis, then talk about significance levels and test statistics, and end with a discussion of one versus two-sided tests. As it's easier to learn material through an example than just pages of theory after theory, we introduce the z-statistic and the z-test, and frame our discussion and examples through this.

Takeaways for the section:

- There is a method for testing the truth of a hypothesis, even when there's randomness involved.
- We assume the hypothesis is true, and then collect data.
- We decide whether or not our assumption is valid by evaluating the likelihood of the data we collect.

We'll extensively use facts about the normal distribution, as well as its cumulative distribution function (often denoted Φ); see Chapter 14 for a review.

22.1.1 Null and Alternative Hypotheses

Suppose McDonald's has come out with a new ad campaign, boasting that the mean time it takes them to fill a typical order is 45 seconds. Clearly every order is different, so there will be some variation around this mean. Being the skeptic you are, you make some observations the next time you visit McDonald's, and in a sample of 20 orders you find the average service time to be 48 seconds with a standard deviation of 8 seconds. Given this data, do you believe McDonald's claim?

There is one major obstacle preventing you from answering this question: the randomness of the sample you drew. Your sample mean suggests that McDonald's is slower than they claim, but it could also be that your sample was slow by chance—perhaps during the course of your observations my kids' baseball teams (Williamstown Rotary and iBerkshires.com) came in to celebrate a win. We need a formal process to determine whether McDonald's is telling the truth, one that takes into account the possibility that your sample is unusual. This process is known as **hypothesis testing**.

There are many circumstances where we want to assess the validity of a claim, from finding out how fast McDonald's can fill an order to determining whether a new drug works better than existing ones. The first step in hypothesis testing is establishing a **null hypothesis**. The null hypothesis is typically a statement contrary to what the experimenter or researcher is attempting to show; we assume the null hypothesis to be true, and use data to try and refute it. In the McDonald's example, since we think the average service time μ is slower than 45 seconds, our null hypothesis (denoted H_0)

might look like this:

Null Hypothesis: H_0 : $\mu \leq 45$.

That is, we hypothesize that the mean service time is *at most* 45 seconds. After developing our null hypothesis, we also have to articulate the **alternative hypothesis**, which is frequently the statement we want to show true. For the McDonald's example, our alternative hypothesis (denoted H_a) would be

Alternative Hypothesis: H_a : $\mu > 45$,

which is what we're expecting to find. Note that the alternative and null hypotheses are complements, meaning that together they encompass every possibility for the value of μ. We couldn't have the following set of hypotheses:

$$H_0 : \mu < 45, \qquad H_a : \mu > 46,$$

because they don't allow for the possibility that $45 \leq \mu \leq 46$. This is an important point to remember: *your null and alternative hypotheses must allow for every possible value of the parameters you're measuring.*

One of the most important aspects of hypothesis testing is the way we phrase our arguments. If we want to show that a drug is effective, we would take our null hypothesis to be, "The drug is ineffective." If the performance of the drug is inconceivable given the assumption that it's ineffective, we reject the null hypothesis and conclude that the drug works. Notice, though, that we would *not* assume that the drug is effective, and revise our assumption only if the drug gives us good reason to. It is much more convincing to say, "This drug has proven itself," than to say, "This drug hasn't screwed up yet."

22.1.2 Significance Levels

Once we have formulated our null and alternative hypotheses, how do we actually test these hypotheses? We assume the null hypothesis to be true, and then look at our data. If the probability of collecting this data under the null hypothesis H_0 is sufficiently small, we reject H_0 in favor of the alternative hypothesis H_a. This form of argumentation might be new to you, so here's an example. Imagine you're a biologist measuring a newly discovered group of crocodiles. Previous studies have shown that the local adult crocodiles are normally distributed with a mean length of 6 feet and a standard deviation of 6 inches (remember there are 12 inches in a foot). Suppose you measure one of these new crocodiles and find it to be 14 feet long. Do you believe that this is one of the local crocodiles? Probably not, because its length is a whopping 16 standard deviations above the mean! What you are implicitly saying is this: "If the height of local adult crocodiles is normally distributed with a mean length of 6 feet and a standard deviation of 6 inches, then the odds of seeing a local crocodile as long as I just did is extraordinarily small." Since the evidence you've collected is so unlikely under the assumption that the crocodile is local, you decide that the crocodile probably is not local (in other words, that your assumption was wrong). Here H_0 would be the crocodile is local, and H_a would be that the crocodile is not. We reject H_0 in favor of H_a because it is unlikely that given H_0 is true, then the crocodile would be 14 feet long.

You might have noticed an issue with the line of reasoning we just gave: *how unlikely is unlikely enough?* What if the crocodile were 7 feet long? In this case the crocodile is only 2 standard deviations above the mean, which is rare but certainly not unheard of.

Critical Region for a Two-Sided Z–Test

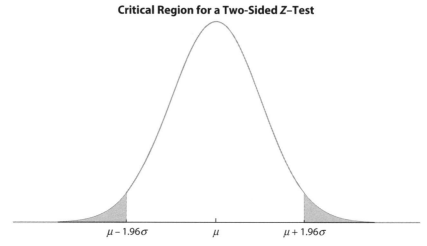

$\mu - 1.96\sigma$ $\qquad$ μ $\qquad$ $\mu + 1.96\sigma$

Figure 22.1. The probability of finding a normally distributed parameter more than 1.96 standard deviations from its mean is 0.05. This therefore determines the critical region for a two-sided z-test with an α-level of 0.05.

To get around this problem, we often establish a **significance level** (significance levels are also known as α-**levels**). A significance level is a limit on how unusual a result we will accept. An α-level of 0.05 means that if our observations from our collected data would occur less than 5% of the time given that the null hypothesis is true, then we will reject the null hypothesis. The advantage of setting an α-level is that it gives a hard and fast cutoff of how unlikely an event we will accept; this is also a disadvantage, as it gives us no flexibility! We'll discuss the blessing and the curse of significance levels later on. If you choose to go this route, you should determine your cutoff value *before* you see the data!

There is an easy way to visualize α-levels. Suppose we've hypothesized that some population parameter we're interested in is distributed $N(\mu, \sigma^2)$, and we want to test this hypothesis at an α-level of 0.05. Using a z-table, we see that if the null hypothesis is true, we will measure the parameter to be between $\mu - 1.96\sigma$ and $\mu + 1.96\sigma$ with probability 0.95. In other words, 95% of the time (or 95 out of 100 experiments) yield a value in this range if the null hypothesis is true. So, if we measure something more than 1.96 standard deviations from the mean, we have witnessed an event that would happen less than 5% of the time under the null hypothesis, and we reject the null hypothesis. We can think of the null hypothesis as establishing a **critical region**, where if the measurement we make lands in the critical region then we reject the null hypothesis. When we're using an α-level of 0.05 and the parameter we're testing is normally distributed, the critical region is everywhere more than 1.96 standard deviations from the mean. See Figure 22.1.

In this book, when we say **significance** we use statistical significance rather than practical significance. Just because something is statistically significant does not mean it is practically significant, or that the difference warrants some sort of action. For example, perhaps a newly developed type of long underwear insulates 2% better than the previously best kind of long underwear, and this difference is statistically significant. This does not mean people will choose to buy the new long underwear over the old one solely based on the difference in insulation, for a 2% improvement is not that much better. Thus, we do not have practical significance. Now suppose we have a 30% improvement in insulation. This is much more likely to be practically significant, especially for consumers in Massachusetts during the winter. However, this 30% is not necessarily statistically significant, meaning we do not know whether the large difference is real. We would need to collect more data to show that indeed the difference exists, which we will discuss in a later section when we talk about sample size and power.

We often have many choices in choosing an α. The most common α-levels are 0.10, 0.05, 0.01, and 0.001. Clearly, the smaller α is, the more difficult it is for the data we have observed to be considered unlikely enough to reject the null hypothesis. Thus, the α we choose should vary based on what we are observing. For example, if we're assessing the effectiveness of a new surgical procedure, we probably want a very low α-level to make sure it is truly significantly different from an older procedure, which most surgeons have likely gotten used to performing, to warrant its adoption. For something such as whether people prefer sprinkles (or jimmies as we say in Massachusetts) on their ice cream, it may be reasonable to choose 0.10 as the α-level. Because it is difficult to decide whether an α-level is too high or too low, 0.05 is arbitrarily considered the standard α-level, which of course should not be the only one we use. *Most importantly, we must select the α-level before observing the data.* The fear is that after observing the data, we will cheat by purposely choosing an α-level that allows us to reject the null hypothesis. For example, if a pharmaceutical company tries to sell a new drug, perhaps it is considered significantly better than the old drug under the α-level of 0.05 but not for an α-level of 0.01. The performance of a drug, especially if used to treat fatal diseases, is very serious and thus should probably be considered under a smaller α-level than 0.05. However, pharmaceutical companies could raise the α-level to 0.05 in order to say that the new drug is significantly better and thus sell more of it.

One more piece of terminology before we move on. For an α-level of 0.05, the 1.96 value we keep mentioning is known as a **critical value**. As you can imagine, the critical value is different for each value of α. We'll give a more precise definition of the critical value in the next section.

22.1.3 Test Statistics

Once we've set our significance level, we're ready to test our hypothesis. All we need to do is formulate a **test statistic**. A test statistic is a measurement we get from the data whose distribution we assume we know (remember we start by assuming the null hypothesis is true). Since we know the distribution of our test statistic, we can determine the probability of measuring a test statistic that large or larger; this is the probability we use to decide whether or not to reject the null hypothesis. One common test statistic is the sample mean. After we identify our test statistic, we need to figure out the distribution it follows. This is the most important question in hypothesis testing, since without knowing the distribution of our test statistic, we will have no way to assess whether our result was unusual or not.

As an example, what kind of distribution should the sample mean follow? To figure that out, we recall the Central Limit Theorem: for a group of n identically and independently distributed random variables X_i with mean μ and variance σ^2, the random variable

$$Y_n = \frac{1}{n}\left(\sum_{i=1}^{n} X_i\right)$$

is normally distributed in the limit as $n \to \infty$. If each X_i represents the measurement of a member of our population, then Y_n is the sample mean, and the X_i satisfy the conditions above. What are the mean and variance of Y_n? By linearity of expectation we see that

$$\mathbb{E}[Y_n] = \mathbb{E}\left[\frac{1}{n}\left(\sum_{i=1}^{n} X_i\right)\right] = \frac{1}{n}\left(\sum_{i=1}^{n}\mathbb{E}[X_i]\right) = \frac{1}{n}\cdot n\mathbb{E}[X] = \mu;$$

the expected value of Y_n is the expected value of X. Furthermore, we see that the variance of Y_n is given by

$$\mathrm{Var}(Y_n) = \mathrm{Var}\left(\frac{1}{n}\left(\sum_{i=1}^{n} X_i\right)\right) = \frac{1}{n^2}\mathrm{Var}\left(\sum_{i=1}^{n} X_i\right). \tag{22.1}$$

Since the X_i's are independent, we know

$$\mathrm{Var}\left(\sum_{i=1}^{n} X_i\right) = \sum_{i=1}^{n}\mathrm{Var}(X_i) = n\mathrm{Var}(X_i),$$

so the variance of Y_n simplifies to

$$\mathrm{Var}(Y_n) = \frac{1}{n^2}\cdot n\mathrm{Var}(X_i) = \frac{\sigma^2}{n}.$$

The wonderful feature about the variance of the sample mean is that it decreases as the sample size increases. To get a feel for why this should be, imagine repeatedly flipping a fair coin (a coin which lands heads or tails with equal probability). Let's call a result "unusual" if we flip fewer than 45% heads or more than 55% heads. With a sample of size 2, there's a 50% chance that we will have an unusual result, since we can only have 2 heads (unusual), 2 tails (unusual), or 1 of each (not unusual). What about a sample of size 100? Now if we flip fewer than 45 heads or more than 55 heads, we will have an unusual result. How many ways can we do this? Recalling our combinatorial knowledge, there are

$$\sum_{i=45}^{i=55}\binom{100}{i} = 923,796,541,447,310,445,480,620,479,776$$

ways to flip between 45 and 55 heads (i.e., not getting an unusual result), and 2^{100} total ways to flip 100 coins. Therefore the probability of having an unusual result is just one

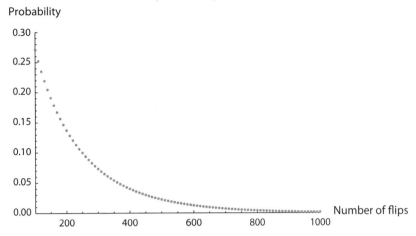

Figure 22.2. As our sample size increases (that is, as we flip more coins), the odds of obtaining an "unusual result" decrease drastically. This is why we expect the variance to decrease as we take larger and larger samples. For more see Exercise 22.8.3.

minus the probability of not having an unusual result:

$$\text{Prob(unusual result)} \;=\; 1 - \frac{1}{2^{100}} \sum_{i=45}^{55} \binom{100}{i} \;=\; 0.271.$$

Repeating the same argument for a sample of size 1000, we find a 0.0014 probability of obtaining an unusual result. As the sample size increases, the total number of possible results increases much faster than the number of ways to get an unusual result, so with a larger sample we expect to get a much more faithful estimate of the true mean (see Figure 22.2). This is an example of a very general principle: **more data is always better**. Unfortunately, to double your accuracy (cut the variance in half) you need to collect four times as much data, which in many real-life situations can be expensive or time-consuming.

We plot the result of some simulations in Figure 22.2, and give some simple code to generate it below.

```
temp = {};
For[n = 100, n <= 1000, n = n + 10, (* will plot every 10th value *)
  temp = AppendTo[
    temp, {n, 1 - Sum[Binomial[n, i], {i, .45 n, .55 n}]/2^n}]]
ListPlot[temp, AxesLabel -> {"Number of flips", "Probability"},
  PlotRange -> {{100, 1000}, {0, .3}},
  PlotLabel -> "Probability of getting an unusual result"]
}
```

Once we have the distribution of the test statistic, hypothesis testing is straightforward. Suppose our null hypothesis is that the mean of a random variable X is equal to μ. If the null hypothesis is true, then for large n the sample mean $\bar{x}$ should be normally distributed with mean μ and variance σ^2/n; in other words, $\bar{X} \sim N(\mu, \sigma^2/n)$. Then our test statistic follows.

> **Test statistic for a sample mean—z-statistic**: Let X be a normal random variable with known variance σ^2 and hypothesized mean μ, and let $x_1, x_2, \ldots, x_n$ be n independent observations drawn from this distribution. Set $\bar{x} = (x_1 + \cdots + x_n)/n$, the sample mean. Then the observed z-test statistic values
>
> $$ z = \frac{\bar{x} - \mu}{\sqrt{\sigma^2/n}} $$
>
> are normally distributed with mean 0 and variance 1 (so $Z \sim \mathrm{N}(0, 1)$). If instead of being normally distributed X is just a nice, well-behaved distribution, then a good rule of thumb is that $\bar{X} = (X_1 + \cdots + X_n)/n$ is nearly normal for if $n \geq 30$. It's very important that the variance is known; if it isn't, more involved tests are needed.

This test is called a **z-test** because our test statistic in normally distributed. This is because a sum of normally distributed random variables is again normally distributed. This mean that the normal distribution is **stable**; we have talked about the benefits of having a stable distribution many times in this book.

For any z value we measure, we can use the standard normal table to find the probability of obtaining a test statistic that far or farther from zero. This probability is called the **p-value** (which stands for probability value). If the p-value is less than the α-level, we reject the null hypothesis in favor of the alternative hypothesis. This also gives us a clearer meaning for the **critical value**. For a given hypothesis test, the critical value is the value such that if our test statistic is larger than the critical value (in the absolute value sense), we reject the null hypothesis.

Now we can finally finish off our McDonald's example. From our discussion above, if we accept the null hypothesis and let $\mu = 45$ and assume $\sigma = 8$ (we'll return to this assumption in §22.3), then in a sample of size 20 we should have

$$ \bar{X} \sim \mathrm{N}\left(45, 8^2/20\right). $$

This means that for our sample of size 20, the wait time should have a mean of 45 seconds and a standard deviation of about 1.79 seconds. We measured $\bar{x} = 48$, so our z value is $(48 - 45)/1.79 = 1.68$. Using a z-table, we see that the odds of measuring a test statistic larger than 1.68 by chance alone is 0.046, or a little over 4%. Since this p-value is less than our significance level of 0.05, we reject the null hypothesis and conclude that McDonald's is indeed slower than they claim. If, however, we chose an alpha level of 0.01 we would have a very different answer; this problem illustrates how susceptible our recommendation is based on our critical value. There are some people who believe it is not the job of the researcher to make conclusions, and their job is only to report the p-values.

22.1.4 One-sided versus Two-sided Tests

While going through that last example you might have wondered, "Why did you only look at the probability of finding a z-score higher than 1.68? Don't we also need to worry about the possibility of measuring a test statistic smaller than the hypothesized mean, too?" This is an important question which we've been a little lax about up to now, but it

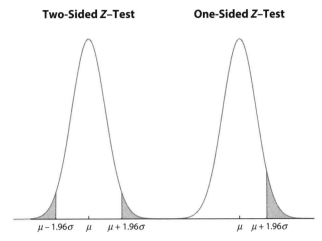

Figure 22.3. The difference between a two-sided z-test and a one-sided z-test. Both tests have an α-level of 0.05.

illuminates the core difference between **one-sided** and **two-sided hypothesis tests**. The McDonald's case is an example of a one-sided test, where we're interested in seeing whether the parameter we're measuring is greater than (or less than) a specific value. To do this, we calculate the probability of observing a test statistic as large or larger than the one we have. With a two-sided test, we're interested in seeing whether the parameter is significantly different from a given value, so we calculate the probability of observing a test statistic as far from the hypothesized mean or further than the one we have. Clearly the p-value we measure depends on which test you're doing. The difference between one- and two-sided tests is depicted in Figure 22.3.

You'll notice that the critical value is smaller for the one-sided test than the two-sided test (your test statistic only needs to be 1.64 standard deviations from the mean instead of 1.96). This is a general property of one- and two-sided tests: two-sided tests require more evidence than one-sided tests.

Why do two-sided tests require more evidence, and if so why don't we always use one-sided tests? The issue is we need information to justify performing a one-sided test. In the McDonald's example we could safely assume that they weren't any faster than they advertised, because if they were they most certainly would have advertised it! However, if we were unable to discard that possibility, then we would have needed to use a two-sided test.

One final word about one-sided tests. In our McDonald's example, our null hypothesis was $\mu \leq 45$, however when we actually performed our test, we just let $\mu = 45$. Why don't we need to worry about the possibility that the mean was, say, 43? Well, imagine we did let $\mu = 43$. Then our test statistic would have become $(48 - 43)/1.79 \approx 2.79$—larger than we had before! For *any* mean less than 45, we would have measured a higher z-value. So $\mu = 45$ is the most difficult case; if we can reject the null hypothesis while letting $\mu = 45$, then we can reject it for any hypothesized mean less than 45. This is a handy property of one-sided tests: you only need to check the most extreme case.

Now that we have laid out the framework for hypothesis testing, let's go through some examples.

Example: Imagine you're measuring the lifetime of lightbulbs, which are known to have a standard deviation of $\sigma = 100$ hours, and you want to test whether their lifetime is significantly different from 2,000 hours. In a random sample of 20 lightbulbs, you find the mean lifetime to be 2050 hours. With an α-level of 0.05, would you reject the hypothesis that the mean lifetime is 2000 hours?

Solution: We've already been given the null hypothesis, namely that $\mu = 2000$. Therefore our alternative hypothesis is

$$H_a : \quad \mu \neq 2000.$$

Notice that this is a two-sided test, since we have no reason to dismiss the possibility that lightbulbs last less than 2000 hours. We now need to calculate our test statistic, which is

$$z = \frac{\text{observed mean} - \text{hypothesized mean}}{\text{standard deviation}}.$$

Be careful! A common mistake people make is to say, "Well the mean I found was 2050 and the hypothesized mean is 2000, and since the standard deviation is 100, my z-score is 1/2. This is too small to conclude that the means are different." However, they're using the wrong standard deviation! The standard deviation of 100 hours is for a sample of *one* lightbulb. For a sample of 20 lightbulbs, you expect to get a more reliable estimate of the mean, and the correct standard deviation to use is $\sigma / \sqrt{n} = 100 / \sqrt{20} \approx 22.36$ hours. Thus our z-score is $(2050 - 2000) / 22.36 \approx 2.24$. Since we're using a two-sided test, our p-value is the probability of measuring a test statistic greater than 2.24 or less than -2.24, which turns out to be 0.025. This is very compelling evidence that the true mean is not 2000 hours.

Example: Suppose an auto insurance company has recently moved from Boston to Seattle, and is trying to get a sense of their new market. In Boston, the percent of their policyholders that filed a claim in a given year was described by a binomial distribution with probability of success (i.e., a driver filing a claim) of $p_b = 0.25$. In their first year in Seattle, the firm finds that 2300 of their 10,000 drivers filed a claim. Using an α-level of 0.05, test whether p_s, the probability of an insured driver in Seattle filing a claim, is less than 0.25.

Solution: What are our null and alternative hypotheses in this case? Since we want to test whether $p_s < 0.25$, our null hypothesis is

$$H_0 : \quad p_s \geq 0.25,$$

and our alternative hypothesis is

$$H_a : \quad p_s < 0.25.$$

Here we're using a one-sided test, so we need to be careful about justifying why we don't consider the possibility that Seattle drivers have more accidents than Boston drivers. Perhaps previous research suggests that this is the case, or maybe Seattle's smaller population density would make us expect fewer accidents. Assuming our one-sided test is reasonable, once we have our hypotheses, we proceed as we typically do. Let's assume the null hypothesis is true, and take $p_s = 0.25$. If this were the case, the number of claims in Seattle should be described by a binomial distribution of size 10,000 and probability of success 0.25. However, we can make our lives a little bit simpler. For large N, a binomial distribution of size N with probability of success p is well approximated by a normal distribution with mean Np and standard deviation $\sqrt{Np(1-p)}$ (this is just a special case of the Central Limit Theorem; we gave the proof when $p = .5$ in §18.3). In most books we often consider any N-value greater than 30 to be large; as we're at 10,000 here, it's fairly safe to replace the binomial with a normal. Therefore, if $p_s = 0.25$, the number of drivers filing claims out of a group of 10,000 should be approximately normally distributed with mean $10000 \cdot 0.25 = 2500$ and standard deviation $\sqrt{10000 \cdot 0.25 \cdot 0.75} \approx 43.3$.

Now that we know what the distribution of drivers filing for claims should look like, we can proceed to calculate our test statistic. We've shown that under the null hypothesis, the number of drivers filing a claim should be normally distributed with mean 2500 and standard deviation 43.3. We had 2300 drivers file a claim. Therefore our test statistic is

$$\frac{2300 - 2500}{43.3} \approx -4.62.$$

This test statistic should follow a standard normal distribution.

The last step is to compute our p-value, which is the probability of observing a test statistic as or more extreme than the one we did. Since we're testing whether $p_s < 0.25$, we only need to find the probability of measuring a test statistic lower than -4.62, which is just $\Phi(-4.62) \approx 1.92 \times 10^{-6}$ (here Φ is the cumulative distribution function of the standard normal). This is much less than our α-level of 0.05, so we reject the null hypothesis and conclude that the probability of an insured driver in Seattle filing for a claim in a given year is less than 0.25.

 In hypothesis testing we need to justify the distribution of the test statistics. When we're testing hypotheses about sample means, we can typically appeal to the Central Limit Theorem to conclude that $\bar{X}$ is normally distributed. However, we need to be careful with this method of argumentation. If our underlying distribution is "nice" and our sample size is large, then it's probably fine to take $\bar{X}$ as normally distributed. But if you're ever in a situation where you're dealing with a weird distribution or have only a few data points, you need to be really careful if you're going to appeal to the CLT (or you might have a Cauchy distribution, in which case the mean does not even exist!). To put it another way: if you're going to use a z-test, you need to be able to provide a good reason for why the data you've collected are normally distributed.

One final point to be aware of about z-tests is that they assume perfect knowledge of the variance. In the McDonald's example we actually used the *sample variance* and assumed it to be the true variance. This is technically incorrect, and we will discuss how to fix this situation later in the chapter.

22.2 On p-values

In the last section we developed a method of hypothesis testing. Within the hypothesis testing framework, the most important measurement we make is the **p-value**. The p-value tells us the probability of collecting data as or more extreme than what we have, *assuming the null hypothesis is true.* In this section we provide some more intuition about *p*-values, and warn against some possible misinterpretations of it.

Takeaways for the section:

- The *p*-value is a conditional probability: the odds of collecting the data you have given that the null hypothesis is true.
- The *p*-value depends on context.
- Different tests of the same data can yield different *p*-values.
- The *p*-value is not the probability that the null hypothesis is true.

22.2.1 Extraordinary Claims and *p*-values

As we outlined in the first section, the *p*-value is a measure of how likely it is for the event we observed to occur by chance alone, assuming the null hypothesis is true. If we observe an event with a very small *p*-value, there are two options available to us: we can either conclude that we have seen a rare event, or conclude that the event we saw was so unlikely given our assumptions that our assumptions were probably wrong (that is, reject the null hypothesis).

One point we'd like to emphasize is that how convincing a *p*-value is depends on context. How can that be? Imagine a musician walks up to you and tells you he has perfect pitch (he can identify any note just by listening to it). Suppose that to test this you play 8 different notes for him and he accurately identifies all 8 of them. Would you believe his claim? Probably, since trained musicians tend to have good ears, and if he were guessing, the odds of him getting all 8 right are exceedingly small. Now imagine a man walks up to you and says he can tell you what note you're going to play next just by thinking hard about it. To test his claim, you have him write down what note he thinks you're going to play, and you then write down the note you actually play. After 8 notes, you reconcile your lists. Suppose he gets all 8 of the notes right. Would you believe his claim? Even though we have the same evidence as in our first example, you would probably be hesitant to believe that this man can predict your actions (and you would probably want to get more data points!) simply because his claim is so incredible. So even though we have identical situations in terms of evidence (that is, if we defined the null hypothesis that these guys were just guessing, the *p*-value would be the same in both cases), we're more likely to believe one *p*-value over the other. As Carl Sagan once said, "Extraordinary claims require extraordinary evidence."

22.2.2 Large *p*-values

Suppose you want to test whether a coin is fair (that is, lands heads or tails with equal probability), and that in flipping the coin 20 times, you recorded 12 heads. If our null hypothesis is that the coin is fair, what is the *p*-value? Assuming the coin is fair, the number of heads flipped H should follow a binomial distribution:

$$\Pr(H = x) = \frac{1}{2^{20}} \binom{20}{k}.$$

The probability of flipping 12 or more heads (or tails), that is, our p-value, is

$$p = \frac{1}{2^{20}} \left(\sum_{k=0}^{8} \binom{20}{k} + \sum_{k=12}^{20} \binom{20}{k} \right) \approx 0.503.$$

This is a larger p-value than we've seen before, and we definitely could not reject our null hypothesis with this kind of data. So what does this mean? Have we proven that the coin is fair? Certainly not! This data would actually be more consistent with the hypothesis that the probability of flipping a head was 0.6. However, this data is not out of line with what we would expect from a fair coin. Since we cannot say that we have proven the coin fair, we say that we **fail to reject the null hypothesis** that the coin is fair. *This is an important part of statistical language—we never "accept" any hypotheses, we merely reject them or fail to reject them.* (This is similar to juries returning "not guilty" verdicts instead of "innocent" ones.) It could be that the coin is not fair, and that with more data points we would see a stronger bias emerge. However, from the sample that we've observed, there's no compelling evidence to make us revise our initial claim.

22.2.3 Misconceptions about p-values

One final point about p-values: the p-value is *not* the probability that the null hypothesis is true. This is a very common mistake, and a reasonable one to make. However, consider the coin example: the coin is either fair or it isn't—there's no randomness to the state of the coin. In this case it doesn't even make sense to talk about the probability of the null hypothesis being true. However, the p-value still makes sense: it's the conditional probability of collecting the evidence we have given that the null hypothesis is true, not the probability that the null hypothesis is true given the data we collected.

  Example: You're an office manager and have purchased 20 brand new photocopiers. The manufacturer tells you that the probability of one of them breaking down in a given year is $p_{ph} = 0.03$. Over the course of the first year, two of the photocopiers break down. Do you now believe the value of p_{ph} the manufacturer told you? At what α level could you reject the manufacturer's claim?

Solution: This is a nice introduction to hypothesis testing with binomial distributions, as our sample is too small to use the normal approximation. We must assume that each photocopier breaking down is independent of whether or not other photocopiers break down in order to use a binomial distribution. As always, we first formulate our null and alternative hypotheses. We suspect that the machines break with frequency greater than 0.03, so let's hypothesize as follows:

$$H_0 : \ p_{ph} \leq 0.03, \quad H_a : \ p_{ph} > 0.03. \tag{22.2}$$

Let's assume the null hypothesis to be true, and take $p_{ph} = 0.03$. If this were the case, then the probability of having x machines break down in a year is

$$\text{Prob}(n \text{ machines break}) = \binom{20}{n} (0.03)^n (1 - 0.03)^{20-n}.$$

The beautiful thing about hypothesis testing with binomial distributions is that we already know our test statistic—it's 2. Our p-value is just the probability that 2 or more machines break in a given year. We could do this the long way by adding the probability that two fail to the probability that three fail, et cetera, or we could just note that

$$\text{Prob}(2 \text{ or more machines break}) = 1 - \text{Prob}(0 \text{ or } 1 \text{ machines break}),$$

which is just $1 - \left((0.97)^{20} + 20 \cdot 0.03 \cdot (0.97)^{19}\right) \approx 0.12$. This is not a very convincing p-value. Yes, having two photocopiers break in a year is rare, but not so rare that you'd begin to wonder whether the manufacturer was lying to you. To be able to reject the null hypothesis, your α-level would have to be at least 0.12, which is too large to be accepted in any formal circumstances.

Example: From 1997–2006, the number of cases of the flu in Hartford—per 10,000 in the population—was modeled by a Poisson distribution with $\lambda = 350$. As a reminder, this means that the probability of seeing k cases of the flu in a particular year is given by

$$\text{Prob}(k \text{ cases of the flu}) = \frac{e^{-350} \cdot 350^k}{k!}.$$

Starting in 2007, a law was passed that increased the number of flu shots given out. In the three years since, the number of cases of the flu per 10,000 people in the population have been 330, 320, and 325 respectively. Assuming the new law was the only meaningful change affecting flu rates (this is a big assumption), do you believe that increasing the availability of flu shots helped reduce the number of cases of the flu in Hartford?

Solution: Let's take our null hypothesis to be that the law did nothing, and that the low totals of the past few years are due to nothing but chance. In this case, the number of cases of the flu will be modeled by a Poisson distribution with $\lambda = 350$. What kind of distribution should we have over the course of three years? As we saw back in §19.1, if X follows a Poisson distribution with parameter λ_1 and Y follows a Poisson distribution with parameter λ_2, then $X + Y$ follows a Poisson distribution with parameter $\lambda_1 + \lambda_2$ (in other words, we again have a **stable** distribution). If we had three Poisson random variables, it would be a Poisson with parameter equal to the sum of the three parameters.

We see that over the course of three years, the number of cases of the flu in Hartford should be modeled by a Poisson distribution with $\lambda = 350 + 350 + 350 = 1050$. We've had 975 cases of the flu over the past three years. The probability of seeing that few or fewer is

$$\sum_{n=0}^{975} \frac{e^{-350} \cdot 350^n}{n!} \approx 0.010,$$

giving a p-value of 0.01. This is pretty compelling evidence that the law was effective.

Remark: We mentioned that assuming we can attribute all the changes in flu cases to this new law is a big assumption. This is because there are many other factors that may have been responsible for lowering the incidence of the flu. Perhaps a recently launched public health program increased the number of Hartfordians who washed their hands consistently, which helped reduce the transmission of germs. Or maybe a few unseasonably warm winters meant people weren't spending as much time cooped up indoors, and thus were less likely to pass germs to each other. For problems like

these, it's always important to recognize the assumptions you're making, and to think about potential factors you've left out. This is important for two reasons: it lets you recognize the shortcomings of your model, and (more importantly) also helps steer you towards important data. For example, if we think weather played a factor in Hartford, maybe we could look at other towns that had warm winters but didn't have a flu vaccine law passed.

22.3 On *t*-tests

At the end of §22.1 we mentioned one potential issue with z-tests: they assume perfect knowledge of the variance. However, there are very few instances where we actually know the variance. Most of the time we need to estimate it from the data itself. In this section we discuss how to find an accurate estimator of the variance, and how to use this estimator in our hypothesis testing framework.

Takeaways for the section:

- The z-test assumes perfect knowledge of the variance, which in many cases we don't have.
- We can get an estimate of the variance from the data, and use that in place of the actual variance.
- Using the sample variance changes the distribution of our test statistic.

22.3.1 Estimating the Sample Variance

In the case that we collect data on a population parameter without any prior information about its variance, the most straightforward way to estimate the variance is to calculate the **sample variance** as follows: find the sample mean $\bar{x}$ as always ($\bar{x} = (x_1 + \cdots + x_n)/n$), and then take the sample variance s^2 to be

$$s^2 := \frac{1}{n-1} \sum_{i=1}^{n} (x_i - \bar{x})^2 = \frac{1}{n-1} \sum_{i=1}^{n} x_i^2 - \frac{n}{n-1}\bar{x}^2.$$

This looks remarkably like our regular formula for variance, except we now have $n-1$ in the denominator. This is due to an important concept in statistics known as **degrees of freedom**. For any given estimate, the degrees of freedom are the number of independent observations that contribute to that estimate. Imagine we know that in Boston the average high temperature in August is 80°F, and we want to get a sense of the variance. If we record the high for the next day to be 85°F, we could estimate that the variance is $(85 - 80)^2 = 25$ (the units would be degrees Fahrenheit^2, of course). Suppose, as is usually the case, we didn't know the mean beforehand. Then if we measure the high temperature to be 85°F, we could use that as our estimate for the mean. However, we can't say anything about the variance, because there's nothing to deviate from the mean. What happens if we try to use the formula above for a sample of size one? You get the indeterminate 0/0—even the equation knows it shouldn't be used in this instance!

What if we made a second observation, and say the high was 88°F? Now our sample mean is 86.5°F, and we can estimate the sample variance. However, since we're using the sample mean to estimate the variance, both observations don't independently contribute to the variance; once you know the mean and the value of one of the highs, you automatically know the value of the second high. Thus only one observation contributes to the sample variance. In general, we use $n - 1$ because we want to measure the amount of variation in our sample, per observation that contributes to the variance.

*Remark: Another way to convince yourself that we should use $n - 1$ instead of n is to show that s^2 is an **unbiased estimator** of σ^2. This means $\mathbb{E}(s^2) = \sigma^2$, so on average our prediction is good.*

22.3.2 From z-tests to t-tests

With the sample variance in hand, we're ready to fix the problem with the z-tests we mentioned earlier. Let's recall that when performing a z-test, we find a test statistic $\bar{x}$ that is normally distributed with mean μ and variance σ^2/n (under the null hypothesis). Given this, we form the test statistic

$$(\bar{x} - \mu) / \sqrt{\sigma^2/n},$$

which has a standard normal distribution. What if we don't know σ^2? It seems reasonable to use our estimate s^2 in its place. That is, we place $\sqrt{s^2/n}$ in the denominator instead of $\sqrt{\sigma^2/n}$. How should this affect the distribution of our test statistic? Since $\mathbb{E}[\bar{x}] = \mu$ and $\mathbb{E}[s^2] = \sigma^2$, it should be close to the z-statistic. However, the distribution should be a little more spread out than the normal distribution. Why is this? When we knew σ^2, there was only one source of randomness—$\bar{x}$. Now we have two measures that can bounce around—$\bar{x}$ and s^2—so it makes sense that our distribution should not be as localized as it was before. To put it another way: we shouldn't have better estimates now that we know less! It turns out that our test statistic does indeed follow a distribution which is symmetric about 0—a distribution called the **t-distribution**.

---

t-distribution: The **t-distribution** is a family of distributions parametrized by their degrees of freedom. Let $X_1, \ldots, X_n$ be independent standard normal random variables, and let S_n^2 be the sample variance random variable. Then the distribution of $\bar{X}/(S_n/\sqrt{n})$ is called the t-distribution with $n - 1$ degrees of freedom, and is denoted T_{n-1}. The closed form expression for the density of a t-distribution with ν degrees of freedom is

$$\frac{\Gamma\left(\frac{\nu+1}{2}\right)}{\sqrt{\nu\pi}\,\Gamma(\nu/2)} \left(1 + \frac{x^2}{\nu}\right)^{-\frac{\nu+1}{2}}.$$

More generally, if the underlying distribution is conjectured to be normally distributed with mean μ (and unknown variance σ^2) then the test statistic is

$$\frac{\bar{x} - \mu}{\sqrt{s^2/n}} \sim t_{n-1};$$

note the variance cannot enter the definition of the test statistic as it is not known and hence is unavailable!

The t–Distribution with Varying Degrees of Freedom

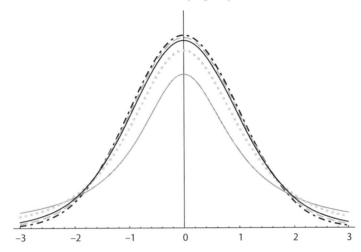

Figure 22.4. The t-distribution with (from shortest to tallest): 1, 3, 9, and 20 degrees of freedom. The tallest curve is the standard normal distribution. Notice that the t-distribution approximates the normal distribution very well once we get to 30 degrees of freedom or so.

The t-distribution looks remarkably like the normal distribution, as shown in Figure 22.4. In fact, in the limit as n goes to infinity, the t-distribution approaches the normal distribution. This makes intuitive sense: we use the t-distribution when we have an underlying normal distribution but need to estimate the variance. When our sample size is large, our estimate of the variance becomes more and more exact, and we don't need to worry as much about the uncertainty in our estimate. A little more formally, using

$$\lim_{n \to \infty} \left(1 + \frac{x}{n} \right)^n = e^{-x},$$

we see

$$\lim_{\nu \to \infty} \left(1 + \frac{x^2}{\nu} \right)^{-\frac{\nu+1}{2}} = \left(\lim_{\nu \to \infty} \left(1 + \frac{x^2}{\nu} \right)^{\nu} \right)^{-1/2} \lim_{\nu \to \infty} \left(1 + \frac{x^2}{\nu} \right)^{-1/2} = e^{-x^2/2}.$$

Once we know how our test statistic is distributed, the t-test works exactly the same way as z-tests. We formulate null and alternative hypotheses, calculate our test statistic, and find the corresponding p-value. The intuition is also exactly the same: if

$$\beta = \frac{\bar{x} - \mu}{\sqrt{s^2/n}}$$

follows a t-distribution, then for any value of β we can calculate how likely it is to measure a test statistic that large or larger. If that probability is sufficiently small, we reject the null hypothesis.

Now that we have the formalism of t-tests at our command, let's return to the McDonald's example we gave at the beginning of the chapter. Instead of taking

$\sigma^2 = 48$ as we did before, we can more accurately use the sample variance $s^2 = 64$, which gives

$$t = \frac{48 - 45}{\sqrt{64/20}} \approx 1.68,$$

as we found earlier. The only difference is that now we expect our test statistic to follow a t-distribution with 19 degrees of freedom. It turns out that for this distribution, the cutoff value for a 0.05 α-level is $t = 1.73$, so now we fail to reject the null hypothesis that $\mu = 45$ at a 5% significance level. Given how close we were to the 5% significance level before, it shouldn't be surprising that a slightly different test yielded a different result, though it might be worrisome!

What happened? Why were we able to reject the null when working with a z-test, but were unable to when using a t-test? There are two factors: we weren't as well informed as we thought we were, and significance levels impose arbitrary cutoffs. When we did our first z-test, we claimed we knew that the standard deviation was 8 seconds. However, in using the t-test we had to admit that we were only estimating the variance, and because of that each of our further estimates became a little more uncertain. The other issue is that we ran up against the 5% cutoff level. Even when using the t-test, our p-value was still 0.055—pretty compelling evidence that McDonald's is slower than they claim, but just over our cutoff. For this reason, as we remarked above many researchers forego rejecting or failing to reject hypotheses, and simply do the analysis and report the p-value.

Example: Suppose you grow tomatoes in your home garden. You've made some observations over the past few years, and have consistently found that the weight of the tomatoes you grow is nearly normally distributed with mean weight 4 ounces. Recently, though, you've seen advertisements for a new fertilizer that claims to increase the size of produce. On an adventurous whim, you decided to test it out. In the next batch of tomatoes you grow there are two 3 ounce tomatoes, four 4 ounce tomatoes, and six 5 ounce tomatoes. From this, can you conclude that the fertilizer increases the yield?

Solution: Since we're interested in seeing whether the fertilizer increases the yield, our null hypothesis is that the fertilizer has no (or perhaps a negative) effect. Denoting the mean tomato size with fertilizer by μ, we have

$$H_0 : \ \mu \leq 4, \qquad H_a : \ \mu > 4. \tag{22.3}$$

As usual, let's assume the null hypothesis is true and let $\mu = 4$. To calculate our test statistic, we need the sample mean and the sample variance. We see the sample mean is

$$\bar{x} = \frac{2 \cdot 3 + 4 \cdot 4 + 6 \cdot 5}{12} = 4.33,$$

and the sample variance is

$$s^2 = \frac{1}{11} \left(2 \cdot (3 - 4.33)^2 + 4 \cdot (4 - 4.33)^2 + 6 \cdot (5 - 4.33)^2\right) \approx 0.61.$$

Our test statistic is therefore

$$\frac{\bar{x} - \mu}{\sqrt{s^2/n}} = \frac{4.33 - 4}{0.225} \approx 1.48.$$

How should this test statistic be distributed? Since we're using the sample variance, our first guess would be a t-distribution. However, to use the t-distribution, we need to have an underlying normal distribution. Do we in this case? Since we said that the distribution of tomato weights was nearly normal, the Central Limit Theorem tells us that a sample of size 12 from this distribution will be very close to normal. Granted our sample size is a little smaller than we would typically like to apply the Central Limit Theorem, but the assumption of normality seems reasonable.

Since $\bar{x}$ is normally distributed, we know

$$\frac{\bar{x} - \mu}{\sqrt{s^2/n}} \sim t_{n-1},$$

so our test statistic should follow a t-distribution with 11 degrees of freedom. For this distribution, a t-score of 1.48 corresponds to a p-value of 0.083. While this isn't concrete evidence that the fertilizer increases the size of the tomatoes, I'd probably keep using it until I'd generated some more conclusive data.

22.4 Problems with Hypothesis Testing

Now that we've developed the basic structure and intuition of hypothesis testing, we have a sobering confession to make: hypothesis testing is not perfect. Since we ultimately appeal to probabilistic arguments, we can't be absolutely certain of our conclusions. There are two types of errors we could possibly make: we could falsely reject a true null hypothesis, or we could fail to reject a false null hypothesis. Statisticians, exhibiting their typical flair for nomenclature, have termed these errors **Type I** and **Type II errors**, respectively.

Takeaways for the section:

- Probabilistic arguments are never conclusive.
- There are two kinds of errors we might make: rejecting a true null hypothesis, or failing to reject a false null hypothesis.
- There is a trade-off between Type I and Type II errors.

22.4.1 Type I Errors

Let's tackle Type I errors first. As we mentioned above, a **Type I error** is the error of rejecting a true null hypothesis. Why would this happen? Suppose we're testing a hypothesis, and the null hypothesis we've identified is in fact true. Our α-level (say it's 0.05) means that if we observe a result that would happen less than 5% of the time by chance alone, we'll reject the null hypothesis. How often will we see a result that would happen at most 5% of the time by chance alone? Exactly 5% of the time! (This isn't rocket science.) More generally, the probability of making a Type I error (conditional on the null hypothesis being true) is exactly the α-level.

22.4.2 Type II Errors

There is another type of error we could potentially make. We could identify the wrong null hypothesis, but fail to reject it. This is known as a **Type II error**. Type II errors are trickier than Type I errors because they depend on which false null you've identified. Why is this the case? If your null hypothesis is way off, it should be relatively easy to reject. However, the closer the null hypothesis comes to the truth, the more difficult it is to reject. Let's do an example.

Example: Imagine you're trying to estimate the average height of adult males at your college. Let's assume that the true population mean is 6 feet (1.83 meters) with a standard deviation of 3 inches (7.62 centimeters), and that you draw a sample of size 20. If your null hypothesis is that the mean height is 5 feet (1.52 meters), what is the chance of making a Type II error if your α level is 0.05?

Solution: For simplicity, let's assume we know that the standard deviation is 3 inches and that the sample mean should be roughly normally distributed. This implies we'll reject H_0 if your sample average is more than 1.96 standard deviations away from the mean. With a sample of size 20, our standard deviation is $s = \sigma/\sqrt{n} = 3/\sqrt{20} \approx 0.67$ inches. Thus we reject H_0 if the sample mean is greater than 5 foot 1.31 inches, or less than 4 foot 10.69 inches. Alternatively, we make a Type II error if our sample mean falls between 4 foot 10.69 inches and 5 foot 1.31 inches. We know that the true distribution for samples of size 20 is normal with mean 6 feet and standard deviation 0.67 inches. Given this, the probability of making a Type II error is about 10^{-57}—no real worries here!

What if we had formulated the more reasonable null hypothesis that the average height is 5 foot 11 inches? The same analysis carries through, except now we would make a Type II error if our sample mean was between 5 foot 9.69 inches and 6 foot 0.31 inches. This type of sample would happen over 67% of the time, so it's very likely that we would make a Type II error. The reason is that our null hypothesis is so close to the truth that we will frequently see data that are consistent with the null. The probability of making a Type II error (conditional on the null hypothesis being false) is called β. We will use this idea to describe power, which we will talk about in a later section.

22.4.3 Error Rates and the Justice System

An important concept related to Type I and Type II errors is the error rate. The **Type I error rate** is the probability of committing a Type I error if the null hypothesis is true. Similarly, the **Type II error rate** is the probability of committing a Type II error when the null hypothesis is false. In our last example, the Type II error rate was 0.67.

One way to think about Type I and Type II errors is in the context of a criminal trial. Jurors formulate the null hypothesis that the defendant is innocent, and then listen to testimony. If the evidence presented seems sufficiently unlikely under the premise of innocence, the juror submits a guilty vote, while if the evidence is not substantial enough the juror enters a not guilty vote (note the guilty/not guilty terminology: in much the same manner that we never accept null hypotheses but only reject or fail to reject them, criminal trials never find someone "innocent," only guilty or not guilty). The possible outcomes are summarized in Table 22.1.

TABLE 22.1.

The possible outcomes of a criminal trial. The null hypothesis is that the defendant is innocent; had it been that he is guilty, the Type I and Type II errors would be flipped.

	Defendant Innocent	*Defendant Guilty*
Convict	Type I Error	Good
Don't Convict	Good	Type II Error

We see that in our trial case, a Type I error means convicting an innocent person, while a Type II error would mean letting a guilty person go free (hence Type I errors are also called *false positives*, while Type II errors are called *false negatives*). Neither of these outcomes seems ideal—is there a way we could lower the incidence of both Type I and Type II errors? At first glance, it might seem not. If a juror decided to require more convincing evidence in the hopes of reducing Type I errors, she would automatically make it harder to convict *anyone*, including a guilty defendant. Thus by trying to reduce the incidence of Type I errors, she necessarily increased the likelihood of committing a Type II error. There is, however, one thing we could do to reduce both types of errors—simply listen to more evidence. The clearer we are about what is going on, the less likely we are to make an error in judgement.

 Example: A local college is testing its students for swine flu, which is known to raise your white blood cell (WBC) count. Suppose the WBC count of healthy people is normally distributed with a mean of 7000 WBCs per microliter and standard deviation of 1000, and that the WBC count of people with swine flu is normally distributed with a mean of 11,000 WBCs per microliter and a standard deviation of 1500. Since the disease is very communicable, the college wants to quarantine people they suspect of having it. If the college quarantines everyone with a WBC count over 9,000, what are the corresponding Type I and Type II error rates? If the college wants to have a Type II error rate of 0.05, where should they set their threshold?

Solution: Let's first consider the case where the college quarantines anyone with a WBC count over 9,000. What is a Type I error in this case? A Type I error is a false positive, so a Type I error would be deciding that a healthy person had swine flu. The odds of this are just the odds that a healthy person has a WBC count over 9,000, which happens with probability $1 - \Phi(2) \approx 0.023$ (remember Φ is the cumulative distribution function for the standard normal).

What about Type II errors? These happen when we determine a sick person is actually healthy, which occurs whenever a sick person has a WBC count less than 9,000. This happens with probability $\Phi(-1.33) \approx 0.091$.

If the college wanted a Type II error rate of 0.05, what should their threshold be? Again, a Type II error will occur whenever a sick person has a WBC count that falls below the threshold. Thus we need to find the z such that $\Phi(z) = 0.05$. Using a standard normal table, we find $z = -1.64$. Thus the threshold should be at $11000 - 1.64 * 1500 = 8540$.

Just for fun, what's the Type I error rate in the case where the threshold is a WBC count of 8540? That's the probability that a healthy person has a WBC count over 8540, which is $1 - \Phi(1.54) \approx 0.062$. Again, we see the general phenomenon that lowering the incidence of Type II errors increases the likelihood of Type I errors.

22.4.4 Power

In the case that the null hypothesis is false, we would ideally want our test to correctly reject it. The **power** of the test allows us to measure how likely we are to successfully reject the null hypothesis given that it is indeed false. Thus, since β is the probability of making a Type II error, or failing to reject a false null hypothesis, then the power of the test is the probability that our test successfully rejects a false null hypothesis, or $1 - \beta$.

Clearly, the lower the power of our test, the more likely we are to make a Type II error. As a result, the first place to look when we fail to reject the null hypothesis is the test's power and ways to increase it. Often, problems emerge from having too small of a sample size. It is very possible that the sample size is too small to be representative of the population we wish to generalize on, resulting in us failing to detect any significant difference when one exists outside of our sample. We will have more data with a larger sample size, meaning we should have more evidence that the null hypothesis is false if it is indeed false. In other words, if the population is on average significantly different, then increasing the sample size, which makes the sample closer to the true population, should increase the likelihood of detecting the difference.

For example, suppose we want to know if 30 passengers on an airplane enjoy a new lunch menu item, fish, more or less than the typically offered meal of steak (assuming no one is vegetarian). The null hypothesis would be that there is no difference in enjoyment. Suppose 25 passengers do enjoy the fish more or less than the steak, but 5 are indifferent. If we have a small sample size of say 4 and happen to select only among the 5 who are indifferent, then we will fail to reject the null hypothesis even though it is clearly false for our population of 30 passengers. If we sample 10 passengers, we will already detect the difference since only 5 are indifferent. Thus, one way to increase power is to increase sample size. By the end of this chapter, we'll know a few ways of increasing power.

22.4.5 Effect Size

As established in the previous section, the power of a test is a quantifier of how far the truth lies from our hypothesized value. This distance between the null value, p_0, and the truth, p, is the effect size. Since we don't know the true value, however, we estimate the effect size as the difference between the null and observed values.

Effect size is crucial to understanding the power of a hypothesis test. A large effect means a smaller Type II error, and thus a larger power. Additionally, a small effect means a larger Type II error and therefore a smaller power. Therefore, knowing the effect size and sample size helps us determine power. The issue here is that, when designing tests, we won't yet know the observed values and therefore can't calculate the effect size. This means we have to try several different effect sizes and look at their consequences. When researchers design a study, they often know what effect size matters to their conclusions and use that to estimate n, the needed sample size.

22.5 Chi-square Distributions, Goodness of Fit

Up to this point, we've only done hypothesis tests on the mean. However, there are tests we can run on other interesting parameters as well. In this section we discuss how to perform hypothesis tests on the variance, and how to test a model itself. Before we can do this, however, we need to introduce a very important distribution known as the χ^2 distribution. We quickly review it below; see Chapter 16 for a more extensive introduction.

Takeaways for the section:

- There is a connection between normal and chi-square random variables.
- This connection leads to sample variances and t-tests.
- We can test goodness of fit of theory and experiment with chi-square random variables.

22.5.1 Chi-square Distributions and Tests of Variance

Suppose X has a standard normal distribution. What distribution does X^2 follow? We can find the density function for X^2 rather easily:

$$\text{Prob}(a \leq X^2 \leq b) \ = \ \text{Prob}(\sqrt{a} \leq X \leq \sqrt{b}) + \text{Prob}(-\sqrt{b} \leq X \leq -\sqrt{a})$$

$$= \ 2 \cdot \text{Prob}(\sqrt{a} \leq X \leq \sqrt{b}) = 2 \int_{\sqrt{a}}^{\sqrt{b}} \frac{1}{\sqrt{2\pi}} e^{-x^2/2} dx.$$

Let's make the substitution $u = x^2$. Then our limits of integration go from a to b and $dx = \frac{du}{2\sqrt{u}}$, yielding

$$\text{Prob}(a \leq X^2 \leq b) \ = \ 2 \int_a^b \frac{1}{\sqrt{2\pi}} e^{-u/2} \frac{du}{2\sqrt{u}} \ = \ \int_a^b \frac{1}{\sqrt{2\pi}} u^{-1/2} e^{-u/2} du.$$

Thus the density function for X^2 is given by $1/\sqrt{2\pi}\, x^{-1/2}e^{-x/2}$. This is one of the most common densities in statistics, and is called the **chi-square distribution with one degree of freedom**. Like the t-distribution, the χ^2 distribution is a family of distributions parametrized by their degrees of freedom. More generally, a χ^2 distribution is defined as follows.

χ^2 **distribution with k degrees of freedom**: Suppose for $1 \leq i \leq k$, X_i are independent, standard normal variables. Then the random variable

$$Y \ = \ X_1^2 + X_2^2 + \cdots + X_k^2$$

follows a χ^2 distribution with k degrees of freedom, often denoted χ_k^2.

Notice that this definition immediately implies that if $Y \sim \chi_k^2$ and $X \sim \chi_l^2$, then $X + Y \sim \chi_{k+l}^2$ so long as X and Y are independent.

The density function for a χ^2 distribution with k degrees of freedom is given by

$$f(x) \ = \ \frac{1}{2^{k/2}\Gamma(k/2)} x^{k/2-1} e^{-x/2},$$

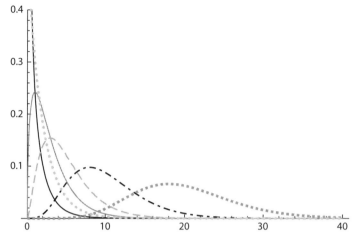

Figure 22.5. Plot of chi-square distributions with $\nu \in \{1, 2, 3, 5, 10, 20\}$; as the degree of freedom increases, the location of the bump moves rightward.

where Γ is the Gamma function. Graphs of the chi-square distribution are shown in Figure 22.5.

So why on earth are we discussing this funny looking distribution? One reason is that the sample variance of a normally distributed random variable is closely related to the χ^2 distribution. Consider a random variable X which is distributed $N\left(\mu, \sigma^2\right)$, and imagine we've drawn a sample of size n from this random variable. We've seen before that our sample mean $\bar{x}$ should be distributed $N\left(\mu, \sigma^2/n\right)$, so

$$\frac{\bar{x} - \mu}{\sqrt{\sigma^2/n}} \sim N(0, 1).$$

Now imagine we need to calculate the sample variance:

$$s^2 = \frac{1}{n-1} \sum_{i=1}^{n} (x_i - \bar{x})^2.$$

One natural question to ask is what kind of distribution the sample variance follows. It can't be a normal distribution, because the sample variance is never negative. However, when we calculate the sample variance, we're squaring a whole bunch of $(x_i - \bar{x})$ terms, so maybe we would expect that the χ^2 distribution should come into play. To get at the question of what distribution the sample variance follows, let's first consider something very close to the true variance:

$$\frac{1}{\sigma^2} \sum_{i=1}^{n} (x_i - \mu)^2 .$$

This is just our regular expression for the variance, except missing a factor of $1/n$ and with an additional $1/\sigma^2$. The reason for these alterations will be made clear in a minute.

By adding 0 in a clever manner, we can rearrange this to

$$\frac{1}{\sigma^2} \sum_{i=1}^{n} (x_i - \mu)^2 = \frac{1}{\sigma^2} \sum_{i=1}^{n} ((x_i - \bar{x}) + (\bar{x} - \mu))^2$$

$$= \frac{1}{\sigma^2} \sum_{i=1}^{n} \left((x_i - \bar{x})^2 + (\bar{x} - \mu)^2 + 2(x_i - \bar{x})(\bar{x} - \mu) \right).$$

Now $\bar{x} - \mu$ is a constant, so $\sum_{i=1}^{n}(\bar{x} - \mu)^2 = n(\bar{x} - \mu)^2$. As $\sum_{i=1}^{n}(x_i - \bar{x}) = 0$, we have $\sum_{i=1}^{n}(x_i - \bar{x})(\bar{x} - \mu) = 0$ as well. Our equation then simplifies to

$$\frac{1}{\sigma^2} \sum_{i=1}^{n} (x_i - \mu)^2 = \frac{1}{\sigma^2} \left(\sum_{i=1}^{n} (x_i - \bar{x})^2 + n (\bar{x} - \mu)^2 \right).$$

One more algebra trick and we're home free:

$$\sum_{i=1}^{n} \left(\frac{x_i - \mu}{\sigma} \right)^2 = \frac{1}{\sigma^2} \sum_{i=1}^{n} (x_i - \bar{x})^2 + \left(\frac{\bar{x} - \mu}{\sqrt{\sigma^2/n}} \right)^2. \tag{22.4}$$

Let's look at this for a second. We know that $(x_i - \mu)/\sigma$ has a standard normal distribution, so by the definition we gave earlier, the left-hand side of this equation is just a χ^2 distribution with n degrees of freedom. Further, we know that $(\bar{x} - \mu)/\sqrt{\sigma^2/n}$ has a standard normal distribution, so the second term on the right-hand side of the equation has a χ^2 distribution with one degree of freedom. Therefore the only term left is the first term on the right-hand side, which you'll notice looks suspiciously like our formula for the sample variance. It's just $(n - 1)s^2/\sigma^2$. So what distribution should it follow? We know that χ^2 distributions have a nice additive property: if $X_1 \sim \chi_k^2$ and $X_2 \sim \chi_l^2$, then $X_1 + X_2 \sim \chi_{k+l}^2$, so long as X_1 and X_2 are independent. Therefore, if the first and second terms of the right-hand side of Equation (22.4) are independent, we have

$$\frac{(n - 1)s^2}{\sigma^2} \sim \chi_{n-1}^2 \tag{22.5}$$

(while they are independent, proving that would take us too far afield for a first course).

So now we see why we've taken so much time to study χ^2 distributions—so long as we're drawing from a normal distribution, the sample variance (more precisely, a multiple of the sample variance) follows a χ^2 distribution! Further, once we know the distribution of a parameter, we can test hypotheses about it. Let's look at a few examples where we test hypotheses about the variance. We can view this as another example of the powerful **Bring It Over Method**.

Example: The Coca-Cola factory is installing a new machine that dispenses Coke into bottles as they move along the bottling apparatus. To ensure that their bottled products are nearly uniform, they want this machine to dispense 12 ounces of Coke on average, with a standard deviation no greater than 0.05 ounces. Suppose that in a sample of 20 bottles, you find the following amount of liquid:

$$
\begin{array}{ccccc}
11.83 & 12.09 & 11.93 & 12.02 & 11.98 \\
11.97 & 12.06 & 12.08 & 12.06 & 12.02 \\
12.01 & 12.10 & 12.04 & 11.98 & 12.04 \\
12.00 & 12.04 & 12.09 & 12.00 & 11.92.
\end{array}
$$

Is this data consistent with what the company wants?

Solution: We have $\bar{x} = 12.01$, so it seems like the machine is dispensing about the right amount of liquid. If we really wanted to, we could test the mean, but for now let's concern ourselves with the variance. The sample variance is $s^2 = 0.0045$, meaning the sample standard deviation is 0.067, which is higher than desired. Is this a significant increase, or just due to chance? Let's take our null hypothesis that $\sigma = 0.05$, meaning $\sigma^2 = 0.0025$. From our test statistic in Equation (22.5), we should have

$$
\frac{(n-1)s^2}{\sigma^2} \sim \chi^2_{19}.
$$

We are using a t-test as 20 observations is not quite large enough to assume the Central Limit Theorem has kicked in, though we are assuming that the individual measurements are normally distributed (with an unknown variance; we are assuming we know the mean).

We get a χ^2 statistic of 34.2. Should we use a one-sided or two-sided test here? Well we're certainly not worried about the variance being too low, so a one-sided test seems reasonable. For a chi-square distribution with 19 degrees of freedom, a test statistic of 34.2 has a one-sided p-value of 0.0174. This is a pretty low p-value, so we'd be concerned that the machine's variance is too high. However, depending on how expensive it is to replace the machine, we might want to collect some more data just to make sure we didn't draw an unusual sample.

Just for fun (and a bit of review), let's test the mean. We want to see whether the machine is dispensing more or less than 12 ounces on average. Taking our null hypothesis to be $\mu = 12$, we know

$$
\frac{\bar{x} - \mu}{\sqrt{s^2/n}} \sim t_{n-1}.
$$

This gives us a test statistic of

$$
\frac{12.01 - 12}{\sqrt{0.0045/20}} \approx 0.67,
$$

which should follow a t-distribution with 19 degrees of freedom. This t-value has a corresponding p-value of about 0.5, clearly is not large enough to reject the null that $\mu = 12$. So while we haven't *proven* that the machine is dispensing more or less than 12 ounces on average, the data we've collected certainly isn't inconsistent with that possibility.

22.5.2 Chi-square Distributions and *t*-distributions

It turns out that χ^2 distributions are related to another distribution we already know and love: the *t*-distribution. In fact, *t*-distributions can be *defined* in terms of χ^2 distributions. The definition goes as follows.

---

t-distribution with k degrees of freedom: Let Z be the standard normal distribution, and Y_n be the χ^2 distribution with n degrees of freedom. The *t*-distribution with n degrees of freedom is defined by

$$\frac{Z}{\sqrt{Y/n}} \sim t_n.$$

This might seem like a weird definition, but we'll see in a minute that it's a very natural one. It comes directly from the situation we found ourselves in a few sections ago: using the sample variance in place of the true variance when performing a *z*-test. Suppose we have a normally distributed random variable X with mean μ and variance σ^2, and from this variable we draw a random sample of size n. Then we know that

$$\frac{\bar{x} - \mu}{\sqrt{\sigma^2/n}} \sim N(0, 1).$$

As we showed earlier, the sample variance s^2 obeys

$$\frac{(n-1)s^2}{\sigma^2} \sim \chi^2_{n-1}.$$

Let's replace σ^2 with s^2 and cleverly multiply by 1:

$$\frac{\bar{x} - \mu}{\sqrt{s^2/n}} = \frac{\bar{x} - \mu}{\sqrt{\frac{(n-1)\sigma^2 \cdot s^2}{(n-1)\sigma^2 \cdot n}}} = \frac{\bar{x} - \mu}{\sqrt{\sigma^2/n}} \cdot \frac{1}{\sqrt{\frac{(n-1)s^2}{(n-1)\sigma^2}}}.$$

Looking at this equation, we notice that $(\bar{x} - \mu)/\sqrt{\sigma^2/n}$ has a standard normal distribution, and $(n-1)s^2/\sigma^2$ has a χ^2 distribution with $n-1$ degrees of freedom. Replacing $(\bar{x} - \mu)/\sqrt{\sigma^2/n}$ with Z and $(n-1)s^2/\sigma^2$ with Y_{n-1}, this becomes

$$\frac{Z}{\sqrt{\frac{Y_{n-1}}{n-1}}},$$

which is exactly the definition we gave above for a *t*-distribution with $n-1$ degrees of freedom.

22.5.3 Goodness of Fit for List Data

One of the most important uses of hypothesis tests is to test models. Since we're always talking about how important it is to justify the choice of distribution, this is a very useful tool. Suppose we're collecting data and there are k possible outcomes (for example, the birth month of everyone in your probability class, which would have 12 possible outcomes). Then we might generate the list of observed data $\{O_1, O_2, \ldots, O_k\}$, where O_i is the number of times we observed the ith outcome. We might also have a model for this data in mind, namely that the ith outcome happens with some probability p_i (e.g., I think 10% of the people in the probability class were born in January, 7% in February, etc.). Then we can test this hypothesis by using the following test statistic:

$$\chi^2 = \sum_{i=1}^{k} \frac{(O_i - E_i)^2}{E_i}, \tag{22.6}$$

where E_i is the number of times we expect to observe event i under the null hypothesis. You'll notice we have suggestively named our test statistic χ^2, and it turns out that this test statistic does follow a χ^2 distribution with $k - 1$ degrees of freedom. Notice that the degrees of freedom are determined by the number of categories we have, not the number of data points.

The general proof of why this test statistic follows a χ^2 distribution is rather advanced, but to get some intuition about the result, let's prove it for the binomial case (that is, when there are only two possible outcomes). Suppose we've gathered n data points and have observed the first outcome O_i times and the second outcome O_2 times. Further suppose we have a model which says the first option should happen with probability p_1, and the second with probability p_2. Notice that we expect to see the first outcome happen np_1 times, and the second outcome np_2 times. This means our test statistic is

$$\chi^2 = \frac{(O_1 - np_1)^2}{np_1} + \frac{(O_2 - np_2)^2}{np_2}.$$

We can simplify this a bit, because we know $p_1 + p_2 = 1$, so $p_2 = 1 - p_1$. Also, $O_1 + O_2 = n$, so $O_2 = n - O_1$. Thus we have

$$\chi^2 = \frac{(O_1 - np_1)^2}{np_1} + \frac{((n - O_1) - n(1 - p_1))^2}{n(1 - p_1)}$$

$$= \frac{(1 - p_1)(O_1 - np_1)^2 + p_1(-O_1 + np_1)^2}{np_1(1 - p_1)}$$

$$= \frac{(O_1 - np_1)^2}{np_1(1 - p_1)} = \left(\frac{O_1 - np_1}{\sqrt{np_1(1 - p_1)}} \right)^2.$$

Since O_1 follows a binomial distribution of size n and probability p_1, the Central Limit Theorem tells us that for large n, $O_1 \approx N(np_1, np_1(1 - p_1))$. Therefore χ^2 is the square of a standard normal distribution, meaning it really does follow a χ^2 distribution with one degree of freedom (when n is large enough to use the Central Limit Theorem). Our proof also helps us see why the number of degrees of freedom is always one less than the number of categories: if there are k possibilities occurring with hypothesized probabilities $\{p_1, p_2, \ldots, p_k\}$, then we can use the restrictions $p_1 + p_2 + \ldots + p_k = 1$ and $O_1 + O_2 + \ldots + O_k = n$ to eliminate one of the categories from our equation, and then rearrange the remaining terms to get a sum of $k - 1$ squares of standard normal distributions (that is, a χ^2 distribution with $k - 1$ degrees of freedom).

TABLE 22.2.

Birth Month	Number of Major Leaguers
January	387
February	329
March	366
April	344
May	336
June	313
July	313
August	503
September	421
October	434
November	398
December	371

Example: Suppose you're investigating the distribution of birth months among major league baseball players. At first glance, you might expect that birth month should have nothing to do with athletic ability, and would take your null hypothesis to be that birth months are equally distributed among baseball players. Following is the data for American Major Leaguers born after 1950 who debuted before 2005 (data from http://www.slate.com/id/2188866).

Given this data, what kind of conclusions can you draw about your hypothesis that birth months should be equally distributed?

Solution: Looking at the data, we notice that there's a considerable bias towards the August through October months. Could this just be chance, or is something going on here? Let's carry out our hypothesis testing as we normally would, and see what happens. Since we're hypothesizing the birth month has no effect, then we would expect that exactly one-twelfth of the 4515 players would be born in each month, or 376.25. Our test statistic of interest is the goodness of fit statistic we just developed, which is given by

$$\chi^2 = \sum_{k=1}^{12} \frac{(\text{Births}_k - 376.25)^2}{376.25} \approx 93.07.$$

Since there are 12 months a player could be born in, our test statistic should follow a χ^2 distribution with 11 degrees of freedom. Before we can find our p-value, however, we need to determine whether we want to use a one-sided or two-sided test. Can it really be the case that this test statistic is "too low"? If the data fit our model perfectly, then every term would be zero, and our test statistic would also be zero. Would this make us want to reject our null hypothesis? Not in the slightest! We'd be jumping with joy to have data that good. So it's only cases when our test statistic is large that we worry about, meaning we'll use a one-sided test. For a one-sided test on a chi-square distribution with 11 degrees of freedom, a test statistic of 93.07 has a p-value of 4.1×10^{-15}. For more on the distribution of birthdays in sports, see Chapter 1.

22.6 Two Sample Tests

One of the most useful applications of hypothesis testing is comparing the means of two (or more!) different samples. These tests are important for any kind of comparative decision making: Did more patients regrow hair with drug A than drug B? Do people prefer Coke or Pepsi? Much as in the one-sample case, the exact test we perform depends on what we know about the variances. There are three possibilities (in order of increasing difficulty): we know the variances; we don't know the variances but have reason to believe they are the same; we don't know the variance and they might be different from each other. We deal with these issues below.

Takeaways for the section:

- There is a test for two populations with known and unequal variances.
- There is a test for two populations with unknown and equal variances. This involves the pooled (sample) variance, which is a weighted estimate of the unknown variance.
- With some approximations we can deal with two populations with unknown and different variances.

22.6.1 Two-sample z-test: Known Variances

Suppose the random variables X and Y have known variances σ_x^2 and σ_y^2, respectively. Let $\bar{X}$ be the distribution of the sample mean of X in a random sample of size n_x, and $\bar{Y}$ the distribution of the sample mean of Y in a random sample of size n_y. We can compare the means of X and Y by forming the random variable $\bar{X} - \bar{Y}$. If X and Y are independent we have

$$\text{Var}(\bar{X} - \bar{Y}) \;=\; \text{Var}(\bar{X}) + \text{Var}(\bar{Y}) \;=\; \frac{\sigma_x^2}{n_x} + \frac{\sigma_y^2}{n_y}.$$

Similarly, the expected value of $\bar{X} - \bar{Y}$ is given by

$$\mathbb{E}(\bar{X} - \bar{Y}) \;=\; \mathbb{E}(\bar{X}) - \mathbb{E}(\bar{Y}) \;=\; \mu_x - \mu_y.$$

If $\bar{X}$ and $\bar{Y}$ are normally distributed, then since sums of normal distributions are still normal (note, yet again, how we use the fact that the normal distribution is a stable distribution), we know

$$\bar{X} - \bar{Y} \;\sim\; \text{N}\left(\mu_x - \mu_y, \sigma_x^2/n_x + \sigma_y^2/n_y\right).$$

We can normalize $\bar{X} - \bar{Y}$ as follows:

$$\frac{(\bar{x} - \bar{y}) - (\mu_x - \mu_y)}{\sqrt{\frac{\sigma_x^2}{n_x} + \frac{\sigma_y^2}{n_y}}} \;\sim\; \text{N}(0, 1). \tag{22.7}$$

Now suppose we wanted to test a hypothesis about the value of $\mu_x - \mu_y$. For simplicity's sake, imagine we wanted to see whether $\mu_x > \mu_y$. We would formulate the null hypothesis that $\mu_x - \mu_y \leq 0$ and take $\mu_x = \mu_y$. We could then sample X and Y, and calculate $\bar{x}$ and $\bar{y}$. Using Equation 22.7 and the fact that under the null hypothesis $\mu_x - \mu_y = 0$, we see that our test statistic is given by

$$z = \frac{\bar{x} - \bar{y}}{\sqrt{\frac{\sigma_x^2}{n_x} + \frac{\sigma_y^2}{n_y}}},$$

which should follow a standard normal distribution. As always, we use this z-value to calculate our p-value, and depending on our significance level will decide whether to reject the null hypothesis or not.

Two sample z-test with known variances: If X and Y are independent random variables with known variances σ_x^2 and σ_y^2, respectively, then

$$\frac{(\bar{x} - \bar{y}) - (\mu_x - \mu_y)}{\sqrt{\frac{\sigma_x^2}{n_x} + \frac{\sigma_y^2}{n_y}}} \sim N(0, 1).$$

Assumptions: This only holds if $\bar{X}$ and $\bar{Y}$ are normally distributed.

To get some experience using a two-sample test, let's look at an example.

Example: You're working for a sleep researcher who wants to look at the impact of sleep on test scores. To do so, the researcher gathers a group of 28 people, and randomly assigns half the group to sleep a full 8 hours before coming in the next morning, while assigning the other half to only sleep 4 hours. The next morning, he administers a test to both groups and records their scores. Suppose he finds the following data.

Given this data, can you conclude at a 5% significance level that sleep helps test performance?

Solution: After collecting this data and calculating the sample mean and variances for the two populations, we formulate the following hypotheses:

$$H_0 : \mu_1 - \mu_2 \leq 0$$

$$H_a : \mu_1 - \mu_2 > 0.$$

That is, we assume that sleep has no benefit, and hope to reject the null hypothesis in favor of the alternative hypothesis that sleep is indeed beneficial. Notice that we're performing a one-sided test, which seems justified because so much evidence suggests that sleep certainly isn't detrimental (save for the students who oversleep and miss the test!). What should our test statistic be in this case? If we assume the null hypothesis, we can take $\mu_1 - \mu_2 = 0$. Since we're interested in testing whether the means are different from each other, it seems natural to consider the random variable $\bar{X} - \bar{Y}$. If we assume

TABLE 22.3.

$X = Sleep\ Group$	$Y = Sleep\text{-}Deprived\ Group$
73	76
95	65
93	74
89	59
79	75
90	76
86	71
91	76
98	74
74	84
91	71
90	77
50	96
70	81
$\mu_1 = 83.5$	$\mu_2 = 75.4$
$s_1^2 = 168.6$	$s_2^2 = 73.32$

both $\bar{X}$ and $\bar{Y}$ are normally distributed, then we know from our discussion above that

$$\frac{\bar{x} - \bar{y}}{\sqrt{\frac{\sigma_x^2}{n_x} + \frac{\sigma_y^2}{n_y}}} \sim N(0, 1).$$

(You might be wondering why we lost the $\mu_x - \mu_y$ term: remember that we're assuming the null hypothesis to be true, which says that $\mu_x - \mu_y = 0$.) You might've spotted an issue here: the formula requires us to use σ_x^2 and σ_y^2, but all we have is s_x^2 and s_y^2! While we will ultimately fix this problem by appealing to a t-distribution (just as we did for one-sample tests), for now let's assume we've gotten the variances right and let $\sigma_x^2 = 168.6$ and $\sigma_y^2 = 73.32$. Then our test statistic is

$$z = \frac{83.5 - 75.4}{\sqrt{\frac{168.6}{14} + \frac{73.32}{14}}} \approx 1.95.$$

This z-score (remember we're doing a one-sided test!) corresponds to a p-value of about 0.0256, which provides pretty solid evidence that sleep is indeed beneficial to test performance.

22.6.2 Two-sample t-test: Unknown but Same Variances

As we saw in the last example, using a z-test for a two-sample hypothesis test requires us to have knowledge about the variances. Typically we don't know the variances beforehand, and estimate them from the sample. We can then use the sample variances in place of the real variances, at the cost of making all of our estimates a little more uncertain. There are actually two cases to consider: when the variances of the samples is the same, and when the variances of the samples might not be the same. The "same variances" case might seem a little artificial, but we begin with it because it contains all the intuition of the general case, and the math is a little more straightforward.

Suppose we know X and Y have the same variance σ^2, but we don't know what it is. In that case, much as with our first introduction to the t-test, we estimate the variance. Before we generated the sample variance s_x^2. However, since we have two variables now, we calculate both s_x^2 and s_y^2. How do we combine them to form one estimate? One possibility would be to disregard s_y^2 altogether and only use s_x^2. This would work, since we know s_x^2 is an unbiased estimator of σ^2. But we can do better than this. Since we've sampled both X and Y, only using s_x^2 would be the equivalent of throwing data away (which is never a good idea!). So how should we combine s_x^2 and s_y^2 to get a better estimate? We could weight the two equally and take our estimate to be $s_p^2 = \frac{1}{2}\left(s_x^2 + s_y^2\right)$ (we call it s_p^2 because we are pooling the variances—this is known as the *pooled variance*). But this also can't be ideal: imagine we had 1000 samples from X and 10 from Y. Would we really want to give s_y^2 half the weight in our estimate? Probably not, since X should be a much better estimate. It turns out the best estimate we can make is a weighted average.

> **Pooled variance for two samples**: If X and Y are independent random variables with common variance σ^2 then the best estimate of the variance is the **pooled variance**, s_p^2, given by
>
> $$s_p^2 = \frac{(n_x - 1) \cdot s_x^2 + (n_y - 1) \cdot s_y^2}{n_x + n_y - 2}.$$

The equation for the pooled variance isn't too complicated, but it does look a bit confusing. Thankfully we can clean it up a little. Let $r = \frac{n_x - 1}{n_x + n_y - 2}$. Notice the numerator is just the number of degrees of freedom for s_x^2, and the denominator is the sum of the degrees of freedom for s_x^2 and s_y^2. The equation then becomes

$$s_p^2 = r \cdot s_x^2 + (1 - r) \cdot s_y^2,$$

so this is indeed a weighted average of s_x^2 and s_y^2 (as $0 \leq r \leq 1$ and $r + (1 - r) = 1$), with the sample variances being weighted by their degrees of freedom.

Now that we have our estimate for the variance, we proceed just as we did in the one-sample case. We replace σ^2 by s_p^2, and our test statistic goes from following a normal distribution to following a t-distribution .

> **Two-sample t-test—same but unknown variance**: If X and Y are independent random variables with common variance σ^2 and estimated pooled variance s_p^2, then
>
> $$\frac{(\bar{x} - \bar{y}) - \left(\mu_x - \mu_y\right)}{\sqrt{\frac{s_p^2}{n_x} + \frac{s_p^2}{n_y}}} \sim t_{n_x + n_y - 2}.$$
>
> Assumptions: We need to know that $\bar{X}$ and $\bar{Y}$ are normally distributed.

Notice that our test statistic follows a t-distribution with $n_x + n_y - 2$ degrees of freedom, which is just the sum of the degrees of freedom for s_x^2 and s_y^2.

 Important: As always, to use the t-test we need to have an underlying normal distribution. In this case, we need to know that

$$\frac{(\bar{x} - \bar{y}) - (\mu_x - \mu_y)}{\sqrt{\frac{\sigma^2}{n_x} + \frac{\sigma^2}{n_y}}} \sim N(0, 1).$$

Thankfully, for most reasonable sample sizes and distributions the Central Limit Theorem assures us that $\bar{X}$ and $\bar{Y}$ will be nearly normal, even if X and Y are not.

Once we have our test statistic, hypothesis testing is business as usual. Let's go through an example to make sure everything's clear.

22.6.3 Unknown and Different Variances

Now that we've discussed what to do for unknown but equal variances, the final (and most general) case to talk about is when we know neither σ_x^2 nor σ_y^2, and σ_x^2 might not equal σ_y^2. In this case, we need to estimate both variances. As usual, we calculate s_x^2 and s_y^2 and we use them in place of σ_x^2 and σ_y^2. In an ideal world, we would now state that

$$\frac{(\bar{x} - \bar{y}) - (\mu_x - \mu_y)}{\sqrt{\frac{s_x^2}{n_x} + \frac{s_y^2}{n_y}}}$$

follows a t-distribution with some easy to calculate number of degrees of freedom. Unfortunately, this is not an ideal world, and the above equation is not true. Why not? You'll remember from earlier that the definition of a t-distribution is

$$\frac{Z}{\sqrt{Y/n}} \sim t_n,$$

where Z is the standard normal random variable, and Y has a χ^2 distribution with n degrees of freedom. In order to prove, in the one-sample case, that replacing σ^2 with s^2 gives us a t-distribution, we had to show that

$$\frac{(n-1)s^2}{\sigma^2} \sim \chi_{n-1}^2.$$

The problem we run into now is this: before, we were able to say that s^2/n was some multiple of a χ^2 distribution. But in the two-sample case, we *cannot* say that $s_x^2/n_x + s_y^2/n_y$ is a multiple of a χ^2 distribution, and so we cannot conclude that the expression above follows a t-distribution. It has *some* distribution, but unfortunately it can't be expressed nicely in terms of the distributions we've met already. So now what? Generally in mathematics, when we cannot find an analytic expression we do the next best thing: approximate. Thankfully in this case there's a well-known and easy to use approximation which states that in the case of unknown variances, the test statistic

$$\frac{(\bar{x} - \bar{y}) - \left(\mu_x - \mu_y\right)}{\sqrt{\frac{s_x^2}{n_x} + \frac{s_y^2}{n_y}}}$$

is approximately distributed t_ν, where ν is the number of degrees of freedom, given by

$$\nu = \frac{\left(\frac{s_x^2}{n_x} + \frac{s_y^2}{n_y}\right)^2}{\frac{s_x^4}{n_x^2(n_x-1)} + \frac{s_y^4}{n_y^2(n_y-1)}}.$$

As a bit of terminology, this approximate test is known as **Welch's t-test**. So we're almost in an ideal world—our test statistic nearly follows a t-distribution. You'll notice, however, that we do not have "some easy to calculate number of degrees of freedom." The equation for ν is a bit daunting, so let's take a look at it. It's not too hard to show that ν is bounded above by $n_x + n_y - 2$, and below by $\min(n_x - 1, n_y - 1)$. This makes sense, since we shouldn't have more degrees of freedom than s_x^2 and s_y^2 have combined, and we shouldn't have fewer than the most uncertain measurement we made. Another limit to consider is when n_x is large (the same argument holds for when n_y is large). When n_x is large, ν approaches $n_y - 1$. This is nice, since when n_x is really big we're not worried about the uncertainty in s_x^2, so our limiting factor becomes the degrees of freedom for s_y^2.

Another question you might have is, "Does this equation always give us an integer number of degrees of freedom, and, if not, what does a non-integer number of degrees of freedom mean?" This equation certainly does not always return an integer! However, we're not too worried if the equation tells us $\nu = 19.394$ because this is an approximation. Remember, the typical interpretation for the degrees of freedom is the number of independent observations that contribute to a measurement (and in that case, 19.934 degrees of freedom makes no sense). But with an approximation like this, that interpretation isn't valid. All the equation is telling us is that this test statistic behaves like a t-distribution with 19.394 "degrees of freedom." Since the formula for the t-distribution uses the degrees of freedom as an input, this is a perfectly well-defined mathematical function.

Once we have our test statistic and know the distribution it follows (approximately), hypothesis testing is the same as always. We summarize two-sample tests in the case of unknown and potentially unequal variances.

Two-sample t-test—unknown, potentially unequal variances: If X and Y are independent random variables with variances σ_x^2 and σ_y^2, respectively, then

$$\frac{(\bar{x} - \bar{y}) - \left(\mu_x - \mu_y\right)}{\sqrt{\frac{s_x^2}{n_x} + \frac{s_y^2}{n_y}}} \approx t_\nu$$

with ν degrees of freedom given by

$$\nu = \frac{\left(\frac{s_x^2}{n_x} + \frac{s_y^2}{n_y}\right)^2}{\frac{s_x^4}{n_x^2(n_x-1)} + \frac{s_y^4}{n_y^2(n_y-1)}}. \tag{22.8}$$

Assumptions: As always, we need $\bar{X}$ and $\bar{Y}$ to be normally distributed.

As an example, let's return to the sleep researcher problem we gave in the first section.

Example: Let's reconsider the sleep example from §22.6.1.

Solution: The analysis we gave there still holds, right up through the p-value we gave. As a reminder, we had

$$\mu_x = 83.5, \quad \mu_y = 75.4, \quad s_x^2 = 168.6, \quad s_y^2 = 73.32,$$

and the null hypothesis that $\mu_x - \mu_y = 0$. Our test statistic is

$$t = \frac{83.5 - 75.4}{\sqrt{\frac{168.6}{14} + \frac{73.32}{14}}} \approx 1.95.$$

Now, however, we also need to calculate the number of degrees of freedom. Using Equation (22.8) to compute the number of degrees of freedom, we see

$$\nu = \frac{\left(\frac{168.6}{14} + \frac{73.32}{14}\right)^2}{\frac{168.6^2}{14^2(14-1)} + \frac{73.32^2}{14^2(14-1)}} \approx 22.5.$$

For a t-distribution with 22.5 degrees of freedom, a t-score of 1.95 corresponds to a p-value of 0.0318. So while our result is a little less certain (with a z-test we had a p-value of 0.025), we can still safely reject the null at a 5% significance level.

22.7 Summary

In this chapter, we discussed the probabilities associated with hypothesis testing in statistics. We learned about one- and two-sided z- and t-tests, which are used for testing the truth of a hypothesis, namely when randomness is involved. A z-test assumes knowledge of the population's standard deviation, while a t-test uses degrees of freedom to adjust the estimated t-distribution based on sample size and various other factors. A one-sided test is used to determine if a result is strictly greater than or less than the null hypothesis value, whereas a two-sided test determines if a result is simply different than the null value (either above or below). All of these tests assume the hypothesis to be true and use collected data to try and disprove it. The probability value (p-value) associated with these tests represents the probability of measuring a value as or more extreme than the observed value, given the null hypothesis. When this value is below a certain threshold, known as the α-level, we can conclude statistically significant evidence against the null and assume the alternative hypothesis.

Additionally, we learned how to compute test statistics, the values used to determine the rarity of observed events, which are quantified as standardized differences from the expected result. We then elaborated more on p-values and the problems, including Type I and II errors, and misconceptions related to them, such as the p-value being the probability that the null hypothesis is true. We also discussed the power of a test and how it relates to effect size, sample size, and Type II error, or β. Lastly, we covered the chi-square distribution and its application in the goodness of fit hypothesis test.

22.8 Exercises

Exercise 22.8.1   *Imagine that systolic blood pressure in healthy people is known to be normally distributed, say $N(110, 100)$. If a patient in a hospital has a systolic blood pressure of 137, find the Z-score for this patient, the associated p-value, and use it to test a hypothesis' about whether this persons blood pressure is "normal" (in the common usage).*

Exercise 22.8.2   *Imagine we have an eight-sided dice which we role twice. Our null hypothesis is that the dice is fair. We roll it twice and the sum of the die is 16. Is this sufficient evidence to say that the die is not fair at a significance level of $\alpha = .05$?*

Exercise 22.8.3   *In §22.1.2 we tossed a fair coin N times and said the result was unusual if there were fewer than 45% or more than 55% heads. Estimate how the probability of obtaining an unusual result decreases with N as $N \to \infty$. To build some intuition, look at a log-log plot of the results from Figure 22.2 (thus the x-axis would be the logarithm of N and the y-axis the logarithm of the probability).*

Exercise 22.8.4   *Plot the number of heads we would need to call the number of heads in n tosses of a fair coin "surprising" if we define surprising as a result extreme enough that we expect to see it only 5% of the time. Plot this number of heads as a percentage of n for n from 5 to 100.*

Exercise 22.8.5   *Define surprising as in the previous exercise. Use a normal approximation for the distribution of heads. Above what n does the number of heads needed for a surprising outcome line up pretty well with the binomial model?*

Exercise 22.8.6   *Show that s^2 is an unbiased estimator for σ^2.*

Exercise 22.8.7   *Prove (22.4):*

$$\sum_{i=1}^{n} \left(\frac{x_i - \mu}{\sigma} \right)^2 = \frac{1}{\sigma^2} \sum_{i=1}^{n} (x_i - \bar{x})^2 + \left(\frac{\bar{x} - \mu}{\sqrt{\sigma^2/n}} \right)^2.$$

Exercise 22.8.8   *We proved that the χ^2-statistic in (22.6) converges to a χ^2 random variable with 1 degree of freedom when $k = 2$; prove it converges to a χ^2 random variable with $k - 1$ degrees of freedom in general.*

Exercise 22.8.9   *Show that as $\nu \to \infty$ the t-distribution approaches the standard normal distribution.*

Exercise 22.8.10 *Imagine you are performing a statistical test to determine if employment has actually increased in the past 3 years, or whether the change in employment can be explained by random fluctuations. What is a Type I error in this case? What is a Type II error?*

Exercise 22.8.11 *Suppose someone is trying to sell you a coin that favors heads. You want such a coin, but you are skeptical. You test out the coin by flipping it 500 times and get a p-value of 0.043. What conclusion can you draw from your finding, and should you buy the coin?*

Exercise 22.8.12 *A recent study testing whether this year's SAT scores have improved on average compared to last year has a p-value of 0.031. Is it reasonable to say that more smart people took the SAT this year?*

Exercise 22.8.13 *What are three ways of increasing power? Explain.*

Exercise 22.8.14 *Your friend Bob is popular on Twitter and is trying to start a YouTube channel. He will devote time to his YouTube career only if at least 40% of his followers subscribe to his channel. After posting a status about his upcoming channel, 3530 of his 8353 followers say they will subscribe. Should Bob start a career on YouTube? Use your knowledge of hypothesis testing to inform Bob of his best course of action.*

Exercise 22.8.15 *Imagine that we planted two types of peppers in the garden, which look very similar. One is a hot pepper and the other is not so we do not want to confuse them, but we unfortunately forgot to label them. There are an equal number of both types. Luckily, the hot peppers are smaller than the sweet peppers. The sizes of both peppers are normally distributed with standard deviation 1 inch, but the mean length of the hot peppers is 3 inches and the mean length of the sweet peppers is 4.5 inches. We randomly pick a pepper. Our null hypothesis is that it is hot. There is a penalty if we misclassify it, since we will put it in the wrong food. If we misclassify it as sweet when it is hot, our utility function is -5. If we misclassify it as hot when it is sweet, our utility is -3. If we correctly classify it, our utility is 10 either way. Set a cutoff that maximizes our utility function for a randomly selected pepper.*

Exercise 22.8.16 *Your friend (who is not the most honest person) claims his die have the following respective probabilities for being a 1, 2, 3, 4, 5, or 6: 10%, 10%, 10%, 20%, 25%, and 25%. In 200 rolls, there were 15 1's, 20 2's, 22 3's, 45 4's, 53 5's, and 45 6's. Should you trust your friend?*

Exercise 22.8.17 *There are two probability classes, each with 30 students. On the first exam, the average score for the first class is 84% and the standard deviation is 9%. The second class has an average of 90% with a standard deviation of 6%. Does the evidence support the claim that the classes have significantly different mean scores? Comment on the conditions for the test.*

Exercise 22.8.18 *Prove that as $\nu \to \infty$ a chi-square distribution with ν degrees of freedom converges to being normally distributed.*

CHAPTER 23 _____

Difference Equations, Markov Processes, and Probability

Domino effect: Once you drop a good idea, the rest will follow.
——LOESJE FOUNDATION, http://www.loesje.org/node/1077

You might not have known this when you purchased this book, but as an added bonus I'm going to share a wonderful strategy to win at roulette. You can make millions with no risk. In fact, as soon as I finish writing this chapter (as I'm so altruistic I want to share this secret with you), I'll be flying back to Vegas to win some more....

Sadly, a lot of people fall for scams like the above. In this chapter we'll talk about what looks like a sure, safe bet, and show why it isn't. What I like about this problem is that it's connected to a lot of great math, and can be understood without a huge amount of mathematical (or gambling) prerequisites. Specifically, we'll see how some-real world problems can be modeled by **recurrence relations**. We'll quickly develop just enough of the theory to solve a few interesting problems, and end the chapter with a short primer on the subject for those who want more. Often these topics are covered in courses on discrete mathematics or differential equations; the reason they fit in this book is that they can be used to compute interesting probabilities.

23.1 From the Fibonacci Numbers to Roulette

The goal of this section is to understand a popular strategy for roulette, and connect it to some mathematics you hopefully have seen before, the Fibonacci numbers.

23.1.1 The Double-plus-one Strategy

To simplify our discussion, we'll talk about an easier version of roulette (see Figure 23.1). We'll assume that every time the wheel spins the ball either lands on a red or a black number, and each outcome happens 50% of the time. The actual game is a bit more complicated, but the strategy we describe below would work in that case too. (The real game usually has 18 red, 18 black, and 2 green places, so that red and black each occur about 47.37% of the time.) To make life easy, we're only allowing

Figure 23.1. A roulette wheel (image from Toni Lozano).

bets on red or black, and we're eliminating the two green numbers (the greens provide a huge advantage to the casino). Say we bet $1 on red (if we bet on black the result is similar). If red comes up we win $1; this means we get back our original dollar plus an additional one. If, however, black comes up then we lose our dollar.

Obviously, our goal is to make money. Here's a famous strategy, called **Double-plus-one**. Bet $1 on red. If it comes up red, great, we're up a dollar. If not, we're down a dollar and now bet $2. If we win, we're now up a grand total of one dollar. What if we lose? If we lose, we're now down $3. In this case, we bet $4. If we win, we're now up a dollar (we lost $3 previously and just won $4 for a profit of a dollar), while if we lose we're down $7, and now we bet $8.

Hopefully the pattern is clear. We keep doubling our bet until we win. When we win, we recoup all our losses and an extra dollar. As *eventually* a red should turn up, *eventually* we should be up a dollar. We then just keep repeating until we've made whatever amount we desire.

What's wrong with this? There are two problems; one requires just some common sense, while the other requires knowing a bit how Vegas works (and why they listen to mathematicians!). The first issue, of course, is that at some point we may need to bet $1,267,650,600,228,229,401,496,703,205,376 (or 2^{100} dollars), and we "may" not have that much money! In order not to worry about such "trivialities", we assume the existence of a rich, but very eccentric, aunt or uncle. This kind family member has unlimited financial reserves, and will advance us whatever amount of money we need to cover our bets, but won't just give us a dollar directly. Why won't they just give us a dollar? That's beyond the scope of this book—we just focus on the mathematics here! The purpose of assuming a rich, eccentric aunt or uncle is to remove the difficulty of needing a large bankroll for the problem and essentially allowing us unlimited betting money, though after we analyze the problem I urge you to modify the argument in the case when you have a fixed, finite amount of money. Also, if we can't even make money in such a favorable case as this....

What's the other problem? This one turns out to be far more serious. We haven't talked too much about how the bets can be done. It turns out that each casino sets both lower and *upper* bounds on how much you can wager on a given spin. For example, the lower limit may be $1 and the upper limit might be $30. If this is the case, if the first five spins are black we're in trouble. If that happens, we've lost $1 + 2 + 4 + 8 + 16 = 31$ dollars. Our method tells us to bet $32, but we can only bet $30, and our system breaks down. We're in even more trouble if we get another black. The problem is that when we win, we win small, but when we lose, we lose *big*.

This should suggest the following natural, and very important, problem: *If we play n times, what's the chance we get 5 or more consecutive blacks?* Interestingly, the same mathematics that we can use to study the Fibonacci numbers can be applied to solve this problem, too. We'll therefore pause and quickly review the Fibonacci numbers, and then return to roulette.

23.1.2 A Quick Review of the Fibonacci Numbers

Let's briefly recall the Fibonacci numbers, though at first there doesn't seem to be any connection. The Fibonacci numbers are the sequence $F_0 = 0$, $F_1 = 1$, $F_2 = 1$, $F_3 = 3$, $F_4 = 5$, $F_5 = 8$, and in general $F_{n+2} = F_{n+1} + F_n$. This is an example of a **linear recurrence relation** (also called a **difference equation**). It's linear as the unknown term depends linearly on previous terms; note we don't have terms multiplying each other, or exponentials of terms. There are many ways to solve this. A great approach is through generating functions (see §19.2 for such a proof), but in the interest of time and to make the exposition self-contained we'll now give the proof by **Divine Inspiration**. Essentially, the way this works is you guess the answer, and see that you're right! Obviously the trouble is that, in general, it's hard to just guess the answer to a difficult math problem! What saves the method is that there's actually a large class of problems where we can just look and rightly guess. For those who want to see more of the general theory, just read on to §23.2.

Let's try $F_n = r^n$ for some r. This is a reasonable guess. It means each term is r times the previous. If we had the simpler relation $G_{n+1} = 2G_n$ then the solution is $G_n = 2^n$, as each term is 2 times the previous. Similarly if we look at $H_{n+1} = 2H_{n-1}$ we see that $H_n = \sqrt{2}^n$. These two observations strongly suggest that the Fibonaccis grow exponentially, which means there are constants C and r such that for large n we have $F_n \approx Cr^n$. As this is such an important point we'll expand on this idea later; my purpose here is to quickly show you how looking at related problems can help you build intuition. Note that if there is such an r we must have $\sqrt{2} \leq r \leq 2$.

If we substitute our guess into the recurrence $F_{n+2} = F_{n+1} + F_n$, we get $r^{n+2} = r^{n+1} + r^n$. This simplifies to $r^2 = r + 1$, or $r^2 - r - 1 = 0$, which by the quadratic formula has two roots: $r_1 = (1 + \sqrt{5})/2 \approx 1.618$ and $r_2 = (1 - \sqrt{5})/2 \approx -.618$. The polynomial $r^2 - r - 1$ is called the **characteristic polynomial of the recurrence relation**.

It turns out that for a linear recurrence relation, any linear combination of solutions of the characteristic polynomial is a solution to the recurrence. In other words, if you plug in $F_{n+2} = c_1 r_1^n + c_2 r_2^n$ for *any* choice of c_1 and c_2, you'll find it solves the recurrence relation because r_1 and r_2 solve the characteristic polynomial; it's a good idea to check this to get a feel for how linearity helps. While we can use any choice of c_1 and c_2, we want our sequence to start off with a 0 when $n = 0$ and a 1 when $n = 1$. In other words, $c_1 + c_2 = 0$ and $c_1 r_1 + c_2 r_2 = 1$. Solving for c_1 and c_2 we find $c_1 = -c_2 = 1/\sqrt{5}$, and we have Binet's formula.

Binet's formula: Let $F_{n+2} = F_{n+1} + F_n$, with $F_0 = 0$ and $F_1 = 1$. Then

$$F_n = \frac{1}{\sqrt{5}} \left(\frac{1 + \sqrt{5}}{2} \right)^n - \frac{1}{\sqrt{5}} \left(\frac{1 - \sqrt{5}}{2} \right)^n.$$

Don't worry if this is a bit incomprehensible right now. We'll talk about recurrence relations in general, and the Fibonacci numbers in particular, in more detail and more leisurely below. Right now, all that matters is you leave this problem knowing that there exists a method to solve linear recurrence relations. Binet's formula is very efficient. It allows us to jump forward and calculate F_{100} without going through all the intermediate terms. While it's of course nice to avoid tedious algebra, if we didn't know the advanced theory we could compute F_{100}, assuming we're very patient. We just keep using the recurrence relation $F_{n+2} = F_{n+1} + F_n$ to find more and more terms, eventually getting $F_{100} = 354, 224, 848, 179, 261, 915, 075$.

It's worth commenting a bit on the Divine Inspiration; what made us think that $a_n = r^n$ would be a good guess? Let's revisit our argument from above. The Fibonacci series is strictly increasing, so $F_{n-2} < F_{n-1} < F_n$. As $F_n = F_{n-1} + F_{n-2}$, we have

$$2F_{n-2} < F_n < 2F_{n-1}.$$

After some algebra, we see $F_n < 2^n$. The lower bound is a bit harder. From $2F_{n-2} < F_n$, we see that every time the index increases by 2, our Fibonacci number at least doubles. Continuing this line backwards, we get

$$F_n > 2F_{n-2} > 2^2 F_{n-4} > 2^3 F_{n-6} > \cdots > 2^{n/2} F_0$$

(at least if n is even); of course, we probably want to stop one step earlier (at F_2 instead of F_0, since $F_0 = 0$). In other words, $F_n > 2^{n/2} = (\sqrt{2})^n$. We've sandwiched the nth Fibonacci number between two exponential bounds; it grows at least as fast as $(\sqrt{2})^n$, and at most as fast as 2^n. It's thus reasonable to *guess* it grows like r^n for some r. (Of course, it might not; it might grow at the rate of $\sqrt{3}^n \log(n)$ for instance.) For large n Binet's formula says F_{n+1} is approximately $\frac{1+\sqrt{5}}{2}$ larger than F_n; note this constant is about 1.61803, sandwiched beautifully between our lower bound of $\sqrt{2} \approx 1.414$ and our upper bound of 2.

23.1.3 Recurrence Relations and Probability

Why are recurrence relations helpful for our roulette problem? Let's try to compute the probability that, in n spins of the wheel, we have at least 5 consecutive blacks. We'll call this probability a_n. It turns out to be easier to compute b_n, the probability that in n spins we *do not* have at least 5 consecutive blacks. Note that a_n is just $1 - b_n$, so if we can find one we can surely find the other. This is a powerful principle in probability, namely

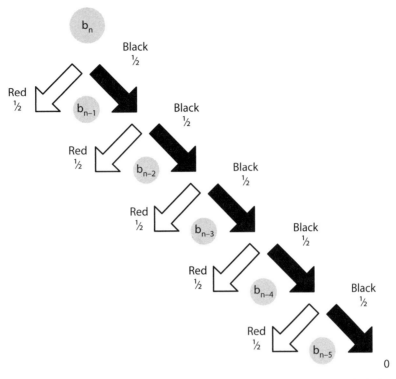

Figure 23.2. Developing the recurrence relation for not having 5 consecutive blacks in n spins of our roulette wheel.

that complementary events have probabilities summing to 1. There are many names for this, including **Law of Total Probability**. Let's use this to get a recurrence relation for b_n. We sketch what happens as we spin in Figure 23.2.

What is b_n? Well, there are two possibilities for the first spin, and each happens with probability 1/2. Half the time we get a red, half the time we get a black. What is the probability we do not have 5 consecutive black spins in n spins, *given that the first spin is a red?* The answer to this question is just b_{n-1}; since the first spin is a red, it can't contribute to 5 consecutive blacks. We now analyze the branch coming from a first spin of black. There are two possibilities for the second spin, again each happening half the time: a red spin, a black spin. If we start off black then red, which happens $\frac{1}{2} \cdot \frac{1}{2} = \frac{1}{4}$ of the time, then the probability that we don't have 5 consecutive blacks is just b_{n-2}.

Continuing along these lines, we find

$$b_n = \frac{1}{2}b_{n-1} + \frac{1}{4}b_{n-2} + \frac{1}{8}b_{n-3} + \frac{1}{16}b_{n-4} + \frac{1}{32}b_{n-5}.$$

Why do we stop here, why aren't there more terms? Well, if we start off with 5 consecutive black spins, then there's no chance that we won't have 5 consecutive black spins! It's precisely for this reason that we're trying to find b_n and not a_n. We now have the recurrence relation. All that remains is to find the initial conditions. This isn't too bad; it's just

$$b_0 = b_1 = b_2 = b_3 = b_4 = 1.$$

Figure 23.3. Finding lower bounds for probability by splitting 100 spins into 20 blocks of 5.

Why are each of these 1? If we have fewer than 5 spins, we can't have at least 5 consecutive blacks! We can either modify the advanced theory or just use the recurrence relation to find the b_n's or the a_n's. After some algebra, we find the a_n's are

$$0, 0, 0, 0, 0, \frac{1}{32}, \frac{3}{64}, \frac{1}{16}, \frac{5}{64}, \frac{3}{32}, \frac{7}{64}, \frac{255}{2048}, \frac{571}{4096}, \dots,$$

or in decimal form,

$$0, 0, 0, 0, 0, 0.03125, 0.046875, 0.0625, 0.078125, 0.09375, 0.109375, 0.124512, \dots.$$

By the time we get to $n = 100$, there is an 81.01% chance that we'll have at least 5 consecutive blacks. At $n = 200$ the probability climbs to 96.59%, while at $n = 400$ it's 99.89%.

23.1.4 Discussion and Generalizations

Our roulette problem has a lot of beautiful features. We can extract a nice mathematical formulation from it which we can solve. Without too much trouble, we can write a simple program to use the recurrence relation and initial conditions to find the probabilities. This illustrates just a small subset of the different types of math that can arise in a probability problem. It also shows the importance of looking at the right object; the recurrence relation is cleaner if we go for the probability of not having 5 consecutive blacks, rather than what we desire (namely the probability of having at least 5 consecutive blacks).

We end with one final feature about this problem. Say we desire the probability of not getting 5 consecutive blacks in 100 spins. We saw we could set up the recurrence relation and find this, but what if we didn't see the recurrence relation? Is there any way to estimate the probability? Estimation is an extremely important skill to develop; good engineers are terrific at this, and with work and practice you can be as well. Let's see what kind of lower bounds we can find for this probability.

Imagine we break our set of 100 spins into 20 sets of 5; see Figure 23.3. In each block of 5, the probability that we don't have 5 consecutive blacks is 31/32, as there is but a one in thirty-two chance that we get black, black, black, black, black. If we want to make sure we don't have 5 consecutive blacks anywhere, we clearly must make sure we don't have 5 consecutive blacks in any of our 20 special blocks of 5; the probability all of these blocks aren't 5 consecutive blacks is just

$$\left(1 - \frac{1}{32}\right)^{100/n} = \left(\frac{31}{32}\right)^{20} \approx 0.529949.$$

This is an upper bound for b_{100}, as it's possible to have 5 consecutive blacks where say three are in one of our special blocks of 5 and two are in an adjacent block.

Thus $b_{100} \le 0.529949$, so $a_{100} \ge 0.470051$, or there is at least a 47% chance we'll get at least 5 consecutive blacks.

I really like the above argument, and strongly urge you to read it a few times until you fully understand what's going on. Note that we've bypassed a lot of advanced mathematics, and quickly obtained a ballpark estimate of the answer. It's really satisfying to take a complicated problem and quickly approximate the answer.

With a little bit of work, we can do better than 47%. Let's again consider 100 spins, and again let's try to find a lower bound for the probability of having at least 5 consecutive blacks. Instead of breaking 100 into 20 disjoint blocks of 5, we now break it into 16 disjoint blocks of length 6, and one block of length 4. For the blocks of length 6, show the probability that there aren't at least 5 consecutive blacks is 61/64, and use this to get a lower bound for a_{100}. If you're patient, you can enumerate the possibilities for 100 split into 10 disjoint blocks of length 10 without too much trouble, and get a really nice lower bound for a_{100}. If you do the computations, you'll find that of the 1024 possible strings of 10 spins, exactly 112 have at least 5 consecutive blacks. This leads to a lower bound of 68.60%, which isn't too far from the actual answer of 81.01%. Also, not surprisingly, we continue to underestimate the true probability.

Our entire analysis was designed to find the probability of getting at least 5 consecutive blacks in n tosses. What if we wanted to know the probability of getting *exactly* 5 consecutive heads somewhere in n tosses? There's a nice way to find this number. We simply find the probability of getting at least 5 consecutive blacks, and subtract off the probability of getting at least 6 consecutive blacks.

23.1.5 Code for Roulette Problem

We end by giving some simple code in Mathematica to attack the roulette problem. The first approach is to directly calculate it by working with the recurrence relation. Let b_n be the probability we do not have 5 consecutive blacks. We can compactly write the recurrence as

$$b_n = \sum_{k=1}^{5} (1/2)^k b_{n-k};$$

note that it's a bit cleaner if we write b_n in terms of the earlier terms, rather than b_{n+1}. We note the first five terms (starting with b_0) are all 1, and then march down.

```
b[0] = 1;
b[1] = 1;
b[2] = 1;
b[3] = 1;
b[4] = 1;
p = 1/2;
For[n = 5, n <= 100, n++,
  b[n] = Sum[p^k b[n - k], {k, 1, 5}]
  ];
```

The above gives the probability we have 5 consecutive blacks in 100 tosses, which is $1 - b_{100}$, is

$$\frac{6418349497949459884697236427 5}{792281625142643375935439503 36}$$

(or about 81.011% for people who don't need this much precision!).

Of course, we can also simulate this probability. We keep track of how many times we have 5 consecutive blacks in our 100 spins with the variable count (so each time it happens we increase our counter by 1). We denote spins that are black with a 1, and use 0 for ones that are red. We let consec denote the number of consecutive blacks obtained (so we start it at 0 and if we spin a black we increase our counter consec by 1, while if we spin a red we reset it to 0). We use a While command below. Thus we continue spinning while we are less than 100 spins *and* we do not have 5 consecutive blacks; once either condition happens we immediately abort the calculation as there is no need to continue—we know at this point if we had 5 consecutive blacks or not.

```
roulette[numspins_, numdo_] := Module[{},
   count = 0;
   For[n = 1, n <= numdo, n++,
    {
     consec = 0;
     spin = 1;
     While[consec < 5 && roll <= numspins,
      {
       toss = If[Random[] <= .5, 1, 0];
       If[toss == 1, consec = consec + 1, consec = 0];
       spin = spin + 1;
       }]; (* end of while loop *)
     If[consec == 5, count = count + 1];
     }]; (* end of n loop *);
   Print["Observe at least 5 heads in a row with prob ",
    100. count/numdo, "%."];
   ];
```

```
Timing[roulette[100, 1000000]]
Observe at least 5 heads in a row with prob 81.044%.
```

We thus see excellent agreement between theory and simulation.

23.2 General Theory of Recurrence Relations

We've just seen the power of recurrence relations in probability. We used them to analyze the roulette problem, and found that what seemed like a surefire method to make money is in fact fatally flawed. As there are many problems where recurrence relations pop up, it's not a bad idea to know more about them. To help, we've collected some facts about them below.

23.2.1 Notation

Before developing the theory, we first set some notation. We'll study **linear recurrence relations**. A linear recurrence relation of depth k is a sequence of numbers $\{a_n\}_{n=0}^{\infty}$ where

$$a_{n+1} = c_1 a_n + c_2 a_{n-1} + \cdots + c_k a_{n-k+1} \qquad (23.1)$$

for some fixed, given real numbers $c_1, c_2, \ldots, c_k$. If we specify the first k terms of the sequence, all remaining terms are uniquely determined. For example, for the Fibonacci numbers we have $k = 2$, $c_1 = c_2 = 1$, $F_0 = 0$, and $F_1 = 1$. Here the recurrence is

$F_{n+1} = F_n + F_{n-1}$. The sequence starts off 0, 1, 1, 2, 3, 5, 8, 13, 21, 34, 55, and so on, where each term (from the third onward) is the sum of the previous two terms.

In some sense, we're done. Once we've specified the recurrence relation and the initial conditions, all subsequent terms are uniquely determined. As this is the case, why should we spend time developing an advanced theory? The main reason is efficiency. We saw in the roulette problem that we might only care about one specific term deep in the sequence; we'd love to be able to jump to it and not have to go through all the previous terms. Related to this, we might be interested in the general behavior of terms in the sequence. Is it possible to say something about their general behavior without computing exactly what they are? For these reasons, there is a real need to find a better approach than just computing term by term.

23.2.2 The Characteristic Equation

In this section it is helpful, though not necessary, to know some linear algebra. We quickly review some ideas from that course that will help put what follows in context; if you haven't taken it yet either skip or skim the next paragraph, or view this as a quick introduction!

In linear algebra you learn an operator T is **linear** if

$$T(a\vec{v} + b\vec{w}) = aT(\vec{v}) + bT(\vec{w})$$

for any constants a, b and any vectors $\vec{v}, \vec{w}$; if T is a matrix we often drop the parentheses and just write $A\vec{v}$. One situation where this is used is in going from a specific solution to a general solution. For example, imagine we are trying to solve $A\vec{x} = b$. We first find a basis for the **null-space** of A (the null-space is the set of all vectors sent to $\vec{0}$ by A); let's say $\vec{v}_1, \ldots, \vec{v}_\ell$ is a basis. Next, we find *one* particular solution to the original problem; this means we find one vector $\vec{v}$ such that $A\vec{v} = \vec{b}$. Then if $\vec{x}$ is a solution to $A\vec{x} = \vec{b}$, there are constants $\alpha_1, \ldots, \alpha_\ell$ such that

$$\vec{x} = \vec{v} + \alpha_1 \vec{v}_1 + \cdots + \alpha_\ell \vec{v}_\ell.$$

What is wonderful about this result is that we have a way to combine solutions to certain problems to find all the solutions to the desired problem; a similar idea works for combining specific solutions of a linear recurrence relation to obtain all the solutions.

Returning to (23.1), let's see how to find a_n as a function of k, the c_i's, and the initial conditions (the values for $a_0, a_1, \ldots, a_{k-1}$). We begin by guessing that $a_n = r^n$ for some constant r; this is the Method of Divine Inspiration we mentioned earlier (we could also use the methods of §19.2 to find the answer via generating functions). It turns out this will always give us a solution to (23.1), though we'll have to do a little work to satisfy the initial conditions.

Plugging $a_n = r^n$ into (23.1) gives

$$r^{n+1} = c_1 r^n + c_2 r^{n-1} + \cdots + c_k r^{n-k+1}. \tag{23.2}$$

Dividing both sides by r^{n-k+1}, Equation (23.2) becomes

$$r^k = c_1 r^{k-1} + c_2 r^{k-2} + \cdots + c_k. \tag{23.3}$$

We call Equation (23.3) the **characteristic polynomial** of the difference equation given by (23.1). Subtracting $c_1 r^{k-1} + c_2 r^{k-2} + \cdots + c_k$ from both sides, we can rewrite (23.3) as

$$r^k - c_1 r^{k-1} - c_2 r^{k-2} - \cdots - c_k = 0. \tag{23.4}$$

Equation (23.4) is a polynomial of degree k, and by the **Fundamental Theorem of Algebra** (see §20.10 for a review of this theorem) has k roots. We call these roots $r_1, r_2, \ldots, r_k$. Note these roots might not be distinct; in fact, if there are repeated roots the analysis is a little harder. For now, we'll assume the roots $r_1, r_2, \ldots, r_k$ are all distinct.

We know $a_n = r_i^n$ is a solution to (23.1) for $1 \leq i \leq k$; each r_i solves the characteristic polynomial, and we created the characteristic polynomial by simple algebraic manipulation of (23.2). Because we're solving a linear difference equation, once we know that each of $r_1^n, r_2^n, \ldots, r_k^n$ is a solution, we know that a linear combination of these solutions also satisfies (23.1) (see Exercise 23.5.2). That is, for *any* constants $\gamma_1, \gamma_2, \ldots, \gamma_k$, we have

$$a_n = \gamma_1 r_1^n + \gamma_2 r_2^n + \cdots + \gamma_k r_k^n \tag{23.5}$$

solves the recurrence. This fact depends on our original recurrence relation being linear. For example, if we had

$$a_{n+1} = n^2 a_n + e^n a_{n-1},$$

Equation (23.5) would not be valid.

Let's prove this in full gory detail for the Fibonacci numbers; the proof in general is similar. For the Fibonacci numbers, we get a characteristic equation of $r^2 - r - 1 = 0$, with roots $r_1 = (1 + \sqrt{5})/2$ and $r_2 = (1 - \sqrt{5})/2$. Knowing that each of these roots solves the characteristic equation, let's look at an arbitrary linear combination $\gamma_1 r_1^n + \gamma_2 r_2^n$ for F_n. We find

$$
\begin{aligned}
& F_{n+1} - F_n - F_{n-1} \\
&= \left(\gamma_1 r_1^{n+1} + \gamma_2 r_2^{n+1} \right) - \left(\gamma_1 r_1^n + \gamma_2 r_2^n \right) - \left(\gamma_1 r_1^{n-1} + \gamma_2 r_2^{n-1} \right) \\
&= \gamma_1 \left(r_1^{n+1} - r_1^n - r_1^{n-1} \right) + \gamma_2 \left(r_2^{n+1} - r_2^n - r_2^{n-1} \right) \\
&= \gamma_1 r_1^{n-1} \left(r_1^2 - r_1 - 1 \right) + \gamma_2 r_2^{n-1} \left(r_2^2 - r_2 - 1 \right) = 0 + 0 = 0.
\end{aligned}
$$

What makes the algebra work is the linearity: sums of solutions are solutions, and a multiple of a solution is a solution.

23.2.3 The Initial Conditions

We've made it about two-thirds of the way to finding a solution to (23.1). We have Equation (23.5) as the general form for the a_n's. In addition, we solved the characteristic polynomial for the roots $r_1, r_2, \ldots, r_k$ (which we assume are distinct). Unfortunately, we're not done yet. We still need to determine the values of $\gamma_1, \gamma_2, \ldots, \gamma_k$ in order to find out what a_n is.

Using our initial conditions, which are the values for $a_0, a_1, \ldots, a_{k-1}$, and our assumption that $a_n = \gamma_1 r_1^n + \cdots + \gamma_k r_k^n$, we can set up the following system of equations:

$$
\begin{aligned}
\gamma_1 + \gamma_2 + \cdots + \gamma_k &= a_0 \\
\gamma_1 r_1 + \gamma_2 r_2 + \cdots + \gamma_k r_k &= a_1 \\
\gamma_1 r_1^2 + \gamma_2 r_2^2 + \cdots + \gamma_k r_k^2 &= a_2 \\
\vdots \ &= \ \vdots \\
\gamma_1 r_1^{k-1} + \gamma_2 r_2^{k-1} + \cdots + \gamma_k r_k^{k-1} &= a_{k-1}.
\end{aligned}
$$

From linear algebra, we know that we can rewrite this system of equations as the product of matrices:

$$
\begin{pmatrix}
1 & 1 & \cdots & 1 \\
r_1 & r_2 & \cdots & r_k \\
r_1^2 & r_2^2 & \cdots & r_k^2 \\
\vdots & \vdots & & \vdots \\
r_1^{k-1} & r_2^{k-1} & \cdots & r_k^{k-1}
\end{pmatrix}
\begin{pmatrix}
\gamma_1 \\ \gamma_2 \\ \gamma_3 \\ \vdots \\ \gamma_k
\end{pmatrix}
=
\begin{pmatrix}
a_0 \\ a_1 \\ a_2 \\ \vdots \\ a_{k-1}
\end{pmatrix}.
\tag{23.6}
$$

It's a wonderful fact that if $r_1, r_2, \ldots, r_k$ are distinct, then our $k \times k$ matrix is invertible. This is a non-trivial fact; for those who are really interested, a proof is given in §23.2.4. In this case, we can solve for the vector of $\gamma_1, \gamma_2, \ldots, \gamma_k$ by multiplying both sides of Equation (23.6) to the left by the inverse of the $k \times k$ matrix:

$$
\begin{pmatrix}
\gamma_1 \\ \gamma_2 \\ \gamma_3 \\ \vdots \\ \gamma_k
\end{pmatrix}
=
\begin{pmatrix}
1 & 1 & \cdots & 1 \\
r_1 & r_2 & \cdots & r_k \\
r_1^2 & r_2^2 & \cdots & r_k^2 \\
\vdots & \vdots & & \vdots \\
r_1^{k-1} & r_2^{k-1} & \cdots & r_k^{k-1}
\end{pmatrix}^{-1}
\begin{pmatrix}
a_0 \\ a_1 \\ a_2 \\ \vdots \\ a_{k-1}
\end{pmatrix}.
\tag{23.7}
$$

Then, Equation (23.7) gives us values for each of $\gamma_1, \gamma_2, \ldots, \gamma_k$. We already solved for $r_1, r_2, \ldots, r_k$ and, according to (23.5), this is all the information we need to find a_n. That is, we substitute the r_i values that we found by solving the characteristic polynomial and the γ_i values we find by Equation (23.7) into Equation (23.5) to solve for a_n.

Let's end by applying this to the Fibonacci numbers. Remember $r_1 = (1 + \sqrt{5})/2$ and $r_2 = (1 - \sqrt{5})/2$, the initial conditions are $F_0 = 0$ and $F_1 = 1$, and $F_n = \gamma_1 r_1^n + \gamma_2 r_2^n$. Our system of equations becomes

$$
\begin{pmatrix} 1 & 1 \\ r_1 & r_2 \end{pmatrix} \begin{pmatrix} \gamma_1 \\ \gamma_2 \end{pmatrix} = \begin{pmatrix} 0 \\ 1 \end{pmatrix}.
$$

The determinant of the matrix is $r_2 - r_1 = -\sqrt{5}$; as this is non-zero, the matrix is invertible. We find

$$
\begin{pmatrix} \gamma_1 \\ \gamma_2 \end{pmatrix} = \frac{-1}{\sqrt{5}} \begin{pmatrix} r_2 & -1 \\ -r_1 & 1 \end{pmatrix} \begin{pmatrix} 0 \\ 1 \end{pmatrix} = \begin{pmatrix} 1/\sqrt{5} \\ -1/\sqrt{5} \end{pmatrix}.
$$

This leads to

$$a_n = \frac{1}{\sqrt{5}} \left(\frac{1+\sqrt{5}}{2} \right)^n - \frac{1}{\sqrt{5}} \left(\frac{1-\sqrt{5}}{2} \right)^n ,$$

and we recover **Binet's formula**. It's a spectacular formula. It allows us to jump to any Fibonacci number without having to compute the intermediate ones. It makes for very efficient computations.

We leave the rest of the roulette problem as an exercise for the interested reader. The difficulty is that the characteristic polynomial has degree 5, and there is no analogue of the quadratic formula. Sadly, this means we can't just write down the roots in terms of the coefficients of the polynomial, but instead have to approximate them. The five roots are approximately $-0.339175 \pm 0.229268i$, $0.0976883 \pm 0.424427i$, and 0.982974.

In analyzing the solutions to recurrence relations, the large n behavior is typically governed by the root whose absolute value is larger. This is because, as n grows, the powers of this root far exceed the powers of the other roots. The only time when it won't control the limiting behavior is if its corresponding coefficient happens to be zero (which only happens for very special, pathological choices of initial conditions).

23.2.4 Proof that Distinct Roots Imply Invertibility

To solve the recurrence relation, we needed the $k \times k$ matrix in (23.7) to be invertible. We need to show that if

$$A = \begin{pmatrix} 1 & 1 & \dots & 1 \\ r_1 & r_2 & \dots & r_k \\ r_1^2 & r_2^2 & \dots & r_k^2 \\ \vdots & & & \vdots \\ r_1^{k-1} & r_2^{k-1} & \dots & r_k^{k-1} \end{pmatrix} ,$$

then A is invertible if and only if the roots are distinct. This is a very special type of matrix, called a **Vandermonde matrix**, and it turns out that a simple matching argument shows that it's invertible if the roots are distinct.

In linear algebra, you learned (or will learn) that a square matrix is invertible if and only if its determinant is non-zero. If two roots are the same, then two columns are the same and the matrix isn't invertible. We see we can therefore restrict ourselves to the case when all roots are distinct.

From linear algebra (basically expand by minors), we know that $\det(A)$ is a function of $r_1, r_2, \dots, r_k$. In addition, we know that, in calculating a determinant of a $k \times k$ matrix, we have $k!$ summands, with each summand a product of k terms. In the product, we always have exactly one element from each row, and exactly one element from each column. We're going to get a massive polynomial in $r_1, r_2, \dots, r_k$. The first question to ask is: what is its degree? Well, the first row is just all 1's, and thus contributes 0 to the degree. The second row gives us an r_i for some i, and this contributes 1 to the degree. For the third row, we get an r_j^2, which contributes 2 to the degree. And so on and so on until the last row, which gives us a factor like r_ℓ^{k-1}, and adds $k-1$ to the degree.

Thus the degree of $\det(A)$ is

$$0 + 1 + 2 + \cdots + (k - 1) = \frac{(k-1)k}{2}$$

(see Appendix A.2.1 for a proof of this sum). We know $\det(A)$ is a polynomial involving $r_1, \ldots, r_k$. We're going to show it's just $\prod_{1 \le i < j \le k} (r_j - r_i)$.

For a minute, let's go back and consider what happens if $r_i = r_j$ for some $i \ne j$. If this is the case, then $\det(A) = 0$ as two columns are equal. As i and j are arbitrary, we see $\det(A)$ must always be divisible by $r_i - r_j$, or $\prod_{1 \le i < j \le k} (r_j - r_i)$ divides $\det(A)$.

Now consider the degree of $\prod_{1 \le i < j \le k} (r_j - r_i)$, which we know to be a factor of $\det(A)$. We have $2 \le j \le k$ and $1 \le i \le j - 1$. Therefore, the degree of this polynomial is

$$\sum_{j=2}^{k} (j - 1) = \sum_{j=1}^{k-1} j = \frac{k(k-1)}{2}.$$

We see that the degree of $\prod_{1 \le i < j \le k} (r_j - r_i)$, which we know to be a factor of $\det(A)$, is the same as the degree of $\det(A)$. This means that

$$\det(A) = \alpha \cdot \prod_{1 \le i < j \le k} (r_j - r_i) \tag{23.8}$$

for some constant α. Then we see from (23.8) that $\det(A)$ can be zero only either α or $\prod_{1 \le i < j \le k} (r_j - r_i)$ is zero. We know that $\prod_{1 \le i < j \le k} (r_j - r_i)$ is zero if $r_i = r_j$, but we've already shown that $\det(A)$ is zero in this case, and we're currently considering the situation in which we have k distinct roots. Thus, we assume that $\prod_{1 \le i < j \le k} (r_j - r_i) \ne 0$, and we must show only that $\alpha \ne 0$. What's really nice is that α is independent of the r_i's, so if we can determine α in one special case, we'll know it in every case.

Let's try $r_i = 10^{10^{i-1}}$. This sequence is growing rapidly. We have $r_1 = 10, r_2 = 10^{10}$, $r_3 = 10^{100}$, and so on. Clearly, r_k will be so large that the determinant cannot vanish (the determinant will be essentially $r_1^0 r_2^1 r_3^2 \cdots r_k^{k-1}$), and therefore we cannot have $\alpha = 0$. Consequently, we see that having k distinct roots $r_1, r_2, \ldots, r_k$ is enough to know that A will be invertible.

It's worth pausing and reflecting on the argument above. By looking at a **special case** we were able to deduce results in general. Specifically, we wanted to show α was not zero. Since we had already isolated all the r_i dependence in the product, all we had to do was find one choice for them such that the determinant was non-zero and then we can deduce $\alpha \ne 0$. Investigating **extreme cases** frequently works, and allows us to transfer knowledge and intuition from one case to all.

23.3 Markov Processes

We end with one final example of how difference equations can be applied to probability. We start with a completely deterministic, incredibly oversimplified situation; after we understand this problem, we'll make the model more reasonable and then explore related applications.

23.3.1 Recurrence Relations and Population Dynamics

Imagine whales mate in pairs, and always give birth to either one or two pairs of whales, with each pair always containing one male and one female. Let's make the following assumptions.

- Whales are born and always die after four full years of life.
- At the start of their second year, each pair of whales gives birth to two pairs of whales.
- At the start of their third year, each pair of whales gives birth to one pair of whales.
- At the start of their fourth year, the whales have had enough and enjoy being grandparents.
- The whales die at the start of their fifth year of life.

We can use recurrence relations to write down a simple formula for how many whales of each type there are each moment in time. Let a_n denote the number of whales born at the start of year n, b_n the number of whales that are 1 year old at the start of year n, c_n the number that are 2 years old at year n, d_n the number that are 3 years old at year n, and finally e_n is the number that are 4 years old at the start of year n. We don't need to worry about whales that are 5 at the start of the year, as they (sadly) immediately die.

Our assumptions imply the following relations.

$$
\begin{aligned}
a_{n+1} &= 2c_{n+1} + d_{n+1} = 2b_n + c_n \\
b_{n+1} &= a_n \\
c_{n+1} &= b_n \\
d_{n+1} &= c_n \\
e_{n+1} &= d_n.
\end{aligned}
$$

Why are these true? The whales born in year $n + 1$ come from whales that are 2 years old at the start of this year, or are 3 years old at the start of the year; however, the whales that are 2 years old at the start of year $n + 1$ were 1 year old at the start of year n. Similarly, the whales that are 4 years old at the start of year $n + 1$ were 3 years old at the start of year n; this gives us the equation $e_{n+1} = d_n$. We may write this as

$$
\begin{pmatrix}
a_{n+1} \\
b_{n+1} \\
c_{n+1} \\
d_{n+1} \\
e_{n+1}
\end{pmatrix}
=
\begin{pmatrix}
2b_n + c_n \\
a_n \\
b_n \\
c_n \\
d_n
\end{pmatrix},
$$

or

$$\begin{pmatrix} a_{n+1} \\ b_{n+1} \\ c_{n+1} \\ d_{n+1} \\ e_{n+1} \end{pmatrix} = \begin{pmatrix} 0 & 2 & 1 & 0 & 0 \\ 1 & 0 & 0 & 0 & 0 \\ 0 & 1 & 0 & 0 & 0 \\ 0 & 0 & 1 & 0 & 0 \\ 0 & 0 & 0 & 1 & 0 \end{pmatrix} \begin{pmatrix} a_n \\ b_n \\ c_n \\ d_n \\ e_n \end{pmatrix} = A \begin{pmatrix} a_n \\ b_n \\ c_n \\ d_n \\ e_n \end{pmatrix}.$$

The matrix A has a very nice structure; these matrices occur all the time in population dynamics, and are called **Leslie matrices**. Iterating, we find

$$\begin{pmatrix} a_{n+1} \\ b_{n+1} \\ c_{n+1} \\ d_{n+1} \\ e_{n+1} \end{pmatrix} = A^{n+1} \begin{pmatrix} a_0 \\ b_0 \\ c_0 \\ d_0 \\ e_0 \end{pmatrix},$$

where the final vector is the number of whales of each age in year 0.

What's so beautiful about this formula is that we can determine the number of whales in later years simply by computing high powers of the matrix A. This is a nice, compact way of writing down the recurrence relation. It generalizes what we did earlier. For example, if we return to the Fibonacci numbers, we have $F_{n+2} = F_{n+1} + F_n$, and we would find

$$\begin{pmatrix} F_{n+2} \\ F_{n+1} \end{pmatrix} = \begin{pmatrix} F_{n+1} + F_n \\ F_{n+1} \end{pmatrix} = \begin{pmatrix} 1 & 1 \\ 1 & 0 \end{pmatrix} \begin{pmatrix} F_{n+1} \\ F_n \end{pmatrix} = B \begin{pmatrix} F_{n+1} \\ F_n \end{pmatrix},$$

which implies

$$\begin{pmatrix} F_{n+2} \\ F_{n+1} \end{pmatrix} = B^{n+1} \begin{pmatrix} F_1 \\ F_0 \end{pmatrix}.$$

Thus the Fibonacci numbers can be attacked using this framework as well.

Why does this belong in a probability course? As it stands, it really doesn't as the process is entirely deterministic (unless we want to ask a question such as, "What is the probability a randomly chosen whale in year n is exactly 3 years old?"). This truly becomes a probability problem when we allow the matrix elements of A to be random variables. Instead of assuming every whale survives each year and dies at the start of its fifth year, let's allow the whales a probability of dying. Let R_i be a random variable denoting the probability a whale that is i years old lives to be $i + 1$. Similarly, let B_i denote the number of pairs of whales birthed by a whale that is exactly i years old. Our matrix A now has elements that are random variables! We now have a product of matrices such as

$$\begin{pmatrix} 0 & B_1 & B_2 & 0 & 0 \\ R_1 & 0 & 0 & 0 & 0 \\ 0 & R_2 & 0 & 0 & 0 \\ 0 & 0 & R_3 & 0 & 0 \\ 0 & 0 & 0 & R_4 & 0 \end{pmatrix}.$$

To understand how the system evolves, we need to understand the behavior of products of matrices with random elements! There's now a plethora of questions we can ask. How does the distribution of the B's and R's affect the long-term behavior of the whale population? Are there certain critical values for these random variables, such that small changes in the means can lead to wildly different behavior? This leads us to **Random Matrix Theory**, a very active area of research. While it would take us too far afield to study these questions further, I wanted you to get a sense of where things go.

23.3.2 General Markov Processes

As Markov processes are so important, it would be criminal to not at least briefly mention and discuss them, and see how some of the problems above fit into this framework. A **Markov process** is essentially a system where all that matters in predicting its behavior at time $n + 1$ is its configuration at time n; in other words, knowing how we got to the state at time n does not provide any additional information in predicting what happens at the next moment. If $X_i = x_i$ denotes the system at time i being in configuration x_i, then we can write this condition as

$$\text{Prob}(X_{n+1} = x_{n+1} | X_n = x_n, \ldots, X_0 = x_0) = \text{Prob}(X_{n+1} = x_{n+1} | X_n = x_n).$$

Recurrence relations are terrific examples if viewed properly. Let's consider the Fibonacci numbers: so $F_{n+1} = F_n + F_{n-1}$ and $F_0 = 0, F_1 = 1$. As our next state depends on our two previous states it doesn't satisfy our definition; however, a trivial modification does. Let

$$v_n = \begin{pmatrix} F_n \\ F_{n-1} \end{pmatrix};$$

then

$$v_{n+1} = \begin{pmatrix} 1 & 0 \\ 1 & 1 \end{pmatrix} v_n.$$

Population problems can often be cast as Markov processes; do so for the one in Exercise 23.5.16. There are wealth of theorems on limiting behavior for Markov processes, and I strongly urge you to do some Web searching and learn more.

23.4 Summary

In this chapter we got a brief introduction to recurrence relations, but even a cursory introduction is enough to see the power and applicability of these methods. I like Loesje's quote: "Domino effect: Once you drop a good idea, the rest will follow". This fits in two ways in recurrence relations. The first is the more obvious. In recurrence relations, once you know the first few terms and you know how things are connected, you can then easily compute later terms. This makes recurrence relations ideally suited for computers.

The second is the importance of a good idea, of the right perspective. We've seen examples of this throughout the book; if you can look at a problem the right way, the algebra often melts away and we can really see what's fundamental. Linear algebra provides a variety of tools to study recurrence relations. It's a great combination when combined with probability, which allows us to take the coefficients to be random variables. By allowing random values, we leave the world of deterministic processes and have a much better chance of building a reasonable model. This is of course just the start of a vast topic.

23.5 Exercises

Exercise 23.5.1   *Consider the following scam: an unscrupulous person is trying to convince you that he is a master at picking stocks. For seven weeks he'll send you a prediction on stock, and will correctly tell you (for free) whether or not it will go up or down. At the end of seven weeks of correct picks, he informs you that his advice now costs $100; should you pay? (Hint: Note that he may be contacting other people as well, and not necessarily giving them the same advice he gives you.)*

Exercise 23.5.2   *Assume there are constants $c_1, \ldots, c_L$ such that for all n*

$$a_{n+1} \;=\; c_1 a_n + c_2 a_{n-1} + \cdots + c_L a_{n-(L-1)}.$$

If $r_1^n, \ldots, r_\ell^n$ are solutions to the recurrence, show that $a_n = \alpha_1 r_1^n + \cdots + \alpha_\ell r_\ell^n$ is also a solution for any $\alpha_1, \ldots, \alpha_\ell$.

Exercise 23.5.3   *Show that for all integer n, Binet's formula returns an integer. (It better, since all the Fibonacci numbers are integers!)*

Exercise 23.5.4   *Using Binet's formula, show for n large that F_{n+1}/F_n approaches the golden mean, $\frac{1+\sqrt{5}}{2}$.*

Exercise 23.5.5   *Use the characteristic polynomial to find a functional equation for $a_{n+1} = 2a_n + a_{n-1}$ with initial conditions $a_0 = 0$ and $a_1 = 1$.*

Exercise 23.5.6   *Find an explicit formula for the recurrence relation $a_{n+1} = 2a_n - a_{n-1}$ and $f_0 = 1$, $f_1 = 2$.*

Exercise 23.5.7   *If the characteristic polynomial of a recurrence relation has a root r of degree d, show that $n^\ell r^n$ is a solution for $\ell \in \{0, 1, \ldots, d-1\}$.*

Exercise 23.5.8   *Find the probability of having a longest run of at least 5 heads in 100 tosses of a fair coin.*

Exercise 23.5.9   *Find the probability of having a longest run of exactly 5 heads in 100 tosses of a fair coin.*

Exercise 23.5.10   *Redo the previous two exercises, but now assume the coin is heads with probability 20/38. We can interpret the answer as showing what would happen to our analysis in roulette if we allow the two greens.*

Exercise 23.5.11   *What is the probability that if we flip a fair coin 100 times we never have four out of five consecutive tosses a tail? What if we toss the coin n times? Can you write down a recurrence relation for this problem?*

Exercise 23.5.12   *Let's modify the roulette problem. We still assume we have a 50% chance of getting a red, but now let's assume we only go bankrupt if there is a set of 6 consecutive spins where at least 5 are black. What is the probability now of going bankrupt in 100 spins?*

Exercise 23.5.13 *Consider the Vandermonde matrices from §23.2.4. Determine α.*

Exercise 23.5.14 *Consider the Vandermonde matrix with parameters $r_1, \ldots, r_n$. Replace the kth row, which was initially $(r_1^{k-1}, \ldots, r_n^{k-1})$ with $(r_1^{2(k-1)}, \ldots, r_n^{2(k-1)})$. Is there still a nice formula for the determinant? If yes, find it.*

Exercise 23.5.15 *Let $A(r_1, \ldots, r_n)$ be a Vandermonde matrix with parameters $r_1, \ldots, r_n$, and $B(s_1, \ldots, s_n)$ a Vandermonde matrix with parameters $s_1, \ldots, s_n$. Find the determinant of $A(r_1, \ldots, r_n)B(s_1, \ldots, s_n)$ in terms of $r_1, \ldots, s_n$.*

Exercise 23.5.16 *Assume initially there are 1 million people in the world, with 700,000 living in Freedonia and 300,000 living in Sylvania. At the end of each year, 80% of the people living in Freedonia stay in Freedonia and the rest move to Sylvania, while 70% of the people living in Sylvania stay in Sylvania (with the remainder moving to Freedonia). If no one is ever born or dies, after 10 years what is the probability a randomly chosen person lives in Freedonia? What about 20 years? What about in the limit as the number of years which passes goes to infinity?*

CHAPTER 24 _____

The Method of Least Squares

The Method of Least Squares is a procedure to determine the best fit line to data; the proof uses calculus and linear algebra. The basic problem is to find the best fit straight line $y = ax + b$ given that, for $n \in \{1, \ldots, N\}$, the pairs (x_n, y_n) are observed. The method easily generalizes to finding the best fit of the form

$$y = a_1 f_1(x) + \cdots + c_K f_K(x);$$

it is not necessary for the functions f_k to be linearly in x—all that is needed is that y is to be a linear combination of these functions. This chapter is thus a *great* application of much of the math you've done in this book and earlier; this is not a collection of cookbook problems to make sure you've mastered the material, but rather one of the most important applications you'll find. To keep this chapter self-contained, we start with a quick review. In addition to developing the theory and doing some examples, we also discuss alternatives one could use for curve fitting. In particular, we'll spend a lot of time trying to figure out what is the right statistic to study, weighing advantages and disadvantages of different alternatives.

24.1 Description of the Problem

Often in the real world one expects to find linear relationships between variables. For example, the force of a spring linearly depends on the displacement of the spring: $y = kx$ (here y is the force, x is the displacement of the spring from rest, and k is the spring constant). To test the proposed relationship, researchers go to the lab and measure what the force is for various displacements. Thus they assemble data of the form (x_n, y_n) for $n \in \{1, \ldots, N\}$; here y_n is the observed force in Newtons when the spring is displaced x_n meters.

Unfortunately, it is extremely unlikely that we will observe a perfect linear relationship. There are two reasons for this. The first is experimental error; the second is that the underlying relationship may not be exactly linear, but rather only approximately linear. (A standard example is the force felt on a falling body. We initially approximate the force as $F = mg$ with g the acceleration due to gravity; however, this is not quite right as there is a resistive force which depends on the velocity.) See Figure 24.1 for a

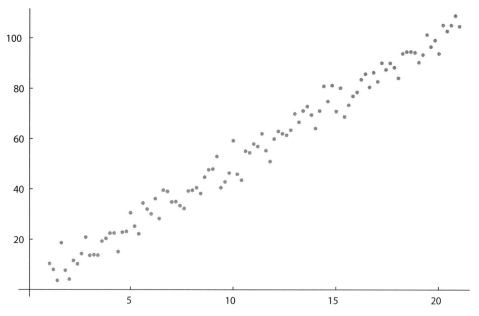

Figure 24.1. 100 "simulated" observations of displacement and force ($k = 5$).

simulated data set of displacements and forces for a spring with spring constant equal to 5.

The Method of Least Squares is a procedure, requiring just some calculus and linear algebra, to determine what the "best fit" line is to the data. Of course, we need to quantify what we mean by "best fit," which will require a brief review of some probability and statistics.

A careful analysis of the proof will show that the method is capable of great generalizations. Instead of finding the best fit line, we could find the best fit given by *any* finite linear combinations of specified functions. Thus the general problem is given functions $f_1, \ldots, f_K$, find values of coefficients $a_1, \ldots, a_K$ such that the *linear* combination

$$y = a_1 f_1(x) + \cdots + a_K f_K(x)$$

is the best approximation to the data.

24.2 Probability and Statistics Review

We give a quick review of the basic elements of probability and statistics which we need for the Method of Least Squares. Given a sequence of data $x_1, \ldots, x_N$, we define the **mean** (or the **expected value**) to be $(x_1 + \cdots + x_N)/N$. We denote this by writing a line above x: thus

$$\bar{x} = \frac{1}{N} \sum_{n=1}^{N} x_n.$$

The mean is the average value of the data.

Consider the following two sequences of data: $\{10, 20, 30, 40, 50\}$ and $\{30, 30, 30, 30, 30\}$. Both sets have the same mean; however, the first data set has greater variation about the mean. This leads to the concept of variance, which is a useful tool to quantify how much a set of data fluctuates about its mean. The **variance**[*] of $\{x_1, \ldots, x_N\}$, denoted by σ_x^2, is

$$\sigma_x^2 = \frac{1}{N} \sum_{n=1}^{N} (x_n - \overline{x})^2;$$

the **standard deviation** σ_x is the square-root of the variance:

$$\sigma_x = \sqrt{\frac{1}{N} \sum_{n=1}^{N} (x_n - \overline{x})^2}.$$

Note that if the x's have units of meters then the variance σ_x^2 has units of meters^2, and the standard deviation σ_x and the mean $\overline{x}$ have units of meters. Thus it is the standard deviation that gives a good measure of the deviations of the x's around their mean, as it has the same units as our quantity of interest.

There are, of course, alternate measures one can use. For example, one could consider

$$\frac{1}{N} \sum_{n=1}^{N} (x_n - \overline{x}).$$

Unfortunately this is a signed quantity, and large positive deviations can cancel with large negatives. In fact, the definition of the mean immediately implies the above is zero! This, then, would be a terrible measure of the variability in data, as it is zero regardless of what the values of the data are.

We can rectify this problem by using absolute values. This leads us to consider

$$\frac{1}{N} \sum_{n=1}^{N} |x_n - \overline{x}|. \tag{24.1}$$

While this has the advantage of avoiding cancelation of errors (as well as having the same units as the x's), the absolute value function is not a good function analytically. It's not differentiable. Thus we'll consider the standard deviation (the square-root of the variance)—this will allow us to use the tools from calculus.

We can now quantify what we mean by "best fit." If we believe $y = ax + b$, then $y - (ax + b)$ should be zero. Thus given observations

$$\{(x_1, y_1), \ldots, (x_N, y_N)\},$$

---

[*]For those who know more advanced statistics, for technical reasons the correct definition of the sample variance is to divide by $N - 1$ and not N.

we look at

$$\{y_1 - (ax_1 + b), \ \ldots, \ y_N - (ax_N + b)\}.$$

The mean should be small (if it is a good fit), and the sum of squares of the terms will measure how good of a fit we have.

We define

$$E(a, b) := \sum_{n=1}^{N} (y_n - (ax_n + b))^2.$$

Large errors are given a higher weight than smaller errors (due to the squaring). Thus our procedure favors many medium sized errors over a few large errors. If we used absolute values to measure the error (see Equation (24.1)), then all errors are weighted equally. We can avoid that problem by taking absolute values; unfortunately the absolute value function is not differentiable, and thus the tools of calculus become inaccessible.

We end this section by expanding on our choice of squaring errors. This is a particular example of one of the greatest challenges in the subject: finding the right statistic to study. We can study anything we want; the questions are when will it be accessible and useful.

Remark 24.2.1 (Choice of how to measure errors): *There are three natural candidates to use in measuring the error between theory and observation:*

$$E_1(a, b) \ = \ \sum_{n=1}^{N} (y_n - (ax_n + b)), \tag{24.2}$$

$$E_2(a, b) \ = \ \sum_{n=1}^{N} |y_n - (ax_n + b)| \tag{24.3}$$

and

$$E_3(a, b) \ = \ \sum_{n=1}^{N} (y_n - (ax_n + b))^2. \tag{24.4}$$

The problem with (24.2) is that the errors are signed quantities, and positive errors can cancel with negative errors. The problem with (24.3) is that the absolute value function is not differentiable, and thus the tools and results of calculus are unavailable. The problem with (24.4) is that errors are not weighted equally: large errors are given significantly more weight than smaller errors. There are thus problems with all three. That said, the problem with (24.4) is not so bad when compared to its advantages, namely that errors cannot cancel and that calculus is available. Thus, most people typically use (24.4) and measure errors by sums of squares.

24.3 The Method of Least Squares

One of the reasons I love the Method of Least Squares so much is that it yields an explicit, closed form solution for the best fit parameters in terms of the observables.

The reason this is possible is that we have the tools of calculus and linear algebra available—there should be some dividends for all our hard work over the years!

Given data $\{(x_1, y_1), \ldots, (x_N, y_N)\}$, we defined the error associated to saying $y = ax + b$ by

$$E(a, b) := \sum_{n=1}^{N} (y_n - (ax_n + b))^2 . \tag{24.5}$$

Note that the error is a function of two variables, the unknown parameters a and b.

The goal is to find values of a and b that minimize the error. In multivariable calculus we learn that this requires us to find the values of (a, b) such that the gradient of E with respect to our variables (which are a and b) vanishes; thus we require

$$\nabla E = \left(\frac{\partial E}{\partial a}, \frac{\partial E}{\partial b} \right) = (0, 0),$$

or

$$\frac{\partial E}{\partial a} = 0, \quad \frac{\partial E}{\partial b} = 0.$$

Note we do not have to worry about boundary points: as $|a|$ and $|b|$ become large, the fit will clearly get worse and worse. Thus we do not need to check on the boundary.

Differentiating $E(a, b)$ yields

$$\frac{\partial E}{\partial a} = \sum_{n=1}^{N} 2 (y_n - (ax_n + b)) \cdot (-x_n)$$

$$\frac{\partial E}{\partial b} = \sum_{n=1}^{N} 2 (y_n - (ax_n + b)) \cdot (-1).$$

Setting $\partial E / \partial a = \partial E / \partial b = 0$ (and dividing by -2) yields

$$\sum_{n=1}^{N} (y_n - (ax_n + b)) \cdot x_n = 0$$

$$\sum_{n=1}^{N} (y_n - (ax_n + b)) = 0.$$

Note we can divide both sides by -2 as it is just a constant; we cannot divide by x_i as that varies with i.

We may rewrite these equations as

$$\left(\sum_{n=1}^{N} x_n^2\right) a + \left(\sum_{n=1}^{N} x_n\right) b = \sum_{n=1}^{N} x_n y_n$$

$$\left(\sum_{n=1}^{N} x_n\right) a + \left(\sum_{n=1}^{N} 1\right) b = \sum_{n=1}^{N} y_n.$$

Thus the values of a and b which minimize the error (defined in (24.5)) satisfy the following matrix equation:

$$\begin{pmatrix} \sum_{n=1}^{N} x_n^2 & \sum_{n=1}^{N} x_n \\ \sum_{n=1}^{N} x_n & \sum_{n=1}^{N} 1 \end{pmatrix} \begin{pmatrix} a \\ b \end{pmatrix} = \begin{pmatrix} \sum_{n=1}^{N} x_n y_n \\ \sum_{n=1}^{N} y_n \end{pmatrix}. \tag{24.6}$$

We need a fact from linear algebra. Recall the inverse of a matrix A is the matrix B such that $AB = BA = I$, where I is the identity matrix. If $A = \begin{pmatrix} \alpha & \beta \\ \gamma & \delta \end{pmatrix}$ is a 2×2 matrix where $\det A = \alpha\delta - \beta\gamma \neq 0$, then A is invertible and

$$A^{-1} = \frac{1}{\alpha\delta - \beta\gamma} \begin{pmatrix} \delta & -\gamma \\ -\beta & \alpha \end{pmatrix}.$$

In other words, $AA^{-1} = \begin{pmatrix} 1 & 0 \\ 0 & 1 \end{pmatrix}$ here. For example, if $A = \begin{pmatrix} 1 & 3 \\ 2 & 7 \end{pmatrix}$ then $\det A = 1$ and $A^{-1} = \begin{pmatrix} 7 & -3 \\ -2 & 1 \end{pmatrix}$; we can check this by noting (through matrix multiplication) that

$$\begin{pmatrix} 1 & 2 \\ 3 & 7 \end{pmatrix} \begin{pmatrix} 7 & -2 \\ -3 & 1 \end{pmatrix} = \begin{pmatrix} 1 & 0 \\ 0 & 1 \end{pmatrix}.$$

We can show the matrix in (24.6) is invertible (so long as at least two of the x_n's are distinct), which implies

$$\begin{pmatrix} a \\ b \end{pmatrix} = \begin{pmatrix} \sum_{n=1}^{N} x_n^2 & \sum_{n=1}^{N} x_n \\ \sum_{n=1}^{N} x_n & \sum_{n=1}^{N} 1 \end{pmatrix}^{-1} \begin{pmatrix} \sum_{n=1}^{N} x_n y_n \\ \sum_{n=1}^{N} y_n \end{pmatrix} \tag{24.7}$$

So, all that's left is to show invertibility. Denote the matrix from (24.6) by M. The determinant of M is

$$\det M = \sum_{n=1}^{N} x_n^2 \cdot \sum_{n=1}^{N} 1 - \sum_{n=1}^{N} x_n \cdot \sum_{n=1}^{N} x_n.$$

As

$$\bar{x} = \frac{1}{N} \sum_{n=1}^{N} x_n,$$

we find that

$$\det M = N \sum_{n=1}^{N} x_n^2 - (N\overline{x})^2$$

$$= N^2 \left(\frac{1}{N} \sum_{n=1}^{N} x_n^2 - \overline{x}^2 \right)$$

$$= N^2 \cdot \frac{1}{N} \sum_{n=1}^{N} (x_n - \overline{x})^2,$$

where the last equality follows from simple algebra. Thus, as long as all the x_n are not equal, $\det M$ will be non-zero and M will be invertible. Using the definition of variance, we notice the above could also be written as

$$\det M = N^2 \sigma_x^2.$$

Thus we find that, so long as the x's are not all equal, the best fit values of a and b are obtained by solving a linear system of equations; the solution is given in (24.7).

We rewrite (24.7) in a simpler form. Using the inverse of the matrix and the definition of the mean and variance, we find

$$\begin{pmatrix} a \\ b \end{pmatrix} = \frac{1}{N^2 \sigma_x^2} \begin{pmatrix} N & -N\overline{x} \\ -N\overline{x} & \sum_{n=1}^{N} x_n^2 \end{pmatrix} \begin{pmatrix} \sum_{n=1}^{N} x_n y_n \\ \sum_{n=1}^{N} y_n \end{pmatrix}. \tag{24.8}$$

Expanding gives

$$a = \frac{N \sum_{n=1}^{N} x_n y_n - N\overline{x} \sum_{n=1}^{N} y_n}{N^2 \sigma_X^2}$$

$$b = \frac{-N\overline{x} \sum_{n=1}^{N} x_n y_n + \sum_{n=1}^{N} x_n^2 \sum_{n=1}^{N} y_n}{N^2 \sigma_X^2}$$

$$\overline{x} = \frac{1}{N} \sum_{n=1}^{N} x_i$$

$$\sigma_x^2 = \frac{1}{N} \sum_{n=1}^{N} (x_i - \overline{x})^2. \tag{24.9}$$

As the formulas for a and b are so important, it is worth giving another expression for them. We also have

$$a = \frac{\sum_{n=1}^{N} 1 \sum_{n=1}^{N} x_n y_n - \sum_{n=1}^{N} x_n \sum_{n=1}^{N} y_n}{\sum_{n=1}^{N} 1 \sum_{n=1}^{N} x_n^2 - \sum_{n=1}^{N} x_n \sum_{n=1}^{N} x_n}$$

$$b = \frac{\sum_{n=1}^{N} x_n \sum_{n=1}^{N} x_n y_n - \sum_{n=1}^{N} x_n^2 \sum_{n=1}^{N} y_n}{\sum_{n=1}^{N} x_n \sum_{n=1}^{N} x_n - \sum_{n=1}^{N} x_n^2 \sum_{n=1}^{N} 1}. \tag{24.10}$$

Remark 24.3.1: *The formulas above for a and b are reasonable, as can be seen by a unit analysis. For example, imagine x is in meters and y is in seconds. Then if y = ax + b we would need b and y to have the same units (namely seconds), and a to have units seconds per meter. If we substitute in the units for the various quantities on the right-hand side of (24.9), we do see a and b have the correct units. While this isn't a proof that we haven't made a mistake, it's a great reassurance. No matter what you're studying, you should always try **unit calculations** such as this.*

There are other, equivalent formulas for a and b; these give the same answer, but arrange the algebra in a slightly different sequence of steps. Essentially what we are doing is the following: imagine we are given

$$4 = 3a + 2b$$
$$5 = 2a + 5b.$$

If we want to solve, we can proceed in two ways. We can use the first equation to solve for b in terms of a and substitute in, or we can multiply the first equation by 5 and the second equation by 2 and subtract; the b terms cancel and we obtain the value of a. Explicitly,

$$20 = 15a + 10b$$
$$10 = 4a + 10b,$$

which yields

$$10 = 11a,$$

or

$$a = 10/11.$$

Remark 24.3.2: *The data plotted in Figure 24.1 was obtained by letting $x_n = 5 + .2n$ and then letting $y_n = 5x_n$ plus an error randomly drawn from a normal distribution with mean zero and standard deviation 4 ($n \in \{1, \ldots, 100\}$). Using these values, we find a best fit line of*

$$y = 4.99x + .48;$$

thus $a = 4.99$ and $b = .48$. As the expected relation is $y = 5x$, we expected a best fit value of a of 5 and b of 0.

While our value for a is very close to the true value, our value of b is significantly off. We deliberately chose data of this nature to indicate the dangers in using the Method of Least Squares. Just because we know 4.99 is the best value for the slope and .48 is the best value for the y-intercept does not mean that these are good estimates of the true values. The theory needs to be supplemented with techniques which provide error estimates. Thus we want to know something like, given this data, there is a 99% chance that the true value of a is in (4.96, 5.02) and the true value of b is in (−.22, 1.18); this is far more useful than just knowing the best fit values.

If instead we used

$$E_{\mathrm{abs}}(a, b) \; = \; \sum_{n=1}^{N} |y_n - (ax_n + b)| \, ,$$

then numerical techniques yield that the best fit value of a is 5.03 and the best fit value of b is less than 10^{-10} in absolute value. The difference between these values and those from the Method of Least Squares is in the best fit value of b (the least important of the two parameters), and is due to the different ways of weighting the errors.

24.4 Exercises

Exercise 24.4.1   *Consider the observed data $(0, 0), (1, 1), (2, 2)$. It should be clear that the best fit line is $y = x$; this leads to zero error in all three systems of measuring error, namely (24.2), (24.3), and (24.4); however, show that if we use (24.2) to measure the error then line $y = 1$ also yields zero error, and clearly this should not be the best fit line!*

Exercise 24.4.2   *Generalize the Method of Least Squares to find the best fit quadratic to $y = ax^2 + bx + c$ (or more generally the best fit degree m polynomial to $y = a_m x^m + a_{m-1} x^{m-1} + \cdots + a_0$).*

While for any real-world problem, direct computation determines whether or not the resulting matrix is invertible, it is nice to be able to prove the determinant is always non-zero for the best fit line (if all the x's are not equal).

Exercise 24.4.3   *If the x's are not all equal, must the determinant be non-zero for the best fit quadratic or the best fit cubic?*

Looking at our proof of the Method of Least Squares, we note that it was not essential that we have $y = ax + b$; we could have had $y = af(x) + bg(x)$, and the arguments would have proceeded similarly. The difference would be that we would now obtain

$$\begin{pmatrix} \sum_{n=1}^{N} f(x_n)^2 & \sum_{n=1}^{N} f(x_n)g(x_n) \\ \sum_{n=1}^{N} f(x_n)g(x_n) & \sum_{n=1}^{N} g(x_n)^2 \end{pmatrix} \begin{pmatrix} a \\ b \end{pmatrix} = \begin{pmatrix} \sum_{n=1}^{N} f(x_n)y_n \\ \sum_{n=1}^{N} g(x_n)y_n \end{pmatrix}. \quad (24.11)$$

Finally, we comment briefly on a very important change of variable that allows us to use the Method of Least Squares in many more situations than one might expect. Consider the case of a researcher trying to prove Newton's Law of Universal Gravity, which says the force felt by two masses m_1 and m_2 has magnitude Gm_1m_2/r^2, where r is the distance between the objects. If we fix the masses, then we expect the magnitude of the force to be inversely proportional to the distance. We may write this as $F = k/r^n$, where we believe $n = 2$ (the value for k depends on G and the product of the masses). Clearly it is n that is the more important parameter here. Unfortunately, as written, we cannot use the Method of Least Squares, as one of the unknown parameters arises non-linearly (as the exponent of the separation).

*We can surmount this problem by **taking a logarithmic transform** of the data. Setting $\mathcal{K} = \log k$, $\mathcal{F} = \log F$, and $\mathcal{R} = \log r$, the relation $F = k/r^n$ becomes $\mathcal{F} = n\mathcal{R} + \mathcal{K}$. We are now in a situation where we can apply the Method of Least Squares. The only difference from the original problem is how we collect and process the data; now* *our data is not the separation between the two masses, but rather the logarithm of the separation. Arguing along these lines, many power relations can be converted to instances where we can use the Method of Least Squares. We thus (finally) fulfill a promise made by many high school math teachers years ago: logarithms can be useful!*

Exercise 24.4.4 *Consider the generalization of the Method of Least Squares given in (24.11). Under what conditions is the matrix invertible?*

Exercise 24.4.5 *The method of proof generalizes further to the case when one expects y is a linear combination of K fixed functions. The functions need not be linear; all that is required is that we have a linear combination, say $a_1 f_1(x) + \cdots + a_K f_K(x)$. One then determines the $a_1, \ldots, a_K$ that minimize the sum of squares of the errors by calculus and linear algebra. Find the matrix equation that the best fit coefficients $(a_1, \ldots, a_K)$ must satisfy.*

Exercise 24.4.6 *Consider the best fit line from the Method of Least Squares, so the best fit values are given by (24.7). Is the point $(\overline{x}, \overline{y})$, where $\overline{x} = \frac{1}{n} \sum_{n=1}^{N} x_n$ and $\overline{y} = \sum_{n=1}^{N} y_n$, on the best fit line? In other words, does the best fit line go through the "average" point?*

Exercise 24.4.7 **(Kepler's Third Law)** *Kepler's Third Law states that if T is the orbital period of a planet traveling in an elliptical orbit about the sun (and no other objects exist), then $T^2 = CL^3$, where L is the length of the semi-major axis. I always found this the hardest of the three laws; how would one be led to the right values of the exponents from observational data? One way is through the Method of Least Squares. Set $\mathcal{T} = \log T$, $\mathcal{L} = \log L$, and $c = \log C$. Then a relationship of the form $T^a = CL^b$ becomes $a\mathcal{T} = b\mathcal{L} + c$, which is amenable to the Method of Least Squares. The semi-major axis of the 8 planets (sadly, Pluto is no longer considered a planet) are Mercury 0.387, Venus 0.723, Earth 1.000, Mars 1.524, Jupiter 5.203, Saturn 9.539, Uranus 19.182, and Neptune 30.06 (the units are astronomical units, where one astronomical unit is $1.496 \cdot 10^8$ km); the orbital periods (in years) are 0.2408467, 0.61519726, 1.0000174, 1.8808476, 11.862615, 29.447498, 84.016846, and 164.79132. Using this data, apply the Method of Least Squares to find the best fit values of a and b in $T^a = CL^b$ (note, of course, you need to use the equation $a\mathcal{T} = b\mathcal{L} + \mathcal{C}$).*

*Actually, as phrased above, the problem is a little indeterminate for the following reason. Imagine we have $T^2 = 5L^3$ or $T^4 = 25L^6$ or $T = \sqrt{5}L^{1.5}$ or even $T^4 = 625L^{12}$. **All of these are the same equation!** In other words, we might as well make our lives easy by taking $a = 1$; there really is no loss in generality in doing this. This is yet another example of how changing our point of view can really help us. At first it looks like this is an exercise involving **three** unknown parameters, a, b, and C; however, **there is absolutely no loss in generality in taking** $a = 1$; thus let us make our lives easier and just look at this special case.*

For your convenience, here are the natural logarithms of the data: the lengths of the semi-major axes are

$$\{-0.949331, -0.324346, 0, 0.421338, 1.64924, 2.25539, 2.95397, 3.4032\}$$

and the natural logarithms of the periods (in years) are

$$\{-1.42359, -0.485812, 0.0000173998, 0.631723, 2.47339, 3.38261, 4.43102, 5.10468\}.$$

*The exercise asks you to find the best fit values of a and b. In some sense this is a bit misleading, as there are infinitely many possible values for the pair (a, b); however, all of these pairs will have the same **ratio** b/a (which Kepler says should be close to 3/2 or 1.50). It is this ratio that is truly important. The content of Kepler's Third Law is that the square of the period is proportional to the cube of the semi-major axis. The key numbers are the powers of the period and the length (the a and the b), not the proportionality constant. This is why I only ask you to find the best fit values of a and b and not C (or $\mathcal{C}$), as C (or $\mathcal{C}$) is not as important. If we take $a = 1$ then the best fit value of $\mathcal{C}$ is 0.000148796, and the best fit value of b is almost 1.50.*

Our notes above have many different formulas to find the best fit values a and b for a relation $y = ax + b$. For us, we have $\mathcal{T} = \frac{b}{a}\mathcal{L} + \frac{\mathcal{C}}{a}$. Thus, for this exercise, the role of a from before is being played by $\frac{b}{a}$ and the role of b from before is being played by $\frac{\mathcal{C}}{a}$. Therefore if we want to find the best fit value for the ratio $\frac{b}{a}$ for this exercise, we just use the first of the two formulas from (24.10).

CHAPTER 25 _____

Two Famous Problems and Some Coding

*There are a handful of famous probability problems which are covered in almost all courses. One of these is the **marriage (or secretary) problem**.*

> **Marriage (or secretary) problem**: *Given a list of n applicants for a job, whom we see one at a time, what strategy maximizes the probability of selecting the best applicant if once we pass on a candidate they are forever unavailable?*

In other words, whenever we meet a candidate we are immediately forced to decide whether or not to offer them the job, and if we decide not to make them an offer they'll be taken by someone else and become unavailable. The modifications to dating and marriage are hopefully obvious, and hence the additional name for this problem. We'll also look at the famous Monty Hall Problem. It's different than the marriage/secretary problem, but as it too involves making a series of choices in a quest for an optimum solution, it seems natural to include it in the same chapter.

There are many more problems which would be fun to include, such as the Envelope Problem as well as Buffon's Needle; unfortunately this book is already quite long and thus I'll encourage you to search these two out.

25.1 The Marriage/Secretary Problem

25.1.1 Assumptions and Strategy

Whenever you're analyzing a difficult problem, it's a good idea to start and carefully enumerate the given conditions. Make sure you're not implicitly assuming anything. Let's do that for the secretary problem (we have to choose a formulation for our discussion, so for definiteness let's do the secretary problem, as I'm worried about lawsuits if you apply this material to your personal life and things don't work out...).

Below are the assumptions in one of the most commonly analyzed versions. If you enjoy this problem I encourage you to explore your own modifications (but again only as an academic exercise, though I did have one student years ago who said he would choose his wife according to what he learned below; he's also still single).

- First, we must either hire or not hire each applicant on the spot. If we choose not to hire them, we cannot hire them later.
- Second, we can determine the relative rank of each applicant. What this means is that we can score each person and compare their score to the scores of earlier applicants. We assume we know how many candidates are in our pool; the problem is a lot harder if this is not known. Note, we are not assuming any knowledge of the distribution of scores (if it were the discrete uniform on n people, for instance, if we had someone ranked at n we would know they're the best!). For simplicity we assume there are no ties.
- Third, the applicants interview for the job in random, independent order with applicants of all qualities equally likely to be in any spot. Another way of stating this is that all $n!$ orderings of the candidates are equally likely to be the order they arrive.
- Fourth, hiring any applicant but the best is a complete failure: there's no difference between hiring the second best applicant and the worst applicant. This is a very harsh condition, and is a strong deviation from the real world. In the real world, at some point the probability of getting no one good becomes so strong that there is a tendency to "settle". Our assumption is not a good approximation to the real world, but it simplifies the mathematics enormously and is a good first model.
- Fifth, if you offer to hire an applicant, the applicant will automatically accept the job. After all, doesn't everyone want to be with you?

We now turn to figuring out what our strategy should be, and analyzing how successful it is. Initially, it is tempting to say this problem is just hopeless. The applicants are coming in for interviews in a random order. We might as well just pick the first applicant, since they are as likely to be the best as any other. If we use this strategy, or more generally if we always pick the ith applicant, then the probability of success is just $1/n$. Note that for n large this strategy has essentially no chance of landing you the best applicant. We must do better!

The following strategy does a terrific job. We choose some number k, which depends on n, and look at the first k candidates one at a time. We will never hire any of them; we're using them to get a feel for the market. We then choose the very next person who scores better than the best we've seen and make them the offer. Not only does this strategy do better than our naive attempt, it does so well that as n tends to infinity we end up with the best person over 30% of the time!

Before turning to the analysis of this approach it's worth commenting on why we chose this as opposed to other options. For example, why wasn't our strategy to take the first person twice as good as the best we've seen (or some other scaling factor), or maybe the first person three points higher? The reason is that we have no idea what distribution the scores come from; it's actually better to forget about the ranking numbers and instead imagine that all we have is a relative ordering of the people. Thus our strategy should only involve the relative comparisons of candidates, and not the comparisons of their numbers.

25.1.2 Probability of Success

Our probability of winning with our strategy is

$$\text{Prob(win)} = \sum_{m=1}^{n} \text{Prob(win|best at } m) \cdot \text{Prob(best at } m).$$

This follows from partitioning; the best person has to be somewhere, and can only be in one spot. The difficulty is figuring out what the probability of winning is given the different location of the best person.

Some of these calculations are easy. If the best candidate is in any of first k spots then we are doomed to fail. It's so bad that not only do we fail to hire the best candidate, but we end up hiring no one! (Although we could tweak the strategy to say that if we haven't made any offers, we hire the last person.) Alright, this handles the analysis of the probability of success when the best candidate is in the first k: it's zero.

Moving on, we can assume the best candidate is somewhere among the last $n - k$ people. For definiteness let's say they're at position $m + 1$, where $k \le m \le n - 1$ (it makes the algebra a little nicer to have them located at $m + 1$ and not m; ah, the benefit of previous experience with this calculation; you could of course do it with them at m). Thus we rewrite the desired probability as

$$\text{Prob(win)} = \sum_{m=k}^{n-1} \text{Prob(win|best at } m + 1) \cdot \text{Prob(best at } m + 1).$$

What is the probability we win *given that* the best candidate is at $m + 1$? This is a conditional probability, and is just k/m; seeing this is the hardest part of the analysis, so let's go through it carefully. Since the *overall* best person is at $m + 1$, we know the best of the first $m + 1$ is also at $m + 1$; it all comes down to where the *second best* person of the first $m + 1$ people is located. Or, since the best is at $m + 1$, it comes down to where is the *best of the first* m located.

If they're one of the first k people then we will choose person $m + 1$. Why? If the second best overall of the first $m + 1$ is in the first k, then that person is better than the person at $k + 1$, or $k + 2$, ..., or m; thus the first person we meet who is better than them is at $m + 1$, and we pick them and win. Conversely, if the best of the first m is *not* in the first k we lose. Why? Now we look at our first k and choose the first person better; sadly we never reach the person at $m + 1$ now as we hit the second best in the first $m + 1$ earlier. *You should **make a picture**.*

So, what's the probability that the best of the first m is in the first k? As each person is equally likely to be in any spot, the probability is just k/m (of the m positions to place them, k of them have the person in the first k spots). Thus Prob(win|best at $m + 1$) $= k/m$.

What is the probability that the best candidate is at m? That's just $1/n$, since the best candidate is equally likely to be in any of the n spots. This means that our overall probability of success is

$$\text{Pr(win)} = \sum_{m=k}^{n-1} \frac{k}{m} \frac{1}{n} = \frac{k}{n} \sum_{m=k}^{n-1} \frac{1}{m}.$$

This is a good start. We now have an expression for our probability of success. For a fixed n and k we can calculate it, and then vary k and see what choice maximizes the sum.

Let's try to approximate this sum. We'll thus assume n and k are both large, which allows us to use results from calculus to optimize our probability. In

$$\Pr(\text{win}) = \frac{k}{n} \sum_{m=k}^{n-1} \frac{1}{m},$$

by **adding zero** we can represent the sum as a difference of two **harmonic series** $\sum_{m=1}^{\ell} \frac{1}{n}$, with ℓ equal to $n - 1$ and $k - 1$. In a calculus class you learn that for large ℓ that sum is approximately $\log(\ell)$. It turns out there's a great approximation to that sum:

$$\sum_{m=1}^{\ell} \frac{1}{n} \approx \log(\ell) + \gamma + \text{Error}_\ell, \tag{25.1}$$

where γ is the **Euler-Mascheroni constant** (about .5772) and $\epsilon_\ell \sim 1/2\ell$ and thus approaches zero rapidly as ℓ grows (see Exercise 25.4.1). We find

$$\Pr(\text{win}) = \frac{k}{n} \sum_{m=k}^{n-1} \frac{1}{m} = \frac{k}{n} \left(\sum_{m=k}^{n-1} \frac{1}{m} + \sum_{m=1}^{k-1} \frac{1}{m} - \sum_{m=1}^{k-1} \frac{1}{m} \right)$$

$$= \frac{k}{n} \left(\sum_{m=1}^{n-1} \frac{1}{m} - \sum_{m=1}^{k-1} \frac{1}{m} \right)$$

$$\approx \frac{k}{n} \left(\log(n - 1) - \log(k - 1) \right)$$

$$\approx \frac{k}{n} \left(\log(n) - \log(k) \right)$$

$$= \frac{k}{n} \cdot \log \left(\frac{n}{k} \right).$$

Now that we have a closed form expression for the probability of success given a k and an n, we just need to find its maximum. Notice that our approximation only depends on the ratio k/n, which suggests we should replace that with some new variable and try to maximize the resulting expression. As we're trying to find a maximum and our probability is a nice, differentiable function of $x = n/k$, it's not surprising that we'll turn to calculus.

Setting $x = n/k$, we have $\Pr(\text{win}) \approx \log(x)/x$. To get a feel for the answer we assume instead that there is no approximation; converting the problem to maximizing $\log(x)/x$ greatly simplifies the algebra, and if n is large it should be very close to the truth.

So, it all reduces to maximizing $\log(x)/x$ for $1 \le k \le n$ (which translates to $n \ge x \ge 1$). In a calculus class you learn that to find candidates for extrema you must check the endpoints and the critical points (remember the critical points are where the first derivative vanishes). The endpoints are easily dispatched, as $x = n$ and $x = 1$ both give small probabilities. (While $x = n$ technically gives an estimate of $\log(n)/n$, which is "large" relative to the value of $1/n$ obtained from always taking a fixed location, remember we assumed n *and* k were both large when we approximated, and thus we shouldn't be too surprised that the probability we get here will be dominated by the probability at the critical point.)

What about the **critical points**, the places where the derivative vanishes? For those we have to solve

$$\frac{d}{dx} \left(\frac{\log(x)}{x} \right) = \frac{\frac{1}{x} \cdot x - \log(x)}{x^2},$$

which equals 0 when the numerator is 0. So $1 - \log(x) = 0$, which implies $x = e$. This is a global maximum and not just a local maximum because the derivative is positive for $x < e$, zero at e, and then negative for $x > e$. Thus the function is increasing up to $x = e$ and decreasing from that point onward, so we must have the global maximum at e. (See Exercise 25.4.2 for another proof.) Note that substituting $x = e$ into $\log(x)/x$ gives a probability of a win of $\log(e)/e = 1/e$, which is significantly larger than $\log(n)/n$ when n is large.

Let's interpret our answer. As $x = n/k$ and the optimal k is when $x \approx e$ (so $k \approx n/e$, or $k/n \approx 1/e$), our optimal strategy has us look through $1/e$ percent of our applicants to get a sense of the market, and then hire the first candidate better than the best of these. If we take this approach, we hire the best candidate approximately an amazing $\log(e)/e = 1/e \approx 36.79\%$ of the time. Not only did our new strategy do enormously better than our original probability $1/n$, it produced a positive probability in the limit as the number of people interviewing goes to infinity! To me this result is absolutely amazing, as the answer it produces is so much greater than what we might initially expect.

While it has been proven that the strategy already discussed leads to the highest probability of success, there are a number of other strategies to consider that make some intuitive sense. For a fuller discussion of these see *Analysis of Heuristic Solutions to the Best Choice Problem* [SSR]. The original secretary problem has also spawned many variants, including trying to minimize the expected rank of the candidate, trying to maximize the expected value of the candidate given they are drawn from a certain distribution, trying to draw a subset of candidates better than all the others, and trying to pick the second best candidate. This last problem was solved by Robert Vanderbei, who gave it the name, **the Postdoc Problem**; for details see http://www.princeton.edu/ rvdb/tex/PostdocProblem/PostdocProb.pdf.

Remark 25.1.1: *We end with a quick warning. If you were reading the above very carefully you might have noticed a small error. Remember that k has to be an integer, yet at the end of the analysis we chose k to be n/e, which clearly is not an integer! Fortunately the problem is easily corrected. Our probability of winning is increasing as k increases from 0 to n/e, then decreases as k continues to n. Thus the optimal k will be either the greatest integer less than n/e, or the smallest integer greater than n/e; we just have to check these two values. In general it's hard to pass from a global optimal choice to a real optimal choice (see Exercise 25.4.18); because of how our function is changing we're quite fortunate here.*

25.1.3 Coding the Secretary Problem

As one of the goals of this chapter is to discuss programming, we end this section with a simple program to test our theoretical conjectures.

```
secretaryproblem[n_, numdo_] := Module[{},
  (* num is num simulations *)
  k = Round[1.0 n/E]; (* takes closest integer to n/e *)
  success = 0; (* number of times win *)
  people = {}; (* makes list of people *)
  For[j = 1, j <= n, j++, people = AppendTo[people, j]];
  For[num = 1, num <= numdo, num++,
   {
    (* randomly order people *)
    order = RandomSample[people];
    (* go through list, see who is best in first k *)
    (* keep going till find someone better; if that is
    n win, else lose *)
    max = 0;
    For[j = 1, j <= k, j++, If[order[[j]] > max, max = order[[j]]]];
    best = 0;
    For[j = k + 1, j <= n, j++,
     If[order[[j]] > max,
       {
        best = order[[j]];
        j = n + 1000; (*
        exit j loop as found someone better than first *)
        }]; (* end of if *)
     ]; (* end of j loop *)
    If[best == n, success = success + 1];
    }]; (* end of num loop *)
  Print["Theory predicts successful ", SetAccuracy[100. / E, 3],
   "%."];
  Print["We were successful ", 100. success / numdo, "%."];
  ];
```

Running the program 10,000 with $n = 1000$ yielded the following times:

```
Theory predicts successful 36.79%.
We were successful 36.78%.
```

A pretty good fit! Can you make the program more efficient? Note that if you want to test and make sure the program is working and doing what you want it to do the above is *not* helpful. There's no way to see if it's getting the right answer or not. We can remedy this by putting in some print statements. I like to add a print variable, and then if we assign the value 1 it prints out some intermediate values, and if we assign the value 0 it doesn't print at all.

```
secretaryproblemdebug[n_, numdo_, print_] := Module[{},
  (* num is num simulations *)
  k = Round[1.0 n/E]; (* takes closest integer to n/e *)
  success = 0; (* number of times win *)
  people = {}; (* makes list of people *)
  If[print == 1,
   Print["Printing results as go along for debugging. When printing
   the best value a value of 0 means the best overall was in the first
   k.\n"];
   ];
```

```
For[j = 1, j <= n, j++, people = AppendTo[people, j]];
For[num = 1, num <= numdo, num++,
 {
  (* randomly order people *)
  order = RandomSample[people];
  (* go through list, see who is best in first k *)
  (* keep going till find someone better; if that is
  n win, else lose *)
  max = 0;
  For[j = 1, j <= k, j++, If[order[[j]] > max, max = order[[j]]]];
  best = 0;
  For[j = k + 1, j <= n, j++,
   If[order[[j]] > max,
    {
     best = order[[j]];
     j = n + 1000; (*
     exit j loop as found someone better than first *)
     }]; (* end of if *)
   ]; (* end of j loop *)
  If[best == n, success = success + 1];

  If[print == 1,
   {
    Print["k = ", k, "; best in first k is ", max,
     "; first better than best in first k is ", best];
    Print["Sorted list is ", order];
    If[best == n, Print["Success!\n"], Print["Failure\n"]];
    }]; (* end of print *)

  }]; (* end of num loop *)
 Print["Theory predicts successful ", SetAccuracy[100. / E, 3],
  "%."];
 Print["We were successful ", 100. success / numdo, "%."];
 ];
```

25.2 Monty Hall Problem

Another famous puzzle is the Monty Hall Problem, a counterintuitive challenge which an introductory probability textbook would not be complete without. While the calculations are simpler than what we saw in the marriage/secretary Problem, the answer is counterintuitive to many at first glance and has generated a lot of discussion. This brain teaser is based on the game show *Let's Make a Deal*, hosted by Monty Hall.

> **Monty Hall Problem**: Imagine you are on a game show. You have a chance to win a car if you choose correctly. There are three closed doors where a car lies behind one and goats lie behind the other two. You pick a door at random, say door number 3. The host then opens another door and reveals a goat, say behind door number 1. You are then asked: do you want to switch doors (to number 2) or keep the door you first picked? Should you switch?

At first glance, this seems simple. As there are now only two doors left and the prize is equally likely to be behind any of the doors, there's just a 50-50 chance at winning

TABLE 25.1.
Analysis of the Monty Hall Problem when we choose door 3.

Behind 1	Behind 2	Behind 3	Result if Stay	Result if Switch
Car	Goat	Goat	Win Goat	Win Car
Goat	Car	Goat	Win Goat	Win Car
Goat	Goat	Car	Win Car	Win Goat

the car and it shouldn't matter if you switch or stay. Do you find this logic convincing? Think about this before you read on. Try your hand at solving this problem, either with pen and paper or by writing some code and simulating a million games and comparing what happens if you switch or don't.

25.2.1 A Simple Solution

A great way to analyze problems like this is to **make a detailed table** of all the possibilities. *We are assuming each configuration is equally likely.* We have to distribute two (identically unappealing) goats and a (highly desirable) car among three doors. There are three ways to do this: the car is behind door 1, or door 2, or door 3. (If you remember your multinomial coefficients, we're looking at how many distinguishable words we can make from CGG.)

In Table 25.1 we show the possible arrangements of two goats and a car and the results after initially picking door number 3 in each case. Note that we might as well assume we've picked door 3 because the problem is **symmetric**. If you're uncomfortable with this, just repeat the analysis but with door 1 chosen, or door 2. You'll see an identical calculation and result.

There's another assumption that's implicit in this problem that we should make explicit: *the host will never open a door revealing the car!* Why? The purpose is to have drama, excitement, suspense. If we choose door number 3 and the host reveals a car behind door number 2, it doesn't matter if we switch or not; we're getting a goat no matter what. We know that, and the audience knows that. There's no suspense, there's no decision to make.

Thus, the host *will not* reveal a car. What does this mean? If we have a goat then in the remaining two doors are a goat and a car. Since the host cannot reveal a car, he'll have to reveal the goat and the remaining door *must* have a car behind it! So, if we have a goat (which happens two out of three times), if we switch we win! What if we have a car behind our door? Then the host is free to open either of the two remaining doors, as both hold a goat. If we switch in this case we always lose, and thus switching leads to a loss one-third of the time (as that's the fraction of times we initially choose the car). We summarize our analysis in Table 25.1.

Notice that if we *stay* with door number 3 we win only one in three times, while if we *switch* we win two out of three times. Thus switching gives us twice as many wins! Also notice how sparse our table is. We're not cluttering it with lots of subcases (we choose door 3 and the host opens door 2, we choose door 3 and the host opens door 1). Sometimes those cases happen, sometimes they don't. It's better to take a higher level view of the problem. It doesn't really matter which door the host opens, it just matters if we switch or stay. It's important to remember which quantities are important in our analysis. The question is not should we switch to door 2 or door 1, but should we switch. We thus have an inkling that it might be better to forget our labels....

25.2.2 An Extreme Case

Often with puzzles it is an effective strategy to check your intuition using **extreme cases**. Using this strategy, let's get an understanding for the case of 3 doors using a case with 1,000,000 doors. Boy would this game show need a lot of goats! We'll assume, however, that math has become so popular and so many people are itching to watch that the networks have the resources to build the superstage with a million doors.

Again, there is a car behind one door, and goats behind all of the others. After the player's initial pick the host opens all but one remaining door. If we do a rigorous analysis as before we'll see that this remaining door has the car behind it 999999/1000000 times. Here's another way to see it. How likely would it be to pick the prize on your first try with one million doors to choose from? Not likely at all! In fact, the odds of that would be one in a million! Thus the probability that the prize is behind one of the remaining 999,999 doors is 999999/1000000, and if we know the car isn't behind 999998 of those doors, this entire probability of 999999/1000000 must collapse to that one remaining door. You should remember that we're not saying a *specific* door has a 999999/1000000 probability of always having a car; we're saying when this process of opening 999998 doors ends the remaining door, whatever it is numbered, has the car 999999/1000000 times.

25.2.3 Coding the Monty Hall Problem

As a major goal of this chapter is to help you master programming, we end by giving a simple program to estimate the probability of winning if you switch, and if you don't, in the Monty Hall Problem.

```
montyhall[num_] := Module[{}, (* num  is number of simulations *)
   switchwin = 0; (* keeps track of how often win when switch *)
   noswitchwin = 0; (* how often won when don't switch *)
   For[n = 1, n <= num, n++,
    {
     (* without loss of generality we may assume
     player chooses door 1 each time *)
     (* randomly choose a door for prize *)
     (* using built in function; could choose
     random number uniformly in [0,1], then
     if at most 1/3 make it door 1... *)
     doorofprize = RandomInteger[{1, 3}];
     (* if prize behind door 1 win if don't switch *)
     (* and lose if switch *)
     If[doorofprize == 1, noswitchwin = noswitchwin + 1];
     (* if prize behind door 2 or 3 win if switch else lose *)
     If[doorofprize == 2 || doorofprize == 3,
      switchwin = switchwin + 1];
     }]; (* end of n loop *)
   Print["Percent of time won when switch: ", 100. switchwin/num,
    "%."];
   Print["Percent of time won when don't switch: ",
    100. noswitchwin/num, "%."];
   ];
```

Running the program 10,000,000 times yielded the following:

```
Percent of time won when switch: 66.6668%.
Percent of time won when don't switch: 33.3332%.
```

25.3 Two Random Programs

We end this chapter by giving code to solve some problems from earlier in the book. These problems were chosen because there's nothing special about them; they're standard, typical examples of problems you should encounter in a probability class.

25.3.1 Sampling with and without Replacement

The code here is from §6.1.3, where we considered the difference in sampling with and without replacement.

```
marblecheck[num_] := Module[{},
    countwith = 0;
    countwithout = 0;
    list = {};
    For[m = 1, m <= 100, m++, list = AppendTo[list, m]];
    p[1] = .1; p[2] = .3; p[3] = .6; p[4] = .9;
    For[n = 1, n <= num, n++ m
      {
        x = Floor[4*Random[]] + 1;
        numgold = 0;
        For[i = 1, i <= 5, i++,
          If[Random[] > p[x], numgold = numgold + 1]];
        If[numgold <= 1, countwith = countwith + 1];

        y = Floor[4*Random[]] + 1;
        numgold = 0;
        templist = RandomSample[list, 5];
        cutoff = Floor[p[y]*100];
        numgold = 0;
        For[m = 1, m <= 5, m++,
          If[templist[[m]] > cutoff, numgold = numgold + 1]];
        If[numgold <= 1, countwithout = countwithout + 1];

      }]; (* end of n loop *)

    Print["Observed probability at least four purple
    (without replacement) is ",
    100 countwithout/num 1.0, "% (32.0597 predicted)."];

    Print["Observed probability of getting at least four
    purple (with replacement)
    is ", 100 countwith/num 1.0, "% (32.1685 predicted)."];
    Print["Did ", num, " iterations."];
    ];
```

This is from §6.3.3; the answer is 37,092,537, and it does get it in a few minutes.

```
solns = 0;
For[x1 = 0, x1 <= 1996/2, x1++,
  For[x2 = 0, x2 <= (1996 - 2 x1)/2, x2++,
    For[x3 = 0, x3 <= (1996 - 2 x1 - 2 x2)/3, x3++,
      If[Mod[1996 - 3 x3 - 2 x2 - 2 x1, 3] == 0, solns = solns + 1];
      ]]];
Print[solns]
```

25.3.2 Expectation

In this problem we toss 4 coins and then retoss *all* the coins if we have fewer than 2 heads (if we have 2 or more heads we keep the outcome). We get $1 for every head when we're done. The question is what is the expected value? It should be 36/16 or 19/8, and the code below supports that answer.

```
retoss[num_] := Module[{},
  winnings = 0;
  For[n = 1, n <= num, n++,
   {
    temp = 0;
    For[i = 1, i <= 4, i++, x[i] = Floor[2*Random[]]];
    If[Sum[x[i], {i, 1, 4}] >= 2, temp = Sum[x[i], {i, 1, 4}],
     {
      For[i = 1, i <= 4, i++, x[i] = Floor[2*Random[]]];
      temp = Sum[x[i], {i, 1, 4}];
      }];
    winnings = winnings + temp;
    }];
  winnings = 1. winnings / num;
  Print["Ran ", num, " simulations."];
  Print["Predict ", 19./8, " and got ", winnings, "."];
  ];
```

After running, 1,000,000 simulations we got 2.37601 (our prediction was 2.375).

25.4 Exercises

The following exercises are all related to the classic secretary problem unless otherwise noted.

Exercise 25.4.1 *Prove the approximation in (25.1) (you might want to Google the Euler-Maclaurin formula).*

Exercise 25.4.2 *Use the second derivative test from calculus to show that $\frac{\log(x)}{x}$ achieves its maximum value at $x = e$.*

Exercise 25.4.3 *Find the smallest N such that for all $n \geq N$, $\log(n)/n \leq 1/e$.*

Exercise 25.4.4 *If we look at the first p percent of the applicants and then take the first person better than the best seen (so $k = pn$), approximately what is the probability we choose the best candidate?*

Exercise 25.4.5 *Find the expected number of applicants interviewed as a fraction of n if the optimal strategy is used for large n.*

Exercise 25.4.6 *Write code that finds the optimal k for small n.*

Exercise 25.4.7 *Plot the maximum probability of picking the best person for each n against the line $1/e$. Comment on the convergence of this probability.*

Exercise 25.4.8 *Does it appear that the probability of picking the best of n people utilizing the optimal strategy is strictly decreasing as n grows?*

Exercise 25.4.9 *Prove that the probability of success utilizing the optimal strategy monotonically decreases with n, that is* Pr(success|n applicants) ≤ Pr(success|n + 1 applicants) *for all n.*

Exercise 25.4.10 *Find the probability no candidate is selected using the optimal strategy.*

Exercise 25.4.11 *Assume we really need to hire someone, so that if we have not yet hired a candidate when we reach the nth person, we hire him or her regardless of relative rank. Find the probability we hire the worst candidate.*

Exercise 25.4.12 *Rigorously prove the strategy analyzed is the optimal strategy.*

Exercise 25.4.13 *If instead of seeking the best candidate, what if we are content to hire either of the top two candidates. What is the optimal strategy now, and what is the probability of success?*

Exercise 25.4.14 *Generalize the previous exercise to hiring any of the top c candidates for some fixed c.*

Exercise 25.4.15 *Generalize the previous exercise to have c grow with n, say $c = n/10$ (someone in the top 10%) or $c = n/2$ (an above average candidate). Can you get a formula for $c = (1 - p)n$ (someone in the top p%)?*

Exercise 25.4.16 *Imagine that the number of candidates n is no longer known, but is known to be drawn from a uniform random variable on $[a, b]$. What is the optimal strategy now, and what is the probability of success?*

Exercise 25.4.17 *Imagine that the number of candidates n is no longer known, but is known to be drawn from a Poisson random variable with known parameter λ. What is the optimal strategy now, and what is the probability of success?*

Exercise 25.4.18 *Imagine we have a knapsack that can hold at most 100 kilograms. There are three items we can pack. The first weighs 51 kilograms and is worth $150 per unit; the second weights 50 kilograms and is worth $100 per unit; the third weighs 50 kilograms and is worth $99 per unit. If x_j represents how much of item j we pack, what are the amounts if we can take any real quantity of each? What if we can only take integer amounts of each?*

APPENDIX A

Proof Techniques

In this chapter we'll discuss how to read a proof, some of the more common ways to prove statements, and highlight a few ways that do not work and should be avoided at all costs! Students are often frustrated when they transition from the more standard courses such as calculus, where there aren't too many theorems and most of the exercises are mechanical and straightforward (if the homework problem is from the integration by parts section, it's pretty clear what you're going to need to do to evaluate the integral), to upper level classes, where frequently the proof of the main theorem of a section is left as an exercise. Even if you happen to be lucky enough to have a book which gives a proof, it's easy to lose the forest in the trees. What this means is that as you're reading the proof you can understand each line in isolation. You can understand how they go from one line to the next; however, it's a complete mystery how the author decided that it would be good to go from *this* line to *that* line, and you're rightly a bit terrified about your turn at proving something, as then you'll be responsible for directing the flow. Learning how to see these paths, learning what's a good next step, is hard, but doing so is essential for your growth in mathematics. The aim below is to describe in detail many of the common methods, in the hope that learning these will help you in following and creating proofs.

We cover the following proof techniques below.

1. Proof by Induction.
2. Proof by Grouping.
3. Proof by Exploiting Symmetry.
4. Proof by Brute Force.
5. Proof by Comparison or Story.
6. Proof by Contradiction.
7. Proof by Exhaustion (also known as Divide and Conquer).
8. Proof by Counterexample.
9. Proof by Generalizing Example.
10. Proof by Pigeon-Hole Principle.
11. Proof by Adding Zero or Multiplying by One.

A.1 How to Read a Proof

Frequently in books you'll find a square, such as □, at the end of the proof. This is meant to alert you that the argument is done, and the claim has been shown. This is done because all too often we're so caught up in following the arguments from line to line that we don't realize we've reached the end! Other people write qed or Q.E.D., which is an abbreviation of the Latin phrase *quod erat demonstrandum*, which means *that which was to be demonstrated*. Sometimes authors also write "*Proof*" at the start of the argument. These are done to help clue you in to what's going on. This helps prevent the proof from blending in with the rest of the text.

Before diving into proof techniques, here's some general advice on reading proofs. On a first pass through a proof don't be too concerned with mastering all the details. Rather, just look for a broad overview of what's happening or being discussed, and don't worry if you're unable to follow the argument from line to line.

Step one is to make sure you understand the conditions and the claim. If you can, take a few examples of objects that satisfy the conditions, and see that the claim is true for them. Sometimes it's particularly good, or at least easy, to try extreme examples. If you have a continuous function, try a constant function. Try a wildly oscillating one like $x^2 \sin(1/x)$, or perhaps one that isn't differentiable at a point, such as $|x|$. Also try a few examples that don't satisfy the assumptions of the theorem. In this case, the claim may or may not hold. Doing a few checks like this can give you a feel for what's going on, and as you do your checks you might start to see what will be needed in the proof. This is especially true when you find examples that don't satisfy the claim, as somehow the assumptions of the theorem must prohibit bad cases like this from happening.

After trying to get a feel of which examples work and which don't, return and think deeply about the assumptions. How are the assumptions used in the argument? When you read the assumptions, your first thoughts should be: okay, so what theorems do I know that require these conditions? For example, if one of your assumptions is that your function is differentiable, maybe you start to think about using the Mean Value Theorem. Or perhaps you've assumed f is a polynomial of degree n; in that case, the Fundamental Theorem of Algebra tells you that f has n complex roots. The more you know, the easier this becomes. I do a lot of work in number theory; if I'm told I have two relatively prime numbers x and y, my first thought is the Euclidean algorithm, which says there are integers a and b such that $ax + by = 1$ (for example, 17 and 11 are relatively prime, and $2 \cdot 17 - 3 \cdot 11 = 1$). Why is this my first thought? Experience—I've done so many problems that I know this is often a great way to start. The more you do, the easier it becomes. Assumptions are sign-posts, they're markers to help direct the flow of the proof. Time spent thinking about them and what they entail is time well spent.

What if the assumptions aren't used? Well, something strange is happening, as why would they be given if they're not used? What's more likely to happen is that sometimes the assumptions aren't truly needed, but are given to allow you access to powerful other theorems to simplify the proof. For example, the proof of the Fundamental Theorem of Calculus only needs our function f to be continuous on a finite interval $[a, b]$; however, whenever I teach calculus I always assume f' exists, is continuous, and is bounded. This is *not* needed, but assuming it simplifies the proof. If you continue deeper in mathematics, you'll revisit old theorems in a quest to have the weakest conditions possible and still get the same result.

Finally, when reading the proof don't worry about understanding every justification. First skim the argument, trying to get a sense of the main ideas. What results were used in the proof? Roughly, why were we able to use these? Sometimes there are lots of technical conditions that need to be met to invoke a theorem. In these cases, a lot of the proof is devoted to showing these conditions are met. When reading the proof for the first time, it's fine to gloss over these parts. Think something like: okay, we need to show the quotient is a finitely generated Abelian group, and the next few lines do that, I'll take their word on it for now. Later, of course, you should go back and try to understand these justifications, but don't obsess too much, as that can lead you to losing the flow of the proof. Often books and papers remove these mini-arguments and isolate them, either before or after the proof, calling them lemmas. A lemma is a smaller result, a building block to the proof of the main claim. Sometimes authors put these first so that by the time you get to the theorem you've seen everything you'll need. Other times these are placed afterwards, to avoid interrupting the flow with technicalities. Each approach is fine.

One last remark about reading proofs. Eventually you'll come across the phrase **without loss of generality**. Typically, this is followed by the author doing one case and saying the other cases follow similarly. If you're new to doing proofs, you should do all the cases in full glory (or is it gory?). These four words can be very dangerous, as sometimes there *are* differences between the various cases, and the only way to be sure is to do each case. If everything checks out and the arguments really are the same, mathematicians will often just give the details for one to save space and time. If you read a proof invoking this claim, it's good practice to fill in the details for the other cases.

Okay, we're now ready to explore different proof methods!

A.2 Proofs by Induction

This section is an expansion of an appendix from [MT-B] on Proofs by Induction. This method is designed to handle the following situation: for each positive integer n we have some statement $P(n)$, and we desire to show that $P(n)$ is true for all n. For example, maybe $P(n)$ is the statement that the sum of the first n odd numbers is always a perfect square. One possibility is to start evaluating it for different choices of n. In this case, we get 1, 4, 9, and 16 for the first four values, and we're feeling confident that the result is true. However, confidence is *not* the same as a proof, and just because it worked for the first few values doesn't mean it'll continue to work. If we eventually find an n such that $P(n)$ fails, then we know the statement is false. What if, however, it always holds for every value we check. Does this mean it must be true? Sadly, it doesn't. It may be we just haven't checked far enough.

For an example of what can go wrong, let's consider a famous polynomial, $f(x) = x^2 + x + 41$. Euler was interested in this polynomial, and you'll see why in a moment. Let's look at some of its values, which we record in Table A.1.

In the interest of space we only recorded a subset of the values of $f(n)$; a little work shows that $f(n)$ is also prime for *all* n up to 38. Based on the data above, it's natural to conjecture that $f(n)$ is *always* prime for any positive integer n. While the data suggests this, testing some values isn't a proof.

How should we proceed? We can take larger and larger values of n and see what happens. In this case, we would find $f(39) = 1601$ is prime, but $f(40) = 1681 = 41^2$ is composite, as is $f(49) = 2491 = 47 \cdot 53$. Here, we were able to go far enough to see

TABLE A.1.
Values of the polynomial $f(x) = x^2 + x + 1$.

n	$f(n)$	Primality of $f(n)$
1	41	prime
2	43	prime
3	47	prime
$\vdots$	$\vdots$	$\vdots$
37	1447	prime
38	1523	prime

TABLE A.2.
Sums of odd integers.

n	Sum of first n odd numbers	Value of the sum
1	1	1
2	1+3	4
3	1+3+5	9
4	$1 + 3 + 5 + 7$	16
$\vdots$	$\vdots$	$\vdots$
100	$1 + 3 + \cdots + 197 + 199$	10000

the pattern break down, and once we have one value that fails we know the claim cannot always hold.

For another example, let's revisit the sum of the first n odd integers. We can make a similar table as before, which we do in Table A.2.

Do you see the pattern? It looks like the sum of the first n odd integers is just n^2. Unlike the previous example, this time our conjecture is true. No matter how far we check, we'll see the pattern hold; however, just observing this equality *is not* a proof.

We need a way to prove statements like this and others. We quickly describe a powerful method, called Proofs by Induction, that works for a variety of problems. The general framework is that we have some statement $P(n)$ which we want to determine whether or not it holds for all positive integers n.

Proof Technique—Proofs by Induction: A statement $P(n)$ is true for all positive integers n if the following two conditions hold.

- **Basis Step**: $P(1)$ is true;

- **Inductive Step**: whenever $P(n)$ is true, $P(n + 1)$ is true.

Proof by Induction is a very useful method for proving results; we'll see many instances of this in this appendix. The reason the method works follows from basic logic. We assume the following two sentences are true:

$$P(1) \text{ is true.}$$

For all $n \geq 1$, $P(n)$ is true implies $P(n+1)$ is true.

Set $n = 1$ in the second statement. As $P(1)$ is true, and $P(1)$ implies $P(2)$, $P(2)$ must be true. Now set $n = 2$ in the second statement. As $P(2)$ is true, and $P(2)$ implies $P(3)$, $P(3)$ must be true. And so on, completing the proof. Verifying the first statement is called the **basis step**, and the second the **inductive step**. In verifying the inductive step, note we assume $P(n)$ is true; this is called the **inductive assumption**. Sometimes instead of starting at $n = 1$ we start at $n = 0$, although in general we could start at any n_0 and then prove for all $n \geq n_0$, $P(n)$ is true.

We give four of the more standard examples of proofs by induction in the next subsections, and one false example; the first example is the most typical. When you have mastered proofs by induction, you might want to return to the following exercise. It's a fun problem involving the Fibonacci numbers.

Exercise A.2.1 (Zeckendorf's Theorem): *Consider the set of distinct Fibonacci numbers:* $\{1, 2, 3, 5, 8, 13, \dots\}$, *where* $F_{n+2} = F_{n+1} + F_n$. *Show every positive integer can be written uniquely as a sum of distinct Fibonacci numbers where we do not allow two consecutive Fibonacci numbers to occur in the decomposition. Equivalently, for any* n *there are choices of* $\epsilon_i(n) \in \{0, 1\}$ *such that*

$$n = \sum_{i=1}^{\ell(n)} \epsilon_i(n) F_i, \quad \epsilon_i(n)\epsilon_{i+1}(n) = 0 \text{ for } i \in \{1, \dots, \ell(n) - 1\}.$$

Does a similar result hold for all recurrence relations? If not, can you find another recurrence relation where such a result holds?

A.2.1 Sums of Integers

Let $P(n)$ be the statement

$$\sum_{k=1}^{n} k = \frac{n(n+1)}{2}.$$

Here we're using summation notation, which is a very compact way of writing expressions. Unwinding, the left-hand side is just

$$\sum_{k=1}^{n} k = 1 + 2 + \dots + k.$$

More generally,

$$\sum_{k=1}^{n} a_k = a_1 + a_2 + \dots + a_n.$$

This is probably the most famous of all examples for proofs by induction. The great Gauss is said to have successfully evaluated this sum when he was five years old. According to the story, his teacher was having a bad day (we all do), and wanted some busywork to occupy the children; he did not count on having a budding master mathematician in the room!

Anyway, let's show that the statement is true by induction. We have two things to check, the basis step (or the base case), and the inductive step (or induction case). Let's go!

Proof: We proceed by induction.

Basis Step: $P(1)$ is clearly true, as both sides equal 1.

Inductive Step: Assuming $P(n)$ is true, we must show $P(n+1)$ is true. By the inductive assumption, $\sum_{k=1}^{n} k = \frac{n(n+1)}{2}$. Thus

$$\sum_{k=1}^{n+1} k = 1 + 2 + \cdots + n + (n+1)$$

$$= (1 + 2 + \cdots + n) + (n+1)$$

$$= \left(\sum_{k=1}^{n} k\right) + (n+1)$$

$$= \frac{n(n+1)}{2} + (n+1)$$

$$= \frac{(n+1)(n+1+1)}{2}.$$

Thus, given $P(n)$ is true, then $P(n+1)$ is true. $\qquad\square$

You might have seen the above example in a calculus class when studying area under curves. This (and the sum in the exercise below) arise in computing the upper and lower sums.

Note how the argument proceeded above. The hard part was showing that if $P(n)$ held then $P(n+1)$ holds too. The way we did this was to look at our expression for $P(n+1)$ and note that there was a $P(n)$ hiding inside it. We then used the fact that $P(n)$ was assumed true to rewrite $P(n+1)$, and then did some simple algebra. Many, many inductions proceed like this. The trick is finding out how to easily work in the induction assumption; however, if you're attempting a proof by induction then you should be on the watch for such an opportunity. The whole point of induction is to build on results for smaller n, so you should try to find the $P(n)$ case lurking in the $P(n+1)$ expression.

 Exercise A.2.2: *Prove*

$$\sum_{k=1}^{n} k^2 = \frac{n(n+1)(2n+1)}{6}.$$

Find a similar formula for the sum of k^3. For the brave, find a similar formula for the sum of k^4. (Hint: The sum of the dth powers of integers up to n is a polynomial in n of degree $d+1$.)

 Exercise A.2.3: *Show the sum of the first n odd numbers is n^2, i.e.,*

$$\sum_{k=1}^{n}(2k-1) = n^2.$$

In the last exercise above, there are two ways to write an odd number. We chose to write the odd numbers as $2k-1$, as this allowed our index k to range from 1 to n (if we want the first n odd numbers). If instead we write odd numbers as $2m+1$, then m would range from 0 to $n-1$ to give the first n odd integers. Either method is fine; the only difference is whether or not you want the index of summation to be nice (from 1 to n) or if you want to avoid the minus sign in the summands.

A.2.2 Divisibility

We now consider a divisibility problem. This is another example of a proof by induction, but the algebra and analysis is a little different, which is why we want to give these arguments too.

 Let $P(n)$ be the statement 133 divides $11^{n+1} + 12^{2n-1}$. We prove this claim by induction.

Proof: We proceed by induction.

Basis Step: A straightforward calculation shows $P(1)$ is true: $11^{1+1} + 12^{2-1} = 121 + 12 = 133$.

Inductive Step: Assume $P(n)$ is true, i.e., 133 divides $11^{n+1} + 12^{2n-1}$. We must show $P(n+1)$ is true, or that 133 divides $11^{(n+1)+1} + 12^{2(n+1)-1}$. But

$$
\begin{aligned}
11^{(n+1)+1} + 12^{2(n+1)-1} &= 11^{n+1+1} + 12^{2n-1+2} \\
&= 11 \cdot 11^{n+1} + 12^2 \cdot 12^{2n-1} \\
&= 11 \cdot 11^{n+1} + (133 + 11)12^{2n-1} \\
&= 11\left(11^{n+1} + 12^{2n-1}\right) + 133 \cdot 12^{2n-1}.
\end{aligned}
$$

By the inductive assumption 133 divides $11^{n+1} + 12^{2n-1}$; therefore, 133 divides $11^{(n+1)+1} + 12^{2(n+1)-1}$, completing the proof. □

The difficulty in this proof was noting that 133 and 11 were lurking together in 144. Specifically, we could write 144 as 133 plus 11. The reason this helps is that the other term is multiplied by 11, and by cleverly regrouping we saw $11^{n+1} + 12^{2n-1}$. It was a very good idea to rewrite 11^{n+2} as $11 \cdot 11^{n+1}$ (and similarly for the expression involving 12), and it was reasonable to try this as we wanted to "see" $P(n)$. In fact, staring at this and thinking back to the sum of integers, we see that *both* proofs had us finding $P(n)$ somewhere in $P(n+1)$. Many induction problems require you to find $P(n)$

lurking in $P(n + 1)$; it's not surprising that this happens, as the whole point of induction arguments is to assume $P(n)$ is true and then show this implies $P(n + 1)$ holds.

 Exercise A.2.4: *Prove 4 divides* $1 + 3^{2n+1}$.

 Exercise A.2.5: *Find a positive integer a such that 5 divides* $1 + 4^{an}$ *for all n, and prove your claim.*

A.2.3 The Binomial Theorem

We end with one more example of a proof by induction, the proof of the Binomial Theorem. This time the result is *clearly* of importance for a probability class. The Binomial Theorem is used all the time; in fact, we even have binomial random variables!

Before stating and proving the result, we first recall the definition and some properties of binomial coefficients.

Definition A.2.6 (Binomial Coefficients): *Let n and k be integers with* $0 \le k \le n$. *We set*

$$\binom{n}{k} = \frac{n!}{k!(n-k)!}.$$

Note that $0! = 1$. *We set* $\binom{n}{k} = 0$ *if* $k > n$.

The combinatorial interpretation of $\binom{n}{k}$ is that this is the number of ways of choosing k people from n when order doesn't matter, and $m!$ is the number of ways of ordering m. It may seem strange to say $0! = 1$, but if we use these interpretations we could read this as saying there are no ways to order an empty set of people. If you don't remember the proofs of these statements, they're given in §A.5 and §A.6.

We're now ready to state the Binomial Theorem.

Theorem A.2.7 (Binomial Theorem): *For all positive integers n we have*

$$(x + y)^n = \sum_{k=0}^{n} \binom{n}{k} x^{n-k} y^k.$$

Proof of the Binomial Theorem: We proceed by induction.

Basis Step: For $n = 1$ we have

$$\sum_{k=0}^{1} \binom{1}{k} x^{1-k} y^k = \binom{1}{0} x + \binom{1}{1} y = (x + y)^1.$$

Inductive Step: Suppose

$$(x + y)^n = \sum_{k=0}^{n} \binom{n}{k} x^{n-k} y^k. \tag{A.1}$$

Then using Lemma A.5.1 we find that

$$
\begin{aligned}
(x+y)^{n+1} &= (x+y)(x+y)^n \\
&= (x+y) \sum_{k=0}^{n} \binom{n}{k} x^{n-k} y^k \\
&= \sum_{k=0}^{n} \binom{n}{k} \left[x^{n+1-k} y^k + \binom{n}{k} x^{n-k} y^{k+1} \right] \\
&= x^{n+1} + \sum_{k=1}^{n} \left[\binom{n}{k} + \binom{n}{k-1} \right] x^{n+1-k} y^k + y^{n+1} \\
&= \sum_{k=0}^{n+1} \binom{n+1}{k} x^{n+1-k} y^k,
\end{aligned}
$$

as $x^{n+1} = \binom{n+1}{0} x^{n+1}$ and $y^{n+1} = \binom{n+1}{n+1} y^{n+1}$. This establishes the induction step, and hence the theorem. □

As always, the hardest part of the proof is figuring out how to use the inductive assumption. The main idea here was to write $(x+y)^{n+1}$ as $(x+y)(x+y)^n$; this is a "natural" thing to do, as we now have a factor of $(x+y)^n$, which by the inductive assumption we know how to handle. Of course, we could also have written it as $(x+y)^n(x+y)$, and the proof would have been similar. This is almost always the goal: find a way to rewrite the expression so you can exploit the inductive assumption. The most troublesome part of this problem is having to adjust the index of summation (if you continue to differential equations, you'll get a lot of practice with this when you do series expansions). A good guideline is to try to make all terms look the same. We thus want the powers of x and y to look the same in each expression, and this helps us figure out how to shift. Typically it's preferable to have the powers of x and y the same and the index of the coefficients different rather than the other way around.

There are other ways to prove the Binomial Theorem; we'll see one in §A.6 where we do proofs by comparison.

A.2.4 Fibonacci Numbers Modulo 2

If it's 10 o'clock now, most people would have no difficulty saying that in 5 hours it'll be 3 o'clock. If we look at what we've just said, are we saying 10 plus 5 is 3? On a clock with twelve hours: yes! The idea of **clock** or **modulo arithmetic** plays a central role in much of number theory, and generalizes nicely. We say **x is congruent to y modulo n** if $x - y$ is divisible by n. Thus, 15 is congruent to 3 modulo 12, as 12 divides 15-3. Similarly we find 67 is equivalent to 7 modulo 12, as $60 = 5 \cdot 12 + 7$. We write 67 modulo $12 = 7$ or $67 = 7$ mod 12.

Let's look at a fun problem involving the Fibonacci numbers. Recall these are defined by $F_0 = 0$, $F_1 = 1$, and $F_{n+2} = F_{n+1} + F_n$. The first few are

$$0, \ 1, \ 1, \ 2, \ 3, \ 5, \ 8, \ 13, \ 21, \ 34, \ 55, \ 89, \ \ldots.$$

Let's look at these numbers modulo 2. A little inspection shows us that x modulo 2 is 0 if x is even (and thus a multiple of 2) and 1 if x is odd. The Fibonacci numbers

modulo 2 are

$$0, 1, 1, 0, 1, 1, 0, 1, 1, 0, 1, 1, \ldots .$$

Looking at this, we see the beginning of a pattern. It seems to be repeating blocks of 0, 1, 1. Does this always continue? It does, and one nice way to prove this is by induction.

Proof that the Fibonacci numbers modulo 2 are the repeating sequence 0, 1, 1, 0, 1, 1, …: We proceed by induction.

Basis Step: Calculating the first few terms verifies that it does start 0, 1, 1, 0, 1, 1.

Inductive Step: The defining property of the Fibonacci numbers is that the two previous terms are added to get the next. It's thus natural to investigate whether or not this holds modulo 2. In other words, is F_{n+2} modulo 2 the same as F_{n+1} modulo 2 plus F_n modulo 2, all of this modulo 2? Unwinding, there are four cases.

- If F_n is even and F_{n-1} is even, is F_{n+2} even?
- If F_n is even and F_{n-1} is odd, is F_{n+2} odd?
- If F_n is odd and F_{n-1} is even, is F_{n+2} odd?
- If F_n is odd and F_{n-1} is odd, is F_{n+2} even?

The four statements are true, and can be verified with a little bit of algebra (at the level of odd plus odd is even, odd plus even is even, even plus even is even). Armed with this, we can now complete the proof. Assume the first k blocks of three are 0, 1, 1; we'll denote this by

$$0, 1, 1, \ldots, 0, 1, 1.$$

Let's look at the next three terms of the Fibonacci numbers modulo 2. The next number is the sum of the two previous modulo 2, so the next number is $1 + 1$ modulo 2, which is zero. Thus our sequence is now

$$0, 1, 1, \ldots, 0, 1, 1, 0.$$

The next term is just $1 + 0$ modulo 2, which is 1, implying our sequence is

$$0, 1, 1, \ldots, 0, 1, 1, 0, 1.$$

The next term is just 1 modulo 2, which is 1 again, giving us

$$0, 1, 1, \ldots, 0, 1, 1, 0, 1, 1.$$

This is exactly what we wanted to prove—we just showed that if the first k blocks of three are 0, 1, 1 then the next block is also 0, 1, 1. This completes the proof. □

There are lots of wonderful patterns that emerge when looking at interesting sequences modulo primes (2 is the smallest prime). We urge you to Google the pattern for Pascal's triangle modulo 2—the resulting pattern is quite surprising!

A.2.5 False Proofs by Induction

After seeing how powerful proofs by induction can be, it's a good idea to be aware of the pitfalls. If you're not careful, you can convince yourself that you've proven many statements that are, in fact, false! Below is a favorite of mine.

Consider the following: Let $P(n)$ be the statement that in any group of n people, everyone has the same name. We give a (false!) proof by induction that $P(n)$ is true for all n!

Proof: We proceed by induction.

Basis Step: Clearly, in any group with just 1 person, every person in the group has the same name.

Inductive Step: Assume $P(n)$ is true, namely, in any group of n people, everyone has the same name. We now prove $P(n + 1)$. Consider a group of $n + 1$ people:

$$\{1, 2, 3, \ldots, n - 1, n, n + 1\}.$$

The first n people form a group of n people; by the inductive assumption, they all have the same name. So, the name of 1 is the same as the name of 2 is the same as the name of 3 ... is the same as the name of n.

Similarly, the last n people form a group of n people; by the inductive assumption they all have the same name. So, the name of 2 is the same as the name of 3 ... is the same as the name of n is the same as the name of $n + 1$. Combining yields everyone has the same name! □

Where is the error? Even Borg drones have different designations (or stormtroopers in *The Force Awakens*, my favorite being JB-007); it's unlikely that everyone reading this book shares my name! Clearly we've done something terribly wrong, but where? Let's go through the above argument slowly and carefully. Rather than trying to follow the proof for an arbitrary n, let's run through it with specific values of n and see what happens.

If $n = 4$, we would have the set $\{1, 2, 3, 4, 5\}$, and the two sets of 4 people would be $\{1, 2, 3, 4\}$ and $\{2, 3, 4, 5\}$. We see that persons 2, 3, and 4 are in both sets, providing the necessary link. If $n = 3$ our set would be $\{1, 2, 3, 4\}$, and the two sets of 3 people would be $\{1, 2, 3\}$ and $\{2, 3, 4\}$. Again we find people in common, providing the necessary link.

What about smaller n? Eventually we reach $n = 1$. Then our set would be $\{1, 2\}$, and the two sets of 1 person would be $\{1\}$ and $\{2\}$; there is no overlap! The error was that we assumed n was "large" in our proof of $P(n) \Rightarrow P(n + 1)$. Yes, in this problem, 2 is large. Terms like large and small are relative. The problem was we accidentally used some facts that only hold for $n \geq 2$. It's very easy to fall into this trap.

Exercise A.2.8: *Similar to the above, give a false proof that any sum of integer squares is an integer square, i.e., $x_1^2 + \cdots + x_n^2 = x^2$. In particular, this would prove all positive integers are squares as $m = 1^2 + \cdots + 1^2$.*

A.3 Proof by Grouping

Our next technique is close to induction. I call it **proof by grouping**. A great example is the rule from calculus that the derivative of a sum is the sum of the derivatives. Most books prove this carefully for a sum of two functions, but then ignore the proof in general. Some care is needed; sadly, the derivative of an infinite sum need not equal the sum of the derivatives; however, if we have a finite sum of differentiable functions, then the derivative of the sum is the sum of the derivative.

We'll give the proof, assuming we know that whenever we have two differentiable functions then the derivative of their sum is the sum of their derivatives. What follows is essentially an induction argument, but I think it's nice to see how we win by cleverly adding parentheses and grouping terms.

Proof: Let

$$g(x) = f_1(x) + f_2(x) + f_3(x) = (f_1(x) + f_2(x)) + f_3(x)$$

be a sum of three differentiable functions; note we've grouped the first two functions together, and written g as a sum of *two* functions (the first is $f_1 + f_2$ and the second is f_3). Taking the derivative, we find

$$\frac{dg}{dx}(x) = \frac{d}{dx}[f_1(x) + f_2(x) + f_3(x)]$$

$$= \frac{d}{dx}[(f_1(x) + f_2(x)) + f_3(x)].$$

We now have the derivative of the sum of two functions, which we know is the sum of the two derivatives. We thus obtain

$$\frac{dg}{dx}(x) = \frac{d}{dx}(f_1(x) + f_2(x)) + \frac{df_3}{dx}(x).$$

We now use the derivative of the sum of two functions is the sum of the derivatives again. We thus obtain

$$\frac{dg}{dx}(x) = \frac{df_1}{dx}(x) + \frac{df_2}{dx}(x) + \frac{df_3}{dx}(x),$$

completing the proof. □

More generally, this type of argument shows the derivative of any finite sum is the sum of the derivatives, extending the common sum rule from calculus. Sadly, most calculus classes gloss over this point, and never remark that you need to be a bit careful as technically we only proved the derivative of a sum of two functions is the sum of the derivatives.

We'll see this method again in §6.2.2, where we meet the multinomial coefficients (a generalization of binomial coefficients), and in §14.3 (when we show sums of normal random variables are normal).

A.4 Proof by Exploiting Symmetries

There are infinitely many similar integrals that calculus professors love to give students. Here's one version: find

$$\int_{-2}^{2} (x^8 - 1701x^6 + 24601) \cos^3 x \sin(x^3 + 2x) \log(x^2 + 4) dx.$$

Good luck finding an anti-derivative for that! Class problems have a huge advantage over the real world: you know there has to be a solution using just the methods you know. Thus, this has to be doable using just Calculus I and II knowledge. The "trick" is to notice that we are *not* being asked to find the anti-derivative. Yes, if we let $f(x)$ be the integrand and $F(x)$ an anti-derivative, then the answer is just $F(2) - F(-2)$. *If* we know an anti-derivative *then* we can evaluate the integral; however, maybe it's possible to evaluate the integral without finding F. It's *helpful* to know F, but it's not always essential. Sometimes all it does is help with the algebra (if you've done multivariable calculus, this is similar to Lagrange multipliers; often we can find the maximum/minimum values without finding the multipliers).

The key observation is to note that it's not an arbitrary integral, but an integral from -2 to 2. Note that this is a *symmetric* region about 0. Further, the integrand is an odd function about 0. Recall that $f(x)$ is an **even function (about the point a)** if $f(a + x) = f(a - x)$, while it is an **odd function (about the point a)** if $f(a + x) = -f(a - x)$. The integral of an odd function about a over a symmetric interval centered at a is zero: this is because the contribution on one side is negated by the contribution on the other side. See Figure A.1 (Top) for an example.

Another nice application of exploiting symmetries is to simplify integrations. The arguments above show that the integral of an odd function over a symmetric region is zero. What if we have an even function about a and we integrate over the interval $[a - b, a + b]$? In that case, the integral is double that of the integral over $[a, a + b]$, as the first half has the same contribution as the second half. See Figure A.1 (Bottom) for an example.

We record our results.

Exploiting Symmetries—integration of odd and even functions: Let $f(x)$ be an odd function about a, and $g(x)$ an even function about a. Then

$$\int_{a-b}^{a+b} f(x)dx = 0, \quad \int_{a-b}^{a+b} g(x)dx = 2\int_{a}^{a+b} g(x)dx.$$

These are two of the most common symmetries worth exploiting, but there are many others, and you should keep your eyes open for them. We'll do one more, which is useful when we prove the cosecant identity of the Gamma function. The following is a gem of mathematics:

$$\sum_{n=1}^{\infty} \frac{1}{n^2} = \frac{\pi^2}{6}.$$

There are lots of proofs of this, and it has probabilistic interpretations (for example, it's the reciprocal of the probability two random numbers are relatively prime). It's often proved in a Fourier Analysis or Complex Analysis course (see [SS1, SS2]).

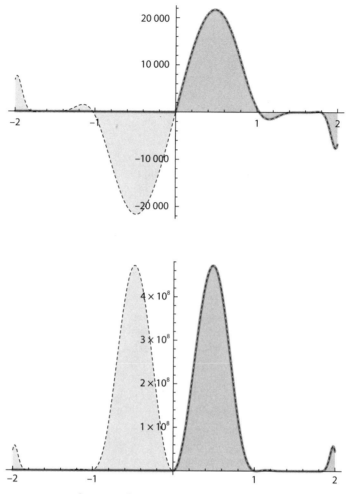

Figure A.1. Let $f(x) = (x^8 - 1701x^6 + 24601)\cos^3 x \sin(x^3 + 2x)\log(x^2 + 4)$. (Top) Area under $f(x)$ (an odd function about 0) from -2 to 2. (Bottom) Area under $f(x)^2$ (an even function about 0) in the symmetric region $[-2, 2]$.

 Let's take this result as a given, and deduce the sum of the reciprocals of the odd squares. We find

$$\sum_{n=1}^{\infty} \frac{1}{n^2} = \sum_{\substack{n=1 \\ n\text{ even}}} \frac{1}{n^2} + \sum_{\substack{n=1 \\ n\text{ odd}}} \frac{1}{n^2}$$

$$= \sum_{n=1}^{\infty} \frac{1}{(2n)^2} + \sum_{n=1}^{\infty} \frac{1}{(2n-1)^2}$$

$$= \frac{1}{4}\sum_{n=1}^{\infty} \frac{1}{n^2} + \sum_{n=1}^{\infty} \frac{1}{(2n-1)^2}$$

$$\frac{3}{4}\sum_{n=1}^{\infty} \frac{1}{n^2} = \sum_{n=1}^{\infty} \frac{1}{(2n-1)^2},$$

which means

$$\sum_{n=1}^{\infty} \frac{1}{(2n-1)^2} = \frac{3}{4} \sum_{n=1}^{\infty} \frac{1}{n^2} = \frac{3}{4} \frac{\pi^2}{6} = \frac{\pi^2}{8}.$$

This is another great example of the powerful consequences if you can **exploit symmetry** properly. The key observation is that the sum over the even terms is just one-fourth of the total sum. We then **brought it over** (we'll see more of this technique in the calculus review problems).

A.5 Proof by Brute Force

There are several ways to attack problems by brute force. They all share a common feature: rolling up your sleeves and diving into the algebra. Sometimes we're lucky and there are only a few items to check, but often there are so many cases that it just isn't feasible. Below we'll give an example to give a flavor of this method.

Recall the lemma on binomial coefficients.

Lemma A.5.1: *For integers $n \geq k \geq 0$ we have*

$$\binom{n}{k} = \binom{n}{n-k}, \quad \binom{n}{k} + \binom{n}{k-1} = \binom{n+1}{k}.$$

Proof of Lemma A.5.1—First Part: The first claim is just the fact that multiplication is commutative:

$$\binom{n}{k} = \frac{n!}{k!(n-k)!} = \frac{n!}{(n-k)!k!} = \binom{n}{n-k}.$$

$\square$

The second claim is more interesting. Following is a "brute force" proof.

Proof of Lemma A.5.1—Second Part: We have

$$\binom{n}{k} + \binom{n}{k-1} = \frac{n!}{k!(n-k)!} + \frac{n!}{(k-1)!(n-k+1)!}$$

$$= \frac{n!}{(k-1)!(n-k)!} \left[\frac{1}{k} + \frac{1}{n-k+1} \right]$$

$$= \frac{n!}{(k-1)!(n-k)!} \left[\frac{n-k+1+k}{k(n-k+1)} \right]$$

$$= \frac{n!}{(k-1)!(n-k)!} \frac{n+1}{k(n-k+1)}$$

$$= \frac{(n+1)!}{k!(n-k+1)!} = \binom{n+1}{k}.$$

$\square$

While the above argument *is* a proof, in some sense it's a terrible one. Yes, it's logically sound, yes, all the steps are correct, yes, it does give us the result; however, after reading it do you have any sense of *why* the result is true? It's just a long list of algebraic manipulations. It's great to be able to do this, but for many problems the algebra will be significantly worse, and it won't be clear at all how to proceed. For this problem, we were lucky. The algebra wasn't too bad, and it was pretty clear what to do: collect common factors and simplify. An alternative algebraic approach to this problem would have been to clear the denominators and then simplify. Is there another way to approach this problem, one which is more enlightening? Fortunately, the answer is a resounding yes, and we give it in the next section.

A.6 Proof by Comparison or Story

We return to the problem from the previous section, where we want to prove $\binom{n}{k} + \binom{n}{k-1} = \binom{n+1}{k}$. We've seen an unenlightening proof; now we'll see a better one that highlights what's really happening. The idea of this method, **Proof by Comparison**, is to compute the desired quantity two different ways. As we're calculating the same thing, these two expressions must be equal. Though the idea is easy to state, in practice it's often very hard to find a viewpoint that leads to an easy calculation. Combinatorial problems are some of the hardest you'll find (both in probability and in mathematics), and you often need a flash of insight (or a lot of experience) to suggest a good way to look at a problem.

Another way to say what we're going to do is that we'll **count the same quantity two different ways**. If we're counting the same quantity two different ways, then the two answers must agree; many identities are derived this way. We're essentially telling a story, with the exciting conclusion that the two main characters are actually one and the same. **Proof by Story** doesn't sound as academic as Proof by Comparison, but that's really what we're doing.

Let's do a simple warm-up example, inspired by the Dr. Seuss story "The Sneetches".

 Proof that $\binom{n}{k} = \binom{n}{n-k}$: Imagine we have a group of n Sneetches, and we want to give some of them stars on their bellies. If exactly k are going to get stars, there are $\binom{n}{k}$ ways to choose k of the n Sneetches to be starred. Alternatively, we could look at this as *excluding* $n - k$ of the Sneetches from getting stars, and there are $\binom{n}{n-k}$ ways to choose $n - k$ Sneetches *not* to be starred. We've counted the same thing two different ways, so $\binom{n}{k} = \binom{n}{n-k}$.

I prefer this proof to the algebraic one, as it illustrates what's really going on and *why* there's an equivalence.

Proof that $\binom{n}{k} + \binom{n}{k-1} = \binom{n+1}{k}$: We find a combinatorial interpretation for all these quantities. Imagine we have $n + 1$ marbles; n of these marbles are red and 1 marble is blue. This is the hardest part of the proof, figuring out what story to tell. While this gets easier with practice, there's at least a reason for doing this. We have $n + 1$ objects, so perhaps n of them are of one type, and 1 is of another.

One half of our story isn't too bad. There are $\binom{n+1}{k}$ ways to choose k marbles from the $n + 1$ marbles when we do not care about order of choice.

How else could we count the number of ways of choosing k marbles from $n + 1$? Well, we could look at how many ways there are to choose k marbles from our $n + 1$ marbles when order doesn't matter, keeping track of whether or not we choose the blue marble. If we don't have the blue marble, then we must choose k marbles from the n red ones; there are $\binom{n}{k}$ ways to do this. If we do have the blue marble (and there is $\binom{1}{1}$ way to do this as we only have one blue marble) then we must choose $k - 1$ red marbles from n red marbles, and there are $\binom{n}{k-1}$ ways of doing this. Collecting, we find our two counts must be equal, so

$$\binom{n}{k} + \binom{n}{k-1} = \binom{n+1}{k}.$$

$\square$

This is a much better proof; it highlights what is going on, and gives a reason for the algebraic miracle.

There's a better way to view the calculation. We should *really* write it as

$$\binom{1}{0}\binom{n}{k} + \binom{1}{1}\binom{n}{k-1} = \binom{n+1}{k}.$$

Looking at it this way, the first factor on the left is the story: we choose 0 of the 1 blue marbles and then k of the n red marbles; the second factor represents choosing 1 of the 1 blue marbles and $k - 1$ of the n red marbles. These two expressions are equal as $\binom{1}{0} = \binom{1}{1} = 1$, but I prefer the second. What's nice now is that a certain symmetry has been restored to both sides of the equation. On both the left- and the right-hand side, the sum of the "top" parts of the terms add up to $n + 1$, and the sum of the "bottom" parts of the terms add up to k. To me, it's a little clearer how we're partitioning, and anything that can highlight what's going on is good! It also decreases the chance that we'll forget a factor, as in other problems these terms won't always be 1.

At the risk of beating the problem to death, it's worth chatting about why we're adding the two terms and not multiplying them. Often in probability we multiply the probabilities of events. Here, what we're doing is partitioning our event, which is choosing k of $n + 1$, into disjoint possibilities (having 0 blue, having 1 blue). For finite sets, the probability of a disjoint union is the sum of the probabilities. This forces us to add the two probabilities together.

Let's do another example. We'll prove $\sum_{k=0}^{n} \binom{n}{k}\binom{n}{n-k} = \binom{2n}{n}$. We'll do this by calculating the same quantity two different ways. This essentially means we need to make up a story, where the expressions above are the quantities involved. Imagine we have n men and n women who want to take a probability class; unfortunately, the classroom is small and only n people can enroll in the class. There are $\binom{2n}{n}$ ways to choose a class of n people from our $2n$ people (n men and n women). That's the right-hand side—what about the left-hand side? Note in any class of n people there must be some number of men and some number of women. If there are k men, there must be $n - k$ women. The number of ways of choosing k men from n men is just $\binom{n}{k}$; similarly there are $\binom{n}{n-k}$ ways to choose $n - k$ women from n women. Thus, the number of ways to have a class of n people with exactly k men is $\binom{n}{k}\binom{n}{n-k}$. There must be some number of men in the class; that number ranges from 0 to n. Thus the total number of possible classes is just $\sum_{k=0}^{n} \binom{n}{k}\binom{n}{n-k}$, which must equal $\binom{2n}{n}$. □

As an aside, since $\binom{n}{n-k} = \binom{n}{k}$, the above implies $\sum_{k=0}^{n} \binom{n}{k}^2 = \binom{2n}{n}$.

Exercise A.6.1: *Find a simple formula involving a triple product of binomial coefficients. You'll have to think a bit and find a good story. (Hint: It isn't $\binom{n}{k}^3$.)*

We end with one last example. Let's see how the Binomial Theorem can be proved in this manner. We want to show

$$(x+y)^n = \sum_{k=0}^{n} \binom{n}{k} x^{n-k} y^k.$$

We have n factors of $x + y$. For each factor, we choose either x or y. We see that $(x + y)^n$ will be a polynomial in x and y, involving terms like $x^j y^k$. What are the possible pairs of (i, j) that work, and what are the coefficients of these terms?

Well, we have n factors and for each factor we *choose* either an x or a y. Thus $j + k$ must equal n, so j must be $n - k$. What about the coefficient of $x^{n-k} y^k$? Every time we choose y from exactly k of the factors (which then forces us to have exactly $n - k$ factors of x), we get a $x^{n-k} y^k$. How many ways are there to choose k of the n factors to be y? Why, this is just the definition of the binomial coefficient, $\binom{n}{k}$, which completes the proof. □

It's worthwhile to see different proofs of the same result, especially if it's an important result. Each of these proofs highlights a different feature. These different approaches will help you not only in understanding the theorem, but in attacking future problems. What can make many math problems seem exceptionally difficult is that it isn't always clear how to start. The more methods you see, the more ideas you have for tackling future problems.

A.7 Proof by Contradiction

Proof by Contradiction is one of my favorite ways of proving statements. Sometimes, instead of trying to directly show that something is true, it's easier to assume it fails and go for a contradiction. Let's look at an example. Remember that a number is **rational** if we can write it as a ratio of two integers (with the denominator non-zero); if we cannot do this, the number is **irrational**.

The square-root of 2 is irrational.

We proceed by assuming it is not irrational, and look for a contradiction; see [MilMo] for a more geometric proof by contradiction. If it isn't irrational then it's rational, and we have $\sqrt{2} = p/q$, and we may assume p and q are relatively prime (this means that no integer 2 or more divides both). If there were a common divisor, we could remove it and get a new fraction p'/q', with $p' < p$ and $q' < q$.

Since we're assuming $\sqrt{2} = p/q$, then $2q^2 = p^2$. We claim that 2 divides p^2. While this appears obvious, this must be proved. It's clearly true if p is even, as an even times any integer is still even. If p is odd, we may write $p = 2m + 1$. Then $p^2 = 4m^2 + 4m + 1 = 2(2m^2 + 2m) + 1$, which is clearly not divisible by 2. Thus p is even, say $p = 2p_1$. Then $2q^2 = p^2$ becomes $2q^2 = 4p_1^2$. We now have $2p_1^2 = q^2$, and a similar argument yields q is even. Hence p and q have a common factor, which contradicts p and q are relatively prime. We were led to this falsehood by assuming $\sqrt{2}$ is rational. Thus that assumption must be false, and $\sqrt{2}$ must be irrational. □

As proofs by counterexample occur so frequently, it's worth doing another example. This one is more involved, and uses some results from analysis and calculus.

 Let $f(x)$ be a continuous function on the real line. If the integral of $f(x)$ vanishes for every interval $[a, b]$ with $a < b$ then $f(x)$ is identically zero.

If we try to prove this directly we might run into some trouble, for we're given information on $f(x)$ over intervals, but must prove something over a point. What if, perhaps, we try to prove by contradiction? We assume for the sake of argument that the result is false: all the hypotheses hold, but there's a counterexample, say f, that is *not* zero everywhere. Now we have something to work with, and we try to show that if such a function existed, then it couldn't possibly satisfy all of our hypotheses. This contradiction means that our initial assumption that there was a counterexample is *false*, and thus the theorem does hold.

Let's try this here. So let's assume we have a continuous function which integrates to zero over any interval, but isn't identically zero. There must be some point, say x_0, where the function isn't zero. Without loss of generality, let's assume our function is positive at the point x_0 (a similar proof works for $f(x_0) < 0$).

Well, let's glean all the information we can out of our hypotheses on f. We assumed f is continuous. So, if we choose any $\epsilon > 0$ then we know there is a δ such that if $|x - x_0| < \delta$, then $|f(x) - f(x_0)| < \epsilon$.

But, we have freedom in choosing ϵ! We know that our f must integrate to zero over any interval, so we have $\int_{x_0-\delta}^{x_0+\delta} f(x)dx = 0$; however, $f(x_0)$ is positive! If ϵ is sufficiently small, by continuity $f(x)$ will be positive around x_0. For example, taking $\epsilon < f(x_0)/20$, we get there is a δ such that $f(x) > 19f(x_0)/20 > 0$.

Now we can get a contradiction. As $f(x) > 19f(x_0)/20$ on this interval but we've assumed the integral on this interval vanishes, standard results from calculus give us

$$0 = \int_{x_0-\delta}^{x_0+\delta} f(x)dx < \int_{x_0-\delta}^{x_0+\delta} \frac{19f(x)}{20}dx = \frac{19f(x_0) \cdot 2\delta}{20},$$

where the first equality follows from our assumption that f integrates to zero over any interval. But $f(x_0)\delta > 0$, and we've reached a contradiction! Basically the above is just a rigorous way of saying that if a continuous function is positive at some point, it's positive in a neighborhood of the point and thus cannot integrate to zero there. □

When you were reading this proof, you probably raised your eyebrows or wondered a bit when numbers like 19/20 entered the proof. Quantities like this are common in analysis. The idea is we want enough control to show that our function is positive in a small interval. We didn't need to take 19/20; many other numbers would've worked too.

A.8 Proof by Exhaustion (or Divide and Conquer)

The more assumptions and hypotheses we have on objects, the more (detailed) theorems and results we should know about them. Often it helps in proving theorems to break the proof up into several cases, covering all possibilities. We call this method **Divide and Conquer**. It's *essential* in using this method that you cover all cases: *make sure you consider all possibilities*. For example, you might do *Case 1: the function is continuous*. Now you have all the theorems about continuity at your disposal. And then *Case 2: the function is not continuous*. Now you have a special point where the function is discontinuous, and theorems and results about such points. The advantage is that, before, you couldn't use either set of results. The disadvantage is that you now have to give two proofs. Often, it's worthwhile having to prove more claims because for each claim you have more at your disposal. Let's do an example.

 For f, g real-valued functions, $|f(x) + g(x)| \leq |f(x)| + |g(x)|$.

If we can show this holds for an arbitrary point x, then we're done. Let's fix an x and investigate.

Case 1: Assume $f(x), g(x) \geq 0$.
    Under this assumption, we have

$$|f(x) + g(x)| = f(x) + g(x) = |f(x)| + |g(x)|,$$

which is what we needed to show.

Case 2: Assume $f(x) \geq 0$, $g(x) < 0$.
    We want to somehow get $f(x) + g(x)$. We can add them together, and get

$$f(x) + g(x) < 0 + f(x) = |f(x)|,$$

but when we take absolute values of both sides, the inequality could change ($-5 < 4$ but $|-5| > |4|$). So, a standard trick is to break this case into *subcases*!

- *Subcase A: Assume $0 \leq f(x) + g(x)$.* Then as $g(x) < 0$, $f(x) + g(x) < f(x)$. So $0 \leq |f(x) + g(x)| < f(x) \leq |f(x)| + |g(x)|$, which is what we needed to show.
- *Subcase B: Assume $f(x) + g(x) < 0$.* Then $0 < -1(f(x) + g(x)) \leq -g(x)$ as $f(x) \geq 0$. So $0 < |f(x) + g(x)| \leq |g(x)| \leq |f(x)| + |g(x)|$, which is what we needed to show.

This completes the analysis of Case 2. Unfortunately, we are not done as Cases 1 and 2 do not exhaust all possibilities.

Case 3: Assume $f(x) < 0$, $g(x) \geq 0$.
    This is proved similarly as in Case 2; essentially just switch the roles of f and g.

Case 4: Assume $f(x) < 0$, $g(x) < 0$. This is proved almost identically as in Case 1. □

Frequently in proofs by exhaustion many of the cases are essentially the same. For example, in the exercise above it doesn't really matter if $f(x) \geq g(x)$ or $g(x) \geq f(x)$, as we can always change the label names of the functions. Because of this, you'll often see proofs using the phrase **without loss of generality**, which means that as it makes no difference in the proof, for definiteness we'll assume a certain ordering or certain values. Be careful, though, as sometimes the different names are important. For example, if we're studying the function $f(x, y) = x^2 y^4 + x^4 y^2$, then once we compute $\partial f / \partial x$ we know $\partial f / \partial y$ by interchanging the roles of x and y. This is not the case for the function $g(x, y) = x^2 y^4 + x^3 y^3$; here there is a real difference between the x-behavior and the y-behavior.

A.9 Proof by Counterexample

One of the most common mistakes students make is to assume that **Proof by Example** is a valid way to prove a relation. This isn't true; just because something sometimes works doesn't mean it will always work. We saw a great example in §A.2 when we looked at Euler's polynomial $x^2 + x + 41$; it was always prime for $n \in \{0, 1, \ldots, 39\}$ but failed to be prime for many n afterwards.

While it's often useful to check a special case and build intuition on how to tackle the general case, checking a few examples isn't a proof. For another example, because $16/64 = 1/4$ and $19/95 = 1/5$, one might think that in dividing two digit numbers if two numbers on a diagonal are the same one just cancels them. Skeptical? Let's test it again. If we look at $49/98$, canceling the 9's gives $4/8$, which simplifies to $1/2$. Convinced? Probably not. A little experimentation brings us to $12/24$. If we really could just cancel the 2's we'd get this equals $1/4$, but it's $1/2$. Of course this is *not* how one divides two digit numbers, but it is interesting to see how many times it works!

However, if we are trying to disprove some statement, this means that if we are able to find just one example where the statement fails under the necessary assumptions of the statement, then we have in fact disproved it, as we have shown that it does not hold for all cases. This is the essence of **Proof by Counterexample**.

Exercise A.9.1: *How many pairs of three digit numbers with the same middle are there such that the ratio of these two numbers is the same as the ratio with the middle digit removed? For example, one pair is* $(561, 462)$, *as* $561/462 = 51/42 = 17/14.$

A.10 Proof by Generalizing Example

Another great way to prove a result is to look at a special case, detect a pattern, and try to generalize what you see. Let's look at an example you may have seen years ago when learning how to multiply and divide. You may remember the rule for divisibility by 3: if the sum of the digits of your number is divisible by three, then so is your number. We check this with 231 (yes, 2+3+1 = 6 which is divisible by 3, as is $231 = 3 \cdot 77$), 9444 (yes, 9+4+4+4 = 21 which is divisible by 3, as is $9444 = 3 \cdot 3148$), and 1717 (no, $1 + 7 + 1 + 7 = 16$ which is not divisible by 3, nor is 1717). Now, while the rule is true, checking a few examples doesn't constitute a proof. We haven't checked *every* number, only three specific numbers. We would have to show that, given an arbitrary number with digits $a_n \ldots a_3 a_2 a_1 a_0$, then if $a_0 + a_1 + \cdots + a_n$ is divisible by 3, so is $a_n \ldots a_3 a_2 a_1 a_0$.

This leads us to proving claims by generalizing an example or known case. Often the way the theorem is stated, it tries to guide you as to what to do. For instance, in the theorem we're trying to prove on divisibility by three, it tells us that divisibility by three is related to the sum of the digits of our number. So, we ask ourselves: how can we get the sum of the digits, given the number $a_n \ldots a_3 a_2 a_1 a_0$?

For example, 314 would be $a_2 a_1 a_0$, with $a_2 = 3$, $a_1 = 1$, $a_0 = 4$, and the sum of digits would be 3+1+4. Well, we might try looking at other ways of writing our number. Often there are different forms that are equivalent, but bring out different properties. For digits, we recall this comes from powers of 10: our number 314 can be written as $314 = 3 \cdot 100 + 1 \cdot 10 + 4 \cdot 1$.

So, notice what happens if we subtract from 314 the sum of its digits:

$$314 - (3 + 1 + 4) = 3 \cdot 100 + 1 \cdot 10 + 4 \cdot 1 - (3 + 1 + 4)$$
$$314 - (3 + 1 + 4) = (3 \cdot 100 - 3) + (1 \cdot 10 - 1) + (4 \cdot 1 - 4)$$
$$314 - (3 + 1 + 4) = (3) \cdot 99 + (1) \cdot 9 + (4) \cdot 0.$$

Ah. Notice that the right-hand side is clearly divisibly by 3, as each term is multiplied by 0 or 9 or 99. If $3 + 1 + 4$ is divisible by 3, when we bring it over to the right-hand side we find 314 equals a number divisible by three! If $3 + 1 + 4$ is not divisible by three, when we bring it over we get 314 equals a number *not* divisible by three!

Now we've done this proof in the special case when our number is 314. There's nothing wrong with first proving something for a specific case or number function, as long as we then generalize. We see that the exact same proof would carry through if instead we considered the number: $a_n \ldots a_3 a_2 a_1 a_0 = a_n \cdot 10^n + \cdots + a_1 \cdot 10^1 + a_0 \cdot 10^0$.

A.11 Dirichlet's Pigeon-Hole Principle

The following seemingly trivial observation appears in a variety of problems and is a very powerful way to prove many claims.

Dirichlet's Pigeon-Hole Principle: Let $A_1, A_2, \ldots, A_n$ be a collection of sets with the property that $A_1 \cup \cdots \cup A_n$ has at least $n + 1$ elements. Then at least one of the sets A_i has at least two elements.

This is called the Pigeon-Hole Principle for the following reason. Imagine we have $n + 1$ pigeons and n boxes, and we put each pigeon in exactly one box. Then at least one box must have two pigeons. If not, then each box has at most 1 pigeon, and as there are n boxes, this can account for at most n pigeons—at least one pigeon is missing! In a more mathematical prose, if we distribute k objects in n boxes and $k > n$, one of the boxes contains at least two objects. The Pigeon-Hole Principle is also known as the **Box Principle**. While there are many applications in number theory, there are a few in probability as well. For example, it's used in the Birthday Problem in Chapter 1 to see that once we have 366 people then we must have at least two sharing a birthday (we assumed no one was born on February 29). Let's do one more example. We'll first give the slick proof, then talk a bit about how to find such arguments.

Let S be any subset of $\{1, 2, \ldots, 2n\}$ *with* $n + 1$ *elements. Then S contains at least two elements* a, b *with a dividing b.*

To see this, we write each element $s \in S$ as $s = 2^\sigma s_0$ with s_0 odd. There are n odd numbers in the set $\{1, 2, \ldots, 2n\}$, and as the set S has $n + 1$ elements, the Pigeon-Hole Principle implies that there are at least two elements a, b with the same odd part. Without loss of generality, we might as well assume $a < b$, and write the numbers as $a = 2^i(2m + 1)$ and $b = 2^j(2m + 1)$. As $a < b$, $i < j$, we see $b = 2^{j-i}a$, proving a does indeed divide b.

The hard part of this problem is figuring out *how* to use the Pigeon-Hole Principle. The phrasing gives us some clues that we *should* use it. We have a collection of objects and we want to show that if we take a large enough subset, then at least two of those have a special relation. The Pigeon-Hole Principle is all about forced relations when we have enough items, so this is a natural approach.

The trick or difficulty is realizing that we should write our numbers as a power of two times an odd number, and then there must be two odd components that are equal. How can we figure out that *this* is what we should try? One way is to take special values of n and look at some sets, and see which elements have things in common. Related to this, try to take sets with just n numbers and see whether or not you can make the claim fail (since we're not taking $n + 1$ objects, it's fine to have the conclusion fail). After some experimentation, you might hit upon looking at the n odd numbers less than $2n$. If $2n = 8$ then this is a good choice, as $\{1, 3, 5, 7\}$ is such that no number divides another in this list; however, if $2n = 10$ we'd have $\{1, 3, 5, 7, 9\}$, and 3 divides 9. This illustrates the dangers of looking at small cases; we might see something that doesn't persist.

Returning to the drawing board, what other good sets are there of $\{1, 2, \ldots, 2n\}$ with n items? Perhaps a good choice is $\{n + 1, n + 2, \ldots, 2n\}$. This set always has n elements, and for all n we never have one element in the list dividing another. This is a great example, and we now know that we can't replace the $n + 1$ in the theorem with n. If the theorem is true, any element x added to $\{n + 1, n + 2, \ldots, 2n\}$ gives two numbers where one divides another. Further, as $x \leq n$, it must be the case that x divides something already in our list. It's not immediately clear, though, what it should divide. If x is large, say $n/2 < x \leq n$, then $2x$ is in our list $\{n + 1, n + 2, \ldots, 2n\}$. If $n/4 < x \leq n/2$, then $4x$ is in our list. This is probably the hardest jump to make, seeing the powers of two come into play. What we're trying to do is gather data and use that to guide us. From here, we somehow have to make the leap to noticing that our special pairs differ by a power of 2.

A.12 Proof by Adding Zero or Multiplying by One

I've saved my personal favorite for last: **adding zero** and **multiplying by one**. At first glance, neither of these seem capable of being that useful. After all, if we multiply by one, we're back where we started. The same goes for adding zero. Neither of these operations changes our expression.

Exactly! These are powerful methods *because* they don't change anything. We can't modify one side of an equality and not the other. We can't discriminate mathematically: whatever we do to one side, we must do to the other. The reason these are useful methods is that we can write 1 or 0 in many different ways, and we don't have to use the same representation on <u>both</u> sides. The point of this is to arrange the algebra in a more illuminating manner, to extract out sub-expressions that we know. Let's do a few examples. I chose these examples from calculus, but all we really need is the definition of the derivative, which states

$$f'(x) = \lim_{h \to 0} \frac{f(x+h) - f(x)}{h} = \lim_{x' \to x} \frac{f(x') - f(x)}{x' - x}$$

(both variants are used below), and that the derivative of x^n is nx^{n-1}. For an example in probability, go to §2.5.2.

Our first example is the proof of the product rule in calculus. Imagine f and g are differentiable functions, and set $A(x) = f(x)g(x)$. It's not unreasonable to hope that there's a nice formula for the derivative of A in terms of f, f', g, and g'. A great way to guess this relationship is to take some special examples. If we try

$$f(x) = x^3 \quad \text{and} \quad g(x) = x^4,$$

then

$$A(x) = x^7 \quad \text{so} \quad A'(x) = 7x^6.$$

At the same time,

$$f'(x) = 3x^2 \quad \text{and} \quad g'(x) = 4x^3.$$

There's only two ways to combine $f(x)$, $f'(x)$, $g(x)$, and $g'(x)$ and get x^6: $f'(x)g(x)$ and $f(x)g'(x)$. (Okay, there are more ways if we allow division; there's only two ways if we restrict ourselves to addition and multiplication.) Interestingly, if we add these together we get $3x^2 \cdot x^4 + x^3 \cdot 4x^3 = 7x^6$, which is just $A'(x)$. This *suggests* that $A'(x) = f'(x)g(x) + f(x)g'(x)$. If we try more and more examples, we'll see this formula keeps working. While this is strong evidence, it's not a proof; however, it *will* suggest the key step in our proof.

From the definition of the derivative and substitution,

$$A'(x) = \lim_{h \to 0} \frac{A(x+h) - A(x)}{h} = \lim_{h \to 0} \frac{f(x+h)g(x+h) - f(x)g(x)}{h}.$$

From our investigations above, we think the answer should be $f'(x)g(x) + f(x)g'(x)$. We can begin to see an $f'(x)$ and a $g'(x)$ lurking above. Imagine the last term were $f(x)g(x+h)$ instead of $f(x)g(x)$. If this were the case, the limit would equal $f'(x)g(x)$ (we pull out the $g(x+h)$, which tends to $g(x)$, and what's left is the definition of $f'(x)$). Similarly, if the first piece were instead $f(x)g(x+h)$, then we'd get $f(x)g'(x)$. What we see is that our expression is *trying* to look like the right things, but we're missing pieces. This can be remedied by adding zero, in the form $f(x)g(x+h) - f(x)g(x+h)$. Let's see what this does. In the algebra below we use the limit of a sum is the sum of the limits and the limit of a product is the product of the limits; we can use these results as

all these limits exist. We find

$$
\begin{aligned}
A'(x) &= \lim_{h \to 0} \frac{f(x+h)g(x+h) - f(x)g(x+h) + f(x)g(x+h) - f(x)g(x)}{h} \\
&= \lim_{h \to 0} \frac{f(x+h) - f(x)}{h} g(x+h) + \lim_{h \to 0} f(x) \frac{g(x+h) - g(x)}{h} \\
&= \lim_{h \to 0} \frac{f(x+h) - f(x)}{h} \lim_{h \to 0} g(x+h) + \lim_{h \to 0} f(x) \lim_{h \to 0} \frac{g(x+h) - g(x)}{h} \\
&= f'(x)g(x) + f(x)g'(x).
\end{aligned}
$$

□

The above proof has a lot of nice features. First off, it's the proof of a result you should know (at least if you've taken a calculus class). Second, we were able to guess the form of the answer by exploring some special cases. Finally, the proof was a natural outgrowth of these cases. We saw terms like $f'(x)g(x)$ and $f(x)g'(x)$ appearing, and thus asked ourselves: *So, what can we do to bring out these terms from what we have?* This led to adding zero in a clever way. It's fine to add zero, as it doesn't change the value. The advantage is we ended up with a new expression where we could now do some great simplifications.

For our second example, we'll look at the chain rule, one of the most dreaded rules from calculus. Now we take $B(x) = f(g(x))$. We assume f and g are differentiable, that $f(g(x))$ is defined, and for convenience we assume $g'(x)$ is continuous and never zero. This assumption isn't needed, but it'll simplify the argument so we make it as our point here is not to prove your old calculus results but rather to highlight the power of multiplying by 1. Since it worked so well last time, let's try to build some intuition from looking at

$$
f(x) = x^3 \quad \text{and} \quad g(x) = x^4.
$$

Again we have

$$
f'(x) = 3x^2 \quad \text{and} \quad g'(x) = 4x^3;
$$

however, we need to remember that we're supposed to evaluate f and f' not at x but at $g(x)$, so the relevant quantities are

$$
\begin{aligned}
B(x) = f(g(x)) &= (x^4)^3 = x^{12} \\
f'(g(x)) &= 3(x^4)^2 = 3x^8 \\
g'(x) &= 4x^3.
\end{aligned}
$$

Since $B'(x) = 12x^{11}$, looking at out building blocks we see that $12x^{11} = 3x^8 \cdot 4x^3$, or in this case we have $B'(x) = f'(g(x)) \cdot g'(x)$. So, just like the product rule, we have a candidate for the derivative. Knowing our goal is a great aid in suggesting the right way to manipulate expressions.

From the definition of the derivative, we have

$$
B'(x) = \lim_{h \to 0} \frac{B(x+h) - B(x)}{h} = \lim_{h \to 0} \frac{f(g(x+h)) - f(g(x))}{h}.
$$

We're searching for $f'(g(x))$ and $g'(x)$. Note the numerator almost looks like the derivative of f at the point $g(x)$; the reason it isn't is that we evaluate f at $g(x+h)$ rather than $g(x)+h$. What if we use the second variant for the definition of the derivative? In that case, $x' = g(x+h)$ tends to x, but the denominator isn't right. It should be $x' - g(x) = g(x+h) - g(x)$, but it's only h. To remedy this, we multiply by 1 in the form of $\frac{g(x+h)-g(x)}{g(x+h)-g(x)}$, and find

$$
\begin{aligned}
B'(x) &= \lim_{h \to 0} \frac{f(g(x+h)) - f(g(x))}{h} \frac{g(x+h) - g(x)}{g(x+h) - g(x)} \\
&= \lim_{h \to 0} \frac{f(g(x+h)) - f(g(x))}{g(x+h) - g(x)} \frac{g(x+h) - g(x)}{h} \\
&= \lim_{h \to 0} \frac{f(g(x+h)) - f(g(x))}{g(x+h) - g(x)} \lim_{h \to 0} \frac{g(x+h) - g(x)}{h} \\
&= f'(g(x)) \cdot g'(x).
\end{aligned}
$$

The last few lines deserve some justification. We're using the second variant of the definition of the derivative. Since the derivative of f exists, $\lim_{x' \to x} \frac{f(x')-f(g(x))}{x'-g(x)}$ equals $f'(g(x))$ for *any* sequence of x' tending to $g(x)$, and thus for the particular sequence where $x' = g(x+h)$.

Where did we use our assumption that $g'(x)$ is continuous and never zero? That assumption implies $g(x) = g(y)$ if and only if $x = y$. It's essential that $g(y) \neq g(x)$ for x and y distinct as otherwise $\frac{g(x+h)-g(x)}{g(x+h)-g(x)}$ could be 0/0.

We end with one last remark on these techniques. It's kind of like drawing auxiliary lines in geometry or trigonometry to highlight relationships. Drawing these lines doesn't change anything, but it often draws our attention to certain aspects of the problem.

APPENDIX B

Analysis Results

Not surprisingly, the tools from calculus (and more generally, real analysis) play a big role in probability. The reason, of course, is that to each random variable we attach a probability distribution. Often that distribution is continuous and even differentiable, and the quantities we want to study can be expressed in terms of our density and its integrals and derivatives.

We quickly review some of the key results from analysis below, and give some idea of how these are used.

B.1 The Intermediate and Mean Value Theorems

This section involves two of the biggest theorems from calculus, the Intermediate and the Mean Value Theorems. We'll use the Intermediate Value Theorem to prove the Mean Value Theorem, which can then be used to approximate numerous probabilities. First, we quickly review some notation. We write (a, b) for the interval $\{x : a < x < b\}$, and call this an **open interval**; by $[a, b]$ we mean $\{x : a \leq x \leq b\}$, and we call this a **closed interval**. We could of course have a half-open interval $[a, b)$ (which is also a half-closed interval!).

> **Theorem B.1.1 (Intermediate Value Theorem (IVT))**: *Let f be a continuous function on $[a, b]$. For all C between $f(a)$ and $f(b)$ there exists $c \in [a, b]$ such that $f(c) = C$. In other words, all intermediate values of a continuous function are obtained.*

If we convert from mathspeak to English, the theorem is a lot clearer, and quite reasonable. One way to do this is with the following example. Imagine we're driving our car. We start off traveling at 20 mph (about 32 kph), and later in the trip we're cruising at 100 mph (about 161 kph). As this example is for math, the police will kindly look the other way this one time. The Intermediate Value Theorem asserts that, at some time in our trip, we must've been traveling 50 mph (about 80 kph). This should be reasonable; we're assuming our speed is given by a nice, continuous function, and thus we can't get from the slow starting speed to the fast final speed without passing through all *intermediate* speeds.

Sketch of the proof: We proceed by **Divide and Conquer**. Without loss of generality, we can assume $f(a) < C < f(b)$, as the proof is trivial if $f(a) = C$ or $f(b) = C$. Many proofs start like this—first get rid of the straightforward cases, and then move on to the heart of the argument.

Let x_1 be the midpoint of $[a, b]$. If $f(x_1) = C$ we're done. If not, there are two cases: either $f(x_1) < C$ or $f(x_1) > C$. If $f(x_1) < C$, we look at the interval $[x_1, b]$. If $f(x_1) > C$ we look at the interval $[a, x_1]$.

In either case, we have a new interval, call it $[a_1, b_1]$, such that $f(a_1) < C < f(b_1)$ and the interval has half the size of $[a, b]$. We continue in this manner, repeatedly taking the midpoint and looking at the appropriate half-interval.

For example, imagine that our function is $f(x) = x^2 + x + 1$, $a = 0$, $b = 1$, and $C = 2$. We have $f(0) = 1$ and $f(1) = 3$. We look at the midpoint and find $f(1/2) = 1.75$, thus our next interval is $[a_1, b_1] = [1/2, 1]$. We continue; we have $f(1/2) = 1.75$, $f(1) = 3$, and at the midpoint $3/4$ we find $f(3/4) = 37/16 = 2.3125$. This means that our next interval is $[a_2, b_2] = [1/2, 3/4]$.

To recap, we have a sequence of intervals

$$[a, b] \supset [a_1, b_1] \supset [a_2, b_2] \supset \cdots$$

such that $f(a_n) \le C \le f(b_n)$, and each a_n and b_n is either an endpoint from the previous interval, or the midpoint of the previous interval. If any of these satisfy $f(x_n) = C$, we're done. If no midpoint works, we divide infinitely often and obtain a sequence of points x_n in intervals $[a_n, b_n]$. This is where rigorous mathematical analysis is required (see, for example, [Rud] for details). In a real analysis class you'll show that

$$\bigcap_{n=1}^{\infty} [a_n, b_n] = [a_1, b_1] \cap [a_2, b_2] \cap [a_3, b_3] \cap \cdots$$

is just a point, say $\{x_0\}$. This is intuitively plausible; at each stage we have an open interval, and we cut its length in half when we go to the next level. Thus the final result cannot have any positive length. It should be non-empty as we have the chain

$$a_1 \le a_2 \le a_3 \le \cdots \le b_3 \le b_2 \le b_1.$$

Let's assume there's a unique point in the intersection. Since f is continuous and $a_n \to x_0$ and $b_n \to x_0$,

$$\lim_{n \to \infty} f(a_n) = f(x_0) = \lim_{n \to \infty} f(b_n);$$

this is just a restatement of what it means for f to be continuous at x_0. But

$$f(a_n) \le C \le f(b_n) \quad \text{and} \quad f(a_n) \le f(x_0) \le f(b_n).$$

This implies that $f(x_0) = C$. Why? They are both "squeezed" to the same thing. Specifically, as $\lim_{n \to \infty} f(a_n) = \lim_{n \to \infty} f(b_n)$, we see that both of these limits equal C as well as $f(x_0)$. Thus, we have found our point! (For the example $f(x) = x^2 + x + 1$ on $[0, 1]$ with $C = 2$, we would find $x_0 = \frac{\sqrt{5}-1}{2} \approx 0.618034$.) $\square$

Theorem B.1.2 (Mean Value Theorem (MVT)): *Let $f(x)$ be differentiable on $[a, b]$. Then there exists $c \in (a, b)$ such that*

$$f(b) - f(a) = f'(c) \cdot (b - a).$$

Let's give an interpretation of the Mean Value Theorem. Let $f(x)$ represent the distance our car has traveled from the starting point at time x. The average speed from a to b is the distance traveled, $f(b) - f(a)$, divided by the elapsed time, $b - a$. As $f'(x)$ represents the speed at time x, the Mean Value Theorem says that there's some intermediate time at which we're traveling at the average speed.

For example, imagine that our average speed is 50 mph (about 80 kph). If our speed is always below 50 mph, there's no way that our average speed could be 50 mph; similarly, if our speed is always above 50 mph, there's no way our average speed could be 50 mph. Thus either our speed is always 50 mph (in which case the conclusion is trivial), or we can deduce that at some point in time we were traveling slower than 50 mph and at another point in time we were traveling faster. We can now use the Intermediate Value Theorem to prove that at some point we must be traveling at 50 mph, as that is an *intermediate* speed. This is essentially the proof; the only difference is that usually in a math book one sees impressive looking math symbols rather than text about cars!

To prove the Mean Value Theorem in familiar math language, it suffices to consider the special case when $f(a) = f(b) = 0$; this case is known as Rolle's Theorem.

Theorem B.1.3 (Rolle's Theorem): *Let f be differentiable on $[a, b]$, and assume $f(a) = f(b) = 0$. Then there exists $c \in (a, b)$ such that $f'(c) = 0$.*

Show the Mean Value Theorem follows from Rolle's Theorem. (Hint: Consider

$$h(x) = f(x) - \frac{f(b) - f(a)}{b - a}(x - a) - f(a).$$

Note $h(a) = f(a) - f(a) = 0$ and $h(b) = f(b) - (f(b) - f(a)) - f(a) = 0$. The conditions of Rolle's Theorem are satisfied for $h(x)$, and

$$h'(c) = f'(c) - \frac{f(b) - f(a)}{b - a}.)$$

Proof of Rolle's Theorem: Step one is to handle some special cases. We'll assume that $f'(a)$ and $f'(b)$ are non-zero. If one of these is zero we sadly aren't quite done, as the theorem asserts there is a c *strictly between* a and b; however, as a similar proof to what we give below handles this case, we leave that case as an exercise to the reader.

Multiplying $f(x)$ by -1 if needed, we may assume $f'(a) > 0$. *For convenience, we assume $f'(x)$ is continuous*. This assumption simplifies the proof, but isn't necessary. As you read the proof below, try to see where we use f' is continuous.

Case 1—$f'(b) < 0$: As $f'(a) > 0$ and $f'(b) < 0$, the Intermediate Value Theorem applied to $f'(x)$ asserts that all intermediate values are attained. As $f'(b) < 0 < f'(a)$, this implies the existence of a $c \in (a, b)$ such that $f'(c) = 0$.

Case 2—$f'(b) > 0$: $f(a) = f(b) = 0$, and the function f is increasing at a and b. If x is real close to a then $f(x) > 0$ if $x > a$. This follows from the fact that

$$f'(a) = \lim_{x \to a} \frac{f(x) - f(a)}{x - a}.$$

As $f'(a) > 0$, the limit is positive. As the denominator is positive for $x > a$, the numerator must be positive. Thus $f(x)$ must be greater than $f(a)$ for such x. Similarly $f'(b) > 0$ implies $f(x) < f(b) = 0$ for x slightly less than b.

Therefore the function $f(x)$ is positive for x slightly greater than a and negative for x slightly less than b. If the first derivative were always positive then $f(x)$ could never be negative as it starts at 0 at a. This can be seen by again using the limit definition of the first derivative to show that if $f'(x) > 0$ then the function is increasing near x. Thus the first derivative cannot always be positive. Either there must be some point $y \in (a, b)$ such that $f'(y) = 0$ (and we're then done) or $f'(y) < 0$. By the Intermediate Value Theorem, as 0 is between $f'(a)$ (which is positive) and $f'(y)$ (which is negative), there's some $c \in (a, y) \subset [a, b]$ such that $f'(c) = 0$. $\square$

Did you see where we used f' was continuous? It happened when we invoked the Intermediate Value Theorem. Whenever you use a theorem, you need to make sure all the conditions are satisfied. To use the IVT, we need our function to be continuous.

B.2 Interchanging Limits, Derivatives, and Integrals

B.2.1 Interchanging Orders: Theorems

For the convenience of the reader we record exact statements of several standard results from advanced calculus that are used at various points of the text. As the Change of Variables Theorem is so important, it gets its own chapter (in the online supplemental chapters).

Theorem B.2.1 (Fubini's Theorem): *Assume f is continuous and*

$$\int_a^b \int_c^d |f(x, y)| dx dy < \infty.$$

Then

$$\int_a^b \left[\int_c^d f(x, y) dy \right] dx = \int_c^d \left[\int_a^b f(x, y) dx \right] dy.$$

Similar statements hold if we instead have

$$\sum_{n=N_0}^{N_1} \int_c^d f(x_n, y)dy, \quad \sum_{n=N_0}^{N_1} \sum_{m=M_0}^{M_1} f(x_n, y_m).$$

For a proof in special cases, see [BL, VG]; an advanced, complete proof is given in [Fol]. See Exercise B.7.2 for an example where the orders of integration cannot be changed.

Theorem B.2.2 (Interchanging Differentiation and Integration): *Let $f(x, t)$ be a continuous function whose partial derivatives with respect to x and with respect to t are continuous in the region $\{(x, t) : x \in [a, b], t \in [c, d]\}$ with a, b, c, d finite. Then*

$$\frac{d}{dx} \int_a^b f(x, t)dt = \int_a^b \frac{\partial f}{\partial x}(x, t)dt.$$

The above theorem holds in greater generality. We can allow the regions to be infinite, at the cost of requiring additional decay in the functions. For a proof and generalizations, see [La2].

Our last result is on interchanging limits and integrals. We state one of the most useful below, though *not* in its most general form (see [Fol] for the more general phrasing and a proof).

Theorem B.2.3 (Dominated Convergence Theorem): *Let $\{f_n\}$ be a sequence of piecewise continuous real-valued functions on $\mathbb{R}$, and assume there is a non-negative, piecewise continuous function g with $|f_n(x)| \le g(x)$ for all n. Assume $\lim_{n\to\infty} f_n(x)$ converges pointwise to a piecewise continuous function f. Then*

$$\lim_{n\to\infty} \int_{-\infty}^{\infty} f_n(x)dx = \int_{-\infty}^{\infty} \lim_{n\to\infty} f_n(x)dx;$$

in other words, we may interchange the limit and the integral.

B.2.2 Interchanging Orders: Examples

The purpose of this section is to give a quick crash course in using analysis to justify certain statements. What follows is essentially independent of the rest of the book. As it's important to know *how* to justify statements (this lessens the chance of accidentally using results that can't be justified!), it's fine to skim or skip what follows.

In general, we need to appeal to some advanced theorems in analysis to interchange the order of operations, such as switching the order of integration or interchanging a sum and a derivative. In the case of the geometric series, however, we can justify interchanging the sum and the derivative without appealing to advanced machinery. The reason is that if we truncate the geometric series

$$\sum_{n=0}^{\infty} x^n = 1 + x + x^2 + x^3 + x^4 + \cdots = \frac{1}{1 - x} \tag{B.1}$$

at any N, the geometric series formula gives us an explicit formula for the sum of the tail:

$$\sum_{n=0}^{\infty} x^n = \left(1 + x + x^2 + \cdots + x^N\right) + \left(x^{N+1} + x^{N+2} + \cdots\right)$$

$$= 1 + x + x^2 + \cdots + x^N + \frac{x^{N+1}}{1-x} = \frac{1}{1-x}.$$

We show that we may interchange differentiation and summation for the geometric series (assuming, of course, that $|x| < 1$). The derivative of the right-hand side (with respect to x) of (B.1) is just $(1-x)^{-2}$. We want to say the derivative of the left-hand side of (B.1) is

$$\sum_{n=0}^{\infty} nx^{n-1},$$

but to do so requires us to justify

$$\frac{d}{dx} \sum_{n=0}^{\infty} x^n = \sum_{n=0}^{\infty} \frac{d}{dx} x^n.$$

A standard way to justify statements like this is as follows. We note that $\sum_{n=0}^{\infty} nx^{n-1}$ converges for $|x| < 1$; if we can show that for any $\epsilon > 0$ that this is within ϵ of $(1-x)^{-2}$, then we will have justified the interchange.

To see this, fix an $\epsilon > 0$. For each N, as discussed above we may write

$$\sum_{n=0}^{\infty} x^n = \sum_{n=0}^{N} x^n + \sum_{n=N+1}^{\infty} x^n$$

$$= \sum_{n=0}^{N} x^n + \frac{x^{N+1}}{1-x} = \frac{1}{1-x}.$$

We can differentiate each side, and we can justify interchanging the differentiation and the summation because we have *finitely many* sums. Specifically, there are only $N + 2$ terms ($N + 1$ from the sum and then one more, $\frac{x^{N+1}}{1-x}$). Therefore we have

$$\frac{d}{dx} \sum_{n=0}^{N} x^n + \frac{d}{dx} \frac{x^{N+1}}{1-x} = \frac{d}{dx} \frac{1}{1-x}$$

$$\sum_{n=0}^{N} nx^{n-1} + \frac{(N+1)x^N(1-x) - x^{N+1}(-1)}{(1-x)^2} = \frac{1}{(1-x)^2}$$

$$\sum_{n=0}^{N} nx^{n-1} + \frac{(N+1)(1-x) + x}{(1-x)^2} x^N = \frac{1}{(1-x)^2}.$$

As $|x| < 1$, given any $\epsilon > 0$ we can find an N_0 such that for all $N \geq N_0$,

$$\left| \frac{(N+1)(1-x) + x}{(1-x)^2} x^N \right| \leq \frac{\epsilon}{2}.$$

Similarly we can find an N_1 such that for all $N \geq N_1$ we have

$$\left| \sum_{n=N+1}^{\infty} n x^{n-1} \right| \leq \frac{\epsilon}{2}.$$

Therefore we have shown that for every $\epsilon > 0$ we have

$$\left| \frac{1}{(1-x)^2} - \sum_{n=0}^{\infty} n x^{n-1} \right| \leq \epsilon,$$

proving the claim. Instead of studying these sums for a specific x, we can consider $x \in [a, b]$ with $-1 < a \leq b < 1$, and N_0, N_1 will just depend on a, b and ϵ.

Exercise B.2.4: *In the argument above, make all dependence explicit; in other words, whenever it says "for sufficiently large," quantify that in terms of ϵ.*

One situation where we cannot interchange differentiation and summation is when we have series that are **conditionally convergent** but not absolutely convergent. This means $\sum a_n$ converges but $\sum |a_n|$ does not. For example, consider

$$\sum_{n=0}^{\infty} \frac{x^n}{n}. \tag{B.2}$$

If $x = -1$ this series conditionally converges but not absolutely; in fact, as

$$-\log(1-x) = x + \frac{x^2}{2} + \frac{x^3}{3} + \cdots = \sum_{n=1}^{\infty} \frac{x^n}{n},$$

then (B.2) with $x = -1$ is just $-\log 2$. What happens if we try to differentiate? We have

$$\frac{d}{dx}[-\log(1-x)] = \frac{d}{dx} \left[\sum_{n=1}^{\infty} \frac{x^n}{n} \right].$$

The left-hand side is easy to differentiate for $x \in [-1, 0]$, giving $\frac{1}{1-x}$. But if we interchange the differentiation and summation we would have

$$\frac{d}{dx} \left[\sum_{n=1}^{\infty} \frac{x^n}{n} \right] = \sum_{n=1}^{\infty} x^{n-1},$$

and this does not converge when $x = -1$ (aside: the sum oscillates between 1 and 0; in some sense it can be interpreted as $\frac{1}{2}$, which is what $\frac{1}{1-x}$ equals when $x = -1$!).

Sometimes, however, conditionally convergent but absolutely divergent series can be managed. Consider

$$\sum_{n=2}^{\infty} \frac{x^n}{n \log n}.$$

This series converges conditionally when $x = -1$ but diverges upon inserting absolute values. If we interchange differentiation and summation we get

$$\sum_{n=2}^{\infty} \frac{x^{n-1}}{\log n},$$

and this sum does converge (conditionally, not absolutely) when $x = -1$.

B.3 Convergence Tests for Series

In calculus classes we learn various tests to determine whether or not a series converges or diverges. There are many reasons for all those hours you spent mastering these, as you're now perfectly prepared to actually *use* these for problems you might care about. In Chapter 19 we'll meet generating functions. These are series that encode a wealth of information about a probability distribution. If these sums converge and are differentiable, then simple differentiation gives us nice formulas for many properties; however, it's sadly not the case that these infinite series always converge. We thus (finally!) see applications of the various series convergence tests from calculus.

As with any result from an earlier course, if you haven't used it in awhile it's easy to be rusty. For completeness we quickly state some of the more popular and powerful tests, and give a few examples illustrating their use. Before doing so, we quickly recall some standard results and notation about series. First, the summation notation:

$$\sum_{n=0}^{N} a_n = a_0 + a_1 + \cdots + a_N;$$

if instead of N we had ∞ as the upper bound the sum would be $a_0 + a_1 + a_2 + \cdots$. For finite N, we have

$$\sum_{n=0}^{N} (a_n + b_n) = \sum_{n=0}^{N} a_n + \sum_{n=0}^{N} b_n;$$

if the two sums on the right are finite then this result also holds if $N = \infty$. If these two sums are infinite, however, things are trickier. The problem is one sum could be ∞ and the other $-\infty$, and $\infty - \infty$ is undefined. (Imagine the examples where $a_n = 2n$ and $b_n = -4n$, and $a_n = 2n$ and $b_n = -n$.) If c is any real number,

$$\sum_{n=0}^{N} c a_n = c \sum_{n=0}^{N} a_n.$$

Root Test: Assume $\lim_{n\to\infty} \sqrt[n]{|a_n|}$ exists, and denote this limit by ρ. Then the series $\sum_{n=0}^{\infty} a_n s^n$ converges for $|s| < 1/\rho$ and diverges for $|s| > 1/\rho$; if $\rho = 0$ we interpret $1/\rho$ as infinity, meaning the series converges for all s. If $\rho = 1$ then there's no information on whether or not it converges or diverges.

Ratio Test: Assume $\lim_{n\to\infty} |a_{n+1}/a_n|$ exists, and denote this limit by ρ. Then the series $\sum_{n=0}^{\infty} a_n s^n$ converges for $|s| < 1/\rho$ and diverges for $|s| > 1/\rho$; if $\rho = 0$ we interpret $1/\rho$ as infinity, meaning the series converges for all s. If $\rho = 1$ then there's no information on whether or not it converges or diverges.

For example, let $a_n = n^2/4^n$. By the ratio test, we have

$$\lim_{n\to\infty} \frac{a_{n+1}}{a_n} = \lim_{n\to\infty} \frac{(n+1)^2/4^{n+1}}{n^2/4^n} = \lim_{n\to\infty} \left(\frac{n+1}{n}\right)^2 \frac{1}{4} = \frac{1}{4}.$$

Thus

$$G(s) = \sum_{n=0}^{\infty} \frac{n^2}{4^n} s^n$$

converges for $|s| < 4$.

Two other tests that are frequently used are the comparison test and the integral test. These can be a little harder to use, as you need to choose a comparison sequence or function, while the ratio and root tests are automatic (simply compute the limit). That said, with experience these become easier to apply.

Comparison Test: Let $\{b_n\}_{n=1}^{\infty}$ be a sequence of non-negative terms (so $b_n \geq 0$). Assume the series $\sum_{n=0}^{\infty} b_n$ converges, and $\{a_n\}_{n=1}^{\infty}$ is another sequence such that $|a_n| \leq b_n$ for all n. Then the series $\sum_{n=0}^{\infty} a_n$ also converges. If instead $\sum_{n=0}^{\infty} b_n$ diverges and $a_n \geq b_n$, then the series $\sum_{n=0}^{\infty} a_n$ also diverges.

Integral Test: Consider a sequence $\{a_n\}_{n=1}^{\infty}$ of non-negative terms. Assume there's some function f such that $f(n) = a_n$ and f is non-increasing. Then the series

$$\sum_{n=1}^{\infty} a_n$$

converges if and only if the integral

$$\int_1^{\infty} f(x)dx$$

converges; thus if the integral diverges the series diverges.

Note: in both these tests, if instead of starting the sums at $n = 0$ we start at $n = N$, the conclusions still hold; this is because the convergence of series depend only on the tails, and we can add or remove finitely many terms without harm.

Let's determine if the series $\sum_{n=1}^{\infty} \frac{1}{2^n + \sqrt{n}}$ converges or diverges. We use the comparison test. The hardest part about using this test is figuring out what to compare our sequence to. If we think it converges we should find a series that converges that is always greater, while if it diverges we should look for a series that is always small and diverges. When n is large, 2^n is larger than $\sqrt{n}$, and thus the denominator essentially looks like 2^n. We thus expect our series to converge by a comparison with the geometric series $b_n = 1/2^n$. Writing down the algebra formally, we would argue that since $2^n + \sqrt{n} \geq 2^n$, we have

$$0 \leq \frac{1}{2^n + \sqrt{n}} \leq \frac{1}{2^n}.$$

Thus the series converges by the comparison test. We can easily modify this to a problem in generating functions. Consider

$$G(s) = \sum_{n=1}^{\infty} \frac{1}{2^n + \sqrt{n}} s^n.$$

A similar argument shows that we can compare this to $\sum_{n=1}^{\infty} (s/2)^n$, which is a geometric series converging for $|s| < 2$. Thus $G(s)$ converges for $|s| < 2$.

Now let's consider $a_n = \frac{1}{n \ln^p n}$ for some $p > 0$. For which p does it converge? Diverge? We know $\sum_{n=1}^{\infty} \frac{1}{n}$ diverges; unfortunately, this is useless for the comparison test as $\frac{1}{n \ln^p n} \leq \frac{1}{n}$ for n large. If we want to show a series diverges by the comparison test, we must compare it to something smaller that diverges, not something larger. It's hard to find a good series to compare this to, and unfortunately the ratio and root tests don't provide any useful information (as the limit in both cases is 1). We are left with trying the integral test.

The first step is to find a strictly decreasing function $f(x)$ that equals $\frac{1}{n \ln^p n}$ when $x = n$ for n large. Looking at what we've written, you should be able to hear the integral test screaming which function to use: $f(x) = \frac{1}{x \ln^p x}$; it's very common in these problems to just replace n with x. Thus the series converges or diverges depending on whether or not

$$\int_{x=\text{BIG}}^{\infty} \frac{1}{x \ln^p x} \, dx$$

converges or diverges; we write "BIG" to indicate that the lower bound doesn't really matter–what matters is the behavior at infinity. We use a u-substitution. This is a *very* natural thing to do. The reason is the derivative of $\ln x$ is $1/x$; looking at our integrand, we see it's begging us to change variables as we have $1/x$. We try $u = \ln x$. This gives $du = dx/x$, and thus our integral becomes

$$\int_{u=\ln(\text{BIG})}^{\infty} u^{-p} du.$$

The integral of u^{-p} is $\frac{u^{1-p}}{p}$ if $p \neq 1$ and $\ln u$ if $p = 1$. Thus the integral converges if $p > 1$ and diverges if $p \leq 1$.

We can turn this into a statement about the generating function

$$G(s) \;=\; \sum_{n=1}^{\infty} \frac{1}{n \ln^p n} s^n.$$

For any choice of p, with some work you can show the sum converges for $|s| \leq 1$ and diverges for $|s| > 1$.

B.4 Big-Oh Notation

The purpose of this section is to introduce some notation to make it easy for us to compare two quantities as some parameter tends to infinity. If the definition seems technical, there's a natural reason: it is! The entire point of this definition is to allow us to carefully discuss and compare two expressions in some limit situation. The point is to bypass handwaving arguments, to avoid using phrases such as "clearly" and "of course." This notation is used throughout analysis whenever one needs to make rigorous comparisons.

As a motivating example, think of the standard normal and the standard exponential. As the first has density function $\frac{1}{\sqrt{2\pi}} \exp(-x^2/2)$ while the second has the density function $\exp(-x)$, "clearly" the standard normal is decaying faster as $x \to \infty$ than the standard exponential. What we want to do now is clarify how much faster the standard normal decays, as well as avoid using the word "clearly."

Definition B.4.1 (Big-Oh Notation): $A(x) = O(B(x))$, read "$A(x)$ is of order (or big-Oh) $B(x)$," means there's a $C > 0$ and an x_0 such that for all $x \geq x_0$, $|A(x)| \leq C\, B(x)$. This is also written $A(x) \ll B(x)$ or $B(x) \gg A(x)$.

Let's unwind this. The part about C is no problem; it's just saying there's some positive constant which will surface later. The purpose of the x_0 constant is to define our universe of discourse. We're saying what happens from some point onward; we're making *no* claims about the behavior for "small" x; all we're saying is that we know what happens as $x \to \infty$. Specifically, for all large x we have $|A(x)|$ is at most $C B(x)$. Frequently the actual value of C doesn't matter; what's important is the growth (or decay) in x. Additionally, in many problems the inequality holds for each and every x, and thus we don't need to worry about x_0.

Sometimes we use big-Oh notation for $x \to 0$ instead of $x \to \infty$; in that case we modify the definition to there's an x_0 such that for all x with $|x| \leq x_0$ we have $|A(x)| \leq C B(x)$.

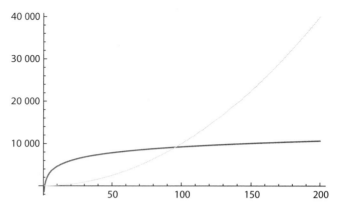

Figure B.1. Plot of $2010 \log x$ versus x^2.

 In Figure B.1 we plot $A(x) = 2010 \log x$ versus $B(x) = x^2$. For small values of x, we see that $A(x)$ is larger; however, as x increases we see eventually $B(x)$ is greater. The reason is that x^2 is growing faster than $\log x$, so in the limit x^2 dominates $\log x$. We can't, however, say that $2010 \log x \le x^2$, though, as this inequality fails for small x. It's only true for x large ($x \ge 100$ suffices). In many problems, we're only interested in making comparisons as our input parameter tends to infinity, and thus such restrictions are fine.

Big-Oh notation is a convenient way to handle lower order terms. For example, if we write $F(x) = x^5 + O(x^2)$, this means that as x tends to infinity, the main term of $F(x)$ grows like x^5, and the correction (or error) terms are at most some constant times x^2.

 Not surprisingly, this is used all the time in Taylor series expansions. Consider the Taylor series expansion for $\cos x$:

$$\cos x = \sum_{n=0}^{\infty} \frac{(-1)^n x^{2n}}{(2n)!} = 1 - \frac{x^2}{2!} + \frac{x^4}{4!} - \cdots.$$

Let's take $x \in [-\pi, \pi]$ near 0, and see how good of a job the various partial Taylor series expansions do of approximating $\cos x$. We have, for instance,

$$\cos x = 1 - \frac{x^2}{2} + O(x^4),$$

and we claim this works for all x. The reason is the error in the approximation is

$$-\frac{x^4}{4!} + \frac{x^6}{6!} - \frac{x^8}{8!} + \cdots.$$

We can trivially bound this by dropping all the minus signs, and thus the error is at most

$$\frac{x^4}{4!} + \frac{x^6}{6!} + \frac{x^8}{8!} + \cdots.$$

How big is this sum? Remember we plan on taking x near 0, so the higher the power of x, the smaller the contribution. Thus the "main" term in the error comes from the $x^4/4!$ piece. Pulling that out, we find the error is at most

$$\frac{x^4}{4!} \left(1 + x^2 + x^4 + \cdots\right).$$

For x close to zero, we clearly have $|x| \le 1/2$ and thus we may use the geometric series formula to evaluate the sum (the ratio is just x^2); note the sum is largest when $|x| = 1/2$ (that's the worst case). We finally see that the error is at most

$$\frac{x^4}{4!} \frac{1}{1 - x^2};$$

if we assume $|x| \le 1/2$ then we finally obtain

$$\left| \cos x - \left(1 - \frac{x^2}{2}\right) \right| \le \frac{4}{3} \frac{x^4}{4!} = \frac{x^4}{18}.$$

In other words, if $|x| \le 1/2$ the error in using the second order Taylor series to approximate $\cos x$ is quite small, as it's at most $x^4/18$. For example, if we take $x = .1$ then we would say $\cos(.1)$ is approximately $1 - \frac{.1^2}{2} = .995$, with an error that is at most $.1^4/18 \approx 5.5556 \cdot 10^{-6}$. The actual value of $\cos(.1)$ (to ten decimal places) is 0.995004165, which means the true error is about $4.16528 \cdot 10^{-6}$. Note the true error is less than our theoretical bound, so it's likely we have done the algebra correctly!

Two very important relations are that x^r grows slower than e^x for any fixed r as $x \to \infty$, and $\log x$ grows slower than x^c for any $c > 0$ as $x \to \infty$. There are many ways to prove these relations. We prove the first one now to highlight the method, and leave the second one for you. Let's consider x^r versus e^x as $x \to \infty$. We want to show $x^r = O(e^x)$. Clearly this is true if r is negative, so we need only look at $r \ge 0$. If r happened to be an integer, we can use L'Hôpital's rule:

$$\lim_{x \to \infty} \frac{x^r}{e^x} = \lim_{x \to \infty} \frac{rx^{r-1}}{e^x} = \lim_{x \to \infty} \frac{r(r-1)x^{r-2}}{e^x} = \cdots = \lim_{x \to \infty} \frac{r!}{e^x} = 0.$$

Why did we assume r was an integer? This is just to make applying L'Hôpital a little cleaner; if r is an integer then after applying L'Hôpital r times the numerator is just $r!$. As this limit is zero, by definition there's some x_0 such that for $x \ge x_0$ we have $x^r/e^x \le 1/2$, which gives $x^r = O(e^x)$ if r is a positive integer. For general r, we can either use L'Hôpital (ending up with a power of x in the denominator of the fraction), or note that $x^r \le x^{\lceil r \rceil}$, where $\lceil r \rceil$ represents the smallest integer at least r.

As $x^r = O(e^x)$ is used in numerous problems, we give one more proof. For convenience, let's assume r is an integer. From the Taylor series expansion of e^x, we know $e^x > x^{r+1}/(r+1)!$ (this is because we're keeping just one term). If $x > (r+1)!$, then

$$x^r < \frac{x^{r+1}}{(r+1)!} < e^x.$$

B.5 The Exponential Function

In this section we study some of the basic properties of the number e. There are many ways to define the number e, the base of the natural logarithm. From the point of view of calculus, the most convenient is through an infinite series:

$$e = \sum_{n=0}^{\infty} \frac{1}{n!}.$$

If we denote the partial sums of the above series by

$$s_m = \sum_{n=0}^{m} \frac{1}{n!},$$

we see e is the limit of the convergent sequence s_m. This representation is one of the main tools in analyzing the nature of e. See Exercise 20.11.38 for a terrific application of the exponential function, where we show how it may be used to derive trig identities. We generalize the above and write the exponential function.

Exponential function: Let x be any real (or complex) number. The exponential function e^x (which for typographical purposes is sometimes written $\exp(x)$ when the argument is complicated) is defined as

$$e^x = \sum_{n=0}^{\infty} \frac{x^n}{n!}.$$

Further, $e^{x+y} = e^x e^y$.

We call the above the exponential function. As remarked, we frequently use the exp notation for typographical purposes; for example, $\exp(-x^2/2)$ is a little easier to read than $e^{-x^2/2}$ or, even worse, $e^{-\frac{x^2}{2}}$!

The series defining the exponential function converges so rapidly that almost any test works. Let's use the ratio test, as it's easy to apply. We have

$$\rho = \lim_{n\to\infty} \frac{|a_{n+1}|}{|a_n|}$$

$$= \lim_{n\to\infty} \frac{|x|^{n+1}/(n+1)!}{|x|^n/n!}$$

$$= \lim_{n\to\infty} \frac{|x|}{n+1} = 0.$$

Thus, the series converges for all x.

This notation is meant to be highly suggestive, and is designed to make you think about raising numbers to powers. We read e^x as e raised to the x power. If asked what is $e^x e^y$, you should immediately answer e^{x+y}; however, it's very important to note that this is *not* obvious and this needs to be proved! Technically e^x, e^y, and e^{x+y}, are three different infinite sums, and we must show the product of the first two equals the third. Of course, if this were not true then our notation would suck (no other word feels right for how horrible our notation would be); unfortunately, math does occasionally have bad notation. I've always hated that cosecant is one over sine and not one over cosine.

 The proof that $e^x e^y = e^{x+y}$ is a nice application of the Binomial Theorem. We have

$$e^x e^y = \sum_{m=0}^{\infty} \frac{x^m}{m!} \sum_{n=0}^{\infty} \frac{y^n}{n!}.$$

Note that we used two different letters for our summations. It's a very common mistake to use the same letter twice; we can't and shouldn't do this. The reason it's wrong to use the same letter is that we have two sums, and each sum has a dummy variable for summation (similar to the dummy variables of integration). Consider for example

$$(1 + 2 + 3) \cdot (1^2 + 2^2 + 3^2) = 84.$$

If we use the same dummy variable, we might be led to the following flawed calculation:

$$\sum_{n=1}^{3} n \sum_{n=1}^{3} n^2 = \sum_{n=1}^{3} n^3 = 1^3 + 2^3 + 3^3 = 36.$$

Using a different letter for each sum minimizes our chance of making such a mistake.

Returning to our analysis of $e^x e^y$, we see we have a sum over terms of the form $\frac{x^m y^n}{m!n!}$, with $m, n \geq 0$. What we will do now is collect all terms where the sum of the power of x plus the power of y is constant. In other words, for a given $k \geq 0$ let's look at all pairs (m, n) with $m + n = k$. We need to introduce one more dummy variable. Let's let ℓ equal the power of x. If the power of x plus the power of y is k, this means that the power of y is $k - \ell$ whenever the power of x is ℓ; furthermore, ℓ ranges from 0 to k (as the powers of x and y are non-negative integers). Collecting, we find

$$e^x e^y = \sum_{k=0}^{\infty} \sum_{\ell=0}^{k} \frac{x^\ell y^{k-\ell}}{\ell!(k-\ell)!}.$$

We now need to do some pattern recognition. Note the denominator looks a lot like a binomial coefficient; it's the bottom of $\binom{k}{\ell}$. This suggests **multiplying by one** (see §A.12 for more examples), in this case $k!/k!$.

Note that the denominator invokes thoughts of binomial coefficients. Specifically, $\binom{k}{\ell} = \frac{k!}{\ell!(k-\ell)!}$. If we multiply by 1 in the form $k!/k!$, we'll see the binomial coefficient emerge:

$$e^x e^y = \sum_{k=0}^{\infty} \sum_{\ell=0}^{k} \frac{1}{k!} \frac{k!}{\ell!(k-\ell)!} x^\ell y^{k-\ell}$$

$$= \sum_{k=0}^{\infty} \frac{1}{k!} \sum_{\ell=0}^{k} \binom{k}{\ell} x^\ell y^{k-\ell}$$

$$= \sum_{k=0}^{\infty} \frac{1}{k!} (x+y)^k$$

$$= \sum_{k=0}^{\infty} \frac{(x+y)^k}{k!} = e^{x+y},$$

where we used the Binomial Theorem to replace the ℓ sum with $(x+y)^\ell$ and we used the series expansion to replace the k sum with e^{x+y}. $\square$

All that matters from the above discussion is that our intuition is correct, and our notation is good. It's also worth noting the power of multiplying by 1. This is one of the hardest math skills to learn, but one of the most important. We can always multiply by 1 (or do something similar, add zero); the trick is finding *good* ways to do this which lead to simpler expressions.

There is another definition of e^x, which also arises in probability. You might remember it from compound interest problems where the money is compounded instantaneously. This definition is very useful in proving the Central Limit Theorem for certain sums of independent random variables.

An alternative definition of e^x is

$$e^x = \lim_{n \to \infty} \left(1 + \frac{x}{n}\right)^n.$$

A nice exercise is to show that this definition agrees with the series expansion.

No introduction to e^x would be complete without a few words about its derivative. Using the series expansion, the natural temptation is to differentiate term by term, which gives

$$\frac{d}{dx} e^x = \frac{d}{dx} \left(1 + x + \frac{x^2}{2!} + \frac{x^3}{3!} + \cdots\right)$$

$$= 1 + \frac{2x}{2!} + \frac{3x^2}{3!} + \frac{4x^3}{4!} + \cdots$$

$$= 1 + x + \frac{x^2}{2!} + \frac{x^3}{3!} + \cdots = e^x.$$

Of course, we need to justify interchanging a sum and a derivative. This is typically done in an advanced analysis course.

 Without using a calculator or computer, determine which is larger: e^π or π^e. (Hint: One approach is to study the function $x^{1/x}$; take the $e\pi$ root of both sides to reduce the problem to comparing $e^{1/e}$ and $\pi^{1/\pi}$. Use calculus to find the maximum value.) One could also study $f(x) = e^x - x^e$ and try to show $f(x) > 0$ when $x > e$; however, it's hard to analyze all the critical points. It's easier to study $g(x) = e^{x/e} - x$, and show $g(x) > 0$ for $x > e$.

B.6 Proof of the Cauchy-Schwarz Inequality

Our last analysis result is the Cauchy-Schwarz inequality, which is very useful in bounding certain integrals.

Lemma B.6.1 (Cauchy-Schwarz Inequality): *For complex-valued functions f and g,*

$$\int_{-\infty}^{\infty} |f(x)g(x)|dx \ \leq \ \left(\int_{-\infty}^{\infty} |f(x)|^2 dx\right)^{1/2} \cdot \left(\int_{-\infty}^{\infty} |g(x)|^2 dx\right)^{1/2}. \tag{B.3}$$

Proof of the Cauchy-Schwarz inequality: For notational simplicity, assume f and g are non-negative functions. Working with $|f|$ and $|g|$ we see there's no harm in the above assumption. As the proof is immediate if either of the integrals on the right-hand side of (B.3) is zero or infinity, we assume both integrals are non-zero and finite. Let

$$h(x) \ = \ f(x) - \lambda g(x), \quad \lambda \ = \ \frac{\int_{-\infty}^{\infty} f(x)g(x)dx}{\int_{-\infty}^{\infty} g(x)^2 dx}.$$

As $\int_{-\infty}^{\infty} h(x)^2 dx \geq 0$ we have

$$0 \ \leq \ \int_{-\infty}^{\infty} (f(x) - \lambda g(x))^2 \, dx$$

$$= \ \int_{-\infty}^{\infty} f(x)^2 dx \ - \ 2\lambda \int_{-\infty}^{\infty} f(x)g(x)dx \ + \ \lambda^2 \int_{-\infty}^{\infty} g(x)^2 dx$$

$$= \ \int_{-\infty}^{\infty} f(x)^2 dx \ - \ 2\frac{\left(\int_{-\infty}^{\infty} f(x)g(x)dx\right)^2}{\int_{-\infty}^{\infty} g(x)^2 dx} \ + \ \frac{\left(\int_{-\infty}^{\infty} f(x)g(x)dx\right)^2}{\int_{-\infty}^{\infty} g(x)^2 dx}$$

$$= \ \int_{-\infty}^{\infty} f(x)^2 dx \ - \ \frac{\left(\int_{-\infty}^{\infty} f(x)g(x)dx\right)^2}{\int_{-\infty}^{\infty} g(x)^2 dx}.$$

This implies

$$\frac{\left(\int_{-\infty}^{\infty} f(x)g(x)dx\right)^2}{\int_{-\infty}^{\infty} g(x)^2 dx} \ \leq \ \int_{-\infty}^{\infty} f(x)^2 dx,$$

or equivalently

$$\left(\int_{-\infty}^{\infty} f(x)g(x)dx \right)^2 \le \int_{-\infty}^{\infty} f(x)^2 dx \cdot \int_{-\infty}^{\infty} g(x)^2 dx.$$

Taking square-roots completes the proof. □

 This proof uses one of the most important identities in all of mathematics: if u is a real number then $u^2 \ge 0$. The clever part is in choosing u. For those loving a challenge, think why this works. Why is this a good choice? A good starting point is to determine when the Cauchy-Schwarz inequality is an equality.

B.7 Exercises

Exercise B.7.1   *In our proof of Rolle's Theorem we assumed $f'(a)$ and $f'(b)$ were non-zero; handle the case when one of these vanish.*

Exercise B.7.2   *One cannot always interchange orders of integration. For simplicity, we give a sequence a_{mn} such that $\sum_m (\sum_n a_{m,n}) \ne \sum_n (\sum_m a_{m,n})$. For $m, n \ge 0$ let*

$$a_{m,n} = \begin{cases} 1 & \text{if } n = m \\ -1 & \text{if } n = m + 1 \\ 0 & \text{otherwise.} \end{cases}$$

Show that the two different orders of summation yield different answers (the reason for this is that the sum of the absolute value of the terms diverges).

Exercise B.7.3   *In justifying interchanging a derivative and a sum we needed the existence of an N_0 such that for all $N \ge N_0$,*

$$\left| \frac{(N+1)(1-x) + x}{(1-x)^2} x^N \right| \le \frac{\epsilon}{2}$$

(where ϵ is a fixed positive number). Find an N_0 that works (your answer should depend on ϵ).

Exercise B.7.4   *Consider the dominated convergence theorem. Show that its conclusion need not hold if there is no non-negative, piecewise continuous function g such that $|f_n(x)| \le g(x)$ for all n.*

APPENDIX C

Countable and Uncountable Sets

Our goal here is to introduce just enough of the theory of countable and uncountable sets for applications in a first course in probability. Briefly, we're going to see that not all infinities are equally infinite! Amazingly, we can have two sets with infinitely many elements and it makes sense to say that one has fewer elements than the other. We'll prove there are levels of infinities. The smallest is called countable (an example is the number of integers); anything larger is called uncountable (an example is the number of real numbers). Countable infinities aren't too bad to handle in mathematics, but uncountable infinities can lead to some very strange situations. Fortunately it's the countable infinities that typically arise in probability (of course, this isn't entirely a coincidence as these are the infinities we can handle!).

This is a very rich and vast subject with numerous uses in mathematics; we cannot, nor do we try to, do it justice in such a brief exposition. We need to have some idea of what countable and uncountable mean for probability. The reason is that there are subtleties that arise when we work with infinite sets, and we need some machinery to avoid error (see for instance §8.4).

C.1 Sizes of Sets

Let's review the standard infinite sets we've encountered over the years. We have the following inclusions: the natural numbers $\mathbb{N} = \{0, 1, 2, 3, \dots\}$ are a subset of the integers $\mathbb{Z} = \{\dots, -1, 0, 1, \dots\}$ are a subset of the rationals $\mathbb{Q} = \{p/q : p, q \in \mathbb{Z}, q \neq 0\}$ are a subset of the real numbers $\mathbb{R}$ are a subset of the complex numbers $\mathbb{C}$. The notation $\mathbb{Z}$ comes from the German zahl (number) and $\mathbb{Q}$ comes from quotient. While all of these sets are infinite, we'll see later that some are more infinite than others.

At the face of it, it seems strange to talk about quantifying infinities. Once two sets have infinitely many elements, aren't they both equally large? Surprisingly, the answer is no; there are very natural demarcations we may make between various sizes of infinity.

On the other hand, perhaps it isn't so strange. If $A \subset B$ and B contains some elements not in A, shouldn't B be larger? While this is true for finite sets, it may fail for infinite sets. For example, consider the closed intervals $A = [0, 1]$ and $B = [0, 2]$. As we remarked, in one sense the second set is larger as the first is a proper subset. In another sense they are the same size as each element $x \in [0, 2]$ can be paired with a unique element $y = x/2 \in [0, 1]$. Using this pairing, we see that given any element in B we can find a unique element in A, and we never have to use the same element of A for two different elements of B. From this point of view, it seems like A and B are the same size!

The above example suggests a way to compare two sets. Before isolating this out as a definition, we first quickly state some needed notation. A function $f : A \to B$ is **one-to-one** (or **injective**) if $f(x) = f(y)$ implies $x = y$; f is **onto** (or **surjective**) if given any $b \in B$ there exists $a \in A$ with $f(a) = b$. A **bijection** is a one-to-one and onto function. Colloquially, a function is one-to-one if distinct inputs go to distinct outputs, and it's surjective if every potential output is hit.

Consider $f : \mathbb{R} \to \mathbb{R}$ given by $f(x) = x^2$. This isn't a bijection, as $f(-1) = f(1)$, or more generally $f(-x) = f(x)$. As the function isn't injective, it can't be a bijection. All isn't lost, however, as it's easy to modify the function and get an injection. Consider $g : [0, \sqrt{2}) \to [0, 2)$ given by $g(x) = x^2$. We'll show this function is a bijection. If $g(x) = g(y)$ then $x^2 = y^2$ which implies $x = \pm y$. As g is defined only on $[0, \sqrt{2})$, we can't have negative input. Thus if $x^2 = y^2$ we see that x and y must be non-negative, which implies a unique solution. All that is left is to show surjectivity, namely that every element in $[0, 2)$ is the image of something in $[0, \sqrt{2})$. Given $y \in [0, 2)$, we take $x = \sqrt{y}$, and then $g(x) = \sqrt{y}^2 = y$.

We now come to the definition of the size of a set, as well as when two sets are the same size.

> We say two sets A and B **have the same cardinality** (i.e., are the same size) if there is a bijection $f : A \to B$. We denote the common cardinality by $|A| = |B|$. If A has finitely many elements (say n elements), then there's a bijection from A to $\{1, \dots, n\}$. We say A is **finite** and $|A| = n < \infty$.

Two finite sets have the same cardinality if and only if they have the same number of elements; prove this. This is one case where our intuition works—if A is a proper subset of B (which means B contains something not in A) and both are finite sets, then $|A| < |B|$.

If A and B are two sets such that there are onto maps $f : A \to B$ and $g : B \to A$, then $|A| = |B|$; prove this.

One last bit of notation. A set A is said to be **infinite** if there's a one-to-one map $f : A \to A$ which isn't onto. In other words, an infinite set has infinitely many elements. Using this definition, we see that the sets $\mathbb{N}$ and $\mathbb{Z}$ are infinite sets.

The definition of an infinite set seems a little strange. Let's look at some examples. Assume A has finitely many elements, say $A = \{a_1, a_2, \dots, a_n\}$. There can't be a one-to-one map on A that isn't onto. Why? Imagine $f : A \to A$ is onto but misses some element; without loss of generality, let's say we never map anything to a_n. As f is one-to-one, each of $a_1, a_2, \dots, a_n$ are mapped to distinct elements; however, as nothing is mapped to a_n our n elements are mapped to $n - 1$ elements. By the Pigeon-Hole Principle (or Dirichlet's Box Principle, see §A.11), two of the a_i's must be mapped to the same element, as we have n objects mapped to $n - 1$ objects. This contradicts our assumption that f is one-to-one, as now two elements are mapped to the same element. We've just shown that no finite set can have an injective map that isn't onto; therefore, if our set A has an injective map that isn't onto then A must be infinite!

As an example of such a situation, imagine A is the set of natural numbers, so $A = \{0, 1, 2, \ldots\}$. We define a map $f : A \to A$ by $f(n) = n + 1$. Clearly this map is one-to-one, and it isn't onto as nothing is mapped to zero.

Here's an example which should seem strange at first. The cardinality of the positive even integers is the same as the cardinality of the positive integers. The bijective map we use is $f(n) = n/2$.

Let's look at this in detail and see why it's so surprising. Let E_N be all positive even integers at most N. The fraction of positive integers less than $2M$ and even is $M/2M = 1/2$, yet the even numbers have the same cardinality as $\mathbb{N}$. If S_N is all perfect squares up to N, one can similarly show the fraction of perfect squares up to N is approximately $1/\sqrt{N}$, which goes to zero as $N \to \infty$. Hence in one sense there are a lot more even numbers or integers than perfect squares, but in another sense these sets are the same size.

C.2 Countable Sets

We are finally ready to define countable sets. These sets have the "smallest" possible infinity as their size.

> A set A is **countable** if there's a bijection between A and the integers $\mathbb{Z}$ (or, as we'll prove later, a bijection between A and the natural numbers $\mathbb{N}$); A is **at most countable** if A is either finite or countable, and A is **uncountable** if A isn't at most countable.

We often build complicated sets out of simpler ones. One of the most important ways to build up sets is through the Cartesian product.

> If A and B are sets, the **Cartesian product** $A \times B$ is $\{(a, b) : a \in A, b \in B\}$.

You can't take a multivariable calculus class without seeing Cartesian products. If we have a real-valued function of two variables, we write this as $f : \mathbb{R}^2 \to \mathbb{R}$. This means we take two real numbers as input and output one real number. The $\mathbb{R}^2 = \mathbb{R} \times \mathbb{R}$ represents the input. If we have a vector valued function, we might have $\overrightarrow{F} : \mathbb{R}^3 \to \mathbb{R}^3$.

We now show that several common sets are countable. Consider the set of whole numbers $\mathbb{W} = \{1, 2, 3, \ldots\}$. Define $f : \mathbb{W} \to \mathbb{N}$ by $f(n) = n - 1$. As f is a bijection, these two sets have the same cardinality. Now let's look at the set of integers $\mathbb{Z} = \{\ldots, -1, 0, 1, \ldots, \}$, and set $f(n) = 2n$ if $n \geq 0$ and $f(n) = -(2n - 1)$ if $n < 0$. We see that f is a bijection from $\mathbb{Z}$ to $\mathbb{N}$, and thus these sets are also the same size.

It's a nice exercise to show that if $f : A \to B$ is a bijection and $g : B \to C$ is a bijection then $h : A \to C$ is a bijection, with $h(a) = g(f(a))$ (so $h = g \circ f$ is the composition of g and f). This means that cardinality obeys some highly desirable properties: if A and B have the same size and B and C have the same size, then A and C have the same size. While this should ring true, it does need a proof. The following result is a very important consequence of the above exercise and these properties of composition.

> To show a set S is countable, it's sufficient to find a bijection from S to either $\mathbb{W}$ or $\mathbb{N}$ or $\mathbb{Z}$.

To continue our exposition, we need the following intuitively plausible result (see any good book on set theory, such as [HJ], for a proof).

> **Theorem C.2.1**: *Let A, B, and C be three sets.*
>
> 1. *If $A \subset B$, then $|A| \leq |B|$.*
> 2. *If $f : A \to C$ is a one-to-one function (not necessarily onto), then $|A| \leq |C|$. Further, if $C \subset A$ then $|A| = |C|$.*
> 3. *(The Cantor-Bernstein Theorem) If $|A| \leq |B|$ and $|B| \leq |A|$, then $|A| = |B|$.*

We are now in the position to show how we can take some countable sets and generate many more countable sets.

> **Theorem C.2.2**: *If A and B are countable then so is $A \cup B$ and $A \times B$.*

Proof: As A and B are countable, we have bijections $f : \mathbb{N} \to A$ and $g : \mathbb{N} \to B$. Thus we can label the elements of A and B by

$$A = \{a_0, a_1, a_2, a_3, \dots\}$$

$$B = \{b_0, b_1, b_2, b_3, \dots\}.$$

Assume $A \cap B$ is empty. Define $h : \mathbb{N} \to A \cup B$ by $h(2n) = a_n$ and $h(2n + 1) = b_n$. As h is a bijection from $\mathbb{N}$ to $A \cup B$, this proves $A \cup B$ is countable. We leave it to you to handle the case when $A \cap B$ isn't empty.

To prove $A \times B$ is countable, consider the following function $h : \mathbb{N} \to A \times B$ (see Figure C.1):

$h(1) = (a_0, b_0)$

$h(2) = (a_1, b_0), h(3) = (a_1, b_1), h(4) = (a_0, b_1)$

$h(5) = (a_2, b_0), h(6) = (a_2, b_1), h(7) = (a_2, b_2), h(8) = (a_1, b_2), h(9) = (a_0, b_2)$

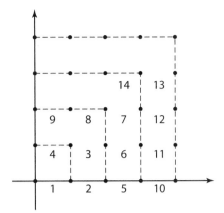

Figure C.1. $A \times B$ is countable.

and so on. For example, at the nth stage we have

$$h(n^2 + 1) = (a_n, b_0), h(n^2 + 2) = (a_n, b_{n-1}), \ldots$$

$$h(n^2 + n + 1) = (a_n, b_n), h(n^2 + n + 2) = (a_{n-1}, b_n), \ldots$$

$$\ldots, h((n + 1)^2) = (a_0, b_n).$$

We're looking at all pairs of integers (a_x, b_y) in the first quadrant (including those on the axes). The above function h starts at $(0, 0)$, and then moves through the first quadrant, hitting each pair once and only once, by going up and over and then restarting on the x-axis. $\qquad\square$

One of the most important consequences of the above theorem is how hard it is to start with countable sets and end up with something that isn't countable!

Corollary C.2.3: *Let $(A_i)_{i \in \mathbb{N}}$ be a collection of sets such that A_i is countable for all $i \in \mathbb{N}$. Then for any n, $A_1 \cup \cdots \cup A_n$ and $A_1 \times \cdots \times A_n$ are countable, where the last set is all n-tuples $(a_1, \ldots, a_n)$, $a_i \in A_i$. Further $\cup_{i=0}^{\infty} A_i$ is countable. If each A_i is at most countable, then $\cup_{i=0}^{\infty} A_i$ is at most countable.*

We now prove a very important and useful fact.

Theorem C.2.4: *The rationals are countable as well!*

Proof: We're back in the realm of possibly surprising results. After all, there are infinitely many rationals between any two integers, and we're now saying that these two sets have the same size!

Clearly the size of the rationals must be at least as large as the size of the integers, as the integers are a proper subset. All we need to do is show that the rational numbers are at most countable and we're done, as the rationals are clearly infinite. The set $\mathbb{Z} \times \mathbb{W}$ is clearly countable, as it's the Cartesian product of two countable sets (the integers and the whole numbers). Given any rational $r = p/q$ we may associate the pair (p, q) with p and q integers and $q > 0$. This gives us a map $f : \mathbb{Q} \to \mathbb{Z} \times \mathbb{W}$ that is one-to-one, though not necessarily onto. Thus $\mathbb{Q}$ is at most countable, which is what we needed to show. □

 Let's stop and think about the consequences of what we've just shown, which are amazing. There is a bijection between the natural numbers and the rational numbers! We can thus write

$$\mathbb{Q} = \{q_1, q_2, q_3, \dots\}$$

and not miss any elements, nor repeat any elements! While we've shown such a function exists, it doesn't mean that it's easy to write down (though for "fun" I urge you to find such a function). Fortunately, there are many instances in mathematics where we only need to know that something exists or can be done, and we don't need to know how to do it! For the rationals, it's almost always sufficient to know that they can be "ordered" somehow without knowing what the actual order is.

As the natural numbers, integers and rationals are countable, by taking each $A_i = \mathbb{N}$, $\mathbb{Z}$, or $\mathbb{Q}$ in Corollary C.2.3 we immediately obtain the following consequence.

> **Corollary C.2.5**: $\mathbb{N}^n$, $\mathbb{Z}^n$, and $\mathbb{Q}^n$ are countable.

For the proof, proceed by induction; for example write $\mathbb{Q}^{n+1}$ as $\mathbb{Q}^n \times \mathbb{Q}$.

C.3 Uncountable Sets

Recall that we've declared a set to be **uncountable** if it's infinite and there's no bijection between it and the rationals (or the integers, or any countable set). We've given many examples of countable sets; in addition to the natural numbers and the integers and the rational numbers, we also have $\mathbb{Q}^n$ for any n. At this point, you should be asking whether or not we need the word uncountable. After all, everything we've seen so far is countable. Uncountable sets do exist, as we can now show.

> **Theorem C.3.1 (Cantor)**: *The set of all real numbers is uncountable. This follows from the fact that if S is the set of all sequences $(y_i)_{i \in \mathbb{N}}$ with $y_i \in \{0, 1\}$ then S is uncountable.*

Proof: Let's show how the uncountability of the real numbers follows from our claim about sequences, as this should surely motivate reading the technical proof. Consider all numbers in the interval $[0, 1]$ whose decimal expansion consists entirely of 0's and 1's. There is a bijection between this subset of $\mathbb{R}$ and the set S, and two different expansions correspond to two different numbers. Once we establish that S is uncountable, we'll know that $\mathbb{R}$ has an uncountable subset, and hence $\mathbb{R}$ is uncountable.

We're *just* left with proving the claim about the subsequences. We proceed by contradiction. Suppose there's a bijection $f : S \to \mathbb{N}$. It's clear that this is equivalent to a listing of the elements of $\mathcal{S}$:

$$x_1 = .x_{11}x_{12}x_{13}x_{14}\cdots$$

$$x_2 = .x_{21}x_{22}x_{23}x_{24}\cdots$$

$$x_3 = .x_{31}x_{32}x_{33}x_{34}\cdots$$

$$\vdots$$

$$x_n = .x_{n1}x_{n2}x_{n3}x_{n4}\cdots x_{nn}\cdots$$

$$\vdots$$

Note we have such a listing as we assumed $\mathcal{S}$ is countable. Define an element $\theta = (\theta_i)_{i\in\mathbb{N}} \in \mathcal{S}$ by $\theta_i = 1 - x_{ii}$. Note θ can't be in the list. It can't be x_N because $1 - x_{NN} \neq x_{NN}$; we've forced it to disagree with x_N in at least one place. As our list was supposed to be a complete enumeration of $\mathcal{S}$, we've reached a contradiction! $\qquad\square$

The above proof is due to Cantor (1873–1874) and is known as **Cantor's Diagonalization Argument**.

As a nice exercise, show $|[0, 1]| = |\mathbb{R}| = |\mathbb{R}^n| = |\mathbb{C}^n|$, and find a set with strictly larger cardinality than $\mathbb{R}$.

We would be remiss if we didn't mention the phrase **power set** here. The power set of A is the set of all subsets of A, and is denoted $\mathcal{P}(A)$. If A is a finite set with n elements, then $\mathcal{P}(A)$ has 2^n elements. One way to see this is to note that each element of A may be in a subset, or it may not. Note the empty set $\varnothing$ and A are elements of the power set of A. For example, if $A = \{a, b\}$ then

$$\mathcal{P}(A) = \{\varnothing, \{a\}, \{b\}, \{a, b\}\}.$$

We can view the set of all countable sequences of 0's and 1's from above as the power set of $\mathbb{N}$. Power sets arise naturally in probability. Unfortunately, we'll see that it isn't possible to assign probabilities to all elements in the power set of A for uncountable A such as $[0, 1]$.

It's an interesting questions as to whether or not there exists an infinite set with size strictly between $\mathbb{Z}$ and $\mathbb{R}$. From our discussion above, the size of $\mathbb{R}$ is the same as the size of the power set of $\mathbb{N}$ or $\mathbb{Z}$ or $\mathbb{Q}$. It's common practice to denote the size of the rationals by $\aleph_0$ (pronounced aleph-naught, and famous for starting the longest song in the English language: Aleph-naught bottles of beer on the wall, ...). Cantor's Diagonalization Argument tells us that $\aleph_0 < 2^{\aleph_0}$. Is there a set whose size is strictly between $\aleph_0$ and $2^{\aleph_0}$? The **Continuum Hypothesis** asserts that no such set exists. By deep work of Kurt Gödel and Paul Cohen, we now know that this hypothesis is *independent* of the other axioms of set theory. This means that if the other axioms of set theory are consistent, then they will be if we assume the Continuum Hypothesis *or* if we assume its negation!

So there you have it. Not all infinite sets have the same size; there are both countable and uncountable sets. As a final remark, let's look at the unit interval $[0, 1]$. We can break that into two disjoint subsets, those elements that are rational and those that are irrational. From our analysis above, we have the rational subset is countable and the irrational subset is uncountable. We'll see in §C.4 that we can use this to quantify the following statement: if we choose a random number in $[0, 1]$ uniformly, then with probability one it's irrational!

C.4 Length of the Rationals

We discuss sizes of subsets of $[0, 1]$. It's natural to define the length of an interval $I = [a, b]$ (or $[a, b)$ and so on) as $b - a$. We denote this by $|I|$, and refer to this as the **length** or **measure** of I. Our definition implies a point a has zero length. What about more exotic sets, such as the rationals and the irrationals? What are the measures of these sets? A proper explanation is given by measure theory (see [La2, Rud]); we introduce enough for our purposes.

Before we discuss the length or size of the rationals in $[0, 1]$, a few preliminaries to set the stage. Let I be a countable union of disjoint intervals $I_n \subset [0, 1)$; thus $I_n \cap I_m$ is empty if $n \neq m$. It's *natural* to say

$$|I| = \sum_n |I_n|.$$

This is true, but only because we are taking a countable union. Consider an uncountable union with $I_x = \{x\}$ for $x \in [0, 1]$. As each singleton $\{x\}$ has length zero, we expect their union to also have length zero; however, their union is $[0, 1]$, which has length 1.

Finally, if $A \subset B$ it's natural to say $|A|$ (the length of A) is at most $|B|$ (the length of B). Note our definition implies $[a, b)$ and $[a, b]$ have the same length.

Our assumptions imply that the rationals in $[0, 1]$ have zero length, hence the irrationals in $[0, 1]$ have length 1. This is so important that we isolate it, to remind ourselves that sets with infinitely many elements can have zero probability! Why are we using the word probability? Well, if we take the uniform distribution on $[0, 1]$, we can ask what is the probability we chose a rational number. This should just be the length of the rationals in $[0, 1]$ divided by the length of $[0, 1]$ (which is of course just 1). If we show that the rationals in $[0, 1]$ have zero length, then we may interpret this as showing that, with probability one, a randomly chosen number in $[0, 1]$ is irrational!

Theorem C.4.1: *The rationals $\mathbb{Q}$ have zero length. This follows from showing the rationals in $[0, 1]$ have zero length.*

Sketch of the proof: It suffices to show $Q = \mathbb{Q} \cap [0, 1]$ has measure zero. To see this, let $n + Q = \mathbb{Q} \cap [n, n + 1]$. Then

$$\mathbb{Q} = \bigcup_{n \in \mathbb{Z}} (n + Q).$$

As each $n + Q$ has the same length as Q, we have a countable union of sets of length zero, and thus the set of all rational numbers has length zero.

To prove $|Q| = 0$ we show that given any $\epsilon > 0$ we can find a countable set of intervals I_n such that

1. $|Q| \subset \cup_n I_n$;

2. $\sum_n |I_n| < \epsilon$.

If we can do this, we're done. Why? This implies that the length of the rationals in $[0, 1]$, whatever it is, is less than ϵ; however, as ϵ was arbitrarily chosen this forces the length of Q to be zero. If it wasn't zero, all we would need to do to get a contradiction is take ϵ less than half the length of Q.

Okay. We're now reduced to showing that we can find a countable set of intervals I_n with the desired properties. As the rationals are countable, we can enumerate Q, say $Q = \{x_n\}_{n=0}^\infty$. For each n let

$$ I_n = \left[x_n - \frac{\epsilon}{4 \cdot 2^n}, \, x_n + \frac{\epsilon}{4 \cdot 2^n} \right], \quad |I_n| = \frac{\epsilon}{2 \cdot 2^n}. $$

Clearly $Q \subset \cup_n I_n$. The intervals I_n are not necessarily disjoint, but

$$ |\cup_n I_n| \leq \sum_n |I_n| = \epsilon, $$

which completes the proof. $\qquad\square$

You can now show that any countable set has measure zero.

C.5 Length of the Cantor Set

It's time to see how misleading our intuition can be with infinite sets. We know there are levels of infinity, with countable sets having fewer elements than uncountable sets. We've just shown that all countable sets have zero length. It's reasonable to posit that all uncountable sets have positive length; unfortunately this isn't true! The standard example is the Cantor set.

The **Cantor set** is a fascinating subset of $[0, 1]$. We construct it in stages. Let $C_0 = [0, 1]$. We remove the middle third of C_0 and obtain $C_1 = [0, \frac{1}{3}] \cup [\frac{2}{3}, 1]$. Note C_1 is a union of two closed intervals (we keep all endpoints). To construct C_2 we remove the middle third of all remaining intervals and obtain

$$ C_2 = \left[0, \frac{1}{9} \right] \cup \left[\frac{2}{9}, \frac{3}{9} \right] \cup \left[\frac{6}{9}, \frac{7}{9} \right] \cup \left[\frac{8}{9}, 1 \right]. $$

We continue this process. Note C_n is the union of 2^n closed intervals, each of size 3^{-n}, and

$$ C_0 \supset C_1 \supset C_2 \supset \cdots . $$

The formal definition follows.

> **Definition C.5.1 (Cantor Set):** *The Cantor set C is defined by*
>
> $$C = \bigcap_{n=1}^{\infty} C_n = \{x \in \mathbb{R} : x \in C_n \text{ for all } n\}.$$

We claim the length of the Cantor set is zero. For each n, the Cantor set is contained in the union of 2^n disjoint sets each of size $1/3^n$. Thus the length of the Cantor set is at most $(2/3)^n$ for any n. Taking n arbitrarily large, we see that the length of the Cantor set is zero. For example, imagine it was not zero; for definiteness, say it's of length 10^{-2011}. All we have to do is choose n so large that $(2/3)^n < 10^{-2011}$; if we take $n > 11,500$ then $(2/3)^n < 10^{-2011}$. This would mean that the Cantor set, whose size we assumed was 10^{-2011}, is a subset of a set whose length is less than 10^{-2011}, a contradiction.

 If x is an endpoint of C_n for some n, then $x \in C$. At first, one might expect that these are the only points, especially as the Cantor set has length zero. With a little work you can show that $1/4$ and $3/4$ are in C, but neither is an endpoint. (Hint: Proceed by induction.) To construct C_{n+1} from C_n, we removed the middle third of intervals. For each sub-interval, what is left looks like the union of two pieces, each one-third the length of the previous. Thus, we have shrinking maps fixing the left and right parts $L, R : \mathbb{R} \to \mathbb{R}$ given by $L(x) = x/3$ and $R(x) = (x+2)/3$, and $C_{n+1} = R(C_n) + L(C_n)$.

 One can show that the Cantor set is also the set of all numbers $x \in [0, 1]$ which have no 1's in their base three expansion. For rationals such as $1/3$, we may write these by using repeating 2's: $1/3 = .02222\ldots$ in base three. By considering base two expansions, you can show there's a one-to-one and onto map from $[0, 1]$ to the Cantor set. This means, of course, that the Cantor set is uncountable, as there's a bijection between it and $[0, 1]$!

To recap: the Cantor set is uncountable and is in a simple correspondence to all of $[0, 1]$, *but* it has length zero! Thus, the notion of "length" is different from the notion of "cardinality": two sets can have the same cardinality but very different lengths.

For more on the Cantor set, including dynamical interpretations, see [Dev, Edg, Fal, SS3].

C.6 Exercises

Exercise C.6.1 *The Hilbert Hotel is managed beautifully by mathematicians. It has infinitely many rooms, and even if all the rooms are occupied it can always accommodate more guests without anyone sharing a room! (a) Show how we can move the guests around to free up a room if a new guest arrives. (b) Show how we can move the guests around to free up rooms for countably many guests who arrive.*

Exercise C.6.2 *Prove two finite sets have the same cardinality if and only if they have the same number of elements.*

Exercise C.6.3 *Prove that if A and B are two sets such that there are onto maps f : A → B and g : B → A, then |A| = |B|.*

Exercise C.6.4 *A real number α is said to be **algebraic** if it is a root of a polynomial of finite degree and integer coefficients; otherwise α is said to be **transcendental**. (a) Prove all rational numbers are algebraic. (b) Prove all square-roots are algebraic. (c) Prove the set of all algebraic numbers is countable. (d) Prove the set of transcendental numbers is uncountable.*

APPENDIX D _____

Complex Analysis and the Central Limit Theorem

In Chapter 20 we gave a proof of the Central Limit Theorem using generating functions; unfortunately that proof isn't complete as it assumed some results from complex analysis. Moreover, we had to assume the moment generating function existed, which isn't always true.

We tried again in Chapter 21; we proved the Central Limit Theorem by using Fourier Analysis. Instead of using the moment generating function, which can fail to even exist, this time we used the Fourier transform (also called the characteristic function), which has the very nice and useful property of actually existing! Unfortunately, here too we needed to appeal to some results from complex analysis.

This leaves us in a quandary, where we have a few options.

1. We can just accept as true some results from complex analysis and move on.
2. We can try and find yet another proof, this time one that doesn't need complex analysis.
3. We can drop everything and take a crash course in complex analysis.

This chapter is for those who like the third option. We'll explain some of the key ideas of complex analysis, in particular we'll show why it's such a different subject than real analysis. Obviously, it helps to have seen real analysis, but if you're comfortable with Taylor series and basic results on convergence you'll be fine.

It turns out that assuming a function of a real variable is differentiable doesn't mean too much, but assume a function of a complex variable is differentiable and all of a sudden doors are opening everywhere with additional, powerful facts that must be true. Obviously this chapter can't replace an entire course, nor is that our goal. We want to show you some of the key ideas of this beautiful subject, and hopefully when you finish reading you'll have a better sense of why the black-box results from complex analysis (Theorems 20.5.3 and 20.5.4) are true.

This chapter is meant to supplement our discussions on moment generating functions and proofs of the Central Limit Theorem. We thus assume the reader is familiar with the notation and concepts from Chapters 19 through 21.

D.1 Warnings from Real Analysis

The following example is one of my favorites from real analysis. It indicates why real analysis is hard, almost surely much harder than you might expect. Consider the function $g : \mathbb{R} \to \mathbb{R}$ given by

$$g(x) = \begin{cases} e^{-1/x^2} & \text{if } x \neq 0 \\ 0 & \text{otherwise.} \end{cases} \tag{D.1}$$

Using the definition of the derivative and L'Hôpital's rule, we can show that g is infinitely differentiable, and all of its derivatives at the origin vanish. For example,

$$g'(0) = \lim_{h \to 0} \frac{e^{-1/h^2} - 0}{h}$$

$$= \lim_{h \to 0} \frac{1/h}{e^{1/h^2}}$$

$$= \lim_{k \to \infty} \frac{k}{e^{k^2}}$$

$$= \lim_{k \to \infty} \frac{1}{2ke^{k^2}} = 0,$$

where we used **L'Hôpital's rule** in the last step ($\lim_{k \to \infty} A(k)/B(k) = \lim_{k \to \infty} A'(k)/B'(k)$ if $\lim_{k \to \infty} A(k) = \lim_{k \to \infty} B(k) = \infty$). (We replaced h with $1/k$ as this allows us to re-express the quantities above in a familiar form, one where we can apply L'Hôpital's rule.) A similar analysis shows that the nth derivative vanishes at the origin for all n, i.e., $g^{(n)}(0) = 0$ for all positive integer n. If we consider the Taylor series for g about 0, we find

$$g(x) = g(0) + g'(0)x + \frac{g''(0)x^2}{2!} + \cdots = \sum_{n=0}^{\infty} \frac{g^{(n)}(0)x^n}{n!} = 0;$$

however, clearly $g(x) \neq 0$ if $x \neq 0$. We are thus in the ridiculous case where the Taylor series (which converges for all x!) only agrees with the function when $x = 0$. This isn't that impressive, as the Taylor series is *forced* to agree with the original function at 0, as both are just $g(0)$.

We can learn a lot from the above example. The first is that it's possible for a Taylor series to converge for all x, but only agree with the function at one point! It's not too impressive to agree at just one point, as by construction the Taylor series *has* to agree at that point of expansion. The second, which is far more important, is that *a Taylor series does not uniquely determine a function!* For example, both $\sin x$ and $\sin x + g(x)$ (with $g(x)$ the function from Equation (D.1)) have the same Taylor series about $x = 0$.

The reason this is so important for us is that we want to understand when a moment generating function uniquely determines a probability distribution. If our distribution was discrete, there was no problem (Theorem 19.6.5). For continuous distributions, however, it's much harder, as we saw in Equation (19.6.5) where we met two densities that had the same moments.

Apparently, we must impose some additional conditions for continuous random variables. For discrete random variables, it was enough to know all the moments; this doesn't suffice for continuous random variables. What should those conditions be?

Recall that if we have a random variable X with density f_X, its kth moment, denoted by μ'_k, is defined by

$$\mu'_k = \int_{-\infty}^{\infty} x^k f_X(x) dx.$$

Let's consider again the pair of functions in Equation (19.6.5). A nice calculus exercise shows that $\mu'_k = e^{k^2/2}$. This means that the moment generating function is

$$M_X(t) = \sum_{k=0}^{\infty} \frac{\mu'_k t^k}{k!} = \sum_{k=0}^{\infty} \frac{e^{k^2/2} t^k}{k!}.$$

For what t does this series converge? Amazingly, this series converges *only* when $t = 0$! To see this, it suffices to show that the terms do not tend to zero. As $k! \leq k^k$, for any fixed t, for k sufficiently large $t^k/k! \geq (t/k)^k$; moreover, $e^{k^2/2} = (e^{k/2})^k$, so the kth term is at least as large as $(e^{k/2}t/k)^k$. For any $t \neq 0$, this clearly does not tend to zero, and thus the moment generating function has a radius of convergence of zero!

This leads us to the following conjecture: *If the moment generating function converges for $|t| < \delta$ for some $\delta > 0$, then it uniquely determines a density.* We'll explore this conjecture below.

D.2 Complex Analysis and Topology Definitions

Our purpose here is to give a flavor of what kind of inputs are needed to ensure that a moment generating function uniquely determines a probability density. We first collect some definitions, and then state some useful results from complex analysis.

Definition D.2.1 (Complex variable, complex function): *Any complex number z can be written as $z = x + iy$, with x and y real and $i = \sqrt{-1}$. We denote the set of all complex numbers by $\mathbb{C}$. A complex function is a map f from $\mathbb{C}$ to $\mathbb{C}$; in other words $f(z) \in \mathbb{C}$. Frequently one writes $x = \Re(z)$ for the **real part**, $y = \Im(z)$ for the **imaginary part**, and $f(z) = u(x, y) + iv(x, y)$ with u and v functions from $\mathbb{R}^2$ to $\mathbb{R}$.*

There are many ways to write complex numbers. The most common is the definition above; however, a polar coordinate approach is sometimes useful. One of the most remarkable relations in all of mathematics is

$$e^{i\theta} = \cos\theta + i\sin\theta.$$

There are several ways to see this, depending on how much math you want to assume. One way is to use the Taylor series expansions for the exponential, sine, and cosine functions. This gives another way of writing complex numbers; instead of $1 + i$ we could write $\sqrt{2}\exp(i\pi/4)$. A particularly interesting choice of θ is π, which gives

$e^{i\pi} = -1$, a beautiful formula involving many of the most important constants in mathematics!

Noting $i^2 = -1$, it isn't too hard to show that

$$(a + ib) + (x + iy) = (a + x) + i(b + y)$$

$$(a + ib) \cdot (x + iy) = (ax - by) + i(ay + bx).$$

The **complex conjugate** of $z = x + iy$ is $\bar{z} := x - iy$, and we define the **absolute value** (or the **modulus** or **magnitude**) of z to be $\sqrt{z\bar{z}}$, and denote this by $|z|$. This is real-valued, and equals $\sqrt{x^2 + y^2}$. If we were to write z as a vector, it would be $z = (x, y)$; note that in this case we see that $|z|$ equals the length of the corresponding vector.

We can write almost anything as an example of a complex function; one possible function is $f(z) = z^2 + |z|$. The question is when is such a function differentiable in z, and what does that differentiability entail? Actually, before we answer this we first need to state what it means for a complex function to be differentiable!

Definition D.2.2 (Differentiable): *We say a complex function f is **(complex)** differentiable at z_0 if it's differentiable with respect to the complex variable z, which means*

$$\lim_{h \to 0} \frac{f(z_0 + h) - f(z_0)}{h}$$

exists, where h tends to zero along any *path in the complex plane. If the limit exists we write $f'(z_0)$ for the limit. If f is differentiable, then $f(x + iy) = u(x, y) + iv(x, y)$ satisfies the **Cauchy-Riemann equations**:*

$$f'(z) = \frac{\partial u}{\partial x} + i\frac{\partial v}{\partial x} = -i\frac{\partial u}{\partial y} + \frac{\partial v}{\partial y}$$

(one direction is easy, arising from sending $h \to 0$ along the paths $\tilde{h}$ and $i\tilde{h}$, with $\tilde{h} \in \mathbb{R}$).

Here's a quick hint to see why differentiability implies the Cauchy-Riemann equations—try and fill in the details. Since the derivative exists at z_0, the key limit is independent of the path we take to the point $x_0 + iy_0$. Consider the path $x + iy_0$ with $x \to x_0$, and the path $x_0 + iy$ with $y \to y_0$, and use results from multivariable calculus on partial derivatives.

Let's explore a bit and see which functions are complex differentiable. We let $h = h_1 + ih_2$ below, with $h \to 0 + 0i$.

- If $f(z) = z$ then

$$\lim_{h \to 0} \frac{f(z + h) - f(z)}{h} = \lim_{h \to 0} \frac{z + h - z}{h} = \lim_{h \to 0} 1 = 1;$$

thus the function is complex differentiable and the derivative is 1.

- If $f(z) = z^2$ then

$$
\begin{aligned}
\lim_{h \to 0} \frac{f(z+h) - f(z)}{h} &= \lim_{h \to 0} \frac{(z+h)^2 - z^2}{h} \\
&= \lim_{h \to 0} \frac{z^2 + 2zh + h^2 - z^2}{h} \\
&= \lim_{h \to 0} \frac{2zh + h^2}{h} \\
&= \lim_{h \to 0} (2z + h) \\
&= \lim_{h \to 0} 2z + \lim_{h \to 0} h \\
&= 2z + 0 = 2z.
\end{aligned}
$$

We're using the following properties of complex numbers: $h/h = 1$ and $2zh + h^2 = (2z+h)h$. Note how similar this is to the real-valued analogue, $f(x) = x^2$.

- If $f(z) = \bar{z}$ then

$$
\lim_{h \to 0} \frac{f(z+h) - f(z)}{h} = \lim_{h \to 0} \frac{\overline{z+h} - \bar{z}}{h}.
$$

Unlike the other limits, this one isn't immediately clear. Let's write $z = x + iy$, $h = h_1 + ih_2$ (and of course $\bar{z} = x - iy$, $\bar{h} = h_1 - ih_2$). The limit is

$$
\lim_{h \to 0} \frac{x - iy + h - ih_2 - (x - iy)}{h_1 + ih_2} = \lim_{h \to 0} \frac{h_1 - ih_2}{h_1 + ih_2}.
$$

This limit does not exist; depending on how $h \to 0$ we obtain different answers. For example, if $h_2 = 0$ (traveling along the x-axis) the limit is just $\lim_{h \to 0} h_1/h_1 = 1$, while if $h_1 = 0$ (traveling along the y-axis) the limit is just $\lim_{h \to 0} -ih_2/ih_2 = -1$. Thus this function isn't complex differentiable anywhere, even though it's a fairly straightforward function to define.

If we continue to argue along these lines, we find that a function is complex differentiable if the x and y dependence is in a very special form, namely everything is a function of $z = x + iy$. In other words, we don't allow our function to depend on $\bar{z} = x - iy$. If we could depend on both, we could isolate out x (which is $z + \bar{z}$) and y (which is $(z - \bar{z})/i$). We can begin to see why being complex differentiable once implies that we're complex differentiable infinitely often, namely because of the very special dependence on x and y. Also, in the plane there's really only two ways to approach a point: from above, or from below. In the complex plane, the situation is strikingly different. There are so many ways we can move in two dimensions, and *each* path must give the same answer if we're to be complex differentiable. This is why differentiability means far more for a complex variable than for a real variable.

To state the needed results from complex analysis, we also require some terminology from point set topology. In particular, many of the theorems below deal with open sets. We briefly review their definition and give some examples.

Definition D.2.3 (Open set, closed set): *A subset U of $\mathbb{C}$ is an **open set** if for any $z_0 \in U$ there's a δ such that whenever $|z - z_0| < \delta$ then $z \in U$ (note δ is allowed to depend on z_0). A set C is **closed** if its **complement**, $\mathbb{C} \setminus C$, is open.*

The following are examples of open sets in $\mathbb{C}$.

1. $U_1 = \{z : |z| < r\}$ for any $r > 0$. This is usually called the **open ball of radius** r centered at the origin.

2. $U_2 = \{z : \Re(z) > 0\}$. To see this is open, if $z_0 \in U_2$ then we can write $z_0 = x_0 + iy_0$, with $x_0 > 0$. Letting $\delta = x_0/2$, for $z = x + iy$ we see that if $|z - z_0| < \delta$ then $|x - x_0| < x_0/2$, which implies $x > x_0/2 > 0$; U_2 is often called the open **right half-plane**.

For examples of closed sets, consider the following:

1. $C_1 = \{z : |z| \leq r\}$. Note that if we take z_0 to be any point on the boundary, then the ball of radius δ centered at z_0 will contain points more than r units from the origin, and thus C_1 isn't open. A little work shows, however, that C_1 is closed (in fact, C_1 is called the **closed ball of radius** r about the origin). We prove it's closed by showing its complement is open. What we need to do is show that, given any point in the complement, there's a small ball about that point entirely contained in the complement. I urge you to draw a picture for the following argument. If $z_0 \in \mathbb{C} \setminus C_1$ then $|z_0| > r$ (as otherwise it would be inside C_1). If we take $\delta < \frac{|z_0| - r}{2}$ then after some algebra we'll find that if $|z - z_0| < \delta$ then $z \in \mathbb{C} \setminus C_1$. Thus $\mathbb{C} \setminus C_1$ is open, so C_1 is closed.

2. $C_2 = \{z : \Re(z) \geq 0\}$. To see this set isn't open, consider any $z_0 = iy$ with $y \in \mathbb{R}$. A similar calculation as the one we did for U_2 or C_1 shows C_2 is closed.

For a set that is neither open nor closed, consider $S = U_1 \cup C_2$.

We now state two of the most important properties a complex function could have. One of the most important results in the subject is that these two seemingly very different properties are actually equivalent!

Definition D.2.4 (Holomorphic, analytic): *Let U be an open subset of $\mathbb{C}$, and let f be a complex function. We say f is **holomorphic** on U if f is differentiable at every point $z \in U$, and we say f is **analytic** on U if f has a series expansion that converges and agrees with f on U. This means that for any $z_0 \in U$, for z close to z_0 we can choose a_n's such that*

$$f(z) = \sum_{n=0}^{\infty} a_n (z - z_0)^n.$$

As alluded to above, saying a function of a complex variable is differentiable turns out to imply *far* more than saying a function of a real variable is differentiable, as the following theorem shows us.

> **Theorem D.2.5**: *Let f be a complex function and U an open set. Then f is holomorphic on U if and only if f is analytic on U, and the series expansion for f is its Taylor series.*

The above theorem is amazing; its result seems to good to be true. Namely, as soon as we know f is differentiable once, it's infinitely (real) differentiable and f agrees with its Taylor series expansion! This is very different than what happens in the case of functions of a real variable. For instance, the function

$$h(x) = x^3 \sin(1/x) \tag{D.2}$$

is differentiable once and only once at $x = 0$, and while the function $g(x)$ from (D.1) is infinitely differentiable, the Taylor series expansion only agrees with $g(x)$ at $x = 0$. Complex analysis is a *very* different subject than real analysis!

The next theorem provides a very nice condition for when a function is identically zero. It involves the notion of a limit or accumulation point, which we define first.

> **Definition D.2.6 (Limit or accumulation point)**: *We say z is a **limit** (or an **accumulation**) **point** of a sequence $\{z_n\}_{n=0}^{\infty}$ if there exists a subsequence $\{z_{n_k}\}_{k=0}^{\infty}$ converging to z.*

Let's do some examples to clarify the definitions.

1. If $z_n = 1/n$, then 0 is a limit point.
2. If $z_n = \cos(\pi n)$ then there are two limit points, namely 1 and -1. (If $z_n = \cos(n)$ then *every* point in $[-1, 1]$ is a limit point of the sequence, though this is harder to show.)
3. If $z_n = (1 + (-1)^n)^n + 1/n$, then 0 is a limit point. We can see this by taking the subsequence $\{z_1, z_3, z_5, z_7, \dots\}$; note the subsequence $\{z_0, z_2, z_4, \dots\}$ diverges to infinity.
4. Let z_n denote the number of distinct prime factors of n. Then every positive integer is a limit point! For example, let's show 5 is a limit point. The first five primes are $2, 3, 5, 7$, and 11; consider $N = 2 \cdot 3 \cdot 5 \cdot 7 \cdot 11 = 2310$. Consider the subsequence $\{z_N, z_{N^2}, z_{N^3}, z_{N^4}, \dots\}$; as N^k has exactly 5 distinct prime factors for each k, 5 is a limit point.
5. If $z_n = n^2$ then there are no limit points, as $\lim_{n \to \infty} z_n = \infty$.
6. Let z_0 be any odd, positive integer, and set

$$z_{n+1} = \begin{cases} 3z_n + 1 & \text{if } z_n \text{ is odd} \\ z_n/2 & \text{if } z_n \text{ is even.} \end{cases}$$

It's *conjectured* that 1 is always a limit point (and if some $z_m = 1$, then the next few terms have to be $4, 2, 1, 4, 2, 1, 4, 2, 1, \dots$, and hence the sequence cycles). This is the famous $3x + 1$ **problem**. Kakutani called it a conspiracy to slow down American mathematics because of the amount of time people spent on this; Erdös said mathematics isn't yet ready for such problems. See [Lag1, Lag2, Lag3] for some nice expositions, but be warned that this problem can be addictive!

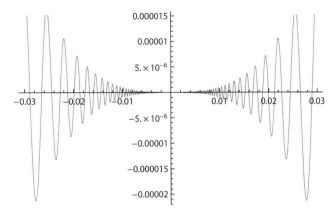

Figure D.1. Plot of $x^3 \sin(1/x)$.

We can now state the theorem which, for us, is the most important result from complex analysis. It's the basis of the black-box results.

Theorem D.2.7: *Let f be an analytic function on an open set U, with infinitely many zeros $z_1, z_2, z_3, \ldots$. If $\lim_{n \to \infty} z_n \in U$, then f is identically zero on U. In other words, if a function is zero along a sequence in U whose accumulation point is also in U, then that function is identically zero in U.*

Note the above is *very* different than what happens in real analysis. Consider again the function from (D.2),

$$h(x) = x^3 \sin(1/x).$$

This function is continuous and differentiable. It's zero whenever $x = 1/\pi n$ with n an integer. If we let $z_n = 1/\pi n$, we see this sequence has 0 as a limit point, and our function is also zero at 0 (see Figure D.1). It's clear, however, that this function is *not* identically zero. Yet again, we see a stark difference between real- and complex-valued functions. As a nice exercise, show that $x^3 \sin(1/x)$ is *not* complex differentiable. It will help if you recall $e^{i\theta} = \cos\theta + i\sin\theta$, or $\sin\theta = (e^{i\theta} - e^{-i\theta})/2$.

D.3 Complex Analysis and Moment Generating Functions

We conclude our technical digression by stating a few more very useful facts. The proof of these requires properties of the **Laplace transform**, which is defined by $(\mathcal{L}f)(s) = \int_0^\infty e^{-sx} f(x)dx$. The reason the Laplace transform plays such an important role in the theory is apparent when we recall the definition of the moment generating function of a random variable X with density f:

$$M_X(t) = \mathbb{E}[e^{tX}] = \int_{-\infty}^{\infty} e^{tx} f(x)dx;$$

in other words, the moment generating function is the Laplace transform of the density evaluated at $s = -t$.

Remember that if F_X and G_Y are the cumulative distribution functions of the random variables X and Y with densities f and g, then

$$F_X(x) = \int_{-\infty}^{x} f(t)dt$$

$$G_Y(y) = \int_{-\infty}^{y} g(v)dv.$$

We remind the reader of the two important results we assumed in the text (Theorems 20.5.3 and 20.5.4), which we restate below. After stating them we discuss their proofs.

Theorem D.3.1: *Assume the moment generating functions $M_X(t)$ and $M_Y(t)$ exist in a neighborhood of zero (i.e., there's some δ such that both functions exist for $|t| < \delta$). If $M_X(t) = M_Y(t)$ in this neighborhood, then $F_X(u) = F_Y(u)$ for all u. As the densities are the derivatives of the cumulative distribution functions, we have $f = g$.*

Theorem D.3.2: *Let $\{X_i\}_{i \in I}$ be a sequence of random variables with moment generating functions $M_{X_i}(t)$. Assume there's a $\delta > 0$ such that when $|t| < \delta$ we have $\lim_{i \to \infty} M_{X_i}(t) = M_X(t)$ for some moment generating function $M_X(t)$, and all moment generating functions converge for $|t| < \delta$. Then there exists a unique cumulative distribution function F whose moments are determined from $M_X(t)$ and for all x where $F_X(x)$ is continuous, $\lim_{i \to \infty} F_{X_i}(x) = F_X(x)$.*

The proof of these theorems follow from results in complex analysis, specifically the Laplace and Fourier inversion formulas. To give an example as to how the results from complex analysis allow us to prove results such as these, we give most of the details in the proof of the next theorem. We *deliberately* do not try and prove the following result in as great generality as possible!

Theorem D.3.3: *Let X and Y be two continuous random variables on $[0, \infty)$ with continuous densities f and g, all of whose moments are finite and agree. Suppose further that:*

1. *There is some $C > 0$ such that for all $c \le C$, $e^{(c+1)t} f(e^t)$ and $e^{(c+1)t} g(e^t)$ are Schwartz functions (see Definition 21.1.3). This isn't a terribly restrictive assumption; f and g need to have decay in order for all moments to exist and be finite. As we're evaluating f and g at e^t and not t, there's enormous decay*

here. The meat of the assumption is that f and g are infinitely differentiable and their derivatives decay.

2. *The (not necessarily integral) moments*

$$\mu'_{r_n}(f) = \int_0^\infty x^{r_n} f(x)dx \quad \text{and} \quad \mu'_{r_n}(g) = \int_0^\infty x^{r_n} g(x)dx$$

agree for some sequence of non-negative real numbers $\{r_n\}_{n=0}^\infty$ which has a finite accumulation point (i.e., $\lim_{n\to\infty} r_n = r < \infty$).

Then $f = g$ (in other words, knowing all these moments uniquely determines the probability density).

Proof: We sketch the proof, which is long and sadly a bit technical. Remember the purpose of this proof is to highlight why our needed results from complex analysis are true. Feel free to skim or skip the proof, but we urge you to read the example at the end of this section, where we return to the two densities that are causing us so much heartache. Let $h(x) = f(x) - g(x)$, and define

$$A(z) = \int_0^\infty x^z h(x)dx.$$

Note that $A(z)$ exists for all z with the real part non-negative. To see this, let $\Re(z)$ denote the real part of z, and let k be the unique non-negative integer with $k \le \Re(z) < k+1$. Then $x^{\Re z} \le x^k + x^{k+1}$, and

$$|A(z)| \le \int_0^\infty x^{\Re(z)} \left[|f(x)| + |g(x)| \right] dx$$

$$\le \int_0^\infty (x^k + x^{k+1}) f(x)dx + \int_0^\infty (x^k + x^{k+1})g(x)dx = 2\mu'_k + 2\mu'_{k+1}.$$

Results from analysis now imply that $A(z)$ exists for all z. The key point is that A is also differentiable. Interchanging the derivative and the integration (which can be justified; see Theorem B.2.2), we find

$$A'(z) = \int_0^\infty x^z (\log x)h(x)dx.$$

To show that $A'(z)$ exists, we just need to show this integral is well-defined. There are only two potential problems with the integral, namely when $x \to \infty$ and when $x \to 0$. For x large, $x^z \log x \le x^{\Re(z)+1}$ and thus the rapid decay of h gives $\left| \int_1^\infty x^z (\log x)h(x)dx \right| < \infty$. For x near 0, $h(x)$ looks like $h(0)$ plus a small error (remember we're assuming f and g are continuous); thus there's a C so that $|h(x)| \le C$ for $|x| \le 1$. Note

$$\lim_{\epsilon \to 0} \int_\epsilon^1 \left| \int_0^\infty x^z (\log x)h(x)dx \right| \le \lim_{\epsilon \to 0} 1 \int_\epsilon^1 1 \cdot (-\log x) \cdot C dx.$$

The anti-derivative of $\log x$ is $x \log x - x$, and $\lim_{\epsilon \to 0}(\epsilon \log \epsilon - \epsilon) = 0$. This is enough to prove that this integral is bounded, and thus from results in analysis we get $A'(z)$ exists.

We (finally!) use our results from complex analysis. As A is differentiable once, it's infinitely differentiable and it equals its Taylor series for z with $\Re(z) > 0$. Therefore A is an analytic function which is zero for a sequence of z_n's with an accumulation point, and thus it's identically zero. This is spectacular—initially we only knew $A(z)$ was zero if z was a positive integer or if z was in the sequence $\{r_n\}$; we now know it's zero for all z with $\Re(z) > 0$. This remarkable conclusion comes from complex analysis; it's here that we use it.

We change variables, and replace x with e^t and dx with $e^t dt$. The range of integration is now $-\infty$ to ∞, and we set $\mathfrak{h}(t)dt = h(e^t)e^t dt$. We now have

$$A(z) = \int_{-\infty}^{\infty} e^{tz} \mathfrak{h}(t)dt = 0.$$

Choosing $z = c + 2\pi i y$ with c less than the C from our hypotheses gives

$$A(c + 2\pi i y) = \int_{-\infty}^{\infty} e^{2\pi i t y} \left[e^{ct} \mathfrak{h}(t) \right] dt = 0.$$

Our assumptions imply that $e^{ct} \mathfrak{h}(t)$ is a Schwartz function, and thus it has a unique inverse Fourier transform. As we know this transform is zero, it implies that $e^{ct} \mathfrak{h}(t) = 0$, or $h(x) = 0$, or $f(x) = g(x)$. $\qquad\square$

We needed the analysis at the end on the inverse Fourier transform as our goal is to show that $f(x) = g(x)$, not that $A(z) = 0$. It seems absurd that $A(z)$ could identically vanish without $f = g$, but we must rigorously show this.

What if we lessen our restrictions on f and g; perhaps one of them isn't continuous? Perhaps there's a unique continuous probability distribution attached to a given sequence of moments such as in the above theorem, but if we allow non-continuous distributions there could be additional possibilities. This topic is beyond the scope of this book, requiring more advanced results from analysis; however, we wanted to point out where the dangers lie, where we need to be careful.

After proving Theorem D.3.3, it's natural to go back to the two densities that are causing so much trouble, namely (see (19.2))

$$f_1(x) = \frac{1}{\sqrt{2\pi x^2}} e^{-(\log^2 x)/2}$$

$$f_2(x) = f_1(x)[1 + \sin(2\pi \log x)].$$

We know these two densities have the same integral moments (their k^{th} moments are $e^{k^2/2}$ for k a non-negative integer). These functions have the correct decay; note

$$e^{(c+1)t} f_1(e^t) = e^{(c+1)t} \cdot \frac{e^{-t^2/2}}{\sqrt{2\pi} e^t},$$

which decays fast enough for any c to satisfy the assumptions of Theorem D.3.3. As these two densities are not the same, *some* condition must be violated. The only condition left to check is whether or not we have a sequence of numbers $\{r_n\}_{n=0}^{\infty}$

with an accumulation point $r > 0$ such that the r_n^{th} moments agree. Using more results from complex analysis (specifically, contour integration), we can calculate the $(a + ib)^{\text{th}}$ moments. We find

$$(a + ib)^{\text{th}} \text{ moment of } f_1 \text{ is } e^{(a+ib)^2/2}$$

and

$$(a + ib)^{\text{th}} \text{ moment of } f_1 \text{ is } e^{(a+ib)^2/2} + \frac{i}{2}\left(e^{(a+i(b-2\pi))^2/2} - e^{(a+i(b+2\pi))^2/2}\right).$$

While these moments agree for $b = 0$ and a a positive integer, there's no sequence of real moments having an accumulation point where they agree. To see this, note that when $b = 0$ the ath moment of f_2 is

$$e^{a^2/2} + e^{(a-2i\pi)^2/2}\left(1 - e^{4ia\pi}\right), \tag{D.3}$$

and this is never zero unless a is a half-integer (i.e., $a = k/2$ for some integer k). In fact, the reason we wrote (D.3) as we did was to highlight the fact that it's only zero when a is a half-integer. Exponentials of real or complex numbers are never zero, and thus the only way this can vanish is if $1 = e^{4ia\pi}$. Recalling that $e^{i\theta} = \cos\theta + i\sin\theta$, we see that the vanishing of the ath moment is equivalent to $1 - \cos(4\pi a) - i\sin(4\pi a) = 0$; the only way this can happen is if $a = k/2$ for some k. If this happens, the cosine term is 1 and the sine term is 0.

D.4 Exercises

Exercise D.4.1 Let $f(x) = x^3 \sin(1/x)$ for $x \neq 0$ and set $f(0) = 0$. *(a) Show that f is differentiable once when viewed as a function of a real variable, but that it is not differentiable twice. (b) Show that f is not differentiable when viewed as a function of a complex variable z; it might be useful to note that $\sin u = (e^{iu} - e^{-iu})/2i$.*

Exercise D.4.2 *If we're told that all the moments of f are finite and f is infinitely differentiable, must there be some C such that for all $c < C$ we have $e^{(c+1)t} f(e^t)$ is a Schwartz function?*

BIBLIOGRAPHY

[Ba]   A. Banner, *The Calculus Lifesaver*, Princeton University Press, Princeton, NJ, 2007.

[BL]   P. Baxandall and H. Liebeck, *Vector Calculus*, Clarendon Press, Oxford, 1986.

[Ben]   F. Benford, *The law of anomalous numbers*, Proceedings of the American Philosophical Society **78** (1938) 551–572.

[BBH]   A. Berger, Leonid A. Bunimovich, and T. Hill, *One-dimensional dynamical systems and Benford's Law*, Trans. Amer. Math. Soc. **357** (2005), no. 1, 197–219.

[Ber]   M. Bernstein, *Games, hats, and codes*, lecture at the SUMS 2005 Conference.

[BD]   P. Bickel and K. Doksum, *Mathematical Statistics: Basic Ideas and Selected Topics*, Holden-Day, San Francisco, 1977.

[Bi]   P. Billingsley, *Probability and Measure*, 3rd edition, Wiley, New York, 1995.

[Bol]   B. Bollobás, *Random Graphs*, Cambridge Studies in Advanced Mathematics, Cambridge University Press, Cambridge, MA, 2001.

[BoDi]   W. Boyce and R. DiPrima, *Elementary Differential Equations and Boundary Value Problems*, 7th edition, John Wiley & Sons, New York, 2000.

[CaBe]   G. Casella and R. Berger, *Statistical Inference*, 2nd edition, Duxbury Advanced Series, Pacific Grove, CA, 2002.

[Chr]   J. Christiansen, *An introduction to the moment problem*, lecture notes.

[Conw]   J. H. Conway, *The weird and wonderful chemistry of audioactive decay*. Pages 173–178 in *Open Problems in Communications and Computation*, ed. T. M. Cover and B. Gopinath, Springer-Verlag, New York, 1987.

[Cor1]   Cornell University, *arXiv*, http://arxiv.org.

[Cor2]   Cornell University, *Project Euclid*, http://projecteuclid.org/.

[CM]   M. Cozzens and S. J. Miller, *The Mathematics of Encryption: An Elementary Introduction*, AMS Mathematical World series **29**, Providence, RI, 2013.

[DN]   H. A. David and H. N. Nagaraja, *Order Statistics*, 3rd edition, Wiley Interscience, Hoboken, NJ, 2003.

[Dev]   R. Devaney, *An Introduction to Chaotic Dynamical Systems*, 2nd edition, Westview Press, Cambridge, MA, 2003.

[DVB]   R. De Veaux, P. Velleman, and D. Bock, *Intro Stats*, 4th edition, Pearson Press, Boston, MA, 2014.

[Du]   R. Durrett, *Probability: Theory and Examples*, 2nd edition, Duxbury Press, 1996.

[Edg]   G. Edgar, *Measure, Topology, and Fractal Geometry*, 2nd edition, Springer-Verlag, 1990.

[Fal]   K. Falconer, *Fractal Geometry: Mathematical Foundations and Applications*, 2nd edition, John Wiley & Sons, New York, 2003.

[Fa1]   E. Fama, *Mandelbrot and Stable Paretian Hypothesis*, The Journal of Business **36** (1963), no. 4, 420–429, http://www.jstor.org/stable/2350971.

[Fa2]   E. Fama, *Random Walks in Stock Market Prices*, Financial Analysts Journal (January-February 1995), 75–80 (reprinted from the September-October 1965 issue, 55–59).

[Fe]   W. Feller, *An Introduction to Probability Theory and Its Applications*, 2nd edition, Vol. II, John Wiley & Sons, New York, 1971.

[Fol]   G. Folland, *Real Analysis: Modern Techniques and Their Applications*, 2nd edition, Pure and Applied Mathematics, Wiley-Interscience, New York, 1999.

[Fr]   J. Franklin, *Mathematical Methods of Economics: Linear and Nonlinear Programming, Fixed-Point Theorem*, Springer-Verlag, New York, 1980.

[GH]   Y. Gerchak and M. Henig, *The basketball shootout: strategy and winning probabilities*, Operations Research Letters **5** (1986), no. 5, 241–244.

[Gl]   M. Gladwell, *Outliers: The Story of Success*, Back Bay Books, Reprint edition (June 7, 2011).

[Hi1]   T. Hill, *The first-digit phenomenon*, American Scientist **86** (1996), 358–363.

[Hi2]   T. Hill, *A statistical derivation of the significant-digit law*, Statistical Science **10** (1996), 354–363.

[HJ]   K. Hrbáček and T. Jech, *Introduction to Set Theory*, Pure and Applied Mathematics, Marcel Dekker, New York, 1984.

[Hu]   W. J. Hurley, *The birthday matching problem when the distribution of birthdays is nonuniform*, Chance Magazine **21** (2008), no. 4, 20–24.

[Knu]   D. Knuth, *The Art of Computer Programming, Volume 2: Seminumerical Algorithms*, 3rd edition, Addison-Wesley, MA, 1997.

[KonMi] A. Kontorovich and S. J. Miller, *Benford's law, values of L-functions and the $3x + 1$ problem*, Acta Arith. **120** (2005), 269–297.

[KonSi] A. Kontorovich and Ya. G. Sinai, *Structure theorem for (d, g, h)-maps*, Bull. Braz. Math. Soc. (N.S.) 33 (2002), no. 2, 213–224.

[Kos]   T. Koshy, *Fibonacci and Lucas Numbers with Applications*, Wiley-Interscience, New York, 2001

[KN]   L. Kuipers and H. Niederreiter, *Uniform Distribution of Sequences*, John Wiley & Sons, New York, 1974.

[Lag1]   J. Lagarias, *The $3x + 1$ problem and its generalizations*, American Mathematical Monthly **92** (1985), no. 1, 3–23.

[Lag2]   J. Lagarias, *The $3x + 1$ problem and its generalizations*. Pages 305–334 in *Organic mathematics (Burnaby, BC, 1995)*, CMS Conf. Proc., vol. 20, AMS, Providence, RI, 1997.

[Lag3]   J. Lagarias, *The Ultimate Challenge: The $3x + 1$ Problem*, American Mathematical Society, Providence, RI, 2010.

[LaSo]   J. Lagarias and K. Soundararajan, *Benford's Law for the $3x + 1$ function*, J. London Math. Soc. (2) **74** (2006), no. 2, 289–303.

[La1]   S. Lang, *Calculus of Several Variables*, Springer-Verlag, New York, 1987.

[La2]   S. Lang, *Undergraduate Analysis*, 2nd edition, Springer-Verlag, New York, 1997.

[La3]   S. Lang, *Complex Analysis*, Graduate Texts in Mathematics, Vol. 103, Springer-Verlag, New York, 1999.

[Le]   L. M. Leemis, *Probability, 56263rd edition*, Amazon, 2010.

[LF]   R. Larson and B. Farber, *Elementary Statistics: Picturing the World*, Prentice-Hall, Englewood Cliffs, NJ, 2003.

[Man]   B. Mandelbrot, *Variation on certain speculative prices*, The Journal of Business **36** (1963), no. 4, 394–419, http://www.jstor.org/stable/2350970.

[ManHu] B. Mandelbrot and R. L. Hudson, *The (Mis)behavior of Markets. A Fractal View of Risk, Ruin, and Reward*, Basic Books, New York, 2004.

[Meh1]   M. Mehta, *On the statistical properties of level spacings in nuclear spectra*, Nucl. Phys. **18** (1960), 395–419.

[Meh2]   M. Mehta, *Random Matrices*, 2nd edition, Academic Press, Boston, 1991.

[Met]   N. Metropolis, *The beginning of the Monte Carlo method*, Los Alamos Science, No. 15, Special Issue (1987), 125–130.

[MU]   N. Metropolis and S. Ulam, *The Monte Carlo method*, J. Amer. Statist. Assoc. **44** (1949), 335–341.

[Mil]   S. J. Miller, *The Pythagorean won-loss formula in baseball*, Chance Magazine **20** (2007), no. 1, 40–48 (an abridged version appeared in the Newsletter of the SABR Statistical Analysis Committee **16** (February 2006), no. 1, 17–22).

[MN]   S. J. Miller and M. Nigrini, *Order Statistics and Shifted Almost Benford Behavior*, preprint.

[MilMo] S. J. Miller and D. Montague, *Rational irrationality proofs*, Mathematics Magazine **85** (2012), no. 2, 110–114.

[MT-B]   S. J. Miller and R. Takloo-Bighash, *An Invitation to Modern Number Theory*, Princeton University Press, Princeton, NJ, 2006.

[MoMc]   D. Moore and G. McCabe, *Introduction to the Practice of Statistics*, W. H. Freeman and Co., London, 2003.

[Ni1]   T. Nicely, *The pentium bug*, http://www.trnicely.net/pentbug/pentbug.html.

[Ni2]   T. Nicely, *Enumeration to 10^{14} of the Twin Primes and Brun's Constant*, Virginia J. Sci. **46** (1996), 195–204.

[Nig1]   M. Nigrini, *Digital Analysis and the Reduction of Auditor Litigation Risk*. Pages 69–81 in *Proceedings of the 1996 Deloitte & Touche/University of Kansas Symposium on Auditing Problems*, ed. M. Ettredge, University of Kansas, Lawrence, KS, 1996.

[Nig2]   M. Nigrini, *The use of Benford's Law as an aid in analytical procedures*, Auditing: A Journal of Practice & Theory **16** (1997), no. 2, 52–67.

[Re]   F. Reif, *Fundamentals of Statistical and Thermal Physics*, McGraw-Hill, New York, 1965.

[Ro]   S. Ross, *A First course in Probability*, 9th edition, Pearson, Essex, UK, 2014.

[Rud]   W. Rudin, *Principles of Mathematical Analysis*, 3rd edition, International Series in Pure and Applied Mathematics, McGraw-Hill, New York, 1976.

[Si]   B. Simon, *The classical moment problem as a self-adjoint finite difference operator*, Adv. Math. **137** (1998), no. 1, 82–203.

[Sl]   N. Sloane, *On-Line Encyclopedia of Integer Sequences*, http://www.research.att. com/~njas/sequences/Seis.html.

[SS1]   E. Stein and R. Shakarchi, *Fourier Analysis: An Introduction*, Princeton University Press, Princeton, NJ, 2003.

[SS2]   E. Stein and R. Shakarchi, *Complex Analysis*, Princeton University Press, Princeton, NJ, 2003.

[SS3]   E. Stein and R. Shakarchi, *Real Analysis: Measure Theory, Integration, and Hilbert Spaces*, Princeton University Press, Princeton, NJ, 2005.

[SSR]   W. E. Stein, D. A. Seale, and A. Rapoport, *Analysis of heuristic solutions to the best choice problem*, European Journal of Operational Research **151** (2003), 140–152.

[St]   G. Strang, *Linear Algebra and Its Applications*, 3rd edition, Wellesley-Cambridge Press, Wellesley, MA, 1998.

[Te]   G. Tenenbaum, *Introduction to Analytic and Probabilistic Number Theory*, Cambridge University Press, Cambridge, 1995.

[VG]   W. Voxman and R. Goetschel, Jr., *Advanced Calculus*, Mercer Dekker, New York, 1981.

[We]   E. Weisstein, *Stirling's Approximation*, MathWorld—A Wolfram Web Resource, http://mathworld.wolfram.com/Stirling'sApproximation.html.

[Wis]   J. Wishart, *The generalized product moment distribution in samples from a normal multivariate population*, Biometrika **20 A** (1928), 32–52.

[Wor]   N. C. Wormald, *Models of random regular graphs*. Pages 239–298 in *Surveys in combinatorics, 1999 (Canterbury)* London Mathematical Society Lecture Note Series, vol. 267, Cambridge University Press, Cambridge, 1999.

[Zy]   A. Zygmund, *Trigonometrical Series*, vols. I and II, Cambridge University Press, Cambridge, 1968.

INDEX

Further Praise for *The Probability Lifesaver*

"*The Probability Lifesaver* contains a lot of explanations and examples and provides step-by-step instructions to how definitions and ideas are formulated. I appreciated that it tries to provide multiple solutions to each problem. Interesting, informative, approachable, and comprehensive, this book was easy to read and would make a good supplement for a first probability course at the undergraduate level."

—Jingchen Hu, *Vassar College*

"Filled with many interesting and contemporary examples, *The Probability Lifesaver* would have undoubtedly helped me while I was taking statistics. Miller offers careful, detailed explanations in simple terms that are easy to understand."

—James Coyle, former student at *Rutgers University*

"In *The Probability Lifesaver*, Miller does more than simply present the theoretical framework of probability. He takes complex concepts and describes them in understandable language, provides realistic applications that highlight the far-extending reaches of probability, and engages the problem-solving intuitions that lie at the heart of mathematics. Lastly, and most importantly, I am reminded throughout this textbook of why I chose to study mathematics: because it's fun!"

—Michael Stone, *Williams College* '16

"*The Probability Lifesaver* motivates introductory probability theory with concrete applications in an approachable and engaging manner. From computing the probability of various poker hands to defining σ-algebras, it strikes a balance between applied computation and mathematical theory that makes it easy to follow while still being mathematically satisfying."

—David Burt, *Williams College* '17

"A balanced mix of theoretical and practical problem-solving approaches in probability—suited for personal study as well as textbook reading in and out of the classroom. After college, while working, I took a probability class remotely and with this book, I was able to follow easily despite being without a TA or easy access to the professor. From research examples to interview questions, it has saved my life more than once."

—Dan Zhao, *Williams College* '14